Student Solutions Guide to Accompany

Intermediate Algebra

Larson/Hostetler

Gerry C. Fitch
Louisiana State University
Baton Rouge, Louisiana

D.C. Heath and Company
Lexington, Massachusetts   Toronto

Address editorial correspondence to:
D.C. Heath and Company
125 Spring Street
Lexington, MA 02173

Published simultaneously in Canada.

Printed in the United States of America.

International Standard Book Number: 0-669-41639-8

10 9 8 7 6 5 4 3 2

Preface

This *Complete Solutions Guide* is a supplement to *Intermediate Algebra,* by Roland E. Larson and Robert P. Hostetler. This guide includes solutions for all exercises in the text, including the chapter reviews, chapter test, and cumulative test. These solutions give step- by-step details of each exercise. I have tries to see that these solutions are correct. Corrections to the solutions or suggestions for improvement are welcome.

Producing this guide has been quite a challenge and a learning experience for me. I would like to thank Ann Marie Jones and D.C. Heath Company for allowing me this experience, I would also like to thank Cathy Cantin, David Heyd, and Meridian Creative Group for their contributions to the production of this guide. Finally, a special word of thanks goes to Karla Neal for her help and to my children, Kara and Parker, for their patience with "mom" during production time and to my new husband, Chuck for his encouragement.

Gerry C. Fitch
Louisiana State University
Baton Rouge, Louisiana 70803

Contents

CHAPTER P
Prerequisites: Fundamental Concepts of Algebra

P.1 Real Numbers: Order and Absolute Value

7. Which of the real numbers in the following set are (a) natural numbers, (b) integers, (c) rational numbers, and
(d) irrational numbers?

Solution

$$\left\{-10, -\sqrt{5}, -\tfrac{2}{3}, -\tfrac{1}{4}, 0, \tfrac{5}{8}, 1, \sqrt{3}, 4, 2\pi, 6\right\}$$

(a) Natural numbers: $\{1, 4, 6\}$

(b) Integers: $\{-10, 0, 1, 4, 6\}$

(c) Rational numbers: $\left\{-10, -\tfrac{2}{3}, -\tfrac{1}{4}, 0, \tfrac{5}{8}, 1, 4, 6\right\}$

(d) Irrational numbers: $\{-\sqrt{5}, \sqrt{3}, 2\pi\}$

9. List the integers between -5.8 and 3.2.

Solution

$\{-5, -4, -3, -2, -1, 0, 1, 2, 3\}$

11. List the odd integers between 0 and 3π.

Solution

$\{1, 3, 5, 7, 9\}$

13. Locate the real numbers (a) 3, (b) $\tfrac{5}{2}$, (c) $-\tfrac{7}{2}$, and (d) -5.2 on the real number line.

Solution

(a) The point representing the real number 3 lies between 2 and 4.

(b) The point representing the real number $\tfrac{5}{2}$ lies between 2 and 3.

(c) The point representing the real number $-\tfrac{7}{2}$ lies between -4 and -3.

(d) The point representing the real number -5.2 lies between -6 and -5.

15. Plot the numbers 2 and 5 and place the correct inequality symbol ($<$ or $>$) between them.

Solution

$2 < 5$

17. Plot the numbers -7 and -2 and place the correct inequality symbol ($<$ or $>$) between them.

Solution

$-7 < -2$

19. Plot the numbers $-\frac{2}{3}$ and $-\frac{10}{3}$ and place the correct inequality symbol ($<$ or $>$) between them.

Solution

$-\frac{2}{3} > -\frac{10}{3}$

21. Using the linegraph in the textbook, approximate the numbers and order them.

Solution

$-1 < 3$

23. Using the linegraph in the textbook, approximate the numbers and order them.

Solution

$-\frac{9}{2} < -2$

25. Find the distance between 4 and 10.

Solution

Distance $= 10 - 4 = 6$

27. Find the distance between 18 and -32.

Solution

Distance $= 18 - (-32) = 18 + 32 = 50$

29. Find the distance between -35 and 0.

Solution

Distance $= 0 - (-35) = 0 + 35 = 35$

31. Evaluate $-|3.5|$.

Solution

$-|3.5| = -3.5$

33. Evaluate $-|-25|$.

Solution

$-|-25| = -25$

35. Place the correct symbol ($<$, $>$, or $=$) between $|-6|$ and $|2|$.

Solution

$|-6| > |2|$ since $6 > 2$.

37. Place the correct symbol ($<$, $>$, or $=$) between $-|-16.8|$ and $-|16.8|$.

Solution

$-|-16.8| = -|16.8|$ since $-16.8 = -16.8$.

39. Find the opposite and absolute value of 14.

Solution

Opposite: -14
Absolute value: 14

41. Find the opposite and absolute value of $-\frac{5}{4}$.

Solution

Opposite: $\frac{5}{4}$
Absolute value: $\frac{5}{4}$

43. Plot -3 and its opposite on the real number line.

Solution

Opposite of -3 is 3.

45. Plot $\frac{5}{3}$ and its opposite on the real number line.

Solution

Opposite of $\frac{5}{3}$ is $-\frac{5}{3}$.

47. Write "x is negative" using inequality notation.

Solution

$x < 0$

49. Write "The price p is less than \$225" using inequality notation.

Solution

$p < 225$

51. Plot the numbers -5 and 2 and place the correct inequality symbol ($<$ or $>$) between them.

Solution

$-5 < 2$

53. Plot the numbers $\frac{1}{3}$ and $\frac{1}{4}$ and place the correct inequality symbol ($<$ or $>$) between them.

Solution

$\frac{1}{3} > \frac{1}{4}$

55. Plot the numbers $-\frac{5}{8}$ and $\frac{1}{2}$ and place the correct inequality symbol ($<$ or $>$) between them.

Solution

$-\frac{5}{8} < \frac{1}{2}$

57. Plot the numbers $-\frac{5}{3}$ and $-\frac{3}{2}$ and place the correct inequality symbol ($<$ or $>$) between them.

Solution

$-\frac{5}{3} < -\frac{3}{2}$

59. Find the distance between -12 and 7.

Solution

Distance $= 7 - (-12) = 7 + 12 = 19$

61. Find the distance between -8 and 0.

Solution

Distance $= 0 - (-8) = 0 + 8 = 8$

63. Find the distance between -6 and -45.

Solution

Distance $= -6 - (-45) = -6 + 45 = 39$

65. Evaluate $|18.6|$.

Solution

$|18.6| = 18.6$

67. Evaluate $-|16|$.

Solution

$-|16| = -16$

69. Evaluate $-|-85|$.

Solution

$-|-85| = -85$

71. Evaluate $|-\pi|$.

Solution

$|-\pi| = \pi$

73. Place the correct symbol ($<$, $>$, or $=$) between $|-2|$ and $|2|$.

Solution

$|-2| = |2|$ since $2 = 2$.

75. Place the correct symbol ($<$, $>$, or $=$) between $-|12.5|$ and $|-25|$.

Solution

$-|12.5| < |-25|$ since $-12.5 < 25$.

77. Find the opposite and absolute value of 34.

Solution

Opposite: -34
Absolute value: 34

79. Find the opposite and absolute value of -160.

Solution

Opposite: 160
Absolute value: 160

81. Find the opposite and absolute value of $-\frac{3}{11}$.

Solution

Opposite: $\frac{3}{11}$
Absolute value: $\frac{3}{11}$

83. Plot the number 7 and its opposite on the real number line.

Solution

$|7| = 7$

85. Plot the number $-\frac{3}{5}$ and its opposite on the real number line.

Solution

$\left|-\frac{3}{5}\right| = \frac{3}{5}$

87. Write "x is nonnegative" using inequality notation.

Solution

$x \geq 0$

89. Write "z is greater than 2 and no more than 10" using inequality notation.

Solution

$10 \geq z > 2$ or $2 < z \leq 10$

91. *True or False?* Determine whether the following statement is true or false.

Every integer is a rational number.

Solution

True

93. *True or False?* Determine whether the following statement is true or false. If the statement is false, give an example of a real number that makes the statement false.

Every rational number is an integer.

Solution

False. $\frac{1}{2}$ is a rational number but is not an integer.

P.2 Operations with Real Numbers

7. Evaluate $13 + 32$.

Solution

$13 + 32 = 45$

9. Evaluate $-13 + 32$.

Solution

$-13 + 32 = +(32 - 13) = 19$

11. Evaluate $-7 - 15$.

Solution

$-7 - 15 = -7 + (-15)$

$\qquad = -(7 + 15) = -22$

13. Evaluate $\frac{3}{4} - \frac{1}{4}$.

Solution

$\frac{3}{4} - \frac{1}{4} = \frac{3 - 1}{4}$

$\qquad = \frac{2}{4} = \frac{1}{2}$

15. Evaluate $\frac{5}{8} + \frac{1}{4} - \frac{5}{6}$.

Solution

$\frac{5}{8} + \frac{1}{4} - \frac{5}{6} = \frac{5(3)}{8(3)} + \frac{1(6)}{4(6)} - \frac{5(4)}{6(4)}$

$\qquad = \frac{15}{24} + \frac{6}{24} - \frac{20}{24}$

$\qquad = \frac{15 + 6 - 20}{24}$

$\qquad = \frac{1}{24}$

17. Evaluate $5\frac{3}{4} + 7\frac{3}{8}$.

Solution

$5\frac{3}{4} + 7\frac{3}{8} = \frac{23}{4} + \frac{59}{8}$

$\qquad = \frac{23(2)}{4(2)} + \frac{59(1)}{8(1)}$

$\qquad = \frac{46 + 59}{8}$

$\qquad = \frac{105}{8} \text{ or } 13\frac{1}{8}$

19. Evaluate $5.8 - 6.2 + 1.1 - 4.7 - 9.2$.

Solution

$5.8 - 6.2 + 1.1 - 4.7 - 9.2 = 5.8 + (-6.2) + 1.1 + (-4.7) + (-9.2)$

$\qquad = -(6.2 - 5.8) + 1.1 + (-4.7) + (-9.2)$

$\qquad = -0.4 + 1.1 + (-4.7) + (-9.2)$

$\qquad = +(1.1 - 0.4) + (-4.7) + (-9.2)$

$\qquad = 0.7 + (-4.7) + (-9.2)$

$\qquad = -(4.7 - 0.7) + (-9.2)$

$\qquad = -4 + (-9.2)$

$\qquad = -(4 + 9.2)$

$\qquad = -13.2$

21. Write $4(5)$ as a repeated addition problem.

Solution

$4(5) = 5 + 5 + 5 + 5$

23. Write $(-15) + (-15) + (-15) + (-15)$ as a multiplication problem.

Solution

$(-15) + (-15) + (-15) + (-15) = 4(-15)$

25. Evaluate the product of $5(-6)$ without using a calculator.

Solution

$5(-6) = -30$

27. Evaluate the product of $\left(-\frac{5}{8}\right)\left(-\frac{4}{5}\right)$ without using a calculator.

Solution

$\left(-\frac{5}{8}\right)\left(-\frac{4}{5}\right) = \frac{1}{2}$

29. Evaluate $\frac{-18}{-3}$ using a calculator. Round the result to two decimal places.

Solution

$\frac{-18}{-3} = 6$

31. Evaluate $-\frac{4}{5} \div \frac{8}{25}$ using a calculator. Round the result to two decimal places.

Solution

$-\frac{4}{5} \div \frac{8}{25} = -\frac{4}{5} \cdot \frac{25}{8} = -2.5$

33. Evaluate $5\frac{3}{4} \div 2\frac{1}{8}$ using a calculator. Round the result to two decimal places.

Solution

$5\frac{3}{4} \div 2\frac{1}{8} = \frac{23}{4} \div \frac{17}{8}$

$= \frac{23}{4} \cdot \frac{8}{17}$

$= \frac{46}{17}$ or 2.71

35. Write $(-3)^4$ as a repeated multiplication problem.

Solution

$(-3)^4 = (-3)(-3)(-3)(-3)$

37. Write $(-5)(-5)(-5)(-5)$ using exponential notation.

Solution

$(-5)(-5)(-5)(-5) = (-5)^4$

39. Evaluate $(-4)^2$.

Solution

$(-4)^2 = (-4)(-4) = 16$

41. Evaluate -4^2.

Solution

$-4^2 = -1(4)(4) = -16$

43. Evaluate $16 - 5(6 - 10)$.

Solution

$16 - 5(6 - 10) = 16 - 5(-4)$

$= 16 + (20)$

$= 36$

45. Evaluate $\dfrac{3^2 - 5}{12} - 3\frac{1}{6}$.

Solution

$\dfrac{3^2 - 5}{12} - 3\frac{1}{6} = \dfrac{9 - 5}{12} - 3\frac{1}{6}$

$= \dfrac{4}{12} - 3\frac{1}{6}$

$= \dfrac{1}{3} - \dfrac{19}{6}$

$= \dfrac{1(2)}{3(2)} - \dfrac{19(1)}{6(1)}$

$= \dfrac{2 - 19}{6}$

$= \dfrac{-17}{6}$ or $-2\frac{5}{6}$

47. Evaluate the following expression using a calculator. Round the result to two decimal places.

$\dfrac{25.5}{6.325}$

Solution

$\dfrac{25.5}{6.325} = 4.031620553 \approx 4.03$

49. Evaluate the following expression using a calculator. Round the result to two decimal places.

$$\frac{(1.5)^{15}}{3}$$

Solution

$$\frac{(1.5)^{15}}{3} = 145.9646301 \approx 145.96$$

53. Evaluate $13 + (-32)$.

Solution

$$13 + (-32) = -(32 - 13) = -19$$

57. Evaluate $4 - 16 + (-8)$.

Solution

$$\begin{aligned}
4 - 16 + (-8) &= 4 + (-16) + (-8) \\
&= -(16 - 4) + (-8) \\
&= -12 + (-8) \\
&= -(12 + 8) \\
&= -20
\end{aligned}$$

61. Evaluate $\frac{3}{5} + \left(-\frac{1}{2}\right)$.

Solution

$$\begin{aligned}
\frac{3}{5} + \left(-\frac{1}{2}\right) &= \frac{3(2)}{5(2)} - \frac{1(5)}{2(5)} \\
&= \frac{6}{10} - \frac{5}{10} \\
&= \frac{6 - 5}{10} \\
&= \frac{1}{10}
\end{aligned}$$

65. Evaluate $|-16.25| - 54.78$.

Solution

$$\begin{aligned}
|-16.25| - 54.78 &= 16.25 + (-54.78) \\
&= -(54.78 - 16.25) \\
&= -38.53
\end{aligned}$$

69. Write $3(-4)$ as a repeated addition problem.

Solution

$$3(-4) = (-4) + (-4) + (-4)$$

51. *Savings Plan* You save $50 a month for 18 years. How much do you set aside during the 18 years?

Solution

$$\$50(12)(18) = \$10,800$$

55. Evaluate $-13 + (-8)$.

Solution

$$-13 + (-8) = -(13 + 8) = -21$$

59. Evaluate $\frac{3}{8} + \frac{7}{8}$.

Solution

$$\begin{aligned}
\frac{3}{8} + \frac{7}{8} &= \frac{3 + 7}{8} \\
&= \frac{10}{8} \\
&= \frac{5}{4}
\end{aligned}$$

63. Evaluate $85 - |-25|$.

Solution

$$85 - |-25| = 85 - 25 = 60$$

67. Evaluate $-\left|-15\frac{2}{3}\right| - 12\frac{1}{3}$.

Solution

$$\begin{aligned}
-\left|-15\tfrac{2}{3}\right| - 12\tfrac{1}{3} &= -15\tfrac{2}{3} + \left(-12\tfrac{1}{3}\right) \\
&= -\left(15\tfrac{2}{3} + 12\tfrac{1}{3}\right) \\
&= -28
\end{aligned}$$

71. Write $\frac{1}{4} + \frac{1}{4} + \frac{1}{4} + \frac{1}{4} + \frac{1}{4} + \frac{1}{4}$ as a multiplication problem.

Solution

$$\frac{1}{4} + \frac{1}{4} + \frac{1}{4} + \frac{1}{4} + \frac{1}{4} + \frac{1}{4} = 6\left(\frac{1}{4}\right)$$

73. Evaluate the product $(-8)(-6)$ without using a calculator.

Solution

$(-8)(-6) = 48$

75. Evaluate the product $6.3(5.1)$ without using a calculator.

Solution

$6.3(5.1) = 32.13$

77. Evaluate the product $\left(\frac{10}{13}\right)\left(-\frac{3}{5}\right)$ without using a calculator.

Solution

$\left(\frac{10}{13}\right)\left(-\frac{3}{5}\right) = -\frac{6}{13}$

79. Evaluate $\frac{-48}{16}$ using a calculator. Round the result to two decimal places.

Solution

$\frac{-48}{16} = -3$

81. Evaluate the following expression using a calculator. Round the result to two decimal places.

$$-\frac{-1/3}{5/6}$$

Solution

$-\dfrac{-1/3}{5/6} = \left(\dfrac{-1}{3} \div \dfrac{5}{6}\right) = -\left(\dfrac{-1}{3} \cdot \dfrac{6}{5}\right) = 0.4$

83. Evaluate $4\frac{1}{8} \div 3\frac{3}{2}$ using a calculator. Round your answer to two decimal places.

Solution

$4\frac{1}{8} \div 3\frac{3}{2} = \frac{33}{8} \div \frac{9}{2}$

$\qquad = \frac{33}{8} \cdot \frac{2}{9}$

$\qquad = \frac{11}{12}$ or 0.92

85. Write 4^3 as a repeated multiplication problem.

Solution

$4^3 = (4)(4)(4)$

87. Write $\left(-\frac{4}{5}\right)^6$ as a repeated multiplication problem.

Solution

$$\left(-\frac{4}{5}\right)^6 = \left(-\frac{4}{5}\right)\left(-\frac{4}{5}\right)\left(-\frac{4}{5}\right)\left(-\frac{4}{5}\right)\left(-\frac{4}{5}\right)\left(-\frac{4}{5}\right)$$

89. Write $\left(\frac{5}{8}\right) \times \left(\frac{5}{8}\right) \times \left(\frac{5}{8}\right) \times \left(\frac{5}{8}\right)$ using exponential notation. In exponential notation, is it necessary to use parentheses to clearly convey the expression's meaning? Explain your reasoning.

Solution

$$\left(\frac{5}{8}\right) \times \left(\frac{5}{8}\right) \times \left(\frac{5}{8}\right) \times \left(\frac{5}{8}\right) = \left(\frac{5}{8}\right)^4$$

Parentheses are necessary to indicate that both the numerator and the denominator are raised to the power.

91. Write $(-4)(-4)(-4)(-4)(-4)(-4)$ using exponential notation. In exponential notation, is it necessary to use parentheses to clearly convey the expression's meaning? Explain your reasoning.

Solution

$(-4)(-4)(-4)(-4)(-4)(-4) = (-4)^6$

Parentheses are necessary to indicate that both the negative sign and the number are raised to the power.

93. Evaluate $(-4)^4$.

Solution

$(-4)^4 = (-4)(-4)(-4)(-4) = 256$

95. Evaluate -4^4.

Solution

$-4^4 = (-1)(4)(4)(4)(4) = -256$

97. Evaluate $\left(\frac{4}{5}\right)^3$.

Solution

$\left(\frac{4}{5}\right)^3 = \frac{64}{125}$

99. Evaluate $5^3 + |-14 + 4|$.

Solution

$5^3 + |-14 + 4| = 125 + |-10|$

$\qquad\qquad\qquad = 125 + 10$

$\qquad\qquad\qquad = 135$

101. Evaluate $45 + 3(16 \div 4)$.

Solution

$$
\begin{aligned}
45 + 3(16 \div 4) &= 45 + 3(4) \\
&= 45 + 12 \\
&= 57
\end{aligned}
$$

103. Evaluate $0.2(6 - 10)^3 + 85$.

Solution

$$
\begin{aligned}
0.2(6 - 10)^3 + 85 &= 0.2(-4)^3 + 85 \\
&= 0.2(-64) + 85 \\
&= -12.8 + 85 \\
&= 72.2
\end{aligned}
$$

105. Evaluate $5.6[13 - 2.5(-6.3)]$ using a calculator. Round the result to two decimal places.

Solution

$$
\begin{aligned}
5.6[13 - 2.5(-6.3)] &= 5.6[13 + 15.75] \\
&= 5.6[28.75] \\
&= 161
\end{aligned}
$$

107. Evaluate $5^6 - 3(400)$ using a calculator. Round the result to two decimal places.

Solution

$$
5^6 - 3(400) = 15{,}625 - 1200 = 14{,}425
$$

109. Evaluate the following expression using a calculator. Round the result to two decimal places.

$$
\frac{500}{(1.055)^{20}}
$$

Solution

$$
\frac{500}{(1.055)^{20}} = \frac{500}{2.9177575} = 171.36448 \approx 171.36
$$

111. *Balance in an Account* During one month, you made the following transactions in your checking account.

Initial Balance:	$2618.68
Deposit:	$1236.45
Withdrawal:	$25.62
Withdrawal:	$455.00
Withdrawal:	$125.00
Withdrawal:	$715.95

Find the balance at the end of the month. (Disregard any interst that may have been earned.)

Solution

$\$2618.68 + \$1236.45 - \$25.62 - \$455.00 - \$125.00 - \$715.95 = \$2533.56$

113. *Circle Graphs*   Find the unknown fractional part of the circle (see figure in textbook). What property of pie graphs did you use to solve the problem?

Solution

$\frac{1}{4} + \frac{2}{9} + \frac{1}{10} + x + \frac{1}{3} = 1$

Thus:

$$x = 1 - \left(\frac{1}{4} + \frac{2}{9} + \frac{1}{10} + \frac{1}{3} \right)$$

$$= 1 - \left(\frac{45}{180} + \frac{40}{180} + \frac{18}{180} + \frac{60}{180} \right) = 1 - \left(\frac{45 + 40 + 18 + 60}{180} \right) = 1 - \frac{163}{180} = \frac{180}{180} - \frac{163}{180} = \frac{17}{180}.$$

The sum of the parts of the circle equals 1.

115. *Organizing Data*   On Monday you purchased $500 worth of stock. The value of the stock during the remainder of the week is shown in the bar graph in the textbook. (a) Use the graph to complete the table shown in the textbook. (b) Find the sum of the daily gains and losses. Interpret the result in the context of the problem. How could you determine this sum from the graph?

Solution

(a)

Day	Daily Gain or Loss
Tuesday	+5
Wednesday	+8
Thursday	−5
Friday	+16

(b) $(+5) + (+8) + (-5) + (+16) = +24 =$ the sum of the daily gains and losses. The sum of the daily gains and losses is equal to the difference of the value of the stock on Friday and the value of the stock on Monday. This sum could be determined from the graph by $524 (value on Friday) - $500 (value on Monday) = $24.

117. *Geometry*   Using the following information, a bale of hay in the form of a rectangular solid has dimensions 14 inches by 18 inches by 42 inches and weighs approximately 50 pounds (see figure in textbook), find the volume of the bale of hay in cubic feet if the volume of a rectangular solid is the product of its length, width, and height. (Use the fact that 1728 cubic inches = 1 cubic foot.)

Solution

$V = l \cdot w \cdot h = 14" \cdot 18" \cdot 42" = 10,584 \text{ in}^3 = 6.125 \text{ ft}^3$

119. *True or False?*   Determine whether the following statement is true or false. Explain your reasoning.

The reciprocal of every nonzero integer is an integer.

Solution

False. If n is a nonzero integer, then its reciprocal is $\frac{1}{n}$ which is a fraction not an integer.

121. Determine whether the following statement is true or false. Explain your reasoning.

If a negative real number is raised to the 12th power, the result will be positive.

Solution

True. Any negative real number raised to an even numbered power will be a positive real number. Any negative real number raised to an odd-numbered power will be a negative real number.

P.3   Properties of Real Numbers

7. Name the property of real numbers that justifies the statement $(10 + 8) + 3 = 10 + (8 + 3)$.

Solution

Associative Property of Addition

9. Name the property of real numbers that justifies the statement $6(-10) = -10(6)$.

Solution

Commutative Property of Multiplication

11. Name the property of real numbers that justifies the statement $2x - 2x = 0$.

Solution

Additive Inverse Property

13. Name the property of real numbers that justifies the statement $1 \cdot (5t) = 5t$.

Solution

Multiplicative Identity Property

15. Name the property of real numbers that justifies the statement $3(2 + x) = 3 \cdot 2 + 3x$.

Solution

Distributive Property

17. Name the property of real numbers that justifies the statement $10(2x) = (10 \cdot 2)x$.

Solution

Associative Property of Multiplication

19. Complete the statement.

Associative Property of Multiplication:

$3(6y) = $ _____

Solution

$3(6y) = (3 \cdot 6)y$

21. Complete the statement.

Distributive Property:

$5(6 + z) = $ _____

Solution

$5(6 + z) = 5 \cdot 6 + 5 \cdot z$

23. Complete the statement.

Multiplicative Identity Property:

$(x + 8) \cdot 1 = $ _____

Solution

$(x + 8) \cdot 1 = (x + 8)$

25. Write (a) the additive inverse and (b) the multiplicative inverse of 10.

Solution

(a) Additive Inverse: -10

(b) Multiplicative Inverse: $\frac{1}{10}$

27. Write (a) the additive inverse and (b) the multiplicative inverse of $x + 1$.

Solution

(a) Additive Inverse: $-(x + 1)$ or $-x - 1$

(b) Multiplicative Inverse: $\dfrac{1}{x + 1}$

29. Rewrite $(x + 5) - 3$ using the Associative Property of Addition or Multiplication.

Solution

$(x + 5) - 3 = x + (5 - 3)$

31. Rewrite $3(4 \cdot 5)$ using the Associative Property of Addition or Multiplication.

Solution

$3(4 \cdot 5) = (3 \cdot 4)5$

33. Rewrite $20(2 + 5)$ using the Distributive Property.

Solution

$20(2 + 5) = 20 \cdot 2 + 20 \cdot 5$

35. Rewrite $(x + 6)(-2)$ using the Distributive Property.

Solution

$(x + 6)(-2) = x \cdot (-2) + 6 \cdot (-2)$

37. *Mental Math* Given $16(1.75) = 16\left(2 - \frac{1}{4}\right)$, use the Distributive Property to perform the required arithmetic mentally. For example, suppose you work in an industry where the wage is \$14 per hour with "time and one-half" for overtime. Thus, your hourly wage for overtime is $14(1.5) = 14\left(1 + \frac{1}{2}\right) = 14 + 7 = \$21$.

Solution

$$16(1.75) = 16\left(2 - \tfrac{1}{4}\right) = 16(2) - 16\left(\tfrac{1}{4}\right)$$
$$= 32 - 4$$
$$= 28$$

39. Identify the property of real numbers that justifies each rewriting of the equation.

$$x + 5 = 3$$
$$(x + 5) + (-5) = 3 + (-5)$$
$$x + (5 + (-5)) = 3 - 5$$
$$x + 0 = -2$$
$$x = -2$$

Solution

$x + 5 = 3$	Given
$(x + 5) + (-5) = 3 + (-5)$	Addition Property of Equality
$x + (5 + (-5)) = 3 - 5$	Associative Property of Addition
$x + 0 = -2$	Additive Inverse Property
$x = -2$	Additive Identity Property

41. Use the figure in the textbook to fill in the blanks. What property is being demonstrated?

_____ (_____ + _____) = _____ + _____

Solution

$a(b + c) = ab + ac$

43. Name the property of real numbers that justifies the statement $3 + (-5) = -5 + 3$.

Solution

Commutative Property of Addition

45. Name the property of real numbers that justifies the statement $5(2a) = (5 \cdot 2)a$.

Solution

Associative Property of Multiplication

47. Name the property of real numbers that justifies the statement $7 \cdot 1 = 7$.

Solution

Multiplicative Identity Property

49. Name the property of real numbers that justifies the statement $3x + 0 = 3x$.

Solution

Additive Identity Property

51. Name the property of real numbers that justifies the statement

$$10x \cdot \frac{1}{10x} = 1.$$

Solution

Multiplicative Inverse Property

53. Name the property of real numbers that justifies the statement $25 - 25 = 0$.

Solution

Additive Inverse Property

55. Name the property of real numbers that justifies the statement $(5 + 10)(8) = 8(5 + 10)$.

Solution

Commutative Property of Multiplication

57. Name the property of real numbers that justifies the statement $3 + (12 - 9) = (3 + 12) - 9$.

Solution

Associative Property of Addition

59. Name the property of real numbers that justifies the statement $(16 + 8) - 5 = 16 + (8 - 5)$.

Solution

Associative Property of Addition

61. Name the property of real numbers that justifies the statement $(6 + x) - m = 6 + (x - m)$.

Solution

Associative Property of Addition

63. Complete the statement.

Commutative Property of Multiplication:

$$15(-3) = \underline{\quad\quad}$$

Solution

$15(-3) = (-3)15$

65. Complete the statement.

Distributive Property:

$$-3(4 - x) = \underline{\quad\quad}$$

Solution

$-3(4 - x) = (-3)(4) + (-3)(-x) = -12 + 3x$

67. Complete the statement.

Commutative Property of Addition:

$$25 + (-x) = \underline{\quad\quad}$$

Solution

$25 + (-x) = (-x) + 25$

69. Write (a) the additive inverse and (b) the multiplicative inverse of -16.

Solution

(a) Additive Inverse: 16

(b) Multiplicative Inverse: $-\frac{1}{16}$

71. Write (a) the additive inverse and (b) the multiplicative inverse of $6z$.

Solution

(a) Additive Inverse: $-6z$

(b) Multiplicative Inverse: $\frac{1}{6z}$

73. Rewrite $3 + (8 - x)$ using the Associative Property of Addition or Multiplication.

Solution

$3 + (8 - x) = (3 + 8) - x = 11 - x$

75. Rewrite $6(2y)$ using the Associative Property of Addition or Multiplication.

Solution

$6(2y) = (6 \cdot 2) \cdot y = 12y$

77. Rewrite $-8(3-5)$ using the Distributive Property.

Solution

$-8(3-5) = -8 \cdot 3 + -8 \cdot -5$

79. Rewrite $5(3x-4)$ using the Distributive Property.

Solution

$5(3x-4) = 5 \cdot 3x - 5 \cdot 4 \text{ or } 5 \cdot 3x + 5 \cdot -4$

81. Identify the property of real numbers that justifies each rewriting of the equation.

$$x - 8 = 20$$
$$(x-8) + 8 = 20 + 8$$
$$x + (-8 + 8) = 28$$
$$x + 0 = 28$$
$$x = 28$$

Solution

$x - 8 = 20$	Given
$(x-8) + 8 = 20 + 8$	Addition Property of Equality
$x + (-8 + 8) = 28$	Associative Property of Addition
$x + 0 = 28$	Additive Inverse Property
$x = 28$	Additive Identity Property

83. Prove that if $ac = bc$ and $c \neq 0$, then $a = b$.

Solution

$ac = bc, \ c \neq 0$	Given
$\dfrac{1}{c}(ac) = \dfrac{1}{c}(bc)$	Multiplication Property of Equality
$\dfrac{1}{c}(ca) = \dfrac{1}{c}(cb)$	Commutative Property of Multiplication
$\left(\dfrac{1}{c} \cdot c\right) a = \left(\dfrac{1}{c} \cdot c\right) b$	Associative Property of Multiplication
$1 \cdot a = 1 \cdot b$	Multiplicative Inverse Property
$a = b$	Multiplicative Identity Property

85. *Interpreting a Graph*   The dividend paid per share by the General Electric Company for the years 1989 through 1993 is approximated by the model

Dividend per share $= -0.93 + 0.66x$.

In this model, the dividend per share is measured in dollars and x represents the earnings per share in dollars (see figure in textbook). (Source: General Electric Company). Use the graph to approximate the increase in the dividend per share if the earnings per share increases by $0.50.

Solution

The increase in the dividend per share is .33 or approximately $\frac{1}{3}$ if the earnings per share increases by $0.50.

87. *Interpreting a Graph* The dividend paid per share by the General Electric Company for the years 1989 through 1993 is approximated by the model

Dividend per share $= -0.93 + 0.66x$.

In this model, the dividend per share is measured in dollars and x represents the earnings per share in dollars (see figure in textbook). (Source: General Electric Company). In 1989, General Electric's earnings per share was \$3.88. Use the model to approximate the dividend per share.

Solution

Dividend per share $= -0.93 + 0.66x$

1989 dividend per share $= -0.93 + 0.66(3.88) = 1.6308$ dollars

89. *Exploration* Suppose you define a new mathematical operation using the symbol $\odot$. This operation is defined as $a \odot b = 2 \cdot a + b$. Give examples to show that neither the Commutative Property nor the Associative Property is true for this operation.

Solution

(a) let $a = 1$ & $b = 2$

$a \odot b \neq b \odot a$

$1 \odot 2 \neq 2 \odot 1$

$2 \cdot 1 + 2 \neq 2 \cdot 2 + 1$

$4 \neq 5$

(b) let $a = 1, b = 2, c = 3$

$a \odot (b \odot c) \neq (a \odot b) \odot c$

$1 \odot (2 \odot 3) \neq (1 \odot 2) \odot 3$

$2 \cdot 1 + (2 \cdot 2 + 3) \neq (2 \cdot 1 + 2) \odot 3$

$2 + 7 \neq 4 \odot 3$

$9 \neq 2 \cdot 4 + 3$

$9 \neq 8 + 3$

$9 \neq 11$

Mid-Chapter Quiz for Chapter P

1. Show each real number -4.5 and -6 as a point on the real number line and place the correct inequality symbol ($<$ or $>$) between them.

Solution

$-4.5 > -6$

2. Show each real number $\frac{3}{4}$ and $\frac{3}{2}$ as a point on the real number line and place the correct inequality symbol ($<$ or $>$) between them.

Solution

$\frac{3}{4} < \frac{3}{2}$

3. Evaluate $|-3.2|$.

Solution

$|-3.2| = 3.2$

4. Evaluate $-|5.75|$.

Solution

$-|5.75| = -5.75$

5. Find the distance between -15 and 7.

Solution

$|-15 - 7| = |-22| = 22$

6. Find the distance between -10.5 and -6.75.

Solution

$|(-10.5) - (-6.75)| = |-10.5 + 6.75|$

$= |-3.75| = 3.75$

7. Evaluate $32 + (-18)$.

Solution

$32 + (-18) = 14$

8. Evaluate $-10 - 12$.

Solution

$-10 - 12 = (-10) + (-12) = -22$

9. Evaluate $\frac{3}{4} + \frac{7}{4}$. Write fractions in reduced form.

Solution

$$\frac{3}{4} + \frac{7}{4} = \frac{3+7}{4} = \frac{10}{4} = \frac{5}{2}$$

10. Evaluate $2 + \frac{2}{3} - \frac{1}{6}$. Write fractions in reduced form.

Solution

$$2 + \frac{2}{3} - \frac{1}{6} = \frac{12}{6} + \frac{4}{6} - \frac{1}{6}$$
$$= \frac{12 + 4 - 1}{6} = \frac{15}{6} = \frac{5}{2}$$

11. Evaluate $\left|4 - 1\frac{3}{8}\right|$. Write fractions in reduced form.

Solution

$$\left|4 - 1\tfrac{3}{8}\right| = \left|\tfrac{32}{8} - \tfrac{11}{8}\right| = \left|\tfrac{21}{8}\right| = \tfrac{21}{8}$$

12. Evaluate $\left(-\frac{4}{5}\right)\left(\frac{15}{32}\right)$. Write fractions in reduced form.

Solution

$$\left(-\frac{4}{5}\right)\left(\frac{15}{32}\right) = \frac{(-4)(15)}{(5)(32)}$$
$$= \frac{\overset{1}{(-4)}\,\overset{3}{(15)}}{\underset{1}{(5)}\,\underset{8}{(32)}} = -\frac{3}{8}$$

13. Evaluate $\frac{7}{12} \div \frac{5}{6}$. Write fractions in reduced form.

Solution

$$\frac{7}{12} \div \frac{5}{6} = \frac{7}{12} \cdot \frac{6}{5}$$
$$= \frac{(7)(6)}{(12)(5)} = \frac{(7)\overset{1}{(6)}}{\underset{2}{(12)}(5)} = \frac{7}{10}$$

14. Evaluate $\left(-\frac{3}{2}\right)^3$. Write fractions in reduced form.

Solution

$$\left(-\frac{3}{2}\right)^3 = \left(-\frac{3}{2}\right)\left(-\frac{3}{2}\right)\left(-\frac{3}{2}\right)$$
$$= \frac{(-3)(-3)(-3)}{(2)(2)(2)} = -\frac{27}{8}$$

15. Evaluate the following expression and round your answer to two decimal places. (A calculator may be useful.)

$$\frac{-130.45}{-45.2}$$

Solution

$$\frac{-130.45}{-45.2} \approx 2.89$$

16. Evaluate $(3.8)^4$ and round your answer to two decimal places. (A calculator may be useful.)

Solution

$(3.8)^4 = (3.8)(3.8)(3.8)(3.8) \approx 208.51$

17. Name the property of real numbers that justifies the following statements.

(a) $8(u - 5) = 8 \cdot u - 8 \cdot 5$

(b) $10x - 10x = 0$

Solution

(a) Distributive Property

(b) Additive Inverse Property

18. Name the property of real numbers that justifies the following statements.

(a) $(7 + y) - z = 7 + (y - z)$

(b) $2x \cdot 1 = 2x$

Solution

(a) Associative Property of Addition

(b) Multiplicative Identity Property

19. You deposit $30 in a retirement account twice each month. How much will you deposit in the account in 5 years?

Solution

$(\$30)(2)(12)(5) = \$3600$

20. Determine the unknown fractional part of the circle graph in the textbook. Explain how you were able to make this determination.

Solution

$$1 = \tfrac{1}{3} + \tfrac{1}{4} + \tfrac{1}{8} + x$$

$$1 - \tfrac{1}{3} - \tfrac{1}{4} - \tfrac{1}{8} = x$$

$$\tfrac{24}{24} - \tfrac{8}{24} - \tfrac{6}{24} - \tfrac{3}{24} = x$$

$$\tfrac{7}{24} = x$$

The sum of the parts of a circle is equal to 1.

P.4 Algebraic Expressions

7. Identify the terms of the expression $10x + 5$.

Solution

$10x, 5$

9. Identify the terms of the expression $\dfrac{3}{t^2} - \dfrac{4}{t} + 6$.

Solution

$\dfrac{3}{t^2}, -\dfrac{4}{t}, 6$

11. Identify the coefficient of $5y^3$.

Solution

The coefficient of $5y^3$ is 5.

13. Identify the rule of algebra that is illustrated by $4 - 3x = 3x + 4$.

Solution

Commutative Property of Addition

15. Identify the rule of algebra that is illustrated by $-5(2x) = (-5 \cdot 2)x$.

Solution

Associative Property of Multiplication

17. Identify the rule of algebra that is illustrated by $(x + 5) \cdot \dfrac{1}{x + 5} = 1$.

Solution

Multiplicative Inverse Property

19. Identify the rule of algebra that is illustrated by $5(y^3 + 3) = 5y^3 + 5 \cdot 3$.

Solution

Distributive Property

21. Identify the rule of algebra that is illustrated by $(16t^4) \cdot 1 = 16t^4$.

Solution

Multiplicative Identity Property

23. Write $(-5x)(-5x)(-5x)(-5x)$ in exponential notation.

Solution

$(-5x)(-5x)(-5x)(-5x) = (-5x)^4$

25. Simplify $x^5 \cdot x^7$.

Solution

$x^5 \cdot x^7 = x^{5+7} = x^{12}$

27. Simplify $(2xy)(3x^2y^3)$.

Solution

$$(2xy)(3x^2y^3) = (2 \cdot 3) \cdot (x \cdot x^2) \cdot (y \cdot y^3)$$
$$= 6 \cdot (x^{1+2}) \cdot (y^{1+3})$$
$$= 6x^3y^4$$

29. Simplify $3x - 2y + 5x + 20y$ by combining like terms.

Solution

$$3x - 2y + 5x + 20y = (3x + 5x) + (-2y + 20y)$$
$$= 8x + 18y$$

31. Simplify $2uv + 5u^2v^2 - uv - (uv)^2$ by combining like terms.

Solution

$$2uv + 5u^2v^2 - uv - (uv)^2 = (2uv - uv) + (5u^2v^2 - u^2v^2)$$
$$= uv + 4u^2v^2$$

33. Use the Distributive Property to rewrite $4(2x^2 + x - 3)$.

Solution

$4(2x^2 + x - 3) = 8x^2 + 4x - 12$

35. Use the Distributive Property to rewrite $4x - 8$.

Solution

$4x - 8 = 4(x - 2)$

37. Simplify $10(x - 3) + 2x - 5$.

Solution

$$10(x - 3) + 2x - 5 = 10x - 30 + 2x - 5$$
$$= (10x + 2x) + (-30 - 5)$$
$$= 12x - 35$$

39. Simplify $2[3(b - 5) - (b^2 + b + 3)]$.

Solution

$$2[3(b - 5) - (b^2 + b + 3)] = 2[3b - 15 - b^2 - b - 3]$$
$$= 6b - 30 - 2b^2 - 2b - 6$$
$$= (-2b^2) + (6b - 2b) + (-30 - 6)$$
$$= -2b^2 + 4b - 36$$

41. Evaluate $-x^2 + 2x + 5$ when $x = -3$.

Solution

$$-(-3)^2 + 2(-3) + 5 = -9 - 6 + 5$$
$$= -15 + 5$$
$$= -10$$

43. Evaluate the expression $5 - 3x$ for the values (a) $x = \frac{2}{3}$, and (b) $x = 5$. If it is not possible, state the reason.

Solution

(a) Substitute: $5 - 3\left(\frac{2}{3}\right)$

    Value of expression: 3

(b) Substitute: $5 - 3(5)$

    Value of expression: -10

45. Evaluate the following expression for the values (a) $x = 0$, $y = 10$, and (b) $x = 4$, $y = 4$. If it is not possible, state the reason.

$$\frac{x}{x - y}$$

Solution

(a) Substitute: $\dfrac{0}{0 - 10}$

Value of expression: 0

(b) Substitute: $\dfrac{4}{4 - 4}$

Value of expression: Undefined

47. *Geometry* Find an expression for the area of the figure. Then evaluate the expression when $b = 15$.

Solution

$A = \frac{1}{2}b(b - 3)$

$A = \frac{1}{2}(15)(15 - 3)$

$\quad = \frac{1}{2}(15)(12)$

$\quad = 90$

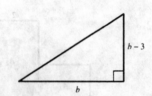

49. Identify the terms of the expression $-3y^2 + 2y - 8$.

Solution

$-3y^2, 2y, -8$

51. Identify the terms of the expression $x^2 - 2.5x - \dfrac{1}{x}$.

Solution

$x^2, -2.5x, -\dfrac{1}{x}$

53. Identify the coefficient of $-\frac{3}{4}t^2$.

Solution

The coefficient of $-\frac{3}{4}t^2$ is $-\frac{3}{4}$.

55. Identify the rule of algebra that is illustrated by $3[2(x + 4)] = (3 \cdot 2)(x + 4)$.

Solution

Associative Property of Multiplication

57. Identify the rule of algebra that is illustrated by

$$(z + 10) \cdot \frac{1}{z + 10} = 1.$$

Solution

Multiplicative Inverse Property

59. Identify the rule of algebra that is illustrated by $(x + 8) - 4 = x + (8 - 4)$.

Solution

Associative Property of Addition

61. Use the Distributive Property to rewrite $5(x + 6) = $ _____ .

Solution

$5(x + 6) = 5x + 30$

63. Use the Distributive Property to rewrite $6x + 6 = $ _____ .

Solution

$6x + 6 = 6(x + 1)$

65. Use the Commutative Property of Multiplication to rewrite $6(xy) = $ _____ .

Solution

$6(xy) = (xy)6$

67. Use the Additive Identity Property to rewrite $3 + 0 = $ _____ .

Solution

$3 + 0 = 3$

69. Use the Additive Inverse Property to rewrite $4 + (-4) =$ _____ .

Solution

$4 + (-4) = 0$

71. Use the Associative Property of Addition to rewrite $(3 + 6) + (-9) =$ _____ .

Solution

$(3 + 6) + (-9) = 3 + (6 - 9)$

73. Rewrite $(-2x)^3$ as a repeated multiplication problem.

Solution

$(-2x)^3 = (-2x)(-2x)(-2x)$

75. *Geometry* Write an expression for the area of the region in the figure. Then simplify the expression.

Solution

Area = Area of Region 1 + Area of Region 2

$= (2 \cdot x) + (2 \cdot 2x)$

$= 2x + 4x$

$= 6x$

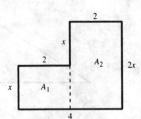

77. Simplify the expression $(-4x)^2$.

Solution

$(-4x)^2 = (-4)^2 \cdot x^2 = 16x^2$

79. Simplify the expression $3^3 y^4 \cdot y^2$.

Solution

$3^3 y^4 \cdot y^2 = 3^3 y^{4+2} = 27y^6$

81. Simplify the expression $(3uv)^2(-6u^3v)$.

Solution

$(3uv)^2(-6u^3v) = (3^2 u^2 v^2)(-6u^3 v)$

$\qquad = (3^2 \cdot -6) \cdot (u^2 \cdot u^3) \cdot (v^2 \cdot v)$

$\qquad = (9 \cdot -6) \cdot (u^{2+3}) \cdot (v^{2+1})$

$\qquad = -54u^5 v^3$

83. Simplify $3x + 4x$ by combining like terms.

Solution

$3x + 4x = (3 + 4)x = 7x$

85. Simplify $9y - 5y + 4y$ by combining like terms.

Solution

$9y - 5y + 4y = (9 - 5 + 4)y = 8y$

87. Simplify the following expression by combining like terms.

$$3\left(\frac{1}{x}\right) - \frac{1}{x} + 8$$

Solution

$$3\left(\frac{1}{x}\right) - \frac{1}{x} + 8 = \left[3\left(\frac{1}{x}\right) - \left(\frac{1}{x}\right)\right] + 8$$

$$= 2\left(\frac{1}{x}\right) + 8$$

89. Use the Distributive Property to rewrite $-3(6y^2 - y - 2)$.

Solution

$-3(6y^2 - y - 2) = -18y^2 + 3y + 6$

91. Use the Distributive Property to rewrite $ab + 2a$.

Solution

$ab + 2a = a(b + 2)$

93. Simplify $-3(y^2 + 3y - 1) + 2(y - 5)$.

Solution

$$
\begin{aligned}
-3(y^2 + 3y - 1) + 2(y - 5) &= -3y^2 - 9y + 3 + 2y - 10 \\
&= (-3y^2) + (-9y + 2y) + (3 - 10) \\
&= -3y^2 - 7y - 7
\end{aligned}
$$

95. Simplify $2x(5x^2) - (x^3 + 5)$.

Solution

$$
\begin{aligned}
2x(5x^2) - (x^3 + 5) &= 10x^3 - x^3 - 5 \\
&= (10x^3 - x^3) - 5 \\
&= 9x^3 - 5
\end{aligned}
$$

97. Simplify $y^2(y + 1) + y(y^2 + 1)$.

Solution

$$
\begin{aligned}
y^2(y + 1) + y(y^2 + 1) &= y^3 + y^2 + y^3 + y \\
&= (y^3 + y^3) + (y^2) + (y) \\
&= 2y^3 + y^2 + y
\end{aligned}
$$

99. Evaluate the expression $10 - |x|$ for the values (a) $x = 3$, and (b) $x = -3$. If it is not possible, state the reason.

Solution

(a) Substitute: $10 - |3|$

Value of expression: 7

(b) Substitute: $10 - |-3|$

Value of expression: 7

101. Evaluate the following expression for the values (a) $x = 0$, and (b) $x = 3$. If it is not possible, state the reason.

$$\frac{x}{x^2 + 1}$$

Solution

(a) Substitute: $\dfrac{0}{0^2 + 1}$

Value of expression: 0

(b) Substitute: $\dfrac{3}{3^2 + 1}$

Value of expression: $\dfrac{3}{10}$

103. Evaluate the expression $3x + 2y$ for the values (a) $x = 1$, $y = 5$, and (b) $x = -6$, $y = -9$. If it is not possible, state the reason.

Solution

(a) Substitute: $3(1) + 2(5)$

Value of expression: 13

(b) Substitute: $3(-6) + 2(-9)$

Value of expression: -36

105. Evaluate the expression rt for the values (a) $r = 40$, $t = 5\frac{1}{4}$, and (b) $r = 35$, $t = 4$. If it is not possible, state the reason.

Solution

(a) Substitute: $40\left(5\frac{1}{4}\right)$

Value of expression: 210

(b) Substitute: $35(4)$

Value of expression: 140

107. *Creating a Table*

(a) Complete the table shown in the textbook by evaluating $2x - 5$.

(b) How much does the value of the expression increase for each one-unit increase in x?

(c) Use the results of (a) and (b) to guess the increase in $\frac{3}{4}x + 5$ for each one-unit increase in x.

Solution

(a)

x	-1	0	1	2	3	4
$2x - 5$	-7	-5	-3	-1	1	3

(b) For each one unit increase in x there is a two unit increase in $2x - 5$.

(c) For each one unit increase in x there is a $\frac{3}{4}$ unit increase in $\frac{3}{4}x + 5$.

109. *Using a Model*   Use the following model, which approximates the annual sales (in millions of dollars) of exercise equipment in the United States from 1983 to 1992 (see figure in textbook).

$$\text{Sales} = 10.38t^2 - 14.97t + 935.42$$

In this formula, $t = 0$ represents 1980. (Source: National Sporting Goods Association). Graphically approximate the sales of exercise equipment in 1988. Then use the model to confirm your estimate algebraically.

Solution

Graphically, the sales in 1988 is approximately \$1500 million.

Let $t = 8$.

$$\text{Sales} = 10.38(8)^2 - 14.97(8) + 935.42$$

$$= 10.38(64) - 14.97(8) + 935.42$$

$$= \$1479.98 \text{ million}$$

111. *Geometry* The area of the trapezoid shown in the figure in the textbook, with parallel bases of lengths b_1 and b_2 and height h, is given by $A = \frac{1}{2}(b_1 + b_2)h$. Use the Distributive Property to show that the area can also be expressed as $A = b_1 h + \frac{1}{2}(b_2 - b_1)h$.

Solution

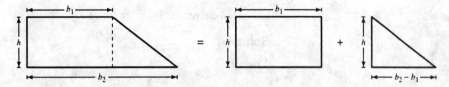

Area of Trapezoid = Area of Rectangle + Area of Triangle

$$= b_1 h + \frac{1}{2}(b_2 - b_1)h$$

$$= h\left[b_1 + \frac{1}{2}(b_2 - b_1)\right]$$

$$= h\left[b_1 + \frac{1}{2}b_2 - \frac{1}{2}b_1\right]$$

$$= h\left[\frac{1}{2}b_1 + \frac{1}{2}b_2\right]$$

$$= \frac{1}{2}h(b_1 + b_2)$$

$$= \frac{h}{2}(b_1 + b_2)$$

113. *Geometry* The roof shown in the figure in the textbook is made up of two trapezoids and two triangles. Find the total area of the roof.

Solution

Total area = 2 [Area Trapezoid] + 2 [Area Triangle]

$$= 2\left[\tfrac{12}{2}(40 + 60)\right] + 2\left[\tfrac{1}{2} \cdot 12 \cdot 20\right]$$

$$= 2[6(100)] + 2[6 \cdot 20]$$

$$= 1200 + 240$$

$$= 1440$$

P.5 | Constructing Algebraic Expressions

7. Translate the following statement into an algebraic expression.

The sum of 8 and a number n

Solution

$8 + n$

9. Translate the following statement into an algebraic expression.

Fifteen decreased by three times a number n

Solution

$15 - 3n$

11. Translate the following statement into an algebraic expression.

The quotient of a number x and 6

Solution

$\dfrac{x}{6}$

13. Translate the following statement into an algebraic expression.

The sum of 3 and four times x, all divided by 8

Solution

$\dfrac{3 + 4x}{8}$

15. Write a verbal description of $x - 5$ without using the variable.

Solution

A number is decreased by 5.

17. Write a verbal description of $z/2$ without using the variable.

Solution

The quotient of a number and two

19. Write a verbal description of $\dfrac{x - 2}{3}$ without using the variable.

Solution

A number is decreased by 2 and the difference is divided by 3.

21. Write an algebraic expression that represents the specified quantity in the following verbal statement and simplify if possible.

The amount of money (in dollars) represented by n quarters

Solution

n = number of quarters

$0.25n$ = amount of money (in dollars)

23. Write an algebraic expression that represents the specified quantity in the following verbal statement and simplify if possible.

The distance traveled in t hours at an average speed of 55 miles per hour

Solution

$55t$ = distance traveled in t hours at 55 mph

25. Write an algebraic expression that represents the specified quantity in the following verbal statement and simplify if possible.

The amount of antifreeze in a cooling system containing y gallons of coolant that is 45% antifreeze

Solution

$0.45y$ = amount of antifreeze in y gallons of 45% antifreeze

27. *Geometry* Write an expression for the area of the region shown in the figure in the textbook.

Solution

$a = s^2$

29. Write an expression for the sum of three consecutive integers, the first of which is n.

Solution

$n + (n + 1) + (n + 2)$ = the sum of three consecutive integers.

31. Translate the following statement into an algebraic expression.

Five more than a number n

Solution

$5 + n$

33. Translate the following statement into an algebraic expression.

Six less than a number n

Solution

$n - 6$

35. Translate the following statement into an algebraic expression.

One-third of a number n

Solution

$\frac{1}{3}n$

37. Translate the following statement into an algebraic expression.

Thirty percent of the list price L

Solution

$0.30L$

39. Translate the following statement into an algebraic expression.

The sum of x and 5 is divided by 10.

Solution

$\frac{x + 5}{10}$

41. Translate the following statement into an algebraic expression.

The product of three and the square of a number x, all decreased by 4

Solution

$3x^2 - 4$

43. Translate the following statement into an algebraic expression.

The absolute value of the difference between a number n and 5

Solution

$|n - 5|$

45. Write a verbal description of $t - 2$ without using the variable.

Solution

A number decreased by 2

47. Write a verbal description of $3x + 2$ without using the variable.

Solution

The sum of three times a number and two

49. Write a verbal description of $8(x - 5)$ without using the variable.

Solution

Eight times the difference of a number and five

51. Write a verbal description of $\frac{y}{8}$ without using the variable.

Solution

The ratio of y to 8

53. Write a verbal description of $\frac{x + 10}{3}$ without using the variable.

Solution

The sum of a number and ten, all divided by three

55. Write a verbal description of $x(x + 7)$ without using the variable.

Solution

Some number times the sum of the same number and seven

57. Write an algebraic expression that represents the specified quantity in the following verbal statement and simplify if possible.

The amount of money (in dollars) represented by x nickels

Solution

$x =$ number of nickels

$0.05x =$ amount of money (in dollars)

59. Write an algebraic expression that represents the specified quantity in the following verbal statement and simplify if possible.

The amount of money (in cents) represented by m nickels and n dimes

Solution

$$m = \text{number of nickels}$$

$$n = \text{number of dimes}$$

$5m + 10n = \text{amount of money (in cents)}$

61. Write an algebraic expression that represents the specified quantity in the following verbal statement and simplify if possible.

The time to travel 100 miles at an average speed of r miles per hour

Solution

$\dfrac{100}{r} = \text{time to travel 100 miles at } r \text{ mph}$

63. Write an algebraic expression that represents the specified quantity in the following verbal statement and simplify if possible.

The amount of wage tax due for a taxable income of I dollars that is taxed at the rate of 1.25%

Solution

$0.0125I = \text{tax on } I \text{ dollars taxed at } 1.25\%$

65. Write an algebraic expression that represents the specified quantity in the following verbal statement and simplify if possible.

The sale price of a coat that has a list price of L dollars if the sale is a "20% off" sale

Solution

$0.80L = \text{sale price of a coat at price } L \text{ at 20% off}$

67. Write an algebraic expression that represents the specified quantity in the following verbal statement and simplify if possible.

The total hourly wage for an employee when the base pay is $8.25 per hour plus 60 cents for each of q units produced per hour

Solution

$8.25 + 0.60q = \text{total wage with base pay \$8.25 and 60 cents for each of } q \text{ units produced}$

69. Write an algebraic expression that represents the specified quantity in the following verbal statement and simplify if possible.

The auto repair bill if the cost for parts was $69.50 and there were t hours of labor at $32 per hour

Solution

$69.50 + 32t = \text{total bill of parts and labor}$

71. Write an algebraic expression that represents the specified quantity in the following verbal statement and simplify if possible.

The length of a rectangle if the length is 6 feet more than the width

Solution

$\text{Length} = w + 6$

73. *Geometry* Write expressions for the perimeter and area of the region shown in the figure in the textbook. Then simplify the expressions.

Solution

$\text{Perimeter} = 2(2w) + 2(w) = 4w + 2w = 6w$

$\text{Area} = 2w \cdot w = 2w^2$

75. *Geometry* Write expressions for the perimeter and area of the region shown in the figure in the textbook. Then simplify the expressions.

Solution

$\text{Perimeter} = 3 + 2x + 6 + x + 3 + x = 4x + 12$

$\text{Area} = (x \cdot 3) + (3 \cdot 2x) = 3x + 6x = 9x$

77. Write an expression for the sum of a number n and three times the number.

Solution

$n + 3n + \text{sum of a number } n \text{ and three times the number}$

79. Write an expression for the sum of two consecutive odd integers, the first of which is $2n + 1$.

Solution

$(2n + 1) + (2n + 3) = $ sum of two consecutive odd integers

83. *Finding a Pattern*   Complete the table shown in the textbook. The third row contains the differences between consecutive entries of the second row. Describe the pattern of the third row.

Solution

n	0	1	2	3	4	5
$5n - 3$	-3	2	7	12	17	22
Differences		5	5	5	5	5

81. *Geometry*   Write an expression that represents the area of the top of the billiard table shown in the figure in the textbook. What is the unit of measure for the area?

Solution

$l \cdot (l - 6) = $ area of billiard table

The unit of measure for the area is meters^2 or square meters.

85. *Finding a Pattern*   Using the results of Exercises 83 and 84, guess the third-row difference that would result in a similar table if the algebraic expression were $an + b$.

Solution

The third row difference for the algebraic expression $an + b$ would be a.

Review Exercises for Chapter P

1. Plot the real numbers $-\frac{1}{8}$ and 3 on a number line and place the correct inequality symbol ($<$ or $>$) between them.

Solution

$-\frac{1}{8} < 3$

5. Evaluate $|-7.2|$.

Solution

$|-7.2| = 7.2$

9. Evaluate $15 + (-4)$. If it is not possible, state the reason. Write all fractions in reduced form.

Solution

$15 + (-4) = 11$

3. Plot the real numbers $-\frac{8}{5}$ and $-\frac{2}{5}$ on a number line and place the correct inequality symbol ($<$ or $>$) between them.

Solution

$-\frac{8}{5} < -\frac{2}{5}$

7. Evaluate $-|-7.2|$.

Solution

$-|-7.2| = -7.2$

11. Evaluate $340 - 115 + 5$. If it is not possible, state the reason. Write all fractions in reduced form.

Solution

$340 - 115 + 5 = 230$

13. Evaluate $|-96| - |134|$. If it is not possible, state the reason. Write all fractions in reduced form.

Solution

$|-96| - |134| = -38$

15. Evaluate $120(-5)(7)$. If it is not possible, state the reason. Write all fractions in reduced form.

Solution

$120(-5)(7) = -4200$

17. Evaluate $\frac{-56}{-4}$. If it is not possible, state the reason. Write all fractions in reduced form.

Solution

$\dfrac{-56}{-4} = 14$

19. Evaluate the following expression. If it is not possible, state the reason. Write all fractions in reduced form.

$$\frac{45 - |-45|}{2}$$

Solution

$\dfrac{45 - |-45|}{2} = 0$

21. Evaluate $\frac{4}{21} + \frac{7}{21}$. If it is not possible, state the reason. Write all fractions in reduced form.

Solution

$\frac{4}{21} + \frac{7}{21} = \frac{11}{21}$

23. Evaluate $-\frac{5}{6} + 1$. If it is not possible, state the reason. Write all fractions in reduced form.

Solution

$-\frac{5}{6} + 1 = \frac{1}{6}$

25. Evaluate $8\frac{3}{4} - 6\frac{5}{8}$. If it is not possible, state the reason. Write all fractions in reduced form.

Solution

$8\frac{3}{4} - 6\frac{5}{8} = \frac{35}{4} - \frac{53}{8} = \frac{70}{8} - \frac{53}{8} = \frac{17}{8}$

27. Evaluate $\frac{3}{8} \cdot \frac{-2}{15}$. If it is not possible, state the reason. Write all fractions in reduced form.

Solution

$$\frac{3}{8} \cdot \frac{-2}{15} = \frac{\overset{1}{\cancel{3}}}{8} \cdot \frac{\overset{1}{\cancel{-2}}}{\underset{5}{\cancel{15}}} = \frac{-1}{20}$$

29. Evaluate $-\frac{7}{15} \div -\frac{7}{30}$. If it is not possible, state the reason. Write all fractions in reduced form.

Solution

$-\dfrac{7}{15} \div -\dfrac{7}{30} = -\dfrac{7}{15} \cdot \dfrac{30}{-7} = 2$

31. Evaluate the following expression. If it is not possible, state the reason. Write all fractions in reduced form.

$$\frac{(3/4) - (1/2)}{5/8}$$

Solution

$\dfrac{(3/4) - (1/2)}{5/8} = \dfrac{1/4}{5/8} = \dfrac{1}{4} \cdot \dfrac{8}{5} = \dfrac{2}{5}$

33. Evaluate the following expression. If it is not possible, state the reason. Write all fractions in reduced form.

$$\frac{6.25}{1.25}$$

Solution

$\dfrac{6.25}{1.25} = 5$

35. Evaluate $(-6)^3$. If it is not possible, state the reason. Write all fractions in reduced form.

Solution

$(-6)^3 = -216$

37. Evaluate the following expression If it is not possible, state the reason. Write all fractions in reduced form.

$$\frac{3}{6^2}$$

Solution

$$\frac{3}{6^2} = \frac{3}{36} = \frac{1}{12}$$

39. Evaluate $120 - (5^2 \cdot 4)$. If it is not possible, state the reason. Write all fractions in reduced form.

Solution

$$120 - (5^2 \cdot 4) = 120 - (25 \cdot 4)$$
$$= 120 - 100$$
$$= 20$$

41. Name the property of real numbers that justifies $13 - 13 = 0$.

Solution

Additive Inverse Property

43. Name the property of real numbers that justifies $7(9 + 3) = 7 \cdot 9 + 7 \cdot 3$.

Solution

Distributive Property

45. Name the property of real numbers that justifies $5 + (4 - y) = (5 + 4) - y$.

Solution

Associative Property of Addition

47. Identify the rule of algebra that is illustrated by $(u - v)(2) = 2(u - v)$.

Solution

Commutative Property of Multiplication

49. Identify the rule of algebra that is illustrated by

$$ab \cdot \frac{1}{ab} = 1.$$

Solution

Multiplicative Inverse Property

51. Expand $\frac{2}{3}(6s - 12t)$ by using the Distributive Property.

Solution

$$\frac{2}{3}(6s - 12t) = 4s - 8t$$

53. Expand $-y(3y - 10)$ by using the Distributive Property.

Solution

$$-y(3y - 10) = -3y^2 + 10y$$

55. Expand $-(-u + 3v)$ by using the Distributive Property.

Solution

$$-(-u + 3v) = u - 3v$$

57. Simplify $5(x - 4) + 10$.

Solution

$$5(x - 4) + 10 = 5x - 20 + 10$$
$$= 5x - 10$$

59. Simplify $3x - (y - 2x)$.

Solution

$$3x - (y - 2x) = 3x - y + 2x$$
$$= 5x - y$$

61. Simplify $-2(1 - 20u) + 2(1 - 10u)$.

Solution

$$-2(1 - 20u) + 2(1 - 10u) = -2 + 40u + 2 - 20u$$
$$= (-2 + 2) + (40u - 20u)$$
$$= 0 + 20u$$
$$= 20u$$

63. Simplify $3[b + 5(b - a)]$.

Solution

$$3[b + 5(b - a)] = 3[b + 5b - 5a]$$
$$= 3b + 15b - 15a$$
$$= 18b - 15a$$

65. Simplify $x^2 \cdot x^3 \cdot x$.

Solution

$$x^2 \cdot x^3 \cdot x = x^{2+3+1} = x^6$$

67. Simplify $(xy)(-3x^2 y^3)$.

Solution

$$(xy)(-3x^2 y^3) = -3 \cdot x^{1+2} \cdot y^{1+3}$$
$$= -3x^3 y^4$$

69. Simplify $(-2a^2)^3 (8a)$.

Solution

$$(-2a^2)^3 (8a) = -8a^6 \cdot 8a$$
$$= (-8 \cdot 8) \cdot (a^{6+1})$$
$$= -64a^7$$

71. Perform the indicated operations and simplify
$6 \cdot 10^3 + 9 \cdot 10^2 + 1 \cdot 10^1$.

Solution

$$6 \cdot 10^3 + 9 \cdot 10^2 + 1 \cdot 10^1 = 6910$$

73. Evaluate $x^2 - 2x - 3$ for the values (a) $x = 3$, and
(b) $x = 0$.

Solution

(a) $x = 3$
Substitute: $3^2 - 2(3) - 3$
Value of expression: 0

(b) $x = 0$
Substitute: $0^2 - 2(0) - 3$
Value of expression: -3

75. Evaluate $3x^2 - y(x + 1)$ for the values (a) $x = 2$, $y = -1$, and (b) $x = -1$, $y = 20$.

Solution

(a) $x = 2$, $y = -1$
Substitute: $3(2)^2 - (-1)(2 + 1)$
Value of expression: 15

(b) $x = -1$, $y = 20$
Substitute: $3(-1)^2 - 20(-1 + 1)$
Value of expression: 3

77. *Geometry* Write expressions for the perimeter and area of the region shown in the figure in the textbook, then simplify.

Solution

Perimeter $= 2(l) + 2(l - 10) = 2l + 2l - 20 = 4l - 20$

Area $= l \cdot (l - 10) = l^2 - 10l$

79. *Geometry* Write expressions for the perimeter and area of the region shown in the figure in the textbook, then simplify.

Solution

Perimeter $= (3b + 2) + (2b + 3) + (4b) = 9b + 5$

Area $= \frac{1}{2} \cdot 4b \cdot 2b = 4b^2$

81. *Graphical Interpretation*   Use the figure in the textbook, which shows the expenditures (in billions of dollars) for advertising for various media in 1991. (Source: McCann-Erickson, Inc.). Determine the combined expenditures for advertising by the five media.

Solution

Combined expenditures $= 30.4 + 6.5 + 27.4 + 8.5 + 9.2 = 82$ billion dollars

83. *Graphical Interpretation*   Use the figure in the textbook, which shows the average weekly overtime for production workers in August for the years 1982, 1986, 1990, and 1994. (Source: U. S. Department of Labor). Determine the increase in average weekly overtime during the four years from 1990 to 1994.

Solution

Increase in overtime $= 4.8 - 3.8 = 1.0$

85. *Total Charge*   You purchase a product and make a down payment of \$239 plus nine monthly payments of \$45 each. What is the total amount you pay for the product?

Solution

Total paid $= \$239 + 9(\$45)$

$\qquad\quad = 239 + 405$

$\qquad\quad = \$644$

87. *Calculator Experiment*   Enter any number between 0 and 1 in a calculator. Take the square root of the number. Then take the square root of the result, and keep continuing this process. What number does the calculator display seem to be approaching?

Solution

The calculator display seems to be approaching the number 1.

89. Translate the following phrase into an algebraic expression. (Let n represent the arbitrary real number.)

Two hundred decreased by three times a number

Solution

$200 - 3n$

91. Translate the following phrase into an algebraic expression. (Let n represent the arbitrary real number.)

The sum of the square of a real number and 49

Solution

$n^2 + 49$

93. Translate the following phrase into an algebraic expression. (Let n represent the arbitrary real number.)

The absolute value of the quotient of a number and 5

Solution

$\left| \dfrac{n}{5} \right|$

95. Write a verbal description of $2y + 7$ without using the variable.

Solution

The sum of twice a number and seven

97. Write a verbal description of the following expression without using the variable.

$$\dfrac{x - 5}{4}$$

Solution

The difference of a number and five, all divided by four

99. Write an algebraic expression that represents the quantit given by the following verbal statement.

The amount of income tax on a taxable income of I dollars when the tax rate is 18%

Solution

$0.18I =$ tax on I dollars at 18%

101. Write an algebraic expression that represents the quantity given by the following verbal statement.

The area of a rectangle whose length is l inches and whose width is five units less than its length

Solution

$l \cdot (l - 5) =$ area of rectangle with length l and width $(l - 5)$

103. Write an algebraic expression that represents the quantity given by the following verbal statement.

The cost of 30 acres of land if the price per acre is p dollars

Solution

$30p =$ cost of 30 acres of land at $30 per acre

Test for Chapter P

1. Place the correct symbol ($<$ or $>$) between the numbers (a) $-\frac{5}{2}$ and $-|-3|$, and (b) $-\frac{2}{3}$ and $-\frac{3}{2}$.

Solution

(a) $-\frac{5}{2} > -|-3|$

(b) $-\frac{2}{3} > -\frac{3}{2}$

2. Find the distance between -6.2 and 5.7.

Solution

$d = |-6.2 - 5.7| = 11.9$

3. Evaluate $-2(225 - 150)$.

Solution

$-2(225 - 150) = -2(75) = -150$

4. Evaluate $\frac{2}{3} + \left(-\frac{7}{6}\right)$.

Solution

$\frac{2}{3} + \left(-\frac{7}{6}\right) = \frac{4}{6} + \left(-\frac{7}{6}\right) = -\frac{3}{6} = -\frac{1}{2}$

5. Evaluate $\left(-\frac{7}{16}\right)\left(-\frac{8}{21}\right)$.

Solution

$\left(-\frac{7}{16}\right)\left(-\frac{8}{21}\right) = \frac{1}{6}$

6. Evaluate $\frac{5}{18} \div \frac{15}{8}$.

Solution

$\frac{5}{18} \div \frac{15}{8} = \frac{5}{18} \cdot \frac{8}{15} = \frac{4}{27}$

7. Evaluate $\left(-\frac{3}{5}\right)^3$.

Solution

$\left(-\frac{3}{5}\right)^3 = \frac{-27}{125}$

8. Evaluate $\frac{4^2 - 6}{5} + 13$.

Solution

$$\frac{4^2 - 6}{5} + 13 = \frac{16 - 6}{5} + 13$$

$$= \frac{10}{5} + 13$$

$$= 2 + 13$$

$$= 15$$

9. Name the property of real numbers demonstrated by:

(a) $(-3 \cdot 5) \cdot 6 = -3(5 \cdot 6)$

(b) $3y \cdot \frac{1}{3y} = 1$

Solution

(a) Associative Property of Multiplication

(b) Multiplicative Inverse Property

10. Rewrite the expression $5(2x - 3)$ using the Distributive Property.

Solution

$5(2x - 3) = 5(2x) - 5(3) = 10x - 15$

11. Simplify $(3x^2 y)(-xy)^2$.

Solution

$(3x^2 y)(-xy)^2 = (3x^2 y)(x^2 y^2) = 3x^4 y^3$

12. Simplify $3x^2 - 2x - 5x^2 + 7x - 1$.

Solution

$3x^2 - 2x - 5x^2 + 7x - 1 = -2x^2 + 5x - 1$

13. Simplify $a(5a - 4) - 2(2a^2 - 2a)$.

Solution

$a(5a - 4) - 2(2a^2 - 2a) = 5a^2 - 4a - 4a^2 + 4a$

$= a^2$

14. Simplify $4t - [3t - (10t + 7)]$.

Solution

$4t - [3t - (10t + 7)] = 4t - [3t - 10t - 7]$

$= 4t - [-7t - 7]$

$= 4t + 7t + 7$

$= 11t + 7$

15. Explain the meaning of "evaluating an expression." Evaluate the expression $4 - (x + 1)^2$ when (a) $x = -1$ and (b) $x = 3$.

Solution

Evaluating an expression means to substitute numerical values for each of the variables in the expression and then to simplify according to the rules for order of operations.

(a) $x = -1 \rightarrow 4 - (-1 + 1)^2$

$4 - (0)^2$

4

(b) $x = 3 \rightarrow 4 - (3 + 1)^2$

$4 - (4)^2$

$4 - 16$

-12

16. Translate the statement "The product of a number n and 5 is decreased by 8" into an algebraic expression.

Solution

$5n - 8$

17. Write algebraic expressions for the perimeter and area of the rectangle shown in the figure in the textbook. Then simplify the expressions.

Solution

Perimeter $= 2(l) + 2(0.6l) = 2l + 1.2l = 3.2l$
Area $= l(0.6l) = 0.6l^2$

18. Write an algebraic expression for the sum of two consecutive even integers, the first of which is $2n$.

Solution

$2n + (2n + 2)$

19. It is necessary to cut a 144-foot rope into nine pieces of equal length. What is the length of each piece?

Solution

Verbal model: 9 · $\boxed{\text{Length of each piece}}$ = $\boxed{\text{Total length}}$

Equation: $9 \cdot n = 144$

$n = 16$

20. A *cord* of wood is a pile 4 feet high, 4 feet wide, and 8 feet long. The volume of a rectangular solid is its length times its width times its height. Find the number of cubic feet in 5 cords of wood.

Solution

Verbal model: $\boxed{\text{Volume of 1 cord}}$ = $\boxed{\text{Length}}$ · $\boxed{\text{Width}}$ · $\boxed{\text{Height}}$

Equation: $V = 4 \cdot 4 \cdot 8$

$V = 128$ cubic feet

Verbal model: $\boxed{\text{Volume of 5 cords}}$ = 5 · $\boxed{\text{Volume of 1 cord}}$

Equation: $V = 5 \cdot 128$

$= 640$ cubic feet

CHAPTER 1
Linear Equations and Inequalities

1.1   Linear Equations

7. Decide whether (a) $x = 0$ and (b) $x = 3$ are solutions of $3x - 7 = 2$.

Solution

(a) $x = 0$

$$3(0) - 7 \overset{?}{=} 2$$

$$-7 \neq 2$$

No

(b) $x = 3$

$$3(3) - 7 \overset{?}{=} 2$$

$$9 - 7 = 2$$

$$2 = 2$$

Yes

9. Decide whether (a) $x = -4$ and (b) $x = 12$ are solutions of $\frac{1}{4}x = 3$.

Solution

(a) $x = -4$

$$\frac{1}{4}(-4) \overset{?}{=} 3$$

$$-1 \neq 3$$

No

(b) $x = 12$

$$\frac{1}{4}(12) \overset{?}{=} 3$$

$$3 = 3$$

Yes

11. Identify $3(x - 1) = 3x$ as a conditional equation, an identity, or an equation with no solution.

Solution

$3(x - 1) = 3x$	Original equation
$3x - 3 = 3x$	Distributive Property
$3x - 3 - 3x = 3x - 3x$	Subtract $3x$ from both sides.
$-3 = 0$	Simplify.

No solution since $-3 \neq 0$.

13. Identify $5(x + 3) = 2x + 3(x + 5)$ as a conditional equation, an identity, or an equation with no solution.

Solution

$5(x + 3) = 2x + 3(x + 5)$	Original equation
$5x + 15 = 2x + 3x + 15$	Distributive Property
$5x + 15 = 5x + 15$	Combine like terms.

Identity since both sides equal.

15. Determine whether $3x + 4 = 10$ is linear. If it is not, state why.

Solution

$3x + 4 = 10$ is linear since variable has exponent 1.

17. Determine whether $\frac{4}{x} - 3 = 5$ is linear. If it is not, state why.

Solution

$\frac{4}{x} - 3 = 5$ is not linear since variable has exponent -1 not 1.

19. Solve $2(x - 4) = 6$ in two ways. Then explain which way you prefer.

Solution

(1)

$2(x - 4) = 6$	Original equation
$2x - 8 = 6$	Distributive Property
$2x - 8 + 8 = 6 + 8$	Add 8 to both sides.
$2x = 14$	Combine like terms.
$\dfrac{2x}{2} = \dfrac{14}{2}$	Divide both sides by 2.
$x = 7$	Simplify.

(2)

$2(x - 4) = 6$	Original equation
$\dfrac{2(x - 4)}{2} = \dfrac{6}{2}$	Divide both sides by 2.
$x - 4 = 3$	Simplify.
$x - 4 + 4 = 3 + 4$	Add 4 to both sides.
$x = 7$	Combine like terms.

21. Solve $\frac{1}{2}(x + 3) = 7$ in two ways. Then explain which way you prefer.

Solution

(1)

$\frac{1}{2}(x + 3) = 7$	Original equation
$\frac{1}{2}x + \frac{3}{2} = 7$	Distributive Property
$\frac{1}{2}x + \frac{3}{2} - \frac{3}{2} = 7 - \frac{3}{2}$	Subtract $\frac{3}{2}$ from both sides.
$\frac{1}{2}x = \frac{11}{2}$	Combine like terms.
$2\left(\frac{1}{2}x\right) = \left(\frac{11}{2}\right)2$	Multiply both sides by 2.
$x = 11$	Simplify.

(2)

$\frac{1}{2}x(x + 3) = 7$	Original equation
$2\left[\frac{1}{2}(x + 3)\right] = [7]2$	Multiply both sides by 2.
$x + 3 = 14$	Simplify.
$x + 3 - 3 = 14 - 3$	Subtract 3 from both sides.
$x = 11$	Combine like terms.

23. Justify each step of the solution shown in the textbook.

Solution

$3x + 15 = 0$	Original equation
$3x + 15 - 15 = 0 - 15$	Subtract 15 from both sides.
$3x = -15$	Combine like terms.
$\dfrac{3x}{3} = \dfrac{-15}{3}$	Divide both sides by 3.
$x = -5$	Simplify.

25. Solve $3x = 12$ and check the result. (If it is not possible, state the reason.)

Solution

$3x = 12$

$\dfrac{3x}{3} = \dfrac{12}{3}$

$x = 4$

Check $3(4) \overset{?}{=} 12$

$12 = 12$

27. Solve $7 - 8x = 13x$ and check the result. (If it is not possible, state the reason.)

Solution

$7 - 8x = 13x$

$7 - 8x + 8x = 13x + 8x$

$7 = 21x$

$\dfrac{7}{21} = \dfrac{21x}{21}$

$\dfrac{1}{3} = x$

Check $7 - 8\left(\dfrac{1}{3}\right) \overset{?}{=} 13\left(\dfrac{1}{3}\right)$

$7 - \dfrac{8}{3} \overset{?}{=} \dfrac{13}{3}$

$\dfrac{21}{3} - \dfrac{8}{3} \overset{?}{=} \dfrac{13}{3}$

$\dfrac{13}{3} = \dfrac{13}{3}$

29. Solve $-8t + 7 = -8t$ and check the result. (If it is not possible, state the reason.)

Solution

$$-8t + 7 = -8t$$

$$-8t + 7 + 8t = -8t + 8t$$

$$7 = 0$$

No solution

31. Solve $8(x - 8) = 24$ and check the result. (If it is not possible, state the reason.)

Solution

$$8(x - 8) = 24 \qquad \textbf{Check} \quad 8(11 - 8) \overset{?}{=} 24$$

$$8x - 64 = 24 \qquad\qquad\qquad 8(3) \overset{?}{=} 24$$

$$8x - 64 + 64 = 24 + 64 \qquad\qquad 24 = 24$$

$$8x = 88$$

$$\frac{8x}{8} = \frac{88}{8}$$

$$x = 11$$

33. Solve $12y = 6(y + 1)$ and check the result. (If it is not possible, state the reason.)

Solution

$$12y = 6(y + 1) \qquad \textbf{Check} \quad 12(1) \overset{?}{=} 6(1 + 1)$$

$$12y = 6y + 6 \qquad\qquad\qquad 12 \overset{?}{=} 6(2)$$

$$12y - 6y = 6y + 6 - 6y \qquad\qquad 12 = 12$$

$$6y = 6$$

$$\frac{6y}{6} = \frac{6}{6}$$

$$y = 1$$

35. Solve $t - \frac{2}{5} = \frac{3}{2}$ and check the result. (If it is not possible, state the reason.)

Solution

$$t - \frac{2}{5} = \frac{3}{2} \qquad \textbf{Check} \quad \frac{19}{10} - \frac{2}{5} \overset{?}{=} \frac{3}{2}$$

$$10\left(t - \frac{2}{5}\right) = \left(\frac{3}{2}\right)10 \qquad\qquad \frac{19}{10} - \frac{4}{10} \overset{?}{=} \frac{15}{10}$$

$$10t - 4 = 15 \qquad\qquad\qquad \frac{15}{10} = \frac{15}{10}$$

$$10t - 4 + 4 = 15 + 4$$

$$10t = 19$$

$$\frac{10t}{10} = \frac{19}{10}$$

$$t = \frac{19}{10}$$

37. Solve $\frac{t}{5} - \frac{t}{2} = 1$ and check the result. (If it is not possible, state the reason.)

Solution

$$\frac{t}{5} - \frac{t}{2} = 1 \qquad \textbf{Check} \quad \frac{\frac{10}{-3}}{5} - \frac{\frac{10}{-3}}{2} \overset{?}{=} 1$$

$$10\left(\frac{t}{5} - \frac{t}{2}\right) = (1)10 \qquad\qquad \frac{10}{-15} + \frac{10}{6} \overset{?}{=} 1$$

$$2t - 5t = 10 \qquad\qquad\qquad -\frac{2}{3} + \frac{5}{3} \overset{?}{=} 1$$

$$-3t = 10 \qquad\qquad\qquad\qquad \frac{3}{3} \overset{?}{=} 1$$

$$\frac{-3t}{-3} = \frac{10}{-3} \qquad\qquad\qquad\qquad 1 = 1$$

$$t = \frac{10}{-3}$$

39. Solve $1.2(x - 3) = 10.8$ and check the result. (If it is not possible, state the reason.)

Solution

$$1.2(x - 3) = 10.8$$

$$1.2x - 3.6 = 10.8$$

$$10(1.2x - 3.6) = (10.8)10$$

$$12x - 36 = 108$$

$$12x - 36 + 36 = 108 + 36$$

$$12x = 144$$

$$\frac{12x}{12} = \frac{144}{12}$$

$$x = 12$$

Check $1.2(12 - 3) \overset{?}{=} 10.8$

$$1.2(9) \overset{?}{=} 10.8$$

$$10.8 = 10.8$$

41. Solve $1.234x + 3 = 7.805$ with the aid of a calculator. Round the solution to two decimal places.

Solution

$$1.234x + 3 = 7.805$$

$$1000(1.234x + 3) = (7.805)1000$$

$$1234x + 3000 = 7805$$

$$1234x + 3000 - 3000 = 7805 - 3000$$

$$1234x = 4805$$

$$\frac{1234x}{1234} = \frac{4805}{1234}$$

$$x = 3.8938412$$

$$x \approx 3.89$$

43. *Writing a Model*   The sum of two consecutive integers is 251. Find the integers.

Solution

let n = first integer

$n + 1$ = second integer

$$n + (n + 1) = 251$$

$$2n + 1 = 251$$

$$2n + 1 - 1 = 251 - 1$$

$$2n = 250$$

$$\frac{2n}{2} = \frac{250}{2}$$

$$n = 125$$

$$n + 1 = 126$$

45. Evaluate $-360 + 120$.

Solution

$$-360 + 120 = -240$$

47. Evaluate $-\frac{4}{15} \times \frac{15}{16}$.

Solution

$$-\frac{4}{15} \times \frac{15}{16} = \frac{(-4)(15)}{(15)(16)} = \frac{(-4)(5)(3)}{(5)(3)(4)(4)} = -\frac{1}{4}$$

49. Evaluate $(12 - 15)^3$.

Solution

$$(12 - 15)^3 = (-3)^3 = -27$$

51. Decide whether (a) $x = 4$ and (b) $x = -4$ are solutions of $x + 8 = 3x$.

Solution

(a) $x = 4$

$$4 + 8 \stackrel{?}{=} 3(4)$$
$$12 = 12$$
Yes

(b) $x = -4$

$$-4 + 8 \stackrel{?}{=} 3(-4)$$
$$4 \neq -12$$
No

53. Decide whether (a) $x = -11$ and (b) $x = 5$ are solutions of $3x + 3 = 2(x - 4)$.

Solution

(a) $x = -11$

$$3(-11) + 3 \stackrel{?}{=} 2(-11 - 4)$$
$$-33 + 3 = 2(-15)$$
$$-30 = -30$$
Yes

(b) $x = 5$

$$3(5) + 3 \stackrel{?}{=} 2(5 - 4)$$
$$15 + 3 = 2(1)$$
$$18 \neq 2$$
No

55. Justify each step of the solution shown in the textbook.

Solution

$-2x + 5 = 12$	Original equation.
$-2x + 5 - 5 = 12 - 5$	Subtract 5 from both sides.
$-2x = 7$	Combine like terms.
$\dfrac{-2x}{-2} = \dfrac{7}{-2}$	Divide both sides by –2.
$x = -\dfrac{7}{2}$	Simplify.

57. Solve $6y = 42$ and check the result. (If it is not possible, state the reason.)

Solution

$$6y = 42 \qquad \text{Check} \quad 6(7) \stackrel{?}{=} 42$$
$$\frac{6y}{6} = \frac{42}{6} \qquad\qquad\qquad 42 = 42$$
$$y = 7$$

59. Solve $6x + 4 = 0$ and check the result. (If it is not possible, state the reason.)

Solution

$$6x + 4 = 0 \qquad \text{Check} \quad 6\left(-\frac{2}{3}\right) + 4 \stackrel{?}{=} 0$$
$$6x + 4 - 4 = 0 - 4$$
$$6x = -4 \qquad\qquad\qquad -4 + 4 \stackrel{?}{=} 0$$
$$\frac{6x}{6} = \frac{-4}{6} \qquad\qquad\qquad 0 = 0$$
$$x = -\frac{4}{6}$$
$$x = -\frac{2}{3}$$

61. Solve $23x - 4 = 42$ and check the result. (If it is not possible, state the reason.)

Solution

$$23x - 4 = 42 \qquad \text{Check} \quad 23(2) - 4 \stackrel{?}{=} 42$$
$$23x - 4 + 4 = 42 + 4 \qquad\qquad 46 - 4 \stackrel{?}{=} 42$$
$$23x = 46 \qquad\qquad\qquad\qquad 42 = 42$$
$$\frac{23x}{23} = \frac{46}{23}$$
$$x = 2$$

63. Solve $4y - 3 = 4y$ and check the result. (If it is not possible, state the reason.)

Solution

$$4y - 3 = 4y$$
$$4y - 3 + 3 = 4y + 3$$
$$4y = 4y + 3$$
$$4y - 4y = 4y + 3 - 4y$$
$$0 = 3$$

No solution

65. Solve $8 - 5t = 20 + t$ and check the result. (If it is not possible, state the reason.)

Solution

$$8 - 5t = 20 + t$$
$$8 - 5t + 5t = 20 + t + 5t$$
$$8 = 20 + 6t$$
$$8 - 20 = 20 + 6t - 20$$
$$-12 = 6t$$
$$\frac{-12}{6} = \frac{6t}{6}$$
$$-2 = t$$

Check $\quad 8 - 5(-2) \stackrel{?}{=} 20 + (-2)$
$$8 + 10 \stackrel{?}{=} 18$$
$$18 = 18$$

67. Solve $15t = 0$ and check the result. (If it is not possible, state the reason.)

Solution

$$15t = 0$$
$$\frac{15t}{15} = \frac{0}{15}$$
$$t = 0$$

Check $\quad 15(0) \stackrel{?}{=} 0$
$$0 = 0$$

69. Solve $5 - (2y - 4) = 15$ and check the result. (If it is not possible, state the reason.)

Solution

$$5 - (2y - 4) = 15$$
$$5 - 2y + 4 = 15$$
$$-2y + 9 = 15$$
$$-2y + 9 - 9 = 15 - 9$$
$$-2y = 6$$
$$\frac{-2y}{-2} = \frac{6}{-2}$$
$$y = -3$$

Check $\quad 5 - [2(-3) - 4] \stackrel{?}{=} 15$
$$5 - [-6 - 4] \stackrel{?}{=} 15$$
$$5 - [-10] \stackrel{?}{=} 15$$
$$5 + 10 \stackrel{?}{=} 15$$
$$15 = 15$$

71. Solve $-4(t + 2) = 0$ and check the result. (If it is not possible, state the reason.)

Solution

$$-4(t + 2) = 0 \qquad \textbf{Check} \quad -4\,[(-2) + 2] \overset{?}{=} 0$$

$$-4t - 8 = 0 \qquad\qquad\qquad -4[0] \overset{?}{=} 0$$

$$-4t - 8 + 8 = 0 + 8 \qquad\qquad\qquad 0 = 0$$

$$-4t = 8$$

$$\frac{-4t}{-4} = \frac{8}{-4}$$

$$t = -2$$

73. Solve $12(x + 3) = 7(x + 3)$ and check the result. (If it is not possible, state the reason.)

Solution

$$12(x + 3) = 7(x + 3) \qquad \textbf{Check} \quad 12\,[(-3) + 3] \overset{?}{=} 7\,[(-3) + 3]$$

$$12x + 36 = 7x + 21 \qquad\qquad\qquad 12[0] \overset{?}{=} 7[0]$$

$$12x + 36 - 7x = 7x + 21 - 7x \qquad\qquad\qquad 0 = 0$$

$$5x + 36 = 21$$

$$5x + 36 - 36 = 21 - 36$$

$$5x = -15$$

$$\frac{5x}{5} = \frac{-15}{5}$$

$$x = -3$$

75. Solve $2(x + 7) - 9 = 5(x - 4)$ and check the result. (If it is not possible, state the reason.)

Solution

$$2(x + 7) - 9 = 5(x - 4) \qquad \textbf{Check} \quad 2\left[\frac{25}{3} + 7\right] - 9 \overset{?}{=} 5\left[\frac{25}{3} - 4\right]$$

$$2x + 14 - 9 = 5x - 20$$

$$2x + 5 = 5x - 20 \qquad\qquad 2\left[\frac{25}{3} + \frac{21}{3}\right] - 9 \overset{?}{=} 5\left[\frac{25}{3} - \frac{12}{3}\right]$$

$$2x + 5 - 2x = 5x - 20 - 2x$$

$$5 = 3x - 20 \qquad\qquad 2\left[\frac{46}{3}\right] - 9 \overset{?}{=} 5\left[\frac{13}{3}\right]$$

$$5 + 20 = 3x - 20 + 20$$

$$25 = 3x \qquad\qquad\qquad \frac{92}{3} - \frac{27}{3} \overset{?}{=} \frac{65}{3}$$

$$\frac{25}{3} = \frac{3x}{3} \qquad\qquad\qquad\qquad \frac{65}{3} = \frac{65}{3}$$

$$\frac{25}{3} = x$$

77. Solve $\dfrac{u}{5} = 10$ and check the result. (If it is not possible, state the reason.)

Solution

$$\frac{u}{5} = 10 \qquad \textbf{Check} \quad \frac{50}{5} \stackrel{?}{=} 10$$

$$5\left(\frac{u}{5}\right) = (10)5 \qquad\qquad\qquad 10 = 10$$

$$u = 50$$

79. Solve $\frac{1}{3}x + 1 = \frac{1}{12}x - 4$ and check the result. (If it is not possible, state the reason.)

Solution

$$\frac{1}{3}x + 1 = \frac{1}{12}x - 4 \qquad \textbf{Check} \quad \frac{1}{3}(-20) + 1 \stackrel{?}{=} \frac{1}{12}(-20) - 4$$

$$12\left(\frac{1}{3}x + 1\right) = \left(\frac{1}{12}x - 4\right)12 \qquad\qquad \frac{-20}{3} + \frac{3}{3} \stackrel{?}{=} \frac{-20}{12} - 4$$

$$4x + 12 = x - 48 \qquad\qquad\qquad \frac{-17}{3} \stackrel{?}{=} \frac{-5}{3} - \frac{12}{3}$$

$$4x + 12 - x = x - 48 - x \qquad\qquad\qquad -\frac{17}{3} = -\frac{17}{3}$$

$$3x + 12 = -48$$

$$3x + 12 - 12 = -48 - 12$$

$$3x = -60$$

$$\frac{3x}{3} = \frac{-60}{3}$$

$$x = -20$$

81. Solve $\dfrac{t + 4}{14} = \dfrac{2}{7}$ and check the result. (If it is not possible, state the reason.)

Solution

$$\frac{t + 4}{14} = \frac{2}{7} \qquad \textbf{Check} \quad \frac{0 + 4}{14} \stackrel{?}{=} \frac{2}{7}$$

$$14\left(\frac{t + 4}{14}\right) = \left(\frac{2}{7}\right)14 \qquad\qquad \frac{4}{14} \stackrel{?}{=} \frac{2}{7}$$

$$t + 4 = 4 \qquad\qquad\qquad \frac{2}{7} = \frac{2}{7}$$

$$t + 4 - 4 = 4 - 4$$

$$t = 0$$

83. Solve $\dfrac{25 - 4u}{3} = \dfrac{5u + 12}{4} + 6$ and check the result. (If it is not possible, state the reason.)

Solution

$$\frac{25 - 4u}{3} = \frac{5u + 12}{4} + 6$$

$$12\left(\frac{25 - 4u}{3}\right) = \left(\frac{5u + 12}{4} + 6\right)12$$

$$4(25 - 4u) = 3(5u + 12) + 72$$

$$100 - 16u = 15u + 36 + 72$$

$$100 - 16u = 15u + 108$$

$$100 - 16u + 16u = 15u + 108 + 16u$$

$$100 = 31u + 108$$

$$100 - 108 = 31u + 108 - 108$$

$$-8 = 31u$$

$$\frac{-8}{31} = \frac{31u}{31}$$

$$-\frac{8}{31} = u$$

Check

$$\frac{25 - 4\left(-\frac{8}{31}\right)}{3} \overset{?}{=} \frac{5\left(-\frac{8}{31}\right) + 12}{4} + 6$$

$$\frac{25 + \frac{32}{31}}{3} \overset{?}{=} \frac{-\frac{40}{31} + 12}{4} + 6$$

$$\frac{25}{3} + \frac{32}{93} \overset{?}{=} \frac{-10}{31} + 3 + 6$$

$$\frac{775}{93} + \frac{32}{93} \overset{?}{=} \frac{-10}{31} + \frac{93}{31} + \frac{186}{31}$$

$$\frac{807}{93} \overset{?}{=} \frac{269}{31}$$

$$\frac{269}{31} = \frac{269}{31}$$

85. Solve $0.3x + 1.5 = 8.4$ and check the result. (If it is not possible, state the reason.)

Solution

$$0.3x + 1.5 = 8.4$$

$$10(0.3x + 1.5) = (8.4)10$$

$$3x + 15 = 84$$

$$3x + 15 - 15 = 84 - 15$$

$$3x = 69$$

$$\frac{3x}{3} = \frac{69}{3}$$

$$x = 23$$

Check

$$0.3(23) + 1.5 \overset{?}{=} 8.4$$

$$6.9 + 1.5 \overset{?}{=} 8.4$$

$$8.4 = 8.4$$

87. Solve $\dfrac{x}{10.625} = 2.850$ with the aid of a calculator. Round the solution to two decimal places.

Solution

$$\frac{x}{10.625} = 2.850$$

$$10.625\left(\frac{x}{10.625}\right) = (2.850)10.625$$

$$x = 30.28125$$

$$x \approx 30.28$$

89. *Finding a Pattern*   The length of a certain rectangle is t times its width. Thus, the perimeter P is given by $P = 2w + 2(tw)$, where w is the width of the rectangle. The perimeter of the rectangle is 1200 meters. (See figure in textbook.)

(a) Complete the table of lengths, widths, and areas of the rectangle (shown in the textbook) for the specified values of t.

(b) Use the table to write a short paragraph describing the relationship among the width, length, and area of a rectangle that has a *fixed* perimeter.

Solution

(a)

t	1	1.5	2	3	4	5
Width	300	240	200	150	120	100
Length	300	360	400	450	480	500
Area	90,000	86,400	80,000	67,500	57,600	50,000

If $t = 1$: $1200 = 2w + 2w$ If $t = 1.5$: $1200 = 2w + 2(1.5w)$

$\qquad\qquad 1200 = 4w$ $\qquad\qquad\qquad 1200 = 5w$

$\qquad\qquad\ \ 300 = w$ $\qquad\qquad\qquad\ \ 240 = w$

$\qquad\qquad\ \ 300 = l$ $\qquad\qquad\qquad\ \ 360 = l$

If $t = 2$: $1200 = 2w + 2(2w)$ If $t = 3$: $1200 = 2w + 2(3w)$

$\qquad\qquad 1200 = 6w$ $\qquad\qquad 1200 = 8w$

$\qquad\qquad\ \ 200 = w$ $\qquad\qquad\ \ 150 = w$

$\qquad\qquad\ \ 400 = l$ $\qquad\qquad\ \ 450 = l$

If $t = 4$: $1200 = 2w + 2(4w)$ If $t = 5$: $1200 = 2w + 2(5w)$

$\qquad\qquad 1200 = 10w$ $\qquad\qquad 1200 = 12w$

$\qquad\qquad\ \ 120 = w$ $\qquad\qquad\ \ 100 = w$

$\qquad\qquad\ \ 480 = l$ $\qquad\qquad\ \ 500 = l$

(b) Since the length is t times the width and the perimeter is fixed, as t gets larger the length gets larger and the area gets smaller. The maximum area occurs when the length and width are equal.

91. *Maximum Height of an Object*   The velocity v of an object projected vertically with an initial velocity of 64 feet per second is given by $v = 64 - 32t$, where t is time in seconds. When does the object reach its maximum height? Explain.

Solution

The object reaches its maximum height when the velocity is zero. The object then returns to the ground.

$$v = 64 - 32t$$

$$0 = 64 - 32t$$

$$0 + 32t = 64 - 32t + 32t$$

$$32t = 64$$

$$\frac{32t}{32} = \frac{64}{32}$$

$$t = 2 \text{ seconds}$$

93. *Creating a Bar Graph*   The total amount of tuition and fees y (in millions of dollars) collected by colleges and universities in the United States from 1985 to 1990 can be approximated by the model

$$y = 8166.4 + 2526.2t, \quad 5 \leq t \leq 10$$

where $t = 5$ represents 1985. Create a bar graph that shows the amounts from 1985 through 1990. What can you conclude? (Source: U.S. National Center for Education Statistics)

Solution

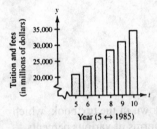

You can conclude that the total amount of tuition and fees collected by colleges and universities in the United States is increasing.

95. *Using a Two-Part Model*   Use the following *two-part* model, which approximates the number of cable television subscribers in the United States from 1970 to 1993.

$$y = 4193 + 1099.4t, \quad 0 \leq t \leq 9$$

$$y = -15,731 + 3210t, \quad 10 \leq t \leq 23$$

In this model, y represents the number of subscribers (in thousands) and t represents the year, with $t = 0$ corresponding to 1970 (see figure in textbook). (Source: Corporation for Public Broadcasting). During which year were there 10,789.4 (thousand) cable television subscribers?

Solution

$$10,789.4 = 4193 + 1099.4t$$

$$10,789.4 - 4193 = 4193 + 1099.4t - 4193$$

$$6596.4 = 1099.4t$$

$$\frac{6596.4}{1099.4} = \frac{1099.4t}{1099.4}$$

$$6 = t \quad \text{(year) 1976}$$

1.2 Linear Equations and Problem Solving

7. *Geometry* Construct a verbal model and write an algebraic equation that represents the following problem. Do not solve the equation.

A "Slow Moving Vehicle" sign has the shape of an equilateral triangle. The sign has a perimeter of 129 centimeters. Find the length of each side. Include a labeled diagram with your model.

Solution

Verbal model: $\boxed{\text{Perimeter}} = \boxed{\text{side}} + \boxed{\text{side}} + \boxed{\text{side}}$

Equation: $129 = x + x + x$

9. Complete the table shown in the textbook, which shows the equivalent forms of various percents.

Solution

Percent: 30%

Parts out of 100: 30

Decimal: 0.30

Fraction: $\frac{30}{100} = \frac{3}{100}$

11. Complete the table shown in the textbook, which shows the equivalent forms of various percents.

Solution

Percent: 7.5%

Parts out of 100: 7.5

Decimal: 0.075

Fraction: $\frac{75}{1000} = \frac{3}{40}$

13. Complete the table shown in the textbook, which shows the equivalent forms of various percents.

Solution

Percent: 12.5%

Parts out of 100: 12.5

Decimal: 0.125

Fraction: $\frac{1}{8}$

15. What is 35% of 250?

Solution

$a = p \cdot b$

$a = (0.35)(250)$

$a = 87.5$

17. 96 is 0.8% of what number?

Solution

$a = p \cdot b$

$96 = (0.008)(b)$

$\dfrac{96}{0.008} = b$

$12{,}000 = b$

19. 1650 is what percent of 5000?

Solution

$a = p \cdot b$

$1650 = (p)(5000)$

$\dfrac{1650}{5000} = p$

$33\% = p$

21. *Monthly Rent* If you spend 15% of your monthly income of $2800 for rent, what is your monthly rent payment?

Solution

Verbal model: | Monthly rent | = | Percent | · | Monthly income |

Equation: $a = p \cdot b$

$a = (0.15)(2800)$

$a = \$420$

23. *Salary Adjustment* During a year of financial difficulties your company reduces your salary by 7%. What percent increase in this reduced salary is required to raise your salary to the amount it was prior to the reduction? Why isn't the percent increase the same as the percent of the reduction?

Solution

Verbal model: | Original salary | − | Reduction | = | Reduced salary |

Equation: $x - 0.07x = 0.93x$

Verbal model: | Reduction | = | Percent | · | Reduced salary |

Equation: $a = p \cdot b$

$0.07x = p \cdot 0.93x$

$\dfrac{0.07x}{0.93x} = p$

$7.53\% = p$

Base number is smaller.

25. Express 36 inches to 48 inches as a ratio. Use the same units in both the numerator and denominator, and simplify.

Solution

$\dfrac{36 \text{ inches}}{48 \text{ inches}} = \dfrac{36}{48} = \dfrac{3}{4}$

27. *Compression Ratio* The compression ratio of a cylinder is the ratio of its expanded volume to its compressed volume (see figure in textbook). The expanded volume of one cylinder of a diesel engine is 425 cubic centimeters, and its compressed volume is 20 cubic centimeters. Find the compression ratio of this engine.

Solution

$\dfrac{\text{Expanded volume}}{\text{Compressed volume}} = \dfrac{425 \text{ cu cm}}{20 \text{ cu cm}} = \dfrac{85}{4}$

29. *Unit Prices* Find the unit price (in dollars per ounce) of a 20-ounce can of pineapple for 95¢.

Solution

$\dfrac{\text{Total price}}{\text{Total units}} = \dfrac{0.95}{20} = \dfrac{95}{2000} = \$0.0475$ per ounce

31. *Unit Prices* Find the unit price (in dollars per ounce) of a 1-pound, 4-ounce loaf of bread for $1.39.

Solution

$\dfrac{\text{Total price}}{\text{Total units}} = \dfrac{1.39}{20} = \dfrac{139}{2000} = \$0.0695$ per ounce

33. *Comparison Shopping* Use unit prices to determine the better buy: (a) a $14\frac{1}{2}$-ounce bag of chips for $2.32 or (b) a 32-ounce bag of chips for $4.85.

Solution

(a) Unit price $= \dfrac{2.32}{14.5} = \$0.16$ per ounce

(b) Unit price $= \dfrac{4.85}{32} = \$0.1515625$ per ounce

The 32-ounce bag is a better buy.

35. *Comparison Shopping* Use unit prices to determine the better buy: (a) an 8-ounce tube of toothpaste for $1.69 or (b) a 12-ounce tube of toothpaste for $2.39.

Solution

(a) Unit price $= \dfrac{1.69}{8} = \$0.21125$ per ounce

(b) Unit price $= \dfrac{2.39}{12} = \$0.1991667$ per ounce

The 12-ounce tube is a better buy.

37. Solve $\dfrac{t}{4} = \dfrac{3}{2}$.

Solution

$$\dfrac{t}{4} = \dfrac{3}{2}$$

$$t = 4 \cdot \dfrac{3}{2}$$

$$t = 6$$

39. Solve $\dfrac{y+1}{10} = \dfrac{y-1}{6}$.

Solution

$$\dfrac{y+1}{10} = \dfrac{y-1}{6}$$

$$6(y+1) = 10(y-1)$$

$$6y + 6 = 10y - 10$$

$$16 = 4y$$

$$4 = y$$

41. *Property Tax* The taxes on property with an assessed value of $75,000 are $1125. Find the taxes on property with an assessed value of $120,000.

Solution

Verbal model: $\boxed{\dfrac{\text{Taxes}}{\text{Assessed value}}} = \boxed{\dfrac{\text{Taxes}}{\text{Assessed value}}}$

Proportion: $\dfrac{x}{120,000} = \dfrac{1125}{75,000}$

$$x = 120,000 \cdot \dfrac{1125}{75,000}$$

$$x = \$1800 \text{ taxes}$$

43. *Geometry* Solve for the length x of the side of the triangle by using the fact that corresponding sides of similar triangles are proportional. (See figure in textbook.)

Solution

Proportion: $\dfrac{x}{6} = \dfrac{2}{4}$

$$x = 6 \cdot \dfrac{2}{4}$$

$$x = 3$$

45. Solve $x + \dfrac{x}{2} = 4$. (If it is not possible, state the reason.)

Solution

$$x + \frac{x}{2} = 4 \qquad \textbf{Check} \qquad \frac{8}{3} + \frac{8/3}{2} \overset{?}{=} 4$$

$$2\left(x + \frac{x}{2}\right) = (4)2 \qquad\qquad \frac{8}{3} + \frac{4}{3} \overset{?}{=} 4$$

$$2x + x = 8 \qquad\qquad\qquad \frac{12}{3} \overset{?}{=} 4$$

$$3x = 8 \qquad\qquad\qquad\qquad 4 = 4$$

$$\frac{3x}{3} = \frac{8}{3}$$

$$x = \frac{8}{3}$$

47. Solve $8(x - 14) = 0$. (If it is not possible, state the reason.)

Solution

$$8(x - 14) = 0 \qquad \textbf{Check} \qquad 8(14 - 14) \overset{?}{=} 0$$

$$\frac{8(x - 14)}{8} = \frac{0}{8} \qquad\qquad\qquad 8(0) \overset{?}{=} 0$$

$$x - 14 = 0 \qquad\qquad\qquad\qquad 0 = 0$$

$$x = 14$$

49. Solve $12(3 - x) = 5 - 7(2x + 1)$. (If it is not possible, state the reason.)

Solution

$$12(3 - x) = 5 - 7(2x + 1) \qquad \textbf{Check} \qquad 12[3 - (-19)] \overset{?}{=} 5 - 7[2(-19) + 1]$$

$$36 - 12x = 5 - 14x - 7 \qquad\qquad\qquad 12[3 + 19] \overset{?}{=} 5 - 7[-38 + 1]$$

$$36 - 12x = -2 - 14x \qquad\qquad\qquad\qquad 12(22) \overset{?}{=} 5 - 7(-37)$$

$$36 + 2x = -2 \qquad\qquad\qquad\qquad\qquad 264 \overset{?}{=} 5 + 259$$

$$2x = -38 \qquad\qquad\qquad\qquad\qquad\qquad 264 = 264$$

$$x = -19$$

51. *Mathematical Modeling* Construct a verbal model and write an algebraic equation that represents the following problem. Do not solve the equation.

Find a number such that the sum of the number and 30 is 82.

Solution

Verbal model: | Number | + | 30 | = | 82 |

Equation: $x + 30 = 82$

53. *Mathematical Modeling* Construct a verbal model and write an algebraic equation that represents the following problem. Do not solve the equation.

> The bill for the repair of an automobile is $380. Included in this bill is a charge of $275 for parts. If the remainder of the bill is for labor, how many hours were spent repairing the car? (The charge for labor is $35 per hour.)

Solution

Verbal model: | Total bill | = | bill for parts | + | bill for labor |

Equation: $\qquad 380 = 275 + 35x$

55. Using the figure in the textbook, what percent of the entire figure is shaded? (Assume that each rectangular portion of the figure has the same area.)

Solution

$\frac{5}{12} = 41\frac{2}{3}\%$

57. What is 8.5% of 816?

Solution

$a = p \cdot b$

$a = (0.085)(816)$

$a = 69.36$

59. What is 0.4% of 150,000?

Solution

$a = p \cdot b$

$a = (0.004)(150{,}000)$

$a = 600$

61. 84 is 24% of what number?

Solution

$a = p \cdot b$

$84 = (0.24)(b)$

$\dfrac{84}{0.24} = b$

$350 = b$

63. 42 is $10\frac{1}{2}\%$ of what number?

Solution

$a = p \cdot b$

$42 = (0.105)(b)$

$\dfrac{42}{0.105} = b$

$400 = b$

65. 2100 is what percent of 1200?

Solution

$a = p \cdot b$

$2100 = (p)(1200)$

$\dfrac{2100}{1200} = p$

$175\% = p$

67. *Pension Funds* Your employer withholds $6\frac{1}{2}\%$ of your gross monthly income for your retirement. Determine the amount withheld each month if your gross monthly income is $3800.

Solution

Verbal model: | Amount withheld | = | Percent | · | Gross monthly income |

Equation: $\qquad a = p \cdot b$

$\qquad a = (0.065)(3800)$

$\qquad a = \$247$

69. *Company Layoff*   Because of slumping sales, a small company laid off 25 of its 160 employees. What percent of the work force was laid off?

Solution

Verbal model: $\boxed{\text{Number laid off}} = \boxed{\text{Percent}} \cdot \boxed{\text{Number of employees}}$

Equation: $a = p \cdot b$

$$25 = (p)(160)$$

$$\frac{25}{160} = p$$

$$15.625\% = p$$

71. *Defective Parts*   A quality control engineer reports that 1.5% of a sample of parts are defective. The engineer found three defective parts. How large was the sample?

Solution

Verbal model: $\boxed{\text{Defective parts}} = \boxed{\text{Percent}} \cdot \boxed{\text{Total parts}}$

Equation: $a = p \cdot b$

$$3 = (0.015)(b)$$

$$\frac{3}{0.015} = b$$

$$200 = b \text{ total parts}$$

73. *Course Grade*   You missed a B by 8 points. Your point total for the course is 372. How many points were possible in the course? (Assume that you needed 80% of the course total for a B.)

Solution

Verbal model: $\boxed{\text{Percent}} \cdot \boxed{\text{Points possible in course}} = \boxed{\text{Total points}} + 8$

Equation: $p \cdot x = t + 8$

$$0.80x = 372 + 8$$

$$0.80 = 380$$

$$x = \frac{380}{0.80} = 475 \text{ points}$$

75. *Graphical Estimation* The bar graph in the textbook shows the per capita food consumption of selected meats for 1970, 1980, and 1991. (Source: U.S. Department of Agriculture). Using this graph, approximate the decrease in the per capita consumption of beef from 1970 to 1991. Use this estimate to approximate the percent decrease.

Solution

Approximate decrease from graph is 16.5 pounds.

Verbal model: $\boxed{\text{Amount decrease}} = \boxed{\text{Percent}} \cdot \boxed{\text{Beef consumption in 1970}}$

Equation: $a = p \cdot b$

$16.5 = p \cdot 80$

$\dfrac{16.5}{80} = p$

$20\% = p$

77. *Graphical Estimation* The bar graph in the textbook shows the per capita food consumption of selected meats for 1970, 1980, and 1991. (Source: U.S. Department of Agriculture). Using this graph, approximate the total number of pounds of pork consumed in 1991 if the population of the United States was approximately 250 million.

Solution

Verbal model: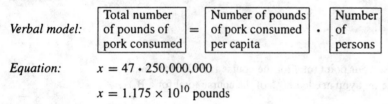

Equation: $x = 47 \cdot 250,000,000$

$x = 1.175 \times 10^{10}$ pounds

79. *Reading a Circle Graph* The expenses for a small company for January are shown in the circle graph in the textbook. What percent of the total monthly expenses is each budget item?

Solution

Verbal model:

$$\boxed{\dfrac{\text{Budget item}}{\text{Total expense}}} = \boxed{\dfrac{\text{Percent}}{100\%}}$$

Proportion: Taxes: $\dfrac{2400}{33{,}700} = \dfrac{x}{100}$

$$x = \dfrac{2400}{33{,}700} \cdot 100$$

$$x = 7.12\%$$

Misc.: $\dfrac{4200}{33{,}700} = \dfrac{x}{100}$

$$x = \dfrac{4200}{33{,}700} \cdot 100$$

$$x = 12.46\%$$

Supplies: $\dfrac{900}{33{,}700} = \dfrac{x}{100}$

$$x = \dfrac{900}{33{,}700} \cdot 100$$

$$x = 2.67\%$$

Rent: $\dfrac{1300}{33{,}700} = \dfrac{x}{100}$

$$x = \dfrac{1300}{33{,}700} \cdot 100$$

$$x = 3.86\%$$

Benefits: $\dfrac{3700}{33{,}700} = \dfrac{x}{100}$

$$x = \dfrac{3700}{33{,}700} \cdot 100$$

$$x = 10.98\%$$

Insurance: $\dfrac{800}{33{,}700} = \dfrac{x}{100}$

$$x = \dfrac{800}{33{,}700} \cdot 100$$

$$x = 2.37\%$$

Utilities: $\dfrac{400}{33{,}700} = \dfrac{x}{100}$

$$x = \dfrac{400}{33{,}700} \cdot 100$$

$$x = 1.19\%$$

Wages: $\dfrac{20{,}000}{33{,}700} = \dfrac{x}{100}$

$$x = \dfrac{20{,}000}{33{,}700} \cdot 100$$

$$x = 59.35\%$$

81. Write 1 pint to 1 gallon as a ratio. Use the same units in both the numerator and denominator, and simplify.

Solution

$$\dfrac{1 \text{ pint}}{1 \text{ gallon}} = \dfrac{1 \text{ pint}}{8 \text{ pints}} = \dfrac{1}{8}$$

83. Write 40 milliliters to 1 liter as a ratio. Use the same units in both the numerator and denominator, and simplify.

Solution

$$\dfrac{40 \text{ milliliters}}{1 \text{ liter}} = \dfrac{0.04 \text{ liter}}{1} = \dfrac{4}{100} = \dfrac{1}{25}$$

85. *State Income Tax* You have \$12.50 of state tax withheld from your paycheck per week when your gross pay is \$625. Find the ratio of tax to gross pay.

Solution

$$\dfrac{\text{Tax}}{\text{Pay}} = \dfrac{12.50}{625} = \dfrac{125}{6250} = \dfrac{1}{50}$$

87. Solve $\dfrac{x}{6} = \dfrac{2}{3}$.

Solution

$$\dfrac{x}{6} = \dfrac{2}{3}$$

$$x = 6 \cdot \dfrac{2}{3}$$

$$x = 4$$

89. Solve $\dfrac{5}{4} = \dfrac{t}{6}$.

Solution

$$\frac{5}{4} = \frac{t}{6}$$

$$t = \frac{5}{4} \cdot 6$$

$$t = \frac{15}{2} = 7\frac{1}{2} = 7.5$$

91. Solve $\dfrac{y+5}{6} = \dfrac{y-2}{4}$.

Solution

$$\frac{y+5}{6} = \frac{y-2}{4}$$

$$4(y+5) = 6(y-2)$$

$$4y + 20 = 6y - 12$$

$$32 = 2y$$

$$16 = y$$

93. *Spring Length* A force of 32 pounds stretches a spring 6 inches. Determine the number of pounds of force required to stretch it 9 inches.

Solution

Verbal model: $\boxed{\dfrac{\text{Pounds}}{\text{Inches}}} = \boxed{\dfrac{\text{Pounds}}{\text{Inches}}}$

Proportion: $\dfrac{x}{9} = \dfrac{32}{6}$

$$x = 9 \cdot \frac{32}{6}$$

$$x = 48 \text{ pounds}$$

95. *Defective Units* A quality control engineer for a manufacturer found one defective unit in a sample of 75. At that rate, what is the expected number of defective units in a shipment of 200,000?

Solution

Verbal model: $\boxed{\dfrac{\text{Defective units}}{\text{Total units}}} = \boxed{\dfrac{\text{Defective units}}{\text{Total units}}}$

Proportion: $\dfrac{x}{200{,}000} = \dfrac{1}{75}$

$$x = 200{,}000 \cdot \frac{1}{75}$$

$$x = 2667 \text{ defective units}$$

97. *Geometry* Solve for the length x of the side of the triangle by using the fact that corresponding sides of similar triangles are proportional. (See figure in textbook.)

Solution

Proportion: $\dfrac{x}{7} = \dfrac{4}{5.5}$

$$x = 7 \cdot \frac{4}{5.5}$$

$$x \approx 5.1$$

99. *Geometry* Find the length of the shadow of a man who is 6 feet tall and is standing 15 feet from a streetlight that is 20 feet high. (See figure in textbook.)

Solution

Proportion: $\dfrac{x}{6} = \dfrac{x+15}{20}$

$$20x = 6x + 90$$

$$14x = 90$$

$$x = \frac{90}{14}$$

$$x = \frac{45}{7} \text{ or } 6\frac{3}{7}$$

1.3	**Business and Scientific Problems**

7. *Using a Circle Graph*   The circle graph in the textbook shows how approximately 17 million barrels of oil per day are consumed in the United States. How many barrels are used in the transportation sector? (Source: Energy Information Administration)

Solution

Verbal model: $\boxed{\text{Total barrels}} \cdot \boxed{\text{Percent used in transportation}} = \boxed{\text{Barrels used in transportation}}$

Equation: $17 \cdot 0.63 = x$

$10.71 \text{ million barrels} = x$

9. Find the markup and the markup rate if the cost is \$45.95 and the selling price is \$64.33. (Assume the markup rate is based on the cost.)

Solution

Verbal model: $\boxed{\text{Selling price}} = \boxed{\text{Cost}} + \boxed{\text{Markup}}$

Equation: $64.33 = 45.95 + x$

$x = 64.33 - 45.95$

$x = \$18.38$

Verbal model: $\boxed{\text{Markup}} = \boxed{\text{Markup rate}} \cdot \boxed{\text{Cost}}$

Equation: $18.38 = x \cdot 45.95$

$\dfrac{18.38}{45.95} = x$

$40\% = x$

11. Find the cost and the markup rate if the selling price is \$26,922.50 and the markup is \$4672.50. (Assume the markup rate is based on the cost.)

Solution

Verbal model: $\boxed{\text{Selling price}} = \boxed{\text{Cost}} + \boxed{\text{Markup}}$

Equation: $26,922.50 = x + 4672.50$

$26,922.50 - 4672.50 = x$

$\$22,250.00 = x$

Verbal model: $\boxed{\text{Markup}} = \boxed{\text{Markup rate}} \cdot \boxed{\text{Cost}}$

Equation: $4672.50 = x \cdot 22,250.00$

$\dfrac{4672.50}{22,250.00} = x$

$21\% = x$

13. Find the discount and the discount rate if the list price is $49.50 and the sale price is $25.74. (Assume the discount rate is based on the list price.)

Solution

Verbal model: $\boxed{\text{Sale price}} = \boxed{\text{List price}} - \boxed{\text{Discount}}$

Equation: $25.74 = 49.50 - x$

$$x = 49.50 - 25.74$$

$$x = \$23.76$$

Verbal model: $\boxed{\text{Discount}} = \boxed{\text{Discount rate}} \cdot \boxed{\text{List price}}$

Equation: $23.76 = x \cdot 49.50$

$$\frac{23.76}{49.50} = x$$

$$48\% = x$$

15. Find the list price and the discount rate if the sale price is $893.10 and the discount is $251.90. (Assume the discount rate is based on the list price.)

Solution

Verbal model: $\boxed{\text{Sale price}} = \boxed{\text{List price}} - \boxed{\text{Discount}}$

Equation: $893.10 = x - 251.90$

$$893.10 + 251.90 = x$$

$$\$1145.00 = x$$

Verbal model: $\boxed{\text{Discount}} = \boxed{\text{Discount rate}} \cdot \boxed{\text{List price}}$

Equation: $251.90 = x \cdot 1145.00$

$$\frac{251.90}{1145.00} = x$$

$$22\% = x$$

17. *Comparison Shopping*   A department store is offering a discount of 20% on a sewing machine with a list price of $279.95. A mail–order catalog has the same machine for $228.95 plus $4.32 for shipping. Which is the better buy?

Solution

Verbal model: $\boxed{\text{Discount}} = \boxed{\text{Discount rate}} \cdot \boxed{\text{List price}}$

Equation: $x = 20\% \cdot 279.95$

$x = \$55.99$

Verbal model: $\boxed{\text{Selling price}} = \boxed{\text{List price}} - \boxed{\text{Discount}}$

Equation: $x = 279.95 - 55.99$

$x = \$223.96$

Verbal model: $\boxed{\text{Selling price}} = \boxed{\text{List price}} + \boxed{\text{Shipping}}$

Equation: $x = 228.95 + 4.32$

$x = \$233.27$

The department store machine is the better buy.

19. *Commission Rate*   Determine the commission rate for an employee who earned $450 in commissions for sales of $5000.

Solution

Verbal model: $\boxed{\text{Commission}} = \boxed{\text{Commission rate}} \cdot \boxed{\text{Sales}}$

Equation: $450 = x \cdot 5000$

$\dfrac{450}{5000} = x$

$9\% = x$

21. *Finding a Pattern*   A rancher must purchase 500 bushels of a feed mixture for cattle and is considering oats and corn, which cost $1.70 per bushel and $3.00 per bushel, respectively. (a) Complete the table shown in the textbook where x is the number of bushels of oats in the mixture. (b) How does the increase in the number of bushels of oats affect the number of bushels of corn in the mixture? (c) How does the increase in the number of bushels of oats affect the price per bushel of the mixture? (d) If there were an equal number of bushels of oats and corn in the mixture, how would the price of the mixture be related to the price of each component?

Solution

(a)

	Oats x	Corn $500 - x$	Price/bushel of mixture
(i)	0	500	3.00
(ii)	100	400	2.74
(iii)	200	300	2.48
(iv)	300	200	2.22
(v)	400	100	1.96
(vi)	500	0	1.70

(i) Price $= 1.70(0) + 3.00(500)$ $= \$1500$

(ii) Price $= 1.70(100) + 3.00(400) = \$1370$

(iii) Price $= 1.70(200) + 3.00(300) = \$1240$

(iv) Price $= 1.70(300) + 3.00(200) = \$1110$

(v) Price $= 1.70(400) + 3.00(100) = \$980$

(vi) Price $= 1.70(500) + 3.00(0)$ $= \$850$

(b) The increase in the number of bushels of oats causes the number of bushels of corn to decrease.

(c) The increase in the number of bushels of oats decreases the price per bushel of the mixture.

(d) If there were an equal number of bushels of oats and corn, the price of the mixture ($2.35/bushel) would be equal to the average of the price of the oats and the corn per bushel [$(1.70 + 3.00)/2 = \$2.35$].

23. *Mixtures*   If the concentration of Solution 1 is 20% and the concentration of Solution 2 is 60%, find the number of units of Solution 1 and Solution 2 needed to obtain 100 gallons of the final solution with a final concentration of 40%.

Solution

Verbal model: $\boxed{\begin{array}{c}\text{Amount of} \\ \text{Solution 1}\end{array}} + \boxed{\begin{array}{c}\text{Amount of} \\ \text{Solution 2}\end{array}} = \boxed{\begin{array}{c}\text{Amount of} \\ \text{final solution}\end{array}}$

Equation: $0.20x + 0.60(100 - x) = 0.40(100)$

$$0.20x + 60 - 0.60x = 40$$

$$-0.40x = -20$$

$$x = 50 \text{ gal at } 20\%$$

$$100 - x = 50 \text{ gal at } 60\%$$

25. *Mixtures* If the concentration of Solution 1 is 15% and the concentration of Solution 2 is 60%, find the number of units of Solution 1 and Solution 2 needed to obtain 24 quarts of the final solution with a final concentration of 45%.

Solution

Verbal model: $\boxed{\begin{array}{c}\text{Amount of}\\\text{Solution 1}\end{array}} + \boxed{\begin{array}{c}\text{Amount of}\\\text{Solution 2}\end{array}} = \boxed{\begin{array}{c}\text{Amount of}\\\text{final solution}\end{array}}$

Equation:

$$0.15x + 0.60(24 - x) = 0.45(24)$$
$$0.15x + 14.4 - 0.60x = 10.8$$
$$-0.45x = -3.6$$
$$x = 8 \text{ qt at } 15\%$$
$$24 - x = 16 \text{ qt at } 60\%$$

27. If the rate r is 650 mi/hr and the time t is $3\frac{1}{2}$ hrs, what is the distance, d?

Solution

Verbal model: $\boxed{\text{Distance}} = \boxed{\text{Rate}} \cdot \boxed{\text{Time}}$

Equation: $d = 650 \cdot 3.5$

$$d = 2275 \text{ mi}$$

29. If the distance d is 1000 km and the rate r is 110 km/hr, what is the time t?

Solution

Verbal model: $\boxed{\text{Distance}} = \boxed{\text{Rate}} \cdot \boxed{\text{Time}}$

Equation: $1000 = 110 \cdot t$

$$\frac{1000}{110} = t$$
$$\frac{100}{11} \text{ hr} = t$$

31. If the distance d is 1000 ft and the time t is $\frac{3}{2}$ sec, what is the rate r?

Solution

Verbal model: $\boxed{\text{Distance}} = \boxed{\text{Rate}} \cdot \boxed{\text{Time}}$

Equation: $1000 = r \cdot \dfrac{3}{2}$

$$\frac{1000}{3/2} = r$$
$$\frac{2000}{3} \text{ ft/sec} = r$$

33. *Average Speed* Determine the time for the space shuttle to travel a distance of 5000 miles when its average speed is 17,000 miles per hour. (See figure in textbook.)

Solution

Verbal model: $\boxed{\text{Distance}} = \boxed{\text{Rate}} \cdot \boxed{\text{Time}}$

Equation: $5000 = 17,000 \cdot t$

$$\frac{5000}{17,000} = t$$
$$\frac{5}{17} \text{ hr} = t$$

35. *Work Rate* Determine the work rate for each task.

(a) A printer can print eight pages per minute.

Solution

(a) Printer's rate = 8 pages per minute

(b) A machine shop can produce 30 units in 8 hours.

(b) Shop's rate = 30 units in 8 hours

$$= \frac{30}{8} \text{ units per hour}$$
$$= \frac{15}{4} \text{ units per hour}$$

37. *Ohm's Law*   Solve for R in $E = IR$.

Solution

$$E = IR$$

$$\frac{E}{I} = R$$

39. *Simple Interest*   Solve for r in $A = P + Prt$.

Solution

$$A = P + Prt$$

$$A - P = Prt$$

$$\frac{A - P}{Pt} = r$$

41. *Geometry*   A rectangular picture frame has a perimeter of 3 feet. The width of the frame is 0.6 times its height. Find the height of the frame.

Solution

Verbal model: Perimeter $= 2$ $\boxed{\text{Width}}$ $+ 2$ $\boxed{\text{Height}}$

Equation:

$$3 = 2(0.6x) + 2(x)$$

$$3 = 1.2x + 2x$$

$$3 = 3.2x$$

$$11.25 \text{ inches} = 0.9375 \text{ feet} = x$$

43. *Simple Interest*   Find the interest on a \$5000 bond that pays an annual rate of $9\frac{1}{2}\%$ for 6 years.

Solution

Common formula: $I = Prt$

Equation: $I = (5000)(0.095)(6)$

$$I = \$2850$$

45. Solve $44 - 16x = 0$.

Solution

$$44 - 16x = 0$$

$$44 - 16x + 16x = 0 + 16x$$

$$44 = 16x$$

$$\frac{44}{16} = \frac{16x}{16}$$

$$\frac{11}{4} = \frac{44}{16} = x$$

47. Solve $3[4 + 5(x - 1)] = 6x + 2$.

Solution

$$3[4 + 5(x - 1)] = 6x + 2$$

$$3[4 + 5x - 5] = 6x + 2$$

$$3[5x - 1] = 6x + 2$$

$$15x - 3 = 6x + 2$$

$$15x - 6x - 3 = 6x + 2 - 6x$$

$$9x - 3 = 2$$

$$9x - 3 + 3 = 2 + 3$$

$$9x = 5$$

$$\frac{9x}{9} = \frac{5}{9}$$

$$x = \frac{5}{9}$$

49. Write "y is no more than 45" using inequality notation.

Solution

$$y \leq 45$$

51. Find the cost and the markup rate if the selling price is $250.80 and the markup is $98.80. (Assume the markup rate is based on the cost.)

Solution

Verbal model: | Selling price | $=$ | Cost | $+$ | Markup |

Equation: $$250.80 = x + 98.80$$

$$250.80 - 98.80 = x$$

$$\$152.00 = x$$

Verbal model: | Markup | $=$ | Markup rate | $\cdot$ | Cost |

Equation: $$98.80 = x \cdot 152.00$$

$$\frac{98.80}{152.00} = x$$

$$65\% = x$$

53. Find the selling price and the markup if the cost is $225.00 and the markup rate is 85.2%. (Assume the markup rate is based on the cost.)

Solution

Verbal model: | Markup | $=$ | Markup rate | $\cdot$ | Cost |

Equation: $$x = 85.2\% \cdot 225.00$$

$$x = \$191.70$$

Verbal model: | Selling price | $=$ | Cost | $+$ | Markup |

Equation: $$x = 225.00 + 191.70$$

$$x = \$416.70$$

55. Find the sale price and the discount rate if the list price is $300.00 and the discount is $189.00. (Assume the discount rate is based on the list price.)

Solution

Verbal model: | Sale price | $=$ | List price | $-$ | Discount |

Equation: $$x = 300.00 - 189.00$$

$$x = \$111.00$$

Verbal model: | Discount | $=$ | Discount rate | $\cdot$ | List price |

Equation: $$\$189.00 = x \cdot 300.00$$

$$\frac{189.00}{300.00} = x$$

$$63\% = x$$

57. Find the sale price and the discount if the list price is $95.00 and the discount rate is 65%. (Assume the discount rate is based on the list price.)

Solution

Verbal model: $\boxed{\text{Discount}} = \boxed{\text{Discount rate}} \cdot \boxed{\text{List price}}$

Equation: $x = 65\% \cdot 95.00$

              $x = \$61.75$

Verbal model: $\boxed{\text{Sale price}} = \boxed{\text{List price}} - \boxed{\text{Discount}}$

Equation: $x = 95.00 - 61.75$

              $x = \$33.25$

59. *Tire Cost*   An auto store gives the list price of a tire as $79.42. During a promotional sale, the store is selling four tires for the price of three. The store needs a markup of 10% during the sale. What is the cost to the store for each tire?

Solution

Verbal model: $\boxed{\text{Cost}} + \boxed{\text{Markup}} = \boxed{\text{Selling price}}$

Equation: $x + 0.10x = 59.565$

            $1.10x = 59.565$

                 $x = \$54.15$

61. *Tip Rate*   A customer left a total of $10 for a meal that cost $8.45. Determine the tip rate.

Solution

Verbal model: $\boxed{\text{Tip}} = \boxed{\text{Total paid}} - \boxed{\text{Meal cost}}$

Equation: $x = 10 - 8.45$

             $x = \$1.55$

Verbal model: $\boxed{\text{Tip}} = \boxed{\text{Tip rate}} \cdot \boxed{\text{Meal cost}}$

Equation: $1.55 = x \cdot 8.45$

        $\dfrac{1.55}{8.45} = x$

     $18.3\% = x$

63. *Long Distance Rates*   The weekday rate for a telephone call is $0.75 for the first minute plus $0.55 for each additional minute. Determine the length of a call that cost $5.15. What would the cost of the call have been if it had been made during the weekend when there is a 60% discount?

Solution

Verbal model: $\boxed{\text{Total cost}} = \boxed{\begin{array}{c}\text{Cost of}\\\text{first minute}\end{array}} + \boxed{\begin{array}{c}\text{Cost of}\\\text{additional minutes}\end{array}}$

Equation: $5.15 = 0.75 + 0.55x$

$4.40 = 0.55x$

$8 = x$

Length of call = 9 minutes

Verbal model: $\boxed{\text{Discount}} = \boxed{\text{Discount rate}} \cdot \boxed{\text{List price}}$

Equation: $x = 60\% \cdot 5.15$

$x = \$3.09$

Verbal model: $\boxed{\text{Selling price}} = \boxed{\text{List price}} - \boxed{\text{Discount}}$

Equation: $x = 5.15 - 3.09$

$x = \$2.06$

65. *Amount Financed*   A customer bought a lawn tractor that cost $4450 plus 6% sales tax. Find the amount of the sales tax and the total bill. Find the amount financed if a down payment of $1000 was made.

Solution

Verbal model: $\boxed{\text{Sales tax}} = \boxed{\text{Sales tax rate}} \cdot \boxed{\text{Cost}}$

Equation: $x = 6\% \cdot 4450$

$x = \$267$

Verbal model: $\boxed{\text{Total bill}} = \boxed{\text{Cost}} + \boxed{\text{Sales tax}}$

Equation: $x = 4450 + 267$

$x = \$4717$

Verbal model: $\boxed{\text{Amount financed}} = \boxed{\text{Total bill}} - \boxed{\text{Downpayment}}$

Equation: $x = 4717 - 1000$

$x = \$3717$

67. *Number of Stamps* You have a set of 70 stamps with a total value of $16.50. If the set includes 15¢ stamps and 30¢ stamps, find the number of each type.

Solution

Verbal model: | Total value | $=$ | Value of 15¢ stamps | $+$ | Value of 30¢ stamps |

Equation:
$$16.50 = 0.15x + 0.30(70 - x)$$
$$16.50 = 0.15x + 21.00 - 0.30x$$
$$-4.50 = -0.15x$$
$$30 = x \quad (15¢ \text{ stamps})$$
$$40 = 70 - x \quad (30¢ \text{ stamps})$$

69. *Opinion Poll* Fourteen hundred people were surveyed in an opinion poll. Political candidates A and B received approximately the same preference, but candidate C was preferred by twice the number of people as candidates A and B. Determine the number in the sample that preferred candidate C.

Solution

Verbal model: | Total people | $=$ | Number preferring A | $+$ | Number preferring B | $+$ | Number preferring C |

Equation:
$$1400 = x + x + 2x$$
$$1400 = 4x$$
$$350 = x$$
$$700 = 2x$$

71. *Antifreeze Coolant* The cooling system on a truck contains 5 gallons of coolant that is 40% antifreeze. How much must be withdrawn and replaced with 100% antifreeze to bring the coolant in the system to 50% antifreeze?

Solution

Verbal model: | Original antifreeze solution | $-$ | Some antifreeze solution | $+$ | Pure antifreeze | $=$ | Final antifreeze solution |

Equation:
$$0.40(5) - 0.40x + 1.00x = 0.50(5)$$
$$2 - 0.40x + 1.00x = 2.5$$
$$0.60x = 0.5$$
$$x = \tfrac{5}{6} \text{ gal}$$

73. *Flying Distance* Two planes leave an airport at approximately the same time and fly in opposite directions. How far apart are the planes after $1\frac{1}{3}$ hours if their speeds are 480 miles per hour and 600 miles per hour?

Solution

Verbal model: | Distance | $=$ | Rate | $\cdot$ | Time |

Equation:
$$x = 480\left(\tfrac{4}{3}\right) + 600\left(\tfrac{4}{3}\right)$$
$$x = 1440 \text{ miles}$$

75. *Geometry*   Using the closed rectangular box shown in the figure in the textbook, find the volume of the box.

Solution

Common formula: $V = lwh$

Equation: $V = 3 \cdot 4 \cdot 2$

$$V = 24$$

77. *Average Wage*   The average hourly wage y for public school bus drivers in the United States from 1980 to 1992 can be modeled by $y = 5.23 + 0.396t$ where $t = 0$ represents 1980 (see figure in textbook). During which year was the average hourly wage $8.40? What was the average annual raise for bus drivers during this 12-year period? (Source: *National Survey of Salaries and Wages in Public Schools*)

Solution

The average hourly wage was $8.40 in 1988. The average annual raise for bus drivers during this 12-year period was $0.396.

79. *Comparing Wage Increases*   Use the information given in Exercises 77 and 78 to determine which of the two groups' average wages was increasing at a greater annual amount during the 12-year period from 1980 to 1992.

Solution

The bus drivers' average salaries were increasing at a greater annual amount during the 12-year period.

81. *Average Speed*   An Olympic runner completes a 5000-meter race in 13 minutes and 20 seconds. What was the average speed of the runner?

Solution

Verbal model: $\boxed{\text{Distance}} = \boxed{\text{Rate}} \cdot \boxed{\text{Time}}$

Equation: $5000 = x \cdot 13\frac{20}{60}$

$$5000 = x \cdot 13\frac{1}{3}$$

$$\frac{5000}{13\frac{1}{3}} = x$$

$$5000 \cdot \frac{3}{40} = x$$

$$375 \text{ m/min} = x$$

$$6.25 \text{ m/sec} = x$$

Mid-Chapter Quiz for Chapter 1

1. Solve $4x + 3 = 11$ and check the result. (If it is not possible, state the reason.)

Solution

$$4x + 3 = 11 \qquad\qquad \textbf{Check} \quad 4(2) + 3 \overset{?}{=} 11$$

$$4x + 3 - 3 = 11 - 3 \qquad\qquad\qquad 8 + 3 \overset{?}{=} 11$$

$$4x = 8 \qquad\qquad\qquad\qquad 11 = 11$$

$$\frac{4x}{4} = \frac{8}{4}$$

$$x = 2$$

2. Solve $-3(z - 2) = 0$ and check the result. (If it is not possible, state the reason.)

Solution

$$-3(z - 2) = 0 \qquad\qquad \textbf{Check} \quad -3(2 - 2) \overset{?}{=} 0$$

$$\frac{-3(z - 2)}{-3} = \frac{0}{-3} \qquad\qquad\qquad -3(0) \overset{?}{=} 0$$

$$z - 2 = 0 \qquad\qquad\qquad\qquad 0 = 0$$

$$z - 2 + 2 = 0 + 2$$

$$z = 2$$

3. Solve $2(y + 3) = 18 - 4y$ and check the result. (If it is not possible, state the reason.)

Solution

$$2(y + 3) = 18 - 4y \qquad\qquad \textbf{Check} \quad 2(2 + 3) \overset{?}{=} 18 - 4(2)$$

$$2y + 6 = 18 - 4y \qquad\qquad\qquad\qquad 2(5) \overset{?}{=} 18 - 8$$

$$2y + 4y + 6 = 18 - 4y + 4y \qquad\qquad\qquad 10 = 10$$

$$6y + 6 - 6 = 18 - 6$$

$$6y = 12$$

$$\frac{6y}{6} = \frac{12}{6}$$

$$y = 2$$

4. Solve $5t + 7 = 7(t + 1) - 2t$ and check the result. (If it is not possible, state the reason.)

Solution

$$5t + 7 = 7(t + 1) - 2t$$

$$5t + 7 = 7t + 7 - 2t$$

$$5t + 7 = 5t + 7$$

Identity

5. Solve $\frac{1}{4}x + 6 = \frac{3}{2}x - 1$ and check the result. (If it is not possible, state the reason.)

Solution

$$\frac{1}{4}x + 6 = \frac{3}{2}x - 1$$

$$4\left(\frac{1}{4}x + 6\right) = 4\left(\frac{3}{2}x - 1\right)$$

$$x + 24 = 6x - 4$$

$$x - x + 24 = 6x - 4 - x$$

$$24 = 5x - 4$$

$$24 + 4 = 5x - 4 + 4$$

$$28 = 5x$$

$$\frac{28}{5} = \frac{5x}{5}$$

$$\frac{28}{5} = x$$

Check

$$\frac{1}{4}\left(\frac{28}{5}\right) + 6 \stackrel{?}{=} \frac{3}{2}\left(\frac{28}{5}\right) - 1$$

$$\frac{7}{5} + \frac{30}{5} \stackrel{?}{=} \frac{42}{5} - \frac{5}{5}$$

$$\frac{37}{5} = \frac{37}{5}$$

6. Solve $\frac{u}{4} + \frac{u}{3} = 1$ and check the result. (If it is not possible, state the reason.)

Solution

$$\frac{u}{4} + \frac{u}{3} = 1$$

$$12\left(\frac{u}{4} + \frac{u}{3}\right) = 12(1)$$

$$3u + 4u = 12$$

$$7u = 12$$

$$\frac{7u}{7} = \frac{12}{7}$$

$$u = \frac{12}{7}$$

Check

$$\frac{12/7}{4} + \frac{12/7}{3} \stackrel{?}{=} 1$$

$$\frac{3}{7} + \frac{4}{7} \stackrel{?}{=} 1$$

$$\frac{7}{7} \stackrel{?}{=} 1$$

$$1 = 1$$

7. Solve $\dfrac{4-x}{5} + 5 = \dfrac{5}{2}$ and check the result. (If it is not possible, state the reason.)

Solution

$$\frac{4-x}{5} + 5 = \frac{5}{2}$$

$$10\left(\frac{4-x}{5} + 5\right) = 10\left(\frac{5}{2}\right)$$

$$2(4-x) + 50 = 25$$

$$8 - 2x + 50 = 25$$

$$-2x + 58 = 25$$

$$-2x + 58 - 58 = 25 - 58$$

$$-2x = -33$$

$$\frac{-2x}{-2} = \frac{-33}{-2}$$

$$x = \frac{33}{2}$$

Check

$$\frac{4 - \frac{33}{2}}{5} + 5 \stackrel{?}{=} \frac{5}{2}$$

$$\frac{\frac{8}{2} - \frac{33}{2}}{5} + 5 \stackrel{?}{=} \frac{5}{2}$$

$$-\frac{25}{2} \cdot \frac{1}{5} + 5 \stackrel{?}{=} \frac{5}{2}$$

$$-\frac{5}{2} + \frac{10}{2} \stackrel{?}{=} \frac{5}{2}$$

$$\frac{5}{2} = \frac{5}{2}$$

8. Solve $0.2x + 0.3 = 1.5$ and check the result. (If it is not possible, state the reason.)

Solution

$$0.2x + 0.3 = 1.5$$

$$10(0.2x + 0.3) = 10(1.5)$$

$$2x + 3 = 15$$

$$2x + 3 - 3 = 15 - 3$$

$$2x = 12$$

$$\frac{2x}{2} = \frac{12}{2}$$

$$x = 6$$

Check

$$0.2(6) + 0.3 \stackrel{?}{=} 1.5$$

$$1.2 + 0.3 \stackrel{?}{=} 1.5$$

$$1.5 = 1.5$$

9. Solve $3x + \frac{11}{12} = \frac{5}{16}$ and round your answer to two decimal places. (A calculator may be helpful.)

Solution

$$3x + \frac{11}{12} = \frac{5}{16}$$

$$3x + \frac{11}{12} - \frac{11}{12} = \frac{5}{16} - \frac{11}{12}$$

$$3x = \frac{15}{48} - \frac{44}{48}$$

$$3x = -\frac{29}{48}$$

$$\frac{3x}{3} = -\frac{29}{48} \div 3$$

$$x = -\frac{29}{48} \cdot \frac{1}{3}$$

$$x = -\frac{29}{144} = -0.20$$

10. Solve $0.42x + 6 = 5.25x - 0.80$ and round your answer to two decimal places. (A calculator may be helpful.)

Solution

$$0.42x + 6 = 5.25x - 0.80$$

$$0.42x + 6 - 5.25x = 5.25x - 0.80 - 5.25x$$

$$-4.83x + 6 = -0.80$$

$$-4.83x + 6 - 6 = -0.80 - 6$$

$$-4.83x = -6.80$$

$$\frac{-4.83x}{-4.83} = \frac{-6.80}{-4.83}$$

$$x = 1.41$$

11. Explain how to write the decimal 0.45 as a fraction and as a percent.

Solution

0.45 is 45 hundredths so $0.45 = \frac{45}{100}$ which reduces to $\frac{9}{20}$ and since percent means hundredths, $0.45 = 45\%$.

12. 500 is 250% of what number?

Solution

$$a = p \cdot b$$

$$500 = (2.50)(b)$$

$$\frac{500}{2.50} = b$$

$$200 = b$$

13. Find the unit price (in dollars per ounce) of a 12-ounce box of cereal that sells for $2.35.

Solution

$$\frac{\text{Total price}}{\text{Total units}} = \frac{2.35}{12} = \frac{235}{1200} = \$0.1958\overline{3} \text{ per ounce}$$

14. A quality control engineer for a manufacturer found one defective unit in a sample of 300. At that rate, what is the expected number of defective units in a shipment of 600,000?

Solution

Verbal model: $\boxed{\dfrac{\text{Defective units}}{\text{Total units}}} = \boxed{\dfrac{\text{Defective units}}{\text{Total units}}}$

Proportion: $\dfrac{x}{600{,}000} = \dfrac{1}{300}$

$$x = 600{,}000 \cdot \frac{1}{300}$$

$$x = 2000 \text{ defective units}$$

15. A store is offering a discount of 25% on a computer with a list price of $1750. A mail-order catalog has the same machine for $1250 plus $24.95 for shipping. Which is the better buy?

Solution

Store computer

Verbal model: $\boxed{\text{Discount}} = \boxed{\text{Discount rate}} \cdot \boxed{\text{List price}}$

Equation: $x = (0.25)(1750)$

$x = \$437.50$

Verbal model: $\boxed{\text{Selling price}} = \boxed{\text{List price}} - \boxed{\text{Discount}}$

Equation: $x = 1750 - 427.50$

$x = \$1312.50$

Mail-order computer

Verbal model: $\boxed{\text{Selling price}} = \boxed{\text{List price}} + \boxed{\text{Shipping}}$

Equation: $x = 1250 + 24.95$

$x = \$1274.95$

The mail-order computer is the better buy.

16. Last week you earned $616. Your regular hourly wage is $12.25 for the first 40 hours, and your overtime hourly wage is $18. How many hours of overtime did you work?

Solution

Verbal model: $\boxed{\text{Total wages}} = \boxed{\text{Regular wages}} + \boxed{\text{Overtime wages}}$

Equation: $616 = 40(12.25) + x(18)$

$616 = 490 + 18x$

$126 = 18x$

$7 = x$

17. Fifty gallons of a 30% acid solution is obtained by combining solutions that are 25% acid and 50% acid. How much of each solution is required?

Solution

Verbal model: $\boxed{\begin{array}{c}\text{Amount of}\\\text{Solution 1}\end{array}} + \boxed{\begin{array}{c}\text{Amount of}\\\text{Solution 2}\end{array}} = \boxed{\begin{array}{c}\text{Amount of}\\\text{final solution}\end{array}}$

Equation: $0.25x + 0.50(50 - x) = 0.30(50)$

$0.25x + 25 - 0.50x = 15$

$25 - 0.25x = 15$

$-0.25x = -10$

$x = 40 \text{ gal at } 25\%$

$50 - x = 10 \text{ gal at } 50\%$

18. On the first part of a 300-mile trip, a sales representative averaged 62 miles per hour. The sales representative averaged 46 miles per hour on the last part of the trip because of the increased volume of traffic. Find the amount of driving time at each speed if the total time was 6 hours.

Solution

Verbal model: $\boxed{\text{Distance}} = \boxed{\text{Rate}} \cdot \boxed{\text{Time}}$

Equation: $300 = 62x + 46(6 - x)$

$$300 = 62x + 276 - 46x$$

$$24 = 16x$$

$$1.5 \text{ hours} = x \text{ (first part of trip)}$$

$$4.5 \text{ hours} = 6 - x \text{ (second part of trip)}$$

19. You can paint a room in 6 hours and your friend can paint it in 8 hours. How long will it take both of you to paint the room?

Solution

Solution

Verbal model: $\boxed{\begin{array}{c}\text{Work} \\ \text{done}\end{array}} = \boxed{\begin{array}{c}\text{Portion done by} \\ \text{first person}\end{array}} + \boxed{\begin{array}{c}\text{Portion done by} \\ \text{second person}\end{array}}$

Equation: $1 = \left(\frac{1}{6}\right)(t) + \left(\frac{1}{8}\right)(t)$

$$1 = \left(\frac{1}{6} + \frac{1}{8}\right)(t)$$

$$1 = \left(\frac{14}{48}\right)(t)$$

$$3.43 \text{ hours} \approx \frac{24}{7} = \frac{48}{14} = t$$

20. The figure in the textbook shows three squares. The perimeters of squares I and II are 20 inches and 32 inches, respectively. Find the area of square III.

Solution

Perimeter of square I = 20   Perimeter of square II = 32   Length of side of square III $= 5 + 8$

$$4s = 20 \qquad\qquad\qquad 4s = 32 \qquad\qquad\qquad\qquad = 13$$

$$s = 5 \qquad\qquad\qquad\quad s = 8$$

$$\text{Area} = s^2$$

$$= 13^2$$

$$= 169 \text{ square inches}$$

1.4 Linear Inequalities

7. Determine whether the following values of x satisfy the inequality $7x - 10 > 0$.

(a) $x = 3$ (b) $x = -2$ (c) $x = \frac{5}{2}$ (d) $x = \frac{1}{2}$

Solution

(a) $7(3) - 10 > 0$

$\quad 21 - 10 > 0$

$\qquad 11 > 0$

$\quad$ True

(b) $7(-2) - 10 > 0$

$\quad -14 - 10 > 0$

$\qquad -24 > 0$

$\quad$ False

(c) $7\left(\frac{5}{2}\right) - 10 > 0$

$\quad \frac{35}{2} - 10 > 0$

$\quad \frac{35}{2} - \frac{20}{2} > 0$

$\qquad \frac{15}{2} > 0$

$\quad$ True

(d) $7\left(\frac{1}{2}\right) - 10 > 0$

$\quad \frac{7}{2} - 10 > 0$

$\quad \frac{7}{2} - \frac{20}{2} > 0$

$\qquad -\frac{13}{2} > 0$

$\quad$ False

9. Determine whether the following values of x satisfy the inequality $0 < \frac{x+5}{6} < 2$.

(a) $x = 10$ (b) $x = 4$ (c) $x = 0$ (d) $x = -6$

Solution

(a) $0 < \frac{10+5}{6} < 2$

$\quad 0 < \frac{15}{6} < 2$

$\quad 0 < 2\frac{3}{6} < 2$

$\quad$ False

(b) $0 < \frac{4+5}{6} < 2$

$\quad 0 < \frac{9}{6} < 2$

$\quad 0 < 1\frac{3}{6} < 2$

$\quad$ True

(c) $0 < \frac{0+5}{6} < 2$

$\quad 0 < \frac{5}{6} < 2$

$\quad$ True

(d) $0 < \frac{-6+5}{6} < 2$

$\quad 0 < \frac{-1}{6} < 2$

$\quad$ False

11. Match the inequality $x \geq 4$ with its graph. [The graphs are labeled (a), (b), (c), and (d) in the textbook.]

Solution

Matches graph (d).

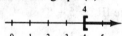

13. Match the inequality $-4 < x < 4$ with its graph. [The graphs are labeled (a), (b), (c), and (d) in the textbook.]

Solution

Matches graph (a).

15. Use a graphing utility to graph the inequality $-5 < x \leq 3$ and write its interval notation.

Solution

$(-5, 3]$

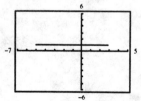

17. Use a graphing utility to graph the inequality $\frac{3}{2} \geq x > 0$ and write its interval notation.

Solution

$\left(0, \frac{3}{2}\right]$

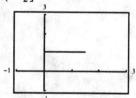

19. Use a graphing utility to graph the inequality $-\frac{15}{4} < x < -\frac{5}{2}$ and write its interval notation.

Solution

$\left(-\frac{15}{4}, -\frac{5}{2}\right)$

21. Solve the inequality $x + 7 \le 9$ and sketch the solution on the real number line.

Solution

$$x + 7 \le 9$$

$$x + 7 - 7 \le 9 - 7$$

$$x \le 2$$

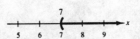

23. Solve the inequality $2x - 5 > 9$ and sketch the solution on the real number line.

Solution

$$2x - 5 > 9$$

$$2x - 5 + 5 > 9 + 5$$

$$2x > 14$$

$$\frac{2x}{2} > \frac{14}{2}$$

$$x > 7$$

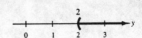

25. Solve the inequality $16 < 4(y + 2) - 5(2 - y)$ and sketch the solution on the real number line.

Solution

$$16 < 4(y + 2) - 5(2 - y)$$

$$16 < 4y + 8 - 10 + 5y$$

$$16 < 9y - 2$$

$$16 + 2 < 9y - 2 + 2$$

$$18 < 9y$$

$$\frac{18}{9} < \frac{9y}{9}$$

$$2 < y$$

27. Solve the following inequality and sketch the solution on the real number line.

$$-3 < \frac{2x - 3}{2} < 3$$

Solution

$$-3 < \frac{2x - 3}{2} < 3$$

$$-6 < 2x - 3 < 6$$

$$-6 + 3 < 2x - 3 + 3 < 6 + 3$$

$$-3 < 2x < 9$$

$$\frac{-3}{2} < \frac{2x}{2} < \frac{9}{2}$$

$$-\frac{3}{2} < x < \frac{9}{2}$$

29. Solve the following inequality and sketch the solution on the real number line.

$$1 > \frac{x - 4}{-3} > -2$$

Solution

$$1 > \frac{x - 4}{-3} > -2$$

$$-3 < x - 4 < 6$$

$$-3 + 4 < x - 4 + 4 < 6 + 4$$

$$1 < x < 10$$

31. Rewrite the statement "x is nonnegative" using inequality notation.

Solution

$x \geq 0$

33. Rewrite the statement "n is no more than -2" using inequality notation.

Solution

$n \leq -2$

35. Write a verbal description of $x \geq \frac{5}{2}$.

Solution

x is at least $\frac{5}{2}$.

37. *Using a Linear Model* A utility company has a fleet of vans. The annual operating cost per van is $C = 0.28m + 2900$ where m is the number of miles traveled by a van in a year. How many miles will yield an annual operating cost that is less than $10,000?

Solution

Verbal model: 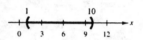 Operating cost $<$ \$10,000

Inequality:

$$0.28m + 2900 < 10{,}000$$

$$0.28m + 2900 - 2900 < 10{,}000 - 2900$$

$$0.28m < 7100$$

$$\frac{0.28m}{0.28} < \frac{7100}{0.28}$$

$$m < 25{,}357.14286$$

$$m \leq 25{,}357$$

39. *Distance* If you live 5 miles from school and your friend lives 3 miles from you, then the distance d that your friend lives from school is in what interval? Use the figure in the textbook to help determine your answer.

Solution

Maximum distance: $5 + 3 = 8$ miles
Minimum distance: $5 - 3 = 2$ miles
$2 \leq d \leq 8$

41. Place the correct inequality symbol ($<$ or $>$) between $-\frac{3}{4}$ and -5.

Solution

$-\frac{3}{4} > -5$

43. Place the correct inequality symbol ($<$ or $>$) between π and -3.

Solution

$\pi > -3$

45. Solve $-2n + 12 = 5$.

Solution

$$-2n + 12 = 5$$
$$-2n + 12 - 12 = 5 - 12$$
$$-2n = -7$$
$$\frac{-2n}{-2} = \frac{-7}{-2}$$
$$n = \frac{7}{2}$$

47. Solve $55 - 4(3 - y) = -5$.

Solution

$$55 - 4(3 - y) = -5$$
$$55 - 12 + 4y = -5$$
$$43 + 4y = -5$$
$$43 + 4y - 43 = -5 - 43$$
$$4y = -48$$
$$\frac{4y}{4} = -\frac{48}{4}$$
$$y = -12$$

49. Match the inequality $-1 < x \leq 2$ with its graph. [The graphs are labeled (a), (b), (c), (d), (e), and (f) in the textbook.]

Solution

Matches graph (f).

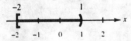

51. Match the inequality $-1 \leq x \leq 1$ with its graph. [The graphs are labeled (a), (b), (c), (d), (e), and (f) in the textbook.]

Solution

Matches graph (a).

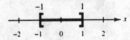

53. Match the inequality $-2 \leq x < 1$ with its graph. [The graphs are labeled (a), (b), (c), (d), (e), and (f) in the textbook.]

Solution

Matches graph (d).

55. Use a graphing utility to graph the inequality $x \leq -2$ and write its interval notation.

Solution

$(-\infty, -2]$

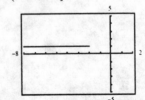

57. Use a graphing utility to graph the inequality $-2 < x \leq 4$ and write its interval notation.

Solution

$(-2, 4]$

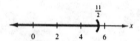

59. Use a graphing utility to graph the inequality $9 \geq x \geq 3$ and write its interval notation.

Solution

$[3, 9]$

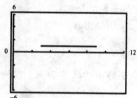

61. Solve the inequality $4x < 22$ and sketch the solution on the real number line.

Solution

$4x < 22$

$\dfrac{4x}{4} < \dfrac{22}{4}$

$x < \dfrac{11}{2}$

63. Solve the inequality $-9x \geq 36$ and sketch the solution on the real number line.

Solution

$-9x \geq 36$

$\dfrac{-9x}{-9} \leq \dfrac{36}{-9}$

$x \leq -4$

65. Solve the inequality $-\frac{3}{4}x < -6$ and sketch the solution on the real number line.

Solution

$-\dfrac{3}{4}x < -6$

$-\dfrac{4}{3} \cdot -\dfrac{3}{4}x > -6 \cdot -\dfrac{4}{3}$

$x > 8$

67. Solve the inequality $5 - x \leq -2$ and sketch the solution on the real number line.

Solution

$5 - x \leq -2$

$5 - x - 5 \leq -2 - 5$

$-x \leq -7$

$-1 \cdot x \geq -7 \cdot -1$

$x \geq 7$

69. Solve the inequality $5 - 3x < 7$ and sketch the solution on the real number line.

Solution

$5 - 3x < 7$

$5 - 3x - 5 < 7 - 5$

$-3x < 2$

$\dfrac{-3x}{-3} > \dfrac{2}{-3}$

$x > -\dfrac{2}{3}$

71. Solve the inequality $3x - 11 > -x + 7$ and sketch the solution on the real number line.

Solution

$$3x - 11 > -x + 7$$
$$3x - 11 + x > -x + 7 + x$$
$$4x - 11 > 7$$
$$4x - 11 + 11 > 7 + 11$$
$$4x > 18$$
$$\frac{4x}{4} > \frac{18}{4}$$
$$x > \frac{9}{2}$$

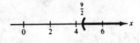

73. Solve the inequality $-3(y + 10) \geq 4(y + 10)$ and sketch the solution on the real number line.

Solution

$$-3(y + 10) \geq 4(y + 10)$$
$$-3y - 30 \geq 4y + 40$$
$$3y - 3y - 30 \geq 3y + 4y + 40$$
$$-30 \geq 7y + 40$$
$$-40 - 30 \geq 7y + 40 - 40$$
$$-70 \geq 7y$$
$$-10 \geq y$$

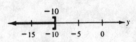

75. Solve the inequality $\dfrac{x}{6} - \dfrac{x}{4} \leq 1$ and sketch the solution on the real number line.

Solution

$$\frac{x}{6} - \frac{x}{4} \leq 1$$
$$2x - 3x \leq 12$$
$$-x \leq 12$$
$$(-1)(-x) \geq 12(-1)$$
$$x \geq -12$$

77. Solve the inequality $0 < 2x - 5 < 9$ and sketch the solution on the real number line.

Solution

$$0 < 2x - 5 < 9$$
$$0 + 5 < 2x - 5 + 5 < 9 + 5$$
$$5 < 2x < 14$$
$$\frac{5}{2} < \frac{2x}{2} < \frac{14}{2}$$
$$\frac{5}{2} < x < 7$$

79. Solve the inequality $\frac{2}{5} < x + 1 < \frac{4}{5}$ and sketch the solution on the real number line.

Solution

$$\frac{2}{5} < x + 1 < \frac{4}{5}$$
$$2 < 5(x + 1) < 4$$
$$2 < 5x + 5 < 4$$
$$2 - 5 < 5x + 5 - 5 < 4 - 5$$
$$-3 < 5x < -1$$
$$\frac{-3}{5} < \frac{5x}{5} < \frac{-1}{5}$$
$$-\frac{3}{5} < x < -\frac{1}{5}$$

81. Rewrite the statement "z is at least 2" using inequality notation.

Solution

$z \geq 2$

83. Write a verbal description of $y \geq 3$.

Solution

y is greater than or equal to 3 or y is at least 3.

85. Write a verbal description of $0 < z \leq \pi$.

Solution

z is greater than 0 and no more than π.

87. *Travel Budget* A student group has $4500 budgeted for a field trip. The cost of transportation for the trip is $1900. To stay within the budget, all other costs C must be no more than what amount?

Solution

Verbal model: | Transportation costs | $+$ | Other costs | $\leq$ | Total money for trip |

Inequality:

$$1900 + C \leq 4500$$
$$1900 + C - 1900 \leq 4500 - 1900$$
$$C \leq 2600$$

89. *Comparing Average Temperatures*   The average temperature in Miami is greater than the average temperature in Washington, D.C. The average temperature in Washington, D.C. is greater than the average temperature in New York. How does the average temperature in Miami compare with that in New York?

Solution

Verbal model: Temp in Miami > Temp in Washington and Temp in Washington > Temp in New York

The average temperature in Miami, therefore, is greater than (>) the average temperature in New York.

91. *Hourly Wage*   You must select one of two plans for payment when working for a company. One plan pays a straight $12.50 per hour. The second pays $8.00 per hour plus $0.75 per unit produced per hour. Write an inequality yielding the number of units that must be produced per hour so that the second plan gives the greater hourly wage. Solve the inequality.

Solution

Verbal model: Second plan > First plan

Equation: $8 + 0.75x > 12.50$

$$0.75x > 4.50$$

$$x > 6 \text{ units produced per hour}$$

93. *Profit*   The revenue and cost for selling x units of a product are given by $R = 89.95x$ and $C = 61x + 875$. To obtain a profit, the revenue must be greater than the cost. For what values of x will this product produce a profit?

Solution

Verbal model: Revenue > Cost

Inequality: $89.95x > 61x + 875$

$$89.95x - 61x > 61x + 875 - 61x$$

$$28.95x > 875$$

$$\frac{28.95x}{28.95} > \frac{875}{28.95}$$

$$x > 30.224525$$

$$x \geq 31$$

95. *Long Distance Charges*   The cost for a long distance telephone call is $0.96 for the first minute and $0.75 for each additional minute. If the total cost of the call cannot exceed $5, find the interval of time that is available for the call.

Solution

Verbal model: $\boxed{\begin{array}{c}\text{Cost of} \\ \text{first minute}\end{array}} + \boxed{\begin{array}{c}\text{Cost of additional} \\ \text{minutes}\end{array}} \leq \boxed{\$5.00}$

Inequality:

$$\$0.96 + \$0.75x \leq \$5.00$$

$$0.96 + 0.75x - 0.96 \leq 5.00 - 0.96$$

$$0.75x \leq 4.04$$

$$\frac{0.75x}{0.75} \leq \frac{4.04}{0.75}$$

$$x \leq 5.386667$$

Since x represents the additional minutes after the first minute, the call must be less than 6.38 minutes. If a portion of a minute is billed as a full minute, then the call must be less than or equal to 6 minutes.

97. *Air Pollutant Emissions*   Use the equation $y = 0.518 - 0.048t$, $3 \leq t \leq 11$, which models the amount of air pollutant emissions of lead in the continental United States from 1983 to 1991 (see figure in textbook). In this model, y represents the amount of pollutant in micrograms per cubic meter of air and t represents the year, with $t = 0$ corresponding to 1980. (Source: U. S. Environmental Protection Agency). During which years between 1983 and 1991 was the air pollutant emission of lead greater than 0.23 micrograms per cubic meter?

Solution

$$0.518 - 0.048t > 0.23$$

$$-0.048t > -0.288$$

$$t < 6$$

Years 1983, 1984, 1985

99. *Geometry*   If a, b, and c are the lengths of the sides of the triangle in the figure, then $a + b$ must be greater than what value? (See figure in textbook.)

Solution

$a + b > c$

101. Five times a number n must be at least 15 and no more than 45. What interval contains this number?

Solution

$$15 \leq 5n \leq 45$$

$$\frac{15}{5} \leq \frac{5n}{5} \leq \frac{45}{5}$$

$$3 \leq n \leq 9$$

$[3, 9]$

103. Determine all real numbers n such that $\frac{1}{4}n$ must be more than 8.

Solution

$$\frac{1}{4}n > 8$$

$$4 \cdot \frac{1}{4}n > 8 \cdot 4$$

$$n > 32$$

$(32, \infty)$

1.5 Absolute Value Equations and Inequalities

7. Determine whether $x = -3$ is a solution of the equation $|4x + 5| = 10$.

Solution

$$|4(-3) + 5| \overset{?}{=} 10$$
$$|-12 + 5| \overset{?}{=} 10$$
$$|-7| \overset{?}{=} 10$$
$$7 \neq 10$$

No

9. Determine whether $w = 4$ is a solution of the equation $|6 - 2w| = 2$.

Solution

$$|6 - 2(4)| \overset{?}{=} 2$$
$$|6 - 8| \overset{?}{=} 2$$
$$|-2| \overset{?}{=} 2$$
$$2 = 2$$

Yes

11. Transform $|x - 10| = 17$ into two linear equations.

Solution

$x - 10 = 17$ or $x - 10 = -17$

13. Transform $|4x + 1| = \frac{1}{2}$ into two linear equations.

Solution

$4x + 1 = \frac{1}{2}$ or $4x + 1 = -\frac{1}{2}$

15. Solve $|t| = 45$. (There may be no solution.)

Solution

$|t| = 45$
$t = 45$ or $t = -45$

17. Solve $|2s + 3| = 25$. (There may be no solution.)

Solution

$$2s + 3 = 25 \quad \text{or} \quad 2s + 3 = -25$$
$$2s = 22 \qquad\qquad 2s = -28$$
$$s = 11 \qquad\qquad s = -14$$

19. Solve $|3x + 4| = -16$. (There may be no solution.)

Solution

No solution

21. Solve $\left|\frac{2}{3}x + 4\right| = 9$. (There may be no solution.)

Solution

$$\frac{2}{3}x + 4 = 9 \quad \text{or} \quad \frac{2}{3}x + 4 = -9$$
$$\frac{2}{3}x = 5 \qquad\qquad \frac{2}{3}x = -13$$
$$2x = 15 \qquad\qquad 2x = -39$$
$$x = \frac{15}{2} \qquad\qquad x = -\frac{39}{2}$$

23. Solve $|x + 2| = |3x - 1|$. (There may be no solution.)

Solution

$$x + 2 = 3x - 1 \quad \text{or} \quad x + 2 = -(3x - 1)$$
$$3 = 2x \qquad\qquad x + 2 = -3x + 1$$
$$\frac{3}{2} = x \qquad\qquad 4x = -1$$
$$\qquad\qquad\qquad x = -\frac{1}{4}$$

25. *Think About It* Write a single equation that is equivalent to the two equations $2x + 3 = 5$ and $2x + 3 = -5$.

Solution

$|2x + 3| = 5$

27. Determine whether the following x-values are solutions of the inequality $|x| < 3$.

(a) $x = 2$ (b) $x = -4$ (c) $x = 4$ (d) $x = -1$

Solution

(a) $|2| < 3$ (b) $|-4| < 3$ (c) $|4| < 3$ (d) $|-1| < 3$

$\quad 2 < 3$ $\quad 4 < 3$ $\quad 4 < 3$ $\quad 1 < 3$

$\quad$ Yes $\quad$ No $\quad$ No $\quad$ Yes

29. Determine whether the following x-values are solutions of the inequality $|x - 7| \geq 3$.

(a) $x = 9$ (b) $x = -4$ (c) $x = 11$ (d) $x = 6$

Solution

(a) $|9 - 7| \geq 3$ (b) $|-4 - 7| \geq 3$ (c) $|11 - 7| \geq 3$ (d) $|6 - 7| \geq 3$

$\quad |2| \geq 3$ $\quad |-11| \geq 3$ $\quad |4| \geq 3$ $\quad |-1| \geq 3$

$\quad 2 \geq 3$ $\quad 11 \geq 3$ $\quad 4 \geq 3$ $\quad 1 \geq 3$

$\quad$ No $\quad$ Yes $\quad$ Yes $\quad$ No

31. Transform $|y + 5| < 3$ into a double inequality or two separate inequalities.

Solution

$\quad |y + 5| < 3$

$\quad -3 < y + 5 < 3$

33. Transform $|7 - 2h| \geq 9$ into a double inequality or two separate inequalities.

Solution

$|7 - 2h| \geq 9$

$\quad 7 - 2h \geq 9 \text{ or } 7 - 2h \leq -9$

35. Solve $|y| < 4$ and sketch the graph of the solution.

Solution

$\quad |y| < 4$

$\quad -4 < y < 4$

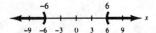

37. Solve $|y + 2| < 4$ and sketch the graph of the solution.

Solution

$\quad |y + 2| < 4$

$\quad -4 < y + 2 < 4$

$\quad\quad -6 < y < 2$

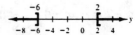

39. Solve $|x| > 6$ and sketch the graph of the solution.

Solution

$|x| > 6$

$\quad x > 6 \text{ or } x < -6$

41. Solve $|y + 2| \geq 4$ and sketch the graph of the solution.

Solution

$\quad y + 2 \leq -4 \quad \text{ or } \quad y + 2 \geq 4$

$\quad\quad y \leq -6 \quad\quad\quad y \geq 2$

43. Solve $|2x + 3| > 9$ and sketch the graph of the solution.

Solution

$$2x + 3 > 9 \quad \text{or} \quad 2x + 3 < -9$$
$$2x > 6 \qquad\qquad 2x < -12$$
$$x > 3 \qquad\qquad x < -6$$

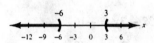

45. Solve the inequality $|0.2x - 3| < 4$. (If it is not possible, state the reason.)

Solution

$$|0.2x - 3| < 4$$
$$-4 < 0.2x - 3 < 4$$
$$-1 < 0.2x < 7$$
$$\frac{-1}{0.2} < x < \frac{7}{0.2}$$
$$-5 < x < 35$$

47. Solve the inequality $\left|\dfrac{z}{10} - 3\right| > 8$. (If it is not possible, state the reason.)

Solution

$$\frac{z}{10} - 3 < -8 \quad \text{or} \quad \frac{z}{10} - 3 > 8$$
$$\frac{z}{10} < -5 \qquad\qquad \frac{z}{10} > 11$$
$$z < -50 \qquad\qquad z > 110$$

49. *Body Temperature* Physicians consider an adult's body temperature to be normal if it is between 97.6°F and 99.6°F. Write an absolute value inequality that describes this normal temperature range.

Solution

$$97.6 < t < 99.6$$
$$97.6 - 98.6 < t - 98.6 < 99.6 - 98.6$$
$$-1 < t - 98.6 < 1$$
$$|t - 98.6| < 1$$

51. What is $7\frac{1}{2}\%$ of 25?

Solution

$$a = p \cdot b$$
$$a = (0.075)(25)$$
$$a = 1.875$$

53. 225 is what percent of 150?

Solution

$$a = p \cdot b$$
$$225 = p \cdot 150$$
$$\frac{225}{150} = p$$
$$150\% = p$$

55. 0.5% of what number is 400?

Solution

$$a = p \cdot b$$
$$400 = (0.005)(b)$$
$$\frac{400}{0.005} = b$$
$$80{,}000 = b$$

57. Solve $|h| = 0$. (There may be no solution.)

Solution

$$h = 0$$

59. Solve $|x - 16| = 5$. (There may be no solution.)

Solution

$$x - 16 = 5 \quad \text{or} \quad x - 16 = -5$$
$$x = 21 \qquad\qquad x = 11$$

61. Solve $|32 - 3y| = 16$. (There may be no solution.)

Solution

$$32 - 3y = 16 \quad \text{or} \quad 32 - 3y = -16$$

$$-3y = -16 \qquad\qquad -3y = -48$$

$$y = \frac{16}{3} \qquad\qquad\qquad y = 16$$

63. Solve $|3x - 2| = -5$. (There may be no solution.)

Solution

No solution

65. Solve $|5x - 3| + 8 = 22$. (There may be no solution.)

Solution

$$|5x - 3| + 8 = 22$$

$$|5x - 3| = 14$$

$$5x - 3 = 14 \quad \text{or} \quad 5x - 3 = -14$$

$$5x = 17 \qquad\qquad 5x = -11$$

$$x = \tfrac{17}{5} \qquad\qquad x = -\tfrac{11}{5}$$

67. Solve $|0.32x - 2| = 4$. (There may be no solution.)

Solution

$$0.32x - 2 = 4 \quad \text{or} \quad 0.32x - 2 = -4$$

$$0.32x = 6 \qquad\qquad 0.32x = -2$$

$$x = \frac{6}{0.32} \qquad\qquad x = \frac{-2}{0.32}$$

$$x = 18.75 \qquad\qquad x = -6.25$$

69. Solve $|x + 8| = |2x + 1|$. (There may be no solution.)

Solution

$$x + 8 = 2x + 1 \quad \text{or} \quad x + 8 = -(2x + 1)$$

$$8 = x + 1 \qquad\qquad x + 8 = -2x - 1$$

$$7 = x \qquad\qquad 3x + 8 = -1$$

$$3x = -9$$

$$x = -3$$

71. Solve $|5x + 4| = |3x + 8|$. (There may be no solution.)

Solution

$$5x + 4 = 3x + 8 \quad \text{or} \quad 5x + 4 = -(3x + 8)$$

$$2x = 4 \qquad\qquad 5x + 4 = -3x - 8$$

$$x = 2 \qquad\qquad 8x = -12$$

$$x = -\tfrac{12}{8} = -\tfrac{3}{2}$$

73. Determine whether the following x-values are solutions of the inequality $|x| \geq 3$.

(a) $x = 2$ (b) $x = -4$ (c) $x = 4$ (d) $x = -1$

Solution

(a) $|2| \geq 3$ (b) $|-4| \geq 3$ (c) $|4| \geq 3$ (d) $|-1| \geq 3$

     $2 \geq 3$ $4 \geq 3$ $4 \geq 3$ $1 \geq 3$

    No   Yes   Yes   No

75. Determine whether the following x-values are solutions of the inequality $|x - 7| < 3$.

(a) $x = 9$ (b) $x = -4$ (c) $x = 11$ (d) $x = 6$

Solution

(a) $|9 - 7| < 3$ (b) $|-4 - 7| < 3$ (c) $|11 - 7| < 3$ (d) $|6 - 7| < 3$

      $|2| < 3$ $|-11| < 3$ $|4| < 3$ $|-1| < 3$

       $2 < 3$ $11 < 3$ $4 < 3$ $1 < 3$

   Yes   No   No   Yes

77. Sketch a graph that represents the statement "All x greater than -2 *and* less than 5."

Solution

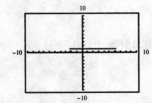

79. Sketch a graph that represents the statement "All x less than or equal to 4 *or* greater than 7."

Solution

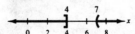

81. Solve $|y - 2| \le 4$. Use a graphing utility to verify the solution.

Solution

$$|y - 2| \le 4$$
$$-4 \le y - 2 \le 4$$
$$-2 \le y \le 6$$

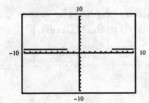

83. Solve $|y| \ge 4$. Use a graphing utility to verify the solution.

Solution

$$y \le -4 \quad \text{or} \quad y \ge 4$$

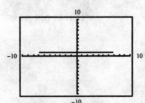

85. Solve $|y - 2| > 4$. Use a graphing utility to verify the solution.

Solution

$$y - 2 < -4 \quad \text{or} \quad y - 2 > 4$$
$$y < -2 \qquad\qquad y > 2$$

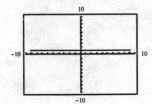

87. Solve $|2x| < 14$. Use a graphing utility to verify the solution.

Solution

$$-14 < 2x < 14$$
$$-7 < x < 7$$

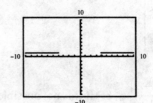

89. Solve $\left|\dfrac{y}{3}\right| \le 3$. Use a graphing utility to verify the solution.

Solution

$$-3 \le \frac{y}{3} \le 3$$
$$-9 \le y \le 9$$

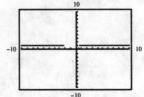

91. Solve $|3x + 2| > 4$. Use a graphing utility to verify the solution.

Solution

$$3x + 2 < -4 \quad \text{or} \quad 3x + 2 > 4$$
$$3x < -6 \qquad\qquad 3x > 2$$
$$x < -2 \qquad\qquad x > \frac{2}{3}$$

93. Solve $|x - 5| + 3 > 5$. Use a graphing utility to verify the solution.

Solution

$$|x - 5| + 3 > 5$$
$$|x - 5| > 2$$
$$x - 5 < -2 \quad \text{or} \quad x - 5 > 2$$
$$x < 3 \qquad\qquad x > 7$$

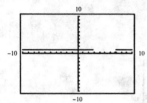

97. Solve the inequality $\dfrac{|x + 2|}{10} \le 8$. (If it is not possible, state the reason.)

Solution

$$\frac{|x + 2|}{10} \le 8$$
$$|x + 2| \le 80$$
$$-80 \le x + 2 \le 80$$
$$-82 \le x \le 78$$

101. Solve the inequality $\dfrac{|s - 3|}{5} > 4$. (If it is not possible, state the reason.)

Solution

$$\frac{|s - 3|}{5} > 4$$
$$|s - 3| > 20$$
$$s - 3 < -20 \quad \text{or} \quad s - 3 > 20$$
$$s < -17 \qquad\qquad s > 23$$

95. *Think About It*   Complete the following statement so that the solution is $0 \le x \le 6$.

$$|x - 3| \le \text{\_\_\_\_}$$

Solution

$|x - 3| \le 3$ since
$$-3 \le x - 3 \le 3$$
$$0 \le \quad x \quad \le 6$$

99. Solve the inequality $|6t + 15| \ge 30$. (If it is not possible, state the reason.)

Solution

$$6t + 15 \le -30 \quad \text{or} \quad 6t + 15 \ge 30$$
$$6t \le -45 \qquad\qquad 6t \ge 15$$
$$t \le -\frac{45}{6} \qquad\qquad t \ge \frac{15}{6}$$
$$t \le -\frac{15}{2} \qquad\qquad t \ge \frac{5}{2}$$

103. Match $|x - 4| \le 4$ with its graph. [The graphs are labeled (a), (b), (c), and (d) in the textbook.]

Solution

Matches graph (d).

$$|x - 4| \le 4$$
$$-4 \le x - 4 \le 4$$
$$0 \le \quad x \quad \le 8$$

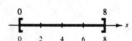

105. Match $\frac{1}{2}|x - 4| > 4$ with its graph. [The graphs are labeled (a), (b), (c), and (d) in the textbook.]

Solution

Matches graph (b).

$\frac{1}{2}|x - 4| > 4$

$|x - 4| > 8$

$x - 4 > 8$ or $x - 4 < -8$

$x > 12$ $x < -4$

107. Write an absolute value inequality that represents the interval shown on the graph in the textbook.

Solution

$[-2, 2]$

$|x| \leq 2$

109. Write an absolute value inequality that represents the interval shown on the graph in the textbook.

Solution

$(7, 13)$

$7 < x < 13$

$-3 < x - 10 < 3$

$|x - 10| < 3$

111. Write an absolute value inequality that represents the interval shown on the graph in the textbook.

Solution

$(16, 22)$

$16 < x < 22$

$-3 < x - 19 < 3$

$|x - 19| < 3$

113. *Temperature* The operating temperature for an electronic device must satisfy the inequality $|t - 72| < 10$ where t is given in degrees Fahrenheit. Sketch the graph of the solution of the inequality.

Solution

$|t - 72| < 10$

$-10 < t - 72 < 10$

$62 < t < 82$

115. Write an absolute value inequality that represents the following verbal statement.

The set of all real numbers x whose distance from 0 is less than 3

Solution

$|x| < 3$

117. Write an absolute value inequality that represents the following verbal statement.

The set of all real numbers x whose distance from 5 is more than 6

Solution

$|x - 5| > 6$

Review Exercises for Chapter 1

1. Determine whether (a) $x = 3$ and (b) $x = 6$ are solutions of $45 - 7x = 3$.

Solution

(a) $45 - 7(3) = 3$

$\qquad 45 - 21 = 3$

$\qquad\qquad 24 = 3$

No

(b) $45 - 7(6) = 3$

$\qquad 45 - 42 = 3$

$\qquad\qquad 3 = 3$

Yes

3. Determine whether (a) $x = \frac{35}{12}$ and (b) $x = -\frac{2}{35}$ are solutions of

$$\frac{x}{7} + \frac{x}{5} = 1.$$

Solution

(a) $\dfrac{\frac{35}{12}}{7} + \dfrac{\frac{35}{12}}{5} = 1$

$\qquad \dfrac{5}{12} + \dfrac{7}{12} = 1$

$\qquad\qquad \dfrac{12}{12} = 1$

$\qquad\qquad\quad 1 = 1$

Yes

(b) $\dfrac{-\frac{2}{35}}{7} + \dfrac{-\frac{2}{35}}{5} = 1$

$\qquad -\dfrac{2}{245} + -\dfrac{2}{175} = 1$

$\qquad \dfrac{-10}{1225} + \dfrac{-14}{1225} = 1$

$\qquad\qquad -\dfrac{24}{1225} = 1$

No

5. Solve $17 - 7x = 3$ and check the result. (There may be no solution.)

Solution

$$17 - 7x = 3$$
$$17 - 7x - 17 = 3 - 17$$
$$-7x = -14$$
$$\frac{-7x}{-7} = \frac{-14}{-7}$$
$$x = 2$$

Check $17 - 7(2) \overset{?}{=} 3$

$\qquad\quad 17 - 14 \overset{?}{=} 3$

$\qquad\qquad\quad 3 = 3$

7. Solve $4y - 6(y - 5) = 2$ and check the result. (There may be no solution.)

Solution

$$4y - 6(y - 5) = 2$$
$$4y - 6y + 30 = 2$$
$$-2y + 30 = 2$$
$$-2y + 30 - 30 = 2 - 30$$
$$-2y = -28$$
$$\frac{-2y}{-2} = \frac{-28}{-2}$$
$$y = 14$$

Check $4(14) - 6(14 - 5) \overset{?}{=} 2$

$\qquad\quad 56 - 6(9) \overset{?}{=} 2$

$\qquad\quad 56 - 54 \overset{?}{=} 2$

$\qquad\qquad\quad 2 = 2$

9. Solve $1.4t + 4.1 = 0.9t$ and check the result. (There may be no solution.)

Solution

$$1.4t + 4.1 = 0.9t$$

$$1.4t + 4.1 - 0.9t = 0.9t - 0.9t$$

$$0.5t + 4.1 = 0$$

$$0.5t + 4.1 - 4.1 = 0 - 4.1$$

$$0.5t = -4.1$$

$$\frac{0.5t}{0.5} = \frac{-4.1}{0.5}$$

$$t = -8.2$$

Check $1.4(-8.2) + 4.1 \overset{?}{=} 0.9(-8.2)$

$$-11.48 + 4.1 \overset{?}{=} -7.38$$

$$-7.38 = -7.38$$

11. Solve $\dfrac{3x}{4} = 4$ and check the result. (There may be no solution.)

Solution

$$\frac{4}{3} \cdot \frac{3x}{4} = 4 \cdot \frac{4}{3}$$

$$x = \frac{16}{3}$$

Check $\dfrac{3\left(\dfrac{16}{3}\right)}{4} \overset{?}{=} 4$

$$\frac{16}{4} \overset{?}{=} 4$$

$$4 = 4$$

13. Solve $\frac{4}{5}x - \frac{1}{10} = \frac{3}{2}$ and check the result. (There may be no solution.)

Solution

$$10\left[\frac{4}{5}x - \frac{1}{10}\right] = \left[\frac{3}{2}\right]10$$

$$8x - 1 = 15$$

$$8x - 1 + 1 = 15 + 1$$

$$8x = 16$$

$$\frac{8x}{8} = \frac{16}{8}$$

$$x = 2$$

Check $\dfrac{4}{5}(2) - \dfrac{1}{10} \overset{?}{=} \dfrac{3}{2}$

$$\frac{8}{5} - \frac{1}{10} \overset{?}{=} \frac{3}{2}$$

$$\frac{16}{10} - \frac{1}{10} \overset{?}{=} \frac{3}{2}$$

$$\frac{15}{10} \overset{?}{=} \frac{3}{2}$$

$$\frac{3}{2} = \frac{3}{2}$$

15. Solve $\dfrac{v - 20}{-8} = 2v$ and check the result. (There may be no solution.)

Solution

$$(-8)\left[\dfrac{v - 20}{-8}\right] = [2v](-8)$$

$$v - 20 = -16v$$

$$v - 20 - v = -16v - v$$

$$-20 = -17v$$

$$\dfrac{-20}{-17} = \dfrac{-17v}{-17}$$

$$\dfrac{20}{17} = v$$

Check

$$\dfrac{\dfrac{20}{17} - 20}{-8} \stackrel{?}{=} 2\left(\dfrac{20}{17}\right)$$

$$\dfrac{\dfrac{20 - 340}{17}}{-8} \stackrel{?}{=} \dfrac{40}{1.7}$$

$$\dfrac{-320}{17} \cdot \dfrac{1}{-8} \stackrel{?}{=} \dfrac{40}{17}$$

$$\dfrac{40}{17} = \dfrac{40}{17}$$

17. Solve $382x - 575 = 715$ using a calculator. Round the result to two decimal places.

Solution

$$382x - 575 = 715$$

$$382x - 575 + 575 = 715 + 575$$

$$382x = 1290$$

$$\dfrac{382x}{382} = \dfrac{1290}{382}$$

$$x = 3.3769634$$

$$x \approx 3.38$$

19. Solve $\dfrac{x}{2.33} = 14.302$ using a calculator. Round the result to two decimal places.

Solution

$$2.33\left[\dfrac{x}{2.33}\right] = [14.302]2.33$$

$$x = 33.32366$$

$$x \approx 33.32$$

21. Complete the table shown in the textbook.

Solution

Percent	Parts out of 100	Decimal	Fraction
87%	87	0.87	$\frac{87}{100}$

23. What is 130% of 50?

Solution

Verbal model: | Compared number | $=$ | Percent | $\cdot$ | Base number |

Equation:

$$a = p \cdot b$$

$$a = 1.30 \cdot 50$$

$$a = 65$$

25. 645 is $21\frac{1}{2}\%$ of what number?

Solution

Verbal model: | Compared number | $=$ | Percent | $\cdot$ | Base number |

Equation:

$$a = p \cdot b$$

$$645 = 0.215 \cdot b$$

$$\dfrac{645}{0.215} = b$$

$$3000 = b$$

27. 250 is what percent of 200?

Solution

Verbal model: | Compared number | $=$ | Percent | $\cdot$ | Base number |

Equation:

$$a = p \cdot b$$

$$250 = p \cdot 200$$

$$\dfrac{250}{200} = p$$

$$1.25 = p \quad \text{or} \quad 125\%$$

29. *Completing a Table*   Complete the table shown in the textbook which shows data on instant replays in the National Football League. (Source: National Football League)

Solution

Year	Reviewed	Reversed	Percent Reversed
1986	374	38	10.16%
1987	490	57	11.63%
1988	537	53	9.87%
1989	492	65	13.21%

1986:

$$38 = p \cdot 374$$

$$\frac{38}{374} = p$$

1987:

$$57 = p \cdot 490$$

$$\frac{57}{490} = p$$

1988:

$$53 = p \cdot 537$$

$$\frac{53}{537} = p$$

1989:

$$65 = p \cdot 492$$

$$\frac{65}{492} = p$$

31. Express 16 feet to 4 yards as a ratio. (Use the same units for the numerator and denominator.)

Solution

$$\frac{16 \text{ feet}}{12 \text{ feet}} = \frac{4}{3}$$

33. Solve $\dfrac{7}{8} = \dfrac{y}{4}$.

Solution

$$8y = 28$$

$$y = \frac{28}{8}$$

$$y = \frac{7}{2}$$

35. *Property Tax*   The tax on property with an assessed value of $80,000 is $1350. Find the tax on property with an assessed value of $110,000.

Solution

$$\frac{1350}{80,000} = \frac{x}{110,000}$$

$$80,000x = 1.485 \times 10^8$$

$$x = \frac{1.485 \times 10^8}{80,000}$$

$$x = \$1856.25$$

37. *Geometry*   You want to measure the height of a flagpole. To do this, you measure the flagpole's shadow and find that it is 30 feet long. You also measure the height of a five-foot lamp post and find its shadow to be 3 feet long. How tall is the flagpole? (See figure in textbook.)

Solution

Verbal model: $\boxed{\dfrac{\text{Flagpole's height}}{\text{Length of flagpole's shadow}}} = \boxed{\dfrac{\text{Lamp post's height}}{\text{Length of lamp post's shadow}}}$

Proportion: $\dfrac{x}{30} = \dfrac{5}{3}$

$$x = 30 \cdot \frac{5}{3} = 50 \text{ feet}$$

39. *Price Increase*   The manufacturer's suggested retail price for a certain truck model is \$25,750. Estimate the price of a comparably equipped truck for the next model year if it is projected that truck prices will increase by $5\frac{1}{2}\%$.

Solution

Verbal model: $\boxed{\text{Increase}} = \boxed{\text{Percent}} \cdot \boxed{\text{Base number}}$

Equation: $x = 0.055 \cdot 25{,}750$

$x = \$1416.25$

Verbal model: $\boxed{\text{New model}} = \boxed{\text{Old price}} + \boxed{\text{Increase}}$

Equation: $x = 25{,}750 + 1416.25$

$x = \$27{,}166.25$

41. *Markup Rate*   A calculator selling for \$175.00 costs the retailer \$95.00. Find the markup rate.

Solution

Verbal model: $\boxed{\text{Markup}} = \boxed{\text{Selling price}} - \boxed{\text{Retailer's cost}}$

Equation: $x = \$175.00 - \$95.00$

$x = \$80.00$

Verbal model: $\boxed{\text{Markup}} = \boxed{\text{Markup rate}} \cdot \boxed{\text{Retailer's cost}}$

Equation: $80.00 = x \cdot 95.00$

$\dfrac{80.00}{95.00} = x$

$84.21\% = x$

43. *Comparison Shopping*   A mail-order catalog has an attaché case with a list price of \$99.97 plus \$4.50 for shipping and handling. A department store has the same case for \$125.95. The department store has a special 20% off sale. Which is the better buy?

Solution

Verbal model: $\boxed{\text{Total price}} = \boxed{\text{List price}} + \boxed{\text{Shipping}}$

Equation: $x = 99.97 + 4.50$

$x = \$104.47$

Verbal model: $\boxed{\text{Discount}} = \boxed{\text{Discount rate}} \cdot \boxed{\text{List price}}$

Equation: $x = 0.20 \cdot 125.95$

$x = \$25.19$

Verbal model: $\boxed{\text{Sale price}} = \boxed{\text{List price}} - \boxed{\text{Discount}}$

Equation: $x = 125.95 - 25.19$

$x = \$100.76$

The department store price is the better buy.

45. *Mixture* Determine the number of liters of a 30% saline solution and the number of liters of a 60% saline solution that are required to make 10 liters of a 50% saline solution.

Solution

Verbal model: | Amount of solution 1 | + | Amount of solution 2 | = | Amount of final solution |

Equation: $0.30x + 0.60(10 - x) = 0.50(10)$

$$0.30x + 6 - 0.60x = 5$$

$$-0.30x = -1$$

$$x = 3.\overline{3} \text{ liters 30\% solution}$$

$$10 - x = 6.\overline{6} \text{ liters 60\% solution}$$

47. *Travel Time* Determine the time for a bus to travel 330 miles if its average speed is 52 miles per hour.

Solution

Verbal model: | Distance | = | Rate | · | Time |

Equation: $330 = 52 \cdot x$

$$\frac{330}{52} = x$$

$$6.35 \text{ hrs} = x$$

49. *Work Rate* Find the time for two people working together to complete a task if it takes them 4.5 hours and 6 hours working individually .

Solution

Verbal model: | Work done | = | Work done by person 1 | + | Work done by person 2 |

Equation: $1 = \frac{1}{4.5}(t) + \frac{1}{6}(t)$

$$27 = 6t + 4.5t$$

$$27 = 10.5t$$

$$\frac{27}{10.5} = t$$

$$2.57 \text{ hrs} = t$$

51. *Simple Interest* Find the total simple interest you will earn on a $1000 corporate bond that matures in 4 years and has an 8.5% interest rate.

Solution

Verbal model: | Interest | = | Principal | · | Rate | · | Time |

Equation: $x = 1000 \cdot 0.085 \cdot 4$

$$x = \$340$$

53. *Simple Interest*   Find the principal required to have an annual interest income of $20,000 if the annual simple interest rate on the principal is 9.5%.

Solution

Verbal model: $\boxed{\text{Interest}} = \boxed{\text{Principal}} \cdot \boxed{\text{Rate}} \cdot \boxed{\text{Time}}$

Equation: $20{,}000 = x \cdot 0.095 \cdot 1$

$$\frac{20{,}000}{0.095} = x$$

$$\$210{,}526.32 = x$$

55. *Simple Interest*   An inheritance of $50,000 is divided between two investments earning 8.5% and 10% simple interest. How much is in each investment if the total interest for 1 year is $4700?

Solution

Verbal model: $\boxed{\text{Interest}} = \boxed{\text{Principal}} \cdot \boxed{\text{Rate}} \cdot \boxed{\text{Time}}$

Equation: $4700 = 0.085x + 0.10(50{,}000 - x)$

$$4700 = 0.085x + 5000 - 0.10x$$

$$-300 = -0.015x$$

$$\frac{-300}{-0.015} = x$$

$$\$20{,}000 = x$$

$$\$30{,}000 = 50{,}000 - x$$

57. *Geometry*   The area of the rectangle shown in the textbook is 48 square inches. Its width is 6 inches. Find the dimensions of the rectangle.

Solution

Verbal model: $\boxed{\text{Area}} = \boxed{\text{Length}} \cdot \boxed{\text{Width}}$

Equation: $48 = x \cdot 6$

$$8 = x$$

59. Solve for x in $2x - 7y + 4 = 0$.

Solution

$$2x - 7y + 4 = 0$$

$$2x = 7y - 4$$

$$x = \tfrac{7}{2}y - 2$$

61. Solve for h in $V = \pi r^2 h$.

Solution

$$V = \pi r^2 h$$

$$\frac{V}{\pi r^2} = h$$

63. Solve $|x - 2| - 2 = 4$.

Solution

$$|x - 2| - 2 = 4$$

$$|x - 2| = 6$$

$$x - 2 = 6 \quad \text{or} \quad x - 2 = -6$$

$$x = 8 \qquad\qquad x = -4$$

65. Solve $|3x - 4| = |x + 2|$.

Solution

$$3x - 4 = x + 2 \quad \text{or} \quad 3x - 4 = -(x + 2)$$

$$2x = 6 \qquad\qquad 3x - 4 = -x - 2$$

$$x = 3 \qquad\qquad 4x = 2$$

$$x = \tfrac{2}{4} = \tfrac{1}{2}$$

67. Solve the inequality $5x + 3 > 18$ and sketch the solution on the real number line.

Solution

$$5x + 3 > 18$$
$$5x > 15$$
$$x > 3$$

69. Solve the inequality $\frac{1}{3} - \frac{1}{2}y < 12$ and sketch the solution on the real number line.

Solution

$$\frac{1}{3} - \frac{1}{2}y < 12$$
$$2 - 3y < 72$$
$$-3y < 70$$
$$y > -\frac{70}{3}$$

71. Solve the inequality $-4 < \frac{x}{5} \le 4$ and sketch the solution on the real number line.

Solution

$$-4 < \frac{x}{5} \le 4$$
$$-20 < x \le 20$$

73. Solve the inequality $5 > \frac{x+1}{-3} > 0$ and sketch the solution on the real number line.

Solution

$$5 > \frac{x+1}{-3} > 0$$
$$-15 < x + 1 < 0$$
$$-16 < x < -1$$

75. Solve the inequality $|2x - 7| < 15$ and sketch the solution on the real number line.

Solution

$$-15 < 2x - 7 < 15$$
$$-8 < 2x < 22$$
$$-4 < x < 11$$

77. Solve the inequality $|x - 4| > 3$ and sketch the solution on the real number line.

Solution

$$x - 4 < -3 \quad \text{or} \quad x - 4 > 3$$
$$x < 1 \qquad\qquad x > 7$$

79. Solve the inequality $|b + 2| - 6 > 1$ and sketch the solution on the real number line.

Solution

$$|b + 2| - 6 > 1$$
$$|b + 2| > 7$$
$$b + 2 < -7 \quad \text{or} \quad b + 2 > 7$$
$$b < -9 \qquad\qquad b > 5$$

81. Write an inequality for the statement "z is no more than 10."

Solution

$$z \le 10$$

83. Write an inequality for the statement "y is at least 7 but less than 14."

Solution

$7 \leq y < 14$

85. Write an absolute value inequality that represents the interval shown on the graph in the textbook.

Solution

$(1, 5)$

$$1 < x < 5$$
$$1 - 3 < x - 3 < 5 - 3$$
$$-2 < x - 3 < 2$$
$$|x - 3| < 2$$

Test for Chapter 1

1. Determine whether (a) $x = -4$ and (b) $x = 2$ are solutions of the equation $3(5 - 2x) - (3x - 1) = -2$.

Solution

(a) $x = -4$

$$3[5 - 2(-4)] - [3(-4) - 1] \overset{?}{=} -2$$
$$3[13] - [-13] = -2$$
$$39 + 13 = -2$$
$$52 \neq -2$$

No

(b) $x = 2$

$$3[5 - 2(2)] - [3(2) - 1] \overset{?}{=} -2$$
$$3[1] - [5] = -2$$
$$-2 = -2$$

Yes

2. Solve $6x - 5 = 19$.

Solution

$$6x - 5 = 19$$
$$6x - 5 + 5 = 19 + 5$$
$$6x = 24$$
$$\frac{6x}{6} = \frac{24}{6}$$
$$x = 4$$

3. Solve $15 - 7(1 - x) = 3(x + 8)$.

Solution

$$15 - 7(1 - x) = 3(x + 8)$$
$$15 - 7 + 7x = 3x + 24$$
$$8 + 7x = 3x + 24$$
$$8 + 7x - 3x = 3x + 24 - 3x$$
$$8 - 8 + 4x = 24 - 8$$
$$4x = 16$$
$$\frac{4x}{4} = \frac{16}{4}$$
$$x = 4$$

4. Solve $\dfrac{2x}{3} = \dfrac{x}{2} + 4$.

Solution

$$6\left[\dfrac{2x}{3} = \dfrac{x}{2} + 4\right]6$$

$$4x = 3x + 24$$

$$4x - 3x = 3x + 24 - 3x$$

$$x = 24$$

5. Solve $\dfrac{t-5}{12} = \dfrac{3}{8}$.

Solution

$$24\left[\dfrac{t-5}{12} = \dfrac{3}{8}\right]24$$

$$2(t - 5) = 3(3)$$

$$2t - 10 = 9$$

$$2t - 10 + 10 = 9 + 10$$

$$2t = 19$$

$$\dfrac{2t}{2} = \dfrac{19}{2}$$

$$t = \dfrac{19}{2}$$

6. What is 27% of 3200?

Solution

$$a = p \cdot b$$

$$x = 0.27 \cdot 3200$$

$$x = 864$$

7. 1200 is what percent of 800?

Solution

$$a = p \cdot b$$

$$1200 = x \cdot 800$$

$$\dfrac{1200}{800} = x$$

$$1.5 = x$$

$$150\% = x$$

8. A store is offering a 20% discount on all items in its inventory. Find the list price on a tractor that has a sale price of $6400.

Solution

Verbal model: | Sale price | = | Percent | · | List Price |

Equation: $6400 = 0.80 \cdot x$

$$\dfrac{6400}{0.80} = x$$

$$\$8000 = x$$

9. Which is the better buy: a 12-ounce can for $2.49 or a 15-ounce can for $2.99? Explain your reasoning.

Solution

$$\dfrac{\text{Total price}}{\text{Total units}} = \dfrac{2.49}{12} = \dfrac{249}{1200} = 0.2075 \text{ per ounce}$$

$$\dfrac{\text{Total price}}{\text{Total units}} = \dfrac{2.99}{15} = \dfrac{299}{1500} = 0.199\overline{3} \text{ per ounce}$$

The 15-ounce can is the better buy.

10. The tax on property with an assessed value of $90,000 is $1200. Estimate the tax on property with an assessed value of $110,000. (Assume the same tax rate.)

Solution

Verbal model: $\boxed{\dfrac{\text{Tax}}{\text{Assessed value}}} = \boxed{\dfrac{\text{Tax}}{\text{Assessed value}}}$

Proportion:

$$\frac{1200}{90,000} = \frac{x}{110,000}$$

$$x = \frac{(1200)(110,000)}{90,000}$$

$$x \approx \$1466.67$$

11. The bill (including parts and labor) for the repair of a home appliance was $165. The cost for parts was $85. How many hours were spent to repair the appliance if the cost of labor was $16 per half hour?

Solution

Verbal model: $\boxed{\text{Total bill}} = \boxed{\text{Cost of parts}} + \boxed{\text{Cost of labor}}$

Equation:

$$165 = 85 + 6x$$

$$80 = 16x$$

$$5 \text{ half hours } = x$$

$$x = 2\tfrac{1}{2} \text{ hours}$$

12. Two solutions—10% concentration and 40% concentration—are mixed to create 100 liters of a 30% solution. Determine the numbers of liters of the 10% solution and the 40% solution that are required.

Solution

Verbal model: $\boxed{\begin{array}{c}\text{Amount of}\\\text{solution 1}\end{array}} + \boxed{\begin{array}{c}\text{Amount of}\\\text{solution 2}\end{array}} = \boxed{\begin{array}{c}\text{Amount of}\\\text{final solution}\end{array}}$

Equation:

$$0.10x + 0.40(100 - x) = 0.30(100)$$

$$0.10x + 40 - 0.40x = 30$$

$$-0.30x = -10$$

$$x = 33\tfrac{1}{3} \text{ liters at } 10\%$$

$$100 - x = 66\tfrac{2}{3} \text{ liters at } 40\%$$

13. Two cars start at a given time and travel in the same direction at average speeds of 40 miles per hour and 55 miles per hour. How much time must elapse before the two cars are 10 miles apart?

Solution

Verbal model: $\boxed{\text{Distance Car 1}} + 10 \text{ miles} = \boxed{\text{Distance Car 2}}$

Equation: $40x + 10 = 55x$

$$10 = 15x$$

$$\frac{10}{15} = x$$

$$\frac{2}{3} \text{ hour} = x$$

14. One number is four times a second. Find the numbers if their difference is 39.

Solution

Verbal model: $\boxed{\text{One number}} - \boxed{\text{Second number}} = \boxed{\text{Difference}}$

Equation: $4x - x = 39$

$$3x = 39$$

$$x = 13$$

$$4x = 52$$

15. Find the principal required to earn \$300 in simple interest in 2 years if the annual interest rate is 7.5%.

Solution

Verbal model: $\boxed{\text{Interest}} = \boxed{\text{Principal}} \cdot \boxed{\text{Rate}} \cdot \boxed{\text{Time}}$

Equation: $300 = x \cdot 0.075 \cdot 2$

$$\$2000 = x$$

16. Solve $1 + 2x > 7 - x$ and sketch its solution.

Solution

$$1 + 2x > 7 - x$$

$$3x > 6$$

$$x > 2$$

17. Solve $0 \leq \dfrac{1 - x}{4} < 2$ and sketch its solution.

Solution

$$0 \leq \frac{1 - x}{4} < 2$$

$$0 \leq 1 - x < 8$$

$$-1 \leq -x < 7$$

$$1 \geq x > -7$$

18. Solve $|x - 3| \le 2$ and sketch its solution.

Solution

$$|x - 3| \le 2$$
$$-2 \le x - 3 \le 2$$
$$1 \le x \le 5$$

19. Solve $|x + 4| > 1$ and sketch its solution.

Solution

$$|x + 4| > 1 \qquad \text{or} \qquad x + 4 < -1$$
$$x + 4 > 1 \qquad\qquad x < -5$$
$$x > -3$$

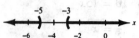

23. Use inequality notation to denote the phrase "t is at least 8."

Solution

$t \ge 8.$

CHAPTER 2
Introduction to Analytic Geometry

2.1   The Rectangular Coordinate System

7. Plot the points $(4, 3)$, $(-5, 3)$, and $(3, -5)$ on a rectangular coordinate system.

Solution

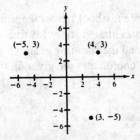

9. Plot the points $\left(\frac{5}{2}, -2\right)$, $\left(-2, \frac{1}{4}\right)$, and $\left(\frac{3}{2}, -\frac{7}{2}\right)$ on a rectangular coordinate system.

Solution

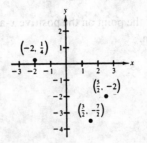

11. Approximate the coordinates of the points shown in the figure.

Solution

$A = (4, -2)$, $B = \left(-3, -\frac{5}{2}\right)$, $C = \left(3, \frac{1}{2}\right)$

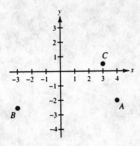

13. *Geometry* Plot the points $(-1, 2)$, $(2, 0)$, $(3, 5)$ and connect them with line segments to form a triangle.

Solution

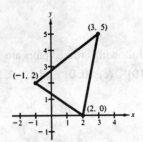

15. *Geometry* Plot the points $(4, 0)$, $(6, -2)$, $(0, -4)$, $(-2, -2)$ and connect them with line segments to form a Parallelogram.

Solution

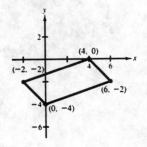

101

17. Determine the quadrant or quadrants in which $(-3, -5)$ is located.

Solution

Quadrant III

19. Determine the quadrant or quadrants in which $(x, 4)$ is located.

Solution

Quadrants I, II

21. Determine the quadrant or quadrants in which (x, y), $xy < 0$ is located.

Solution

Quadrants II, IV

23. Find the coordinates of the point located 5 units to the left of the y-axis and 2 units above the x-axis.

Solution

$(-5, 2)$

25. Find the coordinates of the point on the positive x-axis 10 units from the origin.

Solution

$(10, 0)$

27. *Exam Score* The table (see textbook) gives the time x in hours invested in concentrated study for five different algebra exams and the resulting exam score y. Plot the points whose coordinates are given in the table.

Solution

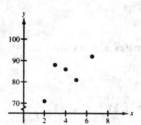

29. Determine whether the following ordered pairs are solutions of $y = 3x + 8$.

(a) $(3, 17)$ (b) $(-1, 10)$ (c) $(0, 0)$ (d) $(-2, 2)$

Solution

(a) $17 = 3(3) + 8$

$17 = 9 + 8$

$17 = 17$

Yes

(b) $10 = 3(-1) + 8$

$10 = -3 + 8$

$10 \neq 5$

No

(c) $0 = 3(0) + 8$

$0 \neq 8$

No

(d) $2 = 3(-2) + 8$

$2 = -6 + 8$

$2 = 2$

Yes

31. Complete the table of values shown in the textbook. Then plot the solution points on a rectangular coordinate system.

Solution

x	-2	0	2	4	6
$y = 5x - 1$	-11	-1	9	19	29

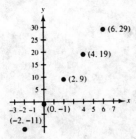

33. Use a graphing utility to complete the table of values shown in the textbook.

Solution

x	-2	0	2	4	6
$y = 4x^2 + x - 2$	12	-2	16	66	148

35. Plot the points $(3, -2)$ and $(3, 5)$ and find the distance between them. State whether the points lie on a horizontal or vertical line.

Solution

$d = |5 - (-2)|$

$\quad = |7|$

$\quad = 7$

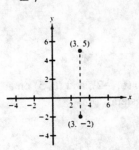

37. Plot the points $\left(\frac{1}{2}, \frac{7}{8}\right)$ and $\left(\frac{11}{2}, \frac{7}{8}\right)$ and find the distance between them. State whether the points lie on a horizontal or vertical line.

Solution

$d = \left|\frac{11}{2} - \frac{1}{2}\right|$

$\quad = |5|$

$\quad = 5$

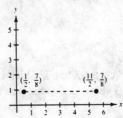

39. Find the distance between the points $(1, 3)$ and $(5, 6)$.

Solution

$d = \sqrt{(1-5)^2 + (3-6)^2} = \sqrt{16+9} = \sqrt{25} = 5$

41. Find the distance between the points $(0, 0)$ and $(12, -9)$.

Solution

$d = \sqrt{(0-12)^2 + [0-(-9)]^2}$

$\quad = \sqrt{144 + 81} = \sqrt{225} = 15$

43. Using the figure in text, find the coordinates of (x, y), the lengths of the vertical and horizontal sides of the right triangle, and the length of the hypotenuse.

Solution

coordinates:

$x = 10, y = 2$

vertical side $= |8 - 2| = 6$

horizontal side $= |10 - 2| = 8$

hypotenuse $= c^2 = 6^2 + 8^2$

$c = \sqrt{36 + 64}$

$\quad = \sqrt{100} = 10$

45. *Making a Conjecture* Plot the points (2, 1), (−3, 5), and (7, −3) on a rectangular coordinate system. Then change the sign of the *x*-coordinate of each point and plot the three new points on the same rectangular coordinate system. What conjecture can you make about the location of a point when the sign of the *x*-coordinate is changed?

Solution

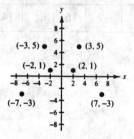

When the sign of the *x*-coordinate is changed, the point is on the opposite side of the *y*-axis as the original point.

47. Solve the inequality $-4 < 10x + 1 < 6$.

Solution

$$-4 < 10x + 1 < 6$$
$$-4 - 1 < 10x + 1 - 1 < 6 - 1$$
$$-5 < 10x < 5$$
$$\frac{-5}{10} < \frac{10x}{10} < \frac{5}{10}$$
$$-\frac{1}{2} < x < \frac{1}{2}$$

49. Solve the inequality $-3 \le -\dfrac{x}{2} \le 3$.

Solution

$$-3 \le -\frac{x}{2} \le 3$$
$$2 \cdot -3 \le 2 \cdot -\frac{x}{2} \le 3 \cdot 2$$
$$-6 \le -x \le 6$$
$$\frac{-6}{-1} \le \frac{-x}{-1} \le \frac{6}{-1}$$
$$6 \ge x \ge -6$$
$$-6 \le x \le 6$$

51. *Price Inflation* A new van costs $32,500, which is about 112% of what it was 3 years ago. What was its price 3 years ago?

Solution

Verbal model: $\boxed{\text{New Price}} = \boxed{\text{Percent}} \cdot \boxed{\text{Old Price}}$

Equation: $32,500 = 1.12 \cdot x$

$$\frac{32,500}{1.12} = x$$
$$\$29,017.86 = x \approx \$29,018$$

53. Plot the points (−8, −2), (6, −2), and (6, 5) on a rectangular coordinate system.

Solution

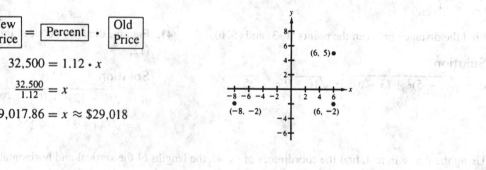

55. Plot the points $\left(\frac{3}{2}, 1\right)$, $(4, -3)$, and $\left(-\frac{4}{3}, \frac{7}{3}\right)$ on a rectangular coordinate system.

Solution

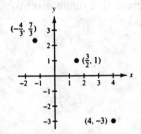

57. Approximate the coordinates of the points. (See figure in textbook.)

Solution

$A = (-2, 4)$, $B = (0, -2)$, $C = (4, -2)$

59. *Geometry* Plot the points $(0, 6)$, $(3, 3)$, $(0, 0)$, and $(-3, 3)$ and connect them with line segments to form a square.

Solution

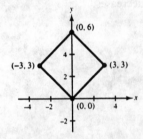

61. Plot the points $(0, 0)$, $(3, 2)$, $(2, 3)$, and $(5, 5)$ and connect them with line segments to form a rhombus.

Solution

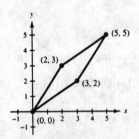

63. Without plotting it, determine the quadrant in which each point is located.

(a) $\left(3, -\frac{5}{8}\right)$ (b) $(-6.2, 8.05)$

Solution

(a) Quadrant IV   (b) Quadrant II

65. Determine the quadrant or quadrants in which $(-3, y)$ is located.

Solution

Quadrants II, III

67. Determine the quadrant or quadrants in which (x, y), $xy > 0$ is located.

Solution

Quadrants I, III

69. Find the coordinates of the point in the following statement.

The point is located three units to the right of the y-axis and four units below the x-axis.

Solution

Point 3 units right of y-axis and 4 units below x-axis $= (3, -4)$

71. Find the coordinates of the point in the following statement.

The coordinates of the point are equal, and it is located in the third quadrant 10 units to the left of the y-axis.

Solution

The coordinates of the point are equal and located in Quadrant III, 10 units left of y-axis $= (-10, -10)$.

73. *Fuel Efficiency* The table shown in the textbook gives the speed x of a car in miles per hour and the approximate fuel efficiency y in miles per gallon. Plot the points whose coordinates are given in the table. Then write a paragraph that summarizes the relationship between x and y.

Solution

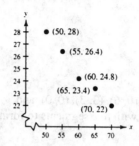

The relationship between x and y is as the value of x increases the value of y decreases.

75. Determine whether the following ordered pairs are solution of $y = \frac{7}{8}x$.

(a) $\left(\frac{8}{7}, 1\right)$ (b) $\left(4, \frac{7}{2}\right)$ (c) $(0, 0)$ (d) $(-16, 14)$

Solution

(a) $1 = \frac{7}{8}\left(\frac{8}{7}\right)$

$1 = 1$

Yes

(b) $\frac{7}{2} = \frac{7}{8}(4)$

$\frac{7}{2} = \frac{7}{2}$

Yes

(c) $0 = \frac{7}{8}(0)$

$0 = 0$

Yes

(d) $14 = \frac{7}{8}(-16)$

$14 \neq -14$

No

77. Determine whether the following ordered pairs are solution of $4y - 2x = -1$.

(a) $(0, 0)$ (b) $\left(\frac{1}{2}, 0\right)$ (c) $\left(-3, -\frac{7}{4}\right)$ (d) $\left(1, -\frac{3}{4}\right)$

Solution

(a) $4(0) - 2(0) = -1$

$0 \neq -1$

No

(b) $4(0) - 2\left(\frac{1}{2}\right) = -1$

$0 - 1 = -1$

$-1 = -1$

Yes

(c) $4\left(-\frac{7}{4}\right) - 2(-3) = -1$

$-7 + 6 = -1$

$-1 = -1$

Yes

(d) $4\left(-\frac{3}{4}\right) - 2(1) = -1$

$-3 - 2 = -1$

$-5 \neq -1$

No

79. Complete the table of values shown in the textbook. Then plot the solution points on a rectangular coordinate system.

Solution

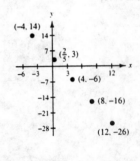

x	-4	$\frac{2}{5}$	4	8	12
$y = -\frac{5}{2}x + 4$	14	3	-6	-16	-26

$y = \dfrac{-5}{2}(-4) + 4$ $y = \dfrac{-5}{2}\left(\dfrac{2}{5}\right) + 4$ $y = \dfrac{-5}{2}(4) + 4$

$\quad = 10 + 4$ $\qquad\quad = -1 + 4$ $\qquad = -10 + 4$

$\quad = 14$ $\qquad\qquad = 3$ $\qquad\quad = -6$

$y = \dfrac{-5}{2}(8) + 4$ $y = \dfrac{-5}{2}(12) + 4$

$\quad = -20 + 4$ $\quad = -30 + 44$

$\quad = -16$ $\qquad = -26$

81. *Numerical Interpretation* The cost c of producing x units is given by $c = 28x + 3000$. Use a table to help write a paragraph that describes the relationship between x and c.

Solution

x	100	150	200	250	300
$c = 28x + 3000$	5800	7200	8600	$10{,}000$	$11{,}400$

$y = 28(100) + 3000$ $y = 28(150) + 3000$ $y = 28(200) + 3000$

$\quad = 2800 + 3000$ $\qquad = 4200 + 3000$ $\qquad = 5600 + 3000$

$\quad = 5800$ $\qquad\quad = 7200$ $\qquad\quad = 8600$

$y = 28(250) + 3000$ $y = 28(300) + 3000$

$\quad = 7000 + 3000$ $\qquad = 8400 + 3000$

$\quad = 10{,}000$ $\qquad = 11{,}400$

For each additional 50 units produced, costs increase by $1400.

83. Plot the points (3, 2) and (10, 2) and find the distance between them. State whether the points lie on a horizontal or vertical line.

Solution

$d = |10 - 3|$

$\quad = |7|$

$\quad = 7$

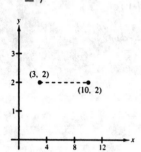

Horizontal line

85. Plot the points $\left(-3, \frac{3}{2}\right)$ and $\left(-3, \frac{9}{4}\right)$ and find the distance between them. State whether the points lie on a horizontal or vertical line.

Solution

$d = \left|\frac{3}{2} - \frac{9}{4}\right|$

$\quad = \left|\frac{6}{4} - \frac{9}{4}\right|$

$\quad = \frac{3}{4}$

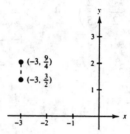

Vertical line

87. Find the coordinates of (x, y), the lengths of the vertical and horizontal sides of the right triangle and the length of the hypotenuse. (See figure in textbook.)

Solution

$x = 4, y = -4$ vertical side $= |4 - (-4)| = 8$ hypotenuse $= c^2 = 8^2 + 7^2$

horizontal side $= |4 - (-3)| = 7$ $c = \sqrt{64 + 49} = \sqrt{113}$

89. Find the distance between the points $(-2, -3)$ and $(4, 1)$.

Solution

$d = \sqrt{(-2 - 4)^2 + (-3 - 1)^2}$

$\quad = \sqrt{36 + 16} = \sqrt{52} = 2\sqrt{13}$

91. Find the distance between the points $(1, 3)$ and $(3, -2)$.

Solution

$d = \sqrt{(1 - 3)^2 + [3 - (-2)]^2} = \sqrt{4 + 25} = \sqrt{29}$

93. Use the Distance Formula to determine whether the three points (2, 3), (2, 6) and (6, 3) lie on a line.

Solution

$d = \sqrt{(2 - 2)^2 + (3 - 6)^2} = \sqrt{0 + 9} = 3$

$d = \sqrt{(2 - 6)^2 + (3 - 3)^2} = \sqrt{16 + 0} = 4$

$d = \sqrt{(2 - 6)^2 + (6 - 3)^2} = \sqrt{16 + 9} = 5$

$3 + 4 \neq 5$

Not collinear

95. Use the Distance Formula to determine whether the three points (8, 3), (5, 2) and (2, 1) lie on a line.

Solution

$d = \sqrt{(8 - 5)^2 + (3 - 2)^2} = \sqrt{9 + 1} = \sqrt{10}$

$d = \sqrt{(8 - 2)^2 + (3 - 1)^2} = \sqrt{36 + 4} = \sqrt{40} = 2\sqrt{10}$

$d = \sqrt{(5 - 2)^2 + (2 - 1)^2} = \sqrt{9 + 1} = \sqrt{10}$

$\sqrt{10} + \sqrt{10} = 2\sqrt{10}$

Collinear

97. *Geometry* Find the perimeter of the triangle having the vertices $(-2, 0)$, $(0, 5)$, and $(1, 0)$.

Solution

$$d = \sqrt{(-2-0)^2 + (0-5)^2} = \sqrt{4+25} = \sqrt{29}$$

$$d = \sqrt{(0-1)^2 + (5-0)^2} = \sqrt{1+25} = \sqrt{26}$$

$$d = \sqrt{(-2-1)^2 + (0-0)^2} = \sqrt{9} = 3$$

$$P = \sqrt{29} + \sqrt{26} + 3$$

99. Plot the points $(-2, 0)$ and $(4, 8)$ and the midpoint of the line segment joining the points. The coordinates of the midpoint of the line segment joining the points (x_1, y_1) and (x_2, y_2) are given by

$$\text{Midpoint} = \left(\frac{x_1 + x_2}{2}, \frac{y_1 + y_2}{2} \right).$$

Solution

$$M = \left(\frac{-2+4}{2}, \frac{0+8}{2} \right) = (1, 4)$$

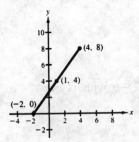

101. Plot the points $(1, 6)$ and $(6, 3)$ and the midpoint of the line segment joining the points. The coordinates of the midpoint of the line segment joining the points (x_1, y_1) and (x_2, y_2) are given by

$$\text{Midpoint} = \left(\frac{x_1 + x_2}{2}, \frac{y_1 + y_2}{2} \right).$$

Solution

$$M = \left(\frac{1+6}{2}, \frac{6+3}{2} \right) = \left(\frac{7}{2}, \frac{9}{2} \right)$$

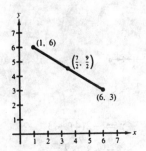

103. *Shifting a Graph* The figure in the textbook is shifted to a new location in the plane. Find the coordinates of the vertices of the figure in its new location.

Solution

$(-2, -1)$ shifted 2 units right and 5 units up $= (0, 4)$

$(-3, -4)$ shifted 2 units right and 5 units up $= (-1, 1)$

$(1, -2)$ shifted 2 units right and 5 units up $= (3, 2)$

105. *Housing Construction* A house is 30 feet wide and the ridge of the roof is 7 feet above the top of the walls (see figure in textbook). The rafters overhang the edges of the walls by 2 feet. How long are the rafters?

Solution

$x^2 = 7^2 + 15^2$

$x^2 = 49 + 225$

$x = \sqrt{274} = 16.55294536$

Rafter $= 2 + x = 18.55294536 \approx 19$ feet

2.2   Graphs of Equations

5. Match the equation $y = 2$ with its graph.

Solution

(e)

7. Match the equation $y = 2 - x$ with its graph.

Solution

(f)

9. Match the equation $y = -x^3$ with its graph.

Solution

(d)

11. Complete the table (see textbook) and use the results to sketch the graph of $2x + y = 3$.

Solution

x	-4	-2	0	2	4
y	11	7	3	-1	-5
(x, y)	$(-4, 11)$	$(-2, 7)$	$(0, 3)$	$(2, -1)$	$(4, -5)$

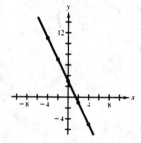

$y = -2x + 3$ $7 = -2x + 3$ $3 = -2x + 3$ $y = -2(2) + 3$ $y = -2(4) + 3$

$y = -2(-4) + 3$ $4 = -2x$ $0 = -2x$ $= -4 + 3$ $= -8 + 3$

$= 8 + 3$ $-2 = x$ $0 = x$ $= -1$ $= -5$

$= 11$

13. Complete the table (see textbook) and use the results to sketch the graph of $y = 4 - x^2$.

Solution

x	± 2	-1	0	2	± 3
y	0	3	4	0	-5
(x, y)	$(2, 0)$ $(-2, 0)$	$(-1, 3)$	$(0, 4)$	$(2, 0)$	$(3, -5)$ $(-3, -5)$

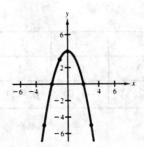

$$0 = 4 - x^2 \qquad y = 4 - (-1)^2 \qquad 4 = 4 - x^2 \qquad y = 4 - 2^2 \qquad -5 = 4 - x^2$$
$$x^2 = 4 \qquad\qquad = 4 - 1 \qquad 0 = -x^2 \qquad\qquad = 4 - 4 \qquad -9 = -x^2$$
$$x = \pm 2 \qquad\qquad = 3 \qquad 0 = x \qquad\qquad = 0 \qquad\qquad 9 = x^2$$
$$\pm 3 = x$$

15. Use a graphing utility to create a table of values. Then sketch the graph of $y = 4 - |x|$.

Solution

x	-4	-2	0	2	4
$y = 4 - |x|$	0	2	4	2	0
Solutions	$(-4, 0)$	$(-2, 2)$	$(0, 4)$	$(1, 2)$	$(4, 0)$

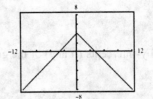

17. Sketch the graph of $y = 3x$.

Solution

x	-2	-1	0	1	2
$y = 3x$	-6	-3	0	3	6
Solutions	$(-2, -6)$	$(-1, -3)$	$(0, 0)$	$(1, 3)$	$(2, 6)$

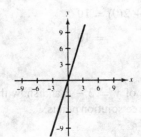

19. Sketch the graph of $y = 2x - 3$.

Solution

x	-2	-1	0	1	2
$y = 2x - 3$	-7	-5	-3	-1	1
Solutions	$(-2, -7)$	$(-1, -5)$	$(0, -3)$	$(1, -1)$	$(2, 1)$

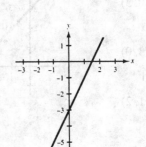

21. Sketch the graph of $y = x^2 - 1$.

Solution

x	-2	-1	0	1	2
$y = x^2 - 1$	3	0	-1	0	3
Solutions	$(-2, 3)$	$(-1, 0)$	$(0, -1)$	$(1, 0)$	$(2, 3)$

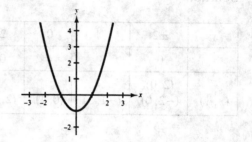

23. Sketch the graph of $y = |x| - 1$.

Solution

x	-2	-1	0	1	2
$y = |x| - 1$	1	0	-1	0	1
Solutions	$(-2, 1)$	$(-1, 0)$	$(0, -1)$	$(1, 0)$	$(2, 1)$

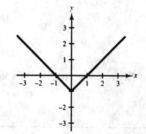

25. Find the x-and y-intercepts (if any) of the graph of $x + 2y = 10$.

Solution

y-intercept: $0 + 2y = 0$

$y = 5$

x-intercept: $x + 2(0) = 10$

$x = 10$

27. Find the x- and y-intercepts (if any) of the graph of $y = (x + 5)(x - 5)$.

Solution

y-intercept: $y = (0 + 5)(0 - 5)$

$y = -25$

x-intercept: $0 = (x + 5)(x - 5)$

$x + 5 = 0$ $x - 5 = 0$

$x = -5$ $x = 5$

29. Sketch the graph of $y = 3 - x$ and show the coordinates of three solution points.

Solution

$y = 3 - 0$

$y = 3$ $(0, 3)$

$0 = 3 - x$

$x = 3$ $(3, 0)$

$y = 3 - 1$

$y = 2$ $(1, 2)$

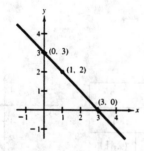

31. Sketch the graph of $y = 4$ and show the coordinates of three solution points.

Solution

No x-intercept

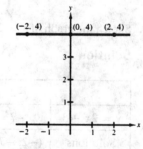

33. Sketch the graph of $4x + y = 3$ and show the coordinates of three solution points.

Solution

$$4x + y = 3$$

$$4(0) + y = 3$$

$$y = 3 \quad (0, 3)$$

$$4x + 0 = 3$$

$$4x = 3$$

$$x = \frac{3}{4} \quad \left(\frac{3}{4}, 0\right)$$

$$4(1) + y = 3$$

$$y = -1 \quad (1, -1)$$

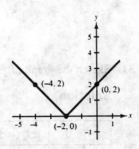

35. Sketch the graph of $y = x^2 - 4$ and show the coordinates of three solution points.

Solution

$$y = 0^2 - 4$$

$$= -4 \quad (0, -4)$$

$$y = 2^2 - 4$$

$$= 0 \quad (2, 0)$$

$$y = (-2)^2 - 4$$

$$= 0 \quad (-2, 0)$$

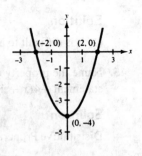

37. Sketch the graph of $y = |x + 2|$ and show the coordinates of three solution points.

Solution

$$y = |0 + 2|$$

$$= 2 \quad (0, 2)$$

$$y = |-2 + 2|$$

$$= 0 \quad (-2, 0)$$

$$y = |-4 + 2|$$

$$= 2 \quad (-4, 2)$$

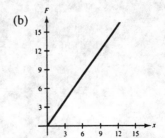

39. *Using a Graph* The force F (in pounds) to stretch a spring x inches from its natural length is given by

$$F = \frac{4}{3}x, \quad 0 \le x \le 12.$$

(a) Use the model to complete the following table.

(b) Sketch the graph of the model.

(c) Use the graph to determine how the length of the spring changes each time the force is doubled. Explain your reasoning.

Solution

(a)

x	0	3	6	9	12
$\frac{4}{3}x$	0	4	8	12	16

(b)

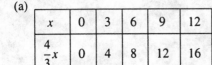

(c) Length doubles

41. *Misleading Graphs* The two graphs represent the *same* data points. (See graphs in textbook.) Which graph is misleading, and why?

Solution

(a) It is difficult to assess the increase in sales.

45. Name the property of real numbers that justifies the statement $-4(x + 10) = -4 \cdot x + (-4)(10)$

Solution

Distributive Property

43. Name the property of real numbers that justifies the statement $8x \cdot \dfrac{1}{8x} = 1$

Solution

Multiplicative Inverse Property

47. *Weight of Sand* The ratio of cement to sand in a 90-pound bag of dry mix is 1 to 4. Find the number of pounds of sand in the bag.

Solution

Verbal Model:

$$\boxed{\frac{\text{cement}}{\text{sand}}} = \boxed{\frac{\text{cement}}{\text{sand}}}$$

Proportion:

$$\frac{1}{4} = \frac{90 - x}{x}$$

$$x = 4(90 - x)$$

$$x = 360 - 4x$$

$$5x = 360$$

$$x = 72 \ \text{pounds sand}$$

49. Sketch the graph of $y = 4 - x$.

Solution

x	0	1	4
y	4	3	0

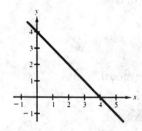

51. Sketch the graph of $y = x^2 - 4$.

Solution

x	0	± 1	± 2
y	-4	-3	0

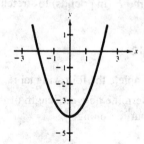

53. Sketch the graph of $y = |x|$.

Solution

x	0	1	-1
y	0	1	1

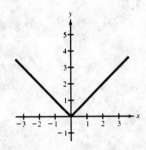

55. Sketch the graph of $y = 2x - 3$ and show the coordinates of three solution points.

Solution

$y = 2(0) - 3$

$y = -3 \quad (0, -3)$

$0 = 2x - 3$

$3 = 2x$

$\dfrac{3}{2} = x \quad \left(\dfrac{3}{2}, 0\right)$

$y = 2(3) - 3$

$y = 3 \quad (3, 3)$

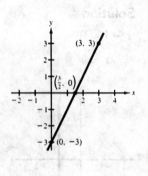

57. Sketch the graph of $2x - 3y = 6$ and show the coordinates of three solution points.

Solution

$2(0) - 3y = 6$

$-3y = 6$

$y = -2 \quad (0, -2)$

$2x - 3(0) = 6$

$2x = 6$

$x = 3 \quad (3, 0)$

$2(1) - 3y = 6$

$-3y = 4$

$y = -\dfrac{4}{3} \quad \left(1, -\dfrac{4}{3}\right)$

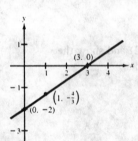

59. Sketch the graph of $x + 5y = 10$ and show the coordinates of three solution points.

Solution

$x + 5y = 10$

$0 + 5y = 10$

$y = 2 \quad (0, 2)$

$x + 5(0) = 10$

$x = 10 \quad (10, 0)$

$5 + 5y = 10$

$5y = 5$

$y = 1 \quad (5, 1)$

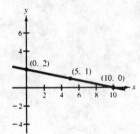

61. Sketch the graph of $x - 1 = 0$ and show the coordinates of three solution points.

Solution

$x - 1 = 0$

$x = 1$

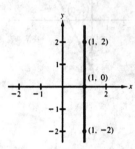

63. Sketch the graph of $y = x^2 - 9$ and show the coordinates of three solution points.

Solution

$0 = x^2 - 9$

$0 = (x - 3)(x + 3)$

$3 = x \quad x = -3 \quad (3, 0)(-3, 0)$

$y = 0^2 - 9$

$y = -9 \quad (0, -9)$

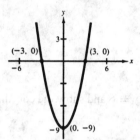

65. Sketch the graph of $y = x(x - 2)$ and show the coordinates of three solution points.

Solution

$y = 0^2 - 2(0)$

$y = 0 \quad (0, 0)$

$0 = x^2 - 2x$

$0 = x(x - 2)$

$0 = x \quad x = 2 \quad (0, 0), (2, 0)$

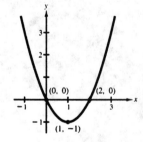

67. Sketch the graph of $y = x^3 - 1$ and show the coordinates of three solution points.

Solution

$y = 0^3 - 1$

$\quad = -1 \quad (0, -1)$

$y = 1^3 - 1$

$\quad = 0 \quad (1, 0)$

$y = (-1)^3 - 1$

$\quad = -2 \quad (-1, -2)$

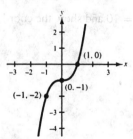

69. Sketch the graph of $y = |x| - 3$ and show the coordinates of three solution points.

Solution

$y = |0| - 3$

$y = -3 \quad (0, -3)$

$0 = |x| - 3$

$3 = |x|$

$3 = x \quad x = -3 \quad (3, 0), (-3, 0)$

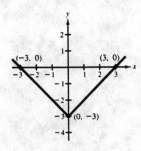

71. Sketch the graph of $y = -|x|$ and show the coordinates of three solution points.

Solution

$y = -|0|$

$\quad = 0 \quad (0, 0)$

$y = -|1|$

$\quad = -1 \quad (1, -1)$

$y = -|-1|$

$\quad = -1 \quad (-1, -1)$

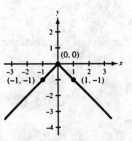

73. *Graphical and Algebraic Solutions* Graphically estimate the intercepts of $y = x^2 + 3$. Then check your results algebraically.

Solution

Estimate y-intercept ≈ 3 **Check** $y = 0^2 + 3 = 3$

   no x-intercepts $0 = x^2 + 3$

                                    $-3 = x^2$ no real solution

75. *Graphical and Algebraic Solutions* Graphically estimate the intercepts of $y = |x - 2|$. Then check your results algebraically.

Solution

Estimate y-intercept ≈ 2 **Check** $y = |0 - 2| = |-2| = 2$

       x-intercept ≈ 2 $0 = |x - 2|$

                                    $0 = x - 2$

                                    $2 = x$

77. *Straight-Line Depreciation*   A manufacturing plant purchases a new molding machine for $225,000. The depreciated value y after t years is given by

$$y = 225,000 - 20,000t, \quad 0 \le t \le 8.$$

Sketch the graph of the model.

Solution

$y = 225,000 - 20,000(0)$

$\quad = 225,000 \quad (0, \ 225,000)$

$y = 225,000 - 20,000(8)$

$\quad = 225,000 - 160,000$

$\quad = 65,000 \quad (8, \ 65,000)$

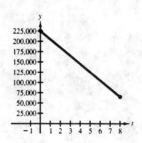

79. Graph the equations on the same set of coordinate axes. What conclusions can you make by comparing the graphs?

(a) $y = x^2$ (b) $y = (x - 2)^2$

(c) $y = (x + 2)^2$ (d) $y = (x + 4)^2$

Solution

(a) $y = x^2$ (b) $y = (x - 2)^2$

(c) $y = (x + 2)^2$ (d) $y = (x + 4)^2$

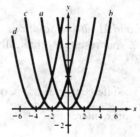

Graph of $y = (x - h)^2$ is graph of $y = x^2$ shifted h units right or left.

2.3   Graphs and Graphing Utilities

5. Use a graphing utility to match the equation $y = 4 + x$ with its graph.

Solution

(c)

7. Use a graphing utility to match the equation $y = 4 - x^2$ with its graph.

Solution

(d)

9. Use a graphing utility to match the equation $y = x^3 + 4$ with its graph.

Solution

(a)

11. Use a graphing utility to graph $y = 2x - 6$. Use a standard setting.

Solution

Keystrokes:

Y= 2 X, T, θ − 6 GRAPH

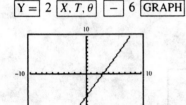

13. Use a graphing utility to graph $y = x^2 - 3$. Use a standard setting.

Solution

Keystrokes:

Y= X, T, θ x^2 − 3 GRAPH

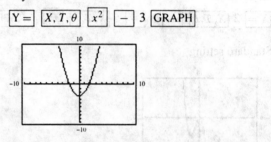

15. Use a graphing utility to graph $y = \sqrt{x + 4}$. Use a standard setting.

Solution

Keystrokes:

Y= √ (X, T, θ + 4) GRAPH

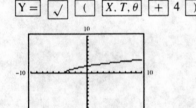

17. Use a graphing utility to graph $y = |x| - 6$. Use a standard setting.

Solution

Keystrokes:

Y= ABS X, T, θ − 6 GRAPH

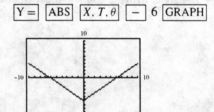

19. Use a graphing utility to graph $y = \frac{1}{2}x - 10$. Begin by using a standard setting. Then graph the equation a second time using the specified setting. Which setting is better? Explain.

Solution

Keystrokes:

Y= (1 ÷ 2) X, T, θ − 10 GRAPH

Standard setting   Specified setting

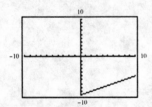   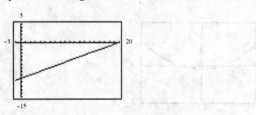

The second setting is better because both intercepts are shown.

21. Use a graphing utility to graph $y = 2x^3 - 5x^2$. Begin by using a standard setting. Then graph the equation a second time using the specified setting. Which setting is better? Explain.

Solution

Keystrokes:

Standard setting   Specified setting

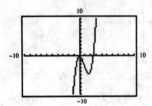   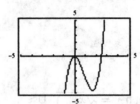

The second setting is better because the intercepts are easier to determine.

23. Find a setting on a graphing utility such that the graph of $y = -3x + 15$ agrees with the graph in the textbook.

Solution

```
RANGE
Xmin=-2
Xmax=8
Xscl=1
Ymin=-1
Ymax=17
Yscl=1
Xres=1
```

25. Find a setting on a graphing utility such that the graph of $y = x^3 - 3x + 2$ agrees with the graph in the textbook.

Solution

```
RANGE
Xmin=-5
Xmax=5
Xscl=1
Ymin=-3
Ymax=6
Yscl=1
Xres=1
```

27. Use the zoom feature of a graphing utility to estimate the lowest point on the graph of
$y = \frac{1}{4}(x^2 - 1)$.

Solution

Keystrokes:

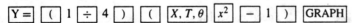

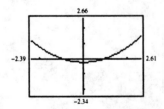

The lowest point is $(0, -.25)$.

29. Use the ZOOM feature of a graphing utility to estimate the lowest point on the graph of $y = x^3(3x + 4)$.

Solution

Keystrokes:

$\boxed{\text{Y}=}$ $\boxed{X, T, \theta}$ $\boxed{\wedge}$ 3 $\boxed{(}$ 3 $\boxed{X, T, \theta}$ $\boxed{+}$ 4 $\boxed{)}$ $\boxed{\text{GRAPH}}$

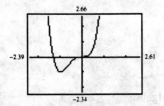

The lowest point is $(-1, -1)$.

31. *Think About It* In Exercises 27 and 29, you graphically estimated the lowest point on a graph. Explain how you could confirm your estimates algebraically with a table.

Solution

Evaluate the expression for values of x near the lowest point on the graph.

33. Use a graphing utility to approximate the x-intercepts of the graph of $y = 0.4x^2 - 0.2x - 3$. (*Hint:* The ZOOM and TRACE features can help you get a better view of the intercepts.)

Solution

Keystrokes:

$\boxed{\text{Y}=}$ 0.4 $\boxed{X, T, \theta}$ $\boxed{x^2}$ $\boxed{-}$ 0.2 $\boxed{X, T, \theta}$ $\boxed{-}$ 3 $\boxed{\text{GRAPH}}$

x-intercepts $= -2.5$ and 3 approximately

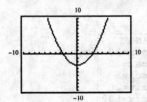

35. Use a graphing utility to approximate the x-intercepts of the graph of $y = x^3 + x - 1$.

Solution

Keystrokes:

$\boxed{\text{Y}=}$ $\boxed{X, T, \theta}$ $\boxed{\wedge}$ 3 $\boxed{+}$ $\boxed{X, T, \theta}$ $\boxed{-}$ 1 $\boxed{\text{GRAPH}}$

x-intercept $= 0.68$, approximately

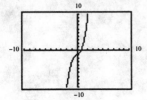

37. Explain how to use a graphing utility to verify that $y_1 = y_2$. Then identify the rule of algebra that is illustrated.

$$y_1 = \frac{1}{2}(x - 4) \qquad y_2 = \frac{1}{2}x - 2$$

Solution

Keystrokes:

$\boxed{Y =}$ .5 $\boxed{(}$ $\boxed{X, T, \theta}$ $\boxed{-}$ 4 $\boxed{)}$ $\boxed{\text{GRAPH}}$

$\boxed{Y =}$ .5 $\boxed{X, T, \theta}$ $\boxed{-}$ 2 $\boxed{\text{GRAPH}}$

y_1 and y_2 are the same line, illustrating the distributive property.

39. Explain how to use a graphing utility to verify that $y_1 = y_2$. Then identify the rule of algebra that is illustrated.

$$y_1 = x + (2x - 1) \qquad y_2 = (x + 2x) - 1$$

Solution

Keystrokes:

$\boxed{Y =}$ $\boxed{X, T, \theta}$ $\boxed{+}$ $\boxed{(}$ 2 $\boxed{X, T, \theta}$ $\boxed{-}$ 1 $\boxed{)}$ $\boxed{\text{GRAPH}}$

$\boxed{Y =}$ $\boxed{(}$ $\boxed{X, T, \theta}$ $\boxed{+}$ 2 $\boxed{X, T, \theta}$ $\boxed{)}$ $\boxed{-}$ 1 $\boxed{\text{GRAPH}}$

y_1 and y_2 are the same line, illustrating the associative property of addition.

41. *Stopping Distance*   Consider the stopping distance y of an automobile (in feet), which is modeled by

$$y = 30.00 + 0.08x^2, \quad 25 \le x \le 75$$

where x is the speed in miles per hour. Find a graphing utility viewing rectangle that gives a good view of the graph of this model.

Solution

Keystrokes:

$\boxed{Y =}$ 30 $\boxed{+}$ 0.08 $\boxed{X, T, \theta}$ $\boxed{x^2}$ $\boxed{\text{GRAPH}}$

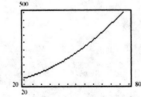
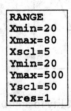

```
RANGE
Xmin=20
Xmax=80
Xscl=5
Ymin=20
Ymax=500
Yscl=50
Xres=1
```

43. Solve $3x + y = 4$ for y in terms of x.

Solution

$$3x + y = 4$$
$$y = -3x + 4$$

45. Solve $x^2 + 3y = 4$ for y in terms of x.

Solution

$$x^2 + 3y = 4$$
$$3y = -x^2 + 4$$
$$y = \frac{-x^2 + 4}{3}$$

47. Use the point-plotting method to sketch the graph of $2x - 7y + 14 = 0$.

Solution

$$2(0) - 7y + 14 = 0$$
$$-7y = -14$$
$$y = 2 \quad (0, 2)$$
$$2x - 7(0) + 14 = 0$$
$$2x = -14$$
$$x = -7 \quad (-7, 0)$$

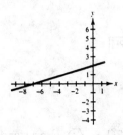

49. Write an algebraic expression that represents the quantity in the verbal statement, and simplify it if possible. The total hourly wage for an employee when the base pay is \$9.35 per hour plus 75 cents for each of q units produced per hour.

Solution

Verbal Model: $\boxed{\text{Total Wage}} = \boxed{\text{Base Pay}} + \boxed{\text{Extra Pay}}$

Equation: Total wage $= \$9.35 + .75q$

51. Use a graphing utility to graph $y = \dfrac{2}{3}x + 1$. Use a standard setting.

Solution

Keystrokes:

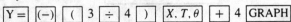

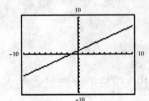

53. Use a graphing utility to graph $y = -\dfrac{3}{4}x + 4$. Use a standard setting.

Solution

Keystrokes:

$\boxed{Y=}\ \boxed{(-)}\ \boxed{(}\ 3\ \boxed{\div}\ 4\ \boxed{)}\ \boxed{X,T,\theta}\ \boxed{+}\ 4\ \boxed{GRAPH}$

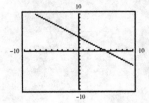

55. Use a graphing utility to graph $y = -x^2 + 4x - 1$. Use a standard setting.

Solution

Keystrokes:

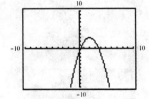

57. Use a graphing utility to graph $y = x^3 - 2$. Use a standard setting.

Solution

Keystrokes:

$\boxed{Y=}$ $\boxed{X, T, \theta}$ $\boxed{\wedge}$ 3 $\boxed{-}$ 2 $\boxed{GRAPH}$

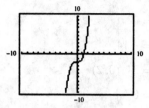

59. Use a graphing utility to graph $y = 2(1 - \sqrt{x})$. Use a standard setting.

Solution

Keystrokes:

$\boxed{Y=}$ 2 $\boxed{(}$ 1 $\boxed{-}$ $\boxed{\sqrt{}}$ $\boxed{X, T, \theta}$ $\boxed{)}$ $\boxed{GRAPH}$

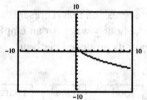

61. Use a graphing utility to graph $y = |x| - 4$. Use a standard setting.

Solution

Keystrokes:

$\boxed{Y=}$ $\boxed{ABS}$ $\boxed{X, T, \theta}$ $\boxed{-}$ 4 $\boxed{GRAPH}$

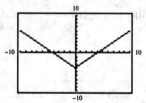

63. Use a graphing utility to graph $y = |x - 4|$. Use a standard setting.

Solution

Keystrokes:

$\boxed{Y=}$ $\boxed{ABS}$ $\boxed{(}$ $\boxed{X, T, \theta}$ $\boxed{-}$ 4 $\boxed{)}$ $\boxed{GRAPH}$

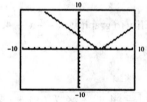

65. Use a graphing utility to graph $y = \frac{1}{6}(2x + 3)$. Begin by using a standard setting. Then graph the equation a second time using the specified setting. Which setting is better? Explain.

Solution

Keystrokes:

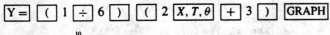

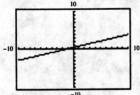

The first setting is better because the second setting does not show entire line and x-intercept.

67. Use a graphing utility to graph $y = 20 - \frac{1}{4}x^2$. Begin by using a standard setting. Then graph the equation a second time using the specified setting. Which setting is better? Explain.

Solution

Keystrokes:

Y= 20 − (1 ÷ 4) X,T,θ x² GRAPH

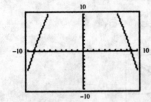

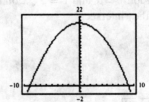

The second setting is better because it shows vertex and intercepts.

69. Use a graphing utility to graph $y = 0.6x^3 - 2x - 1$. Begin by using a standard setting. Then graph the equation a second time using the specified setting. Which setting is better? Explain.

Solution

Keystrokes:

Y= 0.6 X,T,θ ∧ 3 − 2 X,T,θ − 1 GRAPH

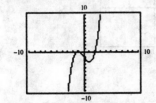

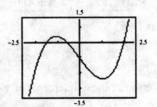

The second setting is better because it shows both negative x-intercepts.

71. Find a setting on a graphing utility such that the graph of $y = 100 - 3x$ agrees with the given graph.

Solution

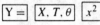

```
RANGE
Xmin=-10
Xmax=40
Xscl=5
Ymin=-10
Ymax=100
Yscl=10
```

73. Find a setting on a graphing utility such that the graph of $y = -\frac{1}{8}x^2 + x$ agrees with the given graph.

Solution

```
RANGE
Xmin=-2
Xmax=10
Xscl=1
Ymin=-5
Ymax=5
Yscl=1
Xres=1
```

75. Use a graphing utility to graph $y = x^2$ and $y = 4x - x^2$ on the same screen. Do the graphs intersect? If so, how many times do they intersect?

Solution

Keystrokes:

$\boxed{Y=}$ $\boxed{X,T,\theta}$ $\boxed{x^2}$

$\boxed{Y=}$ 4 $\boxed{X,T,\theta}$ $\boxed{-}$ $\boxed{X,T,\theta}$ $\boxed{x^2}$ $\boxed{\text{GRAPH}}$

Yes, they intersect in two points.

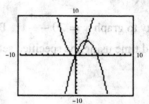

77. Use a graphing utility to graph $y = x$ and $y = 3\sqrt{x}$ on the same screen. Do the graphs intersect? If so, how many times do they intersect?

Solution

Keystrokes:

$\boxed{Y=}$ $\boxed{X,T,\theta}$

$\boxed{Y=}$ 3 $\boxed{\sqrt{}}$ $\boxed{X,T,\theta}$ $\boxed{\text{GRAPH}}$

Yes, they intersect in two points.

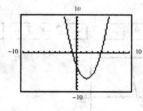

79. Use a graphing utility to estimate the x-intercepts of $y = x^2 - 4x - 3$.

Solution

Keystrokes:

$\boxed{Y=}$ $\boxed{X,T,\theta}$ $\boxed{x^2}$ $\boxed{-}$ 4 $\boxed{X,T,\theta}$ $\boxed{-}$ 3 $\boxed{\text{GRAPH}}$

2 x-intercepts $= -0.65$ and 4.65

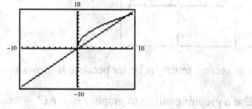

81. Use a graphing utility to estimate the x-intercepts of $y = -x^2 + 12x - 37$.

Solution

Keystrokes:

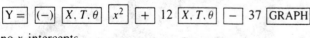

$\boxed{Y=}$ $\boxed{(-)}$ $\boxed{X,T,\theta}$ $\boxed{x^2}$ $\boxed{+}$ 12 $\boxed{X,T,\theta}$ $\boxed{-}$ 37 $\boxed{\text{GRAPH}}$

no x-intercepts

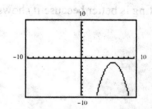

83. Use a graphing utility to find a graph with (2, 0) and (0, 2) as intercepts. Explain your strategy.

Solution

Find the slope: $m = \dfrac{2-0}{0-2} = \dfrac{2}{-2} = -1$.

So, equation is $y = -x + 2$.

85. *Comparing Views* The equation $y = 0.02x^2 + 0.75x - 10$ gives the profit y (in thousands of dollars) for selling x units of a product. Which graphing utility viewing rectangle in the text gives a better view of the graph? Explain your reasoning.

Solution

Keystrokes:

$\boxed{Y=}$ .02 $\boxed{X, T, \theta}$ $\boxed{x^2}$ $\boxed{+}$ .75 $\boxed{X, T, \theta}$ $\boxed{-}$ 10 $\boxed{\text{GRAPH}}$

Rectangle 1

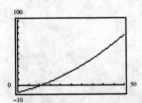

Rectangle 2

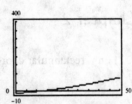

The first viewing rectangle gives a better view because it utilizes the entire calculator display.

87. *Income* The net income per share of common stock for McDonald's Corporation from 1984 through 1993 can be approximated by the model

$$y = (0.081x + 0.648)^2, \quad 4 \leq x \leq 13,$$

where y is the net income per share and x represents the year, with $x = 0$ corresponding to 1980. (Source: McDonald's Corporation)

(a) Use a graphing utility to graph the model.

(b) The actual net income per share of stock in 1993 was $2.91. How closely does the model predict this value?

Solution

(a)

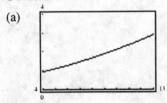

(b) Model gives $2.89 for $x = 13$

89. *Strength of Copper* Use the following model, which relates the percent strength of copper y in terms of its Celsius temperature x.

$$y = \sqrt{10{,}700 - 17.6x}, \quad 100 \le x \le 500$$

The percent strength is relative to 100% at 20°C.

Use a graphing utility to sketch the graph of the equation using the range settings as shown in the textbook.

Solution

Keystrokes:

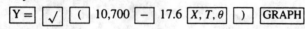

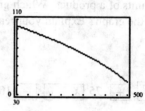

Mid-Chapter Quiz for Chapter 2

1. Plot the points $(-1, 5)$ and $(3, 2)$ on a rectangular coordinate system and find the distance between them.

Solution

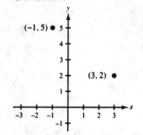

$$d = \sqrt{(-1 - 3)^2 + (5 - 2)^2}$$
$$= \sqrt{16 + 9}$$
$$= \sqrt{25}$$
$$= 5$$

2. Plot the points $(-3, -2)$ and $(2, 10)$ on a rectangular coordinate system and find the distance between them.

Solution

$$d = \sqrt{(-3 - 2)^2 + (-2 - 10)^2}$$
$$= \sqrt{25 + 144}$$
$$= \sqrt{169}$$
$$= 13$$

3. Determine the quadrants in which the point $(x, 4)$ must be located if x is a real number. Explain your reasoning.

Solution

Quadrants I, II. Since x can be any real number and y is 4, the point $(x, 4)$ can only be located in quadrants in which the y coordinate is positive.

4. Find the coordinates of the point that lies 10 units to the right of the y-axis and 3 units below the x-axis.

Solution

$(10, -3)$

5. Determine whether the following ordered pairs are solution points of the equation $4x - 3y = 10$.

Solution

(a) $(2, 1)$ $4(2) - 3(1) \overset{?}{=} 10$

$8 - 3 \overset{?}{=} 10$

$5 \neq 10$ no

(c) $(2.5, 0)$ $4(2.5) - 3(0) \overset{?}{=} 10$

$10 - 0 \overset{?}{=} 10$

$10 = 10$ yes

(b) $(1, -2)$ $4(1) - 3(-2) \overset{?}{=} 10$

$4 + 6 \overset{?}{=} 10$

$10 = 10$ yes

(d) $\left(2, -\frac{2}{3}\right)$ $4(2) - 3\left(-\frac{2}{3}\right) \overset{?}{=} 10$

$8 + 2 \overset{?}{=} 10$

$10 = 10$ yes

6. Find the x- and y-intercepts of the graph of the equation $6x - 8y + 48 = 0$.

Solution

x-intercept $6x - 8(0) + 48 = 0$

$6x = -48$

$x = -8$ $(-8, 0)$

y-intercept $6(0) - 8y + 48 = 0$

$-8y = -48$

$y = 6$ $(0, 6)$

7. Sketch the graph of $y = 2x - 3$ and show the coordinates of three solution points (including intercepts).

Solution

(1) $y = 2(0) - 3$

$= -3$ $(0, -3)$

(2) $0 = 2x - 3$

$3 = 2x$

$\frac{3}{2} = x$ $\left(\frac{3}{2}, 0\right)$

(3) $y = 2(2) - 3$

$= 1$ $(2, 1)$

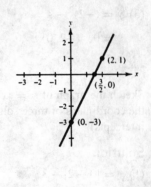

8. Sketch the graph of $y = 5$ and show the coordinates of three solution points (including intercepts).

Solution

$y = 5$ $(0, 5)$

$(4, 5)$

$(-2, 5)$

x can be any real number as long as y is 5.

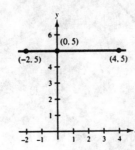

9. Sketch the graph of $3x + y - 6 = 0$ and show the coordinates of three solution points (including intercepts).

Solution

(1) $3(0) + y - 6 = 0$

$y = 6$ $(0, 6)$

(2) $3x + 0 - 6 = 0$

$3x = 6$

$x = 2$ $(2, 0)$

(3) $3(4) + y - 6 = 0$

$y = -6$ $(4, -6)$

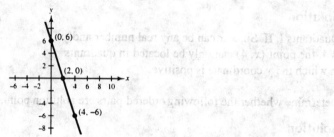

10. Sketch the graph of $y = x^2 - 4$ and show the coordinates of three solution points (including intercepts).

Solution

(1) $y = 0^2 - 4$

$= -4$ $(0, -4)$

(2) $y = 2^2 - 4$

$= 0$ $(2, 0)$

(3) $y = (-2)^2 - 4$

$= 0$ $(-2, 0)$

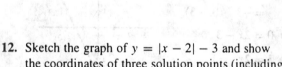

11. Sketch the graph of $y = 6x - x^2$ and show the coordinates of three solution points (including intercepts).

Solution

(1) $y = 6(0) - 0^2$

$= 0$ $(0, 0)$

(2) $y = 6(6) - 6^2$

$= 0$ $(6, 0)$

(3) $y = 6(3) - 3^2$

$= 18 - 9$

$= 9$ $(3, 9)$

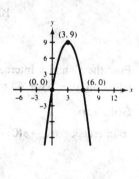

12. Sketch the graph of $y = |x - 2| - 3$ and show the coordinates of three solution points (including intercepts).

Solution

(1) $y = |0 - 2| - 3$

$= -1$ $(0, -1)$

(2) $y = |5 - 2| - 3$

$= 0$ $(5, 0)$

(3) $y = |2 - 2| - 3$

$= -3$ $(2, -3)$

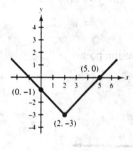

13. Use a graphing utility to graph $y = 7 - 2x$. Use a standard setting.

Solution

Keystrokes:

Y= 7 − 2 X,T,θ GRAPH

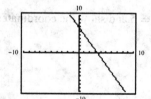

14. Use a graphing utility to graph $y = x^2 + 4x$. Use a standard setting.

Solution

Keystrokes:

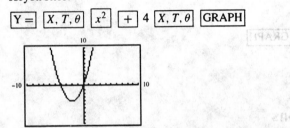

15. Use a graphing utility to graph $y = \sqrt{8 - x}$. Use a standard setting.

Solution

Keystrokes:

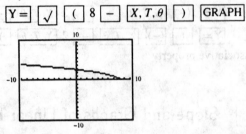

16. Use a graphing utility to graph $y = |x - 2| + |x + 2|$. Use a standard setting.

Solution

Keystrokes:

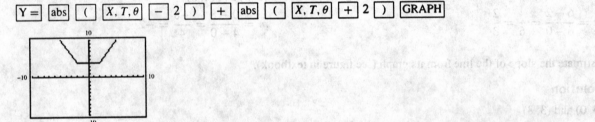

17. Find a setting on a graphing utility such that the graph of $y = -2(x^2 - 8x + 6)$ is approximately the same as the given graph.

Solution

```
RANGE
Xmin=-2
Xmax=8
Xscl=1
Ymin=-4
Ymax=24
Yscl=4
```

18. Find a setting on a graphing utility such that the graph of $y = x^3 - 5x + 5$ is approximately the same as the given graph.

Solution

```
RANGE
Xmin=-4
Xmax=4
Xscl=1
Ymin=-2
Ymax=12
Yscl=2
```

19. Use a graphing utility to approximate the x-intercepts of the graph of $y = x^2 - x - 2.75$.

Solution

Keystrokes:

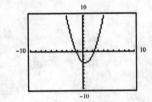

x-intercepts -1.23 and 2.23

20. Use a graphing utility to verify that $y_1 = y_2$ if $y_1 = -x + (2x - 2)$ and $y_2 = (-x + 2x) - 2$. Identify the rule of algebra that is illustrated.

Solution

Keystrokes:

y_1　$\boxed{Y=}$　$\boxed{(-)}$　$\boxed{X,T,\theta}$　$\boxed{+}$　$\boxed{(}$　2　$\boxed{X,T,\theta}$　$\boxed{-}$　2　$\boxed{)}$

y_2　$\boxed{Y=}$　$\boxed{(}$　$\boxed{(-)}$　$\boxed{X,T,\theta}$　$\boxed{+}$　2　$\boxed{X,T,\theta}$　$\boxed{)}$　$\boxed{-}$　2　$\boxed{GRAPH}$

associative property

2.4 | Slope and Graphs of Linear Equations

7. Estimate the slope of the line from its graph (see figure in textbook).

Solution

$(0, 2)$ and $(6, 6)$

$$m = \frac{6-2}{6-0} = \frac{4}{6} = \frac{2}{3}$$

9. Estimate the slope of the line from its graph (see figure in textbook).

Solution

$(0, 8)$ and $(4, 0)$

$$m = \frac{0-8}{4-0} = \frac{-8}{4} = -2$$

11. Estimate the slope of the line from its graph (see figure in textbook).

Solution

$(3, 0)$ and $(3, 8)$

$$m = \frac{8-0}{3-3} = \frac{8}{0} = \text{undefined}$$

13. Plot the points $(0, 0)$ and $(5, -4)$ and find the slope (if possible) of the line passing through them. State whether the line is rising, falling, horizontal, or vertical.

Solution

$$m = \frac{-4-0}{5-0} = \frac{-4}{5} \qquad \text{Line is falling.}$$

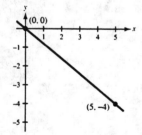

15. Plot the points $(-2, -3)$ and $(6, 1)$ and find the slope (if possible) of the line passing through them. State whether the line is rising, falling, horizontal, or vertical.

Solution

$$m = \frac{1-(-3)}{6-(-2)} = \frac{4}{8} = \frac{1}{2} \qquad \text{Line is rising.}$$

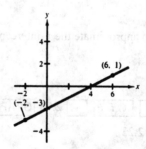

17. Plot the points $(3, -2)$ and $(3, 6)$ and find the slope (if possible) of the line passing through them. State whether the line is rising, falling, horizontal, or vertical.

Solution

$$m = \frac{6 - (-2)}{3 - 3} = \frac{8}{0} = \text{not possible}$$

Line is vertical.

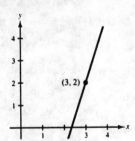

19. Plot the points $\left(\frac{3}{4}, 2\right)$ and $\left(\frac{7}{2}, 0\right)$ and find the slope (if possible) of the line passing through them. State whether the line is rising, falling, horizontal, or vertical.

Solution

$$m = \frac{0 - 2}{\frac{7}{2} - \frac{3}{4}} = \frac{-2}{\frac{14}{4} - \frac{3}{4}} = \frac{-2}{\frac{11}{4}}$$

$$= -2 \cdot \frac{4}{11} = \frac{-8}{11} \qquad \text{Line is falling.}$$

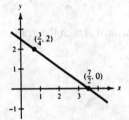

21. A point on a line $(5, 2)$ and the slope of the line $m = 0$ are given. Find two additional points on the line. (There are many correct answers to each problem.)

Solution

$$0 = \frac{y - 2}{x - 5}$$

Horizontal line: $(1, 2)$, $(0, 2)$, $(3, 2)$

Any points with a y-coordinate of 2

23. A point on a line $(3, -4)$ and the slope of the line $m = 3$ are given. Find two additional points on the line. (There are many correct answers to each problem.)

Solution

$$3 = \frac{y + 4}{x - 3}$$

$(4, -1)$, $(5, 2)$

25. Sketch the graph of a line through the point $(3, 2)$ having slope $m = 3$.

Solution

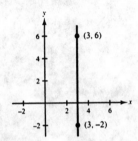

27. Sketch the graph of a line through the point $(3, 2)$ having slope $m = \frac{-1}{3}$.

Solution

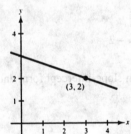

29. Which line is steeper? $m_1 = 3$ or $m_2 = -4$?

Solution

The line with slope $m_2 = -4$ is steeper because $|-4| > |3|$.

31. Plot the x- and y-intercepts and sketch the graph of $2x - y + 4 = 0$.

Solution

$2x - y + 4 = 0$

$2(0) - y + 4 = 0$

$4 = y \quad (0, 4)$

$2x - 0 + 4 = 0$

$2x = -4$

$x = -2 \quad (-2, 0)$

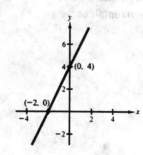

33. Plot the x- and y-intercepts and sketch the graph of $-5x + 2y - 20 = 0$.

Solution

$-5x + 2y - 20 = 0$

$-5(0) + 2y - 20 = 0$

$2y = 20$

$y = 10 \quad (0, 10)$

$y = 10$

$-5x + 2(0) - 20 = 0$

$-5x = 20$

$x = -4 \quad (-4, 0)$

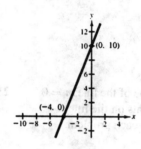

35. Write $3x - y - 2 = 0$ in slope-intercept form and sketch the line. Use a graphing utility to confirm your sketch.

Solution

$3x - y - 2 = 0$

$-y = -3x + 2$

$y = 3x - 2$

slope $= 3 \quad y$-intercept $= -2$

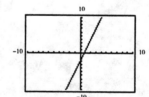

37. Write $3x + 2y - 2 = 0$ in slope-intercept form and sketch the line. Use a graphing utility to confirm your sketch.

Solution

$3x + 2y - 2 = 0$

$2y = -3x + 2$

$y = \dfrac{-3}{2}x + 1$

slope $= \dfrac{-3}{2} \quad y$-intercept $= 1$

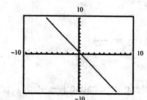

39. Use a graphing utility to graph the pair of equations on the same viewing rectangle. Are the lines parallel, perpendicular, or neither? (Use the *square* setting so the slopes of the lines appear visually correct.)

$$L_1 : y = \frac{1}{2}x - 2 \qquad L_2 : y = \frac{1}{2}x + 3$$

Solution

Keystrokes:

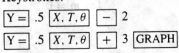

$L_1 \parallel L_2$

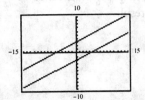

41. Use a graphing utility to graph the pair of equations on the same viewing rectangle. Are the lines parallel, perpendicular, or neither? (Use the *square* setting so the slopes of the lines appear visually correct.)

$$L_1 : y = \frac{3}{4}x - 3 \qquad L_2 : y = -\frac{4}{3}x + 1$$

Keystrokes:

$\boxed{Y=}$ .75 $\boxed{X,T,\theta}$ $\boxed{-}$ 3

$\boxed{Y=}$ $\boxed{(}$ $\boxed{-4}$ $\boxed{\div}$ 3 $\boxed{)}$ $\boxed{X,T,\theta}$ $\boxed{+}$ 1 $\boxed{GRAPH}$

Solution

$L_1 \perp L_2$

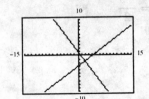

43. *Geometry* The length and width of a rectangular flower garden are 40 feet and 30 feet, respectively. A walkway of width x surrounds the garden.

(a) Write the outside perimeter y of the walkway in terms of x.

(b) Use a graphing utility to graph the equation for the perimeter.

(c) Determine the slope of the graph of part (b). For each additional 1-foot increase in the width of the walkway, determine the increase in its perimeter.

Solution

(a) $y = 2(30 + 2x) + 2(40 + 2x) = 60 + 4x + 80 + 4x = 140 + 8x$

(b) Keystrokes:

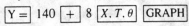

RANGE
Xmin=0
Xmax=50
Xscl=10
Ymin=0
Ymax=700
Yscl=100

(c) slope = 8

8 feet

45. Evaluate $|-15|$.

Solution

$|-15| = 15$

49. *Work Rate* Two people can complete a task in t hours where t must satisfy the equation $\frac{t}{4} + \frac{t}{6} = 1$. Solve this equation and interpret the result.

Solution

$$12\left(\frac{t}{4} + \frac{t}{6} = 1\right)12$$

$$3t + 2t = 12$$

$$5t = 12$$

$$t = \frac{12}{5} = 2\frac{2}{5}$$

Two people working together can complete a task in $2\frac{2}{5}$ hours.

53. Plot the points $(0, 0)$ and $(7, 5)$ and, if possible, find the slope of the line passing through them. State whether the line is rising, falling, horizontal, or vertical.

Solution

$$m = \frac{5-0}{7-0} = \frac{5}{7}$$

Line rises.

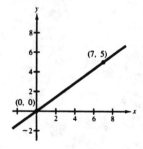

47. Evaluate $-|-6|$.

Solution

$-|-6| = -6$

51. Identify the line that has the specified slope m for each of the following. (See figure in textbook.)

(a) $m = \frac{3}{4}$ (b) $m = 0$

(c) $m = -3$

Solution

(a) $m = \frac{3}{4} \Rightarrow L_3$ (b) $m = 0 \Rightarrow L_2$

(c) $m = -3 \Rightarrow L_1$

55. Plot the points $(0, 12)$ and $(8, 0)$ and, if possible, find the slope of the line passing through them. State whether the line is rising, falling, horizontal, or vertical.

Solution

$$m = \frac{0-12}{8-0} = \frac{-12}{8} = \frac{-3}{2}$$

Line falls.

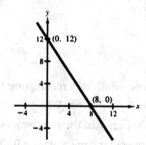

57. Plot the points $(-5, -3)$ and $(-5, 4)$ and, if possible, find the slope of the line passing through them. State whether the line is rising, falling, horizontal, or vertical.

Solution

$$m = \frac{4 - (-3)}{-5 - (-5)} = \frac{7}{0} = \text{undefined}$$

Line is vertical.

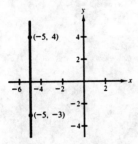

59. Plot the points $(2, -5)$ and $(7, -5)$ and, if possible, find the slope of the line passing through them. State whether the line is rising, falling, horizontal, or vertical.

Solution

$$m = \frac{-5 - (-5)}{7 - 2} = \frac{0}{5} = 0$$

Line is horizontal.

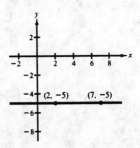

61. Plot the points $\left(\frac{3}{4}, 2\right)$ and $\left(5, -\frac{5}{2}\right)$ and, if possible, find the slope of the line passing through them. State whether the line is rising, falling, horizontal, or vertical.

Solution

$$m = \frac{2 - \frac{-5}{2}}{\frac{3}{4} - 5} \cdot \frac{4}{4} = \frac{8 + 10}{3 - 20} = \frac{18}{-17}$$

Line falls.

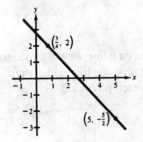

63. Plot the points $(4.2, -1)$ and $(-4.2, 6)$ and, if possible, find the slope of the line passing through them. State whether the line is rising, falling, horizontal, or vertical.

Solution

$$m = \frac{-1 - 6}{4.2 - (-4.2)} = \frac{-7}{8.4} = \frac{-70}{84}$$

$$= \frac{-35}{42} = \frac{-5}{6}$$

Line falls.

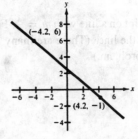

65. Plot the points $(2.5, -2)$ and $(4.75, 5.25)$ and, if possible, find the slope of the line passing through them. State whether the line is rising, falling, horizontal, or vertical.

Solution

$$m = \frac{5.25 - (-2)}{4.75 - 2.5} = \frac{7.25}{2.25} = \frac{725}{225} = \frac{29}{9}$$

Line rises.

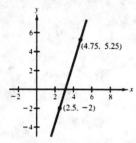

67. Solve for x so that the line through the points $(4, 5)$ and $(x, 7)$ will have $m = -\frac{2}{3}$.

Solution

$$\frac{-2}{3} = \frac{7 - 5}{x - 4}$$

$$-2(x - 4) = 6$$

$$-2x + 8 = 6$$

$$-2x = -2$$

$$x = 1$$

69. Solve for y so that the line through the points $(-3, y)$ and $(9, 3)$ will have $m = \frac{3}{2}$.

Solution

$$\frac{3}{2} = \frac{3 - y}{9 - (-3)}$$

$$3(12) = 2(3 - y)$$

$$36 = 6 - 2y$$

$$30 = -2y$$

$$-15 = y$$

71. The point $(0, 3)$ is a point on a line with $m = -1$. Find two additional points on the line. (There are many correct answers to each problem.)

Solution

$$-1 = \frac{y - 3}{x - 0}$$

$(1, 2)$, $(2, 1)$

73. The point $(-5, 0)$ is a point on a line with $m = \frac{4}{3}$. Find two additional points on the line. (There are many correct answers to each problem.)

Solution

$$\frac{4}{3} = \frac{y - 0}{x + 5}$$

$(-2, 4)$, $(1, 8)$

75. The point $(4, 2)$ is a point on a line with m undefined. Find two additional points on the line. (There are many correct answers to each problem.)

Solution

$(4, 2)$, m is undefined.

Vertical line: $(4, 3)$, $(4, 1)$, $(4, 0)$

Any points with x-coordinate of 4

77. Sketch the graph of a line through the point $(0, 1)$ having the slope $m = 2$.

Solution

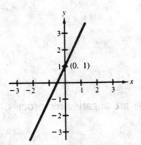

81. Sketch the graph of a line through the point $(0, 1)$ having the slope $m = -\frac{4}{3}$.

Solution

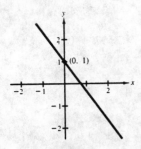

85. Write $x - 4y + 2 = 0$ in slope-intercept form and sketch the line. Use a graphing utility to confirm your sketch.

Solution

$x - 4y + 2 = 0$

$-4y = -x - 2$

$y = \frac{1}{4}x + \frac{1}{2}$

slope $= \frac{1}{4}$ y-intercept $= \frac{1}{2}$

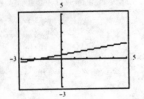

79. Sketch the graph of a line through the point $(0, 1)$ having the slope m is undefined.

Solution

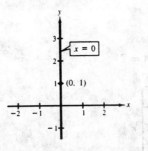

83. Write $x + y = 0$ in slope-intercept form and sketch the line. Use a graphing utility to confirm your sketch.

Solution

$x + y = 0$

$y = -x + 0$

slope $= -1$ y-intercept $= 0$

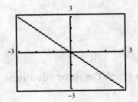

87. Write $\frac{1}{3}x + \frac{1}{2}y = 1$ in slope-intercept form and sketch the line. Use a graphing utility to confirm your sketch.

Solution

$\frac{1}{3}x + \frac{1}{2}y = 1$

$\frac{1}{2}y = -\frac{1}{3}x + 1$

$y = -\frac{2}{3}x + 2$

slope $= -\frac{2}{3}$ y-intercept $= 2$

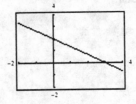

89. Write $y - 2 = 0$ in slope-intercept form and sketch the line. Use a graphing utility to confirm your sketch.

Solution

$y - 2 = 0$

$y = 2$

slope = 0 y-intercept = 2

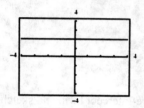

91. Determine whether the lines L_1 and L_2 passing through the points L_1 : $(1, 3)$, $(2, 1)$ and L_2 : $(0, 0)$, $(4, 2)$ are parallel, perpendicular, or neither.

Solution

$m_1 = \dfrac{1-3}{2-1} = \dfrac{-2}{1}$

$m_2 = \dfrac{2-0}{4-0} = \dfrac{2}{4} = \dfrac{1}{2}$

$L_1 \perp L_2$ since m_1 and m_2 are negative reciprocals.

93. Determine whether the lines L_1 and L_2 passing through the points L_1 : $(-2, 0)$, $(4, 4)$ and L_2 : $(1, -2)$, $(4, 0)$ are parallel, perpendicular, or neither.

Solution

$m_1 = \dfrac{4-0}{4+2} = \dfrac{4}{6} = \dfrac{2}{3}$

$m_2 = \dfrac{0+2}{4-1} = \dfrac{2}{3}$

$L_1 \| L_2$ since $m_1 = m_2$.

95. Use a graphing utility to graph the three equations

$y_1 = 3x$

$y_2 = -3x$

$y_3 = \frac{1}{3}x$

on the same viewing rectangle. Describe the relationships among the graphs. (Use the *square* setting so the slopes of the lines appear visually correct.)

Solution

Keystrokes:

$\boxed{Y=}$ 3 $\boxed{X, T, \theta}$

$\boxed{Y=}$ $\boxed{-}$ 3 $\boxed{X, T, \theta}$

$\boxed{Y=}$ $\boxed{(}$ 1 $\boxed{\div}$ 3 $\boxed{)}$ $\boxed{X, T, \theta}$ $\boxed{GRAPH}$

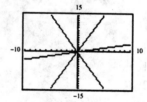

y_2 and y_3 are perpendicular.

97. Use a graphing utility to graph the three equations

$$y_1 = \tfrac{1}{4}x$$
$$y_2 = \tfrac{1}{4}x - 2$$
$$y_3 = \tfrac{1}{4}x + 3$$

on the same viewing rectangle. Describe the relationships among the graphs. (Use the *square* setting so the slopes of the lines appear visually correct.)

Solution

Keystrokes:

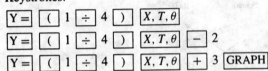

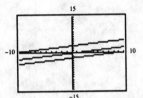

The lines are parallel.

99. *Graphical Estimation*   The graph in the text shows the earnings per share of common stock for Johnson & Johnson for the years 1987 through 1993. Use the slope of each segment to determine the year when earnings (a) decreased most rapidly and (b) increased most rapidly. (Source: Johnson & Johnson 1993 Annual Report)

Solution

(a) 1991:   only segment having a negative slope

(b) 1992:   sharpest rise therefore greatest positive slope

101. *Graphical Estimation*   The graph gives the declared dividend per share of common stock for Emerson Electric Company for the years 1987 through 1993. Use the slope of each segment to determine the year when dividends increased most rapidly. (Source: Emerson Electric Company)

Solution

1989:   sharpest rise therefore greatest positive slope

2.5   Relations and Functions

7. State the domain and range of $\{(-2, 0), (0, 1), (1, 4), (0, -1)\}$. Then draw a graphic representation.

Solution

Domain = $\{-2, 0, 1\}$

Range = $\{-1, 0, 1, 4\}$

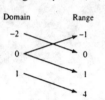

9. State the domain and range of $\{(0, 0), (4, -3), (2, 8), (5, 5), (6, 5)\}$. Then draw a graphic representation.

Solution

Domain = $\{0, 2, 4, 5, 6\}$

Range = $\{-3, 0, 5, 8\}$

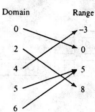

11. Write a set of ordered pairs that represents the rule of correspondence.

In a given week, a salesperson travels a distance d in t hours at an average speed of 50 mph. The travel times for each day are 3 hours, 2 hours, 8 hours, 6 hours, and $\frac{1}{2}$ hour.

Solution

$(3, 150), (2, 100), (8, 400), (6, 300), \left(\frac{1}{2}, 25\right)$

13. Write a set of ordered pairs that represents the rule of correspondence.

The winners of the World Series from 1990 to 1993

Solution

(1990, Cincinnati), (1991, Minnesota), (1992, Toronto), (1993, Toronto)

15. Decide whether the relation is a function.

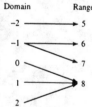

Solution

This relation is not a function because -1 in the domain is paired to 2 numbers (6 and 7) in the range.

17. Decide whether the relation is a function.

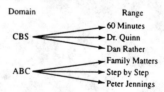

Solution

This relation is not a function because both CBS and ABC in the domain are each paired to 3 different TV shows in the range.

19. Determine which sets of ordered pairs represent functions from A to B.

$A = \{0, 1, 2, 3\}$ and $B = \{-2, -1, 0, 1, 2\}$

(a) $\{(0, 1), (1, -2), (2, 0), (3, 2)\}$

(b) $\{(0, -1), (2, 2), (1, -2), (3, 0), (1, 1)\}$

(c) $\{(0, 0), (1, 0), (2, 0), (3, 0)\}$

(d) $\{(0, 2), (3, 0), (1, 1)\}$

Solution

(a) Yes   (b) No   (c) Yes   (d) No

21. Decide whether $y = 10x + 12$ represents y as a function of x.

Solution

Yes

23. Decide whether $|y - 2| = x$ represents y as a function of x.

Solution

No

25. Using the function $f(x) = 3x + 5$, fill in the blanks and simplify.

(a) $f(2) = 3(\quad) + 5$

(c) $f(k) = 3(\quad) + 5$

(b) $f(-2) = 3(\quad) + 5$

(d) $f(k + 1) = 3(\quad) + 5$

Solution

(a) $f(2) = 3(2) + 5 = 11$

(c) $f(k) = 3(k) + 5 = 3k + 5$

(b) $f(-2) = 3(-2) + 5 = -1$

(d) $f(k + 1) = 3(k + 1) + 5 = 3k + 3 + 5 = 3k + 8$

27. Evaluate the function $f(x) = 12x - 7$ as indicated, and simplify.

(a) $f(3)$

(c) $f(a) + f(1)$

(b) $f\left(\dfrac{3}{2}\right)$

(d) $f(a + 1)$

Solution

(a) $f(3) = 12(3) - 7 = 29$

(c) $f(a) + f(1) = [12(a) - 7] + [12(1) - 7]$

$= 12a - 7 + 5 = 12a - 2$

(b) $f\left(\dfrac{3}{2}\right) = 12\left(\dfrac{3}{2}\right) - 7 = 11$

(d) $f(a + 1) = 12(a + 1) - 7 = 12a + 12 - 7$

$= 12a + 5$

29. Evaluate the function

$$f(x) = \begin{cases} x + 8, & \text{if } x < 0 \\ 10 - 2x, & \text{if } x \geq 0 \end{cases}$$

as indicated, and simplify.

(a) $f(4)$

(c) $f(0)$

(b) $f(-10)$

(d) $f(6) - f(-2)$

Solution

(a) $f(4) = 10 - 2(4) = 10 - 8 = 2$

(c) $f(0) = 10 - 2(0) = 10$

(b) $f(-10) = -10 + 8 = -2$

(d) $f(6) - f(-2) = [10 - 2(6)] - [-2 + 8]$

$= 10 - 12 - 6 = -8$

31. Evaluate the function $f(x) = 2x + 5$ as indicated, and simplify.

(a) $\dfrac{f(x+2) - f(2)}{x}$

(b) $\dfrac{f(x-3) - f(3)}{x}$

Solution

(a) $\dfrac{f(x+2) - f(2)}{x} = \dfrac{[2(x+2)+5] - [2(2)+5]}{x} = \dfrac{2x+4+5-4-5}{x} = \dfrac{2x}{x} = 2$

(b) $\dfrac{f(x-3) - f(3)}{x} = \dfrac{[2(x-3)+5] - [2(3)+5]}{x} = \dfrac{2x-6+5-6-5}{x} = \dfrac{2x-12}{x}$

33. Find the domain of $f(x) = \dfrac{2x}{x-3}$.

Solution

domain $= x \neq 3$

35. Find the domain of $f(x) = \sqrt{2x-1}$.

Solution

domain $= 2x - 1 \geq 0$

$x \geq \dfrac{1}{2}$

37. *Geometry* Express the perimeter P of a square as a function of the length x of one of its sides.

Solution

$P(x) = 4x$

39. *Geometry* An open box is made from a square piece of material 24 inches on a side by cutting equal squares from the corners and turning up the sides (see figure in textbook). Write the volume V of the box as a function of x.

Solution

$V(x) = x(24 - 2x)^2$ or $4x(12 - x)^2$

41. Simplify $4 - 3(2x + 1)$.

Solution

$4 - 3(2x+1) = 4 - 6x - 3$

$= 1 - 6x$

43. Simplify $5(x + 2) - 4(2x - 3)$.

Solution

$5(x+2) - 4(2x-3) = 5x + 10 - 8x + 12$

$= -3x + 22$

45. *Work Rate* You can mow the lawn in 4 hours and your friend can mow it in 5 hours. What fractional part of the lawn can each of you mow in one hour? How long will it take both of you to mow the lawn?

Solution

Your rate $= \dfrac{1}{4}$ Friend's rate $= \dfrac{1}{5}$

Verbal Model: $\boxed{\text{Work}} = \boxed{\text{Rate}} \cdot \boxed{\text{Time}}$

Equation: $1 = \dfrac{1}{4} \cdot t + \dfrac{1}{5} \cdot t$

$20 = 5t + 4t$

$20 = 9t$

$2.2 \text{ hours} \approx \dfrac{20}{9} \text{ hours} = t$

47. Decide whether the relation is a function.

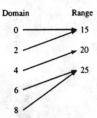

Solution

Yes, this relation is a function as each number in the domain is paired with exactly one number in the range.

51. Decide whether the relation is a function.

Input value	0	1	2	3	4
Output value	0	1	4	9	16

Solution

Yes, this relation is a function as each number in the domain is paired with exactly one number in the range.

55. Use the graph, which shows numbers of high school and college students in the United States. (Source: U.S. National Center for Education Statistics) Is the high school enrollment a function of the year? Is the college enrollment a function of the year? Explain.

Solution

Yes, the high school enrollment is a function of the year as each number in the domain is paired with exactly one number in the range. Yes, the college enrollment is a function of the year as each number in the domain is paired with exactly one number in the range.

59. Decide whether $y = x(x - 10)$ represents y as a function of x.

Solution

Yes

49. Decide whether the relation is a function.

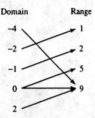

Solution

No, this relation is not a function as 0 in the domain is paired with 2 numbers in the range (5 and 9).

53. Decide whether the relation is a function.

Input value	4	7	9	7	4
Output value	2	4	6	8	10

Solution

No, this relation is not a function as the 4 and the 7 in the domain are each paired with 2 different numbers in the range.

57. Decide whether $3x + 7y - 2 = 0$ represents y as a function of x.

Solution

Yes

61. Determine whether $|y| = 8 - x$ represents y as a function of x.

Solution

No

63. Evaluate $g(x) = \dfrac{1}{2}x^2$ as indicated, and simplify.

(a) $g(4)$

(b) $g\left(\dfrac{2}{3}\right)$

(c) $g(2y)$

(d) $g(4) + g(6)$

Solution

(a) $g(4) = \dfrac{1}{2}(4)^2 = 8$

(b) $g\left(\dfrac{2}{3}\right) = \dfrac{1}{2}\left(\dfrac{2}{3}\right)^2 = \dfrac{1}{2}\left(\dfrac{4}{9}\right) = \dfrac{2}{9}$

(c) $g(2y) = \dfrac{1}{2}(2y)^2 = \dfrac{1}{2}(4y^2) = 2y^2$

(d) $g(4) + g(6) = \dfrac{1}{2}(4)^2 + \dfrac{1}{2}(6)^2 = 8 + 18 = 26$

65. Evaluate $f(x) = \sqrt{x + 5}$ as indicated, and simplify.

(a) $f(-1)$

(b) $f(4)$

(c) $f\left(\dfrac{16}{3}\right)$

(d) $f(5z)$

Solution

(a) $f(-1) = \sqrt{-1 + 5} = 2$

(b) $f(4) = \sqrt{4 + 5} = 3$

(c) $f\left(\dfrac{16}{3}\right) = \sqrt{\dfrac{16}{3} + 5} = \sqrt{\dfrac{31}{3}} = \dfrac{\sqrt{93}}{3}$

(d) $f(5z) = \sqrt{5z + 5}$

67. Evaluate $f(x)\dfrac{3x}{x - 5}$ as indicated, and simplify.

(a) $f(0)$

(b) $f\left(\dfrac{5}{3}\right)$

(c) $f(2) - f(-1)$

(d) $f(x + 4)$

Solution

(a) $f(0) = \dfrac{3(0)}{0 - 5} = 0$

(b) $f\left(\dfrac{5}{3}\right) = \dfrac{3(5/3)}{5/3 - 5} \cdot \dfrac{3}{3} = \dfrac{15}{5 - 15} = \dfrac{15}{-10} = \dfrac{3}{-2}$

(c) $f(2) - f(-1) = \left[\dfrac{3(2)}{2 - 5}\right] - \left[\dfrac{3(-1)}{-1 - 5}\right]$

$= \dfrac{6}{-3} - \dfrac{-3}{-8} = -2 - \dfrac{1}{2} = \dfrac{-5}{2}$

(d) $f(x + 4) = \dfrac{3(x + 4)}{x + 4 - 5} = \dfrac{3x + 12}{x - 1}$

69. Evaluate $g(x) = 1 - x^2$ as indicated, and simplify.

(a) $g(2.2)$

(b) $\dfrac{g(2.2) - g(2)}{0.2}$

Solution

(a) $g(2.2) = 1 - (2.2)^2$

$= 1 - 4.84$

$= -3.84$

(b) $\dfrac{g(2.2) - g(2)}{0.2} = \dfrac{-3.84 - \left(1 - 2^2\right)}{0.2}$

$= \dfrac{-3.84 - (-3)}{0.2}$

$= \dfrac{-0.84}{0.2} = -4.2$

71. Evaluate the function

$$h(x) = \begin{cases} 4 - x^2, & \text{if } x \le 2 \\ x - 2, & \text{if } x > 2 \end{cases}$$

as indicated, and simplify.

(a) $h(2)$

(b) $h\left(-\dfrac{3}{2}\right)$

(c) $h(5)$

(d) $h(-3) + h(7)$

Solution

(a) $h(2) = 4 - 2^2 = 0$

(b) $h\left(-\dfrac{3}{2}\right) = 4 - \left(-\dfrac{3}{2}\right)^2$

$$= 4 - \dfrac{9}{4} = \dfrac{16}{4} - \dfrac{9}{4} = \dfrac{7}{4}$$

(c) $h(5) = 5 - 2 = 3$

(d) $h(-3) + h(7) = [4 - (-3)^2] + [7 - 2]$

$$= 4 - 9 + 5 = 0$$

73. Find the domain and range of
$f : \{(0, 0), (2, 1), (4, 8), (6, 27)\}$.

Solution

domain = $\{0, 2, 4, 6\}$
range = $\{0, 1, 8, 27\}$

75. Find the domain and range of the following.

Circumference of a circle: $C = 2\pi r$

Solution

domain = $r \ge 0$
range = $C \ge 0$

77. Find the domain of $h(x) = \dfrac{9}{x^2 + 1}$

Solution

domain = x is any real number

79. Find the domain of $f(t) = \dfrac{t + 3}{t(t + 2)}$.

Solution

domain = $t \ne 0, -2$

81. Find the domain of $g(x) = \sqrt{x + 4}$.

Solution

domain = $x \ge -4$ since $x + 4 \ge 0$

83. Find the domain of $f(t) = t^2 + 4t - 1$.

Solution

domain = t is any real number

85. *Geometry* Express the volume V of a cube as a function of the length x of one of the edges.

Solution

$V(x) = x^3$

87. *Cost* The inventor of a new game believes that the variable cost for producing the game is \$1.95 per unit and the fixed costs are \$8000. Write the total cost C as a function of x, the number of games produced.

Solution

Verbal Model: $\boxed{\text{Total Cost}} = \boxed{\begin{array}{c}\text{Fixed} \\ \text{Costs}\end{array}} + \boxed{\begin{array}{c}\text{Variable} \\ \text{Costs}\end{array}}$

Equation: $C(x) = 8000 + 1.95x$

89. *Safe Load* A solid rectangular beam has a height of 6 inches and width of 4 inches. The safe load S of the beam with the load at the center is a function of its length L and is approximated by the model $S(L) = \dfrac{128{,}160}{L}$, where S is measured in pounds and L is measured in feet. Find

(a) $S(12)$

(b) $S(16)$.

Solution

(a) $S(12) = \dfrac{128{,}160}{12} = 10{,}680$ pounds

(b) $S(16) = \dfrac{128{,}160}{16} = 8010$ pounds

91. Determine whether the statements use the word *function* in ways that are mathematically correct.

(a) The sales tax on a purchased item is a function of the selling price.

(b) Your score on the next algebra exam is a function of the number of hours you study the night before the exam.

Solution

(a) This is a correct mathematical use of the word function.

(b) This is not a correct mathematical use of the word function.

2.6 Graphs of Functions

7. Use a graphing utility to graph $g(x) = 1 - x^2$ and find its domain and range.

Solution

Keystrokes:

$\boxed{Y=}\ 1\ \boxed{-}\ \boxed{X,T,\theta}\ \boxed{x^2}\ \boxed{GRAPH}$

Domain = reals   Range = $(-\infty, 1]$ or $y \le 1$

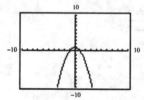

9. Use a graphing utility to graph $f(x) = \sqrt{x - 2}$ and find its domain and range.

Solution

Keystrokes:

$\boxed{Y=}\ \boxed{\sqrt{\ }}\ \boxed{(}\ \boxed{X,T,\theta}\ \boxed{-}\ 2\ \boxed{)}\ \boxed{GRAPH}$

Domain = $[2, \infty)$ or $x \ge 2$ Range $[0, \infty)$ or $y \ge 0$

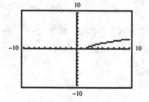

11. Use the Vertical Line Test to determine whether y is a function of x.

$$y = \frac{1}{3}x^3$$

Solution

Yes, $y = \dfrac{1}{3}x^3$ passes the Vertical Line Test and is a function of x.

13. Use the Vertical Line Test to determine whether y is a function of x.

$$y^2 = x$$

Solution

No, $y^2 = x$ does not pass the Vertical Line Test and is not a function of x.

15. *Think About It* Does the graph of $y = \frac{1}{3}x^3$ represent x as a function of y? Explain your reasoning.

Solution

Yes, $y = \frac{1}{3}x^3$ does represent x as a function of y. Solve $y = \frac{1}{3}x^3$ for x.

$$y = \frac{1}{3}x^3$$

$$3y = x^3$$

$$\sqrt[3]{3y} = x$$

$x = \sqrt[3]{3y}$ will pass the Vertical Line Test.

17. Match the function $f(x) = x^2 - 1$ with its graph.

Solution

(b)

19. Match the function $f(x) = 2 - |x|$ with its graph.

Solution

(a)

21. Sketch the graph of $g(x) = \frac{1}{2}x^2$. Then determine its domain and range.

Solution

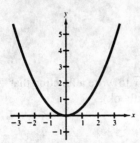

Domain: x is a real

Range: $[0, \infty)$ or $y \geq 0$

23. Sketch the graph of $f(x) = -(x - 1)^2$. Then determine its domain and range.

Solution

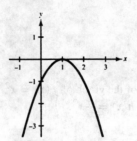

Domain: x is a real

Range: $(-\infty, 0]$ or $y \leq 0$

25. Sketch the graph of $K(s) = |s - 4| + 1$. Then determine its domain and range.

Solution

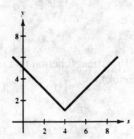

Domain: x is a real

Range: $[1, \infty)$ or $y \geq 1$

27. Sketch the graph of $h(x) = \begin{cases} 2x + 3, & \text{if } x < 0 \\ 3 - x, & \text{if } x \geq 0 \end{cases}$

Solution

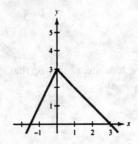

Domain: x is a real

Range: $(-\infty, 3]$ or $y \leq 3$

29. Select the viewing rectangle that shows the most complete graph of $f(x) = -(x^2 - 20x + 50)$.

Solution

(b) shows the most complete graph.

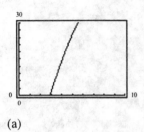

(a)

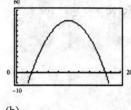

(b)

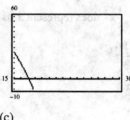

(c)

31. Identify the transformation of the graph of $f(x) = x^2$ and sketch the graph of h if $h(x) = x^2 + 2$.

Solution

vertical shift up 2 units

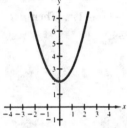

33. Identify the transformation of the graph of $f(x) = x^2$ and sketch the graph of h if $h(x) = (x + 2)^2$.

Solution

horizontal shift left 2 units

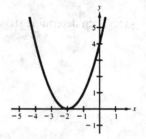

35. Identify the transformation of the graph of $f(x) = x^2$ and sketch the graph of h if $h(x) = -x^2$.

Solution

reflected across x-axis

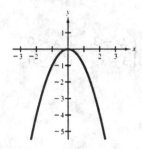

37. Use the graph of $f(x) = \sqrt{x}$ to write a function that represents the graph (see textbook).

Solution

$f(x) = -\sqrt{x}$

39. Use the graph of $f(x) = \sqrt{x}$ to write a function that represents the graph (see textbook).

Solution

$f(x) = \sqrt{x + 2}$

41. Use the graph of $f(x) = \sqrt{x}$ to write a function that represents the graph (see textbook).

Solution

$f(x) = \sqrt{-x}$

43. *Geometry*   The perimeter of a rectangle is 200 meters (see textbook).

  (a) Show that the area of the rectangle is given by $A = x(100 - x)$, where x is its length.

  (b) Use a graphing utility to graph the area function.

  (c) Graphically, what value of x yields the largest value of A? Interpret the results.

Solution

(a)
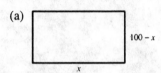

$A = l \cdot w$

$A = x(100 - x)$

let $x =$ length

$100 - x =$ width

$P = 2l + 2w$

$200 = 2l + 2w$

$100 = l + w$

$100 - l = w$

(b)
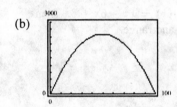

(c) When $x = 50$, the largest value of A is 2500. (50, 2500) is the highest point on the graph of A giving the largest value of the function.

45. Solve the equation $4 - \dfrac{1}{2}x = 6$.

Solution

$4 - \dfrac{1}{2}x = 6$

$-\dfrac{1}{2}x = 2$

$x = -4$

47. Solve the equation $4(x - 3) = 2x$.

Solution

$4(x - 3) = 2x$

$4x - 12 = 2x$

$2x = 12$

$x = 6$

49. *Phone Charges*   The cost for a phone call is \$1.10 for the first minute and \$0.45 for each additional minute. The total cost of a call cannot exceed \$11. Find the interval of time that is available for the call.

Solution

Verbal Model: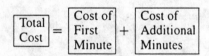

Inequality: $1.10 \le 1.10 + 0.45x \le 11$

$0 \le 0.45x \le 9.90$

$0 \le x \le 22$ minutes

51. Use the Vertical Line Test to determine whether y is a function of x if $y = (x + 2)^2$. (See graph in textbook.)

Solution

y is a function of x by vertical line test.

53. Use the Vertical Line Test to determine whether y is a function of x if $x^2 + y^2 = 16$. (See graph in textbook.)

Solution

y is not a function of x by vertical line test.

55. Sketch the graph of $3x - 5y = 15$. Does the graph represent y as a function of x?

Solution

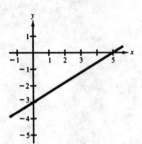

y is a function of x.

59. Sketch the graph of $f(x) = 2x - 7$. Then determine its domain and range.

Solution

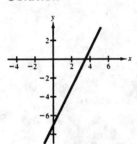

Domain: x is a real

Range: y is a real

63. Sketch the graph of $f(t) = \sqrt{t - 2}$. Then determine its domain and range.

Solution

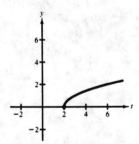

Domain: $[2, \infty)$ or $t \geq 2$

Range: $[0, \infty)$ or $y \geq 0$

57. Sketch the graph of $y^2 = x + 1$. Does the graph represent y as a function of x?

Solution

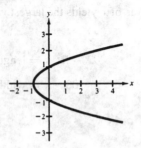

y is not a function of x.

61. Sketch the graph of $h(x) = x^2 - 6x + 8$. Then determine its domain and range.

Solution

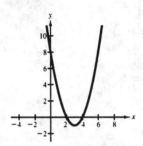

Domain : x is a real

Range: $[-1, \infty)$ or $y \geq -1$

65. Sketch the graph of $g(s) = \dfrac{1}{2}s^3$. Then determine its domain and range.

Solution

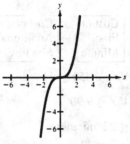

Domain: s is a real

Range: y is a real

67. Sketch the graph of $f(x) = |x + 3|$. Then determine its domain and range.

Solution

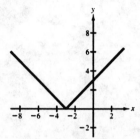

Domain: x is a real

Range: $[0, \infty)$ or $y \geq 0$

71. Sketch the graph of $h(x) = x^3$, $-2 \leq x \leq 2$. Then determine its domain and range.

Solution

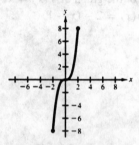

Domain: $-2 \leq x \leq 2$ or $[-2, 2]$

Range: $-8 \leq y \leq 8$ or $[-8, 8]$

69. Sketch the graph of $f(x) = 6 - 3x$, $0 \leq x \leq 2$. Then determine its domain and range.

Solution

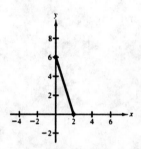

Domain: $0 \leq x \leq 2$ or $[0, 2]$

Range: $0 \leq y \leq 6$ or $[0, 6]$

73. Sketch the graph of

$$f(x) = \begin{cases} x + 6, & \text{if } x < 0 \\ 6 - 2x, & \text{if } x \geq 0 \end{cases}.$$

Then determine its domain and range.

Solution

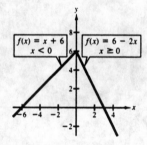

Domain: x is a real

Range: $(-\infty, 6]$ or $y \leq 6$

75. Sketch the graph of

$$h(x) = \begin{cases} 4 - x^2, & \text{if } x \le 2 \\ x - 2, & \text{if } x > 2 \end{cases}.$$

Then determine its domain and range.

Solution

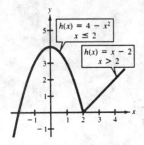

Domain: x is a real

Range: y is a real

77. Identify the transformation of the graph of $f(x) = x^3$ and sketch a graph of h if $h(x) = x^3 + 3$.

Solution

vertical shift up 3 units

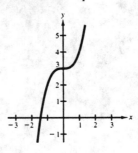

79. Identify the transformation of the graph of $f(x) = x^3$ and sketch a graph of h if $h(x) = (x - 3)^3$.

Solution

horizontal shift right 3 units

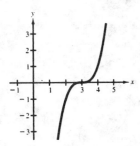

81. Identify the transformation of the graph of $f(x) = x^3$ and sketch a graph of h if $h(x) = 2 - (x - 1)^3$.

Solution

vertical shift up 2 units
reflected across x-axis
horizontal shift right 1 unit

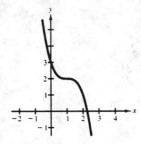

83. Identify the transformation of the graph of $f(x) = |x|$ and use a graphing utility to sketch the graph of h if $h(x) = |x - 5|$.

Solution

Keystrokes:

$\boxed{Y=}$ $\boxed{\text{ABS}}$ $\boxed{(}$ $\boxed{X, T, \theta}$ $\boxed{-}$ 5 $\boxed{)}$ $\boxed{\text{GRAPH}}$

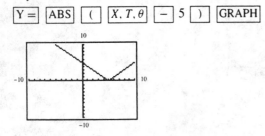

horizontal shift right 5 units

85. Identify the transformation of the graph of $f(x) = |x|$ and use a graphing utility to sketch the graph of h if $h(x) = |x| - 5$.

Solution

Keystrokes:

$\boxed{Y=}$ $\boxed{\text{ABS}}$ $\boxed{X, T, \theta}$ $\boxed{-}$ 5 $\boxed{\text{GRAPH}}$

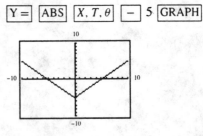

vertical shift down 5 units

87. Identify the transformation of the graph of $f(x) = |x|$ and use a graphing utility to sketch the graph of h if $h(x) = -|x|$.

Solution

Keystrokes:

$\boxed{Y=}$ $\boxed{(-)}$ $\boxed{ABS}$ $\boxed{X, T, \theta}$ $\boxed{GRAPH}$

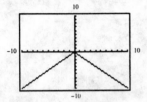

reflected across x-axis

89. Use the graph of $f(x) = x^2$ to write an equation that represents the transformation of f given by the graph (see textbook).

Solution

Graph is shifted 3 units left

$h(x) = (x + 3)^2$

91. Use the graph of $f(x) = x^2$ to write an equation that represents the transformation of f given by the graph (see textbook).

Solution

Graph is reflected across x-axis

$h(x) = -x^2$

93. Use the graph of $f(x) = x^2$ to write an equation that represents the transformation of f given by the graph (see textbook).

Solution

Graph is shifted 3 units left and reflected across x-axis

$h(x) = -(x + 3)^2$

95. Use the graph of $f(x) = x^2$ to write an equation that represents the transformation of f given by the graph (see textbook).

Solution

Graph is reflected across x-axis and shifted up 2 units

$h(x) = -x^2 + 2$

97. *Civilian Population of the United States* For 1950 through 1990, the civilian population, P (in thousands) in the United States can be modeled by

$$P_1(t) = -12t^2 + 2900t + 149,300$$

where $t = 0$ represents 1950. (Source: U.S. Bureau of Census)

(a) Use a graphing utility to graph function over the appropriate domain.

(b) In the transformation of the population function

$$P_2(t) = -12(t + 20)^2 + 2900(t + 20) + 149,300,$$

$t = 0$ corresponds to what calendar year? Explain.

(c) Use a graphing utility to graph P_2 over the appropriate domain.

Solution

(a)

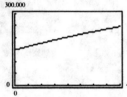

(b) 1970

If $t = 0$ represents 1950, then $t + 20$ represents $1950 + 20 = 1970$.

(c)

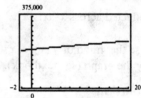

Review Exercises for Chapter 2

1. Plot the points $(0, -3)$, $\left(\frac{5}{2}, 5\right)$, and $(-2, -4)$ on a rectangular coordinate system.

Solution

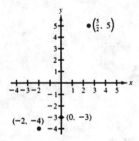

3. Plot the points and verify that the points form the indicated polygon.

Right triangle: $(1, 1)$, $(12, 9)$, $(4, 20)$

Solution

$$d = \sqrt{(1 - 9)^2 + (1 - 12)^2} = \sqrt{64 + 121} = \sqrt{185}$$

$$d = \sqrt{(1 - 4)^2 + (1 - 20)^2} = \sqrt{9 + 361} = \sqrt{370}$$

$$d = \sqrt{(4 - 12)^2 + (20 - 9)^2} = \sqrt{64 + 121} = \sqrt{185}$$

$$\left(\sqrt{185}\right)^2 + \left(\sqrt{185}\right)^2 = \left(\sqrt{370}\right)^2$$

$$185 + 185 = 370$$

$$370 = 370$$

By Pythagorean Theorem, figure is a right triangle.

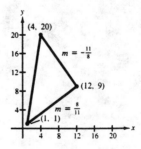

5. Determine the quadrant(s) in which the point $(2, -6)$ is located.

Solution

Quadrant IV

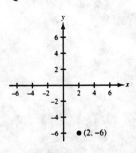

7. Determine the quadrant(s) in which the point $(4, y)$ is located.

Solution

Quadrants I, IV

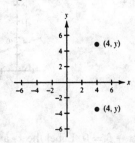

9. Plot the points $(4, 3)$ and $(4, 8)$ and find the distance between them.

Solution

$$d = \sqrt{(4-4)^2 + (3-8)^2}$$
$$= \sqrt{0 + 25}$$
$$= \sqrt{25}$$
$$= 5$$

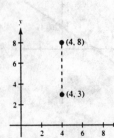

11. Plot the points $(-5, -1)$ and $(1, 2)$ and find the distance between them.

Solution

$$d = \sqrt{(-5-1)^2 + (-1-2)^2}$$
$$= \sqrt{36 + 9}$$
$$= \sqrt{45}$$
$$= 3\sqrt{5}$$

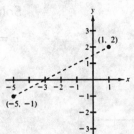

13. Determine whether the ordered pairs are solutions of the equation $y = 4 - \frac{1}{2}x$.

(a) $(4, 2)$ $2 \stackrel{?}{=} 4 - \frac{1}{2}(4)$

$2 \stackrel{?}{=} 4 - 2$

$2 = 2$ yes

(b) $(-1, 5)$ $5 \stackrel{?}{=} 4 - \frac{1}{2}(-1)$

$5 \stackrel{?}{=} 4 + \frac{1}{2}$

$5 \neq 4\frac{1}{2}$ no

(c) $(-4, 0)$ $0 \stackrel{?}{=} 4 - \frac{1}{2}(-4)$

$0 \stackrel{?}{=} 4 + 2$

$0 \neq 6$ no

(d) $(8, 0)$ $0 \stackrel{?}{=} 4 - \frac{1}{2}(8)$

$0 \stackrel{?}{=} 4 - 4$

$0 = 0$ yes

15. Sketch the graph of $y = 6 - \frac{1}{3}x$. Label the x- and y-intercepts.

Solution

$y = 6 - \frac{1}{3}x$

$y = 6 - \frac{1}{3}(0)$

$y = 6 \quad (0, 6)$

$0 = 6 - \frac{1}{3}x$

$\frac{1}{3}x = 6$

$x = 18 \quad (18, 0)$

17. Sketch the graph of $3y - 2x - 3 = 0$. Label the x- and y-intercepts.

Solution

$3y - 2x - 3 = 0$

$3y - 2(0) - 3 = 0$

$3y = 3$

$y = 1 \quad (0, 1)$

$3(0) - 2x - 3 = 0$

$-2x = 3$

$x = -\frac{3}{2} \quad \left(-\frac{3}{2}, 0\right)$

19. Sketch the graph of $x = |y - 3|$. Label the x- and y-intercepts.

Solution

$x = |y - 3|$

$0 = |y - 3|$

$3 = y \quad (0, 3)$

$x = |0 - 3|$

$x = 3 \quad (3, 0)$

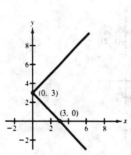

21. Find the slope of the line through the points $(-1, 1)$ and $(6, 3)$.

Solution

$$m = \frac{3 - 1}{6 - (-1)} = \frac{2}{7}$$

23. Find the slope of the line through the points $(-1, 3)$ and $(4, 3)$.

Solution

$$m = \frac{3 - 3}{4 - (-1)} = \frac{0}{5} = 0$$

25. Find the slope of the line through the points $(0, 6)$ and $(8, 0)$.

Solution

$$m = \frac{0 - 6}{8 - 0} = \frac{-6}{8} = \frac{-3}{4} = -\frac{3}{4}$$

27. Find t so that the three points $(-3, -3)$, $(0, t)$, and $(1, 3)$ are collinear. (*Note:* Collinear means that the points lie on the same straight line.)

Solution

$$m = \frac{3 - (-3)}{1 - (-3)} = \frac{6}{4} = \frac{3}{2}$$

$$\frac{3}{2} = \frac{3 - t}{1 - 0}$$

$$3 = 6 - 2t$$

$$-3 = -2t$$

$$\frac{3}{2} = t$$

29. Find two additional points on the line when one point is $(2, -4)$ and the slope is $m = -3$. (There are many solutions to this problem.)

Solution

$$-3 = \frac{y + 4}{x - 2}$$

$(0, 2)$, $(1, -1)$

31. Find two additional points on the line when one point is $(3, 1)$ and the slope is $m = \frac{5}{4}$. (There are many solutions to this problem.)

Solution

$$\frac{5}{4} = \frac{y - 1}{x - 3}$$

$(7, 6)$, $(11, 11)$

33. Find two additional points on the line when one point is $(3, 7)$ and the slope is undefined. (There are many solutions to this problem.)

Solution

Since m is undefined the line is a vertical line so points such as $(3, 0)$, $(3, 1)$, and $(3, -2)$ are on this line.

35. Write the equation $5x - 2y - 4 = 0$ in slope-intercept form and sketch the line.

Solution

$$5x - 2y - 4 = 0$$

$$-2y = -5x + 4$$

$$y = \frac{5}{2}x - 2$$

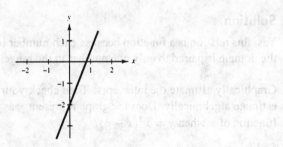

37. Write the equation $x + 2y - 2 = 0$ in slope-intercept form and sketch the line.

Solution

$$x + 2y - 2 = 0$$

$$2y = -x + 2$$

$$y = -\frac{1}{2}x + 1$$

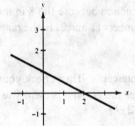

39. Use a graphing utility to graph the equations on the same viewing rectangle. Are the lines parallel, perpendicular, or neither?

$$L_1 : y = \tfrac{3}{2}x + 1 \qquad L_2 : y = \tfrac{2}{3}x - 1$$

Solution

Keystrokes:

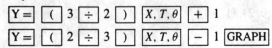

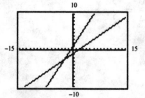

Neither

41. Use a graphing utility to graph the equations on the same viewing rectangle. Are the lines parallel, perpendicular, or neither?

$$L_1 : y = \tfrac{3}{2}x - 2 \qquad L_2 : y = -\tfrac{2}{3}x + 1$$

Solution

Keystrokes:

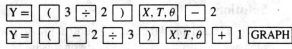

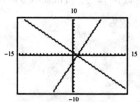

Perpendicular

43. Decide whether the relation is a function.

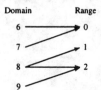

Solution

No, this relation is not a function because the 8 in the domain is paired to two numbers (1 and 2) in the range.

45. Decide whether the relation is a function.

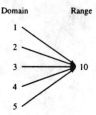

Solution

Yes, this relation is a function because each number in the domain is paired to only one number in the range.

47. Graphically estimate the intercepts. Then check your estimate algebraically. Does the graph represent y as a function of x when $9y^2 = 4x^3$?

Solution

$(0, 0)$, x and y-intercepts

y is not a function of x.

49. Graphically estimate the intercepts. Then check your estimate algebraically. Does the graph represent y as a function of x when $y = x^2(x - 3)$?

Solution

$(0, 0), (3, 0)$ x and y-intercepts

y is a function of x.

51. Evaluate $f(x) = 4 - \frac{5}{2}x$ for the specified value of the independent variable and simplify when possible.

(a) $f(-10)$ (b) $f\left(\frac{2}{5}\right)$ (c) $f(t) + f(-4)$ (d) $f(x + h)$

Solution

(a) $f(-10) = 4 - \frac{5}{2}(-10) = 4 + 25 = 29$ (b) $f\left(\frac{2}{5}\right) = 4 - \frac{5}{2}\left(\frac{2}{5}\right) = 4 - 1 = 3$

(c) $f(t) + f(-4) = \left(4 - \frac{5}{2}t\right) + \left(4 - \frac{5}{2}(-4)\right)$ (d) $f(x + h) = 4 - \frac{5}{2}(x + h) = 4 - \frac{5}{2}x - \frac{5}{2}h$

$$= 4 - \frac{5}{2}t + 4 + 10 = 18 - \frac{5}{2}t$$

53. Evaluate $f(t) = \sqrt{5 - t}$ for the specified value of the independent variable and simplify when possible.

(a) $f(-3)$ (b) $f(5)$ (c) $f\left(\frac{11}{3}\right)$ (d) $f(5z)$

Solution

(a) $f(-3) = \sqrt{5 - (-3)}$ (b) $f(5) = \sqrt{5 - 5}$

$\qquad = \sqrt{8}$ $\qquad = 0$

$\qquad = 2\sqrt{2}$

(c) $f\left(\frac{11}{3}\right) = \sqrt{5 - \frac{11}{3}}$ (d) $f(5z) = \sqrt{5 - 5z}$

$\qquad = \sqrt{\frac{15}{3} - \frac{11}{3}}$

$\qquad = \sqrt{\frac{4}{3} \cdot \frac{3}{3}} = \frac{2\sqrt{3}}{3}$

55. Evaluate

$$f(x) = \begin{cases} -3x, & \text{if } x \leq 0 \\ 1 - x^2, & \text{if } x > 0 \end{cases}$$

for the specified value of the independent variable and simplify when possible.

(a) $f(2)$ (b) $f\left(-\frac{2}{3}\right)$ (c) $f(1)$ (d) $f(4) - f(3)$

Solution

(a) $f(2) = 1 - 2^2$ (b) $f\left(-\frac{2}{3}\right) = -3\left(-\frac{2}{3}\right)$

$\qquad = -3$ $\qquad = 2$

(c) $f(1) = 1 - 1^2$ (d) $f(4) - f(3) = (1 - 4^2) - (1 - 3^2)$

$\qquad = 0$ $\qquad = 1 - 16 - 1 + 9$

$\qquad\qquad\qquad\qquad\qquad\qquad\qquad\qquad = -7$

57. Evaluate $f(x) = 3 - 2x$ for the specified value of the independent variable and simplify when possible.

(a) $\dfrac{f(x + 2) - f(2)}{x}$

(b) $\dfrac{f(x - 3) - f(3)}{x}$

Solution

(a) $\dfrac{f(x + 2) - f(2)}{x} = \dfrac{[3 - 2(x + 2)] - [3 - 2(2)]}{x}$

$= \dfrac{3 - 2x - 4 - 3 + 4}{x}$

$= \dfrac{-2x}{x} = -2$

(b) $\dfrac{f(x - 3) - f(3)}{x} = \dfrac{[3 - 2(x - 3)] - [3 - 2(3)]}{x}$

$= \dfrac{3 - 2x + 6 - 3 + 6}{x}$

$= \dfrac{-2x + 12}{x}$

59. Find the domain of $h(x) = 4x^2 - 7$.

Solution

Domain: Real numbers or $(-\infty, \infty)$

61. Find the domain of $f(x) = \sqrt{5 - 2x}$.

Solution

Domain: $\left(-\infty, \frac{5}{2}\right]$

63. Select the viewing rectangle on a graphing utility that shows the most complete graph of $f(x) = x^4 - 2x^3$.

Solution

(a)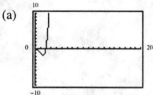

(b)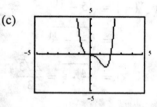

(c)

(c) shows the most complete graph.

65. Sketch the graph of $g(x) = \frac{1}{8}x^2$.

Solution

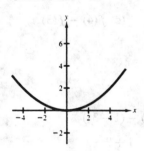

67. Sketch the graph of $y = (x - 2)^2$.

Solution

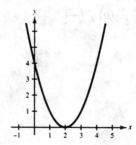

69. Sketch the graph of $y = \frac{1}{2}x(2 - x)$.

Solution

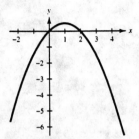

71. Sketch the graph of $y = 8 - 2|x|$.

Solution

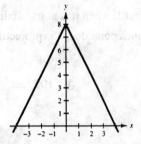

73. Sketch the graph of $g(x) = \frac{1}{4}x^3$.

Solution

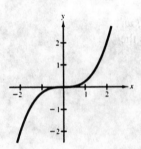

75. Sketch the graph of

$$f(x) = \begin{cases} 2 - (x - 1)^2, & \text{if } x < 1 \\ 2 + (x - 1)^2, & \text{if } x \geq 1 \end{cases}.$$

Solution

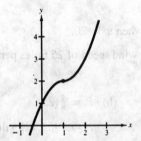

77. Identify the transformation of the graph of $f(x) = x^4$ and sketch the graph of $h(x) = -x^4$.

Solution

The graph of $f(x) = x^4$ is reflected across the x-axis.

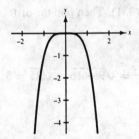

79. Identify the transformation of the graph of $f(x) = x^4$ and sketch the graph of $h(x) = (x - 1)^4$.

Solution

The graph of $f(x) = x^4$ is shifted right one unit.

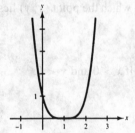

81. *Path of a Projectile*   The height y (in feet) of a projectile is given by $y = -\frac{1}{16}x^2 + 5x$, where x is the horizontal distance (in feet) from where the projectile was launched.

(a) Sketch the path of the projectile.

(b) How high is the projectile when it is at its maximum height?

(c) How far from the launch point does the projectile strike the ground?

Solution

(a)

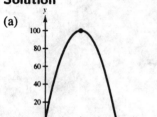

(b) 100 feet   (c) 80 feet

83. *Power Generation*   The power generated by a wind turbine is given by the function $P = kw^3$, where P is the number of kilowatts produced at a wind speed of w miles per hour and k is the constant of proportionality.

(a) Find k if $P = 1000$ when $w = 20$.

(b) Find the output for a wind speed of 25 miles per hour.

Solution

(a) $P = kw^3$

$1000 = k(20)^3$

$\frac{1}{8} = k$

(b) $P = \frac{1}{8}(25)^3$

$P = 1953.125$ kilowatts

Test for Chapter 2

1. Determine the quadrant in which the point (x, y) lies if $x > 0$ and $y < 0$.

Solution

(x, y) lies in Quadrant IV if $x > 0$ and $y < 0$.

2. Plot the points $(0, 5)$ and $(3, 1)$. Then find the distance between them.

Solution

$d = \sqrt{(0 - 3)^2 + (5 - 1)^2} = \sqrt{9 + 16} = \sqrt{25} = 5$

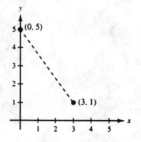

3. Find the x- and y-intercepts of the graph of
   $y = -3(x + 1)$.

   Solution

   (a) $y = -3(0 + 1) = -3$ $(0, -3)$; y-intercept
   (b) $0 = -3(x + 1)$ $x = -1, (-1, 0)$; x-intercept

4. Sketch the graph of $y = |x - 2|$.

   Solution

   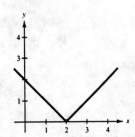

5. Find the slope of the line through $(-4, 7)$ and $(2, 3)$.

   Solution

   $$m = \frac{3 - 7}{2 + 4} = -\frac{4}{6} = -\frac{2}{3}$$

6. Find the slope of the line through $(3, -2)$ and $(3, 6)$.

   Solution

   $$m = \frac{6 + 2}{3 - 3} = \frac{8}{0} = \text{undefined}$$

7. Sketch the line passing through the point $(0, -6)$ with a slope of $m = \frac{4}{3}$.

   Solution

   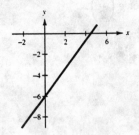

8. Plot the x- and y-intercepts of the graph of $2x + 5y = 10$. Use the result to sketch the graph.

   Solution

   $2x + 5y = 10$

   $2(0) + 5y = 10$

   $5y = 10$

   $y = 2$ $(0, 2)$

   $2x + 5(0) = 10$

   $2x = 10$

   $x = 5$ $(5, 0)$

   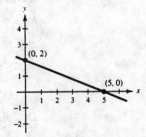

9. Write the equation $5x + 3y - 9 = 0$ in slope-intercept form. Find the slope of a line that is perpendicular to this line.

Solution

$$5x + 3y - 9 = 0$$

$$3y = -5x + 9$$

$$y = \frac{-5}{3}x + 3$$

$$m = \frac{3}{5}$$

10. The graph of $y^2(4 - x) = x^3$ is shown in the textbook. Does the graph represent y as a function of x? Explain your reasoning.

Solution

$y^2(4 - x) = x^3$ is not a function of x, because the graph does not pass the Vertical Line Test.

11. Does $3x^2 - y^2 = 9$ represent y as a function of x? Explain.

Solution

No, y is not a function of x, because the graph does not pass the Vertical Line Test.

12. Does $3x - y = 9$ represent y as a function of x? Explain.

Solution

Yes, y is a function of x, because the graph does not pass the Vertical Line Test.

13. Evaluate $g(x) = x/(x - 3)$ for $g(2)$.

Solution

$$g(2) = \frac{2}{2 - 3}$$

$$= -2$$

14. Evaluate $g(x) = \dfrac{x}{x - 3}$ for $g\left(\dfrac{7}{2}\right)$.

Solution

$$g\left(\frac{7}{2}\right) = \frac{\frac{7}{2}}{\frac{7}{2} - 3}$$

$$= \frac{7}{7 - 6} = 7$$

15. Evaluate $g(x) = \dfrac{x}{x - 3}$ for $g(x + 2)$.

Solution

$$g(x + 2) = \frac{x + 2}{(x + 2) - 3} = \frac{x + 2}{x - 1}$$

16. Find the domain of $h(t) = \sqrt{9 - t}$.

Solution

$$h(t) = \sqrt{9 - t}$$

$$9 - t \geq 0$$

$$-t \geq -9$$

$$t \leq 9$$

Domain: $t \leq 9$ or $(-\infty, 9]$

17. Find the domain of $f(x) = \dfrac{x + 1}{x - 4}$.

Solution

$$f(x) = \frac{x + 1}{x - 4}$$

Domain: $x \neq 4$

18. Sketch the graph of $g(x) = \sqrt{2 - x}$.

Solution

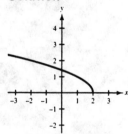

19. Use a graphing utility to graph $y = -0.3x^2 + 2x + 5$. Then use the graph to estimate the intercepts of the graph.

Solution

Keystrokes:

$\boxed{Y=}$ $\boxed{-}$ .3 $\boxed{X, T, \theta}$ $\boxed{x^2}$ $\boxed{+}$ 2 $\boxed{X, T, \theta}$ $\boxed{+}$ 5 $\boxed{GRAPH}$

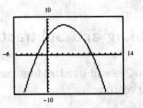

y-intercept $= 5$

x-intercepts ≈ -1.94 and 8.60

20. Use the graph of $y = |x|$ to write an equation for each graph. (See textbook for graphs.)

Solution

(a) $y = |x - 2|$ (b) $y = |x| - 2$ (c) $y = -|x| + 2$ or $2 - |x|$

CHAPTER 3
Polynomials and Factoring

3.1 Adding and Subtracting Polynomials

7. Write $3x^2 + 2 - x$ in standard form, and find its degree and leading coefficient.

Solution

Standard form: $3x^2 - x + 2$
Degree: 2
Leading coefficient: 3

9. Write $5 - 3y^4$ in standard form, and find its degree and leading coefficient.

Solution

Standard form: $-3y^4 + 5$
Degree: 4
Leading Coefficient: -3

11. Determine whether $12 - 5y^2$ is a monomial, binomial, or trinomial.

Solution

$12 - 5y^2$ is a binomial.

13. Determine whether $x^3 + 2x^2 - 4$ is a monomial, binomial, or trinomial.

Solution

$x^3 + 2x^2 - 4$ is a trinomial.

15. Determine whether 5 is a monomial, binomial, or trinomial.

Solution

5 is a monomial.

17. Use a horizontal format to find the sum.

$(2x^2 - 3) + (5x^2 + 6)$

Solution

$(2x^2 - 3) + (5x^2 + 6) = (2x^2 + 5x^2) + (-3 + 6) = 7x^2 + 3$

19. Use a horizontal format to find the sum.

$(2 - 8y) + (-2y^4 + 3y + 2)$

Solution

$(2 - 8y) + (-2y^4 + 3y + 2) = (-2y^4) + (-8y + 3y) + (2 + 2) = -2y^4 - 5y + 4$

21. Use a vertical format to find the sum.

$$5x^2 - 3x + 4$$
$$-3x^2 \qquad - 4$$

Solution

$$5x^2 - 3x + 4$$
$$-3x^2 \qquad - 4$$
$$\overline{\qquad}$$
$$2x^2 - 3x$$

23. Use a vertical format to find the sum.

$(2b - 3) + (b^2 - 2b) + (7 - b^2)$

Solution

$$2b - 3$$
$$b^2 - 2b$$
$$-b^2 \qquad + 7$$
$$\overline{\qquad}$$
$$4$$

25. Use a horizontal format to find the difference.

$$(3x^2 - 2x + 1) - (2x^2 + x - 1)$$

Solution

$$(3x^2 - 2x + 1) - (2x^2 + x - 1) = (3x^2 - 2x + 1) + (-2x^2 - x + 1)$$
$$= (3x^2 - 2x^2) + (-2x - x) + (1 + 1)$$
$$= x^2 - 3x + 2$$

27. Use a horizontal format to find the difference.

$$(8x^3 - 4x^2 + 3x) - \left[(x^3 - 4x^2 + 5) + (x - 5)\right]$$

Solution

$$(8x^3 - 4x^2 + 3x) - \left[(x^3 - 4x^2 + 5) + (x - 5)\right] = (8x^3 - 4x^2 + 3x) - [x^3 - 4x^2 + x]$$
$$= (8x^3 - 4x^2 + 3x) + (-x^3 + 4x^2 - x)$$
$$= (8x^3 - x^3) + (-4x^2 + 4x^2) + (3x - x)$$
$$= 7x^3 + 2x$$

29. Use a vertical format to find the difference.

$$x^2 - x + 3$$
$$\underline{-\qquad x - 2}$$

Solution

$$\begin{array}{l} x^2 - x + 3 \Rightarrow x^2 - x + 3 \\ \underline{- (x - 2) \Rightarrow - x + 2} \\ x^2 - 2x + 5 \end{array}$$

31. Perform the operations.

$$-(2x^3 - 3) + (4x^3 - 2x)$$

Solution

$$-(2x^3 - 3) + (4x^3 - 2x) = -2x^3 + 3 + 4x^3 - 2x$$
$$= (-2x^3 + 4x^3) + (-2x) + (3)$$
$$= 2x^3 - 2x + 3$$

33. Perform the operations.

$$15v - 3(3v - v^2) + 9(8v + 3)$$

Solution

$$15v - 3(3v - v^2) + 9(8v + 3) = 15v - 9v + 3v^2 + 72v + 27$$
$$= (3v^2) + (15v - 9v + 72v) + 27$$
$$= 3v^2 + 78v + 27$$

35. *Graphical Reasoning* Use a graphing utility to graph $y_1 = (x^3 - 3x^2 - 2) - (x^2 + 1)$ and $y_2 = x^3 - 4x^2 - 3$. What conclusion can you make?

Solution

Keystrokes:

y_1 $\boxed{Y=}$ $\boxed{(}$ $\boxed{X, T, \theta}$ $\boxed{\wedge}$ 3 $\boxed{-}$ 3 $\boxed{X, T, \theta}$ $\boxed{x^2}$ $\boxed{-}$ 2 $\boxed{)}$ $\boxed{-}$ $\boxed{(}$ $\boxed{X, T, \theta}$ $\boxed{x^2}$ $\boxed{+}$ 1 $\boxed{)}$ $\boxed{ENTER}$

y_2 $\boxed{Y=}$ $\boxed{X, T, \theta}$ $\boxed{\wedge}$ 3 $\boxed{-}$ 4 $\boxed{X, T, \theta}$ $\boxed{x^2}$ $\boxed{-}$ 3 $\boxed{GRAPH}$

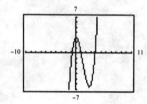

y_1 and y_2 represent equivalent expressions since the graphs of y_1 and y_2 are identical.

37. Evaluate $f(x) = x^3 - 12x$ when the values of the variable are

(a) $x = -2$;   (b) $x = 0$;   (c) $x = 2$;   (d) $x = 4$.

Then use a graphing utility to graph the function and graphically confirm your results.

Solution

Polynomial	Value	Substitute	Simplify
$f(x) = x^3 - 12x$	(a) $x = -2$	$(-2)^3 - 12(-2)$	16
	(b) $x = 0$	$0^3 - 12(0)$	0
	(c) $x = 2$	$2^3 - 12(2)$	-16
	(d) $x = 4$	$4^3 - 12(4)$	16

39. *Free-Falling Object* Find the height of a free-falling object at the specified times using the position polynomial $h = -16t^2 + 32$. Then write a paragraph that describes the path of the object.

Solution

Polynomial	Value	Substitute	Simplify
$h = -16t^2 + 32$	(a) $t = 0$	$-16(0)^2 + 32$	32
	(b) $t = \frac{1}{2}$	$-16\left(\frac{1}{2}\right)^2 + 32$	28
	(c) $t = 1$	$-16(1)^2 + 32$	16
	(d) $t = \frac{3}{2}$	$-16\left(\frac{3}{2}\right)^2 + 32$	-4

At time $t = 0$, the object is at 32 feet and continues to fall.

41. *Profit* A manufacturer can produce and sell x radios per week. The total cost (in dollars) of producing the radios is given by $C = 8x + 15{,}000$ and the total revenue is given by $R = 14x$. Find the profit obtained by selling 5000 radios per week.

Solution

Verbal Model: | Profit | = | Revenue | − | Cost |

Equation:
$$P = R - C$$
$$P = 14x - (8x + 15{,}000)$$
$$P = 6x - 15{,}000$$
$$P = 6(5000) - 15{,}000$$
$$P = \$15{,}000$$

43. Evaluate $5 - |-2|$.

Solution

$$5 - |-2| = 5 - 2 = 3$$

45. Use a graphing utility to graph each of the functions and identify the transformation of $f(x) = x^5$.

(a) $g(x) = x^5 - 2$;

(b) $g(x) = (x - 2)^5$;

(c) $h(x) = -x^5$;

(d) $h(x) = (-x)^5$

Solution

(a) $f(x)$ is shifted 2 units down.

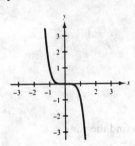

(b) $f(x)$ is shifted 2 units right.

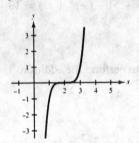

(c) $f(x)$ is reflected across the x-axis.

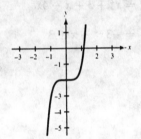

(d) $f(x)$ is reflected across the y-axis.

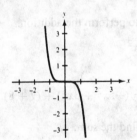

47. Write $10x - 4$ in standard form, and find its degree and leading coefficient.

Solution

Standard form: $10x - 4$
Degree: 1
Leading coefficient: 10

49. Write $-3x^3 - 2x^2 - 3$ in standard form and find its degree and leading coefficient.

Solution

Standard form: $-3x^3 - 2x^2 - 3$
Degree: 3
Leading coefficient: -3

51. Write -4 in standard form, and find its degree and leading coefficient.

Solution

Standard form: -4
Degree: 0
Leading coefficient: -4

53. Write $v_0 t - 16t^2$ (v_0 is constant) in standard form, and find its degree and leading coefficient.

Solution

Standard form: $-16t^2 + v_0 t$
Degree: 2
Leading coefficient: -16

55. Give an example of a polynomial in x that satisfies the given conditions. (*Note:* This problem has many correct answers.)

A monomial of degree 3

Solution

A monomial of degree 3 is any term of form ax^3 where a is any real number.

57. Give an example of a polynomial in x that satisfies the given conditions. (*Note:* This problem has many correct answers.)

A binomial of degree 2 and leading coefficient of 8

Solution

A binomial of degree 2 and leading coefficient of 8 is any binomial beginning $8x^2$ and containing one other term of degree less than 2 such as $8x^2 + 4$ or $8x^2 + x$.

59. Use a horizontal format to perform the addition.

$$(8 - t^4) + (5 + t^4)$$

Solution

$$(8 - t^4) + (5 + t^4) = (8 + 5) + (-t^4 + t^4) = 13$$

61. Use a horizontal format to perform the addition.

$$x^2 + \left[3 + (x - 4x^2)\right]$$

Solution

$$x^2 + \left[3 + (x - 4x^2)\right] = x^2 + 3 + x - 4x^2$$
$$= -3x^2 + x + 3$$

63. Use a horizontal format to perform the addition.

$$(x^2 - 3x + 8) + (2x^2 - 4x) + 3x^2$$

Solution

$$(x^2 - 3x + 8) + (2x^2 - 4x) + 3x^2 = (x^2 + 2x^2 + 3x^2) + (-3x - 4x) + (8)$$
$$= 6x^2 - 7x + 8$$

65. Use a horizontal format to perform the addition.

Add $3x^2 + 8$ to $7 - 5x^2$.

Solution

$$(7 - 5x^2) + (3x^2 + 8) = (-5x^2 + 3x^2) + (7 + 8) = -2x^2 + 15$$

67. Use a vertical format to find the sum.

$$3x^4 - 2x^2 - 9$$
$$\underline{-5x^4 + x^2}$$

Solution

$$3x^4 - 2x^2 - 9$$
$$\underline{-5x^4 + x^2}$$
$$\overline{-2x^4 - x^2 - 9}$$

69. Use a vertical format to find the sum.

$$(5p^2 - 4p + 2) + (-3p^2 + 2p - 7)$$

Solution

$$5p^2 - 4p + 2$$
$$\underline{-3p^2 + 2p - 7}$$
$$\overline{2p^2 - 2p - 5}$$

71. Use a horizontal format to find the difference.

$(4 - y^3) - (4 + y^3)$

Solution

$$(4 - y^3) - (4 + y^3) = (4 - y^3) + (-4 - y^3)$$
$$= (4 - 4) + (-y^3 - y^3)$$
$$= -2y^3$$

73. Use a horizontal format to find the difference.

$(5q^2 - 3q + 5) - (4q^2 - 3q - 10)$

Solution

$$(5q^2 - 3q + 5) - (4q^2 - 3q - 10) = (5q^2 - 3q + 5) + (-4q^2 + 3q + 10)$$
$$= (5q^2 - 4q^2) + (-3q + 3q) + (5 + 10)$$
$$= q^2 + 15$$

75. Use a horizontal format to find the difference.

$(x^3 - 3x) - \left[3x^3 - (x^2 + 5x)\right]$

Solution

$$(x^3 - 3x) - \left[3x^3 - (x^2 + 5x)\right] = (x^3 - 3x) - [3x^3 - x^2 - 5x]$$
$$= x^3 - 3x - 3x^3 + x^2 + 5x$$
$$= -2x^3 + x^2 + 2x$$

77. Use a horizontal format to find the difference.

Subtract $6x^3 - x + 11$ from $10x^3 + 15$.

Solution

$$(10x^3 + 15) - (6x^3 - x + 11) = (10x^3 + 15) + (-6x^3 + x - 11)$$
$$= (10x^3 - 6x^3) + (x) + (15 - 11)$$
$$= 4x^3 + x + 4$$

79. Use a vertical format to find the difference.

$(25 - 15x - 2x^3) - (12 - 13x + 2x^3)$

Solution

$$25 - 15x - 2x^3 \Rightarrow \quad 25 - 15x - 2x^3$$
$$\underline{-(12 - 13x + 2x^3) \Rightarrow -12 + 13x - 2x^3}$$
$$13 - 2x - 4x^3$$

81. Perform the operation.

$(4x^2 + 5x - 6) - (2x^2 - 4x + 5)$

Solution

$$4x^2 + 5x - 6 \Rightarrow \quad 4x^2 + 5x - 6$$
$$\underline{-(2x^2 - 4x + 5) \Rightarrow -2x^2 + 4x - 5}$$
$$2x^2 + 9x - 11$$

83. Perform the operations.

$$(10x^2 - 11) - (-7x^3 - 12x^2 - 15)$$

Solution

$$
\begin{aligned}
(10x^2 - 11) - (-7x^3 - 12x^2 - 15) &= 10x^2 - 11 + 7x^3 + 12x^2 + 15 \\
&= (7x^3) + (10x^2 + 12x^2) + (-11 + 15) \\
&= 7x^3 + 22x^2 + 4
\end{aligned}
$$

85. Perform the operations.

$$5s - [6s - (30s + 8)]$$

Solution

$$
\begin{aligned}
5s - [6s - (30s + 8)] &= 5s - [6s - 30s - 8] \\
&= (5s - 6s + 30s) + (8) \\
&= 29s + 8
\end{aligned}
$$

87. Perform the operations.

$$2(t^2 + 12) - 5(t^2 + 5) + 6(t^2 + 5)$$

Solution

$$
\begin{aligned}
2(t^2 + 12) - 5(t^2 + 5) + 6(t^2 + 5) &= 2t^2 + 24 - 5t^2 - 25 + 6t^2 + 30 \\
&= (2t^2 - 5t^2 + 6t^2) + (24 - 25 + 30) \\
&= 3t^2 + 29
\end{aligned}
$$

89. Evaluate the polynomial $-4x^3 + 16x - 16$ for the values of the variable.

Solution

(a) $x = -1$ $\quad -4(-1)^3 + 16(-1) - 16 = -4(-1) - 16 - 16 = 4 - 16 - 16 = -28$

(b) $x = 0$ $\quad -4(0)^3 + 16(0) - 16 = 0 + 0 - 16 = -16$

(c) $x = 2$ $\quad -4(2)^3 + 16(2) - 16 = -4(8) + 32 - 16 = -32 + 32 - 16 = -16$

(d) $x = \frac{5}{2}$ $\quad -4\left(\frac{5}{2}\right)^3 + 16\left(\frac{5}{2}\right) - 16 = -4\left(\frac{125}{8}\right) + 40 - 16$

$$
\begin{aligned}
&= -\frac{125}{2} + 24 \\
&= -\frac{125}{2} + \frac{48}{2} \\
&= -\frac{77}{2} \text{ or } -38\frac{1}{2} \text{ or } -38.5
\end{aligned}
$$

91. *Free-Falling Object*   Use the position polynomial to find the height of the free-falling object at the specified times. Then write a paragraph that describes the path of the object.

$$h = -16t^2 + 80t + 50.$$

Solution

Polynomial		Value	Substitute	Simplify
$h = -16t^2 + 80t + 50$	(a)	$t = 0$	$-16(0)^2 + 80(0) + 50$	50
	(b)	$t = 2$	$-16(2)^2 + 80(2) + 50$	146
	(c)	$t = 4$	$-16(4)^2 + 80(4) + 50$	114
	(d)	$t = 5$	$-16(5)^2 + 80(5) + 50$	50

At time $t = 0$, the object is at a height of 50 feet. The object moves upward, reaches a maximum height and returns downward. At time $t = 5$, object is again at a height of 50 feet.

93. *Free-Falling Object*   Use the position polynomial $h = -16t^2 + 100$ to determine whether the free-falling object was dropped, thrown upward, or thrown downward. Also determine the height of the object at time $t = 0$.

Solution

The free-falling object was dropped.
$-16(0)^2 + 100 = 100$

95. *Free-Falling Object*   Use the position polynomial $h = -16t^2 - 24t + 50$ to determine whether the free-falling object was dropped, thrown upward, or thrown downward. Also determine the height of the object at time $t = 0$.

Solution

The free-falling object was thrown downward.
$-16(0)^2 - 24(0) + 50 = 50$

97. *Free-Falling Object*   An object is thrown upward from the top of a 200-foot building (see figure in the textbook). The initial velocity is 40 feet per second. Use the positional polynomial

$$h = 16t^2 + 40t + 200$$

to find the height of the object when $t = 1$, $t = 2$, and $t = 3$.

Solution

$h = -16(1)^2 + 40(1) + 200 = 224$

$h = -16(2)^2 + 40(2) + 200 = 216$

$h = -16(3)^2 + 40(3) + 200 = 176$

99. *Geometry*   Find the area of the entire region (see figure in the textbook).

Solution

$$\text{Area of region} = \left(6 \cdot \frac{3}{2}x\right) + \left(6 \cdot \frac{9}{2}x\right) \quad \text{or} \quad 6 \cdot \left[\frac{3}{2}x + \frac{9}{2}x\right]$$

$$= 9x + 27x \quad \text{or} \quad 6\left[\frac{12}{2}x\right]$$

$$= 36x \quad \text{or} \quad 36x$$

101. *Consumption of Milk* The per capita consumption (average consumption per person) of whole milk and lowfat milk in the United States from 1970 to 1991 can be approximated by the following two polynomial models.

$$y = 25.23 - 0.84t + 0.006t^2 \qquad \text{Whole milk}$$

$$y = 4.52 + 0.54t - 0.007t^2 \qquad \text{Lowfat milk}$$

In these models, y represents per capita consumption in gallons and t represents the year, with $t = 0$ corresponding to 1970. (*Source:* U.S. Department of Agriculture.)

(a) Find a polynomial model that represents the per capita consumption of milk (both types) during this time period. Use this model to find the per capita consumption of milk in 1975 and in 1985.

(b) During the given time period, the per capita consumption of whole milk was decreasing and the per capita consumption of lowfat milk was increasing (see figure in the textbook). Use a graphing utility to graph the model for the per capita consumption of milk (both types). Is this consumption increasing or decreasing?

Solution

(a)

Per capita consumption of milk	=	Per capita consumption of whole milk	+	Per capita consumption of lowfat milk

Per capita consumption of milk $= (25.23 - 0.84t + 0.006t^2) + (4.52 + 0.54t - 0.007t^2)$

$$= 29.75 - 0.3t - 0.001t^2$$

Per capita consumption in 1975 $= 29.75 - 0.3(5) - 0.001(5)^2$

$$= 29.75 - 1.5 - 0.025 = 28.225 \text{ gallons}$$

Per capita consumption in 1985 $= 29.75 - 0.3(15) - 0.001(15)^2$

$$= 29.75 - 4.5 - 0.225 = 25.025 \text{ gallons}$$

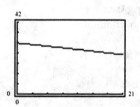

(b) The per capita consumption of milk is decreasing.

3.2 Exponents and Multiplying Polynomials

7. Use $t^3 \cdot t^4$ to illustrate a property of exponents. Use Examples 1 and 2 as models.

Solution

$$t^3 \cdot t^4 = (t \cdot t \cdot t)(t \cdot t \cdot t \cdot t) = t^{3+4} = t^7$$

9. Use $\left(\dfrac{y}{5}\right)^4$ to illustrate a property of exponents. Use Examples 1 and 2 as models.

Solution

$$\left(\frac{y}{5}\right)^4 = \frac{y}{5} \cdot \frac{y}{5} \cdot \frac{y}{5} \cdot \frac{y}{5} = \frac{y \cdot y \cdot y \cdot y}{5 \cdot 5 \cdot 5 \cdot 5} = \frac{y^4}{5^4} = \frac{y^4}{625}$$

11. Use $\dfrac{x^6}{x^4}$ to illustrate a property of exponents. Use Examples 1 and 2 as models.

Solution

$$\frac{x^6}{x^4} = \frac{x \cdot x \cdot x \cdot x \cdot x \cdot x}{x \cdot x \cdot x \cdot x} = x^{6-4} = x^2$$

13. Use $\dfrac{3^7 x^5}{3^3 x^3}$ to illustrate a property of exponents. Use Examples 1 and 2 as models.

Solution

$$\frac{3^7 x^5}{3^3 x^3} = \frac{3 \cdot 3 \cdot 3 \cdot 3 \cdot 3 \cdot 3 \cdot 3 \cdot x \cdot x \cdot x \cdot x \cdot x}{3 \cdot 3 \cdot 3 \quad\quad x \cdot x \cdot x} = 3^{7-3} x^{5-3} = 3^4 x^2 = 81 x^2$$

15. Simplify the expressions.

(a) $-3x^3 \cdot x^5$

(b) $(-3x)^2 \cdot x^5$

Solution

(a) $-3x^3 \cdot x^5 = -3(x^3 \cdot x^5) = -3x^{3+5} = -3x^8$

(b) $(-3x)^2 \cdot x^5 = 9x^2 \cdot x^5 = 9x^{2+5} = 9x^7$

17. Simplify the expressions.

(a) $5x^2 y(-2y)^2(3)$

(b) $5x^2 y(-2y)^3(3)$

Solution

(a) $5x^2 y(-2y)^2(3) = 5x^2 y(4y^2)(3)$

$\quad\quad\quad\quad\quad\quad = 60x^2 y^{1+2}$

$\quad\quad\quad\quad\quad\quad = 60x^2 y^3$

(b) $5x^2 y(-2y)^3(3) = 5x^2 y(-8y^3)(3)$

$\quad\quad\quad\quad\quad\quad = -120x^2 y^{1+3}$

$\quad\quad\quad\quad\quad\quad = -120x^2 y^4$

19. Simplify the expressions.

(a) $\dfrac{12m^5n^6}{9mn^3}$

(b) $\dfrac{-18m^3n^6}{-6mn^3}$

Solution

(a) $\dfrac{12m^5n^6}{9mn^3} = \dfrac{12}{9} \cdot \dfrac{m^5}{m} \cdot \dfrac{n^6}{n^3}$

$= \dfrac{4}{3} \cdot m^{5-1} \cdot n^{6-3}$

$= \dfrac{4}{3}m^4n^3$

(b) $\dfrac{-18m^3n^6}{-6mn^3} = \dfrac{-18}{-6} \cdot \dfrac{m^3}{m} \cdot \dfrac{n^6}{n^3}$

$= 3 \cdot m^{3-1} \cdot n^{6-3}$

$= 3m^2n^3$

21. Simplify the expressions.

(a) $\dfrac{(-5u^3v)^2}{10u^2v}$

(b) $\dfrac{-5(u^3v)^2}{10u^2v}$

Solution

(a) $\dfrac{(-5u^3v)^2}{10u^2v} = \dfrac{(-5)^2 \cdot (u^3)^2 \cdot (v)^2}{10u^2v}$

$= \dfrac{25u^6v^2}{10u^2v}$

$= \dfrac{25}{10} \cdot \dfrac{u^6}{u^2} \cdot \dfrac{v^2}{v}$

$= \dfrac{5}{2} \cdot u^{6-2} \cdot v^{2-1}$

$= \dfrac{5}{2}u^4v$

(b) $\dfrac{-5(u^3v)^2}{10u^2v} = \dfrac{-5 \cdot (u^3)^2 \cdot (v)^2}{10u^2v}$

$= \dfrac{-5u^6v^2}{10u^2v}$

$= \dfrac{-5}{10} \cdot \dfrac{u^6}{u^2} \cdot \dfrac{v^2}{v}$

$= -\dfrac{1}{2} \cdot u^{6-2} \cdot v^{2-1}$

$= \dfrac{-1}{2}u^4v$

23. Simplify $(-2a^2)(-8a)$.

Solution

$(-2a^2)(-8a) = (-2)(-8)a^2 \cdot a = 16a^{2+1} = 16a^3$

25. Simplify $-2x(5 + 3x^2 - 7x^3)$.

Solution

$-2x(5 + 3x^2 - 7x^3) = (-2x)(5) + (-2x)(3x^2) - (-2x)(7x^3) = -10x - 6x^3 + 14x^4$

27. Simplify $5a(a + 2) - 3a(2a - 3)$.

Solution

$5a(a + 2) - 3a(2a - 3) = (5a)(a) + (5a)(2) - (3a)(2a) + (3a)(3)$

$= 5a^2 + 10a - 6a^2 + 9a$

$= -a^2 + 19a$

29. Use a horizontal format to perform the multiplication.

$(u + 5)(2u^2 + 3u - 4)$

Solution

$$\begin{aligned}(u + 5)(2u^2 + 3u - 4) &= (u + 5)(2u^2) + (u + 5)(3u) + (u + 5)(-4) \\ &= 2u^3 + 10u^2 + 3u^2 + 15u - 4u - 20 \\ &= 2u^3 + 13u^2 + 11u - 20\end{aligned}$$

31. Use a horizontal format to perform the multiplication.

$(x - 1)(x^2 - 4x + 6)$

Solution

$$\begin{aligned}(x - 1)(x^2 - 4x + 6) &= (x - 1)(x^2) + (x - 1)(-4x) + (x - 1)(6) \\ &= x^3 - x^2 - 4x^2 + 4x + 6x - 6 \\ &= x^3 - 5x^2 + 10x - 6\end{aligned}$$

33. Use a vertical format to perform the multiplication.

$$\begin{aligned}(7x^2 - 14x + 9) \\ x + 3\end{aligned}$$

Solution

$$\begin{array}{r} 7x^2 - 14x + 9 \\ x + 3 \\ \hline 21x^2 - 42x + 27 \\ 7x^3 - 14x^2 + 9x \\ \hline 7x^3 + 7x^2 - 33x + 27 \end{array}$$

35. Use a vertical format to perform the multiplication.

$(u - 2)(u^2 + u + 3)$

Solution

$$\begin{array}{r} u^2 + u + 3 \\ u - 2 \\ \hline -2u^2 - 2u - 6 \\ u^3 + u^2 + 3u \\ \hline u^3 - u^2 + u - 6 \end{array}$$

37. Expand the product.

$(x + 2)(x - 2)$

Solution

$$\begin{aligned}(x + 2)(x - 2) &= x(x - 2) + 2(x - 2) \\ &= x^2 - 2x + 2x - 4 \\ &= x^2 - 4\end{aligned}$$

39. Expand the product.

$(x + 5)^2$

Solution

$$\begin{aligned}(x + 5)^2 &= (x + 5)(x + 5) \\ &= x(x + 5) + 5(x + 5) \\ &= x^2 + 5x + 5x + 25 \\ &= x^2 + 10x + 25\end{aligned}$$

41. Expand the product.

$$[(x + 2) + y]^2$$

Solution

$$
\begin{aligned}
[(x + 2) + y]^2 &= [(x + 2) + y][(x + 2) + y] \\
&= (x + 2)[(x + 2) + y] + y[(x + 2) + y] \\
&= (x + 2)(x + 2) + y(x + 2) + y(x + 2) + y^2 \\
&= x(x + 2) + 2(x + 2) + y(x + 2) + y(x + 2) + y^2 \\
&= x^2 + 2x + 2x + 4 + yx + 2y + yx + 2y + y^2 \\
&= x^2 + 4x + 4 + 2yx + 4y + y^2
\end{aligned}
$$

43. *Geometry*   Write an expression that represents the area of the shaded portion of the figure as shown in the textbook. Then simplify the expression.

Solution

$$
\begin{aligned}
\text{Area} &= 3x(3x + 10) - x(x + 4) \\
&= 9x^2 + 30x - x^2 - 4x \\
&= 8x^2 + 26x
\end{aligned}
$$

45. *Geometrical Modeling*   Use the area model to write two different expressions for the total area. Then equate the two expressions and name the algebraic property that is illustrated.

Solution

$$
\begin{aligned}
\text{Area} &= l \cdot w \\
&= (x + a)(x + b) \\
&= x^2 + ax + bx + ab
\end{aligned}
$$

$$
\begin{aligned}
\text{Area} &= (x \cdot x) + (x \cdot a) + (x \cdot b) + (a \cdot b) \\
&= x^2 + ax + bx + ab
\end{aligned}
$$

Formula: $(x + a)(x + b) = x^2 + ax + bx + ab.$ Distributive Property

47. Plot the points $(-3, 2)$ and $(5, -4)$ on a rectangular coordinate system and find the slope of the line passing through the points.

Solution

$$m = \frac{2 - (-4)}{-3 - 5} = -\frac{6}{8} = -\frac{3}{4}$$

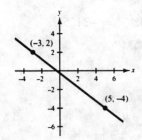

49. Plot the points $\left(\frac{5}{2}, \frac{7}{2}\right)$ and $\left(\frac{7}{3}, -2\right)$ on a rectangular coordinate system and find the slope of the line passing through the points.

Solution

$$m = \frac{-2 - \frac{7}{2}}{\frac{7}{3} - \frac{5}{2}} \cdot \frac{6}{6} = \frac{-12 - 21}{14 - 15}$$

$$= \frac{-33}{-1}$$

$$= 33$$

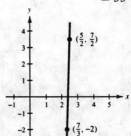

51. *Distance* An automobile is traveling at an average speed of 48 miles per hour. Express the distance d traveled as a function of time t in hours.

Solution

Verbal Model: | Distance | = | Rate | · | Time |

Equation: $d = 48t$

53. Use $(-5x)^5$ to illustrate a property of exponents. Use Examples 1 and 2 as models.

Solution

$$(-5x)^5 = -5x \cdot -5x \cdot -5x \cdot -5x \cdot -5x$$

$$= -5 \cdot -5 \cdot -5 \cdot -5 \cdot -5 \cdot x \cdot x \cdot x \cdot x \cdot x$$

$$= (-5)^5 x^5 = -3125x^5$$

55. Use $\dfrac{3^4 x^5}{3x^2}$ to illustrate a property of exponents. Use Examples 1 and 2 as models.

Solution

$$\frac{3^4 x^5}{3x^2} = 3^{4-1} \cdot x^{5-2} = 3^3 x^3 = 27x^3$$

57. Simplify.

(a) $(-5z)^3$

(b) $(-5z)^2$

Solution

(a) $(-5z)^3 = (-5)^3 z^3 = -125z^3$

(b) $(-5z)^2 = (-5)^2 z^2 = 25z^2$

59. Simplify.

(a) $(-5z)^4$

(b) $-5z^4 \cdot (-5z)^4$

Solution

(a) $(-5z)^4 = (-5)^4 z^4 = 625z^4$

(b) $-5z^4 \cdot (-5z)^4 = -5z^4 \cdot (625z^4)$

$$= -5 \cdot 625 \cdot z^4 \cdot z^4$$

$$= -3125z^{4+4} = -3125z^8$$

61. Simplify.

(a) $\dfrac{28x^2y^3}{21xy^2}$

(b) $\dfrac{21xy^2}{28y}$

Solution

(a) $\dfrac{28x^2y^3}{21xy^2} = \dfrac{28}{21} \cdot \dfrac{x^2}{x} \cdot \dfrac{y^3}{y^2}$

$$= \dfrac{4}{3} \cdot x^{2-1} \cdot y^{3-2}$$

$$= \dfrac{4}{3}xy$$

(b) $\dfrac{21xy^2}{28y} = \dfrac{21}{28} \cdot x \cdot \dfrac{y^2}{y}$

$$= \dfrac{3}{4} \cdot x \cdot y^{2-1}$$

$$= \dfrac{3}{4}xy$$

63. Simplify.

(a) $\left(\dfrac{3x}{4y}\right)^2$

(b) $\left(\dfrac{5u}{3v}\right)^3$

Solution

(a) $\left(\dfrac{3x}{4y}\right)^2 = \dfrac{3^2 \cdot x^2}{4^2 \cdot y^2}$

$$= \dfrac{9x^2}{16y^2}$$

(b) $\left(\dfrac{5u}{3v}\right)^3 = \dfrac{5^3 \cdot u^3}{3^3 \cdot v^3}$

$$= \dfrac{125u^3}{27v^3}$$

65. Simplify $5x(2x)^2$.

Solution

$5x(2x)^2 = 5x \cdot 2^2 x^2 = (5)(2^2)x \cdot x^2 = 20x^{1+2} = 20x^3$

67. Simplify $2y(5 - y)$.

Solution

$2y(5 - y) = (2y)(5) - (2y)(y) = 10y - 2y^2$

69. Simplify $4x(2x^2 - 3x + 5)$.

Solution

$4x(2x^2 - 3x + 5) = (4x)(2x^2) - (4x)(3x) + (4x)(5) = 8x^3 - 12x^2 + 20x$

71. Simplify $-3x(-5x)(5x + 2)$.

Solution

$-3x(-5x)(5x + 2) = 15x^2(5x + 2) = (15x^2)(5x) + (15x^2)(2) = 75x^3 + 30x^2$

73. Simplify $\left(4y - \frac{1}{3}\right)(12y - 9)$.

Solution

$$\left(4y - \tfrac{1}{3}\right)(12y + 9) = 4y(12y + 9) - \tfrac{1}{3}(12y + 9)$$
$$= (4y)(12y) + (4y)(9) - \left(\tfrac{1}{3}\right)(12y) - \left(\tfrac{1}{3}\right)(9)$$
$$= 48y^2 + 36y - 4y - 3 = 48y^2 + 32y - 3$$

75. Simplify $(2x + y)(3x + 2y)$.

Solution

$$(2x + y)(3x + 2y) = 2x(3x + 2y) + y(3x + 2y)$$
$$= (2x)(3x) + (2x)(2y) + (y)(3x) + (y)(2y)$$
$$= 6x^2 + 4xy + 3xy + 2y^2$$
$$= 6x^2 + 7xy + 2y^2$$

77. Simplify $(s - 3t)(s + t) - (s - 3t)(s - t)$.

Solution

$$(s - 3t)(s + t) - (s - 3t)(s - t) = s(s + t) - 3t(s + t) - s(s - t) + 3t(s - t)$$
$$= (s)(s) + (s)(t) - (3t)(s) - (3t)(t) - (s)(s) + (s)(t) + (3t)(s) - (3t)(t)$$
$$= s^2 + st - 3ts - 3t^2 - s^2 + st + 3ts - 3t^2$$
$$= 2st - 6t^2$$

79. Use a horizontal format to perform the multiplication.

$$(x^3 - 3x + 2)(x - 2)$$

Solution

$$(x^3 - 3x + 2)(x - 2) = x^3(x - 2) + (-3x)(x - 2) + 2(x - 2)$$
$$= x^4 - 2x^3 - 3x^2 + 6x + 2x - 4$$
$$= x^4 - 2x^3 - 3x^2 + 8x - 4$$

81. Use a horizontal format to perform the multiplication.

$$(t^2 + t - 2)(t^2 - t + 2)$$

Solution

$$(t^2 + t - 2)(t^2 - t + 2) = t^2(t^2 - t + 2) + t(t^2 - t + 2) - 2(t^2 - t + 2)$$
$$= t^4 - t^3 + 2t^2 + t^3 - t^2 + 2t - 2t^2 + 2t - 4$$
$$= t^4 - t^2 + 4t - 4$$

83. Use a vertical format to perform the multiplication.

$$-x^2 + 2x - 1$$
$$\underline{2x + 1}$$

Solution

$$-x^2 + 2x - 1$$
$$\underline{2x + 1}$$
$$-x^2 + 2x - 1$$
$$\underline{-2x^3 + 4x^2 - 2x}$$
$$-2x^3 + 3x^2 - 1$$

85. Use a vertical format to perform the multiplication.

$$(x^3 + x)(x^2 - 4)$$

Solution

$$x^3 + x$$
$$\underline{x^2 - 4}$$
$$-4x^3 - 4x$$
$$\underline{x^5 + x^3}$$
$$x^5 - 3x^3 - 4x$$

87. Use a special product formula to perform the multiplication.

$$(2 + 7y)(2 - 7y)$$

Solution

$$(2 + 7y)(2 - 7y) = (2)^2 - (7y)^2 = 4 - 49y^2$$

89. Use a special product formula to perform the multiplication.

$$\left(2x - \frac{1}{4}\right)\left(2x + \frac{1}{4}\right)$$

Solution

$$\left(2x - \frac{1}{4}\right)\left(2x + \frac{1}{4}\right) = (2x)^2 - \left(\frac{1}{4}\right)^2 = 4x^2 - \frac{1}{16}$$

91. Use a special product formula to perform the multiplication.

$$(0.2t + 0.5)(0.2t - 0.5)$$

Solution

$$(0.2t + 0.5)(0.2t - 0.5) = (0.2t)^2 - (0.5)^2$$
$$= 0.04t^2 - 0.25$$

93. Use a special product formula to perform the multiplication.

$$(x + 10)^2$$

Solution

$$(x + 10)^2 = (x)^2 - 2(x)(10) + 10^2$$
$$= x^2 - 20x + 100$$

95. Use a special product formula to perform the multiplication.

$$(2x - 7y)^2$$

Solution

$$(2x - 7y)^2 = (2x)^2 - 2(2x)(7y) + (7y)^2$$
$$= 4x^2 - 28xy + 49y^2$$

97. Use a special product formula to perform the multiplication.

$$[u - (v - 3)]^2$$

Solution

$$[u - (v - 3)]^2 = (u)^2 - 2u(v - 3) + (v - 3)^2$$
$$= u^2 - 2uv + 6u + v^2 - 6v + 9$$

99. Simplify $(t + 3)^2 - (t - 3)^2$.

Solution

$$
\begin{aligned}
(t + 3)^2 - (t - 3)^2 &= [t^2 + 6t + 9] - [t^2 - 6t + 9] \\
&= t^2 + 6t + 9 - t^2 + 6t - 9 \\
&= 12t
\end{aligned}
$$

101. Simplify $(x + 3)^3$.

Solution

$$
\begin{aligned}
(x + 3)^3 &= (x + 3)(x + 3)(x + 3) \\
&= (x^2 + 3x + 3x + 9)(x + 3) \\
&= (x^2 + 6x + 9)(x + 3)
\end{aligned}
$$

$$
\begin{array}{r}
x^2 + 6x + 9 \\
x + 3 \\
\hline
3x^2 + 18x + 27 \\
x^3 + 6x^2 + 9x \\
\hline
x^3 + 9x^2 + 27x + 27
\end{array}
$$

103. Simplify $(u + v)^3$.

Solution

$$
\begin{aligned}
(u + v)^3 &= (u + v)(u + v)(u + v) \\
&= (u^2 + uv + uv + v^2)(u + v) \\
&= (u^2 + 2uv + v^2)(u + v)
\end{aligned}
$$

$$
\begin{array}{r}
u^2 + 2uv + v^2 \\
u + v \\
\hline
u^2v + 2uv^2 + v^3 \\
u^3 + 2u^2v + uv^2 \\
\hline
u^3 + 3u^2v + 3uv^2 + v^3
\end{array}
$$

105. *Graphical and Algebraic Reasoning* Use a graphing utility to graph the expressions

$$y_1 = (x + 1)(x^2 - x + 2) \quad \text{and} \quad y_2 = x^3 + x + 2$$

What conclusion can you make? Verify your conclusion algebraically.

Solution

Keystrokes:

y_1 | Y= | (| X,T,θ | + | 1 |) | (| X,T,θ | x^2 | − | X,T,θ | + | 2 |) | ENTER

y_2 | Y= | X,T,θ | ∧ | 3 | + | X,T,θ | + | 2 | GRAPH

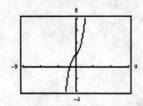

$y_1 = y_2$ because $(x + 1)(x^2 - x + 2) = x^3 - x^2 + 2x + x^2 - x + 2 = x^3 + x + 2$

107. Use a graphing utility to graph the expressions

$$y_1 = (2x - 3)(x + 2) \quad \text{and} \quad y_2 = 2x^2 + x - 6$$

What conclusion can you make? Verify your conclusion algebraically.

Solution

Keystrokes:

y_1 [Y=] [(] [2] [X,T,θ] [−] [3] [)] [(] [X,T,θ] [+] [2] [)] [ENTER]

y_2 [Y=] [2] [X,T,θ] [x²] [+] [X,T,θ] [−] [6] [GRAPH]

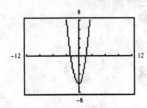

$y_1 = y_2$ because $(2x - 3)(x + 2) = 2x^2 + 4x - 3x - 6 = 2x^2 + x - 6$

109. *Geometry* Find the area of the shaded region (see figure in textbook).

Solution

Verbal model:

Area of Shaded Region	=	Area of Outside Rectangle	−	Area of Inside Rectangle

Equation:

$$A = l \cdot w - l \cdot w$$

$$A = 3x \cdot x - (x - 3)2$$

$$A = 3x^2 - 2x + 6$$

111. Given the function $f(x) = x^2 - 2x$, find and simplify

(a) $f(t - 3)$ and (b) $f(2 + h) - f(2)$.

Solution

(a) $f(t - 3) = (t - 3)^2 - 2(t - 3)$

$$= t^2 - 6t + 9 - 2t + 6$$

$$= t^2 - 8t + 15$$

(b) $f(2 + h) - f(2) = \left[(2 + h)^2 - 2(2 + h)\right] - \left[2^2 - 2(2)\right]$

$$= (4 + 4h + h^2 - 4 - 2h) - (0)$$

$$= 2h + h^2$$

113. *Geometry* The length of a rectangle is $1\frac{1}{2}$ times its width w. Find (a) the perimeter and (b) the area of the rectangle.

Solution

(a) $P = 2l + 2w$

$P = 2\left(\dfrac{3}{2}w\right) + 2w$

$\quad = 3w + 2w$

$P = 5w$

(b) $A = l \cdot w$

$A = \left(\dfrac{3}{2}w\right)(w)$

$A = \dfrac{3}{2}w^2$

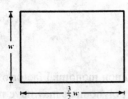

115. *Compound Interest* After two years, an investment of $1000 compounded annually at an interest rate r will yield an amount $\$1000(1 + r)^2$. Find this product.

Solution

$\begin{aligned}
\text{Interest} &= 1000(1 + r)^2 \\
&= 1000(1 + r)(1 + r) \\
&= 1000(1 + 2r + r^2) \\
&= 1000 + 2000r + 1000r^2
\end{aligned}$

117. *Finding a Pattern* Perform the multiplications.

(a) $(x - 1)(x + 1)$

(b) $(x - 1)(x^2 + x + 1)$

(c) $(x - 1)(x^3 + x^2 + x + 1)$

(d) From the pattern formed by these products, can you predict the result of
$(x - 1)(x^4 + x^3 + x^2 + x + 1)$?

Solution

(a) $(x - 1)(x + 1) = x^2 - 1$

(b) $(x - 1)(x^2 + x + 1) = x^3 + x^2 + x - x^2 - x - 1 = x^3 - 1$

(c) $(x - 1)(x^3 + x^2 + x + 1) = x^4 + x^3 + x^2 + x - x^3 - x^2 - x - 1 = x^4 - 1$

(d) $(x - 1)(x^4 + x^3 + x^2 + x + 1) = x^5 - 1$

3.3 Factoring Polynomials: An Introduction

7. Find the greatest common factor of 48, 90.

Solution

$$48 = 2^4 \cdot 3$$

$$90 = 5 \cdot 2 \cdot 3^2$$

$$\text{GCF} = 2 \cdot 3 = 6$$

9. Find the greatest common factor of $28b^2$, $14b^3$, $42b^5$.

Solution

$$28b^2 = 7 \cdot 2^2 \cdot b \cdot b$$

$$14b^3 = 7 \cdot 2 \cdot b \cdot b \cdot b$$

$$42b^5 = 7 \cdot 2 \cdot 3 \cdot b \cdot b \cdot b \cdot b \cdot b$$

$$\text{GCF} = 7 \cdot 2 \cdot b \cdot b$$

$$= 14b^2$$

11. Factor out the greatest common monomial factor of $8z - 8$. (Some of the polynomials may have no common monomial factor other than 1 and -1.)

Solution

$$8z - 8 = 8(z - 1)$$

13. Factor out the greatest common monomial factor of $21u^2 - 14u$. (Some of the polynomials may have no common monomial factor other than 1 and -1.)

Solution

$$21u^2 - 14u = 7u(3u - 2)$$

15. Factor out the greatest common monomial factor of $15xy^2 - 3x^2y + 9xy$. (Some of the polynomials may have no common monomial factor other than 1 and -1.)

Solution

$$15xy^2 - 3x^2y + 9xy = 3xy(5y - x + 3)$$

17. Factor a negative real number out of $10 - x$ and then write the polynomial factor in standard form.

Solution

$$10 - x = -1(-10 + x) = -1(x - 10)$$

19. Factor a negative real number out of $y - 3y - 2y^2$ and then write the polynomial factor in standard form.

Solution

$$y - 3y - 2y^2 = -1(-y + 3y + 2y^2)$$

$$= -1(2y^2 + 3y - y)$$

21. Factor $\dfrac{3}{2}x + \dfrac{5}{4} = \dfrac{1}{4}(\qquad)$.

Solution

$$\frac{3}{2}x + \frac{5}{4} = \frac{1}{4}(6x + 5)$$

23. Factor $\dfrac{5}{8}x + \dfrac{5}{16}y = \dfrac{5}{16}(\qquad)$.

Solution

$$\frac{5}{8}x + \frac{5}{16}y = \frac{5}{16}(2x + y)$$

25. Factor $2y(y - 3) + 5(y - 3)$ by factoring out the greatest common factor.

Solution

$$2y(y - 3) + 5(y - 3) = (y - 3)(2y + 5)$$

27. Factor $a(a + 6) - a^2(a + 6)$ by factoring out the greatest common factor.

Solution

$$a(a + 6) - a^2(a + 6) = (a + 6)(a - a^2)$$

$$= (a + 6)a(1 - a)$$

29. Factor $x^2 + 25x + x + 25$ by grouping.

Solution

$$x^2 + 25x + x + 25 = (x^2 + 25x) + (x + 25)$$

$$= x(x + 25) + 1(x + 25)$$

$$= (x + 25)(x + 1)$$

31. Factor $a^3 - 4a^2 + 2a - 8$ by grouping.

Solution

$$a^3 - 4a^2 + 2a - 8 = (a^3 - 4a^2) + (2a - 8)$$
$$= a^2(a - 4) + 2(a - 4)$$
$$= (a - 4)(a^2 + 2)$$

35. Factor the difference of two squares.

$$4z^2 - y^2$$

Solution

$$4z^2 - y^2 = (2z - y)(2z + y)$$

39. Factor the difference of cubes.

$$x^3 - 8$$

Solution

$$x^3 - 8 = x^3 - 2^3$$
$$= (x - 2)(x^2 + 2x + 4)$$

43. Factor $3x^4 - 300x^2$ completely.

Solution

$$3x^4 - 300x^2 = 3x^2(x^2 - 100)$$
$$= 3x^2(x - 10)(x + 10)$$

47. *Revenue* The revenue from selling x units of a product at a price of p dollars per unit is given by $R = xp$. For a particular commodity, the revenue is $R = 800x - 0.25x^2$. Factor the expression and determine an expression that gives the price in terms of x.

Solution

$$R = 800x - 0.25x^2$$
$$R = 0.25x(3200 - x)$$
$$R = x(800 - 0.25x) \quad \text{so}$$
$$p = 800 - 0.25x$$

33. Factor the difference of two squares.

$$x^2 - 64$$

Solution

$$x^2 - 64 = x^2 - 8^2$$
$$= (x - 8)(x + 8)$$

37. Factor the difference of two squares.

$$(x - 1)^2 - 16$$

Solution

$$(x - 1)^2 - 16 = [(x - 1) - 4][(x - 1) + 4]$$
$$= (x - 5)(x + 3)$$

41. Factor the sum of cubes.

$$x^3 + 64y^3$$

Solution

$$x^3 + 64y^3 = x^3 + (4y)^3$$
$$= (x + 4y)(x^2 - 4xy + 16y^2)$$

45. Factor $2a^4 - 32$ completely.

Solution

$$2a^4 - 32 = 2(a^4 - 16)$$
$$= 2(a^2 - 4)(a^2 + 4)$$
$$= 2(a - 2)(a + 2)(a^2 + 4)$$

49. Simplify $\dfrac{3}{8}x - \dfrac{1}{12}x + 8$ by combining like terms.

Solution

$$\frac{3}{8}x - \frac{1}{12}x + 8 = \left(\frac{3}{8} - \frac{1}{12}\right)x + 8$$
$$= \left(\frac{9}{24} - \frac{2}{24}\right)x + 8$$
$$= \frac{7}{24}x + 8$$

51. Simplify $3x^2 - 5x + 3 + 28x - 33x^2$ by combining like terms.

Solution

$$3x^2 - 5x + 3 + 28x - 33x^2 = (3 - 33)x^2 + (-5 + 28)x + 3$$
$$= -30x^2 + 23x + 3$$

53. Write an absolute value inequality that represents the verbal statement: "the set of all real numbers x whose distance from 0 is less than 5."

Solution

$|x| < 5$

55. Find the greatest common factor of $3x^2$, $12x$.

Solution

$3x^2 = 3 \cdot x \cdot x$

$12x = 2^2 \cdot 3 \cdot x$

$\text{GCF} = 3x$

57. Find the greatest common factor of $30z^2$, $-12z^3$.

Solution

$30z^2 = 2 \cdot 3 \cdot 5 \cdot z \cdot z$

$-12z^3 = -1 \cdot 2^2 \cdot 3 \cdot z \cdot z \cdot z$

$\text{GCF} = 2 \cdot 3 \cdot z \cdot z$

$\quad\quad = 6z^2$

59. Find the greatest common factor of $42(x + 8)^2$, $63(x + 8)^3$.

Solution

$42(x + 8)^2 = 7 \cdot 3 \cdot 2 \cdot (x + 8)^2$

$63(x + 8)^3 = 7 \cdot 3^2 \cdot (x + 8)^3$

$\quad\quad \text{GCF} = 7 \cdot 3(x + 8)^2$

$\quad\quad\quad = 21(x + 8)^2$

61. Factor out the greatest common monomial factor of $4u + 10$. (Some of the polynomials may have no common monomial factor other than 1 and -1.)

Solution

$4u + 10 = 2(2u + 5)$

63. Factor out greatest common monomial factor of $24x^2 - 18$. (Some of the polynomials may have no common monomial factor other than 1 and -1.)

Solution

$24x^2 - 18 = 6(4x^2 - 3)$

65. Factor out the greatest common monomial factor of $2x^2 + x$. (Some of the polynomials may have no common monomial factor other than 1 and -1.)

Solution

$2x^2 + x = x(2x + 1)$

67. Factor out the greatest common monomial factor of $11u^2 + 9$. (Some of the polynomials may have no common monomial factor other than 1 and -1.)

Solution

$11u^2 + 9$ is prime

69. Factor out the greatest common monomial factor of $3x^2y^2 - 15y$. (Some of the polynomials may have no common monomial factor other than 1 and -1.)

Solution

$3x^2y^2 - 15y = 3y(x^2y - 5)$

71. Factor out the greatest common monomial factor of $28x^2 + 16x - 8$. (Some of the polynomials may have no common monomial factor other than 1 and -1.)

Solution

$28x^2 + 16x - 8 = 4(7x^2 + 4x - 2)$

73. Factor out the greatest common monomial factor of $14x^4 + 21x^3 + 9x^2$. (Some of the polynomials may have no common monomial factor other than 1 and -1.)

Solution

$14x^4 + 21x^3 + 9x^2 = x^2(14x^2 + 21x + 9)$

75. Factor a negative real number out of $7 - 14x$ and then write the polynomial factor in standard form.

Solution

$7 - 14x = -7(-1 + 2x) = -7(2x - 1)$

77. Factor a negative real number out of $5 + 4x - x^2$ and then write the polynomial factor in standard form.

Solution

$5 + 4x - x^2 = -1(-5 - 4x + x^2) = -1(x^2 - 4x - 5)$

79. Factor $\dfrac{2}{3}t + \dfrac{3}{4} = \dfrac{1}{12}(\quad\quad)$.

Solution

$\dfrac{2}{3}t + \dfrac{3}{4} = \dfrac{1}{12}(8t + 9)$

81. Factor $2y - \dfrac{3}{5} = \dfrac{1}{5}($).

Solution

$2y - \dfrac{3}{5} = \dfrac{1}{5}(10y - 3)$

83. Factor $5x(x + 2) - 3(x + 2)$ by factoring out the greatest common factor.

Solution

$5x(x + 2) - 3(x + 2) = (x + 2)(5x - 3)$

85. Factor $(10 + b)(c + 7) - (10 + b)(c + 7)^2$ by factoring out the greatest common factor.

Solution

$$(10 + b)(c + 7) - (10 + b)(c + 7)^2 = (10 + b)\left[(c + 7) - (c + 7)^2\right]$$
$$= (10 + b)(c + 7)[1 - (c + 7)]$$
$$= (10 + b)(c + 7)(1 - c - 7)$$
$$= (10 + b)(c + 7)(-c - 6)$$
$$= (10 + b)(c + 7)(c + 6)(-1)$$

87. Factor $y^2 - 6y + 2y - 12$ by grouping.

Solution

$$y^2 - 6y + 2y - 12 = (y^2 - 6y) + (2y - 12)$$
$$= y(y - 6) + 2(y - 6)$$
$$= (y - 6)(y + 2)$$

89. Factor $x^3 + 2x^2 + x + 2$ by grouping.

Solution

$$x^3 + 2x^2 + x + 2 = (x^3 + 2x^2) + (x + 2)$$
$$= x^2(x + 2) + 1(x + 2)$$
$$= (x + 2)(x^2 + 1)$$

91. Factor $z^4 + 3z^3 - 2z - 6$ by grouping.

Solution

$$z^4 + 3z^3 - 2z - 6 = (z^4 + 3z^3) + (-2z - 6)$$
$$= z^3(z + 3) - 2(z + 3)$$
$$= (z + 3)(z^3 - 2)$$

93. Factor the difference of two squares.

$16y^2 - 9$

Solution

$$16y^2 - 9 = (4y)^2 - 3^2$$
$$= (4y - 3)(4y + 3)$$

95. Factor the difference of two squares.

$225 - 9y^2$

Solution

$$225 - 9y^2 = 15^2 - (3y)^2$$
$$= (15 - 3y)(15 + 3y)$$

97. Factor the difference of two squares.

$u^2 - \dfrac{1}{16}$

Solution

$$u^2 - \dfrac{1}{16} = u^2 - \left(\dfrac{1}{4}\right)^2$$
$$= \left(u - \dfrac{1}{4}\right)\left(u + \dfrac{1}{4}\right)$$

99. Factor the difference of two squares.

$81 - (z + 5)^2$

Solution

$81 - (z + 5)^2 = 9^2 - (z + 5)^2 = [9 - (z + 5)][9 + (z + 5)] = [9 - z - 5][9 + z + 5] = (4 - z)(14 + z)$

101. Factor the sum of cubes.

$$a^3 + b^3$$

Solution

$$a^3 + b^3 = (a + b)(a^2 - ab + b^2)$$

103. Factor the difference of cubes.

$$8t^3 - 27$$

Solution

$$8t^3 - 27 = (2t)^3 - 3^3$$
$$= (2t - 3)(4t^2 + 6t + 9)$$

105. Factor the sum of cubes.

$$27u^3 + 1$$

Solution

$$27u^3 + 1 = (3u)^3 + 1^3$$
$$= (3u + 1)(9u^2 - 3u + 1)$$

107. Factor $8 - 50x^2$ completely.

Solution

$$8 - 50x^2 = 2(4 - 25x^2)$$
$$= 2\left[2^2 - (5x)^2\right]$$
$$= 2[2 - 5x][2 + 5x]$$

109. Factor $y^4 - 81$ completely.

Solution

$$y^4 - 81 = (y^2)^2 - 9^2$$
$$= (y^2 - 9)(y^2 + 9)$$
$$= (y - 3)(y + 3)(y^2 + 9)$$

111. Factor $2x^3 - 54$ completely.

Solution

$$2x^3 - 54 = 2(x^3 - 27)$$
$$= 2(x^3 - 3^3)$$
$$= 2(x - 3)(x^2 + 3x + 9)$$

113. *Graphical Reasoning* Use a graphing utility to compare the two graphs. What can you conclude? $y_1 = 3x - 6$, $y_2 = 3(x - 2)$

Solution

Keystrokes:

y_1 $\boxed{\text{Y} =}$ 3 $\boxed{X, T, \theta}$ $\boxed{-}$ 6 $\boxed{\text{ENTER}}$

y_2 $\boxed{\text{Y} =}$ 3 $\boxed{(}$ $\boxed{X, T, \theta}$ $\boxed{-}$ 2 $\boxed{)}$ $\boxed{\text{GRAPH}}$

$y_1 = y_2$

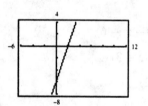

115. *Graphical Reasoning* Use a graphing utility to compare the two graphs. What can you conclude? $y_1 = x^2 - 4$, $y_2 = (x + 2)(x - 2)$

Solution

Keystrokes:

y_1 $\boxed{\text{Y} =}$ $\boxed{X, T, \theta}$ $\boxed{x^2}$ $\boxed{-}$ 4 $\boxed{\text{ENTER}}$

y_2 $\boxed{\text{Y} =}$ $\boxed{(}$ $\boxed{X, T, \theta}$ $\boxed{+}$ 2 $\boxed{)}$ $\boxed{(}$ $\boxed{X, T, \theta}$ $\boxed{-}$ 2 $\boxed{)}$ $\boxed{\text{GRAPH}}$

$y_1 = y_2$

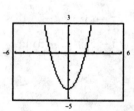

117. *Mental Math* Evaluate $79 \cdot 81$ mentally using the sample as a model.

$$48 \cdot 52 = (50 - 2)(50 + 2) = 50^2 - 2^2 = 2496$$

Solution

$$79 \cdot 81 = (80 - 1)(80 + 1) = 80^2 - 1^2 = 6399$$

119. *Think About It*   Show all the different groupings that can be used to completely factor $3x^3 + 4x^2 - 3x - 4$. Carry out the various factorizations to show that they yield the same result.

Solution

$$3x^3 + 4x^2 - 3x - 4 = (3x^3 + 4x^2) + (-3x - 4) \qquad = (3x^3 - 3x) + (4x^2 - 4)$$
$$= x^2(3x + 4) - 1(3x + 4) \qquad \text{or} \qquad = 3x(x^2 - 1) + 4(x^2 - 1)$$
$$= (x^2 - 1)(3x + 4) \qquad = (x^2 - 1)(3x + 4)$$
$$= (x - 1)(x + 1)(3x + 4) \qquad = (x - 1)(x + 1)(3x + 4)$$

121. *Simple Interest*   The total amount of money from a principal of P invested at $r\%$ simple interest for t years is given by $P + Prt$. Factor this expression.

Solution

$$P + Prt = P(1 + rt)$$

123. *Geometry*   The area of a rectangle of length l is given by $45l - l^2$. Factor this expression to determine the width of the rectangle.

Solution

$$A = 45l - l^2$$
$$= l(45 - l) \quad \text{so}$$
$$w = 45 - l$$

125. *Product Design*   A washer on the drive train of a car has an inside radius of r centimeters and an outside radius of R centimeters (see figure in textbook). Find the area of one of the flat surfaces of the washer and express the area in factored form.

Solution

$$A = \pi R^2 - \pi r^2$$
$$= \pi(R^2 - r^2)$$
$$= \pi(R - r)(R + r)$$

Mid-Chapter Quiz for Chapter 3

1. Determine the degree and leading coefficient of $3 - 2x + 4x^3 - 2x^4$.

Solution

degree = 4
leading coefficient = -2

2. Explain why $2x - 3x^{1/2} + 5$ is not a polynomial.

Solution

$2x - 3x^{1/2} + 5$ is not a polynomial because the term $-3x^{1/2}$ has degree $\frac{1}{2}$. The degree of the variable x is not an integer.

3. Add $2t^3 + 3t^2 - 2$ to $t^3 + 9$ and simplify.

Solution

$$(2t^3 + 3t^2 - 2) + (t^3 + 9) = 3t^3 + 3t^2 + 7$$

4. Perform the operation $(3 - 7y) + (7y^2 + 2y - 3)$ and simplify.

Solution

$$(3 - 7y) + (7y^2 + 2y - 3) = 7y^2 - 5y$$

5. Perform the operation $(7x^3 - 3x^2 + 1) - (x^2 - 2x^3)$ and simplify.

Solution

$$(7x^3 - 3x^2 + 1) - (x^2 - 2x^3) = 7x^3 - 3x^2 + 1 - x^2 + 2x^3$$
$$= 9x^3 - 4x^2 + 1$$

6. Perform the operation $(5 - u) - 2\left[3 - (u^2 + 1)\right]$ and simplify.

Solution

$$\begin{aligned}
(5 - u) - 2\left[3 - (u^2 + 1)\right] &= (5 - u) - 2[3 - u^2 - 1] \\
&= (5 - u) - 2[2 - u^2] \\
&= 5 - u - 4 + 2u^2 \\
&= 2u^2 - u + 1
\end{aligned}$$

7. Perform the operation $(-5n^2)(-2n^3)$ and simplify.

Solution

$(-5n^2)(-2n^3) = 10n^5$

8. Perform the operation $\dfrac{6x^7}{(-2x^2)^3}$ and simplify.

Solution

$$\begin{aligned}
\frac{6x^7}{(-2x^2)^3} &= \frac{6x^7}{-8x^6} \\
&= -\frac{3x}{4}
\end{aligned}$$

9. Perform the operation $7y(4 - 3y)$ and simplify.

Solution

$7y(4 - 3y) = 28y - 21y^2$

10. Perform the operation $2z(z + 5) - 7(z + 5)$ and simplify.

Solution

$$\begin{aligned}
2z(z + 5) - 7(z + 5) &= 2z^2 + 10z - 7z - 35 \\
&= 2z^2 + 3z - 35
\end{aligned}$$

11. Perform the operation $(6r + 5)(6r - 5)$ and simplify.

Solution

$(6r + 5)(6r - 5) = 36r^2 - 25$

12. Perform the operation $(2x - 3)^2$ and simplify.

Solution

$$\begin{aligned}
(2x - 3)^2 &= (2x - 3)(2x - 3) \\
&= 4x^2 - 12x + 9
\end{aligned}$$

13. Perform the operation $(x + 1)(x^2 - x + 1)$ and simplify.

Solution

$$\begin{aligned}
(x + 1)(x^2 - x + 1) &= x^3 - x^2 + x + x^2 - x + 1 \\
&= x^3 + 1
\end{aligned}$$

14. Perform the operation $(v - 3)^2 - (v + 3)^2$ and simplify.

Solution

$$\begin{aligned}
(v - 3)^2 - (v + 3)^2 &= (v^2 - 6v + 9) - (v^2 + 6v + 9) \\
&= v^2 - 6v + 9 - v^2 - 6v - 9 \\
&= -12v
\end{aligned}$$

15. Factor $28a^2 - 21a$ completely.

Solution

$28a^2 - 21a = 7a(4a - 3)$

16. Factor $25 - 4x^2$ completely.

Solution

$25 - 4x^2 = (5 - 2x)(5 + 2x)$

17. Factor $z^3 + 3z^2 - 9z - 27$ completely.

Solution

$$\begin{aligned}
z^3 + 3z^2 - 9z - 27 &= z^2(z + 3) - 9(z + 3) \\
&= (z + 3)(z^2 - 9) \\
&= (z + 3)(z + 3)(z - 3)
\end{aligned}$$

18. Factor $4y^3 - 32$ completely.

Solution

$$\begin{aligned}
4y^3 - 32 &= 4(y^3 - 8) \\
&= 4[y^3 - 2^3] \\
&= 4(y - 2)(y^2 + 2y + 4)
\end{aligned}$$

19. Find all possible products of the form $(5x + m)(2x + n)$ such that $mn = 10$.

Solution

$(5x + 10)(2x + 1)$ $(5x - 10)(2x - 1)$

$(5x + 1)(2x + 10)$ $(5x - 1)(2x - 10)$

$(5x + 2)(2x + 5)$ $(5x - 2)(2x - 5)$

$(5x + 5)(2x + 2)$ $(5x - 5)(2x - 2)$

20. Find the area of the shaded portion of the figure (see the textbook).

Solution

Verbal model:

Area of shaded region	=	Area of large triangle	−	Area of small triangle

Equation:

$$A = \frac{1}{2}(x + 2)^2 - \frac{1}{2}x^2$$

$$= \frac{1}{2}(x^2 + 4x + 4) - \frac{1}{2}x^2$$

$$= \frac{1}{2}x^2 + 2x + 2 - \frac{1}{2}x^2$$

$$= 2x + 2$$

3.4   Factoring Trinomials

7. Factor the perfect square trinomial.

$$x^2 + 4x + 4$$

Solution

$$x^2 + 4x + 4 = x^2 + 2(2x) + 2^2 = (x + 2)^2$$

9. Factor the perfect square trinomial.

$$25y^2 - 10y + 1$$

Solution

$$25y^2 - 10y + 1 = (5y)^2 - 2(5y) + 1 = (5y - 1)^2$$

11. Find all values of b for which $x^2 + bx + 81$ is a perfect square trinomial.

Solution

$$x^2 + bx + 81 = x^2 + bx + 9^2$$

(a) $b = 18$

$$x^2 + 18x + 9^2 = x^2 + 2(9x) + 9^2$$

$$= (x + 9)^2$$

(b) $b = -18$

$$x^2 - 18x + 9^2 = x^2 - 2(9x) + 9^2$$

$$= (x - 9)^2$$

13. Find all values of b for which $4x^2 + bx + 9$ is a perfect square trinomial.

Solution

$$4x^2 + bx + 9 = (2x)^2 + bx + 3^2$$

(a) $b = 12$

$$(2x)^2 + 12x + 3^2 = (2x)^2 + 2(2x)(3) + 3^2$$

$$= (2x + 3)^2$$

(b) $b = -12$

$$(2x)^2 - 12x + 3^2 = (2x)^2 - 2(2x)(3) + 3^2$$

$$= (2x - 3)^2$$

15. Find a real number c such that $x^2 + 8x + c$ is a perfect square trinomial.

Solution

$c = 16$

$$x^2 + 8x + c = x^2 + 2(4x) + c$$
$$= x^2 + 2(4x) + 4^2$$
$$= (x + 4)^2$$

17. Find a real number c such that $y^2 - 6y + c$ is a perfect square trinomial.

Solution

$c = 9$

$$y^2 - 6y + c = y^2 - 2(3y) + c$$
$$= y^2 - 2(3y) + 3^2$$
$$= (y - 3)^2$$

19. Find the missing factor for
$x^2 + 5x + 4 = (x + 4)(\quad)$.

Solution

$x^2 + 5x + 4 = (x + 4)(x + 1)$

21. Find the missing factor for
$y^2 - y - 20 = (y + 4)(\quad)$.

Solution

$y^2 - y - 20 = (y + 4)(y - 5)$

23. Factor $x^2 + 4x + 3$.

Solution

$x^2 + 4x + 3 = (x + 3)(x + 1)$

25. Factor $t^2 - 4t - 21$.

Solution

$t^2 - 4t - 21 = (t - 7)(t + 3)$

27. Factor $x^2 - 2xy - 35y^2$.

Solution

$x^2 - 2xy - 35y^2 = (x - 7y)(x + 5y)$

29. Find all values of b for which $x^2 + bx = 35$ can be factored.

Solution

$b = 36:\quad x^2 + 36x + 35 = (x + 35)(x + 1)$
$b = -36:\quad x^2 - 36x + 35 = (x - 35)(x - 1)$
$b = 12:\quad x^2 + 12x + 35 = (x + 7)(x + 5)$
$b = -12:\quad x^2 - 12x + 35 = (x - 7)(x - 5)$

31. Find two values of c for which $x^2 + 6x + c$ can be factored.

Solution

$c = 5:\quad x^2 + 6x + 5 = (x + 5)(x + 1)$
$c = 8:\quad x^2 + 6x + 8 = (x + 4)(x + 2)$
$c = 9:\quad x^2 + 6x + 9 = (x + 3)(x + 3)$

Also note that if $c = $ a negative number, there are many possibilities for c such as the following.

$c = -7:\quad x^2 + 6x - 7 = (x + 7)(x - 1)$
$c = -16:\quad x^2 + 6x - 16 = (x + 8)(x - 2)$
$c = -27:\quad x^2 + 6x - 27 = (x + 9)(x - 3)$

33. Find the missing factor for $5x^2 + 18x + 9 = (x + 3)(\quad)$.

Solution

$5x^2 + 18x + 9 = (x + 3)(5x + 3)$

35. Factor $5x^2 + 7x + 2$, if possible.

Solution

$5x^2 + 7x + 2 = (5x + 2)(x + 1)$

37. Factor $6x^2 - 11x + 3$, if possible.

Solution

$6x^2 - 11x + 3 = (3x - 1)(2x - 3)$

39. Factor $3t^2 - 4t - 10$.

Solution

$3t^2 - 4t - 10 = $ prime

41. Factor $6u^2 - 5uv - 4v^2$, if possible.

Solution

$6u^2 - 5uv - 4v^2 = (3u - 4v)(2u + v)$

43. Factor $-2x^2 - x + 6$, if possible.

Solution

$$-2x^2 - x + 6 = -1(2x^2 + x - 6)$$
$$= -1(2x - 3)(x + 2)$$

45. Factor $1 - 11x - 60x^2$, if possible.

Solution

$$1 - 11x - 60x^2 = -60x^2 - 11x + 1$$
$$= -1(60x^2 + 11x - 1)$$
$$= -1(15x - 1)(4x + 1)$$

47. Factor $3x^2 + 10x + 8$ by grouping.

Solution

$$3x^2 + 10x + 8 = 3x^2 + 6x + 4x + 8$$
$$= (3x^2 + 6x) + (4x + 8)$$
$$= 3x(x + 2) + 4(x + 2)$$
$$= (3x + 4)(x + 2)$$

49. Factor $15x^2 - 11x + 2$ by grouping.

Solution

$$15x^2 - 11x + 2 = 15x^2 - 6x - 5x + 2$$
$$= (15x^2 - 6x) + (-5x + 2)$$
$$= 3x(5x - 2) - 1(5x - 2)$$
$$= (3x - 1)(5x - 2)$$

51. Find the missing coordinate of the solution point.

$$y = \frac{3}{5}x + 4 \quad (15, \quad)$$

Solution

$$y = \frac{3}{5}(15) + 4$$
$$y = 9 + 4$$
$$y = 13 \quad (15, 13)$$

53. Find the missing coordinate of the solution point.

$$y = 5.5 - 0.95x \quad (\quad , -1)$$

Solution

$$-1 = 5.5 - 0.95x$$
$$0.95x = 5.5 + 1$$
$$0.95x = 6.5$$
$$x = \frac{6.5}{0.95}$$
$$x = 6.84 \quad (6.84, -1)$$

55. *Recipe Proportions* Two and one-half cups of flour are required to make one batch of cookies. How many cups are required to make $3\frac{1}{2}$ batches?

Solution

Verbal Model:
$$\frac{\text{Cups of Flour}}{\text{Batches of Cookies}} = \frac{\text{Cups of Flour}}{\text{Batches of Cookies}}$$

Proportion:
$$\frac{2\frac{1}{2}}{1} = \frac{x}{3\frac{1}{2}}$$

$$x = 2\frac{1}{2} \cdot 3\frac{1}{2} = \frac{5}{2} \cdot \frac{7}{2} = \frac{35}{4} = 8\frac{3}{4}$$

57. Factor the perfect square trinomial.

$$a^2 - 12a + 36$$

Solution

$$a^2 - 12a + 36 = a^2 - 2(6a) + 6^2 = (a - 6)^2$$

59. Factor the perfect square trinomial.

$$b^2 + 4b + 4$$

Solution

$$b^2 + 4b + 4 = (b)^2 + 2(2b) + 4 = (b + 2)^2$$

61. Factor the following perfect square trinomial.

$$u^2 + 8uv + 16v^2$$

Solution

$$u^2 + 8uv + 16v^2 = u^2 + 2(4uv) + (4v)^2$$
$$= (u + 4v)^2$$

63. Fill in the missing number for $x^2 + 12x + 50 = (x + 6)^2 + (\quad)$.

Solution

$$x^2 + 12x + 50 = (x + 6)^2 + 14$$

65. Find the missing factor for $x^2 - 2x - 24 = (x + 4)(\qquad)$.

Solution

$$x^2 - 2x - 24 = (x + 4)(x - 6)$$

67. Find the missing factor for $z^2 - 6z + 8 = (z - 4)(\qquad)$.

Solution

$$z^2 - 6z + 8 = (z - 4)(z - 2)$$

69. Factor $x^2 - 5x + 6$.

Solution

$$x^2 - 5x + 6 = (x - 3)(x - 2)$$

71. Factor $y^2 + 7y - 30$.

Solution

$$y^2 + 7y - 30 = (y + 10)(y - 3)$$

73. Factor $x^2 - 20x + 96$.

Solution

$$x^2 - 20x + 96 = (x - 12)(x - 8)$$

75. Factor $x^2 + 30x + 216$.

Solution

$$x^2 + 30x + 216 = (x + 12)(x + 18)$$

77. Find all values of b for which the trinomial $x^2 + bx + 18$ can be factored.

Solution

$b = 19:$ $x^2 + 19x + 18 = (x + 18)(x + 1)$
$b = -19:$ $x^2 - 19x + 18 = (x - 18)(x - 1)$
$b = 9:$ $x^2 + 9x + 18 = (x + 6)(x + 3)$
$b = -9:$ $x^2 - 9x + 18 = (x - 6)(x - 3)$
$b = 11:$ $x^2 + 11x + 18 = (x + 9)(x + 2)$
$b = -11:$ $x^2 - 11x + 18 = (x - 9)(x - 2)$

79. Find all values of b for which the trinomial $x^2 + bx - 21$ can be factored.

Solution

$b = 20:$ $x^2 + 20x - 21 = (x + 21)(x - 1)$
$b = -20:$ $x^2 - 20x - 21 = (x - 21)(x + 1)$
$b = 4:$ $x^2 + 4x - 21 = (x + 7)(x - 3)$
$b = -4:$ $x^2 - 4x - 21 = (x - 7)(x + 3)$

81. Find two values of c for which the trinomial $x^2 - 3x + c$ can be factored.

Solution

$c = 2:$ $x^2 - 3x + 2 = (x - 2)(x - 1)$
$c = -4:$ $x^2 - 3x - 4 = (x - 4)(x + 1)$
$c = -10:$ $x^2 - 3x - 10 = (x - 5)(x + 2)$
$c = -18:$ $x^2 - 3x - 18 = (x - 6)(x + 3)$

There are more possibilities.

83. Find the missing factor for
$5a^2 + 12a - 9 = (a + 3)(\qquad)$.

Solution

$5a^2 + 12a - 9 = (a + 3)(5a - 3)$

85. Find the missing factor for
$2y^2 - 3y - 27 = (y + 3)(\qquad)$.

Solution

$2y^2 - 3y - 27 = (y + 3)(2y - 9)$

87. Factor $3x^2 + 4x + 1$, if possible. (The trinomial may be prime.)

Solution

$3x^2 + 4x + 1 = (3x + 1)(x + 1)$

89. Factor $2x^2 - 9x + 9$, if possible. (The trinomial may be prime.)

Solution

$2x^2 - 9x + 9 = (2x - 3)(x - 3)$

91. Factor $6b^2 + 19b - 7$, if possible. (The trinomial may be prime.)

Solution

$6b^2 + 19b - 7 = (3b - 1)(2b + 7)$

93. Factor $24x^2 - 14xy - 3y^2$, if possible. (The trinomial may be prime.)

Solution

$24x^2 - 14xy - 3y^2 = (6x + y)(4x - 3y)$

95. Factor $3y^2 + 4y + 12$, if possible. (The trinomial may be prime.)

Solution

$3y^2 + 4y + 12 = $ prime

97. Factor $6x^2 + x - 2$ by grouping.

Solution

$$6x^2 + x - 2 = 6x^2 + 4x - 3x - 2$$
$$= (6x^2 + 4x) + (-3x - 2)$$
$$= 2x(3x + 2) - 1(3x + 2)$$
$$= (2x - 1)(3x + 2)$$

99. Factor $3x^4 - 12x^3$ completely.

Solution

$3x^4 - 12x^3 = 3x^3(x - 4)$

101. Factor $10t^3 + 2t^2 - 36t$ completely.

Solution

$$10t^3 + 2t^2 - 36t = 2t(5t^2 + t - 18)$$
$$= 2t(5t - 9)(t + 2)$$

103. Factor $36 - (z + 3)^2$ completely.

Solution

$$36 - (z + 3)^2 = [6 - (z + 3)][6 + (z + 3)]$$
$$= [6 - z - 3][6 + z + 3]$$
$$= (3 - z)(9 + z)$$

105. Factor $54x^3 - 2$ completely.

Solution

$$54x^3 - 2 = 2(27x^3 - 1)$$
$$= 2(3x - 1)(9x^2 + 3x + 1)$$

107. Use a graphing utility to graph the two equations on the same screen. What can you conclude?

$$y_1 = x^2 + 6x + 9, \; y_2 = (x + 3)^2$$

Solution

Keystrokes:

y_1 [Y=] [X, T, θ] [x^2] [+] 6 [X, T, θ] [+] 9 [ENTER]

y_2 [Y=] [(] [X, T, θ] [+] 3 [)] [x^2] [GRAPH]

$y_1 = y_2$

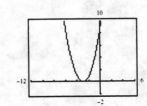

109. Use a graphing utility to graph the two equations on the same screen. What can you conclude?

$$y_1 = x^2 + 2x - 3, \quad y_2 = (x-1)(x+3)$$

Solution

Keystrokes:

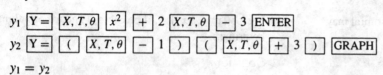

$y_1 = y_2$

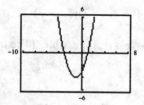

111. *Mental Math* Use mental math to evaluate 52^2.

Sample: $29^2 = (30-1)^2$

$$= 30^2 - 2(30)(1) + 1^2$$

$$= 900 - 60 + 1$$

$$= 841$$

Solution

$$52^2 = (50+2)^2$$

$$= 50^2 + 2(50)(2) + 2^2$$

$$= 2500 + 200 + 4$$

$$= 2704$$

113. *Geometric Factoring Models* Factor $x^2 + 4x + 3$ and represent the result with a geometric factoring model. (See the sample in the textbook.)

Solution

$$x^2 + 4x + 3 = (x+3)(x+1)$$

115. *Geometric Factoring Models* Match the geometric factoring model with the correct factoring formula.

Solution

(c)

117. *Geometric Factoring Models* Match the geometric factoring model with the correct factoring formula.

Solution

(b)

3.5 Solving Polynomial Equations

5. Use the Zero-Factor Property to solve $2x(x-8) = 0$.

Solution

$2x(x-8) = 0$

$2x = 0 \qquad x - 8 = 0$

$x = 0 \qquad x = 8$

7. Use the Zero-Factor Property to solve $25(a+4)(a-2) = 0$.

Solution

$25(a+4)(a-2) = 0$

$a + 4 = 0 \qquad a - 2 = 0$

$a = -4 \qquad a = 2$

9. Use the Zero-Factor Property to solve $(x-3)(2x+1)(x+4)=0$.

Solution

$(x-3)(2x+1)(x+4)=0$

$x-3=0 \qquad 2x+1=0 \qquad x+4=0$

$\qquad x=3 \qquad\qquad x=-\dfrac{1}{2} \qquad x=-4$

11. Solve $5y-y^2=0$ by factoring.

Solution

$5y-y^2=0$

$y(5-y)=0$

$y=0 \quad 5-y=0$

$\qquad\qquad 5=y$

13. Solve $x^2-25=0$ by factoring.

Solution

$x^2-25=0$

$(x+5)(x-5)=0$

$x+5=0 \qquad x-5=0$

$\quad x=-5 \qquad\quad x=5$

15. Solve $(x+2)^2-9=0$ by factoring.

Solution

$(x+2)^2-9=0$

$[(x+2)-3][(x+2)+3]=0$

$(x-1)(x+5)=0$

$(x-1)=0 \qquad (x+5)=0$

$\quad x=1 \qquad\qquad x=-5$

17. Solve $m^2-8m+16=0$ by factoring.

Solution

$m^2-8m+16=0$

$(m-4)(m-4)=0$

$m-4=0 \qquad m-4=0$

$\quad m=4 \qquad\quad m=4$

19. Solve $x(x-3)=10$ by factoring.

Solution

$x(x-3)=10$

$x^2-3x-10=0$

$(x-5)(x+2)=0$

$x-5=0 \quad x+2=0$

$\quad x=5 \qquad\quad x=-2$

21. Solve $x(x+2)-10(x+2)=0$ by factoring.

Solution

$x(x+2)-10(x+2)=0$

$(x-10)(x+2)=0$

$x-10=0 \qquad x+2=0$

$\quad x=10 \qquad\qquad x=-2$

23. Solve $7+13x-2x^2=0$ by factoring.

Solution

$7+13x-2x^2=0$

$(7-x)(1+2x)=0$

$7-x=0 \qquad 1+2x=0$

$\quad -x=-7 \qquad\quad 2x=-1$

$\quad x=7 \qquad\qquad\quad x=-\dfrac{1}{2}$

25. Solve $x^3-19x^2+84x=0$ by factoring.

Solution

$x^3-19x^2+84x=0$

$x(x^2-19x+84)=0$

$x(x-12)(x-7)=0$

$x=0 \quad x-12=0 \quad x-7=0$

$x=0 \qquad x=12 \qquad\quad x=7$

27. Solve $z^2(z+2)-4(z+2)=0$ by factoring.

Solution

$z^2(z+2)-4(z+2)=0$

$(z+2)(z^2-4)=0$

$(z+2)(z-2)(z+2)=0$

$z+2=0 \qquad z-2=0 \qquad z+2=0$

$\quad z=-2 \qquad\quad z=2 \qquad\quad z=-2$

29. Solve $c^3 - 3c^2 - 9c + 27 = 0$ by factoring.

Solution

$$c^3 - 3c^2 - 9c + 27 = 0$$

$$c^2(c - 3) - 9(c - 3) = 0$$

$$(c - 3)(c^2 - 9) = 0$$

$$(c - 3)(c - 3)(c + 3) = 0$$

$$c - 3 = 0 \quad c - 3 = 0 \quad c + 3 = 0$$

$$c = 3 \qquad c = 3 \qquad c = -3$$

31. *Graphical Reasoning* Explain how the x-intercepts of the graph of $y = x^3 - 9$ correspond to the solutions of the polynomial equation when $y = 0$.

Solution

From the graph, the x-intercepts are $(-3, 0)$ and $(3, 0)$. The solutions of the equation $0 = x^2 - 9$ are 3 and -3.

$$0 = (x - 3)(x + 3)$$

$$0 = x - 3 \qquad 0 = x + 3$$

$$3 = x \qquad -3 = x$$

33. *Graphical Reasoning* Explain how the x-intercepts of the graph of $y = x^2 - 2x - 3$ correspond to the solutions of the polynomial equation when $y = 0$.

Solution

From the graph, the x-intercepts are $(-1, 0)$ and $(3, 0)$. The solutions of the equation $x^2 - 2x - 3 = 0$ are -1 and 3.

$$(x - 3)(x + 1) = 0$$

$$(x - 3) = 0 \qquad (x + 1) = 0$$

$$x = 3 \qquad x = -1$$

35. Use a graphing utility to solve $x^2 - 6x = 0$ graphically.

Solution

Keystrokes:

$\boxed{Y=}\ \boxed{X,T,\theta}\ \boxed{x^2}\ \boxed{-}\ 6\ \boxed{X,T,\theta}\ \boxed{\text{GRAPH}}$

The x-intercepts are 0 and 6, so the solutions are 0 and 6.

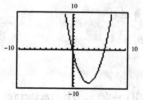

37. Use a graphing utility to solve $2x^2 + 5x - 12 = 0$ graphically.

Solution

Keystrokes:

$\boxed{Y=}\ 2\ \boxed{X,T,\theta}\ \boxed{x^2}\ \boxed{+}\ 5\ \boxed{X,T,\theta}\ \boxed{-}\ 12\ \boxed{\text{GRAPH}}$

The x-intercepts are -4 and $\dfrac{3}{2}$, so the solutions are -4 and $\dfrac{3}{2}$.

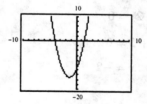

39. *Geometry* An open box is to be made from a piece of material that is 5 meters long and 4 meters wide. The box is to be made by cutting squares of dimension x from the corners and turning up the sides. (See figure in textbook.) The volume V of a rectangular solid is the product of its length, width, and height.

(a) Show that the volume is given by $V = (5 - 2x)(4 - 2x)x$.

(b) Determine the values of x for which $V = 0$. Determine an appropriate domain for the function V in the context of this problem.

(c) Complete the table (in text).

(d) Determine x when $V = 3$. Verify the result algebraically.

(e) Use a graphing utility to graph the volume function. Use the graph to approximate the value of x that yields the box of greatest volume.

Solution

(a) volume V = length · width · height

$$= (5 - 2x)(4 - 2x)x$$

(b) $0 = (5 - 2x)(4 - 2x)x$ Domain: Each side must be positive.

$5 - 2x = 0$ $4 - 2x = 0$ $x = 0$ $x > 0$ $5 - 2x > 0$ $4 - 2x > 0$ so $0 < x < 2$

$x = \dfrac{5}{2}$ $x = 2$ $x = 0$ $x < \dfrac{5}{2}$ $x < 2$

(c)

x	0.25	0.50	0.75	1.00	1.25	1.50	1.75
V	3.94	6	6.56	6	4.69	3	1.31

(d) If $V = 3$, then $x = 1.5$.

$3 = [5 - 2(1.5)]\,[4 - 2(1.5)]\,(1.5)$

$3 = (5 - 3)(4 - 3)(1.5)$

$3 = (2)(1)(1.5)$

$3 = 3$

(e) Keystrokes:

$\boxed{Y=}$ $\boxed{(}$ $\boxed{5}$ $\boxed{-}$ $\boxed{2}$ $\boxed{X,T,\theta}$ $\boxed{)}$ $\boxed{(}$ $\boxed{4}$ $\boxed{-}$ $\boxed{2}$ $\boxed{X,T,\theta}$ $\boxed{)}$ $\boxed{X,T,\theta}$ $\boxed{GRAPH}$

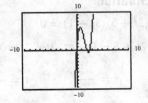

$x = 0.74$ yields the box of greatest volume.

41. Find two consecutive positive integers whose product is 132.

Solution

Verbal Model: $\boxed{\text{First Integer}} \cdot \boxed{\text{Second Integer}} = \boxed{132}$

Equation:
$$x \cdot (x + 1) = 132$$
$$x^2 + x - 132 = 0$$
$$(x + 12)(x - 11) = 0$$

$$x + 12 = 0 \qquad x - 11 = 0$$
$$x = -12 \qquad x = 11 \quad \text{1st integer}$$
$$\text{reject} \qquad x + 1 = 12 \quad \text{2nd integer}$$

43. Simplify $2(x + 5) - 3 - (2x - 3)$.

Solution
$$2(x + 5) - 3 - (2x - 3) = 2x + 10 - 3 - 2x + 3$$
$$= 10$$

45. Simplify $4 - 2[3 + 4(x + 1)]$.

Solution
$$4 - 2[3 + 4(x + 1)] = 4 - 2[3 + 4x + 4]$$
$$= 4 - 2[7 + 4x]$$
$$= 4 - 14 - 8x$$
$$= -10 - 8x$$

47. *Simple Interest* Evaluate the expression Prt when $P = \$2000$, $r = 0.07$, and $t = 10$.

Solution
$$Prt = (2000)(0.07)(10)$$
$$= \$1400$$

49. Use the Zero-Factor Property to solve $(y - 3)(y + 10) = 0$.

Solution
$$(y - 3)(y + 10) = 0$$
$$y - 3 = 0 \quad y + 10 = 0$$
$$y = 3 \qquad y = -10$$

51. Use the Zero-Factor Property to solve $(2t + 5)(3t + 1) = 0$.

Solution
$$(2t + 5)(3t + 1) = 0$$
$$2t + 5 = 0 \qquad 3t + 1 = 0$$
$$t = -\frac{5}{2} \qquad t = -\frac{1}{3}$$

53. Use the Zero-Factor Property to solve $4x(2x - 3)(2x + 25) = 0$.

Solution
$$4x(2x - 3)(2x + 25) = 0$$
$$4x = 0 \quad 2x - 3 = 0 \quad 2x + 25 = 0$$
$$x = 0 \qquad x = \frac{3}{2} \qquad x = -\frac{25}{2}$$

55. Solve $3y^2 - 48 = 0$ by factoring.

Solution
$$3y^2 - 48 = 0$$
$$3(y^2 - 16) = 0$$
$$3(y + 4)(y - 4) = 0$$
$$y + 4 = 0 \qquad y - 4 = 0$$
$$y = -4 \qquad y = 4$$

57. Solve $(t - 2)^2 - 16 = 0$ by factoring.

Solution
$$(t - 2)^2 - 16 = 0$$
$$(t - 2 + 4)(t - 2 - 4) = 0$$
$$(t + 2)(t - 6) = 0$$
$$t + 2 = 0 \qquad t - 6 = 0$$
$$t = -2 \qquad t = 6$$

59. Solve $9x^2 + 15x = 0$ by factoring.

Solution

$9x^2 + 15x = 0$

$3x(3x + 5) = 0$

$3x = 0 \quad 3x + 5 = 0$

$x = 0 \qquad x = -\dfrac{5}{3}$

61. Solve $x^2 - 3x - 10 = 0$ by factoring.

Solution

$x^2 - 3x - 10 = 0$

$(x - 5)(x + 2) = 0$

$x - 5 = 0 \quad x + 2 = 0$

$x = 5 \qquad x = -2$

63. Solve $x^2 + 16x + 64 = 0$ by factoring.

Solution

$x^2 + 16x + 64 = 0$

$(x + 8)(x + 8) = 0$

$x + 8 = 0 \qquad x + 8 = 0$

$x = -8 \qquad x = -8$

65. Solve $4z^2 - 12z + 9 = 0$ by factoring.

Solution

$z^2 - 12z + 9 = 0$

$(2z - 3)(2z - 3) = 0$

$2z - 3 = 0 \qquad 2z - 3 = 0$

$z = \dfrac{3}{2} \qquad z = \dfrac{3}{2}$

67. Solve $14x^2 + 9x + 1 = 0$ by factoring.

Solution

$14x^2 + 9x + 1 = 0$

$(7x + 1)(2x + 1) = 0$

$7x + 1 = 0 \qquad 2x + 1 = 0$

$x = -\dfrac{1}{7} \qquad x = -\dfrac{1}{2}$

69. Solve $y(y + 6) = 72$ by factoring.

Solution

$y(y + 6) = 72$

$y^2 + 6y - 72 = 0$

$(y + 12)(y - 6) = 0$

$y + 12 = 0 \qquad y - 6 = 0$

$y = -12 \qquad y = 6$

71. Solve $t(2t - 3) = 35$ by factoring.

Solution

$t(2t - 3) = 35$

$2t^2 - 3t - 35 = 0$

$(2t + 7)(t - 5) = 0$

$2t + 7 = 0 \qquad t - 5 = 0$

$t = -\dfrac{7}{2} \qquad t = 5$

73. Solve $(a + 2)(a + 5) = 10$ by factoring.

Solution

$(a + 2)(a + 5) = 10$

$a^2 + 7a + 10 - 10 = 0$

$a^2 + 7a = 0$

$a(a + 7) = 0$

$a = 0 \quad a + 7 = 0$

$a = 0 \qquad a = -7$

75. Solve $(x - 4)(x + 5) = 10$ by factoring.

Solution

$$(x - 4)(x + 5) = 10$$
$$x^2 + x - 20 - 10 = 0$$
$$x^2 + x - 30 = 0$$
$$(x + 6)(x - 5) = 0$$
$$x + 6 = 0 \qquad x - 5 = 0$$
$$x = -6 \qquad x = 5$$

77. Solve $x(x + 10) - 2(x + 10) = 0$ by factoring.

Solution

$$x(x + 10) - 2(x + 10) = 0$$
$$(x + 10)(x - 2) = 0$$
$$x + 10 = 0 \qquad x - 2 = 0$$
$$x = -10 \qquad x = 2$$

79. Solve $6t^3 - t^2 - t = 0$ by factoring.

Solution

$$6t^3 - t^2 - t = 0$$
$$t(6t^2 - t - 1) = 0$$
$$t(3t + 1)(2t - 1) = 0$$
$$t = 0 \qquad 3t + 1 = 0 \qquad 2t - 1 = 0$$
$$t = 0 \qquad\qquad t = -\frac{1}{3} \qquad t = \frac{1}{2}$$

81. Solve $x^2(x - 25) - 16(x - 25) = 0$ by factoring.

Solution

$$x^2(x - 25) - 16(x - 25) = 0$$
$$(x - 25)(x^2 - 16) = 0$$
$$(x - 25)(x - 4)(x + 4) = 0$$
$$x - 25 = 0 \qquad x - 4 = 0 \qquad x + 4 = 0$$
$$x = 25 \qquad\quad x = 4 \qquad\quad x = -4$$

83. Solve $a^3 + 2a^2 - 9a - 18 = 0$ by factoring.

Solution

$$a^3 + 2a^2 - 9a - 18 = 0$$
$$(a^3 + 2a^2) + (-9a - 18) = 0$$
$$a^2(a + 2) - 9(a + 2) = 0$$
$$(a + 2)(a^2 - 9) = 0$$
$$(a + 2)(a - 3)(a + 3) = 0$$
$$a + 2 = 0 \qquad a - 3 = 0 \qquad a + 3 = 0$$
$$a = -2 \qquad\quad a = 3 \qquad\quad a = -3$$

85. Use a graphing utility to solve $x^2 - 8x + 12 = 0$ graphically.

Solution

Keystrokes:

$\boxed{Y=}$ $\boxed{X, T, \theta}$ $\boxed{x^2}$ $\boxed{-}$ 8 $\boxed{X, T, \theta}$ $\boxed{+}$ 12 $\boxed{\text{GRAPH}}$

The x-intercepts are 2 and 6, so the solutions are 2 and 6.

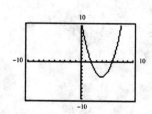

87. Use a graphing utility to solve $x^3 - 4x = 0$ graphically.

Solution

Keystrokes:

$\boxed{Y=}$ $\boxed{X,T,\theta}$ $\boxed{\wedge}$ 3 $\boxed{-}$ 4 $\boxed{X,T,\theta}$ $\boxed{\text{GRAPH}}$

The x-intercepts are $-2, 0,$ and 2, so the solutions are $-2, 0,$ and 2.

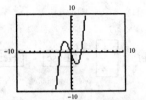

89. The sum of a positive number and its square is 240. Find the number.

Solution

Verbal Model: $\boxed{\text{Number}}$ + $\boxed{\text{Its Square}}$ = $\boxed{240}$

Equation:
$$x + x^2 = 240$$
$$x^2 + x - 240 = 0$$
$$(x + 16)(x - 15) = 0$$
$$x + 16 = 0 \qquad x - 15 = 0$$
$$x = -16 \qquad x = 15$$
$$\text{reject}$$

91. *Geometry* The rectangular floor of a storage shed has an area of 330 square feet. The length of the floor is 7 feet more than its width (see figure in textbook). Find the dimensions of the floor.

Solution

Verbal Model: $\boxed{\text{Length}}$ · $\boxed{\text{Width}}$ = $\boxed{\text{Area}}$

Equation:
$$(x + 7) \cdot x = 330$$
$$x^2 + 7x = 330$$
$$x^2 + 7x - 330 = 0$$
$$(x + 22)(x - 15) = 0$$
$$x + 22 = 0 \qquad x - 15 = 0$$
$$x = -22 \qquad x = 15 \ \text{width}$$
$$\text{reject} \qquad x + 7 = 22 \ \text{length}$$

93. *Geometry*   The triangular cross section of a machined part must have an area of 48 square inches (see figure in textbook). Find the base and height of the triangle if the height is $1\frac{1}{2}$ times the base.

Solution

Verbal Model: $\boxed{\frac{1}{2}} \cdot \boxed{\text{Base}} \cdot \boxed{\text{Height}} = \boxed{\text{Area}}$

Equation:

$$\frac{1}{2} \cdot x \cdot \frac{3}{2}x = 48$$

$$\frac{3}{4}x^2 - 48 = 0$$

$$3x^2 - 192 = 0$$

$$3(x^2 - 64) = 0$$

$$(x + 8)(x - 8) = 0$$

$$x + 8 = 0 \qquad x - 8 = 0$$

$$x = -8 \qquad x = 8 \text{ base}$$

$$\text{reject} \qquad \frac{3}{2}x = 12 \text{ height}$$

95. *Geometry*   An open box with a square base is to be constructed from 880 square inches of material (see figure in textbook). What should the dimensions of the base be if the height of the box is to be 6 inches? (*Hint:* The surface area is given by $S = x^2 + 4xh$.)

Solution

$$S = x^2 + 4xh$$

$$880 = x^2 + 4x(6)$$

$$0 = x^2 + 24x - 880$$

$$0 = (x - 20)(x + 44)$$

$$0 = x - 20 \qquad 0 = x + 44$$

$$20 = x \qquad -44 = x$$

$$20'' \times 20'' \qquad \text{reject}$$

97. *Break-Even Analysis*   The revenue R from the sale of x units of a product is given by $R = 90x - x^2$. The cost of producing x units of the product is given by $C = 200 + 60x$. How many units must the company produce and sell in order to break-even?

Solution

Verbal Model: $\boxed{\text{Revenue}} = \boxed{\text{Cost}}$

Equation:

$$90x - x^2 = 200 + 60x$$

$$0 = x^2 - 30x + 200$$

$$0 = (x - 20)(x - 10)$$

$$x - 20 = 0 \qquad x - 10 = 0$$

$$x = 20 \qquad x = 10$$

99. *Numerical Reasoning*

(a) Complete the table. For the following items, use the product $P = (x + 5)(x - 4)$.

x	3	4	5	6	7	8
P						

(b) Use the table to determine how P changes for each one-unit increase in x.

(c) Use the values of P in the table to solve the equation $(x + 5)(x - 4) = 70$.

Solution

(a)

x	3	4	5	6	7	8
P	-8	0	10	22	36	52

$$P = (3 + 5)(3 - 4) \qquad P = (4 + 5)(4 - 4)$$
$$\quad = (8)(-1) \qquad\qquad\quad = (9)(0)$$
$$\quad = -8 \qquad\qquad\qquad\quad = 0$$
$$P = (5 + 5)(5 - 4) \qquad P = (6 + 5)(6 - 4)$$
$$\quad = (10)(1) \qquad\qquad\quad = (11)(2)$$
$$\quad = 10 \qquad\qquad\qquad\quad = 22$$
$$P = (7 + 5)(7 - 4) \qquad P = (8 + 5)(8 - 4)$$
$$\quad = (12)(3) \qquad\qquad\quad = (13)(4)$$
$$\quad = 36 \qquad\qquad\qquad\quad = 52$$

(b) P changes for each one unit increase in x by an even positive integer (by 8, 10, 12, 14, etc.)

(c) $(x + 5)(x - 4) = 70 \quad x = 9 \quad (9 + 5)(9 - 4) = 70$

101. Let a be a nonzero real number. Find the solutions of $ax^2 - ax = 0$.

Solution

$$ax^2 - ax = 0$$
$$ax(x - 1) = 0$$
$$ax = 0 \qquad x - 1 = 0$$
$$x = \frac{0}{a} \qquad\quad x = 1$$
$$\qquad\qquad\qquad x = 1$$
$$x = 0$$

103. *Exploration* Solve the equations using either method (or both methods) described in Exercise 102 (in the textbook).

(a) $3(x + 6)^2 - 10(x + 6) - 8 = 0$ (b) $8(x + 2)^2 - 18(x + 2) + 9 = 0$

Solution

(a) $3(x + 6)^2 - 10(x + 6) - 8 = 0$

  let $u = (x + 6)$

  $3u^2 - 10u - 8 = 0$

  $(3u + 2)(u - 4) = 0$

  $3u + 2 = 0 \qquad u - 4 = 0$

  $u = -\dfrac{2}{3} \qquad u = 4$

  $\qquad\qquad\qquad x + 6 = 4$

  $x + 6 = -\dfrac{2}{3} \qquad x = -2$

  $x = -\dfrac{20}{3}$

(b) $8(x + 2)^2 - 18(x + 2) + 9 = 0$

  let $u = (x + 2)$

  $8u^2 - 18u + 9 = 0$

  $(4u - 3)(2u - 3) = 0$

  $4u - 3 = 0 \qquad 2u - 3 = 0$

  $u = \dfrac{3}{4} \qquad u = \dfrac{3}{2}$

  $x + 2 = \dfrac{3}{4} \qquad x + 2 = \dfrac{3}{2}$

  $x = -\dfrac{5}{4} \qquad x = -\dfrac{1}{2}$

Review Exercises for Chapter 3

1. State why $y^2 - \dfrac{2}{y}$ is not a polynomial.

Solution

$y^2 - \dfrac{2}{y}$ is not a polynomial because the variable cannot be in the denominator of a term.

3. State why $4 + 2|x^3|$ is not a polynomial.

Solution

$4 + 2|x^3|$ is not a polynomial because the variable cannot be within absolute value signs.

5. For $-4(x - 2) = -4x - 8$, correct the error.

Solution

$-4(x - 2) = -4x - 8$ Negative four times negative two equals

$-4(x - 2) = -4x + 8$ positive eight not negative eight.

7. For $(x + 3)^2 = x^2 + 9$, correct the error.

Solution

$(x + 3)^2 = x^2 + 9$ To square a binomial, FOIL must be used.

$(x + 3)^2 = (x + 3)(x + 3)$

$\qquad\quad = x^2 + 6x + 9$

9. Simplify the expression.

$$5x + 3x^2 + x - 4x^2$$

Solution

$$5x + 3x^2 + x - 4x^2 = (5x + x) + (3x^2 - 4x^2)$$
$$= 6x - x^2$$

11. Simplify the expression.

$$3t - 2(t^2 - t - 5)$$

Solution

$$3t - 2(t^2 - t - 5) = 3t - 2t^2 + 2t + 10$$
$$= (-2t^2) + (3t + 2t) + (10)$$
$$= -2t^2 + 5t + 10$$

13. Simplify the expression.

$$(-x^3 - 3x) - 4(2x^3 - 3x + 1)$$

Solution

$$(-x^3 - 3x) - 4(2x^3 - 3x + 1) = -x^3 - 3x - 8x^3 + 12x - 4$$
$$= (-x^3 - 8x^3) + (-3x + 12x) + (-4)$$
$$= -9x^3 + 9x - 4$$

15. Simplify the expression.

$$3y^2 - \left[2y - 3(y^2 + 5)\right]$$

Solution

$$3y^2 - \left[2y - 3(y^2 + 5)\right] = 3y^2 - [2y - 3y^2 - 15]$$
$$= 3y^2 - 2y + 3y^2 + 15$$
$$= (3y^2 + 3y^2) - 2y + 15$$
$$= 6y^2 - 2y + 15$$

17. Simplify the expression.

$$(-6z)^3$$

Solution

$$(-6z)^3 = (-6z)(-6z)(-6z) = -216z^3$$

19. Simplify the expression.

$$-(u^2v)^2(-4u^3v)$$

Solution

$$-(u^2v)^2(-4u^3v) = -(u^4v^2)(-4u^3v)$$
$$= 4u^{4+3}v^{2+1}$$
$$= 4u^7v^3$$

21. Simplify the expression.

$$\frac{120u^5v^3}{15u^3v}$$

Solution

$$\frac{120u^5v^3}{15u^3v} = \frac{120}{15} \cdot \frac{u^5}{u^3} \cdot \frac{v^3}{v}$$
$$= 8u^2v^2$$

23. Simplify the expression.

$$\frac{72x^4}{6x^2}$$

Solution

$$\frac{72x^4}{6x^2} = 12x^{4-2}$$
$$= 12x^2$$

25. Simplify the expression.

$$(-2x)^3(x + 4)$$

Solution

$$(-2x)^3(x + 4) = -8x^3(x + 4)$$
$$= -8x^4 - 32x^3$$

27. Simplify the expression.

$$(5x + 3)(3x - 4)$$

Solution

$$(5x + 3)(3x - 4) = 5x(3x - 4) + 3(3x - 4) = 15x^2 - 20x + 9x - 12 = 15x^2 - 11x - 12$$

29. Simplify the expression.

$$(2x^2 - 3x + 2)(2x + 3)$$

Solution

$$(2x^2 - 3x + 2)(2x + 3) = 2x^2(2x + 3) - 3x(2x + 3) + 2(2x + 3)$$
$$= 4x^3 + 6x^2 - 6x^2 - 9x + 4x + 6$$
$$= 4x^3 + (6x^2 - 6x^2) + (-9x + 4x) + 6$$
$$= 4x^3 - 5x + 6$$

31. Simplify the expression.

$$2u(u - 7) - (u + 1)(u - 7)$$

Solution

$$2u(u - 7) - (u + 1)(u - 7) = 2u(u - 7) - u(u - 7) - 1(u - 7)$$
$$= 2u^2 - 14u - u^2 + 7u - u + 7$$
$$= (2u^2 - u^2) + (-14u + 7u - u) + 7$$
$$= u^2 - 8u + 7$$

33. Find the product.

$$(4x - 7)^2$$

Solution

$$(4x - 7)^2 = (4x)^2 - 2(4x)(7) + (-7)^2$$
$$= 16x^2 - 56x + 49$$

35. Find the product.

$$(5u - 8)(5u + 8)$$

Solution

$$(5u - 8)(5u + 8) = (5u)^2 - 8^2$$
$$= 25u^2 - 64$$

37. Find the product.

$$(2z + 3)(3z - 5)$$

Solution

$$(2z + 3)(3z - 5) = (2z)(3z) - (2z)(5) + (3)(3z) + (3)(-5) = 6z^2 - 10z + 9z - 15 = 6z^2 - z - 15$$

39. Find the product.

$$[(u - 3) + v][(u - 3) - v]$$

Solution

$$[(u - 3) + v][(u - 3) - v] = (u - 3)^2 - v^2 = u^2 - 2(u)(3) + (-3)^2 - v^2 = u^2 - 6u + 9 - v^2$$

41. Factor out the greatest common factor from $6x^2 + 15x^3$.

Solution

$6x^2 + 15x^3 = 3x^2(2 + 5x)$

43. Factor out the greatest common factor from $28(x + 5) - 70(x + 5)^2$.

Solution

$$28(x + 5) - 70(x + 5)^2 = 14(x + 5)[2 - 5(x + 5)]$$
$$= 14(x + 5)(2 - 5x - 25)$$
$$= 14(x + 5)(-5x - 23)$$
$$= -14(x + 5)(5x + 23)$$

45. Factor $v^3 - 2v^2 - v + 2$ by grouping.

Solution

$$v^3 - 2v^2 - v + 2 = v^2(v - 2) - 1(v - 2)$$
$$= (v - 2)(v^2 - 1)$$
$$= (v - 2)(v - 1)(v + 1)$$

47. Factor $t^3 + 3t^2 - 9t - 9$ by grouping.

Solution

$$t^3 + 3t^2 - 9t - 9 = t^2(t + 3) - 9(t + 1)$$

will not factor further

prime

49. Factor the difference of two squares.

$9a^2 - 100$

Solution

$9a^2 - 100 = (3a - 10)(3a + 10)$

51. Factor the difference of two squares.

$(u + 6)^2 - 81$

Solution

$$(u + 6)^2 - 81 = (u + 6 - 9)(u + 6 + 9)$$
$$= (u - 3)(u + 15)$$

53. Factor the difference of cubes.

$u^3 - 1$

Solution

$u^3 - 1 = (u - 1)(u^2 + u + 1)$

55. Factor the sum of cubes.

$8x^3 + 27$

Solution

$$8x^3 + 27 = (2x)^3 + (3)^3$$
$$= (2x + 3)(4x^2 - 6x + 9)$$

57. Factor the perfect square trinomial.

$x^2 - 40x + 400$

Solution

$x^2 - 40x + 400 = (x - 20)(x - 20)$ or $(x - 20)^2$

59. Factor the perfect square trinomial.

$4s^2 + 40s + 100$

Solution

$$4s^2 + 40s + 100 = 4(s^2 + 10s + 25)$$
$$= 4(s + 5)(s + 5) \text{ or } 4(s + 5)^2$$

61. Factor $x^2 + 2x - 35$.

Solution

$x^2 + 2x - 35 = (x + 7)(x - 5)$

63. Factor $18x^2 + 27x + 10$

Solution

$18x^2 + 27x + 10 = (3x + 2)(6x + 5)$

65. Factor $4a - 64a^3$ completely.

Solution

$$4a - 64a^3 = 4a(1 - 16a^2)$$
$$= 4a(1 - 4a)(1 + 4a)$$

67. Factor $8x(2x - 3) - 4(2x - 3)$ completely.

Solution

$$8x(2x - 3) - 4(2x - 3) = (2x - 3)(8x - 4)$$
$$= (2x - 3)4(2x - 1)$$

69. Factor $8x^3 + 1$ completely.

Solution

$8x^3 + 1 = (2x + 1)(4x^2 - 2x + 1)$

71. Factor $4u^2 - 28u + 49$ completely.

Solution

$4u^2 - 28u + 49 = (2u - 7)(2u - 7)$ or $(2u - 7)^2$

73. Factor $\frac{1}{4}x^2 + x - 1$ completely.

Solution

$\frac{1}{4}x^2 + x - 1$ is prime

75. Factor $4t^2 - t + 13$ completely.

Solution

$4t^2 - t + 13$ is prime

77. Use a graphing utility to graph $y_1 = 1 - (x - 2)^2$ and $y_2 = -(x - 1)(x - 3)$ on the same screen. What can you conclude?

Solution

Keystrokes:

y_1 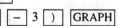 $\boxed{\text{Y}=}$ 1 $\boxed{-}$ $\boxed{(}$ $\boxed{X, T, \theta}$ $\boxed{-}$ 2 $\boxed{)}$ $\boxed{x^2}$ $\boxed{\text{ENTER}}$

y_2 $\boxed{\text{Y}=}$ $\boxed{(-)}$ $\boxed{(}$ $\boxed{X, T, \theta}$ $\boxed{-}$ 1 $\boxed{)}$ $\boxed{(}$ $\boxed{X, T, \theta}$ $\boxed{-}$ 3 $\boxed{)}$ $\boxed{\text{GRAPH}}$

$y_1 = y_2$

79. Find all values of b for which $x^2 + bx + 5$ is factorable.

Solution

$b = 6:$ $x^2 + 6x + 5 = (x + 5)(x + 1)$

$b = -6:$ $x^2 - 6x + 5 = (x - 5)(x - 1)$

81. Find two values of c for which $x^2 + 7x + c$ is factorable.

Solution

$c = 6:$ $x^2 + 7x + 6 \;\; = (x + 6)(x + 1)$

$c = 10:$ $x^2 + 7x + 10 = (x + 5)(x + 2)$

$c = 12:$ $x^2 + 7x + 12 = (x + 4)(x + 3)$

If c is a negative number, there are many possibilities such as the following.

$c = -8:$ $x^2 + 7x - 8 \;\; = (x + 8)(x - 1)$

$c = -18:$ $x^2 + 7x - 18 = (x + 9)(x - 2)$

83. Solve $10x(x - 3) = 0$.

Solution

$10x(x - 3) = 0$

$10x = 0 \quad x - 3 = 0$

$x = 0 \qquad x = 3$

85. Solve $v^2 - 100 = 0$.

Solution

$$v^2 - 100 = 0$$

$$(v - 10)(v + 10) = 0$$

$$v - 10 = 0 \qquad v + 10 = 0$$

$$v = 10 \qquad v = -10$$

87. Solve $x^2 - 25x = -150$.

Solution

$$x^2 - 25x = -150$$
$$x^2 - 25x + 150 = 0$$
$$(x - 15)(x - 10) = 0$$
$$x - 15 = 0 \qquad x - 10 = 0$$
$$x = 15 \qquad x = 10$$

89. Solve $3s^2 - 2s - 8 = 0$.

Solution

$$3s^2 - 2s - 8 = 0$$
$$(3s + 4)(s - 2) = 0$$
$$3s + 4 = 0 \qquad s - 2 = 0$$
$$s = -\frac{4}{3} \qquad s = 2$$

91. Solve $z(5 - z) + 36 = 0$.

Solution

$$z(5 - z) + 36 = 0$$
$$5z - z^2 + 36 = 0$$
$$z^2 - 5z - 36 = 0$$
$$(z - 9)(z + 4) = 0$$
$$z - 9 = 0 \qquad z + 4 = 0$$
$$z = 9 \qquad z = -4$$

93. Use a graphing utility to solve $y = x^2 - 10x + 21$. Explain your strategy.

Solution

Keystrokes:

$\boxed{Y=}$ $\boxed{X,T,\theta}$ $\boxed{x^2}$ $\boxed{-}$ 10 $\boxed{X,T,\theta}$ $\boxed{+}$ 21 $\boxed{GRAPH}$

The x-intercepts are 3 and 7, so the solutions are 3 and 7.

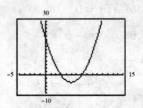

95. *Probability* The probability of three successes in five trials of an experiment is given by $10p^3(1 - p)^2$. Find this product.

Solution

$$10p^3(1 - p)^2 = 10p^3(1 - p)(1 - p)$$
$$= 10p^3(1 - 2p + p^2)$$
$$= 10p^3 - 20p^4 + 10p^5$$

97. *Geometry* The figure (in the textbook) shows a square with sides of length x, within which is a smaller square with sides of length y.

(a) Remove the small square from the larger square. What is the area of the remaining figure?

(b) After removing the small square, slide and rotate the remaining top rectangle so that it fits against the right side of the figure. What are the dimensions of the resulting rectangle and what special product formula have you demonstrated geometrically?

Solution

(a) $x^2 - y^2$

(b) Resulting rectangle $= (x + y)(x - y)$.
The special product formula $x^2 - y^2 = (x + y)(x - y)$ has been demonstrated geometrically.

99. Find two consecutive positive odd integers whose product is 195. Find the two integers.

Solution

Verbal Model: $\boxed{\text{First odd integer}} \cdot \boxed{\text{Second odd integer}} = \boxed{195}$

Equation:

$$(2n + 1)(2n + 3) = 195$$

$$4n^2 + 8n + 3 = 195$$

$$4n^2 + 8n - 192 = 0$$

$$4(n^2 + 2n - 48) = 0$$

$$(n + 8)(n - 6) = 0$$

$$n + 8 = 0 \qquad n - 6 = 0$$

$$n = -8 \qquad n = 6$$

$$\text{reject} \qquad 2n + 1 = 13$$

$$2n + 3 = 15$$

101. *Geometry* The perimeter of a rectangular storage lot at a car dealership is 800 feet. The lot is surrounded by fencing that costs $15 per foot for the front and $10 per foot for the remaining three sides. Find the dimensions of the parking lot if the total cost of the fencing is $9500.

Solution

Verbal Model: $\boxed{\begin{array}{c}\text{Cost of}\\\text{fencing}\\\text{2 widths}\end{array}} + \boxed{\begin{array}{c}\text{Cost of}\\\text{fencing}\\\text{front}\end{array}} + \boxed{\begin{array}{c}\text{Cost of}\\\text{fencing}\\\text{back}\end{array}} = \boxed{9500}$

Equation:

$$\$10(2x) + \$15(400 - x) + \$10(400 - x) = 9500$$

$$20x + 6000 - 15x + 4000 - 10x = 9500$$

$$-5x = -500$$

$$x = 100 \text{ ft}$$

$$400 - x = 300 \text{ ft}$$

Test for Chapter 3

1. Determine the degree and the leading coefficient of the polynomial $-5.2x^3 + 3x^2 - 8$.

Solution

$-5.2x^3 + 3x^2 - 8$

Degree $= 3$ Leading coefficient $= -5.2$

2. Explain why $\dfrac{4}{x^2 + 2}$ is not a polynomial.

Solution

$\dfrac{4}{x^2 + 2}$ is not a polynomial because the variable appears in the denominator.

3. (a) Add $(5a^2 - 3a + 4) + (a^2 - 4)$ and simplify.

Solution

(a) $(5a^2 - 3a + 4) + (a^2 - 4) = 6a^2 - 3a$

(b) Subtract $(16 - y^2) - (16 + 2y + y^2)$ and simplify.

(b) $(16 - y^2) - (16 + 2y + y^2)$

$\qquad = 16 - y^2 - 16 - 2y - y^2$

$\qquad = -2y^2 - 2y$

4. (a) Add $-2(2x^4 - 5) + 4x(x^3 + 2x - 1)$ and simplify.

Solution

(a) $-2(2x^4 - 5) + 4x(x^3 + 2x - 1)$

$\qquad = 4x^4 + 10 + 4x^4 + 8x^2 - 4x$

$\qquad = 8x^2 - 4x + 10$

(b) Subtract $4t - [3t - (10t + 7)]$ and simplify.

(b) $4t - [3t - (10t + 7)] = 4t - [3t - 10t - 7]$

$\qquad\qquad\qquad\qquad\quad = 4t - 3t + 10t + 7$

$\qquad\qquad\qquad\qquad\quad = 11t + 7$

5. (a) Multiply $(-2u^2v)^3(3v^2)$ and simplify.

Solution

(a) $(-2u^2v)^3(3v^2) = (-8u^6v^3)(3v^2)$

$\qquad\qquad\qquad = -24u^6v^5$

(b) Multiply $3(5x)(2xy)^2$ and simplify.

(b) $3(5x)(2xy)^2 = 3(5x)(4x^2y^2)$

$\qquad\qquad\quad = 60x^3y^2$

6. (a) Multiply $2y\left(\dfrac{y}{4}\right)^2$ and simplify.

Solution

(a) $2y\left(\dfrac{y}{4}\right)^2 = 2y\left(\dfrac{y^2}{16}\right)$

$\qquad\qquad = \dfrac{y^3}{8}$

(b) Divide $\dfrac{(-3x^2y)^4}{6x^2}$ and simplify.

(b) $\dfrac{(-3x^2y)^4}{6x^2} = \dfrac{81x^8y^4}{6x^2}$

$\qquad\qquad = \dfrac{27x^6y^4}{2}$

7. (a) Multiply $-3x(x - 4)$ and simplify.

Solution

(a) $-3x(x - 4) = -3x^2 + 12x$

(b) Multiply $(2x - 3y)(x + 5y)$ and simplify.

(b) $(2x - 3y)(x + 5y) = 2x^2 + 7xy - 15y^2$

8. (a) Multiply $(x - 1)[2x + (x - 3)]$ and simplify.

(b) Multiply $(2s - 3)(3s^2 - 4s + 7)$ and simplify.

Solution

(a) $(x - 1)[2x + (x - 3)] = (x - 1)(3x - 3) = 3x^2 - 6x + 3$

(b) $(2s - 3)(3s^2 - 4s + 7) = 6s^3 - 8s^2 + 14s - 9s^2 + 12s - 21$

$$= 6s^3 - 17s^2 + 26s - 21$$

9. (a) Multiply $(4x - 3)^2$ and simplify.

(b) Multiply $[4 - (a + b)][4 + (a + b)]$ and simplify.

Solution

(a) $(4x - 3)^2 = 16x^2 - 24x + 9$

(b) $[4 - (a + b)][4 + (a + b)] = 16 - (a + b)^2$

$$= 16 - (a^2 + 2ab + b^2)$$

$$= 16 - a^2 - 2ab - b^2$$

10. Factor $18y^2 - 12y$ completely.

Solution

$18y^2 - 12y = 6y(3y - 2)$

11. Factor $v^2 - \dfrac{16}{9}$ completely.

Solution

$$v^2 - \frac{16}{9} = \left(v - \frac{4}{3}\right)\left(v + \frac{4}{3}\right)$$

12. Factor $x^3 - 3x^2 - 4x + 12$ completely.

Solution

$$x^3 - 3x^2 - 4x + 12 = x^2(x - 3) - 4(x - 3)$$

$$= (x - 3)(x^2 - 4)$$

$$= (x - 3)(x - 2)(x + 2)$$

13. Factor $9u^2 - 6u + 1$ completely.

Solution

$9u^2 - 6u + 1 = (3u - 1)(3u - 1)$ or $(3u - 1)^2$

14. Factor $6x^2 - 26x - 20$ completely.

Solution

$$6x^2 - 26x - 20 = 2(3x^2 - 13x - 10)$$

$$= 2(2x + 2)(x - 5)$$

15. Factor $x^3 + 27$ completely.

Solution

$$x^3 + 27 = (x + 3)\left(x^2 - 3x + 9\right)$$

16. Solve $(y + 2)^2 - 9 = 0$.

Solution

$$(y + 2)^2 - 9 = 0$$

$$[(y + 2) - 3][(y + 2) + 3] = 0$$

$$y - 1 = 0 \qquad y + 5 = 0$$

$$y = 1 \qquad y = -5$$

17. Solve $12 + 5y - 3y^2 = 0$.

Solution

$$12 + 5y - 3y^2 = 0$$

$$(3 - y)(4 + 3y) = 0$$

$$3 - y = 0 \qquad 4 + 3y = 0$$

$$3 = y \qquad -\frac{4}{3} = y$$

18. Find the area of the shaded region shown in the figure in the textbook.

Solution

$$\text{Area} = 2x(x + 15) - x(x + 4)$$
$$\text{Shaded region} = 2x^2 + 30x - x^2 - 4x$$
$$= x^2 + 26x$$

19. The length of a rectangle is $1\frac{1}{2}$ times its width. Find the dimensions of the rectangle if its area is 54 square centimeters.

Solution

Verbal Model: $\boxed{\text{Area rectangle}} = \boxed{\text{Length}} \cdot \boxed{\text{Width}}$

Equation: $54 = \dfrac{3}{2}w \cdot w$

$$108 = 3w^2$$
$$36 = w^2$$
$$6 = \text{width}$$
$$9 = \text{length}$$

20. The height of a free-falling object is given by $h = -16t^2 - 40t + 144$, where h is measured in feet and t is measured in seconds. The object is projected downward when $t = 0$. How long does it take to hit the ground?

Solution

$$0 = -16t^2 - 40t + 144$$
$$0 = 2t^2 + 5t - 18$$
$$0 = (2t + 9)(t - 2)$$

$$2t + 9 = 0 \qquad t - 2 = 0$$

$$t = -\frac{9}{2} \qquad t = 2 \text{ sec}$$

reject

Cumulative Test for Chapters P–3

1. Evaluate $-\dfrac{8}{45} \div \dfrac{12}{25}$.

Solution

$$-\frac{8}{45} \div \frac{12}{25} = -\frac{8}{45} \cdot \frac{25}{12}$$
$$= -\frac{10}{27}$$

2. Write an algebraic expression for the statement, "The number n is tripled and the product is decreased by 8."

Solution

$3n - 8$.

3. (a) Multiply $(2a^2b)^3(-ab^2)^2$ and simplify.

Solution

(a) $(2a^2b)^3(-ab^2)^2 = (8a^6b^3)(a^2b^4)$
$$= 8a^8b^7$$

(b) Subtract $3x(x^2 - 2) - x(x^2 + 5)$ and simplify.

(b) $3x(x^2 - 2) - x(x^2 + 5) = 3x^3 - 6x - x^3 - 5x$
$$= 2x^3 - 11x$$

4. (a) Multiply $t(3t - 1) - 2t(t + 4)$ and simplify.

Solution

(a) $t(3t - 1) - 2t(t + 4) = 3t^2 - t - 2t^2 - 8t$
$$= t^2 - 9t$$

(b) Expand $[2 + (x - y)]^2$ and simplify.

(b) $[2 + (x - y)]^2 = 4 + 4(x - y) + (x - y)^2$
$$= 4 + 4x - 4y + x^2 - 2xy + y^2$$

5. (a) Solve $12 - 5(3 - x) = x + 3$.

Solution

(a) $12 - 5(3 - x) = x + 3$
$12 - 15 + 5x = x + 3$
$-3 + 5x = x + 3$
$-3 + 5x - x = x + 3 - x$
$3 - 3 + 4x = 3 + 3$
$4x = 6$
$\dfrac{4x}{4} = \dfrac{6}{4}$
$x = \dfrac{3}{2}$

(b) Solve $1 - \dfrac{x + 2}{4} = \dfrac{7}{8}$.

(b) $\qquad 1 - \dfrac{x + 2}{4} = \dfrac{7}{8}$
$8\left[1 - \dfrac{x + 2}{4}\right] = \left[\dfrac{7}{8}\right]8$
$8 - 2(x + 2) = 7$
$8 - 2x - 4 = 7$
$4 - 2x = 7$
$4 - 4 - 2x = 7 - 4$
$-2x = 3$
$\dfrac{-2x}{-2} = -\dfrac{3}{2}$
$x = -\dfrac{3}{2}$

6. (a) Solve $y^2 - 64 = 0$.

Solution

(a) $\qquad y^2 - 64 = 0$

$\qquad (y - 8)(y + 8) = 0$

$\qquad y - 8 = 0 \quad y + 8 = 0$

$\qquad\qquad y = 8 \qquad\quad y = -8$

(b) Solve $2t^2 - 5t - 3 = 0$.

(b) $\qquad 2t^2 - 5t - 3 = 0$

$\qquad (2t + 1)(t - 3) = 0$

$\qquad 2t + 1 = 0 \qquad t - 3 = 0$

$\qquad\qquad t = -\dfrac{1}{2} \qquad\quad t = 3$

7. Your annual automobile premium is $1225. Because of a driving violation, your premium is increased 15%. What is your new premium?

Solution

Verbal Model: $\boxed{\text{Total annual premium}} = \boxed{\text{Annual premium}} + \boxed{\text{Surcharge}}$

Equation: $\qquad x = \$1225 + 0.15(1225)$

$\qquad\qquad x = 1225 + 183.75$

$\qquad\qquad x = \$1408.75$

8. The triangles in the textbook are similar. Solve for x by using the fact that corresponding sides of similar triangles are proportional.

Solution

$\dfrac{9}{4.5} = \dfrac{13}{x}$

$9x = 13(4.5)$

$x = \dfrac{13(4.5)}{9}$

$x = 6.5$

9. Solve $|x - 2| \geq 3$ and sketch its solution.

Solution

$x - 2 \leq -3 \quad$ or $\quad x - 2 \geq 3$

$\qquad x \leq -1 \qquad\qquad\quad x \geq 5$

10. The revenue from selling x units of a product is $R = 12.90x$. The cost of producing x units is $C = 8.50x + 450$. To obtain a profit, the revenue must be greater than the cost. For what values of x will this product produce a profit? Explain your reasoning.

Solution

Verbal Model: $\boxed{\text{Revenue}} > \boxed{\text{Cost}}$

Equation: $\qquad 12.90x > 8.50x + 450$

$\qquad\qquad 4.4x > 450$

$\qquad\qquad x > 102.27273$

$\qquad\qquad x > 103$

11. Determine whether the equation $x - y^3 = 0$ represents y as a function of x.

Solution

$x - y^3 = 0$ does represent y as a function of x.

12. Find the domain of the function $f(x) = \sqrt{x - 2}$.

Solution

$f(x) = \sqrt{x - 2} \quad D = x \geq 2 \quad x - 2 \geq 0$

$\qquad\qquad\qquad\qquad\qquad\qquad x \geq 2$

13. Given $f(x) = x^2 - 3x$, find (a) $f(4)$ and (b) $f(c + 3)$.

Solution

(a) $f(4) = 4^2 - 3(4) = 16 - 12 = 4$

(b) $f(c + 3) = (c + 3)^2 - 3(c + 3)$

$\qquad = c^2 + 6c + 9 - 3c - 9$

$\qquad = c^2 + 3c$

14. Find the slope of the line passing through $(-4, 0)$ and $(4, 6)$. Then, find the distance between the points.

Solution

$m = \dfrac{6 - 0}{4 + 4} = \dfrac{6}{8} = \dfrac{3}{4}$

$d = \sqrt{(-4 - 4)^2 + (0 - 6)^2}$

$\quad = \sqrt{64 + 36}$

$\quad = \sqrt{100}$

$\quad = 10$

15. Factor $y^3 - 3y^2 - 9y + 27$ by grouping.

Solution

$y^3 - 3y^2 - 9y + 27 = y^2(y - 3) - 9(y - 3)$

$\qquad = (y - 3)(y^2 - 9)$

$\qquad = (y - 3)(y - 3)(y + 3)$

16. Factor $3x^2 - 8x - 35$.

Solution

$3x^2 - 8x - 35 = (3x + 7)(x - 5)$

17. Graph $4x + 3y - 12 = 0$.

Solution

$4x + 3y - 12 = 0$

$4(0) + 3y - 12 = 0 \qquad\qquad 4x + 3(0) - 12 = 0$

$\qquad 3y = 12 \qquad\qquad\qquad\qquad 4x = 12$

$\qquad\quad y = 4 \quad (0, 4) \qquad\qquad\qquad x = 3 \quad (3, 0)$

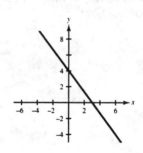

18. Graph $y = 1 - (x - 2)^2$.

Solution

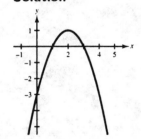

19. Graph $y = |x + 3|$.

Solution

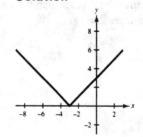

20. Graph $y = x^3 - 1$.

Solution

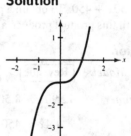

CHAPTER 4
Rational Expressions and Equations

4.1   Simplifying Rational Expressions

7. Find the domain of $\dfrac{5}{x-8}$.

Solution

$x - 8 \neq 0$

$x \neq 8$

$D = (-\infty, 8) \cup (8, \infty)$

9. Find the domain of $\dfrac{x}{x^2 + 4}$.

Solution

$x^2 + 4 \neq 0$

$D = (-\infty, \infty)$

11. Find the domain of $\dfrac{y+5}{y^2 - 3y}$.

Solution

$y^2 - 3y \neq 0$

$y(y - 3) \neq 0$

$y \neq 0 \quad y \neq 3$

$D = (-\infty, 0) \cup (0, 3) \cup (3, \infty)$

13. Evaluate $f(x) = \dfrac{4x}{x+3}$ as indicated. If it is not possible, state the reason.

Solution

(a) $f(1) = \dfrac{4(1)}{1+3} = \dfrac{4}{4} = 1$

(b) $f(-2) = \dfrac{4(-2)}{-2+3} = \dfrac{-8}{1} = -8$

(c) $f(-3) = \dfrac{4(-3)}{-3+3}$

$= \dfrac{-12}{0}$ not possible; undefined

(d) $f(0) = \dfrac{4(0)}{0+3} = \dfrac{0}{3} = 0$

15. Evaluate $h(s) = \dfrac{s^2}{s^2 - s - 2}$ as indicated. If it is not possible, state the reason.

Solution

(a) $h(10) = \dfrac{10^2}{10^2 - 10 - 2} = \dfrac{100}{88} = \dfrac{25}{22}$

(b) $h(0) = \dfrac{0^2}{0^2 - 0 - 2} = \dfrac{0}{-2} = 0$

(c) $h(-1) = \dfrac{(-1)^2}{(-1)^2 - (-1) - 2} = \dfrac{1}{1 + 1 - 2}$

$= \dfrac{1}{0} = $ not possible; undefined

(d) $h(2) = \dfrac{2^2}{2^2 - 2 - 2} = \dfrac{4}{4 - 2 - 2}$

$= \dfrac{4}{0} = $ not possible; undefined

17. *Inventory Cost*   The inventory cost I when x units of a product are ordered from a supplier is given by $I = (0.25x + 2000)/x$. Describe the domain.

Solution

$x =$ units of a product

$D = \{1, 2, 3, 4, \ldots\}$

19. Complete the statement.

$$\frac{5}{6} = \frac{5(\quad)}{6(x + 3)}, \quad x \neq -3$$

Solution

$$\frac{5}{6} = \frac{5(x + 3)}{6(x + 3)}, \quad x \neq -3$$

21. Complete the statement.

$$\frac{8x}{x - 5} = \frac{8x(\quad)}{x^2 - 3x - 10}, \quad x \neq -2$$

Solution

$$\frac{8x}{x - 5} = \frac{8x(x + 2)}{x^2 - 3x - 10}, \quad x \neq -2$$

23. Simplify $\dfrac{5x}{25}$.

Solution

$$\frac{5x}{25} = \frac{5x}{5 \cdot 5}$$

$$= \frac{x}{5}$$

25. Simplify $\dfrac{18x^2 y}{15xy^4}$.

Solution

$$\frac{18x^2 y}{15xy^4} = \frac{3 \cdot 6 \cdot x \cdot x \cdot y}{3 \cdot 5 \cdot x \cdot y \cdot y^3}$$

$$= \frac{6x}{5y^3}$$

27. Simplify $\dfrac{3xy^2}{xy^2 + x}$.

Solution

$$\frac{3xy^2}{xy^2 + x} = \frac{3xy^2}{x(y^2 + 1)}$$

$$= \frac{3y^2}{y^2 + 1}$$

29. Simplify $\dfrac{y^2 - 64}{5(3y + 24)}$.

Solution

$$\frac{y^2 - 64}{5(3y + 24)} = \frac{(y - 8)(y + 8)}{15(y + 8)}$$

$$= \frac{y - 8}{15}$$

31. Simplify $\dfrac{3 - x}{2x^2 - 3x - 9}$.

Solution

$$\frac{3 - x}{2x^2 - 3x - 9} = \frac{-1(x - 3)}{(2x + 3)(x - 3)}$$

$$= -\frac{1}{2x + 3}$$

33. Simplify $\dfrac{3m^2 - 12n^2}{m^2 + 4mn + 4n^2}$.

Solution

$$\frac{3m^2 - 12n^2}{m^2 + 4mn + 4n^2} = \frac{3(m^2 - 4n^2)}{(m + 2n)(m + 2n)} = \frac{3(m - 2n)(m + 2n)}{(m + 2n)(m + 2n)} = \frac{3(m - 2n)}{m + 2n}$$

35. *Using a Table* Complete the table (see textbook). What can you conclude?

Solution

x	−2	−1	0	1	2	3	4
$\dfrac{x^2 - x - 2}{x - 2}$	−1	0	1	2	Undefined	4	5
$x + 1$	−1	0	1	2	3	4	5

Domain of $\dfrac{x^2 - x - 2}{x - 2}$ is $(-\infty, 2) \cup (2, \infty)$.

Domain of $x + 1$ is $(-\infty, \infty)$.

The two expressions are equal for all replacements of the variable x except 2.

37. *Geometry* Find the ratio of the area of the shaded portion to the total area of the figure (see textbook).

Solution

$$\frac{\text{Area of shaded portion}}{\text{Area of total figure}} = \frac{x \cdot \dfrac{x}{3}}{x(x + 3)} = \frac{\dfrac{x}{3}}{x + 3} = \frac{x}{3(x + 3)}$$

39. *Creating and Using a Model* Use the following polynomial models, which give the projected cost of Medicare and the U.S. population age 65 and older, for the years 1990 and 1995 (see figure in textbook).

$$C = 107.1 + 12.64t + 0.54t^2 \qquad \text{Cost of Medicare}$$

$$P = 31.6 + 0.51t - 0.14t^2 \qquad \text{Population}$$

In these models, C represents the total annual cost of Medicare (in billions of dollars), P represents the U.S. population (in millions) age 65 and older, and t represents the year, with $t = 0$ corresponding to 1990. (*Sources:* Congressional Budget Office and U.S. Bureau of Census.)

Find a rational model that represents the average cost of Medicare *per person age 65 or older* during the years 1990 to 1995.

Solution

$$\text{Average cost of Medicare per person aged 65 and older} = \frac{107.1 + 12.64t + 0.54t^2}{31.6 + 0.51t - 0.14t^2}$$

41. Simplify $-6x(10 - 7x)$.

Solution

$$-6x(10 - 7x) = -60x + 42x^2$$

43. Simplify $(11 - x)(11 + x)$.

Solution

$$(11 - x)(11 + x) = 121 - x^2$$

45. Write $\left(\dfrac{y}{3}\right)^4$ as a repeated multiplication.

Solution

$$\left(\frac{y}{3}\right)^4 = \left(\frac{y}{3}\right)\left(\frac{y}{3}\right)\left(\frac{y}{3}\right)\left(\frac{y}{3}\right)$$

$$= \frac{y^4}{81}$$

47. Find the domain of $\dfrac{7x}{x+4}$.

Solution

$$x + 4 \neq 0$$

$$x \neq -4$$

$$D = (-\infty, -4) \cup (-4, \infty)$$

49. Find the domain of $\dfrac{4}{x^2 + 9}$.

Solution

$$x^2 + 9 \neq 0$$

$$D = (-\infty, \infty)$$

51. Find the domain of $\dfrac{5t}{t^2 - 16}$.

Solution

$$t^2 - 16 \neq 0$$

$$(t - 4)(t + 4) \neq 0$$

$$t \neq 4 \quad t \neq -4$$

$$D = (-\infty, -4) \cup (-4, 4) \cup (4, \infty)$$

53. Find the domain of $\dfrac{u^2}{u^2 - 2u - 5}$.

Solution

$$u^2 - 2u - 5 \neq 0$$

$$D = (-\infty, \infty)$$

55. *Geometry* A rectangle of length x inches has an area of 500 square inches. The perimeter of the rectangle is given by $P = 2\left(x + \dfrac{500}{x}\right)$. Describe the domain.

Solution

Since length must be positive, $x \geq 0$. Since $\dfrac{500}{x}$ must be defined, $x \neq 0$. Therefore, the domain is $x > 0$ or $(0, \infty)$.

57. Complete the statement.

$$\frac{x}{2} = \frac{3x(x + 16)^2}{2(\qquad)}, \quad x \neq -16.$$

Solution

$$\frac{x}{2} = \frac{3x(x + 16)^2}{2(3(x + 16)^2)}, \quad x \neq -16.$$

59. Complete the statement.

$$\frac{x + 5}{3x} = \frac{(x + 5)(\qquad)}{3x^2(x - 2)}, \quad x \neq 2$$

Solution

$$\frac{x + 5}{3x} = \frac{(x + 5)(x(x - 2))}{3x^2(x - 2)}, \quad x \neq 2$$

61. Simplify $\dfrac{12y^2}{2y}$.

Solution

$$\frac{12y^2}{2y} = \frac{2 \cdot 6 \cdot y \cdot y}{2 \cdot y}$$

$$= 6y$$

63. Simplify $\dfrac{x^2(x - 8)}{x(x - 8)}$.

Solution

$$\frac{x^2(x - 8)}{x(x - 8)} = \frac{x \cdot x(x - 8)}{x(x - 8)}$$

$$= x$$

65. Simplify $\dfrac{2x-3}{4x-6}$.

Solution

$$\dfrac{2x-3}{4x-6} = \dfrac{2x-3}{2(2x-3)}$$

$$= \dfrac{1}{2}$$

67. Simplify $\dfrac{5-x}{3x-15}$.

Solution

$$\dfrac{5-x}{3x-15} = \dfrac{-1(x-5)}{3(x-5)}$$

$$= -\dfrac{1}{3}$$

69. Simplify $\dfrac{a+3}{a^2+6a+9}$.

Solution

$$\dfrac{a+3}{a^2+6a+9} = \dfrac{a+3}{(a+3)(a+3)}$$

$$= \dfrac{1}{a+3}$$

71. Simplify $\dfrac{x^2-7x}{x^2-14x+49}$.

Solution

$$\dfrac{x^2-7x}{x^2-14x+49} = \dfrac{x(x-7)}{(x-7)(x-7)}$$

$$= \dfrac{x}{x-7}$$

73. Simplify $\dfrac{y^3-4y}{y^2+4y-12}$.

Solution

$$\dfrac{y^3-4y}{y^2+4y-12} = \dfrac{y(y^2-4)}{(y+6)(y-2)}$$

$$= \dfrac{y(y-2)(y+2)}{(y+6)(y-2)}$$

$$= \dfrac{y(y+2)}{y+6}$$

75. Simplify $\dfrac{15x^2+7x-4}{15x^2+x-2}$.

Solution

$$\dfrac{15x^2+7x-4}{15x^2+x-2} = \dfrac{(5x+4)(3x-1)}{(5x+2)(3x-1)}$$

$$= \dfrac{5x+4}{5x+2}$$

77. Simplify $\dfrac{5xy+3x^2y^2}{xy^3}$.

Solution

$$\dfrac{5xy+3x^2y^2}{xy^3} = \dfrac{xy(5+3xy)}{xy \cdot y^2}$$

$$= \dfrac{5+3xy}{y^2}$$

79. Simplify $\dfrac{u^2-4v^2}{u^2+uv-2v^2}$.

Solution

$$\dfrac{u^2-4v^2}{u^2+uv-2v^2} = \dfrac{(u-2v)(u+2v)}{(u-v)(u+2v)}$$

$$= \dfrac{u-2v}{u-v}$$

81. *Think About It* Explain how you can show that the two expressions are *not* equivalent.

$$\dfrac{x-4}{4} \neq x-1$$

Solution

$$\dfrac{x-4}{4} \neq x-1 \qquad \text{Choose a value such as 10 for}$$
$$x \text{ and evaluate both sides.}$$

$$\dfrac{10-4}{4} \neq 10-1$$

$$\dfrac{6}{4} \neq 9$$

83. Write $\dfrac{x^{2n}-4}{x^n+2}$ in reduced form. (Assume n is a positive integer.)

Solution

$$\dfrac{x^{2n}-4}{x^n+2} = \dfrac{(x^n+2)(x^n-2)}{x^n+2}$$

$$= x^n-2$$

85. *Average Cost* A machine shop has a setup cost of $2500 for the production of a new product. The cost of labor and material used to produce each unit is $9.25.

(a) Write a rational expression that gives the average cost per unit when x units are produced.

(b) Determine the domain of the expression in part (a).

(c) Find the average cost per unit when $x = 100$ units are produced.

Solution

(a) $\dfrac{2500 + 9.25x}{x}$ = average cost per unit

(b) Domain = $\{1, 2, 3, 4, \ldots\}$

(c) $\dfrac{2500 + 9.25(100)}{100}$ = $34.25

87. *Geometry* One swimming pool is circular and another is rectangular. The rectangular pool's width is three times its depth, and its length is 6 feet more than its width. The circular pool has a diameter that is twice the width of the rectangular pool, and it is 2 feet deeper. Find the ratio of the volume of the circular pool to the volume of the rectangular pool.

Solution

$$\frac{\text{Circular pool volume}}{\text{Rectangular pool volume}} = \frac{\pi(3d)^2(d+2)}{d(3d)(3d+6)} = \frac{\pi(3d)^2(d+2)}{3d^2 \cdot 3(d+2)} = \frac{\pi(3d)^2(d+2)}{(3d)^2(d+2)} = \pi$$

4.2 Multiplying and Dividing Rational Expressions

5. Evaluate $\dfrac{x - 10}{4x}$ for each value of x. If it is not possible, state the reason.

Solution

(a) $x = 10$ $\quad \dfrac{10 - 10}{4(10)} = \dfrac{0}{40} = 0$

(b) $x = 0$ $\quad \dfrac{0 - 10}{4(0)} = \dfrac{-10}{0} =$ undefined

(c) $x = -2$ $\quad \dfrac{-2 - 10}{4(-2)} = \dfrac{-12}{-8} = \dfrac{3}{2}$

(d) $x = 12$ $\quad \dfrac{12 - 10}{4(12)} = \dfrac{2}{48} = \dfrac{1}{24}$

7. Complete the statement.

$$\frac{7}{3y} = \frac{7x^2}{3y(\quad)}, \quad x \neq 0$$

Solution

$$\frac{7}{3y} = \frac{7x^2}{3y(x^2)}, \quad x \neq 0$$

9. Complete the statement.

$$\frac{13x}{x - 2} = \frac{13x(\qquad)}{4 - x^2}, \quad x \neq -2$$

Solution

$$\frac{13x}{x - 2} = \frac{13x((-1)(2 + x))}{4 - x^2}, \quad x \neq -2$$

11. Multiply and simplify.

$$\frac{7x^2}{3} \cdot \frac{9}{14x}$$

Solution

$$\frac{7x^2}{3} \cdot \frac{9}{14x} = \frac{7x \cdot x \cdot 3 \cdot 3}{3 \cdot 7 \cdot 2 \cdot x}$$

$$= \frac{3x}{2}$$

13. Multiply and simplify.

$$\frac{8}{3 + 4x} \cdot (9 + 12x)$$

Solution

$$\frac{8}{3 + 4x} \cdot (9 + 12x) = \frac{8 \cdot 3(3 + 4x)}{3 + 4x}$$

$$= 24$$

15. Multiply and simplify.

$$\frac{(2x-3)(x+8)}{x^3} \cdot \frac{x}{3-2x}$$

Solution

$$\frac{(2x-3)(x+8)}{x^3} \cdot \frac{x}{3-2x} = \frac{(2x-3)(x+8)x}{x \cdot x^2 \cdot -1(2x-3)} = \frac{x+8}{-x^2}$$

17. Multiply and simplify.

$$\frac{xu-yu+xv-yv}{xu+yu-xv-yv} \cdot \frac{xu+yu+xv+yv}{xu-yu-xv+yv}$$

Solution

$$\frac{xu-yu+xv-yv}{xu+yu-xv-yv} \cdot \frac{xu+yu+xv+yv}{xu-yu-xv+yv} = \frac{u(x-y)+v(x-y)}{u(x+y)-v(x+y)} \cdot \frac{u(x+y)+v(x+y)}{u(x-y)-v(x-y)}$$

$$= \frac{(x-y)(u+v)}{(x+y)(u-v)} \cdot \frac{(x+y)(u+v)}{(x-y)(u-v)}$$

$$= \frac{(x-y)(u+v)(x+y)(u+v)}{(x+y)(u-v)(x-y)(u-v)}$$

$$= \frac{(u+v)(u+v)}{(u-v)(u-v)} \quad \text{or} \quad \frac{(u+v)^2}{(u-v)^2}$$

19. Divide and simplify.

$$\frac{3x}{4} \div \frac{x^2}{2}$$

Solution

$$\frac{3x}{4} \div \frac{x^2}{2} = \frac{3x}{4} \cdot \frac{2}{x^2}$$

$$= \frac{3}{2x}$$

21. Divide and simplify.

$$\frac{3(a+b)}{4} \div \frac{(a+b)^2}{2}$$

Solution

$$\frac{3(a+b)}{4} \div \frac{(a+b)^2}{2} = \frac{3(a+b)}{4} \cdot \frac{2}{(a+b)^2}$$

$$= \frac{3(a+b) \cdot 2}{2 \cdot 2 \cdot (a+b)(a+b)}$$

$$= \frac{3}{2(a+b)}$$

23. Divide and simplify.

$$\frac{16x^2+8x+1}{3x^2+8x-3} \div \frac{4x^2-3x-1}{x^2+6x+9}$$

Solution

$$\frac{16x^2+8x+1}{3x^2+8x-3} \div \frac{4x^2-3x-1}{x^2+6x+9} = \frac{16x^2+8x+1}{3x^2+8x-3} \cdot \frac{x^2+6x+9}{4x^2-3x-1} = \frac{(4x+1)(4x+1)}{(3x-1)(x+3)} \cdot \frac{(x+3)(x+3)}{(4x+1)(x-1)}$$

$$= \frac{(4x+1)(4x+1)(x+3)(x+3)}{(3x-1)(x+3)(4x+1)(x-1)} = \frac{(4x+1)(x+3)}{(3x-1)(x-1)}$$

25. Divide and simplify.

$$\frac{\left(\dfrac{x^2}{12}\right)}{\left(\dfrac{5x}{18}\right)}$$

Solution

$$\frac{\left(\dfrac{x^2}{12}\right)}{\left(\dfrac{5x}{18}\right)} = \frac{x^2}{12} \div \frac{5x}{18} = \frac{x^2}{12} \cdot \frac{18}{5x} = \frac{x^2 \cdot 3 \cdot 3 \cdot 2}{2 \cdot 2 \cdot 3 \cdot 5 \cdot x} = \frac{3x}{10}$$

27. Perform the operations and simplify.

$$\left[\frac{x^2}{9} \cdot \frac{3(x+4)}{x^2+2x}\right] \div \frac{x}{x+2}$$

Solution

$$\left[\frac{x^2}{9} \cdot \frac{3(x+4)}{x^2+2x}\right] \div \frac{x}{x+2} = \frac{x^2}{9} \cdot \frac{3(x+4)}{x(x+2)} \cdot \frac{x+2}{x} = \frac{x+4}{3}$$

29. Perform the operations and simplify.

$$\left[\frac{xy+y}{4x} \div (3x+3)\right] \div \frac{y}{3x}$$

Solution

$$\left[\frac{xy+y}{4x} \div (3x+3)\right] \div \frac{y}{3x} = \frac{y(x+1)}{4x} \cdot \frac{1}{3(x+1)} \cdot \frac{3x}{y} = \frac{1}{4}$$

31. *Graphical Reasoning*   Use a graphing utility to graph the two equations in the same viewing rectangle. Use the graphs to verify that the expressions are equivalent.

$$y_1 = \frac{3x+2}{x} \cdot \frac{x^2}{9x^2-4}, \quad y_2 = \frac{x}{3x-2}$$

Solution

Keystrokes:

y_1 [Y=] [(] [(] [3] [X,T,θ] [+] [2] [)] [÷] [X,T,θ] [)] [×]

[(] [X,T,θ] [x²] [÷] [(] [9] [X,T,θ] [x²] [−] [4] [)] [)] [ENTER]

y_2 [Y=] [X,T,θ] [÷] [(] [3] [X,T,θ] [−] [2] [)] [GRAPH]

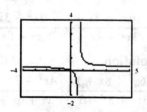

33. *Graphical Reasoning* Use a graphing utility to graph the two equations in the same viewing rectangle. Use the graphs to verify that the expressions are equivalent.

$$y_1 = \frac{3x + 15}{x^4} \div \frac{x + 5}{x^2}, \quad y_2 = \frac{3}{x^2}$$

Solution

Keystrokes:

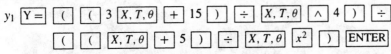

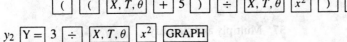

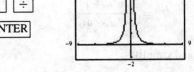

35. *Geometry* Write an expression for the area of the shaded region (see figure in textbook). Then simplify the expression.

Solution

$$\text{Area} = \left(\frac{2w + 3}{3}\right)\left(\frac{w}{2}\right) = \frac{(2w + 3)w}{6}$$

37. Factor $3x^2 - 21x$ completely.

Solution

$$3x^2 - 21x = 3x(x - 7)$$

39. Factor $4t^2 - 169$ completely.

Solution

$$4t^2 - 169 = (2t - 13)(2t + 13)$$

41. *Simple Interest* You borrow \$12,000 for 6 months. At simple interest of 12%, how much will you owe?

Solution

Verbal Model: $\boxed{\text{Interest}} = \boxed{\text{Principal}} \cdot \boxed{\text{Rate}} \cdot \boxed{\text{Time}}$

Equation: $\quad i = \$12,000 \cdot 0.12 \cdot \frac{1}{2}$

$$i = \$720$$

43. Complete the statement.

$$\frac{3x}{x - 4} = \frac{3x(x + 2)^2}{(x - 4)(\quad\quad)}, \quad x \neq -2$$

Solution

$$\frac{3x}{x - 4} = \frac{3x(x + 2)^2}{(x - 4)(x + 2)^2}, \quad x \neq -2$$

45. Complete the statement.

$$\frac{3u}{7v} = \frac{3u(\quad\quad)}{7v(u + 1)}, \quad u \neq -1$$

Solution

$$\frac{3u}{7v} = \frac{3u(u + 1)}{7v(u + 1)}, \quad u \neq -1$$

47. Multiply and simplify.

$$\frac{8s^3}{9s} \cdot \frac{6s^2}{32s}$$

Solution

$$\frac{8s^3}{9s} \cdot \frac{6s^2}{32s} = \frac{8s^3 \cdot 3 \cdot 2s \cdot s}{3 \cdot 3 \cdot s \cdot 8 \cdot 2 \cdot 2 \cdot s}$$

$$= \frac{s^3}{6}$$

49. Multiply and simplify.

$$16u^4 \cdot \frac{12}{8u^2}$$

Solution

$$16u^4 \cdot \frac{12}{8u^2} = \frac{8 \cdot 2 \cdot u^2 \cdot u^2 \cdot 12}{8 \cdot u^2}$$

$$= 24u^2$$

51. Multiply and simplify.

$$\frac{8u^2v}{3u+v} \cdot \frac{u+v}{12u}$$

Solution

$$\frac{8u^2v}{3u+v} \cdot \frac{u+v}{12u} = \frac{4 \cdot 2 \cdot u \cdot u \cdot v(u+v)}{(3u+v) \cdot 4 \cdot 3 \cdot u}$$

$$= \frac{2uv(u+v)}{3(3u+v)}$$

55. Multiply and simplify.

$$\frac{6r}{r-2} \cdot \frac{r^2-4}{33r^2}$$

Solution

$$\frac{6r}{r-2} \cdot \frac{r^2-4}{33r^2} = \frac{3 \cdot 2 \cdot r(r-2)(r+2)}{(r-2) \cdot 3 \cdot 11 \cdot r \cdot r}$$

$$= \frac{2(r+2)}{11r}$$

59. Multiply and simplify.

$$\frac{2t^2-t-15}{t+2} \cdot \frac{t^2-t-6}{t^2-6t+9}$$

Solution

$$\frac{2t^2-t-15}{t+2} \cdot \frac{t^2-t-6}{t^2-6t+9} = \frac{(2t+5)(t-3)(t-3)(t+2)}{(t+2)(t-3)(t-3)} = 2t+5$$

61. Multiply and simplify.

$$\frac{x+5}{x-5} \cdot \frac{2x^2-9x-5}{3x^2+x-2} \cdot \frac{x^2-1}{x^2+7x+10}$$

Solution

$$\frac{x+5}{x-5} \cdot \frac{2x^2-9x-5}{3x^2+x-2} \cdot \frac{x^2-1}{x^2+7x+10} = \frac{x+5}{x-5} \cdot \frac{(2x+1)(x-5)}{(3x-2)(x+1)} \cdot \frac{(x-1)(x+1)}{(x+5)(x+2)}$$

$$= \frac{(x+5)(2x+1)(x-5)(x-1)(x+1)}{(x-5)(3x-2)(x+1)(x+5)(x+2)}$$

$$= \frac{(2x+1)(x-1)}{(3x-2)(x+2)}$$

53. Multiply and simplify.

$$\frac{12-r}{3} \cdot \frac{3}{r-12}$$

Solution

$$\frac{12-r}{3} \cdot \frac{3}{r-12} = \frac{-1(r-12) \cdot 3}{3(r-12)} = -1$$

57. Multiply and simplify.

$$(u-2v)^2 \cdot \frac{u+2v}{u-2v}$$

Solution

$$(u-2v)^2 \cdot \frac{u+2v}{u-2v} = \frac{(u-2v)(u-2v)(u+2v)}{u-2v}$$

$$= (u-2v)(u+2v)$$

63. Divide and simplify.

$$\frac{7xy^2}{10u^2v} \div \frac{21x^3}{45uv}$$

Solution

$$\frac{7xy^2}{10u^2v} \div \frac{21x^3}{45uv} = \frac{7xy^2}{10u^2v} \cdot \frac{45uv}{21x^3}$$

$$= \frac{7xy^2 \cdot 3 \cdot 3 \cdot 5 \cdot u \cdot v}{5 \cdot 2 \cdot u \cdot u \cdot v \cdot 7 \cdot 3x \cdot x^2}$$

$$= \frac{3y^2}{2ux^2}$$

65. Divide and simplify.

$$\frac{(x^3y)^2}{(x+2y)^2} \div \frac{x^2y}{(x+2y)^3}$$

Solution

$$\frac{(x^3y)^2}{(x+2y)^2} \div \frac{x^2y}{(x+2y)^3} = \frac{(x^3y)^2}{(x+2y)^2} \cdot \frac{(x+2y)^3}{x^2y}$$

$$= \frac{(x^3y)(x^3y)(x+2y)^2(x+2y)}{(x+2y)^2x^2y}$$

$$= \frac{(x^3y)(x^2 \cdot xy)(x+2y)}{x^2y}$$

$$= x^4y(x+2y)$$

67. Divide and simplify.

$$\frac{\left(\dfrac{25x^2}{x-5}\right)}{\left(\dfrac{10x}{5-x}\right)}$$

Solution

$$\frac{\left(\dfrac{25x^2}{x-5}\right)}{\left(\dfrac{10x}{5-x}\right)} = \frac{25x^2}{x-5} \div \frac{10x}{5-x} = \frac{25x^2}{x-5} \cdot \frac{5-x}{10x} = \frac{5 \cdot 5 \cdot x \cdot x \cdot (-1)(x-5)}{(x-5) \cdot 5 \cdot 2 \cdot x} = -\frac{5x}{2}$$

69. Divide and simplify.

$$\frac{x(x + 3) - 2(x + 3)}{x^2 - 4} \div \frac{x}{x^2 + 4x + 4}$$

Solution

$$\frac{x(x + 3) - 2(x + 3)}{x^2 - 4} \div \frac{x}{x^2 + 4x + 4} = \frac{x(x + 3) - 2(x + 3)}{x^2 - 4} \cdot \frac{x^2 + 4x + 4}{x}$$

$$= \frac{(x + 3)(x - 2)}{(x - 2)(x + 2)} \cdot \frac{(x + 2)(x + 2)}{x}$$

$$= \frac{(x + 3)(x - 2)(x + 2)(x + 2)}{(x - 2)(x + 2)x}$$

$$= \frac{(x + 3)(x + 2)}{x}$$

71. Perform the operations and simplify your answer.

$$\frac{2x^2 + 5x - 25}{3x^2 + 5x + 2} \cdot \frac{3x^2 + 2x}{x + 5} \div \left(\frac{x}{x + 1}\right)^2$$

Solution

$$\frac{2x^2 + 5x - 25}{3x^2 + 5x + 2} \cdot \frac{3x^2 + 2x}{x + 5} \div \left(\frac{x}{x + 1}\right)^2 = \frac{(2x - 5)(x + 5)}{(3x + 2)(x + 1)} \cdot \frac{x(3x + 2)}{x + 5} \cdot \left(\frac{x + 1}{x}\right)^2$$

$$= \frac{(2x - 5)(x + 5)x(3x + 2)(x + 1)(x + 1)}{(3x + 2)(x + 1)(x + 5)x \cdot x}$$

$$= \frac{(2x - 5)(x + 1)}{x}$$

73. *Photocopy Rate* A photocopier produces copies at a rate of 20 pages per minute.

(a) Determine the time required to copy one page.

(b) Determine the time required to copy x pages.

(c) Determine the time required to copy 35 pages.

Solution

(a) $\dfrac{20 \text{ pages}}{1 \text{ minute}} = \dfrac{20 \text{ pages}}{60 \text{ seconds}} = \dfrac{1 \text{ page}}{3 \text{ seconds}}$ $t = 3$ seconds or $\dfrac{1}{20}$ minutes

(b) $\dfrac{3 \text{ seconds}}{1 \text{ page}} \cdot x \text{ pages} = 3x$ seconds or $\dfrac{x}{20}$ minutes

(c) $\dfrac{3 \text{ seconds}}{1 \text{ page}} \cdot 35 \text{ pages} = 3 \cdot 35$ seconds $= 105$ seconds or $\dfrac{7}{4}$ minutes

75. *Probability* Consider an experiment in which a marble is tossed into a rectangular box whose dimensions are $3x - 2$ by x inches (see figure in textbook). The probability that the marble will come to rest in the unshaded portion of the box is equal to the ratio of the unshaded area to the total area of the figure. Find the probability.

Solution

$$\frac{\text{Unshaded Area}}{\text{Total Area}} = \frac{\frac{x}{2} \cdot \frac{x}{2}}{x(3x - 2)} = \left[\frac{x}{2} \cdot \frac{x}{2}\right] \div [x(3x - 2)]$$

$$= \frac{x}{2} \cdot \frac{x}{2} \cdot \frac{1}{x(3x - 2)}$$

$$= \frac{x}{4(3x - 2)}$$

4.3 Adding and Subtracting Rational Expressions

5. Combine and simplify.

$$\frac{5}{8} + \frac{7}{8}$$

Solution

$$\frac{5}{8} + \frac{7}{8} = \frac{5 + 7}{8} = \frac{12}{8} = \frac{3}{2}$$

7. Combine and simplify.

$$\frac{x}{9} - \frac{x + 2}{9}$$

Solution

$$\frac{x}{9} - \frac{x + 2}{9} = \frac{x - (x + 2)}{9}$$

$$= \frac{x - x - 2}{9} = -\frac{2}{9}$$

9. Combine and simplify.

$$\frac{5x - 1}{x + 4} + \frac{5 - 4x}{x + 4}$$

Solution

$$\frac{5x - 1}{x + 4} + \frac{5 - 4x}{x + 4} = \frac{5x - 1 + 5 - 4x}{x + 4}$$

$$= \frac{x + 4}{x + 4}$$

$$= 1, \ x \neq -4$$

11. Find the least common multiple of the expressions.

$$5x^2, \ 20x^3$$

Solution

$$5x^2 = 5 \cdot x \cdot x$$

$$20x^3 = 5 \cdot 2 \cdot 2 \cdot x \cdot x \cdot x$$

$$\text{LCM} = 20x^3$$

13. Find the least common multiple of the expressions.

$$15x^2, \ 3(x + 5)$$

Solution

$$15x^2 = 5 \cdot 3 \cdot x \cdot x$$

$$3(x + 5) = 3 \cdot (x + 5)$$

$$\text{LCM} = 15x^2(x + 5)$$

15. Find the least common multiple of the expressions.

$$6(x^2 - 4), \ 2x(x + 2)$$

Solution

$$6(x^2 - 4) = 6(x - 2)(x + 2)$$

$$2x(x + 2) = 2 \cdot x \cdot (x + 2)$$

$$\text{LCM} = 6x(x - 2)(x + 2)$$

17. Find the least common denominator of the following two fractions and rewrite each fraction using the least common denominator.

$$\frac{n+8}{3n-12}, \quad \frac{10}{6n^2}$$

Solution

$$\frac{n+8}{3n-12} = \frac{n+8}{3(n-4)} = \frac{n+8(2n^2)}{3(n-4)(2n^2)} = \frac{2n^2(n+8)}{6n^2(n-4)}$$

$$\frac{10}{6n^2} = \frac{10}{3 \cdot 2n^2} = \frac{10(n-4)}{3 \cdot 2n^2(n-4)} = \frac{10(n-4)}{6n^2(n-4)}$$

$$\text{LCD} = 6n^2(n-4)$$

19. Find the least common denominator of the following two fractions and rewrite each fraction using the least common denominator.

$$\frac{x-8}{x^2-25}, \quad \frac{9x}{x^2-10x+25}$$

Solution

$$\frac{x-8}{x^2-25} = \frac{x-8}{(x-5)(x+5)} = \frac{(x-8)(x-5)}{(x-5)(x+5)(x-5)} = \frac{(x-8)(x-5)}{(x-5)^2(x+5)}$$

$$\frac{9x}{x^2-10x+25} = \frac{9x}{(x-5)^2} = \frac{9x(x+5)}{(x-5)^2(x+5)} = \frac{9x(x+5)}{(x-5)^2(x+5)}$$

$$\text{LCD} = (x-5)^2(x+5)$$

21. Perform the operation and simplify.

$$\frac{5}{4x} - \frac{3}{5}$$

Solution

$$\frac{5}{4x} - \frac{3}{5} = \frac{5(5)}{4x(5)} - \frac{3(4x)}{5(4x)}$$

$$= \frac{25}{20x} - \frac{12x}{20x}$$

$$= \frac{25-12x}{20x}$$

23. Perform the operation and simplify.

$$\frac{20}{x-4} + \frac{20}{4-x}$$

Solution

$$\frac{20}{x-4} + \frac{20}{4-x} = \frac{20(1)}{(x-4)(1)} + \frac{20(-1)}{(4-x)(-1)}$$

$$= \frac{20}{x-4} - \frac{20}{x-4}$$

$$= \frac{20-20}{x-4} = 0$$

25. Perform the operation and simplify.

$$25 + \frac{10}{x+4}$$

Solution

$$25 + \frac{10}{x+4} = \frac{25(x+4)}{1(x+4)} + \frac{10(1)}{(x+4)(1)} = \frac{25(x+4)}{x+4} + \frac{10}{x+4} = \frac{25x+100+10}{x+4} = \frac{25x+110}{x+4}$$

27. Perform the operation and simplify.

$$\frac{x}{x^2 - 9} + \frac{3}{x(x-3)}$$

Solution

$$\frac{x}{x^2 - 9} + \frac{3}{x(x-3)} = \frac{x(x)}{(x-3)(x+3)} + \frac{3(x+3)}{x(x-3)(x+3)}$$

$$= \frac{x^2}{x(x-3)(x+3)} + \frac{3(x+3)}{x(x-3)(x+3)}$$

$$= \frac{x^2 + 3x + 9}{x(x-3)(x+3)}$$

29. Perform the operation and simplify.

$$\frac{3u}{u^2 - 2uv + v^2} + \frac{2}{u-v}$$

Solution

$$\frac{3u}{u^2 - 2uv + v^2} + \frac{2}{u-v} = \frac{3u(1)}{(u-v)^2(1)} + \frac{2(u-v)}{(u-v)(u-v)}$$

$$= \frac{3u}{(u-v)^2} + \frac{2u - 2v}{(u-v)^2}$$

$$= \frac{3u + 2u - 2v}{(u-v)^2} = \frac{5u - 2v}{(u-v)^2}$$

31. Perform the operations and simplify.

$$\frac{x+2}{x-1} - \frac{2}{x+6} - \frac{14}{x^2 + 5x - 6}$$

Solution

$$\frac{x+2}{x-1} - \frac{2}{x+6} - \frac{14}{x^2 + 5x - 6} = \frac{(x+2)(x+6)}{(x-1)(x+6)} - \frac{2(x-1)}{(x+6)(x-1)} - \frac{14(1)}{(x+6)(x-1)(1)}$$

$$= \frac{x^2 + 8x + 12}{(x-1)(x+6)} - \frac{2x - 2}{(x+6)(x-1)} - \frac{14}{(x+6)(x-1)}$$

$$= \frac{x^2 + 8x + 12 - 2x + 2 - 14}{(x-1)(x+6)}$$

$$= \frac{x^2 + 6x}{(x-1)(x+6)}$$

$$= \frac{x(x+6)}{(x-1)(x+6)} = \frac{x}{x-1}$$

33. Use a graphing utility to graph the two equations on the same screen. Use the graphs to verify that the expressions are equivalent. Verify the results algebraically.

$$y_1 = \frac{2}{x} + \frac{4}{x(x-2)}, \quad y_2 = \frac{2x}{x(x-2)}$$

Solution

Keystrokes:

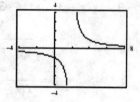

y_1 $\boxed{Y=}$ $\boxed{(}$ $\boxed{2}$ $\boxed{\div}$ $\boxed{X,T,\theta}$ $\boxed{)}$ $\boxed{+}$ $\boxed{(}$ $\boxed{4}$ $\boxed{\div}$ $\boxed{(}$ $\boxed{X,T,\theta}$

$\boxed{(}$ $\boxed{X,T,\theta}$ $\boxed{-}$ $\boxed{2}$ $\boxed{)}$ $\boxed{)}$ $\boxed{)}$ $\boxed{ENTER}$

y_2 $\boxed{Y=}$ $\boxed{2}$ $\boxed{X,T,\theta}$ $\boxed{\div}$ $\boxed{(}$ $\boxed{X,T,\theta}$ $\boxed{(}$ $\boxed{X,T,\theta}$ $\boxed{-}$ $\boxed{2}$ $\boxed{)}$ $\boxed{)}$ $\boxed{GRAPH}$

$$\frac{2}{x} + \frac{4}{x(x-2)} = \frac{2(x-2)}{x(x-2)} + \frac{4}{x(x-2)}$$

$$= \frac{2x-4+4}{x(x-2)}$$

$$= \frac{2x}{x(x-2)}$$

$y_1 = y_2$

35. Simplify the complex fraction.

$$\frac{\dfrac{1}{2}}{\left(3+\dfrac{1}{x}\right)}$$

Solution

$$\frac{\dfrac{1}{2}}{\left(3+\dfrac{1}{x}\right)} = \frac{\dfrac{1}{2}}{\left(3+\dfrac{1}{x}\right)} \cdot \frac{2x}{2x}$$

$$= \frac{\dfrac{1}{2} \cdot 2x}{3(2x)+\dfrac{1}{x}(2x)}$$

$$= \frac{x}{6x+2}$$

37. Simplify the complex fraction.

$$\frac{\left(3+\dfrac{9}{x-3}\right)}{\left(4+\dfrac{12}{x-3}\right)}$$

Solution

$$\frac{\left(3+\dfrac{9}{x-3}\right)}{\left(4+\dfrac{12}{x-3}\right)} = \frac{\left(3+\dfrac{9}{x-3}\right)}{\left(4+\dfrac{12}{x-3}\right)} \cdot \frac{x-3}{x-3}$$

$$= \frac{3(x-3)+\dfrac{9}{x-3}(x-3)}{4(x-3)+\dfrac{12}{x-3}(x-3)}$$

$$= \frac{3x-9+9}{4x-12+12} = \frac{3x}{4x} = \frac{3}{4}$$

39. Simplify the complex fraction.

$$\frac{\left(\dfrac{y}{x} - \dfrac{x}{y}\right)}{\left(\dfrac{x+y}{xy}\right)}$$

Solution

$$\frac{\left(\dfrac{y}{x} - \dfrac{x}{y}\right)}{\left(\dfrac{x+y}{xy}\right)} = \frac{\left(\dfrac{y}{x} - \dfrac{x}{y}\right)}{\left(\dfrac{x+y}{xy}\right)} \cdot \frac{xy}{xy}$$

$$= \frac{\dfrac{y}{x}(xy) - \dfrac{x}{y}(xy)}{\left(\dfrac{x+y}{xy}\right)xy} = \frac{y^2 - x^2}{x+y}$$

$$= \frac{(y-x)(y+x)}{x+y} = y - x$$

41. *Parallel Resistance*   When two resistors are connected in parallel (see figure in textbook), the total resistance is

$$\frac{1}{\left(\dfrac{1}{R_1} + \dfrac{1}{R_2}\right)}.$$

Simplify this complex fraction.

Solution

$$\frac{1}{\left(\dfrac{1}{R_1} + \dfrac{1}{R_2}\right)} = \frac{1}{\left(\dfrac{1}{R_1} + \dfrac{1}{R_2}\right)} \cdot \frac{R_1 R_2}{R_1 R_2}$$

$$= \frac{R_1 R_2}{\dfrac{1}{R_1}(R_1 R_2) + \dfrac{1}{R_2}(R_1 R_2)}$$

$$= \frac{R_1 R_2}{R_2 + R_1}$$

43. *Interpreting a Table*   Use a graphing utility to complete the table. Comment on the domain and equivalence of the expressions.

Solution

x	-3	-2	-1	0	1	2	3
$\dfrac{\left(1 - \dfrac{1}{x}\right)}{\left(1 - \dfrac{1}{x^2}\right)}$	$\dfrac{3}{2}$	2	undef.	undef.	undef.	$\dfrac{2}{3}$	$\dfrac{3}{4}$
$\dfrac{x}{x+1}$	$\dfrac{3}{2}$	2	undef.	0	$\dfrac{1}{2}$	$\dfrac{2}{3}$	$\dfrac{3}{4}$

Keystrokes:

y_1 $\boxed{Y=}$ $\boxed{(}$ 1 $\boxed{-}$ 1 $\boxed{\div}$ $\boxed{X,T,\theta}$ $\boxed{)}$ $\boxed{\div}$

  $\boxed{(}$ 1 $\boxed{-}$ 1 $\boxed{\div}$ $\boxed{X,T,\theta}$ $\boxed{x^2}$ $\boxed{)}$ $\boxed{ENTER}$

y_2 $\boxed{Y=}$ $\boxed{X,T,\theta}$ $\boxed{\div}$ $\boxed{(}$ $\boxed{X,T,\theta}$ $\boxed{+}$ 1 $\boxed{)}$ $\boxed{GRAPH}$

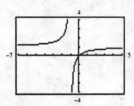

Zero is not in the domain of $\dfrac{1 - \dfrac{1}{x}}{1 - \dfrac{1}{x^2}}$ but is in the domain of

$\dfrac{x}{x+1}$.

45. Simplify $-(-3x^2)^3(2x^4)$.

Solution

$-(-3x^2)^3(2x^4) = -(-27x^6)(2x^4) = 54x^{10}$

47. Simplify $(a^2 + b^2)^0$.

Solution

$(a^2 + b^2)^0 = 1$

49. Simplify $\left(\dfrac{5}{x^2}\right)^2$.

Solution

$$\left(\dfrac{5}{x^2}\right)^2 = \dfrac{25}{x^4}$$

51. *Monthly Wage* A company offers two wage plans. One plan pays a straight $2500 per month. The second pays $1500 per month plus a commission of 4% on gross sales. Let x represent the gross sales and write an inequality that represents gross sales in which the second plan gives the greater monthly wage. Solve the inequality.

Solution

1st plan wages $= \$2500$

2nd plan wages $= \$1500 + 0.04x$

$$\$2500 < \$1500 + 0.04x$$

$$1000 < 0.04x$$

$$\$25,000 < x$$

53. Combine and simplify.

$$\dfrac{4-y}{4} + \dfrac{3y}{4}$$

Solution

$$\dfrac{4-y}{4} + \dfrac{3y}{4} = \dfrac{4-y+3y}{4} = \dfrac{4+2y}{4}$$
$$= \dfrac{2(2+y)}{4} = \dfrac{2+y}{2}$$

55. Combine and simplify.

$$\dfrac{2}{3a} - \dfrac{11}{3a}$$

Solution

$$\dfrac{2}{3a} - \dfrac{11}{3a} = \dfrac{2-11}{3a} = \dfrac{-9}{3a} = \dfrac{-3}{a}$$

57. Combine and simplify.

$$\dfrac{2x+5}{3} + \dfrac{1-x}{3}$$

Solution

$$\dfrac{2x+5}{3} + \dfrac{1-x}{3} = \dfrac{2x+5+1-x}{3} = \dfrac{x+6}{3}$$

59. Combine and simplify.

$$\dfrac{3y}{3} - \dfrac{3y-3}{3} - \dfrac{7}{3}$$

Solution

$$\dfrac{3y}{3} - \dfrac{3y-3}{3} - \dfrac{7}{3} = \dfrac{3y-(3y-3)-7}{3}$$
$$= \dfrac{3y-3y+3-7}{3} = -\dfrac{4}{3}$$

61. Combine and simplify.

$$\dfrac{-16u}{9} - \dfrac{27-16u}{9} + \dfrac{2}{9}$$

Solution

$$\dfrac{-16u}{9} - \dfrac{27-16u}{9} + \dfrac{2}{9} = \dfrac{-16u-(27-16u)+2}{9} = \dfrac{-16u-27+16u+2}{9} = -\dfrac{25}{9}$$

63. Find the least common multiple of the expressions.

$9y^3$, $12y$

Solution

$9y^3 = 3 \cdot 3 \cdot y \cdot y \cdot y$

$12y = 2 \cdot 2 \cdot 3 \cdot y$

$\text{LCM} = 3 \cdot 3 \cdot 2 \cdot 2 \cdot y \cdot y \cdot y$

$\qquad = 36y^3$

65. Find the least common multiple of the expressions.

$6x^2$, $15x(x-1)$

Solution

$6x^2 = 2 \cdot 3 \cdot x \cdot x$

$15x(x-1) = 5 \cdot 3 \cdot x \cdot (x-1)$

$\text{LCM} = 2 \cdot 3 \cdot 5 \cdot x \cdot x \cdot (x-1)$

$\qquad = 30x^2(x-1)$

67. Find the least common denominator of the two fractions and rewrite each fraction using the least common denominator.

$$\frac{v}{2v^2 + 2v}, \quad \frac{4}{3v^2}$$

Solution

$$\frac{v}{2v^2 + 2v} = \frac{v}{2v(v+1)} = \frac{v(3v)}{2v(v+1)(3v)} = \frac{3v^2}{6v^2(v+1)}$$

$$\frac{4}{3v^2} = \frac{4(2(v+1))}{3v^2(2(v+1))} = \frac{8v + 8}{6v^2(v+1)}$$

$$\text{LCD} = 6v^2(v+1)$$

69. Find the least common denominator of the two fractions and rewrite each fraction using the least common denominator.

$$\frac{4x}{(x+5)^2}, \quad \frac{x-2}{x^2 - 25}$$

Solution

$$\frac{4x}{(x+5)^2} = \frac{4x(x-5)}{(x+5)^2(x-5)} = \frac{4x^2 - 20x}{(x+5)^2(x-5)}$$

$$\frac{x-2}{x^2 - 25} = \frac{x-2}{(x-5)(x+5)} = \frac{(x-2)(x+5)}{(x-5)(x+5)(x+5)} = \frac{x^2 + 3x - 10}{(x+5)^2(x-5)}$$

$$\text{LCD} = (x+5)^2(x-5)$$

71. Perform the operation and simplify.

$$\frac{7}{a} + \frac{14}{a^2}$$

Solution

$$\frac{7}{a} + \frac{14}{a^2} = \frac{7(a)}{a(a)} + \frac{14(1)}{a^2(1)} = \frac{7a}{a^2} + \frac{14}{a^2}$$

$$= \frac{7a + 14}{a^2}$$

73. Perform the operation and simplify.

$$\frac{3x}{x - 8} - \frac{6}{8 - x}$$

Solution

$$\frac{3x}{x - 8} - \frac{6}{8 - x} = \frac{3x(1)}{(x-8)(1)} - \frac{6(-1)}{(8-x)(-1)}$$

$$= \frac{3x}{x - 8} + \frac{6}{x - 8}$$

$$= \frac{3x + 6}{x - 8}$$

75. Perform the operation and simplify.

$$\frac{3x}{3x-2} + \frac{2}{2-3x}$$

Solution

$$\frac{3x}{3x-2} + \frac{2}{2-3x} = \frac{3x(1)}{3x-2(1)} + \frac{2(-1)}{(2-3x)(-1)}$$

$$= \frac{3x}{3x-2} + \frac{-2}{3x-2}$$

$$= \frac{3x-2}{3x-2} = 1$$

77. Perform the operation and simplify.

$$-\frac{1}{6x} + \frac{1}{6(x-3)}$$

Solution

$$-\frac{1}{6x} + \frac{1}{6(x-3)} = \frac{1(x-3)}{6x(x-3)} + \frac{1(x)}{6(x-3)x}$$

$$= \frac{-(x-3)}{6x(x-3)} + \frac{x}{6x(x-3)}$$

$$= \frac{-x+3+x}{6x(x-3)}$$

$$= \frac{3}{6x(x-3)} = \frac{1}{2x(x-3)}$$

79. Perform the operation and simplify.

$$\frac{x}{x+3} - \frac{5}{x-2}$$

Solution

$$\frac{x}{x+3} - \frac{5}{x-2} = \frac{x(x-2)}{(x+3)(x-2)} - \frac{5(x+3)}{(x-2)(x+3)}$$

$$= \frac{x(x-2)}{(x+3)(x-2)} - \frac{5(x+3)}{(x-2)(x+3)}$$

$$= \frac{x^2 - 2x - 5x - 15}{(x+3)(x-2)}$$

$$= \frac{x^2 - 7x - 15}{(x+3)(x-2)}$$

81. Perform the operation and simplify.

$$\frac{3}{x+1} - \frac{2}{x}$$

Solution

$$\frac{3}{x+1} - \frac{2}{x} = \frac{3x}{(x+1)x} - \frac{2(x+1)}{x(x+1)}$$

$$= \frac{3x}{x(x+1)} - \frac{2(x+1)}{x(x+1)}$$

$$= \frac{3x - 2x - 2}{x(x+1)}$$

$$= \frac{x-2}{x(x+1)}$$

83. Perform the operation and simplify.

$$\frac{3}{x-5} + \frac{2}{x+5}$$

Solution

$$\frac{3}{x-5} + \frac{2}{x+5} = \frac{3(x+5)}{(x-5)(x+5)} + \frac{2(x-5)}{(x+5)(x-5)}$$

$$= \frac{3(x+5)}{(x-5)(x+5)} + \frac{2(x-5)}{(x+5)(x-5)}$$

$$= \frac{3x + 15 + 2x - 10}{(x-5)(x+5)}$$

$$= \frac{5x+5}{(x-5)(x+5)}$$

85. Perform the operation and simplify.

$$\frac{4}{x^2} - \frac{4}{x^2+1}$$

Solution

$$\frac{4}{x^2} - \frac{4}{x^2+1} = \frac{4(x^2+1)}{x^2(x^2+1)} - \frac{4x^2}{(x^2+1)x^2}$$

$$= \frac{4(x^2+1)}{x^2(x^2+1)} - \frac{4x^2}{x^2(x^2+1)}$$

$$= \frac{4x^2 + 4 - 4x^2}{x^2(x^2+1)}$$

$$= \frac{4}{x^2(x^2+1)}$$

87. Perform the operation and simplify.

$$\frac{4}{x-4} + \frac{16}{(x-4)^2}$$

Solution

$$\frac{4}{x-4} + \frac{16}{(x-4)^2} = \frac{4(x-4)}{(x-4)(x-4)} + \frac{16(1)}{(x-4)^2(1)}$$

$$= \frac{4x-16}{(x-4)^2} + \frac{16}{(x-4)^2}$$

$$= \frac{4x-16+16}{(x-4)^2}$$

$$= \frac{4x}{(x-4)^2}$$

89. Perform the operation and simplify.

$$\frac{y}{x^2+xy} - \frac{x}{xy+y^2}$$

Solution

$$\frac{y}{x^2+xy} - \frac{x}{xy+y^2} = \frac{y}{x(x+y)} - \frac{x}{y(x+y)}$$

$$= \frac{y(y)}{x(x+y)(y)} - \frac{x(x)}{y(x+y)(x)}$$

$$= \frac{y^2}{xy(x+y)} - \frac{x^2}{xy(x+y)}$$

$$= \frac{y^2-x^2}{xy(x+y)}$$

$$= \frac{(y-x)(x+x)}{xy(x+y)} = \frac{y-x}{xy}$$

91. Perform the operations and simplify.

$$\frac{4}{x} - \frac{2}{x^2} + \frac{4}{x+3}$$

Solution

$$\frac{4}{x} - \frac{2}{x^2} + \frac{4}{x+3} = \frac{4x(x+3)}{x(x)(x+3)} - \frac{2(x+3)}{x^2(x+3)} + \frac{4(x^2)}{(x+3)x^2}$$

$$= \frac{4x^2+12x}{x^2(x+3)} - \frac{2x+6}{x^2(x+3)} + \frac{4x^2}{x^2(x+3)}$$

$$= \frac{4x^2+12x-2x-6+4x^2}{x^2(x+3)}$$

$$= \frac{8x^2+10x-6}{x^2(x+3)}$$

93. Simplify the complex fraction.

$$\frac{\left(16x - \dfrac{1}{x}\right)}{\left(\dfrac{1}{x} - 4\right)}$$

Solution

$$\frac{\left(16x - \dfrac{1}{x}\right)}{\left(\dfrac{1}{x} - 4\right)} = \frac{\left(16x - \dfrac{1}{x}\right)}{\left(\dfrac{1}{x} - 4\right)} \cdot \frac{x}{x} = \frac{16x(x) - \dfrac{1}{x}(x)}{\dfrac{1}{x}(x) - 4(x)}$$

$$= \frac{16x^2 - 1}{1 - 4x} = \frac{(4x-1)(4x+1)}{-1(4x-1)}$$

$$= \frac{4x+1}{-1} = -4x - 1$$

95. Simplify the complex fraction.

$$\frac{\left(\dfrac{3}{x^2} + \dfrac{1}{x}\right)}{\left(2 - \dfrac{4}{5x}\right)}$$

Solution

$$\frac{\left(\dfrac{3}{x^2} + \dfrac{1}{x}\right)}{\left(2 - \dfrac{4}{5x}\right)} = \frac{\left(\dfrac{3}{x^2} + \dfrac{1}{x}\right)}{\left(2 - \dfrac{4}{5x}\right)} \cdot \frac{5x^2}{5x^2}$$

$$= \frac{15 + 5x}{10x^2 - 4x} = \frac{5(3+x)}{2x(5x-2)}$$

97. Simplify the complex fraction.

$$\dfrac{\left(1 - \dfrac{1}{y^2}\right)}{\left(1 - \dfrac{4}{y} + \dfrac{3}{y^2}\right)}$$

Solution

$$\dfrac{\left(1 - \dfrac{1}{y^2}\right)}{\left(1 - \dfrac{4}{y} + \dfrac{3}{y^2}\right)} = \dfrac{\left(1 - \dfrac{1}{y^2}\right)}{\left(1 - \dfrac{4}{y} + \dfrac{3}{y^2}\right)} \cdot \dfrac{y^2}{y^2}$$

$$= \dfrac{1(y^2) - \dfrac{1}{y^2}(y^2)}{1(y^2) - \dfrac{4}{y}(y^2) + \dfrac{3}{y^2}(y^2)}$$

$$= \dfrac{y^2 - 1}{y^2 - 4y + 3}$$

$$= \dfrac{(y-1)(y+1)}{(y-1)(y-3)} = \dfrac{y+1}{y-3}$$

99. Simplify the complex fraction.

$$\dfrac{\left(\dfrac{x}{x-3} - \dfrac{2}{3}\right)}{\left(\dfrac{10}{3x} + \dfrac{x^2}{x-3}\right)}$$

Solution

$$\dfrac{\left(\dfrac{x}{x-3} - \dfrac{2}{3}\right)}{\left(\dfrac{10}{3x} + \dfrac{x^2}{x-3}\right)} = \dfrac{\left(\dfrac{x}{x-3} - \dfrac{2}{3}\right)}{\left(\dfrac{10}{3x} + \dfrac{x^2}{x-3}\right)} \cdot \dfrac{3x(x-3)}{3x(x-3)}$$

$$= \dfrac{3x^2 - 2x(x-3)}{10(x-3) + 3x^2}$$

$$= \dfrac{3x^2 - 2x^2 + 6x}{10x - 30 + 3x^3}$$

$$= \dfrac{x^2 + 6x}{3x^3 + 10x - 30}$$

101. *Difference Quotient*   Use the function $f(x) = \dfrac{1}{x}$ to find and simplify the expression for $\dfrac{f(2+h) - f(2)}{h}$. This expression is called a difference quotient and is used in calculus.

Solution

$$\dfrac{f(2+h) - f(2)}{h} = \dfrac{\dfrac{1}{2+h} - \dfrac{1}{2}}{h} = \dfrac{\dfrac{1}{2+h} - \dfrac{1}{2}}{h} \cdot \dfrac{2(2+h)}{2(2+h)}$$

$$= \dfrac{2 - (2+h)}{2h(2+h)} = \dfrac{2 - 2 - h}{2h(2+h)}$$

$$= \dfrac{-h}{2h(2+h)} = \dfrac{-1}{2(2+h)}$$

103. *Work Rate*   After two workers work together for t hours on a common task, the fractional parts of the job done by the two workers are $t/4$ and $t/6$. What fractional part of the task has been completed?

Solution

$$\dfrac{t}{4} + \dfrac{t}{6} = \dfrac{t(3)}{4(3)} + \dfrac{t(2)}{6(2)} = \dfrac{3t}{12} + \dfrac{2t}{12} = \dfrac{5t}{12}$$

105. *Average of Two Numbers* Determine the average of the two real numbers $x/4$ and $x/6$.

Solution

$$\frac{\frac{x}{4} + \frac{x}{6}}{2} = \frac{\left(\frac{x}{4} + \frac{x}{6}\right)}{2} \cdot \frac{12}{12}$$

$$= \frac{\frac{x}{4}(12) + \frac{x}{6}(12)}{2(12)}$$

$$= \frac{3x + 2x}{24} = \frac{5x}{24}$$

107. *Equal Parts* Find three real numbers that divide the real number line (see figure in textbook) between $x/6$ and $x/2$ into four equal parts.

Solution

$$\frac{\frac{x}{2} - \frac{x}{6}}{4} \cdot \frac{6}{6} = \frac{3x - x}{24} = \frac{2x}{24} = \frac{x}{12}$$

Thus,

$$x_1 = \frac{x}{6} + \frac{x}{12} = \frac{3x}{12} = \frac{x}{4}$$

$$x_2 = \frac{x}{4} + \frac{x}{12} = \frac{4x}{12} = \frac{x}{3}$$

$$x_3 = \frac{x}{3} + \frac{x}{12} = \frac{5x}{12}$$

109. *Monthly Payment* The approximate annual interest rate r of a monthly installment loan is given by

$$r = \frac{\left[\dfrac{24(NM - P)}{N}\right]}{\left(P + \dfrac{MN}{12}\right)},$$

where N is the total number of payments, M is the monthly payment, and P is the amount financed.

(a) Approximate the annual rate for a 4-year car loan of $10,000 that has monthly payments of $300.

(b) Simplify the expression for the annual interest rate r, and then rework part (a).

Solution

(a) $r = \dfrac{\left[\dfrac{24(48(300) - 10,000)}{48}\right]}{\left(10,000 + \dfrac{300(48)}{12}\right)} = 19.6\%$

(b) $r = \dfrac{\dfrac{24(NM - P)}{N}}{P + \dfrac{MN}{12}} \cdot \dfrac{12N}{12N} = \dfrac{288(NM - P)}{12NP + MN^2}$

$r = \dfrac{288\left[(48)(300) - 10,000\right]}{12(48)(10,000) + (300)(48)^2}$

$r = 19.6\%$

Mid-Chapter Quiz for Chapter 4

1. Determine the domain of $\dfrac{y+2}{y(y-4)}$.

Solution

$y(y-4) \neq 0$

$y \neq 0 \quad y-4 \neq 0$

$\qquad y \neq 4 \quad D = (-\infty, 0) \cup (0, 4) \cup (4, \infty)$

2. Evaluate $h(x) = (x^2 - 9)/(x^2 - x - 2)$ as indicated. If it is not possible, state the reason.

(a) $h(-3)$ (b) $h(0)$

(c) $h(-1)$ (d) $h(5)$

Solution

(a) $h(-3) = \dfrac{(-3)^2 - 9}{(-3)^2 - (-3) - 2} = \dfrac{9-9}{9+3-2}$

$\qquad = \dfrac{0}{10} = 0$

(b) $h(0) = \dfrac{0^2 - 9}{0^2 - 0 - 2} = \dfrac{-9}{-2}$

$\qquad = \dfrac{9}{2}$

(c) $h(-1) = \dfrac{(-1)^2 - 9}{(-1)^2 - (-1) - 2} = \dfrac{1-9}{1+1-2}$

$\qquad = \dfrac{-8}{0} =$ undefined

(d) $h(5) = \dfrac{5^2 - 9}{5^2 - 5 - 2} = \dfrac{25-9}{25-5-2}$

$\qquad = \dfrac{16}{18} = \dfrac{8}{9}$

3. Write $\dfrac{9y^2}{6y}$ in reduced form.

Solution

$\dfrac{9y^2}{6y} = \dfrac{3y}{2}$

4. Write $\dfrac{8u^3 v^2}{36uv^3}$ in reduced form.

Solution

$\dfrac{8u^3 v^2}{36uv^3} = \dfrac{2u^2}{9v}$

5. Write $\dfrac{4x^2 - 1}{x - 2x^2}$ in reduced form.

Solution

$\dfrac{4x^2 - 1}{x - 2x^2} = \dfrac{(2x-1)(2x+1)}{x(1-2x)}$

$\qquad = \dfrac{(2x-1)(2x+1)}{-x(2x-1)}$

$\qquad = \dfrac{2x+1}{-x}$

6. Write $\dfrac{(z+3)^2}{2z^2 + 5z - 3}$ in reduced form.

Solution

$\dfrac{(z+3)^2}{2z^2 + 5z - 3} = \dfrac{(z+3)(z+3)}{(2z-1)(z+3)}$

$\qquad = \dfrac{z+3}{2z-1}$

7. Write $\dfrac{7ab + 3a^2 b^2}{a^2 b}$ in reduced form.

Solution

$\dfrac{7ab + 3a^2 b^2}{a^2 b} = \dfrac{ab(7 + 3ab)}{a^2 b}$

$\qquad = \dfrac{7 + 3ab}{a}$

8. Write $\dfrac{2mn^2 - n^3}{2m^2 + mn - n^2}$ in reduced form.

Solution

$\dfrac{2mn^2 - n^3}{2m^2 + mn - n^2} = \dfrac{n^2(2m - n)}{(2m - n)(m + n)}$

$\qquad = \dfrac{n^2}{m + n}$

9. Perform the operation and simplify.

$$\frac{11t^2}{6} \cdot \frac{9}{33t}$$

Solution

$$\frac{11t^2}{6} \cdot \frac{9}{33t} = \frac{11t^2(9)}{6(33t)} = \frac{t}{2}$$

10. Perform the operation and simplify.

$$(x^2 + 2x) \cdot \frac{5}{x^2 - 4}$$

Solution

$$(x^2 + 2x) \cdot \frac{5}{x^2 - 4} = \frac{x(x + 2)5}{(x - 2)(x + 2)}$$

$$= \frac{5x}{x - 2}$$

11. Perform the operation and simplify.

$$\frac{4}{3(x - 1)} \cdot \frac{12x}{6(x^2 + 2x - 3)}$$

Solution

$$\frac{4}{3(x - 1)} \cdot \frac{12x}{6(x^2 + 2x - 3)} = \frac{4(12x)}{3(x - 1)6(x + 3)(x - 1)} = \frac{8x}{3(x - 1)^2(x + 3)}$$

12. Perform the operation and simplify.

$$\frac{5u}{3(u + v)} \cdot \frac{2(u^2 - v^2)}{3v} \div \frac{25u^2}{18(u - v)}$$

Solution

$$\frac{5u}{3(u + v)} \cdot \frac{2(u^2 - v^2)}{3v} \div \frac{25u^2}{18(u - v)} = \frac{5u \cdot 2(u - v)(u + v) \cdot 18(u - v)}{3(u + v)(3v)(25u^2)}$$

$$= \frac{4(u - v)^2}{5uv}$$

13. Perform the operation and simplify.

$$\frac{\left(\dfrac{9t^2}{3 - t}\right)}{\left(\dfrac{6t}{t - 3}\right)}$$

Solution

$$\frac{\dfrac{9t^2}{3 - t}}{\dfrac{6t}{t - 3}} \cdot \frac{t - 3}{t - 3} = \frac{-9t^2}{6t} = -\frac{3t}{2}$$

14. Perform the operation and simplify.

$$\frac{\left(\dfrac{10}{x^2 + 2x}\right)}{\left(\dfrac{15}{x^2 + 3x + 2}\right)}$$

Solution

$$\frac{\dfrac{10}{x^2 + 2x}}{\dfrac{15}{x^2 + 3x + 2}} = \frac{\dfrac{10}{x(x + 2)}}{\dfrac{15}{(x + 2)(x + 1)}} \cdot \frac{x(x + 2)(x + 1)}{x(x + 2)(x + 1)}$$

$$= \frac{10(x + 1)}{15x} = \frac{2(x + 1)}{3x}$$

15. Perform the operation and simplify.

$$\frac{4x}{x+5} - \frac{3x}{4}$$

Solution

$$\frac{4x}{x+5} - \frac{3x}{4} = \frac{4x(4)}{4(x+5)} - \frac{3x(x+5)}{4(x+5)}$$

$$= \frac{16x - 3x^2 - 15x}{4(x+5)}$$

$$= \frac{-3x^2 + x}{4(x+5)} = \frac{x(1-3x)}{4(x+5)}$$

16. Perform the operation and simplify.

$$4 + \frac{x}{x^2-4} - \frac{2}{x^2}$$

Solution

$$4 + \frac{x}{x^2-4} - \frac{2}{x^2} = \frac{4x^2(x^2-4)}{x^2(x^2-4)} + \frac{x(x^2)}{x^2(x^2-4)} - \frac{2(x^2-4)}{x^2(x^2-4)}$$

$$= \frac{4x^4 - 16x^2 + x^3 - 2x^2 + 8}{x^2(x^2-4)}$$

$$= \frac{4x^4 + x^3 - 18x^2 + 8}{x^2(x^2-4)}$$

17. Perform the operation and simplify.

$$\frac{\left(1 - \dfrac{2}{x}\right)}{\left(\dfrac{3}{x} - \dfrac{4}{5}\right)}$$

Solution

$$\frac{\left(1 - \dfrac{2}{x}\right)}{\left(\dfrac{3}{x} - \dfrac{4}{5}\right)} = \frac{\left(1 - \dfrac{2}{x}\right)}{\left(\dfrac{3}{x} - \dfrac{4}{5}\right)} \cdot \frac{5x}{5x}$$

$$= \frac{5x - 10}{15 - 4x}$$

18. Perform the operation and simplify.

$$\frac{\left(\dfrac{3}{x} + \dfrac{x}{3}\right)}{\left(\dfrac{x+3}{6x}\right)}$$

Solution

$$\frac{\left(\dfrac{3}{x} + \dfrac{x}{3}\right)}{\left(\dfrac{x+3}{6x}\right)} = \frac{\dfrac{3}{x} + \dfrac{x}{3}}{\dfrac{x+3}{6x}} \cdot \frac{6x}{6x}$$

$$= \frac{18 + 2x^2}{x+3} = \frac{2(9 + x^2)}{x+3}$$

19. You start a business with a setup cost of $6000. The cost of material for producing each unit of your product is $10.50.

(a) Write an algebraic fraction that gives the average cost per unit when x units are produced. Explain your reasoning.

(b) Find the average cost per unit when $x = 500$ units are produced.

Solution

(a) *Verbal Model:* $\boxed{\begin{array}{c}\text{Average}\\\text{Cost}\end{array}} = \boxed{\begin{array}{c}\text{Total}\\\text{Cost}\end{array}} \div \boxed{\begin{array}{c}\text{Number of}\\\text{Units}\end{array}}$

  Equation: $\text{Average cost} = \dfrac{6000 + 10.50x}{x}$

(b) Average cost when $x = 500$ units are produced $= \dfrac{6000 + 10.50(500)}{500} = \$22.50$

20. Find the ratio of the shaded portion of the figure in the textbook to the total area of the figure.

Solution

$$\frac{\text{Shaded Portion}}{\text{Total Area}} = \frac{\left[\dfrac{1}{2}(x + 4)(x + 2)\right] - \left[\dfrac{1}{2}\left(\dfrac{x(x+2)}{x+4}\right)x\right]}{\left[\dfrac{1}{2}(x+4)(x+2)\right]}$$

$$= \frac{\dfrac{1}{2}(x+2)\left[(x+4) - \dfrac{x^2}{x+4}\right]}{\dfrac{1}{2}(x+4)(x+2)}$$

$$= \frac{\dfrac{(x+4)^2 - x^2}{(x+4)}}{x+4}$$

$$= \frac{x^2 + 8x + 16 - x^2}{(x+4)} \cdot \frac{1}{x+4}$$

$$= \frac{8x + 16}{(x+4)^2}$$

$$= \frac{8(x+2)}{(x+4)^2}$$

4.4 Dividing Polynomials

7. Perform the division of a polynomial by a monomial and check your result.

$$\frac{50z^3 + 30z}{-5z}$$

Solution

$$\frac{50z^3 + 30z}{-5z} = \frac{50z^3}{-5z} + \frac{30z}{-5z} = -10z^2 - 6$$

9. Perform the division of a polynomial by a monomial and check your result.

$$(5x^2y - 8xy + 7xy^2) \div 2xy$$

Solution

$$(5x^2y - 8xy + 7xy^2) \div 2xy = \frac{5x^2y - 8xy + 7xy^2}{2xy}$$

$$= \frac{5x^2y}{2xy} - \frac{8xy}{2xy} + \frac{7xy^2}{2xy}$$

$$= \frac{5x}{2} - 4 + \frac{7}{2}y$$

11. Perform the division and check your result.

$$\frac{x^2 - 8x + 15}{x - 3}$$

Solution

$$\frac{x^2 - 8x + 15}{x - 3} = x - 3 \overline{)\,x^2 - 8x + 15}$$

$$\begin{array}{r} x - 5 \\ x - 3 \overline{)\,x^2 - 8x + 15} \\ \underline{x^2 - 3x} \\ -5x + 15 \\ \underline{-5x + 15} \end{array}$$

13. Perform the division and check your result.

Divide $21 - 4x - x^2$ by $3 - x$.

Solution

$$\begin{array}{r} x + 7 \\ -x + 3 \overline{)\,-x^2 - 4x + 21} \\ \underline{-x^2 + 3x} \\ -7x + 21 \\ \underline{-7x + 21} \end{array}$$

15. Perform the division and check your result.

$$\frac{x^3 - 2x^2 + 4x - 8}{x - 2}$$

Solution

$$\begin{array}{r} x^2 \quad\ + 4 \\ x - 2 \overline{)\,x^3 - 2x^2 + 4x - 8} \\ \underline{x^3 - 2x^2} \\ 4x - 8 \\ \underline{4x - 8} \end{array}$$

17. Perform the division and check your result.

$$\frac{x^2 + 16}{x + 4}$$

Solution

$$\begin{array}{r} x - 4 \quad + \dfrac{32}{x + 4} \\ x + 4 \overline{)\,x^2 + 0x + 16} \\ \underline{x^2 + 4x} \\ -4x + 16 \\ \underline{-4x - 16} \\ 32 \end{array}$$

19. Perform the division and check your result.

$$\frac{6z^2 + 7z}{5z - 1}$$

Solution

$$
\begin{array}{r}
\frac{6}{5}z + \frac{41}{25} \quad + \dfrac{\frac{41}{25}}{5z - 1} \\
5z - 1 \overline{)\, 6z^2 + \;\; 7z} \\
6z^2 - \frac{6}{5}z \\
\hline
\frac{41}{5}z \\
\frac{41}{5}z - \frac{41}{25} \\
\hline
\frac{41}{25}
\end{array}
$$

21. Perform the division and check your result.

$$x^5 \div (x^2 + 1)$$

Solution

$$
\begin{array}{r}
x^3 - \;\; x \qquad + \dfrac{x}{x^2 + 1} \\
x^2 + 1 \overline{)\, x^5 \qquad\qquad} \\
x^5 + \;\; x^3 \\
\hline
-x^3 \\
-x^3 - x \\
\hline
x
\end{array}
$$

23. Perform the division and check your result.

$$(x^3 + 4x^2 + 7x + 6) \div (x^2 + 2x + 3)$$

Solution

$$
\begin{array}{r}
x + 2 \\
x^2 + 2x + 3 \overline{)\, x^3 + 4x^2 + 7x + 6} \\
x^3 + 2x^2 + 3x \\
\hline
2x^2 + 4x + 6 \\
2x^2 + 4x + 6
\end{array}
$$

25. Use synthetic division to perform the division.

$$\frac{x^3 + 3x^2 - 1}{x + 4}$$

Solution

$$\frac{x^3 + 3x^2 - 1}{x + 4}$$

$$
\begin{array}{r|rrrr}
-4 & 1 & 3 & 0 & -1 \\
   &   & -4 & 4 & -16 \\
\hline
   & 1 & -1 & 4 & -17
\end{array}
$$

$$\frac{x^3 + 3x^2 - 1}{x + 4} = x^2 - x + 4 + \frac{-17}{x + 4}$$

27. Use synthetic division to perform the division.

$$\frac{0.1x^2 + 0.8x + 1}{x - 0.2}$$

Solution

$$\frac{0.1x^2 + 0.8x + 1}{x - 0.2}$$

$$
\begin{array}{r|rrr}
0.2 & 0.1 & 0.8 & 1 \\
    &   & 0.02 & 0.164 \\
\hline
    & 0.1 & 0.82 & 1.164
\end{array}
$$

$$\frac{0.1x^2 + 0.8x + 1}{x - 0.2} = 0.1x + 0.82 + \frac{1.164}{x - 0.2}$$

29. Use synthetic division to perform the division. Use the result to factor the dividend.

$$\frac{x^2 - 15x + 56}{x - 8}$$

Solution

$$\frac{x^2 - 15x + 56}{x - 8}$$

$$
\begin{array}{r|rrr}
8 & 1 & -15 & 56 \\
  &   & 8 & -56 \\
\hline
  & 1 & -7 & 0
\end{array}
$$

$$x^2 - 15x + 56 = (x - 7)(x - 8)$$

31. Use synthetic division to perform the division. Use the result to factor the dividend.

$$\frac{x^4 - x^3 - 3x^2 + 4x - 1}{x - 1}$$

Solution

$$\frac{x^4 - x^3 - 3x^2 + 4x - 1}{x - 1}$$

$$
\begin{array}{r|rrrrr}
1 & 1 & -1 & -3 & 4 & -1 \\
  &   & 1 & 0 & -3 & 1 \\
\hline
  & 1 & 0 & -3 & 1 & 0
\end{array}
$$

$$x^4 - x^3 - 3x^2 + 4x - 1 = (x^3 - 3x + 1)(x - 1)$$

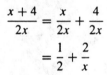 **33.** Use a graphing utility to graph the two equations on the same screen. Use the graphs to verify that the expressions are equivalent. Verify the results algebraically.

$$y_1 = \frac{x + 4}{2x}, \quad y_2 = \frac{1}{2} + \frac{2}{x}$$

Solution

Keystrokes:

y_1 [Y=] [(] [X, T, θ] [+] [4] [)] [÷] [2] [X, T, θ] [ENTER]

y_2 [Y=] [(] [1] [÷] [2] [)] [+] [(] [2] [÷] [X, T, θ] [)] [GRAPH]

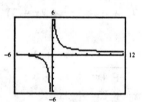

$$\frac{x + 4}{2x} = \frac{x}{2x} + \frac{4}{2x}$$

$$= \frac{1}{2} + \frac{2}{x}$$

So $y_1 = y_2$

35. *Finding a Pattern* Complete the table for the given polynomial, $x^3 - x^2 - 2x$. What conclusion can you draw as you compare the polynomial values with the remainders? (Use synthetic division to find the remainders.)

Solution

x	Polynomial Value	Divisor	Remainder
-2	-8	$x + 2$	-8
-1	0	$x + 1$	0
0	0	x	0
$\frac{1}{2}$	$-\frac{9}{8}$	$x - \frac{1}{2}$	$-\frac{9}{8}$
1	-2	$x - 1$	-2
2	0	$x - 2$	0

$$f(-1) = (-1)^3 - (-1)^2 - 2(-1)$$
$$= -1 - 1 + 2$$
$$= 0$$

$$\begin{array}{r|rrrr} -1 & 1 & -1 & -2 & 0 \\ & & -1 & 2 & 0 \\ \hline & 1 & -2 & 0 & 0 \end{array}$$

$$f(0) = 0^3 - 0^2 - 2(0)$$
$$= 0$$

$$\begin{array}{r|rrrr} 0 & 1 & -1 & -2 & 0 \\ & & 0 & 0 & 0 \\ \hline & 1 & -1 & -2 & 0 \end{array}$$

$$f\left(\tfrac{1}{2}\right) = \left(\tfrac{1}{2}\right)^3 - \left(\tfrac{1}{2}\right)^2 - 2\left(\tfrac{1}{2}\right)$$
$$= \tfrac{1}{8} - \tfrac{1}{4} - 1$$
$$= \tfrac{1}{8} - \tfrac{2}{8} - \tfrac{8}{8}$$
$$= -\tfrac{9}{8}$$

$$\begin{array}{r|rrrr} \tfrac{1}{2} & 1 & -1 & -2 & 0 \\ & & \tfrac{1}{2} & -\tfrac{1}{4} & -\tfrac{9}{8} \\ \hline & 1 & -\tfrac{1}{2} & -\tfrac{9}{4} & -\tfrac{9}{8} \end{array}$$

$$f(1) = 1^3 - 1^2 - 2(1)$$
$$= 1 - 1 - 2$$
$$= -2$$

$$\begin{array}{r|rrrr} 1 & 1 & -1 & -2 & 0 \\ & & 1 & 0 & -2 \\ \hline & 1 & 0 & -2 & -2 \end{array}$$

$$f(2) = 2^3 - 2^2 - 2(2)$$
$$= 8 - 4 - 4$$
$$= 0$$

$$\begin{array}{r|rrrr} 2 & 1 & -1 & -2 & 0 \\ & & 2 & 2 & 0 \\ \hline & 1 & 1 & 0 & 0 \end{array}$$

The polynomial values equal the remainders.

37. *Geometry* You are given $V = x^3 + 18x^2 + 80x + 96$ for the volume of the solid shown (see figure in textbook). Find an expression for the missing dimension.

Solution

Volume = Area of triangle · Height (of prism)

$$\text{Area of triangle} = \frac{\text{Volume}}{\text{Height (of prism)}}$$

$$= \frac{x^3 + 18x^2 + 80x + 96}{x + 12}$$

$$= x^2 + 6x + 8$$

$$\text{Area of triangle} = \tfrac{1}{2} \cdot \text{base} \cdot \text{height}$$

$$\text{height} = \frac{2 \text{ Area of triangle}}{\text{base}}$$

$$= \frac{2\left(x^2 + 6x + 8\right)}{x + 2}$$

$$= 2x + 8 \quad \text{or} \quad 2(x + 4)$$

39. Multiply $(x + 1)^2$.

Solution

$$(x + 1)^2 = (x + 1)(x + 1)$$

$$= x^2 + 2x + 1$$

41. Multiply $(4 - 5z)(4 + 5z)$.

Solution

$$(4 - 5z)(4 + 5z) = 16 - 25z^2$$

43. *Geometry* The base of a triangle is $5x$ and its height is $2x + 9$. Find the area A of the triangle.

Solution

$$\text{Area} = \frac{1}{2}bh = \frac{1}{2}(5x)(2x + 9) = \frac{5}{2}x(2x + 9)$$

45. Perform the division of a polynomial by a monomial and check your result.

$$\frac{6z + 10}{2}$$

Solution

$$\frac{6z + 10}{2} = \frac{6z}{2} + \frac{10}{2} = 3z + 5$$

47. Perform the division of a polynomial by a monomial and check your result.

$$\frac{10z^2 + 4z - 12}{4}$$

Solution

$$\frac{10z^2 + 4z - 12}{4} = \frac{10z^2}{4} + \frac{4z}{4} - \frac{12}{4}$$

$$= \frac{5z^2}{2} + z - 3$$

49. Perform the division of a polynomial by a monomial and check your result.

$$(7x^3 - 2x^2) \div x$$

Solution

$$(7x^3 - 2x^2) \div x = \frac{7x^3 - 2x^2}{x}$$

$$= \frac{7x^3}{x} - \frac{2x^2}{x} = 7x^2 - 2x$$

51. Perform the division of a polynomial by a monomial and check your result.

$$\frac{8z^3 + 3z^2 - 2z}{2z}$$

Solution

$$\frac{8z^3 + 3z^2 - 2z}{2z} = \frac{8z^3}{2z} + \frac{3z^2}{2z} - \frac{2z}{2z}$$

$$= 4z^2 + \frac{3}{2}z - 1$$

53. Perform the division of a polynomial by a monomial and check your result.

$$\frac{m^4 + 2m^2 - 7}{m}$$

Solution

$$\frac{m^4 + 2m^2 - 7}{m} = \frac{m^4}{m} + \frac{2m^2}{m} - \frac{7}{m}$$

$$= m^3 + 2m - \frac{7}{m}$$

55. Perform the division and check your result.

$$\frac{4(x+5)^2 + 8(x+5)}{x+5}$$

Solution

$$\frac{f(x+5)^2 + 8(x+5)}{x+5} = \frac{4(x+5)^2}{x+5} + \frac{8(x+5)}{x+5}$$

$$= 4(x+5) + 8$$

57. Perform the division and check your result.

$$(x^2 - 15x + 50) \div (x+5)$$

Solution

$$(x^2 - 15x + 50) \div (x+5) = x + 5 \overline{\smash{\big)}\begin{array}{r} x + 10 \\ x^2 + 15x + 50 \end{array}}$$
$$\begin{array}{r} x^2 + 5x \\ \hline 10x + 50 \\ 10x + 50 \\ \hline \end{array}$$

59. Perform the division and check your result.

Divide $2y^2 + 7y + 3$ by $2y + 1$.

Solution

$$2y + 1 \overline{\smash{\big)}\begin{array}{r} y + 3 \\ 2y^2 + 7y + 3 \end{array}}$$
$$\begin{array}{r} 2y^2 + y \\ \hline 6y + 3 \\ 6y + 3 \\ \hline \end{array}$$

61. Perform the division and check your result.

$$\frac{12t^2 - 40t + 25}{2t - 5}$$

Solution

$$2t - 5 \overline{\smash{\big)}\begin{array}{r} 6t - 5 \\ 12t^2 - 40t + 25 \end{array}}$$
$$\begin{array}{r} 12t^2 - 30t \\ \hline -10t + 25 \\ -10t + 25 \\ \hline \end{array}$$

63. Perform the division and check your result.

$$\frac{16x^2 - 1}{4x + 1}$$

Solution

$$4x + 1 \overline{\smash{\big)}\begin{array}{r} 4x - 1 \\ 16x^2 + 0x - 1 \end{array}}$$
$$\begin{array}{r} 16x^2 + 4x \\ \hline -4x - 1 \\ -4x - 1 \\ \hline \end{array}$$

65. Perform the division and check your result.

$$\frac{x^3 + 125}{x + 5}$$

Solution

$$x + 5 \overline{\smash{\big)}\begin{array}{r} x^2 - 5x + 25 \\ x^3 + 0x^2 + 0x + 125 \end{array}}$$
$$\begin{array}{r} x^3 + 5x^2 \\ \hline -5x^2 + 0x \\ -5x^2 - 25x \\ \hline 25x + 125 \\ 25x + 125 \\ \hline \end{array}$$

67. Perform the division and check your result.

$$\frac{2x + 9}{x + 2}$$

Solution

$$x + 2 \overline{)\, 2x + 9} \quad 2 \quad + \frac{5}{x + 2}$$
$$\underline{\quad 2x + 4 \quad}$$
$$\qquad\qquad 5$$

69. Perform the division and check your result.

$$\frac{5x^2 + 2x + 3}{x + 2}$$

Solution

$$x + 2 \overline{)\, 5x^2 + 2x + 3} \quad 5x - 8 \quad + \frac{19}{x + 2}$$
$$\underline{\quad 5x^2 + 10x \quad}$$
$$\qquad\qquad -8x + 3$$
$$\qquad\qquad \underline{-8x - 16}$$
$$\qquad\qquad\qquad\qquad 19$$

71. Perform the division and check your result.

$$\frac{12x^2 - 17x - 5}{3x + 2}$$

Solution

$$3x + 2 \overline{)\, 12x^2 - 17x - 5} \quad 4x - \frac{25}{3} + \frac{\frac{35}{3}}{3x + 2}$$
$$\underline{\quad 12x^2 + 8x \quad}$$
$$\qquad\qquad -25x - 5$$
$$\qquad\qquad \underline{-25x - \dfrac{50}{3}}$$
$$\qquad\qquad\qquad\qquad \frac{35}{3}$$

73. Perform the division and check your result.

$$\frac{2x^3 - 5x^2 + x - 6}{x - 3}$$

Solution

$$x - 3 \overline{)\, 2x^3 - 5x^2 + x - 6} \quad 2x^2 + x + 4 \quad + \frac{6}{x - 3}$$
$$\underline{\quad 2x^3 - 6x^2 \quad}$$
$$\qquad\qquad x^2 + x$$
$$\qquad\qquad \underline{x^2 - 3x}$$
$$\qquad\qquad\qquad 4x - 6$$
$$\qquad\qquad\qquad \underline{4x - 12}$$
$$\qquad\qquad\qquad\qquad 6$$

75. Perform the division and check your result.

$$\frac{x^6 - 1}{x - 1}$$

Solution

$$
\begin{array}{r}
x^5 + x^4 + x^3 + x^2 + x + 1 \\
x - 1 \overline{\smash{)}\, x^6 \hspace{5.5cm} -1} \\
\underline{x^6 - x^5} \\
x^5 \\
\underline{x^5 - x^4} \\
x^4 \\
\underline{x^4 - x^3} \\
x^3 \\
\underline{x^3 - x^2} \\
x^2 \\
\underline{x^2 - x} \\
x - 1 \\
\underline{x - 1} \\
0
\end{array}
$$

77. *Think About It* Perform the division assuming that n is a positive integer.

$$\frac{x^{3n} + 3x^{2n} + 6x^n + 8}{x^n + 2}$$

Solution

$$
\begin{array}{r}
x^{2n} + x^n + 4 \\
x^n + 2 \overline{\smash{)}\, x^{3n} + 3x^{2n} + 6x^n + 8} \\
\underline{x^{3n} + 2x^{2n}} \\
x^{2n} + 6x^n \\
\underline{x^{2n} + 2x^n} \\
4x^n + 8 \\
\underline{4x^n + 8}
\end{array}
$$

79. Use synthetic division to perform the division.

$$\frac{x^4 - 4x^3 + x + 10}{x - 2}$$

Solution

$$\frac{x^4 - 4x^3 + x + 10}{x - 2}$$

$$
\begin{array}{r|rrrrr}
2 & 1 & -4 & 0 & 1 & 10 \\
  &   & 2 & -4 & -8 & -14 \\
\hline
  & 1 & -2 & -4 & -7 & -4
\end{array}
$$

$$\frac{x^4 - 4x^3 + x + 10}{x - 2} = x^3 - 2x^2 - 4x - 7 + \frac{-4}{x - 2}$$

81. Use synthetic division to perform the division.

$$\frac{5x^3 + 12}{x + 5}$$

Solution

$$\frac{5x^3 + 12}{x + 5}$$

$$
\begin{array}{r|rrrr}
-5 & 5 & 0 & 0 & 12 \\
   &   & -25 & 125 & -625 \\
\hline
   & 5 & -25 & 125 & -613
\end{array}
$$

$$\frac{5x^3 + 12}{x + 5} = 5x^2 - 25x + 125 + \frac{-613}{x + 5}$$

83. Use synthetic division to perform the division. Use the result to factor the dividend.

$$\frac{2a^2 + 13a - 45}{a + 9}$$

Solution

$$\frac{2a^2 + 13a - 45}{a + 9}$$

$$
\begin{array}{r|rrr}
-9 & 2 & 13 & -45 \\
   &   & -18 & 45 \\
\hline
   & 2 & -5 & 0 \\
\end{array}
$$

$$2a^2 + 13a - 45 = (2a - 5)(a + 9)$$

85. Use synthetic division to perform the division. Use the result to factor the dividend.

$$\frac{15x^2 - 2x - 8}{x - \dfrac{4}{5}}$$

Solution

$$\frac{15x^2 - 2x - 8}{x - \dfrac{4}{5}}$$

$$
\begin{array}{r|rrr}
\dfrac{4}{5} & 15 & -2 & -8 \\
   &   & 12 & 8 \\
\hline
   & 15 & 10 & 0 \\
\end{array}
$$

$$15x^2 - 2x - 8 = (15x + 10)\left(x - \frac{4}{5}\right)$$

87. Use synthetic division to perform the division. Use the result to factor the dividend.

$$\frac{2t^3 + 15t^2 + 19t - 30}{t + 5}$$

Solution

$$\frac{2t^3 + 15t^2 + 19t - 30}{t + 5}$$

$$
\begin{array}{r|rrrr}
-5 & 2 & 15 & 19 & -30 \\
   &   & -10 & -25 & 30 \\
\hline
   & 2 & 5 & -6 & 0 \\
\end{array}
$$

$$2t^3 + 15t^2 + 19t - 30 = (2t^2 + 5t - 6)(t + 5)$$

89. *Think About It*   Find the constant c so that the denominator will divide evenly into the numerator.

$$\frac{x^3 - 2x^2 - 4x + c}{x - 2}$$

Solution

$$\frac{x^3 - 2x^2 - 4x + c}{x - 2}$$

$$
\begin{array}{r|rrrr}
2 & 1 & 2 & -4 & c \\
  &   & 2 & 8 & 8 \\
\hline
  & 1 & 4 & 4 & 0 \\
\end{array}
$$

$$c + 8 = 0$$

$$c = -8$$

91. *Geometry* The rectangle's area is $2x^3 + 3x^2 - 6x - 9$ (see figure in textbook). Find its width if its length is $2x + 3$.

Solution

$$\text{Area} = \text{Length} \cdot \text{Width}, \quad \text{so} \quad \text{Width} = \frac{\text{Area}}{\text{Length}}$$

$$\text{Length} = \frac{2x^3 + 3x^2 - 6x - 9}{2x + 3} = x^2 - 3$$

$$
\begin{array}{r}
x^2 \qquad - 3 \\
2x + 3 \overline{)\ 2x^3 + 3x^2 - 6x - 9} \\
\underline{2x^3 + 3x^2} \qquad\qquad \\
-6x - 9 \\
\underline{-6x - 9} \\
\end{array}
$$

4.5   Graphing Rational Functions

5. Given $f(x) = \dfrac{4}{x - 1}$

(a) complete each table, (b) determine the vertical and horizontal asymptotes of the graph, and (c) find the domain of the function.

Solution

x	0	0.5	0.9	0.99	0.999
y	-4	-8	-40	-400	-4000

x	2	1.5	1.1	1.01	1.001
y	4	8	40	400	4000

x	2	5	10	100	1000
y	4	1	0.44444	0.0404	0.004

7. Find the domain of the function and identify any horizontal and vertical asymptotes.

$$f(x) = \frac{5}{x^2}$$

Solution

Domain: $x^2 \neq 0 \quad (-\infty, 0) \cup (0, \infty)$

$$x \neq 0$$

Vertical asymptote: $x = 0$

Horizontal asymptote:

$y = 0$ since the degree of the numerator is less than the degree of the denominator.

9. Find the domain of the function and identify any horizontal and vertical asymptotes.

$$g(t) = \frac{3}{t^2 + 1}$$

Solution

Domain: $t^2 + 1 \neq 0 \quad (-\infty, \infty)$

no real solution

Vertical asymptote: none

Horizontal asymptote:

$y = 0$ since the degree of the numerator is less than the degree of the denominator.

11. Find the domain of the function and identify any horizontal and vertical asymptotes.

$$y = \frac{5x^2}{x^2 - 1}$$

Solution

Domain: $\qquad x^2 - 1 \neq 0 \quad (-\infty, -1) \cup (-1, 1) \cup (1, \infty)$

$\qquad\qquad (x - 1)(x + 1) \neq 0$

$\qquad\qquad\quad x \neq 1 \quad x \neq -1$

Vertical asymptotes: $x = 1, x = -1$

Horizontal asymptote: $y = 5$ since the degree of the numerator is equal to the degree of the denominator and the leading coefficient of the numerator is 5 and the leading coefficient of the denominator is 1.

13. Match the function with its graph.

$$f(x) = \frac{3}{x + 2}$$

Solution

(d)

15. Match the function with its graph.

$$f(x) = \frac{3x^2}{x + 2}$$

Solution

(a)

17. Sketch the graph of $f(x) = \dfrac{1}{x - 2}$. As sketching aids, check for intercepts, vertical asymptotes, and horizontal asymptotes.

Solution

y-intercept: $f(0) = \dfrac{1}{0 - 2} = -\dfrac{1}{2}$

x-intercept: none, numerator is never zero

Vertical asymptote: $x - 2 = 0$

$$x = 2$$

Horizontal asymptote: $y = 0$ since the degree of the numerator is less than the degree of the denominator

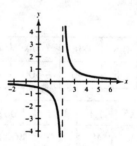

19. Sketch the graph of $g(x) = \dfrac{1}{2 - x}$. As sketching aids, check for intercepts, vertical asymptotes, and horizontal asymptotes.

Solution

y-intercept: $g(0) = \dfrac{1}{2 - 0} = \dfrac{1}{2}$

x-intercept: none, numerator is never zero

Vertical asymptote: $2 - x = 0$

$$x = 2$$

Horizontal asymptote: $y = 0$ since the degree of the numerator is less than the degree of the denominator.

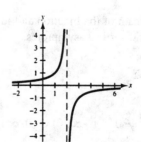

21. Sketch the graph of $y = \dfrac{2x + 4}{x}$. As sketching aids, check for intercepts, vertical asymptotes, and horizontal asymptotes.

Solution

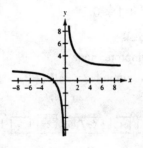

y-intercept: none, denominator cannot be zero

x-intercept: $2x + 4 = 0$

$\qquad\qquad\quad x = -2$

Vertical asymptote: $x = 0$

Horizontal asymptote: $y = 2$ since the degree of the numerator is equal to the degree of the denominator

23. Sketch the graph of $y = \dfrac{2x^2}{x^2 + 1}$. As sketching aids, check for intercepts, vertical asymptotes, and horizontal asymptotes.

Solution

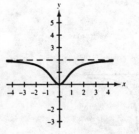

y-intercept: $y = \dfrac{2(0)^2}{0^2 + 1} = 0$

x-intercept: $x = 0$

Vertical asymptote: none, $x^2 + 1 = 0$ has no real solutions

Horizontal asymptote: $y = 2$ since the degree of the numerator is equal to the degree of the denominator

25. Use a graphing utility to graph the function $h(x) = \dfrac{x - 3}{x - 1}$. Give the domain of the function and identify any horizontal or vertical asymptotes.

Solution

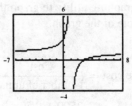

Domain: $x - 1 \neq 0$

$\qquad\qquad x \neq 1$

Vertical asymptote: $x = 1$

Horizontal asymptote: $y = 1$

Keystrokes:

27. Use a graphing utility to graph the function $f(t) = \dfrac{6}{t^2 + 1}$.

Solution

Domain: $t^2 + 1 \neq 0$

       Reals

Vertical asymptote: none

Horizontal asymptote: $y = 0$

Keystrokes:

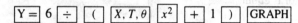

$\boxed{Y=}$ 6 $\boxed{\div}$ $\boxed{(}$ $\boxed{X, T, \theta}$ $\boxed{x^2}$ $\boxed{+}$ 1 $\boxed{)}$ $\boxed{\text{GRAPH}}$

29. Use the graph of $f(x) = \dfrac{1}{x}$ to sketch $g(x) = -\dfrac{1}{x}$.

Solution

Reflect $f(x)$ about the x-axis.

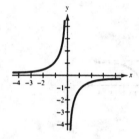

31. Use the graph of $f(x) = \dfrac{1}{x}$ to sketch $g(x) = \dfrac{1}{x-2}$.

Solution

Shift graph 2 units to the right.

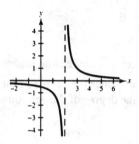

33. *Average Cost* The cost of producing x units is $C = 2500 + 0.50x$, $0 < x$.

(a) Write the average cost $\overline{A}$ as a function of x.

(b) Find the average costs of producing $x = 1000$ and $x = 10{,}000$ units.

(c) Use a graphing utility to graph the average cost function. Determine the horizontal asymptote of the graph.

Solution

(a) Average cost $= \dfrac{\text{Cost}}{\text{Number of units}}$

       $\overline{A} = \dfrac{2500 + 0.50x}{x}$, $0 < x$

(b) $\overline{A} = \dfrac{2500 + 0.50(1000)}{1000} = \$3$

    $\overline{A} = \dfrac{2500 + 0.50(10{,}000)}{10{,}000} = \$0.75$

(c) Keystrokes:

$\boxed{Y=}$ $\boxed{(}$ 2500 $\boxed{+}$ .5 $\boxed{X, T, \theta}$ $\boxed{)}$ $\boxed{\div}$ $\boxed{X, T, \theta}$ $\boxed{\text{GRAPH}}$

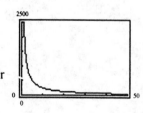

Horizontal asymptote

$\overline{A} = \$0.50$ since the degree of the numerator is equal to the degree of the denominator and the leading coefficient of the numerator is 0.50 and the leading coefficient of the denominator is 1.

35. Solve the inequality $2x - 12 \geq 0$ and sketch the graph of the solution on the real number line.

Solution

$$2x - 12 \geq 0$$

$$2x \geq 12$$

$$x \geq 6$$

37. Solve the inequality $|x - 3| < 2$ and sketch the graph of the solution on the real number line.

Solution

$$|x - 3| < 2$$

$$-2 < x - 3 < 2$$

$$1 < x < 5$$

39. Determine all real numbers n such that $\frac{1}{3}n$ must be at least 10 and no more than 50.

Solution

$$10 \leq \frac{1}{3}n \leq 50$$

$$30 \leq n \leq 150$$

41. Find the domain of the function $f(x) = 2 + \dfrac{1}{x - 3}$ and identify any horizontal and vertical asymptotes.

Solution

Domain: $x - 3 \neq 0 \quad (-\infty, 3) \cup (3, \infty)$

$$x \neq 3$$

Vertical asymptote: $x = 3$

Horizontal asymptote: $y = 2$

43. Find the domain of the function $f(x) = \dfrac{3x}{x^2 - 9}$ and identify any horizontal and vertical asymptotes.

Solution

Domain: $\qquad x^2 - 9 \neq 0$

$$(x - 3)(x + 3) \neq 0 \quad (-\infty, -3) \cup (-3, 3) \cup (3, \infty)$$

$$x \neq 3 \quad x \neq -3$$

Vertical asymptotes: $x = 3, x = -3$

Horizontal asymptote: $y = 0$ since the degree of the numerator is less than the degree of the denominator.

45. Find the domain of the function $f(x) = \dfrac{x}{x + 8}$ and identify any horizontal and vertical asymptotes.

Solution

Domain: $x + 8 \neq 0$

$$x \neq -8 \quad (-\infty, -8) \cup (-8, \infty)$$

Vertical asymptote: $x = -8$

Horizontal asymptote: $y = 1$ since the degree of the numerator is equal to the degree of the denominator and the leading coefficients are 1.

47. Find the domain of the function $g(t) = \dfrac{3}{t(t - 1)}$ and identify any horizontal and vertical asymptotes.

Solution

$$t(t - 1) \neq 0$$

$$t \neq 0 \quad t - 1 \neq 0$$

$$t \neq 1 \quad (-\infty, 0) \cup (0, 1) \cup (1, \infty)$$

Vertical asymptotes: $t = 0, t = 1$

Horizontal asymptote: $y = 0$ since the degree of the numerator is less than the degree of the denominator

49. Find the domain of the function $y = \dfrac{2x^2}{x^2 + 1}$ and identify any horizontal and vertical asymptotes.

Solution

Domain: $x^2 + 1 \neq 0 \quad (-\infty, \infty)$

        no real solution

Vertical asymptote: none

Horizontal asymptote: $y = 2$ since the degree of the numerator is equal to the degree of the denominator and the leading coefficient of the numerator is 2 and the leading coefficient of the denominator is 1.

51. Sketch the graph of $g(x) = \dfrac{5}{x}$. As sketching aids, check for intercepts, vertical asymptotes, and horizontal asymptotes. Use a graphing utility to verify your graph.

Solution

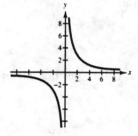

y-intercept: $g(0) = \dfrac{5}{0} =$ undefined, none

x-intercept: none, numerator is never zero

Vertical asymptote: $x = 0$

Horizontal asymptote: $y = 0$ since the degree of the numerator is less than the degree of the denominator

53. Sketch the graph of $f(x) = \dfrac{5}{x^2}$. As sketching aids, check for intercepts, vertical asymptotes, and horizontal asymptotes. Use a graphing utility to verify your graph.

Solution

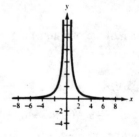

y-intercept: $f(0) = \dfrac{5}{0^2} =$ undefined, none

x-intercept: none, numerator is never zero

Vertical asymptote: $x = 0$

Horizontal asymptote: $y = 0$ since the degree of the numerator is less than the degree of the denominator

55. Sketch the graph of $g(t) = 3 - \dfrac{2}{t}$. As sketching aids, check for intercepts, vertical asymptotes, and horizontal asymptotes. Use a graphing utility to verify your graph.

Solution

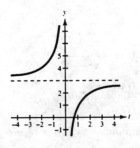

y-intercept: $g(0) = 3 - \dfrac{2}{0} =$ undefined, none

x-intercept: $0 = 3 - \dfrac{2}{t}$

$\qquad\qquad 0 = 3t - 2$

$\qquad\qquad 2 = 3t$

$\qquad\qquad \dfrac{2}{3} = t$

Vertical asymptote: $t = 0$

Horizontal asymptote: $y = 3$

57. Sketch the graph of $y = \dfrac{3x}{x+4}$. As sketching aids, check for intercepts, vertical asymptotes, and horizontal asymptotes. Use a graphing utility to verify your graph.

Solution

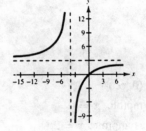

y-intercept: $y = \dfrac{3(0)}{0+4} = 0$

x-intercept: $0 = \dfrac{3x}{x+4}$

$\qquad\qquad 0 = 3x$

$\qquad\qquad 0 = x$

Vertical asymptote: $x + 4 = 0$

$\qquad\qquad\qquad\qquad x = -4$

Horizontal asymptote: $y = 3$ since the degree of the numerator is equal to the degree of the denominator and the leading coefficient of numerator is 3 and the leading coefficient of denominator is 1.

59. Sketch the graph of $y = \dfrac{4}{x^2+1}$. As sketching aids, check for intercepts, vertical asymptotes, and horizontal asymptotes. Use a graphing utility to verify your graph.

Solution

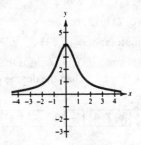

y-intercept: $y = \dfrac{4}{0^2+1} = 4$

x-intercept: none, numerator is never zero

Vertical asymptote: none, $x^2 + 1 \neq 0$

$\qquad\qquad\qquad\qquad$ no real solution

Horizontal asymptote: $y = 0$, since the degree of the numerator is less than the degree of the denominator

61. Sketch the graph of $y = -\dfrac{x}{x^2 - 4}$. As sketching aids, check for intercepts, vertical asymptotes, and horizontal asymptotes. Use a graphing utility to verify your graph.

Solution

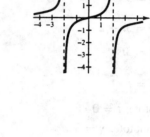

y-intercept: $y = \dfrac{-0}{0^2 - 4} = 0$

x-intercept: $0 = -\dfrac{x}{x^2 - 4}$

$$0 = -x$$

$$0 = x$$

Vertical asymptote: $x = 2, \quad x = -2$

$$x^2 - 4 = 0$$

$$(x - 2)(x + 2) = 0$$

$$x = 2 \quad x = -2$$

Horizontal asymptote: $y = 0$, since the degree of the numerator is less
    than the degree of the denominator

63. Use a graphing utility to graph the function $y = \dfrac{2(x^2 + 1)}{x^2}$. Give its domain.

Solution

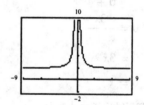

Domain: $x^2 \neq 0$

$$x \neq 0$$

$(-\infty, 0) \cup (0, \infty)$

Vertical asymptote: $x = 0$

Horizontal asymptote: $y = 2$

Keystrokes:

$\boxed{Y=}\ \boxed{(}\ \boxed{2}\ \boxed{(}\ \boxed{X,T,\theta}\ \boxed{x^2}\ \boxed{+}\ \boxed{1}\ \boxed{)}\ \boxed{)}\ \boxed{\div}\ \boxed{X,T,\theta}\ \boxed{x^2}\ \boxed{GRAPH}$

65. Use a graphing utility to graph the function $y = \dfrac{3}{x} + \dfrac{1}{x - 2}$. Give its domain.

Solution

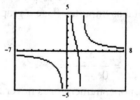

Domain: $x \neq 0 \quad x - 2 \neq 0$

$$x \neq 2$$

$(-\infty, 0) \cup (0, 2) \cup (2, \infty)$

Vertical asymptote: $x = 0, x = 2$

Horizontal asymptote: $y = 0$, since the degree of the numerator is less
    than the degree of the denominator

Keystrokes:

$\boxed{Y=}\ \boxed{3}\ \boxed{\div}\ \boxed{X,T,\theta}\ \boxed{+}\ \boxed{1}\ \boxed{\div}\ \boxed{(}\ \boxed{X,T,\theta}\ \boxed{-}\ \boxed{2}\ \boxed{)}\ \boxed{GRAPH}$

67. Use the graph of $f(x) = \dfrac{4}{x^2}$ to sketch the graph of

$$g(x) = 2 + \frac{4}{x^2}.$$

Solution

Shift $f(x)$ up 2 units

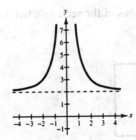

69. Use the graph of $f(x) = \dfrac{4}{x^2}$ to sketch the graph of

$$g(x) = -\frac{4}{(x-2)^2}.$$

Solution

Shift $f(x)$ right 2 units

Reflect across x-axis

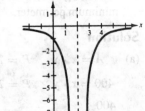

71. *Think About It* Use a graphing utility to graph the function. Explain why there is no vertical asymptote when a superficial examination of the function may indicate that there should be one.

$$g(x) = \frac{4 - 2x}{x - 2}$$

Solution

Reduce $g(x)$ to lowest terms

$$g(x) = \frac{2(2 - x)}{x - 2} = -2$$

Keystrokes:

| Y= | (| 4 | − | 2 | X,T,θ |) | ÷ | (| X,T,θ | − | 2 |) | GRAPH |

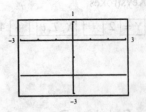

73. *Medicine* The concentration of a certain chemical in the bloodstream t hours after injection into the muscle tissue is given by

$$C = \frac{2t}{4t^2 + 25}, \quad 0 \le t$$

(a) Determine the horizontal asymptote of the function and interpret its meaning in the context of the problem.

(b) Graph the function on a graphing utility. Approximate the time when the concentration is the greatest.

Solution

(a) $C = 0$ is the horizontal asymptote, since the degree of the numerator is less than the degree of the denominator. The meaning in the context of the problem is that the chemical is eliminated from the body.

(b) Keystrokes:

| Y= | 2 | X,T,θ | ÷ | (| 4 | X,T,θ | x² | + | 25 |) | GRAPH |

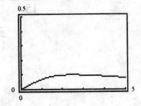

Maximum occurs when $t \approx 2.5$

75. *Geometry* A rectangular region of length x and width y has an area of 400 square meters.

(a) Verify that the perimeter P is given by $P = 2\left(x + \dfrac{400}{x}\right)$.

(b) Determine the domain of the function within the physical constraints of the problem.

(c) Sketch a graph of the function and approximate the dimensions of the rectangle that has a minimum perimeter.

Solution

(a)
$$A = x \cdot y \qquad P = 2l + 2w$$
$$400 = x \cdot y \qquad P = 2(l + w)$$
$$\frac{400}{x} = y \qquad P = 2\left(x + \frac{400}{x}\right)$$

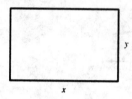

(b) Domain: $x > 0$

(c) Minimum perimeter: 20 units × 20 units

Keystrokes:

$\boxed{Y=}$ 2 $\boxed{(}$ $\boxed{X, T, \theta}$ $\boxed{+}$ 400 $\boxed{\div}$ $\boxed{X, T, \theta}$ $\boxed{)}$ $\boxed{\text{GRAPH}}$

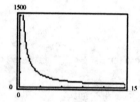

4.6 Solving Rational Equations

7. Determine whether the values of x are solutions to $\dfrac{x}{3} - \dfrac{x}{5} = \dfrac{4}{3}$.

(a) $x = 0$ (b) $x = -1$ (c) $x = \dfrac{1}{8}$ (d) $x = 10$

Solution

(a) $x = 0$

$$\frac{0}{3} - \frac{0}{5} \overset{?}{=} \frac{4}{3}$$
$$0 \neq \frac{4}{3}$$

No

(b) $x = -1$

$$\frac{-1}{3} - \frac{-1}{5} \overset{?}{=} \frac{4}{3}$$
$$\frac{-5}{15} - \frac{-3}{15} \overset{?}{=} \frac{20}{15}$$
$$\frac{-5}{15} + \frac{3}{15} \overset{?}{=} \frac{20}{15}$$
$$\frac{-2}{15} \neq \frac{20}{15}$$

No

(c) $x = \dfrac{1}{8}$

$$\frac{1/8}{3} - \frac{1/8}{5} \overset{?}{=} \frac{4}{3}$$
$$\frac{1}{24} - \frac{1}{40} \overset{?}{=} \frac{4}{3}$$
$$\frac{5}{120} - \frac{3}{120} \overset{?}{=} \frac{160}{120}$$
$$\frac{2}{120} \neq \frac{160}{120}$$

No

(d) $x = 10$

$$\frac{10}{3} - \frac{10}{5} \overset{?}{=} \frac{4}{3}$$
$$\frac{50}{15} - \frac{30}{15} \overset{?}{=} \frac{20}{15}$$
$$\frac{20}{15} = \frac{20}{15}$$

Yes

9. Determine whether the values of x are solutions to $\dfrac{x}{4} + \dfrac{3}{4x} = 1$.

(a) $x = -1$ (b) $x = 1$ (c) $x = 3$ (d) $x = 2$

Solution

(a) $x = -1$

$$\frac{-1}{4} + \frac{3}{4(-1)} \overset{?}{=} 1$$

$$\frac{-1}{4} + \frac{-3}{4} \overset{?}{=} 1$$

$$-1 \neq 1$$

No

(b) $x = 1$

$$\frac{1}{4} + \frac{3}{4(1)} \overset{?}{=} 1$$

$$\frac{1}{4} + \frac{3}{4} \overset{?}{=} 1$$

$$1 = 1$$

Yes

(c) $x = 3$

$$\frac{3}{4} + \frac{3}{4(3)} \overset{?}{=} 1$$

$$\frac{3}{4} + \frac{3}{12} \overset{?}{=} 1$$

$$\frac{3}{4} + \frac{1}{4} \overset{?}{=} 1$$

$$1 = 1$$

Yes

(d) $x = 2$

$$\frac{2}{4} + \frac{3}{4(2)} \overset{?}{=} 1$$

$$\frac{4}{8} + \frac{3}{8} \overset{?}{=} 1$$

$$\frac{7}{8} \neq 1$$

No

11. Solve $\dfrac{x}{4} = \dfrac{3}{8}$.

Solution

$$\frac{x}{4} = \frac{3}{8}$$

$$8\left(\frac{x}{4}\right) = \left(\frac{3}{8}\right)8$$

$$2x = 3$$

$$x = \frac{3}{2}$$

Check: $\dfrac{\frac{3}{2}}{4} \overset{?}{=} \dfrac{3}{8}$

$$\frac{3}{8} = \frac{3}{8}$$

13. Solve $\dfrac{h}{5} - \dfrac{h+2}{9} = \dfrac{2}{3}$.

Solution

$$\frac{h}{5} - \frac{h+2}{9} = \frac{2}{3}$$

$$45\left(\frac{h}{5} - \frac{h+2}{9}\right) = \left(\frac{2}{3}\right)45$$

$$9h - 5(h+2) = 30$$

$$9h - 5h - 10 = 30$$

$$4h - 10 = 30$$

$$4h = 40$$

$$h = 10$$

Check: $\dfrac{10}{5} - \dfrac{10+2}{9} \overset{?}{=} \dfrac{2}{3}$

$$2 - \frac{12}{9} = \frac{2}{3}$$

$$2 - \frac{4}{3} = \frac{2}{3}$$

$$\frac{2}{3} = \frac{2}{3}$$

15. Solve $\dfrac{7}{x} = 21$.

Solution

$$\dfrac{7}{x} = 21 \qquad \textbf{Check:} \quad \dfrac{7}{\frac{1}{3}} \stackrel{?}{=} 2$$

$$x\left(\dfrac{7}{x}\right) = (21)x \qquad\qquad 21 = 21$$

$$7 = 21x$$

$$\dfrac{7}{21} = x$$

$$\dfrac{1}{3} = x$$

17. Solve $\dfrac{12}{y+5} + \dfrac{1}{2} = 2$.

Solution

$$\dfrac{12}{y+5} + \dfrac{1}{2} = 2$$

$$2(y+5)\left(\dfrac{12}{y+5} + \dfrac{1}{2}\right) = (2)2(y+5)$$

$$24 + y + 5 = 4(y+5)$$

$$y + 29 = 4y + 20$$

$$9 = 3y$$

$$3 = y$$

$$\textbf{Check:} \quad \dfrac{12}{3+5} + \dfrac{1}{2} \stackrel{?}{=} 2$$

$$\dfrac{3}{2} + \dfrac{1}{2} = 2$$

$$\dfrac{4}{2} = 2$$

$$2 = 2$$

19. Solve $\dfrac{4}{2x+3} + \dfrac{17}{5(2x+3)} = 3$.

Solution

$$\dfrac{4}{2x+3} + \dfrac{17}{5(2x+3)} = 3$$

$$5(2x+3)\left(\dfrac{4}{2x+3} + \dfrac{17}{5(2x+3)}\right) = (3)5(2x+3)$$

$$20 + 17 = 15(2x+3)$$

$$37 = 30x + 45$$

$$-8 = 30x$$

$$-\dfrac{8}{30} = x$$

$$-\dfrac{4}{15} = x$$

$$\textbf{Check:} \quad \dfrac{4}{2\left(-\dfrac{4}{15}\right)+3} + \dfrac{17}{5\left[2\left(-\dfrac{4}{15}\right)+3\right]} \stackrel{?}{=} 3$$

$$\dfrac{4}{-\dfrac{8}{15}+\dfrac{45}{15}} + \dfrac{17}{5\left[-\dfrac{8}{15}+\dfrac{45}{15}\right]} = 3$$

$$\dfrac{4}{\dfrac{37}{15}} + \dfrac{17}{5\left(\dfrac{37}{15}\right)} = 3$$

$$\dfrac{60}{37} + \dfrac{51}{37} = 3$$

$$\dfrac{111}{37} = 3$$

$$3 = 3$$

21. Solve $\dfrac{x}{x+4} + \dfrac{4}{x+4} + 2 = 0$.

Solution

$$\frac{x}{x+4} + \frac{4}{x+4} + 2 = 0 \qquad \textbf{Check:} \quad \frac{-4}{-4+4} + \frac{4}{-4+4} + 2 \overset{?}{=} 0$$

$$(x+4)\left(\frac{x}{x+4} + \frac{4}{x+4} + 2\right) = (0)(x+4) \qquad\qquad \frac{-4}{0} + \frac{4}{0} + 2 \neq 0$$

$$x + 4 + 2(x+4) = 0$$

$$x + 4 + 2x + 8 = 0$$

$$3x = -12$$

$$x = -4$$

Division by zero is undefined. Solution is extraneous, so equation has no solution.

23. Solve $\dfrac{32}{t} = 2t$.

Solution

$$\frac{32}{t} = 2t \qquad\qquad \textbf{Check:} \quad \frac{32}{4} \overset{?}{=} 2(4)$$

$$t\left(\frac{32}{t}\right) = (2t)t \qquad\qquad\qquad 8 = 8$$

$$32 = 2t^2 \qquad\qquad\qquad\qquad \frac{32}{-4} = 2(-4)$$

$$16 = t^2 \qquad\qquad\qquad\qquad -8 = -8$$

$$0 = t^2 - 16$$

$$0 = (t-4)(t+4)$$

$$t = 4 \quad t = -4$$

25. Solve $\dfrac{1}{x-1} + \dfrac{3}{x+1} = 2$.

Solution

$$\frac{1}{x-1} + \frac{3}{x+1} = 2 \qquad \textbf{Check:} \quad \frac{1}{0-1} + \frac{3}{0+1} \overset{?}{=} 2$$

$$(x-1)(x+1)\left(\frac{1}{x-1} + \frac{3}{x+1}\right) = (2)(x-1)(x+1) \qquad\qquad -1 + 3 = 2$$

$$x + 1 + 3(x-1) = 2(x^2 - 1) \qquad\qquad\qquad 2 = 2$$

$$x + 1 + 3x - 3 = 2x^2 - 2 \qquad\qquad\qquad \frac{1}{2-1} + \frac{3}{2+1} \overset{?}{=} 2$$

$$4x - 2 = 2x^2 - 2 \qquad\qquad\qquad\qquad 1 + 1 = 2$$

$$0 = 2x^2 - 4x \qquad\qquad\qquad\qquad 2 = 2$$

$$0 = 2x(x-2)$$

$$x = 0 \quad x = 2$$

27. Solve $\dfrac{x}{2} = \dfrac{2 - \dfrac{3}{x}}{1 - \dfrac{1}{x}}$.

Solution

$$\frac{x}{2} = \frac{2 - \dfrac{3}{x}}{1 - \dfrac{1}{x}}$$

$$\frac{x}{2} = \frac{2 - \dfrac{3}{x}}{1 - \dfrac{1}{x}} \cdot \frac{x}{x}$$

$$2(x - 1)\left(\frac{x}{2}\right) = \left(\frac{2x - 3}{x - 1}\right)2(x - 1)$$

$$x(x - 1) = (2x - 3)2$$

$$x^2 - x = 4x - 6$$

$$x^2 - 5x + 6 = 0$$

$$(x - 3)(x - 2) = 0$$

$$x = 3 \quad x = 2$$

Check:

$$\frac{3}{2} \overset{?}{=} \frac{2 - (3/3)}{1 - (1/3)} \qquad \frac{2}{2} \overset{?}{=} \frac{2 - (3/2)}{1 - (1/2)}$$

$$\frac{3}{2} = \frac{1}{(2/3)} \qquad 1 = \frac{(1/2)}{(1/2)}$$

$$\frac{3}{2} = \frac{3}{2} \qquad 1 = 1$$

29. (a) Use the graph (in the textbook) to determine any x-intercepts of the equation, and (b) set $y = 0$ and solve the resulting equation to confirm your result.

$$y = \frac{x + 2}{x - 2}$$

Solution

x-intercepts: $(0, -2)$

$$0 = \frac{x + 2}{x - 2}$$

$$(x - 2)\left(0 = \frac{x + 2}{x - 2}\right)(x - 2)$$

$$0 = x + 2$$

$$-2 = x$$

31. (a) Use the graph (in the textbook) to determine any x-intercepts of the equation, and (b) set $y = 0$ and solve the resulting equation to confirm your result.

$$y = x - \frac{1}{x}$$

Solution

x-intercepts: $(-1, 0)$ and $(1, 0)$

$$0 = x - \frac{1}{x}$$

$$x\left(0 = x - \frac{1}{x}\right)x$$

$$0 = x^2 - 1$$

$$0 = (x - 1)(x + 1)$$

$$x - 1 = 0 \quad x + 1 = 0$$

$$x = 1 \qquad x = -1$$

33. *Graphical Reasoning*　(a) Use a graphing utility to graph the equation and determine any
x-intercepts of the equation, and (b) set $y = 0$ and solve the resulting rational equation to
confirm the result of part (a).

$$y = \frac{1}{x} + \frac{4}{x-5}$$

Solution

Keystrokes:

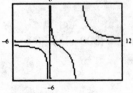

x-intercept: $(1, 0)$

$$0 = \frac{1}{x} + \frac{4}{x-5}$$

$$x(x-5)(0) = \left(\frac{1}{x} + \frac{4}{x-5}\right)x(x-5)$$

$$0 = x - 5 + 4x$$

$$5 = 5x$$

$$1 = x$$

35. *Graphical Reasoning*　(a) Use a graphing utility to graph the equation and determine any
x-intercepts of the equation, and (b) set $y = 0$ and solve the resulting rational equation to
confirm the first result.

$$y = (x + 1) - \frac{6}{x}$$

Solution

Keystrokes:

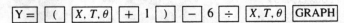

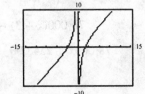

x-intercepts: $(-3, 0)$ and $(2, 0)$

$$0 = (x + 1) - \frac{6}{x}$$

$$x(0) = \left[(x + 1) - \frac{6}{x}\right]x$$

$$0 = x^2 + x - 6$$

$$0 = (x + 3)(x - 2)$$

$$x + 3 = 0 \qquad x - 2 = 0$$

$$x = -3 \qquad x = 2$$

37. *Wind Speed*   A plane with a speed of 300 miles per hour in still air travels 680 miles with a tail wind in the same time it could travel 520 miles with a head wind of equal speed. Find the speed of the wind.

Solution

Verbal Model: $\boxed{\text{Distance}} \div \boxed{\text{Rate}} = \boxed{\text{Time}}$

Equation:

$$\frac{680}{300 + x} = \frac{520}{300 - x}$$

$$(300 + x)(300 - x)\left(\frac{680}{300 + x}\right) = \left(\frac{520}{300 - x}\right)(300 + x)(300 - x)$$

$$680(300 - x) = 520(300 + x)$$

$$204{,}000 - 680x = 156{,}000 + 520x$$

$$-1200x = -48{,}000$$

$$x = 40 \text{ miles per hour}$$

39. *Partnership Costs*   Some partners buy a piece of property for $78,000 by sharing the cost equally. To ease the financial burden, they look for three additional partners to reduce the cost per person by $1300. How many partners are presently in the group?

Solution

Verbal Model: $\boxed{\begin{array}{c}\text{Cost per person}\\\text{original group}\end{array}} - \boxed{\begin{array}{c}\text{Cost per person}\\\text{new group}\end{array}} = \boxed{1300}$

Equation:

$$\frac{78{,}000}{x} - \frac{78{,}000}{x + 3} = 1300$$

$$x(x + 3)\left(\frac{78{,}000}{x} - \frac{78{,}000}{x + 3}\right) = (1300)x(x + 3)$$

$$78{,}000(x + 3) - 78{,}000x = 1300x(x + 3)$$

$$78{,}000x + 234{,}000 - 78{,}000x = 1300x^2 + 3900x$$

$$0 = 1300x^2 + 3900x - 234{,}000$$

$$0 = x^2 + 3x - 180$$

$$0 = (x + 15)(x - 12)$$

$$x = -15 \quad x = 12$$

They would like to have $12 + 3$ or 15 persons in the group.

41. *Swimming Pool* The flow rate of one pipe is $1\frac{1}{4}$ times that of a second pipe. A swimming pool can be filled in five hours using both pipes. Find the time required to fill the pool using only the pipe with the lower flow rate.

Solution

Verbal Model: $\boxed{\text{Rate Pipe 1}} + \boxed{\text{Rate Pipe 2}} = \boxed{\text{Rate Together}}$

Equation:
$$\frac{1}{x} + \frac{1}{\frac{5}{4}x} = \frac{1}{5}$$

$$5x\left(\frac{1}{x} + \frac{4}{5x}\right) = \left(\frac{1}{5}\right)5x$$

$$5 + 4 = x$$

$$9 \text{ hours} = x$$

$$11\frac{1}{4} \text{ hours } = \frac{45}{4} = \frac{5}{4}x$$

43. *Using a Model* Use the following model, which approximates the total revenue y (in billions of dollars) for the car and truck rental industry in the United States from 1985 to 1991.

$$y = 43.31 - \frac{275.25}{t} + \frac{654.53}{t^2}, \quad 5 \le t \le 11$$

In this model, $t = 0$ represents 1980. (Source: *Current Business Reports*)

(a) Use the bar graph (in the textbook) to *graphically* determine the year the total revenue first exceeded $20 billion.

(b) Use the model to *algebraically* confirm your answer to part (a).

(c) What would you estimate the revenue to be in 1996? Explain your reasoning.

Solution

(a) 1989

(b) In 1989, $t = 9$. Substitute 9 for t in model

$$y = 43.31 - \frac{275.25}{9} + \frac{654.53}{9^2}$$

$$y = 43.31 - 30.58333 + 8.080617281$$

$$y = \$20.81 \text{ billion}$$

(c) In 1996, $t = 16$. Substitute 16 for t in model.

$$y = 43.31 - \frac{275.25}{16} + \frac{654.53}{16^2}$$

$$y = 43.31 - 17.203125 + 2.556757813$$

$$y = \$28.66 \text{ billion}$$

45. Solve $125 - 50x = 0$. Show how to use a graphing utility to check your solution.

Solution

$$125 - 50x = 0$$

$$125 = 50x$$

$$\frac{125}{50} = x$$

$$\frac{25}{10} = x$$

$$\frac{5}{2} = x$$

Keystrokes:

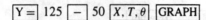

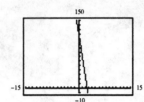

Find x-intercept $x = \dfrac{5}{2}$.

47. Solve $x^2 + x - 42 = 0$. Show how to use a graphing utility to check your solution.

Solution

$$x^2 + x - 42 = 0$$

$$(x + 7)(x - 6) = 0$$

$$x + 7 = 0 \qquad x - 6 = 0$$

$$x = -7 \qquad x = 6$$

Keystrokes:

$\boxed{Y=}$ $\boxed{X, T, \theta}$ $\boxed{x^2}$ $\boxed{+}$ $\boxed{X, T, \theta}$ $\boxed{-}$ 42 $\boxed{GRAPH}$

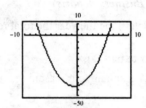

Find x-intercepts $x = -7$ and $x = 6$

49. Find two consecutive positive even integers whose product is 624.

Solution

Verbal Model: $\boxed{\begin{array}{c}\text{1st even} \\ \text{integer}\end{array}} \cdot \boxed{\begin{array}{c}\text{2nd even} \\ \text{integer}\end{array}} = \boxed{624}$

Equation:

$$x \cdot (x + 2) = 624$$

$$x^2 + 2x - 624 = 0$$

$$(x - 24)(x + 26) = 0$$

$$x - 24 = 0 \qquad x + 26 = 0$$

$$x = 24 \qquad x = -26$$

$$x + 2 = 26 \qquad \text{not positive}$$

51. Solve $\dfrac{t}{2} = \dfrac{1}{8}$.

Solution

$$\frac{t}{2} = \frac{1}{8}$$

$$8\left(\frac{t}{2}\right) = \left(\frac{1}{8}\right)8$$

$$4t = 1$$

$$t = \frac{1}{4}$$

53. Solve $\dfrac{z+2}{3} = \dfrac{z}{12}$.

Solution

$$\frac{z+2}{3} = \frac{z}{12}$$

$$12\left(\frac{z+2}{3}\right) = \left(\frac{z}{12}\right)12$$

$$4(z+2) = z$$

$$4z + 8 = z$$

$$3z = -8$$

$$z = -\frac{8}{3}$$

55. Solve $\dfrac{4t}{3} = 15 - \dfrac{t}{6}$.

Solution

$$\frac{4t}{3} = 15 - \frac{t}{6}$$

$$6\left(\frac{4t}{3}\right) = \left(15 - \frac{t}{6}\right)6$$

$$8t = 90 - t$$

$$9t = 90$$

$$t = 10$$

57. Solve $\dfrac{9}{25 - y} = -\dfrac{1}{4}$.

Solution

$$\frac{9}{25 - y} = -\frac{1}{4}$$

$$4(25 - y)\left(\frac{9}{25 - y}\right) = \left(-\frac{1}{4}\right)4(25 - y)$$

$$36 = -(25 - y)$$

$$36 = -25 + y$$

$$61 = y$$

Check:

$$\frac{9}{25 - 61} \overset{?}{=} -\frac{1}{4}$$

$$-\frac{9}{36} = -\frac{1}{4}$$

$$-\frac{1}{4} = -\frac{1}{4}$$

59. Solve $5 - \dfrac{12}{a} = \dfrac{5}{3}$.

Solution

$$5 - \frac{12}{a} = \frac{5}{3}$$

$$3a\left(5 - \frac{12}{a}\right) = \left(\frac{5}{3}\right)3a$$

$$15a - 36 = 5a$$

$$10a = 36$$

$$a = \frac{36}{10}$$

$$a = \frac{18}{5}$$

Check:

$$5 - \frac{12}{\frac{18}{5}} \overset{?}{=} \frac{5}{3}$$

$$5 - \frac{60}{18} = \frac{5}{3}$$

$$\frac{15}{3} - \frac{10}{3} = \frac{5}{3}$$

$$\frac{5}{3} = \frac{5}{3}$$

61. Solve $\dfrac{5}{x} = \dfrac{25}{3(x+2)}$.

Solution

$$\frac{5}{x} = \frac{25}{3(x+2)}$$

$$3x(x+2)\left(\frac{5}{x}\right) = \left(\frac{25}{3(x+2)}\right)3x(x+2)$$

$$15(x+2) = 25x$$

$$15x + 30 = 25x$$

$$30 = 10x$$

$$3 = x$$

Check:

$$\frac{5}{3} \overset{?}{=} \frac{25}{3(3+2)}$$

$$\frac{5}{3} = \frac{25}{15}$$

$$\frac{5}{3} = \frac{5}{3}$$

63. Solve $\dfrac{8}{3x+5} = \dfrac{1}{x+2}$.

Solution

$$\frac{8}{3x+5} = \frac{1}{x+2}$$

$$(3x+5)(x+2)\left(\frac{8}{3x+5}\right) = \left(\frac{1}{x+2}\right)(3x+5)(x+2)$$

$$8(x+2) = 3x+5$$

$$8x + 16 = 3x + 5$$

$$5x = -11$$

$$x = -\frac{11}{5}$$

Check:

$$\frac{8}{3\left(-\dfrac{11}{5}\right)+5} \overset{?}{=} \frac{1}{-\dfrac{11}{5}+2}$$

$$\frac{8}{-\dfrac{33}{5}+\dfrac{25}{5}} = \frac{1}{-\dfrac{11}{5}+\dfrac{10}{5}}$$

$$\frac{8}{-\dfrac{8}{5}} = \frac{1}{-\dfrac{1}{5}}$$

$$-5 = -5$$

65. Solve $\dfrac{3}{x+2} - \dfrac{1}{x} = \dfrac{1}{5x}$.

Solution

$$\frac{3}{x+2} - \frac{1}{x} = \frac{1}{5x}$$

$$5x(x+2)\left(\frac{3}{x+2} - \frac{1}{x}\right) = \left(\frac{1}{5x}\right)5x(x+2)$$

$$15x - 5(x+2) = x + 2$$

$$15x - 5x - 10 = x + 2$$

$$10x - 10 = x + 2$$

$$9x = 12$$

$$x = \frac{12}{9}$$

$$x = \frac{4}{3}$$

Check:

$$\frac{1}{\dfrac{4}{3}+2} - \frac{1}{\dfrac{4}{3}} \overset{?}{=} \frac{1}{5\left(\dfrac{4}{3}\right)}$$

$$\frac{3}{\dfrac{10}{3}} - \frac{1}{\dfrac{4}{3}} = \frac{1}{\dfrac{20}{3}}$$

$$\frac{9}{10} - \frac{3}{4} = \frac{3}{20}$$

$$\frac{18}{20} - \frac{15}{20} = \frac{3}{20}$$

$$\frac{3}{20} = \frac{3}{20}$$

67. Solve $\dfrac{10}{x(x-2)} + \dfrac{4}{x} = \dfrac{5}{x-2}$.

Solution

$$\frac{10}{x(x-2)} + \frac{4}{x} = \frac{5}{x-2}$$

$$x(x-2)\left(\frac{10}{x(x-2)} + \frac{4}{x}\right) = \left(\frac{5}{x-2}\right)x(x-2)$$

$$10 + 4(x-2) = 5x$$

$$10 + 4x - 8 = 5x$$

$$2 = x$$

Check:

$$\frac{10}{2(2-2)} + \frac{4}{2} \overset{?}{=} \frac{5}{2-2}$$

$$\frac{10}{0} + \frac{4}{2} \neq \frac{5}{0}$$

Division by zero is undefined. Solution is extraneous, so equation has no solution.

69. Solve $\dfrac{10}{x+3} + \dfrac{10}{3} = 6$.

Solution

$$\frac{10}{x+3} + \frac{10}{3} = 6$$

$$3(x+3)\left(\frac{10}{x+3} + \frac{10}{3}\right) = (6)3(x+3)$$

$$30 + 10(x+3) = 18(x+3)$$

$$30 + 10x + 30 = 18x + 54$$

$$10x + 60 = 18x + 54$$

$$6 = 8x$$

$$\frac{6}{8} = x$$

$$\frac{3}{4} = x$$

Check:

$$\frac{10}{\dfrac{3}{4}+3} + \frac{10}{3} \overset{?}{=} 6$$

$$\frac{10}{\dfrac{15}{4}} + \frac{10}{3} = 6$$

$$\frac{40}{15} + \frac{10}{3} = 6$$

$$\frac{8}{3} + \frac{10}{3} = 6$$

$$\frac{18}{3} = 6$$

$$6 = 6$$

71. Solve $\dfrac{1}{x-5} + \dfrac{1}{x+5} = \dfrac{x+3}{x^2-25}$.

Solution

$$\frac{1}{x-5} + \frac{1}{x+5} = \frac{x+3}{x^2-25}$$

$$(x-5)(x+5)\left(\frac{1}{x-5} + \frac{1}{x+5}\right) = \left(\frac{x+3}{(x-5)(x+5)}\right)(x-5)(x+5)$$

$$x+5+x-5 = x+3$$

$$2x = x+3$$

$$x = 3$$

Check:

$$\frac{1}{3-5} + \frac{1}{3+5} \overset{?}{=} \frac{3+3}{3^2-25}$$

$$-\frac{1}{2} + \frac{1}{8} = -\frac{6}{16}$$

$$-\frac{4}{8} + \frac{1}{8} = -\frac{3}{8}$$

$$-\frac{3}{8} = -\frac{3}{8}$$

73. Solve $\dfrac{1}{2} = \dfrac{18}{x^2}$.

Solution

$$\dfrac{1}{2} = \dfrac{18}{x^2}$$

$$2x^2\left(\dfrac{1}{2}\right) = \left(\dfrac{18}{x^2}\right)2x^2$$

$$x^2 = 36$$

$$x^2 - 36 = 0$$

$$(x - 6)(x + 6) = 0$$

$$x = 6 \quad x = -6$$

Check: $\dfrac{1}{2} \overset{?}{=} \dfrac{18}{6^2}$ $\quad \dfrac{1}{2} \overset{?}{=} \dfrac{18}{(-6)^2}$

$\dfrac{1}{2} = \dfrac{18}{36}$ $\qquad \dfrac{1}{2} = \dfrac{18}{36}$

$\dfrac{1}{2} = \dfrac{1}{2}$ $\qquad \dfrac{1}{2} = \dfrac{1}{2}$

75. Solve $x + 1 = \dfrac{72}{x}$.

Solution

$$x + 1 = \dfrac{72}{x}$$

$$x(x + 1) = \left(\dfrac{72}{x}\right)x$$

$$x^2 + x = 72$$

$$x^2 + x - 72 = 0$$

$$(x + 9)(x - 8) = 0$$

$$x = -9 \quad x = 8$$

Check: $-9 + 1 \overset{?}{=} \dfrac{72}{-9}$

$-8 = -8$

$8 + 1 \overset{?}{=} \dfrac{72}{8}$

$9 = 9$

77. Solve $1 = \dfrac{16}{y} - \dfrac{39}{y^2}$.

Solution

$$1 = \dfrac{16}{y} - \dfrac{39}{y^2}$$

$$y^2(1) = \left[\dfrac{16}{y} - \dfrac{39}{y^2}\right]y^2$$

$$y^2 = 16y - 39$$

$$y^2 - 16y + 39 = 0$$

$$(y - 13)(y - 3) = 0$$

$$y = 13 \quad y = 3$$

Check: $1 \overset{?}{=} \dfrac{16}{13} - \dfrac{39}{13^2}$

$1 = \dfrac{16}{13} - \dfrac{3}{13}$

$1 = 1$

$1 \overset{?}{=} \dfrac{16}{3} - \dfrac{39}{3^2}$

$1 = \dfrac{16}{3} - \dfrac{13}{3}$

$1 = 1$

79. Solve $\dfrac{2x}{5} = \dfrac{x^2 - 5x}{5x}$.

Solution

$$5x\left(\frac{2x}{5} \quad = \frac{x^2 - 5x}{5x}\right)5x$$

$$2x^2 = x^2 - 5x$$

$$x^2 + 5x = 0$$

$$x(x + 5) = 0$$

$$x = 0 \quad x + 5 = 0$$

$$x = -5$$

Check: $x = 0$

$$\frac{2(0)}{5} \stackrel{?}{=} \frac{0^2 - 5(0)}{5(0)}$$

$$0 \neq \text{undefined}$$

so $x = 0$ is extraneous.

$x = -5$

$$\frac{2(-5)}{5} \stackrel{?}{=} \frac{(-5)^2 - 5(-5)}{5(-5)}$$

$$\frac{-10}{5} = \frac{25 + 25}{-25}$$

$$-2 = -2$$

81. Solve $\dfrac{2(x + 7)}{x + 4} - 2 = \dfrac{2x + 20}{2x + 8}$.

Solution

$$\frac{2(x + 7)}{x + 4} - 2 = \frac{2x + 20}{2x + 8}$$

$$2(x + 4)\left(\frac{2(x + 7)}{x + 4} - 2\right) = \left(\frac{2x + 20}{2(x + 4)}\right)2(x + 4)$$

$$4(x + 7) - 2 \cdot 2(x + 4) = 2x + 20$$

$$4x + 28 - 4x - 16 = 2x + 20$$

$$12 = 2x + 20$$

$$-8 = 2x$$

$$-4 = x$$

Check:

$$\frac{2[-4 + 7]}{-4 + 4} - 2 \stackrel{?}{=} \frac{2(-4) + 20}{2(-4) + 8}$$

$$\frac{6}{0} - 2 \neq \frac{12}{0}$$

Division by zero is undefined. Solution is extraneous, so equation has no solution.

83. Solve $x - \dfrac{24}{x} = 5$.

Solution

$$x - \frac{24}{x} = 5$$

$$x\left(x - \frac{24}{x}\right) = (5)x$$

$$x^2 - 24 = 5x$$

$$x^2 - 5x - 24 = 0$$

$$(x - 8)(x + 3) = 0$$

$$x = 8 \quad x = -3$$

Check:

$$8 - \frac{24}{8} \stackrel{?}{=} 5 \qquad -3 - \frac{24}{-3} \stackrel{?}{=} 5$$

$$8 - 3 = 5 \qquad\quad -3 + 8 = 5$$

$$5 = 5 \qquad\qquad\quad 5 = 5$$

85. *Graphical Reasoning* (a) Use a graphing utility to determine any x-intercepts of the equation, and (b) set $y = 0$ and solve the resulting rational equation to confirm the results of part (a).

$$y = \frac{x - 4}{x + 5}$$

Solution

(a) Keystrokes:

x-intercepts: $x = 4$

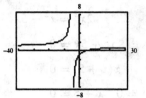

(b) $0 = \dfrac{x - 4}{x + 5}$

$0 = x - 4$

$4 = x$

87. *Graphical Reasoning* (a) Use a graphing utility to determine any x-intercepts of the equation, and (b) set $y = 0$ and solve the resulting rational equation to confirm the results of part (a).

$$y = (x - 1) - \frac{12}{x}$$

Solution

(a) Keystrokes:

Y= (X,T,θ − 1) − 12 ÷ X,T,θ GRAPH

x-intercepts: $x = -3$ and $x = 4$

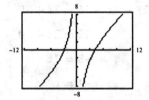

(b) $0 = (x - 1) - \dfrac{12}{x}$

$0 = x^2 - x - 12$

$0 = (x - 4)(x + 3)$

$x - 4 = 0 \qquad x + 3 = 0$

$x = 4 \qquad\quad x = -3$

89. What number can be added to its reciprocal to obtain $\dfrac{65}{8}$?

Solution

Verbal Model: $\boxed{\text{Number}} + \boxed{\text{Reciprocal}} = \boxed{\dfrac{65}{8}}$

Equation:

$$x + \frac{1}{x} = \frac{65}{8}$$

$$8x\left(x + \frac{1}{x}\right) = \left(\frac{65}{8}\right)8x$$

$$8x^2 + 8 = 65x$$

$$8x^2 - 65x + 8 = 0$$

$$(8x - 1)(x - 8) = 0$$

$$x = \frac{1}{8} \quad x = 8$$

91. *Pollution Removal* The cost C in dollars of removing $p\%$ of the air pollution in the stack emission of a utility company is modeled by

$$C = \frac{120{,}000p}{100 - p}.$$

(a) Use a graphing utility to graph the model. Use the result to graphically estimate the percent of stack emission that can be removed for $680{,}000.

(b) Determine the percentage of the stack emission that can be removed for $680{,}000.

Solution

(a) Keystrokes:

$\boxed{Y=}$ $\boxed{(}$ 120,000 $\boxed{X, T, \theta}$ $\boxed{)}$ $\boxed{\div}$ $\boxed{(}$ 100 $\boxed{-}$ $\boxed{X, T, \theta}$ $\boxed{)}$ $\boxed{\text{GRAPH}}$

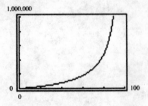

(b) *Verbal Model:* $\boxed{\text{Cost}} = \boxed{\dfrac{120{,}000p}{100 - p}}$

Equation:

$$680{,}000 = \frac{120{,}000p}{100 - p}$$

$$100 - p)(680{,}000) = \left(\frac{120{,}000p}{100 - p}\right)(100 - p)$$

$$68{,}000{,}000 - 680{,}000p = 120{,}000p$$

$$68{,}000{,}000 = 800{,}000p$$

$$85\% = p$$

93. *Comparing Two Speeds*   One person runs 2 miles per hour faster than a second person. The first person runs 5 miles in the same time the second runs 4 miles. Find the speed of each.

Solution

Verbal Model:

$$\boxed{\frac{\text{Distance person 1}}{\text{Rate person 1}}} = \boxed{\frac{\text{Distance person 2}}{\text{Rate person 2}}}$$

Equation:

$$\frac{5}{x+2} = \frac{4}{x}$$

$$x(x+2)\left(\frac{5}{x+2}\right) = \left(\frac{4}{x}\right)x(x+2)$$

$$5x = 4(x+2)$$

$$5x = 4x + 8$$

$$x = 8 \text{ mph person 2}$$

$$x + 2 = 10 \text{ mph person 1}$$

95. *Speed*   A boat travels at a speed of 20 miles per hour in still water. It travels 48 miles upstream, and then returns to the starting point, in a total of 5 hours. Find the speed of the current.

Solution

Verbal Model:

$$\boxed{\begin{array}{c}\text{Time traveled}\\\text{upstream}\end{array}} + \boxed{\begin{array}{c}\text{Time traveled}\\\text{downstream}\end{array}} = \boxed{\begin{array}{c}\text{Total}\\\text{time}\end{array}}$$

Equation:

$$\frac{48}{20-x} + \frac{48}{20+x} = 5$$

$$(20-x)(20+x)\left(\frac{48}{20-x} + \frac{48}{20+x}\right) = (5)(20-x)(20+x)$$

$$48(20+x) + 48(20-x) = 5(400 - x^2)$$

$$960 + 48x + 960 - 48x = 2000 - 5x^2$$

$$1920 = 2000 - 5x^2$$

$$5x^2 = 80$$

$$x^2 = 16$$

$$x = 4 \text{ mph}$$

97. *Partnership Costs*   A group plans to start a new business that will require $240,000 for start-up capital. The partners in the group will share the cost equally. If two additional people join the group, the cost per person will decrease by $4000. How many partners are currently in the group?

Solution

Verbal Model: $\boxed{\dfrac{\text{Cost per person}}{\text{Current group}}} - \boxed{\dfrac{\text{Cost per person}}{\text{New group}}} = \boxed{4000}$

Equation:

$$\frac{240{,}000}{x} - \frac{240{,}000}{x+2} = 4000$$

$$x(x+2)\left(\frac{240{,}000}{x} - \frac{240{,}000}{x+2}\right) = (4000)x(x+2)$$

$$240{,}000(x+2) - 240{,}000x = 4000(x^2 + 2x)$$

$$240{,}000x + 480{,}000 - 240{,}000x = 4000x^2 + 8000x$$

$$0 = 4000x^2 + 8000x - 480{,}000$$

$$0 = x^2 + 2x - 120$$

$$0 = (x+12)(x-10)$$

$$x + 12 = 0 \qquad x - 10 = 0$$

$$x = -12 \qquad x = 10 \text{ people}$$

99. *Using a Table*   Complete the table (see textbook) by finding the time for two individuals to complete a task. The first two columns in the table give the times required for two individuals to complete the task working alone. (Assume that when they work together their individual rates do not change.)

Solution

$$\frac{1}{6} + \frac{1}{6} = \frac{1}{t}$$

$$t + t = 6$$

$$2t = 6$$

$$t = 3 \text{ hours}$$

$$\frac{1}{3} + \frac{1}{5} = \frac{1}{t}$$

$$5t + 3t = 15$$

$$8t = 15$$

$$t = \frac{15}{8} \text{ minutes}$$

$$\frac{1}{5} + \frac{1}{2\frac{1}{2}} = \frac{1}{t}$$

$$\frac{1}{5} + \frac{1}{\frac{5}{2}} = \frac{1}{t}$$

$$\frac{1}{5} + \frac{2}{5} = \frac{1}{t}$$

$$t + 2t = 5$$

$$3t = 5$$

$$t = \frac{5}{3} \text{ hours}$$

Person #1	Person #2	Together
6 hours	6 hours	3 hours
3 minutes	5 minutes	$\dfrac{15}{8}$ minutes
5 hours	$2\frac{1}{2}$ hours	$\dfrac{5}{3}$ hours

101. *Work Rate* One landscaper works $1\frac{1}{2}$ times as fast as a second landscaper. Find their individual rates if it takes them 9 hours working together to complete a certain job.

Solution

Verbal Model: | Rate Person 1 | $+$ | Rate Person 2 | $=$ | Rate Together |

Equation:
$$\frac{1}{x} + \frac{1}{\frac{3}{2}x} = \frac{1}{9}$$

$$9x\left(\frac{1}{x} + \frac{2}{3x}\right) = \left(\frac{1}{9}\right)9x$$

$$9 + 6 = x$$

$$15 \text{ hours} = x$$

$$22\frac{1}{2} = \frac{45}{2} \text{ hours} = \frac{3}{2}x$$

$$15 \text{ hours} = x$$

Review Exercises for Chapter 4

1. Find the domain of $\dfrac{3y}{y-8}$.

Solution

$y - 8 \neq 0$

$y \neq 8$

$D = (-\infty, 8) \cup (8, \infty)$

3. Find the domain of $\dfrac{u}{u^2 - 7u + 6}$.

Solution

$u^2 - 7u + 6 \neq 0$

$(u - 6)(u - 1) \neq 0$

$u \neq 6 \quad u \neq 1$

$D = (-\infty, 1) \cup (1, 6) \cup (6, \infty)$

5. Simplify $\dfrac{6x^4y^2}{15xy^2}$.

Solution

$$\frac{6x^4y^2}{15xy^2} = \frac{2 \cdot 3x \cdot x^3 \cdot y^2}{5 \cdot 3x \cdot y^2}$$

$$= \frac{2x^3}{5}$$

7. Simplify $\dfrac{5b - 15}{30b - 120}$.

Solution

$$\frac{5b - 15}{30b - 120} = \frac{5(b - 3)}{30(b - 4)}$$

$$= \frac{5(b - 3)}{5 \cdot 6(b - 4)}$$

$$= \frac{b - 3}{6(b - 4)}$$

9. Simplify $\dfrac{9x - 9y}{y - x}$.

Solution

$$\frac{9x - 9y}{y - x} = \frac{9(x - y)}{-1(x - y)} = -9$$

11. Simplify $\dfrac{x^2 - 5x}{2x^2 - 50}$.

Solution

$$\frac{x^2 - 5x}{2x^2 - 50} = \frac{x(x - 5)}{2(x^2 - 25)}$$

$$= \frac{x(x - 5)}{2(x - 5)(x + 5)}$$

$$= \frac{x}{2(x + 5)}$$

13. Match $f(x) = \dfrac{5}{x-6}$ with its graph.

Solution

(b)

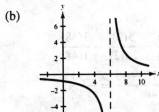

15. Match $f(x) = \dfrac{6x}{x-5}$ with its graph.

Solution

(a)

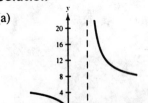

17. Perform the operation(s) and simplify your answer.

$$\frac{7}{8} \cdot \frac{2x}{y} \cdot \frac{y^2}{14x^2}$$

Solution

$$\frac{7}{8} \cdot \frac{2x}{y} \cdot \frac{y^2}{14x^2} = \frac{7 \cdot 2 \cdot x \cdot y \cdot y}{2 \cdot 2 \cdot 2 \cdot y \cdot 7 \cdot 2 \cdot x \cdot x}$$

$$= \frac{y}{8x}$$

19. Perform the operation and simplify your answer.

$$\frac{60z}{z+6} \cdot \frac{z^2-36}{5}$$

Solution

$$\frac{60z}{z+6} \cdot \frac{z^2-36}{5} = \frac{5 \cdot 12z(z-6)(z+6)}{(z+6)5}$$

$$= 12z(z-6)$$

21. Perform the operation and simplify your answer.

$$\frac{u}{u-3} \cdot \frac{3u-u^2}{4u^2}$$

Solution

$$\frac{u}{u-3} \cdot \frac{3u-u^2}{4u^2} = \frac{u}{u-3} \cdot \frac{-u(u-3)}{4u^2}$$

$$= -\frac{1}{4}$$

23. Perform the operation and simplify your answer.

$$\frac{6/x}{2/x^3}$$

Solution

$$\frac{6/x}{2/x^3} = \frac{6}{x} \div \frac{2}{x^3} = \frac{3 \cdot 2}{x} \cdot \frac{x \cdot x^2}{2} = 3x^2$$

25. Perform the operation and simplify your answer.

$$25y^2 \div \frac{xy}{5}$$

Solution

$$25y^2 \div \frac{xy}{5} = 25y \cdot y \cdot \frac{5}{xy} = \frac{125y}{x}$$

27. Perform the operation and simplify your answer.

$$\frac{x^2-7x}{x+1} \div \frac{x^2-14x+49}{x^2-1}$$

Solution

$$\frac{x^2-7x}{x+1} \div \frac{x^2-14x+49}{x^2-1}$$

$$= \frac{x(x-7)}{x+1} \cdot \frac{(x-1)(x+1)}{(x-7)(x-7)}$$

$$= \frac{x(x-1)}{x-7}$$

29. Perform the operation and simplify your answer.

$$\frac{4}{9} - \frac{11}{9}$$

Solution

$$\frac{4}{9} - \frac{11}{9} = \frac{4 - 11}{9} = -\frac{7}{9}$$

31. Perform the operations and simplify your answer.

$$\frac{15}{16} - \frac{5}{24} - 1$$

Solution

$$\frac{15}{16} - \frac{5}{24} - 1 = \frac{15(3)}{16(3)} - \frac{5(2)}{24(2)} - \frac{1(48)}{1(48)}$$

$$= \frac{45 - 10 - 48}{48} = -\frac{13}{48}$$

33. Perform the operation and simplify your answer.

$$\frac{1}{x + 5} + \frac{3}{x - 12}$$

Solution

$$\frac{1}{x + 5} + \frac{3}{x - 12} = \frac{1}{x + 5}\left(\frac{x - 12}{x - 12}\right) + \frac{3}{x - 12}\left(\frac{x + 5}{x + 5}\right)$$

$$= \frac{x - 12}{(x + 5)(x - 12)} + \frac{3(x + 5)}{(x - 12)(x + 5)}$$

$$= \frac{x - 12 + 3x + 15}{(x + 5)(x - 12)} = \frac{4x + 3}{(x + 5)(x - 12)}$$

35. Perform the operations and simplify your answer.

$$5x + \frac{2}{x - 3} - \frac{3}{x + 2}$$

Solution

$$5x + \frac{2}{x - 3} - \frac{3}{x + 2} = \frac{5x(x - 3)(x + 2)}{(x - 3)(x + 2)} + \frac{2}{(x - 3)}\left(\frac{x + 2}{x + 2}\right) - \frac{3}{(x + 2)}\left(\frac{x - 3}{x - 3}\right)$$

$$= \frac{5x^3 - 5x^2 - 30x + 2x + 4 - 3x + 9}{(x - 3)(x + 2)}$$

$$= \frac{5x^3 - 5x^2 - 31x + 13}{(x - 3)(x + 2)}$$

37. Perform the operation and simplify your answer.

$$\frac{6}{x} - \frac{6x - 1}{x^2 + 4}$$

Solution

$$\frac{6}{x} - \frac{6x - 1}{x^2 + 4} = \frac{6(x^2 + 4)}{x(x^2 + 4)} - \frac{6x - 1(x)}{x^2 + 4(x)}$$

$$= \frac{6x^2 + 24 - 6x^2 + x}{x(x^2 + 4)} = \frac{24 + x}{x(x^2 + 4)}$$

39. Perform the operations and simplify your answer.

$$\frac{5}{x+3} - \frac{4x}{(x+3)^2} - \frac{1}{x-3}$$

Solution

$$\frac{5}{x+3} - \frac{4x}{(x+3)^2} - \frac{1}{x-3} = \frac{5}{x+3}\left(\frac{(x+3)(x-3)}{(x+3)(x-3)}\right) - \frac{4x}{(x+3)^2}\left(\frac{x-3}{x-3}\right) - \frac{1}{x-3}\left(\frac{(x+3)^2}{(x+3)^2}\right)$$

$$= \frac{5x^2 - 45 - 4x^2 + 12x - x^2 - 6x - 9}{(x+3)^2(x-3)}$$

$$= \frac{6x - 54}{(x+3)^2(x-3)}$$

41. Perform the operation and simplify your answer.

$$\frac{\left(\dfrac{6x^2}{x^2+2x-35}\right)}{\left(\dfrac{x^3}{x^2-25}\right)}$$

Solution

$$\frac{\left(\dfrac{6x^2}{x^2+2x-35}\right)}{\left(\dfrac{x^3}{x^2-25}\right)} = \frac{\dfrac{6x^2}{(x+7)(x-5)}}{\dfrac{x^3}{(x-5)(x+5)}} \cdot \frac{(x+7)(x-5)(x+5)}{(x+7)(x-5)(x+5)}$$

$$= \frac{6x^2(x+5)}{x^3(x+7)} = \frac{6(x+5)}{x(x+7)}$$

43. Perform the operation and simplify your answer.

$$\frac{3t}{\left(5 - \dfrac{2}{t}\right)}$$

Solution

$$\frac{3t}{\left(5 - \dfrac{2}{t}\right)} \cdot \frac{t}{t} = \frac{3t^2}{5t - 2}$$

45. Perform the operation and simplify your answer.

$$\frac{\left(\dfrac{1}{a^2 - 16} - \dfrac{1}{a}\right)}{\left(\dfrac{1}{a^2 + 4a} + 4\right)}$$

Solution

$$\frac{\left(\dfrac{1}{a^2 - 16} - \dfrac{1}{a}\right)}{\left(\dfrac{1}{a^2 + 4a} + 4\right)} \cdot \frac{a(a-4)(a+4)}{a(a-4)(a+4)} = \frac{a - (a-4)(a+4)}{a - 4 + 4a(a-4)(a+4)} = \frac{a - (a^2 - 16)}{a - 4 + 4a(a^2 - 16)}$$

$$= \frac{a - a^2 + 16}{a - 4 + 4a^3 - 64a} = \frac{-a^2 + a + 16}{4a^3 - 63a - 4}$$

47. Perform the operation and simplify your answer.

$$(4x^3 - x) \div 2x$$

Solution

$$(4x^3 - x) \div 2x = \frac{4x^3 - x}{2x}$$

$$= \frac{4x^3}{2x} - \frac{x}{2x} = 2x^2 - \frac{1}{2}$$

49. Perform the indicated division.

$$\frac{6x^3 + 2x^2 - 4x + 2}{3x - 1}$$

Solution

$$\require{enclose}
\begin{array}{r}
2x^2 + \dfrac{4}{3}x - \dfrac{8}{9} + \dfrac{\frac{10}{9}}{3x-1} \\[4pt]
3x - 1 \enclose{longdiv}{6x^3 + 2x^2 - 4x + 2} \\
\end{array}$$

$$
\begin{array}{r}
6x^3 - 2x^2 \\ \hline
4x^2 - 4x \\
4x^2 - \dfrac{4}{3}x \\ \hline
-\dfrac{8}{3}x + 2 \\
-\dfrac{8}{3}x + \dfrac{8}{9} \\ \hline
\dfrac{10}{9}
\end{array}
$$

51. Perform the operation and simplify your answer.

$$\frac{x^4 - 3x^2 + 2}{x^2 - 1}$$

Solution

$$\require{enclose}
\begin{array}{r}
x^2 - 2 \\
x^2 - 1 \enclose{longdiv}{x^4 - 3x^2 + 2} \\
x^4 - x^2 \\ \hline
-2x^2 + 2 \\
-2x^2 + 2 \\ \hline
\end{array}$$

53. Use synthetic division to perform the division.

$$\frac{x^3 + 7x^2 + 3x - 14}{x + 2}$$

Solution

$$\frac{x^3 + 7x^2 + 3x - 14}{x + 2} = x^2 + 5x - 7$$

$$
\begin{array}{r|rrrr}
-2 & 1 & 7 & 3 & -14 \\
   &   & -2 & -10 & 14 \\
\hline
   & 1 & 5 & -7 & 0
\end{array}
$$

55. Use synthetic division to perform the division.

$$(x^4 - 3x^2 - 25) \div (x - 3)$$

Solution

$$(x^4 - 3x^2 - 25) \div (x - 3) = x^3 + 3x^2 + 6x + 18 + \frac{29}{x - 3}$$

$$
\begin{array}{r|rrrrr}
3 & 1 & 0 & -3 & 0 & -25 \\
  &   & 3 & 9 & 18 & 54 \\
\hline
  & 1 & 3 & 6 & 18 & 29
\end{array}
$$

57. Use a graphing utility to graph the equations on the same screen. Use the graphs to verify that the expressions are equivalent. Verify the results algebraically.

$$y_1 = \frac{x^2 + 6x + 9}{x^2} \cdot \frac{x^2 - 3x}{x + 3}, \quad y_2 = \frac{x^2 - 9}{x}$$

Solution

Keystrokes:

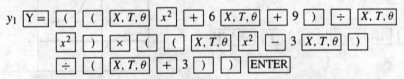

y_1 Y= ((X,T,θ x^2 + 6 X,T,θ + 9) ÷ X,T,θ
x^2) × ((X,T,θ x^2 − 3 X,T,θ)
÷ (X,T,θ + 3)) ENTER

y_2 Y= (X,T,θ x^2 − 9) ÷ X,T,θ GRAPH

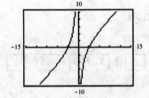

$$\frac{x^2 + 6x + 9}{x^2} \cdot \frac{x^2 - 3x}{x + 3} = \frac{(x + 3)(x + 3)x(x - 3)}{x^2(x + 3)} = \frac{x^2 - 9}{x}$$

59. Use a graphing utility to graph the equations on the same screen. Use the graphs to verify that the expressions are equivalent. Verify the results algebraically.

$$y_1 = \frac{\left(\dfrac{1}{x} - \dfrac{1}{2}\right)}{2x}, \quad y_2 = \frac{2 - x}{4x^2}$$

Solution

Keystrokes:

y_1 Y= ((1 ÷ X,T,θ) − (1 ÷ 2))
÷ 2 X,T,θ ENTER

y_2 Y= (2 − X,T,θ) ÷ 4 X,T,θ x^2 GRAPH

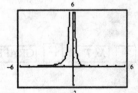

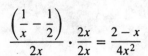

$$\frac{\left(\dfrac{1}{x} - \dfrac{1}{2}\right)}{2x} \cdot \frac{2x}{2x} = \frac{2 - x}{4x^2}$$

61. Use a graphing utility to graph $g(x) = \dfrac{2 + x}{1 - x}$.

Solution

Keystrokes:

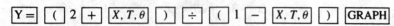 GRAPH

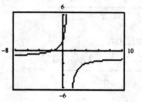

63. Use a graphing utility to graph $f(x) = \dfrac{x}{x^2 + 1}$.

Solution

Keystrokes:

Y= X,T,θ ÷ (X,T,θ x^2 + 1) GRAPH

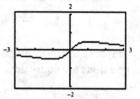

65. Use a graphing utility to graph $P(x) = \dfrac{3x + 6}{x - 2}$.

Solution

Keystrokes:

Y= (3 X,T,θ + 6) ÷ (X,T,θ − 2) GRAPH

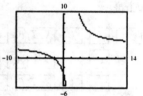

67. Use a graphing utility to graph $h(x) = \dfrac{4}{(x - 1)^2}$.

Solution

Keystrokes:

Y= 4 ÷ (X,T,θ − 1) x^2 GRAPH

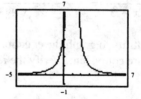

69. Use a graphing utility to graph $f(x) = \dfrac{-5}{x^2}$.

Solution

Keystrokes:

Y= (-) 5 ÷ X,T,θ x^2 GRAPH

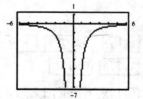

71. Use a graphing utility to graph $y = \dfrac{x}{x^2 - 1}$.

Solution

Keystrokes:

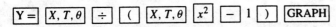

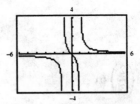

73. Use a graphing utility to graph $y = \dfrac{2x^2}{x^2 - 4}$.

Solution

Keystrokes:

$\boxed{Y=}\ 2\ \boxed{X, T, \theta}\ \boxed{x^2}\ \boxed{\div}\ \boxed{(}\ \boxed{X, T, \theta}\ \boxed{x^2}\ \boxed{-}\ 4\ \boxed{)}\ \boxed{GRAPH}$

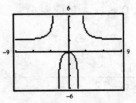

75. Solve $\dfrac{3x}{8} = -15$.

Solution

$$\frac{3x}{8} = -15$$

$$8\left(\frac{3}{8}x\right) = (-15)8$$

$$3x = -120$$

$$x = -40$$

77. Solve $3\left(8 - \dfrac{12}{t}\right) = 0$.

Solution

$$t\left(8 - \frac{12}{t}\right) = (0)t$$

$$8t - 12 = 0$$

$$8t = 12$$

$$t = \frac{12}{8}$$

$$t = \frac{3}{2}$$

79. Solve $\dfrac{2}{y} - \dfrac{1}{3y} = \dfrac{1}{3}$.

Solution

$$\frac{2}{y} - \frac{1}{3y} = \frac{1}{3}$$

$$3y\left(\frac{2}{y} - \frac{1}{3y}\right) = \left(\frac{1}{3}\right)3y$$

$$6 - 1 = y$$

$$5 = y$$

81. Solve $r = 2 + \dfrac{24}{r}$.

Solution

$$r = 2 + \frac{24}{r}$$

$$r(r) = \left(2 + \frac{24}{r}\right)r$$

$$r^2 = 2r + 24$$

$$r^2 - 2r - 24 = 0$$

$$(r - 6)(r + 4) = 0$$

$$r = 6 \quad r = -4$$

83. Solve $\dfrac{2}{x} - \dfrac{x}{6} = \dfrac{2}{3}$.

Solution

$$\frac{2}{x} - \frac{x}{6} = \frac{2}{3}$$

$$6x\left(\frac{2}{x} - \frac{x}{6}\right) = \left(\frac{2}{3}\right)6x$$

$$12 - x^2 = 4x$$

$$0 = x^2 + 4x - 12$$

$$0 = (x-2)(x+6)$$

$$x = 2 \quad x = -6$$

85. Solve $\dfrac{12}{x^2 + x - 12} - \dfrac{1}{x-3} = -1$.

Solution

$$\frac{12}{x^2 + x - 12} - \frac{1}{x-3} = -1$$

$$(x-3)(x+4)\left(\frac{12}{x^2 + x - 12} - \frac{1}{x-3}\right) = (-1)(x-3)(x+4)$$

$$12 - (x+4) = -1(x^2 + x - 12)$$

$$12 - x - 4 = -x^2 - x + 12$$

$$(x^2 - 4) = 0$$

$$(x-2)(x+2) = 0$$

$$x - 2 = 0 \quad x + 2 = 0$$

$$x = 2 \quad\quad x = -2$$

87. Solve $\dfrac{5}{x^2 - 4} - \dfrac{6}{x-2} = -5$.

Solution

$$\frac{5}{x^2 - 4} - \frac{6}{x-2} = -5$$

$$(x-2)(x+2)\left(\frac{5}{x^2 - 4} - \frac{6}{x-2}\right) = (-5)(x-2)(x+2)$$

$$5 - 6(x+2) = -5(x^2 - 4)$$

$$5 - 6x - 12 = -5x^2 + 20$$

$$5x^2 - 6x - 27 = 0$$

$$(5x + 9)(x - 3) = 0$$

$$5x + 9 = 0 \quad x - 3 = 0$$

$$x = -\frac{9}{5} \quad\quad x = 3$$

89. (a) Use a graphing utility to determine any x-intercepts of the graph of $y = \dfrac{1}{x} - \dfrac{1}{2x+3}$,

and (b) set $y = 0$ and solve the resulting rational equation to confirm the results of part (a).

Solution

Keystrokes:

x-intercepts: $x = -3$

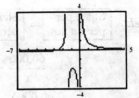

$$0 = \frac{1}{x} - \frac{1}{2x - 3}$$

$$0 = 2x + 3 - x$$

$$0 = x + 3$$

$$-3 = x$$

91. *Average Speed* Suppose you drove 56 miles on a service call for your company. On the return trip, which takes 10 minutes less than the original trip, your average speed was 8 miles per hour faster. What is your average speed on the return trip?

Solution

Verbal Model: $\boxed{\text{Distance}} = \boxed{\text{Rate}} \times \boxed{\text{Time}}$

Equation:
$$\frac{56}{x} = \frac{56}{x+8} + \frac{1}{6}$$

$$6x(x+8)\left(\frac{56}{x}\right) = \left(\frac{56}{x+8} + \frac{1}{6}\right) 6x(x+8)$$

$$336(x+8) = 336x + x(x+8)$$

$$336x + 2688 = 336x + x^2 + 8x$$

$$0 = x^2 + 8x - 2688$$

$$0 = (x - 48)(x + 56)$$

$$x = 48 \quad x = -56$$

$$x + 8 = 56$$

93. *Forming a Partnership*   A group of people agree to share equally in the cost of a $60,000 piece of machinery. If they could find two more people to join the group, each person's share of the cost would decrease by $5000. How many people are presently in the group?

Solution

Verbal Model: $\boxed{\text{Share per person now}} = \boxed{\text{Share per person later}} + \boxed{5000}$

Equation:

$$\frac{60,000}{x} = \frac{60,000}{x+2} + 5000$$

$$x(x+2)\left(\frac{60,000}{x}\right) = \left(\frac{60,000}{x+2} + 5000\right)x(x+2)$$

$$60,000(x+2) = 60,000 + 5000x(x+2)$$

$$60,000x + 120,000 = 60,000x + 5000x^2 + 10,000x$$

$$0 = 5000x^2 + 10,000x - 120,000$$

$$0 = x^2 + 2x - 24$$

$$0 = (x+6)(x-4)$$

$$x = -6 \quad x = 4$$

Test for Chapter 4

1. Find the domain of the rational expression $\dfrac{3y}{y^2 - 25}$.

Solution

$$y^2 - 25 \neq 0$$

$$(y-5)(y+5) \neq 0$$

$$y \neq 5, -5$$

$$D = (-\infty, -5) \cup (-5, 5) \cup (5, \infty)$$

2. Find the least common denominator of

$$\frac{3}{x^2}, \frac{x}{x-3}, \frac{2x}{x^3(x+3)}, \text{ and } \frac{10}{x^2 + 6x + 9}.$$

Solution

$$LCD = x^3(x-3)(x+3)^2$$

3. (a) Simplify $\dfrac{2-x}{3x-6}$.

Solution

(a) $\dfrac{2-x}{3x-6} = \dfrac{2-x}{-3(-x+2)} = -\dfrac{1}{3}$

(b) Simplify $\dfrac{2a^2 - 5a - 12}{5a - 20}$.

(b) $\dfrac{2a^2 - 5a - 12}{5a - 20} = \dfrac{(2a+3)(a-4)}{5(a-4)} = \dfrac{2a+3}{5}$

4. Perform the operation and simplify.

$$\frac{4z^3}{5} \cdot \frac{25}{12z^2}.$$

Solution

$$\frac{4z^3}{5} \cdot \frac{25}{12z^2} = \frac{4 \cdot z^2 \cdot z \cdot 5 \cdot 5}{5 \cdot 4 \cdot 3 \cdot z^2} = \frac{5z}{3}$$

5. Perform the operation and simplify.

$$\frac{y^2 + 8y + 16}{2(y - 2)} \cdot \frac{8y - 16}{(y + 4)^3}.$$

Solution

$$\frac{y^2 + 8y + 16}{2(y - 2)} \cdot \frac{8y - 16}{(y + 4)^3} = \frac{(y + 4)^2 \cdot 8(y - 2)}{2(y - 2)(y + 4)^2(y + 4)}$$

$$= \frac{4}{y + 4}$$

6. Perform the operation and simplify.

$$(4x^2 - 9) \cdot \frac{2x + 3}{2x^2 - x - 3}.$$

Solution

$$(4x^2 - 9) \cdot \frac{2x + 3}{2x^2 - x - 3} = \frac{(2x - 3)(2x + 3)(2x + 3)}{(2x - 3)(x + 1)}$$

$$= \frac{(2x + 3)^2}{x + 1}$$

7. Perform the operation and simplify.

$$\frac{(2xy^2)^3}{15} \div \frac{12x^3}{21}.$$

Solution

$$\frac{(2xy^2)^3}{15} \div \frac{12x^3}{21} = \frac{(2xy^2)^3}{15} \cdot \frac{21}{12x^3}$$

$$= \frac{8x^3y^6 \cdot 7 \cdot 3}{5 \cdot 3 \cdot 4 \cdot 3x^3}$$

$$= \frac{14y^6}{15}$$

8. Perform the operation and simplify.

$$\frac{\left(\dfrac{3x}{x + 2}\right)}{\left(\dfrac{12}{x^3 + 2x^2}\right)}.$$

Solution

$$\frac{\left(\dfrac{3x}{x + 2}\right)}{\left(\dfrac{12}{x^3 + 2x^2}\right)} = \frac{3x}{x + 2} \div \frac{12}{x^3 + 2x^2}$$

$$= \frac{3x}{x + 2} \cdot \frac{x^2(x + 2)}{12} = \frac{x^3}{4}$$

9. Perform the operation and simplify.

$$\frac{\left(9x - \dfrac{1}{x}\right)}{\left(\dfrac{1}{x} - 3\right)}.$$

Solution

$$\frac{\left(9x - \dfrac{1}{x}\right)}{\left(\dfrac{1}{x} - 3\right)} = \frac{\left(9x - \dfrac{1}{x}\right)}{\left(\dfrac{1}{x} - 3\right)} \cdot \frac{x}{x}$$

$$= \frac{9x(x) - \dfrac{1}{x}(x)}{\dfrac{1}{x}(x) - 3(x)} = \frac{9x^2 - 1}{1 - 3x}$$

$$= \frac{(3x - 1)(3x + 1)}{-1(-1 + 3x)} = -(3x + 1)$$

10. Perform the operation and simplify.

$$2x + \frac{1 - 4x^2}{x + 1}.$$

Solution

$$2x + \frac{1 - 4x^2}{x + 1} = 2x\left(\frac{x + 1}{x + 1}\right) + \frac{1 - 4x^2}{x + 1}$$

$$= \frac{2x^2 + 2x}{x + 1} + \frac{1 - 4x^2}{x + 1}$$

$$= \frac{-2x^2 + 2x + 1}{x + 1}$$

11. Perform the operation and simplify.

$$\frac{5x}{x + 2} - \frac{2}{x^2 - x - 6}.$$

Solution

$$\frac{5x}{x + 2} - \frac{2}{x^2 - x - 6}$$

$$= \frac{5x}{x + 2} - \frac{2}{(x - 3)(x + 2)}$$

$$= \frac{5x}{x + 2}\left(\frac{x - 3}{x - 3}\right) - \frac{2}{(x - 3)(x + 2)}$$

$$= \frac{5x^2 - 15x - 2}{(x + 2)(x - 3)}$$

12. Perform the operations and simplify.

$$\frac{3}{x} - \frac{5}{x^2} + \frac{2x}{x^2 + 2x + 1}.$$

Solution

$$\frac{3}{x} - \frac{5}{x^2} + \frac{2x}{x^2 + 2x + 1} = \frac{3}{x} - \frac{5}{x^2} + \frac{2x}{(x + 1)^2}$$

$$= \frac{3}{x}\left(\frac{x(x + 1)^2}{x(x + 1)^2}\right) - \frac{5}{x^2}\left(\frac{x(x + 1)^2}{x(x + 1)^2}\right) + \frac{2x}{(x + 1)^2}\left(\frac{x^2}{x^2}\right)$$

$$= \frac{3x(x^2 + 2x + 1) - 5(x^2 + 2x + 1) + 2x^3}{x^2(x + 1)^2}$$

$$= \frac{3x^3 + 6x^2 + 3x - 5x^2 - 10x - 5 + 2x^3}{x^2(x + 1)^2}$$

$$= \frac{5x^3 + x^2 - 7x - 5}{x^2(x + 1)^2}$$

13. Perform the operation and simplify.

$$\frac{4}{x + 1} + \frac{4x}{x + 1}.$$

Solution

$$\frac{4}{x + 1} + \frac{4x}{x + 1} = \frac{4 + 4x}{x + 1}$$

$$= \frac{4(1 + x)}{x + 1}$$

$$= 4$$

14. Perform the operation and simplify.

$$\frac{t^4 + t^2 - 6t}{t^2 - 2}.$$

Solution

$$\frac{t^4 + t^2 - 6t}{t^2 - 2} = t^2 - 2 \overline{\smash{\big)}\ \begin{aligned} t^2 \qquad\qquad\quad + 3 \\ t^4 + 0t^3 + t^2 - 6t + 0 \end{aligned}} - \frac{6t - 6}{t^2 - 2}$$

$$\begin{aligned}
\underline{t^4 \qquad\quad - 2t^2} \\
3t^2 - 6t \\
\underline{3t^2 \qquad - 6} \\
-6t + 6
\end{aligned}$$

15. Perform the operation and simplify.

$$\frac{2x^4 - 15x^2 - 7}{x - 3}.$$

Solution

$$\frac{2x^4 - 15x^2 - 7}{x - 3}$$

$$\begin{array}{r|rrrrr}
3 & 2 & 0 & -15 & 0 & -7 \\
  &   & 6 & 18 & 9 & 27 \\
\hline
  & 2 & 6 & 3 & 9 & 20
\end{array}$$

$$\frac{2x^4 - 15x^2 - 7}{x - 3} = 2x^3 + 6x^2 + 3x + 9 + \frac{20}{x - 3}$$

16. Sketch the graph of the function. Describe how to find the asymptotes of the function.

(a) $f(x) = \dfrac{3}{x-3}$

(b) $g(x) = \dfrac{3x}{x-3}$

Solution

(a) $f(x) = \dfrac{3}{x-3}$

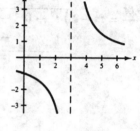

x-intercept: none, numerator is never zero

y-intercept: $f(0) = \dfrac{3}{0-3} = -1$

Vertical asymptote: $x - 3 = 0$

$$x = 3$$

Horizontal asymptote: $y = 0$ since the degree of the numerator is less than the degree of the denominator

(b) $f(x) = \dfrac{3x}{x-3}$

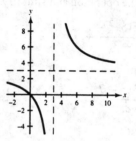

x-intercept: $x = 0$

y-intercept: $y = 0$

Vertical asymptote: $x - 3 = 0$

$$x = 3$$

Horizontal asymptote: $y = 3$ since the degree of the numerator is equal to the degree of the denominator

17. Solve $\dfrac{3}{h+2} = \dfrac{1}{8}$.

Solution

$\dfrac{3}{h+2} = \dfrac{1}{8}$

$3(8) = h + 2$ **Check:** $\dfrac{3}{22+2} = \dfrac{1}{8}$

     $24 = h + 2$ $\dfrac{3}{24} = \dfrac{1}{8}$

     $22 = h$ $\dfrac{1}{8} = \dfrac{1}{8}$

18. Solve $\dfrac{2}{x+5} - \dfrac{3}{x+3} = \dfrac{1}{x}$.

Solution

$$\frac{2}{x+5} - \frac{3}{x+3} = \frac{1}{x}$$

$$2x(x+3) - 3x(x+5) = (x+5)(x+3)$$

$$2x^2 + 6x - 3x^2 - 15x = x^2 + 3x + 5x + 15$$

$$-2x^2 - 17x - 15 = 0$$

$$(-2x - 15)(x + 1) = 0$$

$$-2x - 15 = 0 \qquad x + 1 = 0$$

$$-2x = 15 \qquad\quad x = -1$$

$$x = -\frac{15}{2}$$

Check:

$$\frac{2}{-\dfrac{15}{2}+5} - \frac{3}{-\dfrac{15}{2}+3} = \frac{1}{-\dfrac{15}{2}}$$

$$-\frac{12}{15} + \frac{10}{15} = -\frac{2}{15}$$

$$-\frac{2}{15} = -\frac{2}{15}$$

$$\frac{2}{-1+5} - \frac{3}{-1+3} \stackrel{?}{=} -\frac{1}{1}$$

$$\frac{2}{4} - \frac{3}{2} = -1$$

$$\frac{1}{2} - \frac{3}{2} = -1$$

$$-1 = -1$$

19. Solve $\dfrac{1}{x+1} + \dfrac{1}{x-1} = \dfrac{2}{x^2-1}$.

Solution

$$\frac{1}{x+1} + \frac{1}{x-1} = \frac{2}{x^2-1}$$

$$x - 1 + x + 1 = 2 \qquad \textbf{Check:} \quad \frac{1}{1+1} + \frac{1}{1-1} \neq \frac{2}{1-1}$$

$$2x = 2$$

$$x = 1$$

Division by zero is undefined. Solution is extraneous, so equation has no solution.

20. One painter works $1\frac{1}{2}$ times as fast as another. Find their individual rates for painting a room if it takes them 4 hours working together.

Solution

Verbal Model: $\boxed{\begin{array}{c}\text{Rate}\\\text{Painter 1}\end{array}} + \boxed{\begin{array}{c}\text{Rate}\\\text{Painter 2}\end{array}} = \boxed{\begin{array}{c}\text{Rate}\\\text{together}\end{array}}$

Equation:

$$\frac{1}{x} + \frac{1}{\frac{3}{2}x} = \frac{1}{4}$$

$$12x\left(\frac{1}{x} + \frac{2}{3x}\right) = \frac{1}{4}(12x)$$

$$12 + 8 = 3x$$

$$20 = 3x$$

$$\frac{20}{3} = 6\tfrac{2}{3} \text{ hours} = x$$

$$10 \text{ hours} = \frac{3}{2}x$$

CHAPTER 5
Radicals and Complex Numbers

5.1 Integer Exponents and Scientific Notation

7. Evaluate 5^{-2}.

Solution

$$5^{-2} = \frac{1}{5^2} = \frac{1}{25}$$

9. Evaluate $\dfrac{1}{(-2)^{-5}}$.

Solution

$$\frac{1}{(-2)^{-5}} = \frac{1}{\left(-\frac{1}{2}\right)^5}$$

$$= \frac{1}{-\frac{1}{32}} = -32$$

11. Evaluate $\left(\frac{2}{3}\right)^{-1}$.

Solution

$$\left(\frac{2}{3}\right)^{-1} = \frac{3}{2}$$

13. Evaluate $27 \cdot 3^{-3}$.

Solution

$$27 \cdot 3^{-3} = 3^3 \cdot 3^{-3}$$

$$= 3^{3+(-3)}$$

$$= 3^0$$

$$= 1$$

15. Evaluate $\dfrac{10^3}{10^{-2}}$.

Solution

$$\frac{10^3}{10^{-2}} = 10^{3-(-2)}$$

$$= 10^{3+2}$$

$$= 10^5$$

$$= 100,000$$

17. Evaluate $\left(4^2 \cdot 4^{-1}\right)^{-2}$.

Solution

$$\left(4^2 \cdot 4^{-1}\right)^{-2} = \left(4^{2+(-1)}\right)^{-2}$$

$$= \left(4^1\right)^{-2}$$

$$= 4^{-2}$$

$$= \frac{1}{4^2}$$

$$= \frac{1}{16}$$

19. Evaluate $\left(5^0 - 4^{-2}\right)^{-1}$.

Solution

$$\left(5^0 - 4^{-2}\right)^{-1} = \left(1 - \frac{1}{4^2}\right)^{-1} = \left(\frac{16}{16} - \frac{1}{16}\right)^{-1} = \left(\frac{15}{16}\right)^{-1} = \frac{16}{15}$$

21. Rewrite $y^4 \cdot y^{-2}$ using only positive exponents, and simplify. (Assume any variables in the expression are nonzero.)

Solution

$y^4 \cdot y^{-2} = y^{4+(-2)} = y^2$

23. Rewrite $\dfrac{(4t)^0}{t^{-2}}$ using only positive exponents, and simplify. (Assume any variables in the expression are nonzero.)

Solution

$\dfrac{(4t)^0}{t^{-2}} = \dfrac{1}{t^{-2}} = t^2$

25. Rewrite $\left(-3x^{-3}y^2\right)\left(4x^2y^{-5}\right)$ using only positive exponents, and simplify. (Assume any variables in the expression are nonzero.)

Solution

$\left(-3x^{-3}y^2\right)\left(4x^2y^{-5}\right) = -3 \cdot 4 \cdot x^{-3+2} \cdot y^{2+(-5)} = -12x^{-1}y^{-3} = \dfrac{-12}{xy^3}$

27. Rewrite $\left(3x^2y^{-2}\right)^{-2}$ using only positive exponents, and simplify. (Assume any variables in the expression are nonzero.)

Solution

$\left(3x^2y^{-2}\right)^{-2} = 3^{-2}x^{-4}y^4 = \dfrac{1y^4}{9x^4}$

29. Rewrite $\dfrac{6^2x^3y^{-3}}{12x^{-2}y}$ using only positive exponents, and simplify. (Assume any variables in the expression are nonzero.)

Solution

$\dfrac{6^2x^3y^{-3}}{12x^{-2}y} = \dfrac{6^2x^{3-(-2)}y^{-3-1}}{6 \cdot 2} = \dfrac{6x^5y^{-4}}{2} = \dfrac{3x^5}{y^4}$

31. Rewrite $\left(\dfrac{3u^2v^{-1}}{3^3u^{-1}v^3}\right)^{-2}$ using only positive exponents, and simplify. (Assume any variables in the expression are nonzero.)

Solution

$\left(\dfrac{3u^2v^{-1}}{3^3u^{-1}v^3}\right)^{-2} = \left(\dfrac{3u^{2-(-1)}v^{-1-3}}{3^3}\right)^{-2} = \left(\dfrac{u^3v^{-4}}{3^2}\right)^{-2} = \left(\dfrac{3^2}{u^3v^{-4}}\right)^2 = \dfrac{3^4}{u^6v^{-8}} = \dfrac{81v^8}{u^6}$

33. Rewrite $\dfrac{a+b}{ba^{-1}-ab^{-1}}$ using only positive exponents, and simplify. (Assume any variables in the expression are nonzero.)

Solution

$\dfrac{a+b}{ba^{-1}-ab^{-1}} = \dfrac{a+b}{\dfrac{b}{a}-\dfrac{a}{b}} \cdot \dfrac{ab}{ab} = \dfrac{a^2b+ab^2}{b^2-a^2} = \dfrac{ab(a+b)}{(b-a)(b+a)} = \dfrac{ab}{b-a}$

35. Write in scientific notation.

Land Area of Earth: 57,500,000 square miles

Solution

$57{,}500{,}000 = 5.75 \times 10^7$

37. Write in scientific notation.

Light year: 9,461,000,000,000,000 kilometers

Solution

$9{,}461{,}000{,}000{,}000{,}000 = 9.461 \times 10^{15}$

39. Write in scientific notation.

Relative Density of Hydrogen: 0.0000899

Solution

$0.0000899 = 8.99 \times 10^{-5}$

41. Write in decimal notation.

1993 Pepsi Co Beverage Sales: $\$2.82 \times 10^{10}$

Solution

$\$2.82 \times 10^{10} = \$28{,}200{,}000{,}000$

43. Write in decimal notation.

Temperature of Sum: 1.3×10^7 degrees Celsius

Solution

$1.3 \times 10^7 = 13,000,000$

45. Write in decimal notation.

Charge of Electron: 4.8×10^{10} electrostatic unit

Solution

$4.8 \times 10^{-10} = 0.00000000048$

47. Evaluate without a calculator.
Write your answer in scientific notation.

$\left(6.5 \times 10^6\right)\left(2 \times 10^4\right)$

Solution

$$\left(6.5 \times 10^6\right)\left(2 \times 10^4\right) = (6.5)(2) \times 10^{6+4}$$
$$= 13.0 \times 10^{10}$$
$$= 1.3 \times 10^{11}$$

49. Evaluate without a calculator.
Write your answer in scientific notation.

$\dfrac{3.6 \times 10^{12}}{6 \times 10^5}$

Solution

$$\frac{3.6 \times 10^{12}}{6 \times 10^5} = \frac{3.6}{6} \times 10^{12-5}$$
$$= 0.6 \times 10^7$$
$$= 6.0 \times 10^6$$

51. Evaluate with a calculator

$\dfrac{\left(3.82 \times 10^5\right)^2}{\left(8.5 \times 10^4\right)\left(5.2 \times 10^{-3}\right)}$

Solution

$$\frac{\left(3.82 \times 10^5\right)^2}{\left(8.5 \times 10^4\right)\left(5.2 \times 10^{-3}\right)} = \frac{14.5924 \times 10^{10}}{44.2 \times 10^1}$$
$$= .3301447964 \times 10^9$$
$$= 3.301447964 \times 10^8$$

53. *Mass of Earth and Sun* The masses of the earth and the sun are approximately 5.975×10^{24} and 1.99×10^{30} kilograms, respectively. The mass of the sun is approximately how many times that of the earth?

Solution

$$\frac{1.99 \times 10^{30}}{5.975 \times 10^{24}} = \frac{1.99}{5.975} \times 10^6 = 0.3330544 \times 10^6 = 333054.4$$

55. Factor $x^2 - 3x + 2$.

Solution

$x^2 - 3x + 2 = (x - 2)(x - 1)$

57. Factor $11x^2 + 6x - 5$.

Solution

$11x^2 + 6x - 5 = (11x - 5)(x + 1)$

59. *Comparing Memberships* The current membership of a public television station is 8415, which is 110% of what it was a year ago. How many members did the station have last year?

Solution

Verbal Model: $a = p \cdot b$

Equation: $8415 = 1.10 \cdot x$

$$7650 = x$$

61. Evaluate -10^{-3}.

Solution

$$-10^{-3} = -\frac{1}{10^3} = -\frac{1}{1,000}$$

63. Evaluate $(-3)^{-5}$.

Solution

$$(-3)^{-5} = \frac{1}{(-3)^5} = \frac{1}{-243} = -\frac{1}{243}$$

65. Evaluate $\frac{1}{4^{-3}}$.

Solution

$$\frac{1}{4^{-3}} = \frac{1}{\left(\frac{1}{4}\right)^3} = 4^3 = 64$$

67. Evaluate $\left(\frac{3}{16}\right)^0$.

Solution

$$\left(\frac{3}{16}\right)^0 = 1$$

69. Evaluate $\frac{3^4}{3^{-2}}$.

Solution

$$\frac{3^4}{3^{-2}} = 3^{4-(-2)}$$
$$= 3^{4+2}$$
$$= 3^6$$
$$= 729$$

71. Evaluate $\left(2^{-3}\right)^2$.

Solution

$$\left(2^{-3}\right)^2 = 2^{-6}$$
$$= \frac{1}{2^6}$$
$$= \frac{1}{64}$$

73. Evaluate $2^{-3} + 2^{-4}$.

Solution

$$2^{-3} + 2^{-4} = \frac{1}{2^3} + \frac{1}{2^4}$$
$$= \frac{1}{8} + \frac{1}{16}$$
$$= \frac{2}{16} + \frac{1}{16}$$
$$= \frac{3}{16}$$

75. Evaluate $\left(\frac{3}{4} + \frac{5}{8}\right)^{-2}$.

Solution

$$\left(\frac{3}{4} + \frac{5}{8}\right)^{-2} = \left(\frac{6}{8} + \frac{5}{8}\right)^{-2}$$
$$= \left(\frac{11}{8}\right)^{-2}$$
$$= \left(\frac{8}{11}\right)^2$$
$$= \frac{64}{121}$$

77. Rewrite $z^5 \cdot x^{-3}$ using only positive exponents, and simplify. (Assume any variables in the expression are nonzero.)

Solution

$$z^5 \cdot z^{-3} = z^{5+(-3)} = z^2$$

79. Rewrite $\frac{1}{x^{-6}}$ using only positive exponents, and simplify. (Assume any variables in the expression are nonzero.)

Solution

$$\frac{1}{x^{-6}} = x^6$$

81. Rewrite $\dfrac{a^{-6}}{a^{-7}}$ using only positive exponents, and simplify. (Assume any variables in the expression are nonzero.)

Solution

$\dfrac{a^{-6}}{a^{-7}} = a^{(-6)-(-7)} = a^{-6+7} = a$

83. Rewrite $\left(2x^2\right)^{-2}$ using only positive exponents, and simplify. (Assume any variables in the expression are nonzero.)

Solution

$\left(2x^2\right)^{-2} = \dfrac{1}{\left(2x^2\right)^2} = \dfrac{1}{4x^4}$

85. Rewrite $\left(\dfrac{x}{10}\right)^{-1}$ using only positive exponents, and simplify. (Assume any variables in the expression are nonzero.)

Solution

$\left(\dfrac{x}{10}\right)^{-1} = \dfrac{10}{x}$

87. Rewrite $\left(\dfrac{a^{-2}}{b^{-2}}\right)\left(\dfrac{b}{a}\right)^3$ using only positive exponents, and simplify. (Assume any variables in the expression are nonzero.)

Solution

$\left(\dfrac{a^{-2}}{b^{-2}}\right)\left(\dfrac{b}{a}\right)^3 = \left(\dfrac{b^2}{a^2}\right)\left(\dfrac{b^3}{a^3}\right) = \dfrac{b^5}{a^5}$

89. Rewrite $\left[\left(x^{-4}y^{-6}\right)^{-1}\right]^2$ using only positive exponents, and simplify. (Assume any variables in the expression are nonzero.)

Solution

$\left[\left(x^{-4}y^{-6}\right)^{-1}\right]^2 = \left(x^4y^6\right)^2 = x^8y^{12}$

91. Rewrite $\left(u + v^{-2}\right)^{-1}$ using only positive exponents, and simplify. (Assume any variables in the expression are nonzero.)

Solution

$\left(u + v^{-2}\right)^{-1} = \dfrac{1}{u + v^{-2}} = \dfrac{1}{u + \dfrac{1}{v^2}} \cdot \dfrac{v^2}{v^2} = \dfrac{v^2}{uv^2 + 1}$

93. Rewrite $\left[\left(2x^{-3}y^{-2}\right)^2\right]^{-2}$ using only positive exponents, and simplify. (Assume any variables in the expression are nonzero.)

Solution

$\left[\left(2x^{-3}y^{-2}\right)^2\right]^{-2} = \left[\left(2x^{-3}y^{-2}\right)^{-4}\right] = 2^{-4}x^{12}y^8 = \dfrac{x^{12}y^8}{2^4} = \dfrac{x^{12}y^8}{16}$

95. Write 3,600,000 in scientific notation.

Solution

$3,600,000 = 3.6 \times 10^6$

97. Write 0.0000000381 in scientific notation.

Solution

$0.0000000381 = 3.81 \times 10^{-8}$

99. Write 6×10^7 in decimal notation.

Solution

$6 \times 10^7 = 60,000,000$

101. Write 1.359×10^{-7} in decimal notation.

Solution

$1.359 \times 10^{-7} = 0.0000001359$

103. Evaluate $\left(2 \times 10^9\right)\left(3.4 \times 10^{-4}\right)$ without a calculator. Write your answer in scientific notation.

Solution

$\left(2 \times 10^9\right)\left(3.4 \times 10^{-4}\right) = (2)(3.4)\left(10^5\right) = 6.8 \times 10^5$

105. Evaluate $\dfrac{3.6 \times 10^9}{9 \times 10^5}$ without a calculator. Write your answer in scientific notation.

Solution

$$\frac{3.6 \times 10^9}{9 \times 10^5} = \frac{3.6}{9} \times 10^4 = 0.4 \times 10^4 = 4.0 \times 10^3$$

107. Evaluate $(4,500,000)(2,000,000,000)$ without a calculator. Write your answer in scientific notation.

Solution

$$(4,500,000)(2,000,000,000) = \left(4.5 \times 10^6\right)\left(2 \times 10^9\right) = (4.5)(2) \times 10^{15}$$
$$= 9 \times 10^{15}$$

109. Evaluate $\dfrac{1.357 \times 10^{12}}{\left(4.2 \times 10^2\right)\left(6.87 \times 10^{-3}\right)}$ using a calculator. Round your answer to two decimal places.

Solution

$$\frac{1.357 \times 10^{12}}{\left(4.2 \times 10^2\right)\left(6.87 \times 10^{-3}\right)} = \frac{1.357}{(4.2)(6.87)} \times 10^{13} = 0.0470299 \times 10^{13}$$
$$= 4.70299 \times 10^{11} \approx 4.70 \times 10^{11}$$

111. Evaluate $\dfrac{(0.0000565)(2,850,000,000,000)}{0.00465}$ using a calculator. Round your answer to two decimal places.

Solution

$$\frac{(0.0000565)(2,850,000,000,000)}{0.00465} = \frac{\left(5.65 \times 10^{-5}\right)\left(2.85 \times 10^{12}\right)}{4.65 \times 10^{-3}}$$
$$= \frac{(5.65)(2.85)}{4.65} \times 10^{10}$$
$$= 3.4629032 \times 10^{10} \approx 3.46 \times 10^{10}$$

113. *Distance to the Sun* The distance from the earth to the sun is approximately 93 million miles. Write this distance in scientific notation.

Solution

$$93,000,000 = 9.3 \times 10^7$$

115. *Light-Year* One light year (the distance light can travel in 1 year) is approximately 9.45×10^{15} meters. Approximate the time for light to travel from the sun to the earth if that distance is approximately 1.49×10^{11} meters.

Solution

$$\frac{1.49 \times 10^{11}}{9.45 \times 10^{15}} = \frac{1.49}{9.45} \times 10^{-4} = 0.157672 \times 10^{-4} = 0.0000157672$$

117. *Federal Debt* In 1993, the population of the United States was 257 million, and the federal debt was 4410 billion dollars. Use these two numbers to determine the amount each person would have had to pay (per capita debt) to remove the debt. (Source: U.S. Bureaus of Census)

Solution

$$\frac{4410 \text{ billion}}{257 \text{ million}} = \frac{4,410,000,000,000}{257,000,000}$$

$$= \frac{4.41 \times 10^{12}}{2.57 \times 10^{8}}$$

$$= 1.715953307 \times 10^{4}$$

$$= 17159.53307$$

$$= \$17,159.53$$

5.2   Rational Exponents and Radicals

7. Complete the statement.

Because $7^2 = 49$,   is a square root of 49.

Solution

Because $7^2 = 49$, 7 is a square root of 49.

9. Complete the statement.

Because $4.2^3 = 74.088$, 4.2 is a

Solution

Because $4.2^3 = 74.088$, 4.2 is a cube root of 74.088.

11. Complete the statement.

Because $45^2 = 2025$, 45 is called the   of 2025.

Solution

Because $45^2 = 2025$, 45 is called the square root of 2025.

13. Evaluate $\sqrt{81}$ without a calculator.

If not possible, state the reason.

Solution

$\sqrt{81} = 9$

15. Evaluate $\sqrt{\dfrac{9}{16}}$ without a calculator.

If not possible, state the reason.

Solution

$\sqrt{\dfrac{9}{16}} = \dfrac{3}{4}$

17. Evaluate $\sqrt{-64}$ without a calculator.

If not possible, state the reason.

Solution

$\sqrt{-64}$ is not a real number

19. Evaluate $-\sqrt{\dfrac{4}{9}}$ without a calculator.

If not possible, state the reason.

Solution

$-\sqrt{\dfrac{4}{9}} = -\dfrac{2}{3}$

21. Evaluate $\sqrt[3]{125}$ without a calculator.

If not possible, state the reason.

Solution

$\sqrt[3]{125} = 5$

23. Evaluate $\sqrt[4]{81}$ without a calculator.

If not possible, state the reason.

Solution

$\sqrt[4]{81} = 3$

25. Evaluate $\sqrt[5]{-0.00243}$ without a calculator. If not possible, state the reason.

Solution

$\sqrt[5]{-0.00243} = -0.3$

29. Evaluate $32^{-2/5}$ without a calculator.

Solution

$32^{-2/5} = \dfrac{1}{\left(\sqrt[5]{32}\right)^2} = \dfrac{1}{2^2} = \dfrac{1}{4}$

33. Fill in the missing description.

Radical Form   Rational Exponent Form
$\sqrt[3]{27^2} = 9$

Solution

Radical Form   Rational Exponent Form
$\sqrt[3]{27^2} = 9$ $\qquad\qquad$ $27^{2/3} = 9$

37. Use a calculator to evaluate $\dfrac{8 - \sqrt{35}}{2}$. Round the result to four decimal places.

Solution

$\dfrac{8 - \sqrt{35}}{2} = 1.0420$

41. Simplify $\sqrt[3]{t^6}$.

Solution

$\sqrt[3]{t^6} = t^{6/3} = t^2$

45. Simplify $\dfrac{2^{1/5}}{2^{6/5}}$.

Solution

$\dfrac{2^{1/5}}{2^{6/5}} = 2^{1/5 - 6/5}$

$= 2^{-5/5}$

$= 2^{-1} = \dfrac{1}{2}$

27. Evaluate $25^{1/2}$ without a calculator.

Solution

$25^{1/2} = \sqrt{25} = 5$

31. Fill in the missing description.

Radical Form   Rational Exponent Form
$\sqrt{16} = 4$

Solution

Radical Form   Rational Exponent Form
$\sqrt{16} = 4$ $\qquad\qquad$ $16^{1/2} = 4$

35. Fill in the missing description.

Radical Form   Rational Exponent Form
$\qquad\qquad$ $256^{3/4} = 64$

Solution

Radical Form   Rational Exponent Form
$\sqrt[4]{256^3} = 64$ $\qquad$ $256^{3/4} = 64$

39. Use a calculator to evaluate $\sqrt[4]{342}$. Round the result to four decimal places.

Solution

$\sqrt[4]{342} = 4.3004$

43. Simplify $3^{1/4} \cdot 3^{3/4}$.

Solution

$3^{1/4} \cdot 3^{3/4} = 3^{1/4 + 3/4}$

$= 3^{4/4}$

$= 3^1$

47. Simplify $x^{2/3} \cdot x^{7/3}$.

Solution

$x^{2/3} \cdot x^{7/3} = x^{2/3 + 7/3}$

$= x^{9/3}$

$= x^3$

49. Simplify $\left(\dfrac{x^{1/4}}{x^{1/6}}\right)^3$.

Solution

$$\left(\frac{x^{1/4}}{x^{1/6}}\right)^3 = \left(x^{1/4-1/6}\right)^3$$

$$= \left(x^{3/12-2/12}\right)^3$$

$$= \left(x^{1/12}\right)^3$$

$$= x^{3/12}$$

$$= x^{1/4}$$

51. *Mathematical Modeling* Use the formula for the *declining balances method*

$$r = 1 - \left(\frac{s}{c}\right)^{1/n}$$

to find the depreciation rate r. In the formula, n is the useful life of the item (in years), s is the salvage value (in dollars), and c is the original cost (in dollars). A truck whose original cost is \$75,000 is depreciated over an 8-year period, as shown in the graph in the textbook.

Solution

$$r = 1 - \left(\frac{25,000}{75,000}\right)^{1/8}$$

$$= 1 - \left(\frac{1}{3}\right)^{1/8}$$

$$= 0.128$$

53. Determine the domain of the function $f(x) = 3\sqrt{x}$.

Solution

$f(x) = 3\sqrt{x},\ x \geq 0$,
so Domain $= [0, \infty)$

55. Evaluate $-\dfrac{13}{35} \cdot -\dfrac{25}{104}$.

Solution

$$-\frac{13}{35} \cdot -\frac{25}{104} = \frac{13 \cdot 5 \cdot 5}{7 \cdot 5 \cdot 2 \cdot 2 \cdot 2 \cdot 13} = \frac{5}{56}$$

57. Evaluate $\dfrac{14}{3} \div \dfrac{42}{45}$.

Solution

$$\frac{14}{3} \div \frac{42}{45} = \frac{14}{3} \cdot \frac{45}{42} = \frac{7 \cdot 2 \cdot 3 \cdot 3 \cdot 5}{3 \cdot 7 \cdot 3 \cdot 2} = 5$$

59. Use the function $f(x) = x^2 - 3$ to find and simplify the expression for $\dfrac{f(2+h) - f(2)}{h}$.

Solution

$$\frac{f(2+h) - f(2)}{h} = \frac{\left[(2+h)^2 - 3\right] - \left[2^2 - 3\right]}{h}$$

$$= \frac{4 + 4h + h^2 - 3 - 4 + 3}{h}$$

$$= \frac{4h + h^2}{h}$$

$$= \frac{h(4+h)}{h}$$

$$= 4 + h$$

61. Evaluate $\sqrt{64}$ without a calculator. If not possible, state the reason.

Solution

$\sqrt{64} = 8$

63. Evaluate $\sqrt{-100}$ without a calculator. If not possible, state the reason.

Solution

$\sqrt{-100}$ is not a real number.

65. Evaluate $\sqrt{0.16}$ without a calculator. If not possible, state the reason.

Solution

$\sqrt{0.16} = 0.4$

67. Evaluate $\sqrt{49 - 4(2)(-15)}$ without a calculator. If not possible, state the reason.

Solution

$\sqrt{49 - 4(2)(-15)} = \sqrt{49 + 120} = \sqrt{169} = 13$

69. Evaluate $\sqrt[3]{1000}$ without a calculator. If not possible, state the reason.

Solution

$\sqrt[3]{1000} = 10$

71. Evaluate $\sqrt[3]{-\dfrac{1}{64}}$ without a calculator. If not possible, state the reason.

Solution

$\sqrt[3]{-\dfrac{1}{64}} = -\dfrac{1}{4}$

73. Evaluate $-\sqrt[4]{-625}$ without a calculator. If not possible, state the reason.

Solution

$-\sqrt[4]{-625}$ is not a real number.

75. Determine whether $\sqrt{6}$ is a rational or irrational number.

Solution

$\sqrt{6}$ is not rational.

77. Determine whether $\sqrt{900}$ is a rational or irrational number.

Solution

$\sqrt{900}$ is rational.

79. Evaluate $49^{1/2}$ without a calculator.

Solution

$49^{1/2} = \sqrt{49} = 7$

81. Evaluate $16^{3/4}$ without a calculator.

Solution

$16^{3/4} = (\sqrt[4]{16})^3 = 2^3 = 8$

83. Evaluate $\left(\dfrac{8}{27}\right)^{2/3}$ without a calculator.

Solution

$\left(\dfrac{8}{27}\right)^{2/3} = \left(\sqrt[3]{\dfrac{8}{27}}\right)^2 = \left(\dfrac{2}{3}\right)^2 = \dfrac{4}{9}$

85. Evaluate $\left(\dfrac{121}{9}\right)^{-1/2}$ without a calculator.

Solution

$\left(\dfrac{121}{9}\right)^{-1/2} = \left(\dfrac{9}{121}\right)^{1/2} = \sqrt{\dfrac{9}{121}} = \dfrac{3}{11}$

87. Use a calculator to evaluate $\sqrt{73}$. If it is not possible, state the reason. Round the result to four decimal places.

Solution

$\sqrt{73} = 8.5440$

89. Use a calculator to evaluate $\dfrac{3 - \sqrt{17}}{9}$. If it is not possible, state the reason. Round the result to four decimal places.

Solution

$\dfrac{3 - \sqrt{17}}{9} = -0.1248$

91. Use a calculator to evaluate $1698^{-3/4}$. If it is not possible, state the reason. Round the result to four decimal places.

Solution

$1698^{-3/4} = 0.0038$

93. Use a calculator to evaluate $\sqrt[3]{545^2}$. If it is not possible, state the reason. Round the result to four decimal places.

Solution

$\sqrt[3]{545^2} = 66.7213$

95. Simplify $\sqrt{t^2}$.

Solution

$\sqrt{t^2} = |t|$

97. Simplify $\sqrt[3]{y^9}$.

Solution

$\sqrt[3]{y^9} = y^3$

99. Simplify $\left(\dfrac{2}{3}\right)^{5/3} \cdot \left(\dfrac{2}{3}\right)^{1/3}$.

Solution

$$\left(\frac{2}{3}\right)^{5/3} \cdot \left(\frac{2}{3}\right)^{1/3} = \left(\frac{2}{3}\right)^{5/3+1/3}$$
$$= \left(\frac{2}{3}\right)^{6/3}$$
$$= \left(\frac{2}{3}\right)^{2}$$
$$= \frac{4}{9}$$

101. Simplify $\left(3x^{-1/3}y^{3/4}\right)^2$.

Solution

$$\left(3x^{-1/3}y^{3/4}\right)^2 = 3^2 x^{-2/3} y^{3/2}$$
$$= \frac{9y^{3/2}}{x^{2/3}}$$

103. Simplify $\dfrac{18y^{4/3}z^{-1/3}}{24y^{-2/3}z}$.

Solution

$$\frac{18y^{4/3}z^{-1/3}}{24y^{-2/3}z} = \frac{6 \cdot 3y^{4/3-(-2/3)}z^{-1/3-1}}{6 \cdot 4}$$
$$= \frac{3y^{6/3}z^{-4/3}}{4}$$
$$= \frac{3y^2}{4z^{4/3}}$$

105. Simplify $\left(c^{3/2}\right)^{1/3}$.

Solution

$$\left(c^{3/2}\right)^{1/3} = c^{3/2 \cdot 1/3}$$
$$= c^{1/2}$$

107. Simplify $\sqrt{\sqrt[4]{y}}$.

Solution

$$\sqrt{\sqrt[4]{y}} = \left(y^{1/4}\right)^{1/2}$$
$$= y^{1/4 \cdot 1/2}$$
$$= y^{1/8}$$
$$= \sqrt[8]{y}$$

109. Multiply and simplify.

$x^{1/2}(2x - 3)$

Solution

$x^{1/2}(2x - 3) = 2x^{3/2} - 3x^{1/2}$

111. Multiply and simplify.

$y^{-1/3}\left(y^{1/3} + 5y^{4/3}\right)$

Solution

$$y^{-1/3}\left(y^{1/3} + 5y^{4/3}\right) = y^0 + 5y^{3/3}$$
$$= 1 + 5y$$

113. Determine the doamin of $g(x) = \dfrac{2}{\sqrt[4]{x}}$.

Solution

The domain of $g(x) = \dfrac{2}{\sqrt[4]{x}}$ is the set of all nonnegative real numbers or $[0, \infty)$.

115. Use a graphing utility to graph the function $y = \dfrac{5}{\sqrt[4]{x^3}}$. Check the domain of the function algebraically. Did the graphing utility skip part of the domain? If so, complete the graph by hand.

Solution

Keystrokes:

$\boxed{Y =}$ 5 $\boxed{\div}$ 4 $\boxed{\text{Math 5}}$ $\boxed{X, T, \theta}$ $\boxed{\text{Math 3}}$ $\boxed{\text{GRAPH}}$

Domain = $(0, \infty)$

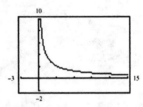

Domain is $(0, \infty)$ so graphing utility did complete the graph.

117. Use a graphing utility to graph the function $g(x) = 2x^{3/5}$. Check the domain of the function algebraically. Did the graphing utility skip part of the domain? If so, complete the graph by hand.

Solution

Keystrokes:

$\boxed{Y =}$ 2 $\boxed{X, T, \theta}$ $\boxed{\wedge}$ $\boxed{(}$ 3 $\boxed{\div}$ 5 $\boxed{)}$ $\boxed{\text{GRAPH}}$

Domain = $(-\infty, \infty)$

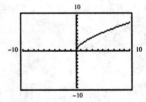

Domain is $(-\infty, \infty)$ so graphing utility did not complete the graph. It should be

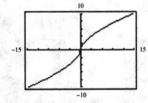

119. *Perfect Squares* Find all possible "last digits" of perfect squares. (For instance, the last digit of 81 is 1 and the last digit of 64 is 4.) Is it possible that 4,322,788,987 is a perfect square?

Solution

(a) "Last digits" $--1$(Perfect square81)   (b) No

$--4$(Perfect square64)

$--5$(Perfect square25)

$--6$(Perfect square36)

$--9$(Perfect square49)

$--0$(Perfect square100)

121. *Velocity of a Stream* A stream of water moving at the rate of v feet per second can carry particles of size $0.03\sqrt{v}$. Find the particle size that can be carried by a stream flowing at the rate of $\frac{3}{4}$ foot per second.

Solution

diameter $= 0.03\sqrt{v}$ inches

diameter $= 0.03\sqrt{\frac{3}{4}}$

diameter $= 0.0259808$ inches

5.3 Simplifying and Combining Radicals

7. Write $\sqrt[3]{11} \cdot \sqrt[3]{10}$ as a single radical.

Solution

$\sqrt[3]{11} \cdot \sqrt[3]{10} = \sqrt[3]{110}$

9. Write $\dfrac{\sqrt{15}}{\sqrt{31}}$ as a single radical.

Solution

$\dfrac{\sqrt{15}}{\sqrt{31}} = \sqrt{\dfrac{15}{31}}$

11. Write $\sqrt{9 \cdot 35}$ as a product or quotient of radicals and simplify.

Solution

$\sqrt{9 \cdot 35} = 3\sqrt{35}$

13. Write $\sqrt[3]{\dfrac{1000}{11}}$ as a product or quotient of radicals and simplify.

Solution

$\sqrt[3]{\dfrac{1000}{11}} = \dfrac{10}{\sqrt[3]{11}} \cdot \dfrac{\sqrt[3]{11^2}}{\sqrt[3]{11^2}} = \dfrac{10\sqrt[3]{121}}{11}$

15. Simplify $\sqrt{20}$.

Solution

$\sqrt{20} = \sqrt{4 \cdot 5} = \sqrt{2^2 \cdot 5} = 2\sqrt{5}$

17. Simplify $\sqrt[3]{24}$.

Solution

$\sqrt[3]{24} = \sqrt[3]{8 \cdot 3} = \sqrt[3]{2^3 \cdot 3} = 2\sqrt[3]{3}$

19. Simplify $\sqrt{\dfrac{15}{4}}$.

Solution

$\sqrt{\dfrac{15}{4}} = \dfrac{\sqrt{15}}{2}$

21. Simplify $\sqrt[5]{\dfrac{15}{243}}$.

Solution

$\sqrt[5]{\dfrac{15}{243}} = \dfrac{\sqrt[5]{15}}{3}$

23. Simplify $\sqrt{9x^5}$.

Solution

$$\sqrt{9x^5} = \sqrt{3^2 x^4 \cdot x}$$
$$= 3 \cdot x^2 \cdot \sqrt{x}$$
$$= 3x^2 \sqrt{x}$$

25. Simplify $\sqrt[4]{3x^4 y^2}$.

Solution

$$\sqrt[4]{3x^4 y^2} = \sqrt[4]{x^4} \cdot \sqrt[4]{3y^2}$$
$$= |x| \sqrt[4]{3y^2}$$

27. Simplify $\sqrt[5]{\dfrac{32x^2}{y^5}}$.

Solution

$$\sqrt[5]{\frac{32x^2}{y^5}} = \sqrt[5]{\frac{2^5 x^2}{y^5}}$$
$$= \frac{2}{y} \sqrt[5]{x^2}$$

29. Simplify $\sqrt{\dfrac{32a^4}{b^2}}$.

Solution

$$\sqrt{\frac{32a^4}{b^2}} = \frac{\sqrt{16 \cdot 2 \cdot a^4}}{\sqrt{b^2}} = \frac{4a^2 \sqrt{2}}{|b|}$$

31. Rationalize the denominator and simplify further, if possible.

$$\sqrt{\frac{1}{3}}$$

Solution

$$\sqrt{\frac{1}{3}} = \frac{1}{\sqrt{3}} \cdot \frac{\sqrt{3}}{\sqrt{3}} = \frac{\sqrt{3}}{3}$$

33. Rationalize the denominator and simplify further, if possible.

$$\sqrt[4]{\frac{5}{4}}$$

Solution

$$\sqrt[4]{\frac{5}{4}} = \frac{\sqrt[4]{5}}{\sqrt[4]{2^2}} \cdot \frac{\sqrt[4]{2^2}}{\sqrt[4]{2^2}} = \frac{\sqrt[4]{5 \cdot 2^2}}{\sqrt[4]{2^4}} = \frac{\sqrt[4]{20}}{2}$$

35. Rationalize the denominator and simplify further, if possible.

$$\frac{1}{\sqrt{y}}.$$

Solution

$$\frac{1}{\sqrt{y}} = \frac{1}{\sqrt{y}} \cdot \frac{\sqrt{y}}{\sqrt{y}} = \frac{\sqrt{y}}{\sqrt{y^2}} = \frac{\sqrt{y}}{|y|}$$

37. Rationalize the denominator and simplify further, if possible.

$$\sqrt[3]{\frac{2x}{3y}}$$

Solution

$$\sqrt[3]{\frac{2x}{3y}} = \frac{\sqrt[3]{2x}}{\sqrt[3]{3y}} \cdot \frac{\sqrt[3]{3^2 y^2}}{\sqrt[3]{3^2 y^2}} = \frac{\sqrt[3]{2x \cdot 3^2 y^2}}{\sqrt[3]{3^3 y^3}} = \frac{\sqrt[3]{18xy^2}}{3y}$$

39. Combine the radical expressions, if possible.

$$3\sqrt{2} - \sqrt{2}$$

Solution

$$3\sqrt{2} - \sqrt{2} = 2\sqrt{2}$$

41. Combine the radical expressions, if possible.

$$2\sqrt[3]{54} + 12\sqrt[3]{16}$$

Solution

$$2\sqrt[3]{54} + 12\sqrt[3]{16} = 2\sqrt[3]{27 \cdot 2} + 12\sqrt[3]{8 \cdot 2}$$
$$= 6\sqrt[3]{2} + 24\sqrt[3]{2} = 30\sqrt[3]{2}$$

43. Combine the radical expressions, if possible.

$$\sqrt{25y} + \sqrt{64y}$$

Solution

$$\sqrt{25y} + \sqrt{64y} = 5\sqrt{y} + 8\sqrt{y} = 13\sqrt{y}$$

45. Use a graphing utility to graph $y_1 = \sqrt{\dfrac{3}{x}}$ and $y_2 = \dfrac{\sqrt{3x}}{x}$ on the same screen. Use the graphs to verify that the expressions are equivalent. Verify the results algebraically.

Solution

Keystrokes:

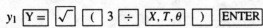

y_1 [Y=] [√] [(] 3 [÷] [X,T,θ] [)] [ENTER]

y_2 [Y=] [√] [(] 3 [X,T,θ] [)] [÷] [X,T,θ] [GRAPH]

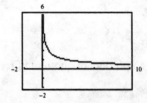

$y_1 = y_2$ since the graphs are the same. $\sqrt{\dfrac{3}{x}} = \sqrt{\dfrac{3}{x}} \cdot \dfrac{\sqrt{x}}{\sqrt{x}} = \dfrac{\sqrt{3x}}{x}$

47. *Geometry* Find the length of the hypotenuse of the right triangle.

Solution

$c = \sqrt{a^2 + b^2}$

$ = \sqrt{6^2 + 3^2}$

$ = \sqrt{36 + 9} = \sqrt{45} = \sqrt{9 \cdot 5} = 3\sqrt{5}$

49. *Vibrating String* The frequency f in cycles per second of a vibrating string is given by

$$f = \frac{1}{100}\sqrt{\frac{400 \times 10^6}{5}}.$$

Use a calculator to approximate this number. (Round your answer to two decimal places.)

Solution

$$f = \frac{1}{100}\sqrt{\frac{400 \times 10^6}{5}} = 8.9443 \times 10^1 = 89.443 \approx 89.44$$

51. Simplify $(x^2 - 3xy)^0$.

Solution

$(x^2 - 3xy)^0 = 1$

53. Simplify $\dfrac{64r^2s^4}{16rs^2}$.

Solution

$\dfrac{64r^2s^4}{16rs^2} = 4r^{2-1}s^{4-2} = 4rs^2$

55. *Geometry* The height of a triangle is 12 inches less than its base. The area is 110 square inches. Find the height and base.

Solution

Formula: $A = \dfrac{1}{2}b \cdot h$

Equation: $110 = \dfrac{1}{2} \cdot x \cdot (x - 12)$ 　　 $x - 22 = 0$ 　　 $x + 10 = 0$

$ 220 = x^2 - 12x$ 　　　　　 $x = 22$ inches 　　 $x = -10$

$ 0 = x^2 - 12x - 220$ 　　 $x - 12 = 10$ inches

$ 0 = (x - 22)(x + 10)$

57. Write $\sqrt{3} \cdot \sqrt{10}$ as a single radical.

Solution

$\sqrt{3} \cdot \sqrt{10} = \sqrt{30}$

59. Write $\dfrac{\sqrt[5]{152}}{\sqrt[5]{3}}$ as a single radical.

Solution

$\dfrac{\sqrt[5]{152}}{\sqrt[5]{3}} = \sqrt[5]{\dfrac{152}{3}}$

61. Write $\sqrt[4]{81 \cdot 11}$ as a product or quotient of radicals and simplify.

Solution

$\sqrt[4]{81 \cdot 11} = 3\sqrt[4]{11}$

63. Write $\sqrt{\dfrac{35}{9}}$ as a product or quotient of radicals and simplify.

Solution

$\sqrt{\dfrac{35}{9}} = \dfrac{\sqrt{35}}{3}$

65. Simplify $\sqrt{27}$.

Solution

$\sqrt{27} = \sqrt{9 \cdot 3} = \sqrt{3^2 \cdot 3} = 3\sqrt{3}$

67. Simplify $\sqrt[4]{30,000}$.

Solution

$\sqrt[4]{30,000} = \sqrt[4]{10,000 \cdot 3} = \sqrt[4]{10^4 \cdot 3} = 10\sqrt[4]{3}$

69. Simplify $\sqrt{\dfrac{15}{49}}$.

Solution

$\sqrt{\dfrac{15}{49}} = \dfrac{\sqrt{15}}{\sqrt{49}} = \dfrac{\sqrt{15}}{7}$

71. Simplify $\sqrt[3]{\dfrac{35}{64}}$.

Solution

$\sqrt[3]{\dfrac{35}{64}} = \dfrac{\sqrt[3]{35}}{4}$

73. Simplify $\sqrt{4 \times 10^{-4}}$.

Solution

$\sqrt{4 \times 10^{-4}} = \sqrt{4} \times \sqrt{10^{-4}} = 2 \times 10^{-2} = 0.02$

75. Simplify $\sqrt[3]{2.4 \times 10^6}$.

Solution

$\sqrt[3]{2.4 \times 10^6} = \sqrt[3]{2.4} \times \sqrt[3]{10^6} = \sqrt[3]{8} \times \sqrt[3]{0.3} \times \sqrt[3]{10^6}$

$= 2 \times \sqrt[3]{0.3} \times 10^2$

$= 200\sqrt[3]{0.3}$

77. Simplify $\sqrt{48y^4}$.

Solution

$\sqrt{48y^4} = \sqrt{16 \cdot 3 \cdot y^4} = 4y^2\sqrt{3}$

79. Simplify $\sqrt[3]{x^4 y^3}$.

Solution

$\sqrt[3]{x^4 y^3} = \sqrt[3]{x^3 \cdot x \cdot y^3} = xy\sqrt[3]{x}$

81. Simplify $\sqrt[5]{32x^5 y^6}$.

Solution

$\sqrt[5]{32x^5 y^6} = \sqrt[5]{2^5 \cdot x^5 \cdot y^5 \cdot y} = 2xy\sqrt[5]{y}$

83. Simplify $\sqrt{\dfrac{13}{25}}$.

Solution

$\sqrt{\dfrac{13}{25}} = \dfrac{\sqrt{13}}{5}$

85. Simplify $\sqrt[3]{\dfrac{54a^4}{b^9}}$.

Solution

$$\sqrt[3]{\dfrac{54a^4}{b^9}} = \sqrt[3]{\dfrac{3^3 \cdot 2 \cdot a^3 \cdot a}{b^9}} = \dfrac{3a}{b^3}\sqrt[3]{2a}$$

87. Rationalize the denominator and simplify further, if possible.

$$\dfrac{12}{\sqrt{3}}$$

Solution

$$\dfrac{12}{\sqrt{3}} = \dfrac{12}{\sqrt{3}} \cdot \dfrac{\sqrt{3}}{\sqrt{3}} = \dfrac{12\sqrt{3}}{3} = 4\sqrt{3}$$

89. Rationalize the denominator and simplify further, if possible.

$$\dfrac{6}{\sqrt[3]{32}}$$

Solution

$$\dfrac{6}{\sqrt[3]{32}} = \dfrac{6}{\sqrt[3]{2^3 \cdot 2^2}} = \dfrac{6}{2\sqrt[3]{2^2}} \cdot \dfrac{\sqrt[3]{2}}{\sqrt[3]{2}} = \dfrac{6\sqrt[3]{2}}{2\sqrt[3]{2^3}} = \dfrac{6\sqrt[3]{2}}{4} = \dfrac{3\sqrt[3]{2}}{2}$$

91. Rationalize the denominator and simplify further, if possible.

$$\sqrt{\dfrac{4}{x}}$$

Solution

$$\sqrt{\dfrac{4}{x}} = \dfrac{\sqrt{4}}{\sqrt{x}} = \dfrac{2}{\sqrt{x}} \cdot \dfrac{\sqrt{x}}{\sqrt{x}} = \dfrac{2\sqrt{x}}{x}$$

93. Rationalize the denominator and simplify further, if possible.

$$\dfrac{6}{\sqrt{3b^3}}$$

Solution

$$\dfrac{6}{\sqrt{3b^3}} = \dfrac{6}{b\sqrt{3b}} \cdot \dfrac{\sqrt{3b}}{\sqrt{3b}} = \dfrac{6\sqrt{3b}}{3b^2} = \dfrac{2\sqrt{3b}}{b^2}$$

95. Rationalize the denominator and simplify further, if possible.

$$\dfrac{a^3}{\sqrt[3]{ab^2}}$$

Solution

$$\dfrac{a^3}{\sqrt[3]{ab^2}} = \dfrac{a^3}{\sqrt[3]{ab^2}} \cdot \dfrac{\sqrt[3]{a^2b}}{\sqrt[3]{a^2b}} = \dfrac{a^3\sqrt[3]{a^2b}}{\sqrt[3]{a^3b^3}} = \dfrac{a^3\sqrt[3]{a^2b}}{ab} = \dfrac{a^2\sqrt[3]{a^2b}}{b}$$

97. Combine the radical expressions, if possible.

$$\sqrt[4]{3} - 5\sqrt[4]{7} - 12\sqrt[4]{3}$$

Solution

$$\sqrt[4]{3} - 5\sqrt[4]{7} - 12\sqrt[4]{3} = -11\sqrt[4]{3} - 5\sqrt[4]{7}$$

99. Combine the radical expressions, if possible.

$$12\sqrt{8} - 3\sqrt[3]{8}$$

Solution

$$12\sqrt{8} - 3\sqrt[3]{8} = 12\sqrt{4 \cdot 2} - 3 \cdot 2 = 24\sqrt{2} - 6$$

101. Combine the radical expressions, if possible.

$$5\sqrt{9x} - 3\sqrt{x}$$

Solution

$$5\sqrt{9x} - 3\sqrt{x} = 15\sqrt{x} - 3\sqrt{x} = 12\sqrt{x}$$

103. Combine the radical expressions, if possible.

$$10\sqrt[3]{z} - \sqrt[3]{z^4}$$

Solution

$$10\sqrt[3]{z} - \sqrt[3]{z^4} = 10\sqrt[3]{z} - \sqrt[3]{z^3 \cdot z} = 10\sqrt[3]{z} - z\sqrt[3]{z}$$
$$= (10 - z)\sqrt[3]{z}$$

105. Perform the subtraction and simplify your answer.

$$\sqrt{5} - \frac{3}{\sqrt{5}}$$

Solution

$$\sqrt{5} - \frac{3}{\sqrt{5}} = \sqrt{5} - \left(\frac{3}{\sqrt{5}} \cdot \frac{\sqrt{5}}{\sqrt{5}}\right) = \sqrt{5} - \frac{3\sqrt{5}}{5}$$

$$= \left(1 - \frac{3}{5}\right)\sqrt{5}$$

$$= \frac{2}{5}\sqrt{5}$$

107. Perform the subtraction and simplify your answer.

$$\sqrt{20} - \sqrt{\frac{1}{5}}$$

Solution

$$\sqrt{20} - \sqrt{\frac{1}{5}} = \sqrt{4 \cdot 5} - \sqrt{\frac{1}{5}} \cdot \frac{\sqrt{5}}{\sqrt{5}} = 2\sqrt{5} - \frac{\sqrt{5}}{5}$$

$$= \left(2 - \frac{1}{5}\right)\sqrt{5}$$

$$= \frac{9}{5}\sqrt{5}$$

109. Place the correct symbol ($<$, $>$, or $=$) between the numbers.

$$\sqrt{7} + \sqrt{18} \qquad \sqrt{7 + 18}$$

Solution

$$\sqrt{7} + \sqrt{18} > \sqrt{7 + 18}$$

111. Place the correct symbol ($<$, $>$, or $=$) between the numbers

$$5 \qquad \sqrt{3^2 + 2^2}$$

Solution

$$5 > \sqrt{3^2 + 2^2}$$

113. *Geometry* Find the length of the hypotenuse of the right triangle.

Solution

$$c = \sqrt{a^2 + b^2}$$

$$= \sqrt{9^2 + 6^2}$$

$$= \sqrt{81 + 36}$$

$$= \sqrt{117}$$

$$= 3\sqrt{13}$$

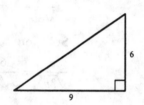

115. *Period of a Pendulum* The period T in seconds of a pendulum (see figure in text) is given by $T = 2\pi\sqrt{\dfrac{L}{32}}$, where L is the length of the pendulum in feet. Find the period of a pendulum whose length is 4 feet. (Round your answer to two decimal places.)

Solution

$$T = 2\pi\sqrt{\frac{4}{32}}$$

$$= 2\pi\sqrt{\frac{1}{8}}$$

$$= 2.2214415$$

$$\approx 2.22$$

117. *Calculator Experiment* Enter any positive real number in your calculator and find its square root. Then repeatedly take the square root of the result.

$$\sqrt{x}, \sqrt{\sqrt{x}}, \sqrt{\sqrt{\sqrt{x}}}, \dots$$

What real number does the display appear to be approaching?

Solution

The display appears to be approaching 1.

Mid-Chapter Quiz for Chapter 4

1. Evaluate -12^{-2}.

Solution

$$-12^{-2} = \frac{1}{-12^2} = \frac{1}{-144} = -\frac{1}{144}$$

2. Evaluate $\left(\frac{3}{4}\right)^{-3}$.

Solution

$$\left(\frac{3}{4}\right)^{-3} = \left(\frac{4}{3}\right)^{3} = \frac{64}{27}$$

3. Evaluate $\sqrt{\frac{25}{9}}$.

Solution

$$\sqrt{\frac{25}{9}} = \frac{\sqrt{25}}{\sqrt{9}} = \frac{5}{3}$$

4. Evaluate $(-64)^{1/3}$.

Solution

$$(-64)^{1/3} = \sqrt[3]{-64} = -4$$

5. Rewrite $\left(t^3\right)^{-1/2}\left(3t^3\right)$ using only positive exponents, and simplify. (Assume any variables in the expression are nonzero.)

Solution

$$\left(t^3\right)^{-1/2}\left(3t^3\right) = t^{-3/2} \cdot 3t^3$$
$$= 3t^{-3/2+3}$$
$$= 3t^{3/2}$$

6. Rewrite $\left(3x^2y^{-1}\right)\left(4x^{-2}y\right)^{-2}$ using only positive exponents, and simplify. (Assume any variables in the expression are nonzero.)

Solution

$$\left(3x^2y^{-1}\right)\left(4x^{-2}y\right)^{-2} = 3x^2y^{-1} \cdot 4^{-2}x^4y^{-2}$$
$$= 3 \cdot \frac{1}{4^2} \cdot x^{2+4}y^{-1+-2}$$
$$= \frac{3}{16}x^6y^{-3} = \frac{3x^6}{16y^3}$$

7. Rewrite $\dfrac{10u^{-2}}{15u}$ using only positive exponents, and simplify. (Assume any variables in the expression are nonzero.)

Solution

$$\frac{10u^{-2}}{15u} = \frac{10}{15}u^{-2-1} = \frac{2}{3}u^{-3} = \frac{2}{3u^3}$$

8. Rewrite $\dfrac{(10x)^0}{\left(x^2+4\right)^{-1}}$ using only positive exponents, and simplify. (Assume any variables in the expression are nonzero.)

Solution

$$\frac{(10x)^0}{\left(x^2+4\right)^{-1}} = \frac{1}{\left(x^2+4\right)^{-1}} = x^2+4$$

9. Write each number in scientific notation:

    (a) 13,400,000;   (b) 0.00075.

Solution

(a) $13,400,000 = 1.34 \times 10^7$

(b) $0.00075 = 7.5 \times 10^{-4}$

10. Evaluate the expression without using a calculator.

    (a) $\left(3 \times 10^3\right)^4$ (b) $\dfrac{3.2 \times 10^4}{16 \times 10^7}$

Solution

(a) $\left(3 \times 10^3\right)^4 = 3^4 \times 10^{12} = 81 \times 10^{12}$

$$= 8.1 \times 10^{13}$$

(b) $\dfrac{3.2 \times 10^4}{16 \times 10^7} = \dfrac{3.2}{16} \times 10^{-3} = 0.2 \times 10^{-3}$

$$= 2 \times 10^{-4}$$

11. Simplify the expression.

(a) $\sqrt{150}$ (b) $\sqrt[3]{54}$

Solution

(a) $\sqrt{150} = \sqrt{25 \cdot 6} = 5\sqrt{6}$

(b) $\sqrt[3]{54} = \sqrt[3]{27 \cdot 2} = 3\sqrt[3]{2}$

12. Simplify the expression.

(a) $\sqrt{27x^2}$ (b) $\sqrt[4]{81x^6}$

Solution

(a) $\sqrt{27x^2} = \sqrt{9 \cdot 3 \cdot x^2} = 3|x|\sqrt{3}$

(b) $\sqrt[4]{81x^6} = \sqrt[4]{81 \cdot x^4 \cdot x^2} = 3|x|\sqrt[4]{x^2}$
$$= 3|x|\sqrt{x}$$

13. Simplify the expression.

(a) $\sqrt[4]{\dfrac{5}{16}}$ (b) $\sqrt{\dfrac{24}{49}}$

Solution

(a) $\sqrt[4]{\dfrac{5}{16}} = \dfrac{\sqrt[4]{5}}{\sqrt[4]{16}} = \dfrac{\sqrt[4]{5}}{2}$

(b) $\sqrt{\dfrac{24}{49}} = \dfrac{\sqrt{4 \cdot 6}}{\sqrt{49}} = \dfrac{2\sqrt{6}}{7}$

14. Simplify the expression.

(a) $\sqrt{\dfrac{40u^3}{9}}$ (b) $\sqrt[3]{\dfrac{16}{u^6}}$

Solution

(a) $\sqrt{\dfrac{40u^3}{9}} = \dfrac{\sqrt{4 \cdot 10 \cdot u^2 \cdot u}}{\sqrt{9}} = \dfrac{2|u|\sqrt{10u}}{3}$

(b) $\sqrt[3]{\dfrac{16}{u^6}} = \dfrac{\sqrt[3]{16}}{\sqrt[3]{u^6}} = \dfrac{\sqrt[3]{16}}{u^2}$

15. Rationalize the denominator and simplify.

(a) $\sqrt{\dfrac{2}{3}}$ (b) $\dfrac{24}{\sqrt{12}}$

Solution

(a) $\sqrt{\dfrac{2}{3}} = \dfrac{\sqrt{2}}{\sqrt{3}} \cdot \dfrac{\sqrt{3}}{\sqrt{3}} = \dfrac{\sqrt{6}}{3}$

(b) $\dfrac{24}{\sqrt{12}} = \dfrac{24}{\sqrt{4 \cdot 3}} = \dfrac{24}{2\sqrt{3}} \cdot \dfrac{\sqrt{3}}{\sqrt{3}}$
$$= \dfrac{24\sqrt{3}}{6}$$
$$= 4\sqrt{3}$$

16. Rationalize the denominator and simplify.

(a) $\dfrac{10}{\sqrt{5x}}$ (b) $\sqrt[3]{\dfrac{3}{2a}}$

Solution

(a) $\dfrac{10}{\sqrt{5x}} = \dfrac{10}{\sqrt{5x}} \cdot \dfrac{\sqrt{5x}}{\sqrt{5x}} = \dfrac{10\sqrt{5x}}{5x} = \dfrac{2\sqrt{5x}}{x}$

(b) $\sqrt[3]{\dfrac{3}{2a}} = \dfrac{\sqrt[3]{3}}{\sqrt[3]{2a}} \cdot \dfrac{\sqrt[3]{4a^2}}{\sqrt[3]{4a^2}} = \dfrac{\sqrt[3]{12a^2}}{2a}$

17. Combine the radical expressions, if possible.

$\sqrt{200y} - 3\sqrt{8y}$

Solution

$\sqrt{200y} - 3\sqrt{8y} = \sqrt{100 \cdot 2y} - 3\sqrt{4 \cdot 2y}$
$$= 10\sqrt{2y} - 6\sqrt{2y}$$
$$= 4\sqrt{2y}$$

18. Combine the radical expressions, if possible.

$6x\sqrt[3]{5x^2} + 2\sqrt[3]{40x^4}$

Solution

$6x\sqrt[3]{5x^2} + 2\sqrt[3]{40x^4} = 6x\sqrt[3]{5x^2} + 2\sqrt[3]{8 \cdot 5 \cdot x^3 \cdot x}$
$$= 6x\sqrt[3]{5x^2} + 4x\sqrt[3]{5x}$$

19. Explain why $\sqrt{5^2 + 12^2} \neq 17$. Determine the correct value of the radical.

Solution

$$\sqrt{5^2 + 12^2} = \sqrt{25 + 144}$$
$$= \sqrt{169}$$
$$= 13$$

20. The four corners are cut from an $8\frac{1}{2}$-inch-by-11-inch sheet of paper as shown in the figure. Find the perimeter of the remaining piece of paper.

Solution

$$C = \sqrt{2^2 + 2^2}$$
$$= \sqrt{4 + 4} = \sqrt{8}$$

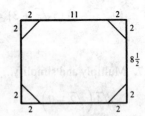

Equation: $P = 2(7) + 2\left(4\frac{1}{2}\right) + 4\left(\sqrt{8}\right)$
$$= 14 + 9 + 8\sqrt{2}$$
$$= 23 + 8\sqrt{2}$$

5.4 | Multiplying and Dividing Radical Expressions

5. Multiply and simplify.

$$\sqrt{3} \cdot \sqrt{6}$$

Solution

$$\sqrt{3} \cdot \sqrt{6} = \sqrt{3 \cdot 6} = \sqrt{18} = \sqrt{9 \cdot 2} = 3\sqrt{2}$$

7. Multiply and simplify.

$$\sqrt{5}\left(2 - \sqrt{3}\right)$$

Solution

$$\sqrt{5}\left(2 - \sqrt{3}\right) = 2\sqrt{5} - \sqrt{5}\sqrt{3} = 2\sqrt{5} - \sqrt{15}$$

9. Multiply and simplify.

$$\left(\sqrt{3} + 2\right)\left(\sqrt{3} - 2\right)$$

Solution

$$\left(\sqrt{3} + 2\right)\left(\sqrt{3} - 2\right) = \left(\sqrt{3}\right)^2 - 2^2$$
$$= 3 - 4$$
$$= -1$$

11. Multiply and simplify.

$$\left(\sqrt{20} + 2\right)^2$$

Solution

$$\left(\sqrt{20} + 2\right)^2 = \left(\sqrt{20}\right)^2 + 2 \cdot \sqrt{20} \cdot 2 + 2^2$$
$$= 20 + 4\sqrt{20} + 4$$
$$= 24 + 4\sqrt{4 \cdot 5}$$
$$= 24 + 8\sqrt{5}$$

13. Multiply and simplify.

$$\sqrt{y}\left(\sqrt{y} + 4\right)$$

Solution

$$\sqrt{y}\left(\sqrt{y} + 4\right) = \left(\sqrt{y}\right)^2 + 4\sqrt{y}$$
$$= y + 4\sqrt{y}$$

15. Multiply and simplify.

$$\left(9\sqrt{x}+2\right)\left(5\sqrt{x}-3\right)$$

Solution

$$\left(9\sqrt{x}+2\right)\left(5\sqrt{x}-3\right) = \left(9\sqrt{x}\right)\left(5\sqrt{x}\right) - 27\sqrt{x} + 10\sqrt{x} - 6$$
$$= 45x - 17\sqrt{x} - 6$$

17. Multiply and simplify.

$$\sqrt[3]{4}\left(\sqrt[3]{2}-7\right)$$

Solution

$$\sqrt[3]{4}\left(\sqrt[3]{2}-7\right) = \sqrt[3]{4}\sqrt[3]{2} - 7\sqrt[3]{4}$$
$$= \sqrt[3]{8} - 7\sqrt[3]{4}$$
$$= 2 - 7\sqrt[3]{4}$$

19. Complete the statement.

$$5\sqrt{3}+15\sqrt{3} = 5(\quad)$$

Solution

$$5\sqrt{3}+15\sqrt{3} = 5\left(\sqrt{3}+3\sqrt{3}\right)$$

21. Complete the statement.

$$4\sqrt{12} - 2x\sqrt{27} = 2\sqrt{3}(\quad)$$

Solution

$$4\sqrt{12} - 2x\sqrt{27} = 4\sqrt{4\cdot 3} - 2x\sqrt{9\cdot 3}$$
$$= 8\sqrt{3} - 6x\sqrt{3}$$
$$= 2\sqrt{3}(4 - 3x)$$

23. Complete the statement.

$$6u^2 + \sqrt{18u^3} = 3u(\quad)$$

Solution

$$6u^2 + \sqrt{18u^3} = 6u^2 + \sqrt{9\cdot 2u^2\cdot u}$$
$$= 6u^2 + 3u\sqrt{2u}$$
$$= 3u\left(2u + \sqrt{2u}\right)$$

25. Find the conjugate of $\sqrt{11} - \sqrt{3}$. Then multiply the number by its conjugate.

Solution

$$\sqrt{11} - \sqrt{3}, \text{conjugate} = \sqrt{11} + \sqrt{3}$$
$$\text{product} = \left(\sqrt{11} - \sqrt{3}\right)\left(\sqrt{11} + \sqrt{3}\right)$$
$$= \left(\sqrt{11}\right)^2 - \left(\sqrt{3}\right)^2$$
$$= 11 - 3 = 8$$

27. Find the conjugate of $\sqrt{x} - 3$. Then multiply the number by its conjugate.

Solution

$$\sqrt{x} - 3, \text{conjugate} = \sqrt{x} + 3$$
$$\text{product} = \left(\sqrt{x} - 3\right)\left(\sqrt{x} + 3\right)$$
$$= \left(\sqrt{x}\right)^2 - 3^2$$
$$= x - 9$$

29. Rationalize the denominator of $\dfrac{6}{\sqrt{22} - 2}$.

Solution

$$\frac{6}{\sqrt{22} - 2} = \frac{6}{\sqrt{22} - 2} \cdot \frac{\sqrt{22} + 2}{\sqrt{22} + 2} = \frac{6\left(\sqrt{22} + 2\right)}{\left(\sqrt{22}\right)^2 - 2^2} = \frac{6\left(\sqrt{22} + 2\right)}{22 - 4} = \frac{6\left(\sqrt{22} + 2\right)}{18} = \frac{\sqrt{22} + 2}{3}$$

31. Rationalize the denominator of $\dfrac{8}{\sqrt{7}+3}$.

Solution

$$\frac{8}{\sqrt{7}+3} = \frac{8}{\sqrt{7}+3} \cdot \frac{\sqrt{7}-3}{\sqrt{7}-3} = \frac{8\left(\sqrt{7}-3\right)}{\left(\sqrt{7}\right)^2 - 3^2} = \frac{8\left(\sqrt{7}-3\right)}{7-9} = \frac{8\left(\sqrt{7}-3\right)}{-2} = -4\left(\sqrt{7}-3\right) = -4\sqrt{7}+12$$

33. Rationalize the denominator of $\dfrac{2t^2}{\sqrt{5t}-\sqrt{t}}$.

Solution

$$\frac{2t^2}{\sqrt{5t}-\sqrt{t}} = \frac{2t^2}{\sqrt{5t}-\sqrt{t}} \cdot \frac{\sqrt{5t}+\sqrt{t}}{\sqrt{5t}+\sqrt{t}} = \frac{2t^2\left(\sqrt{5t}+\sqrt{t}\right)}{\left(\sqrt{5t}\right)^2 - \left(\sqrt{t}\right)^2} = \frac{2t^2\left(\sqrt{5t}+\sqrt{t}\right)}{5t-t} = \frac{2t^2\left(\sqrt{5t}+\sqrt{t}\right)}{4t} = \frac{t\sqrt{5t}+t\sqrt{t}}{2}$$

35. Rationalize the denominator of $\dfrac{\sqrt{u+v}}{\sqrt{u-v}-\sqrt{u}}$.

Solution

$$\frac{\sqrt{u+v}}{\sqrt{u-v}-\sqrt{u}} = \frac{\sqrt{u+v}}{\sqrt{u-v}-\sqrt{u}} \cdot \frac{\sqrt{u-v}+\sqrt{u}}{\sqrt{u-v}+\sqrt{u}} = \frac{\sqrt{u+v}\left(\sqrt{u-v}+\sqrt{u}\right)}{u-v-u} = \frac{\sqrt{u+v}\left(\sqrt{u-v}+\sqrt{u}\right)}{-v}$$

37. Evaluate $f(x) = x^2 - 2x - 1$.

(a) $f\left(1+\sqrt{2}\right)$

(b) $f\left(\sqrt{4}\right)$

Solution

(a) $f\left(1+\sqrt{2}\right) = \left(1+\sqrt{2}\right)^2 - 2\left(1+\sqrt{2}\right) - 1$

$= 1 + 2\sqrt{2} + 2 - 2 - 2\sqrt{2} - 1$

$= 0$

(b) $f\left(\sqrt{4}\right) = \left(\sqrt{4}\right)^2 - 2\sqrt{4} - 1$

$= 4 - 4 - 1$

$= -1$

39. Use a graphing utility to graph the two equations on the same screen. Use the graphs to verify that the expressions are equivalent. Verify the results algebraically.

$$y_1 = \frac{10}{\sqrt{x}+1}, \quad y_2 = \frac{10\left(\sqrt{x}-1\right)}{x-1}$$

Solution

Keystrokes:

y_1 [Y=] 10 [÷] [(] [√] [X, T, θ] [+] 1 [)] [ENTER]

y_2 [Y=] [(] 10 [(] [√] [X, T, θ] [−] 1 [)] [)] [÷]

[(] [X, T, θ] [−] 1 [)] [GRAPH]

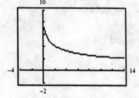

$y_1 = y_2$

$$\frac{10}{\sqrt{x}+1} = \frac{10}{\sqrt{x}+1} \cdot \frac{\sqrt{x}-1}{\sqrt{x}-1} = \frac{10\left(\sqrt{x}-1\right)}{x-1}$$

41. *Strength of a Wooden Beam* The rectangular cross section of a wooden beam cut from a log of diameter 24 inches (see figure in text) will have maximum strength if its width w and height h are given by

$$w = 8\sqrt{3} \quad \text{and} \quad h = \sqrt{24^2 - \left(8\sqrt{3}\right)^2}.$$

Find the area of the rectangular cross section and express the area in simplest form.

Solution

$$\text{Area} = h \cdot w$$

$$= \sqrt{24^2 - \left(8\sqrt{3}\right)^2} \cdot 8\sqrt{3}$$

$$= \sqrt{576 - 192} \cdot 8\sqrt{3}$$

$$= \sqrt{384} \cdot 8\sqrt{3}$$

$$= 8\sqrt{1152}$$

$$= 8\sqrt{2^7 \cdot 3^2}$$

$$= 8 \cdot 2^3 \cdot 3\sqrt{2}$$

$$= 192\sqrt{2}$$

43. Solve $\dfrac{4}{x} - \dfrac{2}{3} = 0$.

Solution

$$3x\left[\frac{4}{x} - \frac{2}{3} = 0\right]3x$$

$$12 - 2x = 0$$

$$12 = 2x$$

$$6 = x$$

45. Solve $3x^2 - 13x - 10 = 0$.

Solution

$$3x^2 - 13x - 10 = 0$$

$$(3x + 2)(x - 5) = 0$$

$$3x + 2 = 0 \qquad x - 5 = 0$$

$$x = -\frac{2}{3} \qquad x = 5$$

47. Use a graphing utility to identify the transformation of the graph of $f(x) = \sqrt{x}$.

$$h(x) = 4 + \sqrt{x}$$

Solution

Keystrokes:

$\boxed{Y=}$ 4 $\boxed{+}$ $\boxed{\sqrt{}}$ $\boxed{X, T, \theta}$ $\boxed{\text{GRAPH}}$

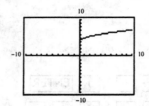

Vertical shift 4 units upward

49. Use a graphing utility to identify the transformation of the graph of $f(x) = \sqrt{x}$.

$$h(x) = -\sqrt{x}$$

Solution

Keystrokes:

$\boxed{Y=}$ $\boxed{(-)}$ $\boxed{\sqrt{}}$ $\boxed{X, T, \theta}$ $\boxed{\text{GRAPH}}$

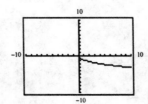

Reflection in the x-axis

51. Multiply and simplify.

$$\sqrt{2} \cdot \sqrt{8}$$

Solution

$$\sqrt{2} \cdot \sqrt{8} = \sqrt{2 \cdot 8}$$
$$= \sqrt{16}$$
$$= 4$$

53. Multiply and simplify.

$$\sqrt{2}\left(\sqrt{20} + 8\right)$$

Solution

$$\sqrt{2}\left(\sqrt{20} + 8\right) = \sqrt{2}\sqrt{20} + 8\sqrt{2}$$
$$= \sqrt{40} + 8\sqrt{2}$$
$$= 2\sqrt{10} + 8\sqrt{2}$$

55. Multiply and simplify.

$$\left(3 - \sqrt{5}\right)\left(3 + \sqrt{5}\right)$$

Solution

$$\left(3 - \sqrt{5}\right)\left(3 + \sqrt{5}\right) = 3^2 - \left(\sqrt{5}\right)^2$$
$$= 9 - 5$$
$$= 4$$

57. Multiply and simplify.

$$\left(2\sqrt{2} + \sqrt{4}\right)\left(2\sqrt{2} - \sqrt{4}\right)$$

Solution

$$\left(2\sqrt{2} + \sqrt{4}\right)\left(2\sqrt{2} - \sqrt{4}\right) = \left(2\sqrt{2}\right)^2 - \left(\sqrt{4}\right)^2$$
$$= 4 \cdot 2 - 4$$
$$= 8 - 4 = 4$$

59. Multiply and simplify.

$$\left(\sqrt{5} + 3\right)\left(\sqrt{3} - 5\right)$$

Solution

$$\left(\sqrt{5} + 3\right)\left(\sqrt{3} - 5\right) = \sqrt{15} - 5\sqrt{5} + 3\sqrt{3} - 15$$

61. Multiply and simplify.

$$\left(10 + \sqrt{2x}\right)^2$$

Solution

$$\left(10 + \sqrt{2x}\right)^2 = 10^2 + 2 \cdot 10 \cdot \sqrt{2x} + \left(\sqrt{2x}\right)^2$$
$$= 100 + 20\sqrt{2x} + 2x$$

63. Multiply and simplify.

$$\left(3\sqrt{x} - 5\right)\left(3\sqrt{x} + 5\right)$$

Solution

$$\left(3\sqrt{x} - 5\right)\left(3\sqrt{x} + 5\right) = \left(3\sqrt{x}\right)^2 - 5^2$$
$$= 9x - 25$$

65. Multiply and simplify.

$$\left(\sqrt{x} + \sqrt{y}\right)\left(\sqrt{x} - \sqrt{y}\right)$$

Solution

$$\left(\sqrt{x} + \sqrt{y}\right)\left(\sqrt{x} - \sqrt{y}\right) = \left(\sqrt{x}\right)^2 - \left(\sqrt{y}\right)^2$$
$$= x - y$$

67. Multiply and simplify.

$$\left(\sqrt[3]{2x} + 5\right)^2$$

Solution

$$\left(\sqrt[3]{2x} + 5\right)^2 = \left(\sqrt[3]{2x}\right)^2 + 2 \cdot 5\sqrt[3]{2x} + 5^2$$
$$= \sqrt[3]{(2x)^2} + 10\sqrt[3]{2x} + 25$$
$$= \sqrt[3]{4x^2} + 10\sqrt[3]{2x} + 25$$

71. Simplify $\dfrac{-2y + \sqrt{12y^3}}{8y}$.

Solution

$$\frac{-2y + \sqrt{12y^3}}{8y} = \frac{-2y + 2y\sqrt{3y}}{8y}$$
$$= \frac{2y\left(-1 + \sqrt{3y}\right)}{8y}$$
$$= \frac{-1 + \sqrt{3y}}{4}$$

75. Find the conjugate of $\sqrt{2u} - \sqrt{3}$. Then multiply the number by its conjugate.

Solution

$$\sqrt{2u} - \sqrt{3}, \text{conjugate} = \sqrt{2u} + \sqrt{3}$$
$$\text{product} = \left(\sqrt{2u} - \sqrt{3}\right)\left(\sqrt{2u} + \sqrt{3}\right)$$
$$= \left(\sqrt{2u}\right)^2 - \left(\sqrt{3}\right)^2$$
$$= 2u - 3$$

69. Simplify $\dfrac{4 - 8\sqrt{x}}{12}$.

Solution

$$\frac{4 - 8\sqrt{x}}{12} = \frac{4\left(1 - 2\sqrt{x}\right)}{12}$$
$$= \frac{1 - 2\sqrt{x}}{3}$$

73. Find the conjugate of $2 + \sqrt{5}$. Then multiply the number by its conjugate.

Solution

$$2 + \sqrt{5}, \text{conjugate} = 2 - \sqrt{5}$$
$$\text{product} = \left(2 + \sqrt{5}\right)\left(2 - \sqrt{5}\right)$$
$$= 2^2 - \left(\sqrt{5}\right)^2$$
$$= 4 - 5 = -1$$

77. Rationalize the denominator of $\dfrac{2}{6 + \sqrt{2}}$.

Solution

$$\frac{2}{6 + \sqrt{2}} = \frac{2}{6 + \sqrt{2}} \cdot \frac{6 - \sqrt{2}}{6 - \sqrt{2}} = \frac{2\left(6 - \sqrt{2}\right)}{36 - 2}$$
$$= \frac{2\left(6 - \sqrt{2}\right)}{34}$$
$$= \frac{6 - \sqrt{2}}{17}$$

79. Rationalize the denominator of $\left(\sqrt{7} + 2\right) \div \left(\sqrt{7} - 2\right)$.

Solution

$$\left(\sqrt{7} + 2\right) \div \left(\sqrt{7} - 2\right) = \frac{\sqrt{7} + 2}{\sqrt{7} - 2} \cdot \frac{\sqrt{7} + 2}{\sqrt{7} + 2} = \frac{\left(\sqrt{7}\right)^2 + 2\sqrt{7} + 2\sqrt{7} + 4}{\left(\sqrt{7}\right)^2 - 2^2} = \frac{7 + 4\sqrt{7} + 4}{7 - 4} = \frac{11 + 4\sqrt{7}}{3}$$

81. Rationalize the denominator of $\dfrac{3x}{\sqrt{15} - \sqrt{3}}$.

Solution

$$\frac{3x}{\sqrt{15} - \sqrt{3}} = \frac{3x}{\sqrt{15} - \sqrt{3}} \cdot \frac{\sqrt{15} + \sqrt{3}}{\sqrt{15} + \sqrt{3}} = \frac{3x\left(\sqrt{15} + \sqrt{3}\right)}{\left(\sqrt{15}\right)^2 - \left(\sqrt{3}\right)^2} = \frac{3x\left(\sqrt{15} + \sqrt{3}\right)}{15 - 3}$$

$$= \frac{3x\left(\sqrt{15} + \sqrt{3}\right)}{12} = \frac{x\sqrt{15} + x\sqrt{3}}{4}$$

83. Rationalize the denominator of $\left(\sqrt{x} - 5\right) \div \left(2\sqrt{x} - 1\right)$.

Solution

$$\left(\sqrt{x} - 5\right) \div \left(2\sqrt{x} - 1\right) = \frac{\sqrt{x} - 5}{2\sqrt{x} - 1} \cdot \frac{2\sqrt{x} + 1}{2\sqrt{x} + 1} = \frac{2x + \sqrt{x} - 10\sqrt{x} - 5}{\left(2\sqrt{x}\right)^2 - 1^2} = \frac{2x - 9\sqrt{x} - 5}{4x - 1}$$

85. Evaluate the function $f(x) = x^2 - 6x + 1$ as indicated.

(a) $f\left(2 - \sqrt{3}\right)$

(b) $f\left(3 - 2\sqrt{2}\right)$

Solution

(a) $f\left(2 - \sqrt{3}\right) = \left(2 - \sqrt{3}\right)^2 - 6\left(2 - \sqrt{3}\right) + 1$

$= 4 - 4\sqrt{3} + 3 - 12 + 6\sqrt{3} + 1$

$= 2\sqrt{3} - 4$

(b) $f\left(3 - 2\sqrt{2}\right) = \left(3 - 2\sqrt{2}\right)^2 - 6\left(3 - 2\sqrt{2}\right) + 1$

$= 9 - 12\sqrt{2} + 8 - 18 + 12\sqrt{2} + 1$

$= 0$

87. Use a graphing utility to graph the two equations on the same screen. Use the graphs to verify that the expressions are equivalent. Verify the results algebraically.

$$y_1 = \frac{4x}{\sqrt{x} + 4}, \quad y_2 = \frac{4x\left(\sqrt{x} - 4\right)}{x - 16}$$

Solution

Keystrokes:

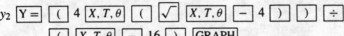

y_1 [Y=] 4 [X,T,θ] [÷] [(] [√] [X,T,θ] [+] 4 [)] [ENTER]

y_2 [Y=] [(] 4 [X,T,θ] [(] [√] [X,T,θ] [−] 4 [)] [)] [÷]

[(] [X,T,θ] [−] 16 [)] [GRAPH]

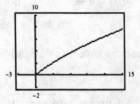

$y_1 = y_2$

$$y_1 = \frac{4x}{\sqrt{x} + 4} = \frac{4x}{\sqrt{x} + 4} \cdot \frac{\sqrt{x} - 4}{\sqrt{x} - 4} = \frac{4x\left(\sqrt{x} - 4\right)}{x - 16}$$

89. *Geometry* The areas of the circles in the figure in the textbook are 15 square centimeters and 20 square centimeters. Find the ratio of the radius of the small circle to the radius of the large circle.

Solution

$$\boxed{\begin{array}{c}\text{radius of}\\\text{small circle}\end{array}} : \boxed{\begin{array}{c}\text{radius of}\\\text{large circle}\end{array}} = \sqrt{\frac{15}{\pi}} : \sqrt{\frac{20}{\pi}}$$

$$= \frac{\sqrt{\dfrac{15}{\pi}}}{\sqrt{\dfrac{20}{\pi}}} = \sqrt{\dfrac{\dfrac{15}{\pi}}{\dfrac{20}{\pi}}} = \sqrt{\frac{15}{20}} = \sqrt{\frac{3}{4}} = \frac{\sqrt{3}}{\sqrt{4}} = \frac{\sqrt{3}}{2}$$

5.5 Solving Equations Involving Radicals

5. Determine whether each value of x is a solution of the equation.

$$\sqrt{x} - 10 = 0$$

Solution

(a) $x = -4$ $\sqrt{-4} - 10 \neq 0$ not a solution

(b) $x = -100$ $\sqrt{-100} - 10 \neq 0$ not a solution

(c) $x = \sqrt{10}$ $\sqrt{\sqrt{10}} - 10 \neq 0$ not a solution

(d) $x = 100$ $\sqrt{100} - 10 = 0$ a solution

7. Determine whether each value of x is a solution of the equation.

$$\sqrt[3]{x - 4} = 4$$

Solution

(a) $x = -60$ $\sqrt[3]{-60 - 4} \neq 4$ not a solution

(b) $x = 68$ $\sqrt[3]{68 - 4} = 4$ a solution

(c) $x = 20$ $\sqrt[3]{20 - 4} \neq 4$ not a solution

(d) $x = 0$ $\sqrt[3]{0 - 4} \neq 4$ not a solution

9. Solve $\sqrt{x} = 20$.

(Some of the equations have no solution.)

Solution

$$\sqrt{x} = 20 \qquad \textbf{Check:} \quad \sqrt{400} \stackrel{?}{=} 20$$

$$\left(\sqrt{x}\right)^2 = 20^2 \qquad\qquad\qquad 20 = 20$$

$$x = 400$$

11. Solve $\sqrt{u} + 13 = 0$.

(Some of the equations have no solution.)

Solution

$$\sqrt{u} + 13 = 0$$

$$\sqrt{u} = -13$$

$$\left(\sqrt{u}\right)^2 = (-13)^2$$

$$u = 169$$

Check: $\sqrt{169} + 13 \overset{?}{=} 0$

$$13 + 13 \neq 0$$

$$\varnothing$$

13. Solve $\sqrt{3y + 5} - 3 = 4$.

(Some of the equations have no solution.)

Solution

$$\sqrt{3y + 5} - 3 = 4$$

$$\sqrt{3y + 5} = 7$$

$$\left(\sqrt{3y + 5}\right)^2 = 7^2$$

$$3y + 5 = 49$$

$$3y = 44$$

$$y = \frac{44}{3}$$

Check: $\sqrt{3\left(\dfrac{44}{3}\right) + 5} - 3 \overset{?}{=} 4$

$$\sqrt{49} - 3 = 4$$

$$7 - 3 = 4$$

$$4 = 4$$

15. Solve $\sqrt{x^2 + 5} = x + 3$.

(Some of the equations have no solution.)

Solution

$$\sqrt{x^2 + 5} = x + 3$$

$$\left(\sqrt{x^2 + 5}\right)^2 = (x + 3)^2$$

$$x^2 + 5 = x^2 + 6x + 9$$

$$-4 = 6x$$

$$-\frac{4}{6} = x$$

$$-\frac{2}{3} = x$$

Check: $\sqrt{\left(-\dfrac{2}{3}\right)^2 + 5} \overset{?}{=} -\dfrac{2}{3} + 3$

$$\sqrt{\frac{4}{9} + \frac{45}{9}} = -\frac{2}{3} + \frac{9}{3}$$

$$\sqrt{\frac{49}{9}} = \frac{7}{3}$$

$$\frac{7}{3} = \frac{7}{3}$$

17. Solve $\sqrt{3y-5} - 3\sqrt{y} = 0$.

(Some of the equations have no solution.)

Solution

$$\sqrt{3y-5} - 3\sqrt{y} = 0$$

$$\sqrt{3y-5} = 3\sqrt{y}$$

$$\left(\sqrt{3y-5}\right)^2 = \left(3\sqrt{y}\right)^2$$

$$3y - 5 = 9y$$

$$-5 = 6y$$

$$-\frac{5}{6} = y$$

Check: $\sqrt{3\left(-\dfrac{5}{6}\right) - 5} - 3\sqrt{-\dfrac{5}{6}} \overset{?}{=} 0$

$\emptyset$

19. Solve $\sqrt[3]{3x-4} = \sqrt[3]{x+10}$.

(Some of the equations have no solution.)

Solution

$$\sqrt[3]{3x-4} = \sqrt[3]{x+10}$$

$$\left(\sqrt[3]{3x-4}\right)^3 = \left(\sqrt[3]{x+10}\right)^3$$

$$3x - 4 = x + 10$$

$$2x = 14$$

$$x = 7$$

Check: $\sqrt[3]{3(7)-4} \overset{?}{=} \sqrt[3]{7+10}$

$\sqrt[3]{17} = \sqrt[3]{17}$

21. Solve $\sqrt{8x+1} = x + 2$.

(Some of the equations have no solution.)

Solution

$$\sqrt{8x+1} = x + 2$$

$$\left(\sqrt{8x+1}\right)^2 = (x+2)^2$$

$$8x + 1 = x^2 + 4x + 4$$

$$0 = x^2 - 4x + 3$$

$$0 = (x-3)(x-1)$$

$$3 = x \qquad x = 1$$

Check: $\sqrt{8(3)+1} \overset{?}{=} 3 + 2$

$\sqrt{25} = 5$

$5 = 5$

$\sqrt{8(1)+1} \overset{?}{=} 1 + 2$

$\sqrt{9} = 3$

$3 = 3$

23. *Graphical Reasoning* Use a graphing utility to graph both sides of the equation on the same screen. Then use the graphs to approximate the solution. Check your approximations algebraically.

$$\sqrt{x} = 2(2 - x)$$

Solution

Keystrokes:

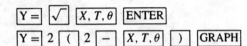

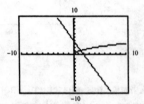

Solution approximation: $x \approx 1.407$

Check algebraically: $\sqrt{1.407} \stackrel{?}{=} 2(2 - 1.407)$

$$1.186 = 1.186$$

25. Use a graphing utility to graph both sides of the equation on the same screen. Then use the graphs to approximate the solution. Check your approximations algebraically.

$$\sqrt{x + 3} = 5 - \sqrt{x}$$

Solution

Keystrokes:

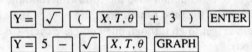

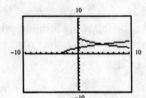

Solution approximation: $x \approx 5$

Solve algebraically:
$$\sqrt{x + 3} = 5 - \sqrt{x}$$
$$\left(\sqrt{x + 3}\right)^2 = \left(5 - \sqrt{x}\right)^2$$
$$x + 3 = 25 - 10\sqrt{x} + x$$
$$-22 = -10\sqrt{x}$$
$$2.2 = \sqrt{x}$$
$$(2.2)^2 = \left(\sqrt{x}\right)^2$$
$$4.84 = x$$

27. *Geometry* Using the figure in the textbook, find the length x of the side labeled x. Round to two decimal places.

Solution

$$x^2 = 8^2 + 6^2$$
$$x = \sqrt{64 + 36}$$
$$x = \sqrt{100}$$
$$x = 10$$

29. *Geometry* Using the figure in the textbook, find the length x of the side labeled x. Round to two decimal places.

Solution

$$7^2 = x^2 + 5.5^2$$
$$49 = x^2 + 30.25$$
$$18.75 = x^2$$
$$\sqrt{18.75} = x$$
$$4.33 = x$$

31. *Drawing a Diagram* A basketball court is 50 feet wide and 94 feet long. Draw a diagram and find the length of the diagonal of the court.

Solution

$$c = \sqrt{94^2 + 50^2}$$

$$= \sqrt{8836 + 2500}$$

$$= \sqrt{11336}$$

$$\approx 106.47 \text{ ft}$$

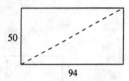

33. *Drawing a Diagram* A house has a basement floor with dimensions 26 feet by 32 feet. The gas hot water heater and furnace are diagonally across the basement from where the natural gas line enters the house. Draw a diagram showing the gas line and find the length of the gas line across the basement.

Solution

$$c = \sqrt{32^2 + 26^2}$$

$$= \sqrt{1024 + 676}$$

$$= \sqrt{1700}$$

$$\approx 41.23$$

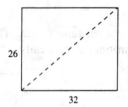

35. *Height of an Object* Use the following formula, which gives the time t in seconds for a free-falling object to fall d feet.

$$t = \sqrt{\frac{d}{16}}$$

A construction worker drops a nail and observes it strike a water puddle after approximately 2 seconds. Estimate the height of the worker.

Solution

$$2 = \sqrt{\frac{d}{16}}$$

$$2^2 = \left(\sqrt{\frac{d}{16}}\right)^2$$

$$4 = \frac{d}{16}$$

$$64 = d$$

37. *Length of a Pendulum* Use the following formula, which gives the time t in seconds for a pendulum of length L feet to go through one complete cycle (its period).

$$t = 2\pi\sqrt{\frac{L}{32}}$$

How long is the pendulum of a grandfather clock with a period of 1.5 seconds (see figure in textbook)?

Solution

$$1.5 = 2\pi\sqrt{\frac{L}{32}}$$

$$\left(\frac{1.5}{2\pi}\right)^2 = \left(\sqrt{\frac{L}{32}}\right)^2$$

$$\frac{2.25}{4\pi^2} = \frac{L}{32}$$

$$\frac{2.25}{4\pi^2}(32) = L$$

$$1.82 = L$$

39. Simplify the fraction.

$$\frac{-16x^2}{12x}$$

Solution

$$\frac{-16x^2}{12x} = \frac{-4x}{3} = -\frac{4x}{3}$$

41. Simplify the fraction.

$$\frac{6u^2v^{-3}}{27uv^3}$$

Solution

$$\frac{6u^2v^{-3}}{27uv^3} = \frac{2uv^{-6}}{9} = \frac{2u}{9v^6}$$

43. Simplify the fraction.

$$\left(\frac{3x^2}{2y^{-1}}\right)^{-2}$$

Solution

$$\left(\frac{3x^2}{2y^{-1}}\right)^{-2} = \left(\frac{2y^{-1}}{3x^2}\right)^2 = \frac{4y^{-2}}{9x^4} = \frac{4}{9x^4 y^2}$$

45. *Comparing Prices* A department store is offering a discount of 20% on a sewing machine with a list price of $239.95. A mail-order catalog has the same machine for $188.95 plus $4.32 for shipping. Which is the better bargin?

Solution

Verbal model: | Discount | = | List Price | · | Discount Rate |

Equation: $\quad x = \$239.95 \cdot 0.20$

$\qquad\quad = \$47.99$

Verbal model: | Selling price | = | List Price | − | Discount |

Equation: $\quad x = \$239.95 - \$47.99$

$\qquad\quad = \$191.96$

Verbal model: | Catalog price | = | List Price | + | Shipping |

Equation: $\quad x = \$188.95 - \$4.32$

$\qquad\quad = \$193.27$

Better bargain is the store price at $191.96.

47. Solve $\sqrt{y} - 7 = 0$.

(Some of the equations have no solution.)

Solution

$$\sqrt{y} - 7 = 0 \qquad \text{Check:} \quad \sqrt{49} - 7 \overset{?}{=} 0$$

$$\sqrt{y} = 7 \qquad\qquad\qquad\quad 7 - 7 = 0$$

$$\left(\sqrt{y}\right)^2 = 7^2 \qquad\qquad\qquad 0 = 0$$

$$y = 49$$

49. Solve $\sqrt{a + 100} = 25$.

(Some of the equations have no solution.)

Solution

$$\sqrt{a + 100} = 25 \qquad \text{Check:} \quad \sqrt{525 + 100} \overset{?}{=} 25$$

$$\left(\sqrt{a + 100}\right)^2 = 25^2 \qquad\qquad\qquad \sqrt{625} = 25$$

$$a + 100 = 625 \qquad\qquad\qquad\quad 25 = 25$$

$$a = 525$$

51. Solve $\sqrt{10x} = 30$.

(Some of the equations have no solution.)

Solution

$$\sqrt{10x} = 30 \qquad \text{Check:} \quad \sqrt{10 \cdot 90} \overset{?}{=} 30$$

$$\left(\sqrt{10x}\right)^2 = 30^2 \qquad\qquad \sqrt{900} = 30$$

$$10x = 900 \qquad\qquad\qquad 30 = 30$$

$$x = 90$$

53. Solve $5\sqrt{x + 2} = 8$.

(Some of the equations have no solution.)

Solution

$$5\sqrt{x + 2} = 8 \qquad \text{Check:} \quad 5\sqrt{\dfrac{14}{25} + 2} \overset{?}{=} 8$$

$$\left(5\sqrt{x + 2}\right)^2 = 8^2 \qquad\qquad 5\sqrt{\dfrac{64}{25}} = 8$$

$$25(x + 2) = 64 \qquad\qquad\qquad 5 \cdot \dfrac{8}{5} = 8$$

$$25x + 50 = 64 \qquad\qquad\qquad 8 = 8$$

$$25x = 14$$

$$x = \dfrac{14}{25}$$

55. Solve $\sqrt{3x + 2} + 5 = 0$.

(Some of the equations have no solution.)

Solution

$$\sqrt{3x + 2} + 5 = 0 \qquad \text{Check:} \quad \sqrt{3\left(\dfrac{23}{3}\right) + 2} + 5 \overset{?}{=} 0$$

$$\sqrt{3x + 2} = -5 \qquad\qquad\qquad \sqrt{23 + 2} + 5 = 0$$

$$\left(\sqrt{3x + 2}\right)^2 = (-5)^2 \qquad\qquad \sqrt{25} + 5 = 0$$

$$3x + 2 = 25 \qquad\qquad\qquad 5 + 5 = 0$$

$$3x = 23 \qquad\qquad\qquad 10 \neq 0$$

$$x = \dfrac{23}{3}$$

No solution

57. Solve $\sqrt{x+3} = \sqrt{2x-1}$.

(Some of the equations have no solution.)

Solution

$$\sqrt{x+3} = \sqrt{2x-1}$$

$$\left(\sqrt{x+3}\right)^2 = \left(\sqrt{2x-1}\right)^2$$

$$x+3 = 2x-1$$

$$4 = x$$

Check: $\sqrt{4+3} \stackrel{?}{=} \sqrt{2(4)-1}$

$$\sqrt{7} = \sqrt{7}$$

59. Solve $\sqrt[3]{2x+15} - \sqrt[3]{x} = 0$.

(Some of the equations have no solution.)

Solution

$$\sqrt[3]{2x+15} - \sqrt[3]{x} = 0$$

$$\sqrt[3]{2x+15} = \sqrt[3]{x}$$

$$\left(\sqrt[3]{2x+15}\right)^3 = \left(\sqrt[3]{x}\right)^3$$

$$2x+15 = x$$

$$x = -15$$

Check: $\sqrt[3]{2(-15)+15} - \sqrt[3]{-15} \stackrel{?}{=} 0$

$$\sqrt[3]{-15} - \sqrt[3]{-15} = 0$$

$$0 = 0$$

61. Solve $(x+4)^{2/3} = 4$.

(Some of the equations have no solution.)

Solution

$$(x+4)^{2/3} = 4$$

$$\sqrt[3]{(x+4)^2} = 4$$

$$\left(\sqrt[3]{(x+4)^2}\right)^3 = (4)^3$$

$$(x+4)^2 = 64$$

$$x+4 = \pm\sqrt{64}$$

$$x = -4 \pm 8$$

$$= 4, -12$$

Check: $(4+4)^{2/3} \stackrel{?}{=} 4$

$$8^{2/3} = 4$$

$$2^2 = 4$$

$$(-12+4)^{2/3} \stackrel{?}{=} 4$$

$$(-8)^{2/3} = 4$$

$$(-2)^2 = 4$$

63. Solve $\sqrt{2x} = x - 4$.

(Some of the equations have no solution.)

Solution

$$\sqrt{2x} = x - 4$$

$$\left(\sqrt{2x}\right)^2 = (x-4)^2$$

$$2x = x^2 - 8x + 16$$

$$0 = x^2 - 10x + 16$$

$$0 = (x-8)(x-2)$$

$$8 = x \quad x = 2$$

Check: $\sqrt{2(8)} \stackrel{?}{=} 8 - 4$

$$\sqrt{16} = 4$$

$$4 = 4$$

$$\sqrt{2(2)} \stackrel{?}{=} 2 - 4$$

$$\sqrt{4} = -2$$

$$2 \neq -2$$

65. *Graphical Reasoning* Use a graphing utility to graph both sides of the equation on the same screen. Then use the graphs to approximate the solution. Check your approximations algebraically.

$$\sqrt{x^2 + 1} = 5 - 2x$$

Solution

Keystrokes:

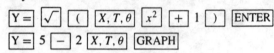

Approximate solution: $x \approx 1.569$

Check algebraically: $\sqrt{1.569^2 + 1} \stackrel{?}{=} 5 - 2(1.569)$

$$1.86 = 1.86$$

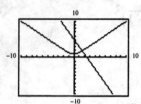

67. Use a graphing utility to graph both sides of the equation on the same screen. Then use the graphs to approximate the solution. Check your approximations algebraically.

$$4\sqrt[3]{x} = 7 - x$$

Solution

Keystrokes:

| Y = | 4 | Math 4 | X, T, θ | ENTER |

| Y = | 7 | − | X, T, θ | GRAPH |

Approximate solution: $x \approx 2$

Check algebraically: $4\sqrt[3]{2} \stackrel{?}{=} 7 - 2$

$$5 = 5$$

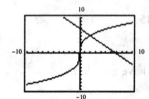

69. *Geometry* Using the figure in the textbook, find the length of the side labeled x. Round to two decimal places.

Solution

$$9^2 = x^2 + 7^2$$

$$81 = x^2 + 49$$

$$32 = x^2$$

$$\sqrt{32} = x$$

$$5.66 = x$$

71. *Geometry* Find the length of the side labeled x. Round to two decimal places.

Solution

$$c^2 = a^2 + b^2$$

$$20^2 = x^2 + 5^2$$

$$400 = x^2 + 25$$

$$375 = x^2$$

$$\sqrt{375} = x$$

$$19.36 =$$

73. Match the function with its graph (see the textbook).

$$f(x) = \sqrt[3]{x} - 1$$

Solution

(c) graph is shifted down 1 unit

75. Match the function with its graph (see the textbook).

$$f(x) = \sqrt[3]{x + 3} + 1$$

Solution

(d) graph is shifted left 3 units and upward 1 unit

77. Match the function with its graph (see the textbook).

$$f(x) = \sqrt{x} - 1$$

Solution

(f) graph is shifted down 1 unit

79. *Length of a Ramp* A ramp is 20 feet long and rests on a porch that is 4 feet high (see figure in textbook). Find the distance between the porch and the base of the ramp.

Solution

$$c^2 = a^2 + b^2$$

$$20^2 = x^2 + 4^2$$

$$400 = x^2 + 16$$

$$384 = x^2$$

$$\sqrt{384} = x$$

$$19.596 \approx x$$

81. *Free Falling Object* Use the equation for the velocity of a free-falling object, $v = \sqrt{2gh}$, as described in Example 9 in the textbook.

An object is dropped from a height of 50 feet. Find the velocity of the object when it strikes the ground.

Solution

$$v = \sqrt{2(32)50}$$

$$v = \sqrt{3200}$$

$$v = 56.57$$

83. *Free Falling Object* Use the equation for the velocity of a free-falling object, $v = \sqrt{2gh}$, as described in Example 9 in the textbook.

An object that was dropped strikes the ground with a velocity of 60 feet per second. Find the height from which the object was dropped.

Solution

$$60 = \sqrt{2(32)h}$$

$$60^2 = \left(\sqrt{64h}\right)^2$$

$$3600 = 64h$$

$$\frac{3600}{64} = h$$

$$56.25 = h$$

85. *Demand for a Product* The demand equation for a certain product is

$$p = 50 - \sqrt{0.8(x-1)}$$

where x is the number of units demanded per day and p is the price per unit. Find the demand if the price is $30.02.

Solution

$$30.02 = 50 - \sqrt{0.8(x-1)}$$

$$\sqrt{0.8(x-1)} = 19.98$$

$$\left(\sqrt{0.8(x-1)}\right)^2 = (19.98)^2$$

$$0.8(x-1) = 399.2004$$

$$0.8x - 0.8 = 399.2004$$

$$0.8x = 400.0004$$

$$x = 500.0005$$

$$\approx 500 \text{ units}$$

87. *Geometry* Write a function that gives the radius r of a circle in terms of the circle's area A. Use a graphing utility to graph this function.

Solution

Area of circle $= \pi r^2$

$$A = \pi r^2$$

$$\frac{A}{\pi} = r^2$$

$$\sqrt{\frac{A}{\pi}} = r$$

Keystrokes:

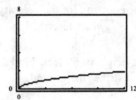

89. *Reading a Graph* An airline offers daily flights between Chicago and Denver. The total monthly cost of the flights is $C = \sqrt{0.2x + 1}$, $0 \le x$, where C is measured in millions of dollars and x is measured in thousands of passengers (see figure in textbook). The total cost of the flights for a certain month is 2.5 million dollars. Approximately how many passengers flew that month?

Solution

Looking at the graph, we see that there are approximately 30,000 passengers.

Check

$$2.5 = \sqrt{0.2x + 1}$$

$$2.5^2 = \left(\sqrt{0.2x + 1}\right)^2$$

$$6.25 = 0.2x + 1$$

$$5.25 = 0.2x$$

$$\frac{5.25}{0.2} = x$$

$$26.25 \text{ thousand} = \text{ number of passengers}$$

$$26,250 = \text{ number of passengers}$$

5.6 Complex Numbers

7. Write $\sqrt{-4}$ in i-form.

Solution

$$\sqrt{-4} = \sqrt{-1 \cdot 4}$$
$$= \sqrt{-1} \cdot \sqrt{4}$$
$$= 2i$$

9. Write $\sqrt{-27}$ in i-form.

Solution

$$\sqrt{-27} = \sqrt{-1 \cdot 9 \cdot 3}$$
$$= \sqrt{-1} \cdot \sqrt{9} \cdot \sqrt{3}$$
$$= 3i\sqrt{3}$$

11. Write $\sqrt{-7}$ in i-form.

Solution

$$\sqrt{-7} = \sqrt{7 \cdot -1}$$
$$= \sqrt{7} \cdot \sqrt{-1}$$
$$= i\sqrt{7}$$

13. Perform the operation and write your answer in standard form.

$$\sqrt{-16} + 6i$$

Solution

$$\sqrt{-16} + 6i = 4i + 6i$$
$$= (4 + 6)i$$
$$= 10i$$

15. Perform the operation and write your answer in standard form.

$$\sqrt{-50} - \sqrt{-8}$$

Solution

$$\sqrt{-50} - \sqrt{-8} = 5i\sqrt{2} - 2i\sqrt{2}$$
$$= \left(5\sqrt{2} - 2\sqrt{2}\right)i$$
$$= 3\sqrt{2}i$$

17. Perform the operation and write your answer in standard form.

$$\sqrt{-8}\sqrt{-2}$$

Solution

$$\sqrt{-8}\sqrt{-2} = \left(2i\sqrt{2}\right)\left(i\sqrt{2}\right)$$
$$= 2 \cdot 2 \cdot i^2$$
$$= 4(-1) = -4$$

19. Perform the operations and write your answer in standard form.

$$\sqrt{-5}\left(\sqrt{-3} - \sqrt{-2}\right)$$

Solution

$$\sqrt{-5}\left(\sqrt{-3} - \sqrt{-2}\right) = i\sqrt{5}\left(i\sqrt{3} - i\sqrt{2}\right)$$
$$= \sqrt{15}i^2 - \sqrt{10}i^2$$
$$= -\sqrt{15} - \sqrt{10}$$

21. Perform the operations and write your answer in standard form.

$$(4 - 3i) + (6 + 7i)$$

Solution

$$(4 - 3i) + (6 + 7i) = (4 + 6) + (-3 + 7)i$$
$$= 10 + 4i$$

23. Perform the operations and write your answer in standard form.

$$13i - (14 - 7i)$$

Solution

$$13i - (14 - 7i) = (-14) + (13 + 7)i$$
$$= -14 + 20i$$

25. Perform the operations and write your answer in standard form.

$$15i - (3 - 25i) + \sqrt{-81}$$

Solution

$$15i - (3 - 25i) + \sqrt{-81} = 15i - 3 + 25i + 9i$$
$$= -3 + (15 + 25 + 9)i$$
$$= -3 + 49i$$

27. Perform the operation and write your answer in standard form.

$$(3i)(12i)$$

Solution

$$(3i)(12i) = 36i^2$$
$$= -36$$

29. Perform the operations and write your answer in standard form.

$$(4 + 3i)(-7 + 4i)$$

Solution

$$\begin{aligned}
(4 + 3i)(-7 + 4i) &= -28 + 16i - 21i + 12i^2 \\
&= -28 - 12 - 5i \\
&= -40 - 5i
\end{aligned}$$

31. Perform the operations and write your answer in standard form.

$$\left(-3 - \sqrt{-12}\right)\left(4 - \sqrt{-12}\right)$$

Solution

$$\begin{aligned}
\left(-3 - \sqrt{-12}\right)\left(4 - \sqrt{-12}\right) &= \left(-3 - 2i\sqrt{3}\right)\left(4 - 2i\sqrt{3}\right) \\
&= -12 + 6i\sqrt{3} - 8i\sqrt{3} + \left(2i\sqrt{3}\right)^2 \\
&= -12 - 2i\sqrt{3} + 12i^2 \\
&= -24 - 2i\sqrt{3}
\end{aligned}$$

33. Multiply the number $-2 - 8i$ by its complex conjugate.

Solution

$-2 - 8i$, conjugate $= -2 + 8i$

$$\begin{aligned}
\text{product} &= (-2 - 8i)(-2 + 8i) \\
&= (-2)^2 - (8i)^2 \\
&= 4 - 64i^2 = 4 + 64 = 68
\end{aligned}$$

35. Multiply the number $1 + \sqrt{-3}$ by its complex conjugate.

Solution

$1 + \sqrt{-3} = 1 + i\sqrt{3}$, conjugate $= 1 - i\sqrt{3}$

$$\begin{aligned}
\text{product} &= \left(1 + i\sqrt{3}\right)\left(1 - i\sqrt{3}\right) \\
&= 1^2 - \left(i\sqrt{3}\right)^2 \\
&= 1 - 3i^2 \\
&= 1 + 3 \\
&= 4
\end{aligned}$$

37. Perform the division and write your answer in standard form.

$$\frac{-12}{2+7i}$$

Solution

$$\frac{-12}{2+7i} = \frac{-12}{2+7i} \cdot \frac{2-7i}{2-7i} = \frac{-12(2-7i)}{4+49}$$

$$= \frac{-12(2-7i)}{53}$$

$$= \frac{-24+84i}{53}$$

$$= \frac{-24}{53} + \frac{84}{53}i$$

39. Perform the division and write your answer in standard form.

$$\frac{20}{2i}$$

Solution

$$\frac{20}{2i} = \frac{10}{i} \cdot \frac{-i}{-i} = \frac{-10i}{1}$$

$$= 0 - 10i$$

41. Simplify $\sqrt{128} + 3\sqrt{50}$.

Solution

$$\sqrt{128} + 3\sqrt{50} = \sqrt{64 \cdot 2} + 3\sqrt{25 \cdot 2}$$

$$= 8\sqrt{2} + 15\sqrt{2}$$

$$= 23\sqrt{2}$$

43. Simplify $\dfrac{8}{\sqrt{10}}$.

Solution

$$\frac{8}{\sqrt{10}} = \frac{8}{\sqrt{10}} \cdot \frac{\sqrt{10}}{\sqrt{10}}$$

$$= \frac{8\sqrt{10}}{10}$$

$$= \frac{4\sqrt{10}}{5}$$

45. *Real Estate Taxes* The tax on a property with an assessed value of \$145,000 is \$2400. Find the tax on a property with an assessed value of \$90,000.

Solution

Verbal model:

$$\boxed{\frac{\text{Tax}}{\text{Property Value}}} = \boxed{\frac{\text{Tax}}{\text{Property Value}}}$$

Proportion:

$$\frac{2400}{145,000} = \frac{x}{90,000}$$

$$x = \frac{(90,000)(2400)}{(145,000)}$$

$$= \$1489.65$$

47. Write $-\sqrt{-144}$ in i-form.

Solution

$$-\sqrt{-144} = -\sqrt{144 \cdot -1}$$

$$= -\sqrt{144} \cdot \sqrt{-1}$$

$$= -12i$$

49. Write $\sqrt{-\dfrac{4}{25}}$ in i-form.

Solution

$$\sqrt{\frac{-4}{25}} = \sqrt{\frac{4}{25} \cdot -1}$$

$$= \sqrt{\frac{4}{25}} \cdot \sqrt{-1}$$

$$= \frac{2}{5}i$$

51. Write $\sqrt{-0.09}$ in i-form.

Solution

$$\sqrt{-0.09} = \sqrt{0.09 \cdot -1}$$
$$= \sqrt{0.09} \cdot \sqrt{-1}$$
$$= 0.3i$$

53. Write $\sqrt{-8}$ in i-form.

Solution

$$\sqrt{-8} = \sqrt{4 \cdot 2 \cdot -1}$$
$$= \sqrt{4} \cdot \sqrt{2} \cdot \sqrt{-1}$$
$$= 2i\sqrt{2}$$

55. Perform the operation and write your answer in standard form.

$$\sqrt{-18}\sqrt{-3}$$

Solution

$$\sqrt{-18}\sqrt{-3} = \left(3i\sqrt{2}\right)\left(i\sqrt{3}\right)$$
$$= 3\sqrt{6} \cdot i^2$$
$$= -3\sqrt{6}$$

57. Perform the operations and write your answer in standard form.

$$\sqrt{-3}\left(\sqrt{-3} + \sqrt{-4}\right)$$

Solution

$$\sqrt{-3}\left(\sqrt{-3} + \sqrt{-4}\right) = i\sqrt{3}\left(i\sqrt{3} + 2i\right)$$
$$= \left(i\sqrt{3}\right)^2 + 2\sqrt{3}i^2$$
$$= -3 - 2\sqrt{3}$$

59. Determine a and b for $5 - 4i = (a+3) + (b-1)i$.

Solution

$$5 - 4i = (a+3) + (b-1)i$$
$$a + 3 = 5 \qquad b - 1 = -4$$
$$a = 2 \qquad b = -3$$

61. Determine a and b for $-4 - \sqrt{-8} = a + bi$.

Solution

$$-4 - \sqrt{-8} = a + bi$$
$$-4 - 2i\sqrt{2} = a + bi$$
$$-4 = a \qquad -2i\sqrt{2} = bi$$
$$-2\sqrt{2} = b$$

63. Perform the operations and write your answer in standard form.

$$(-4 - 7i) + (-10 - 33i)$$

Solution

$$(-4 - 7i) + (-10 - 33i) = (-4 - 10) + (-7 - 33)i$$
$$= -14 - 46i$$

65. Perform the operations and write your answer in standard form.

$$(15 + 10i) - (2 + 10i)$$

Solution

$$(15 + 10i) - (2 + 10i) = (15 - 2) + (10 - 10)i$$
$$= 13 + 0i = 13$$

67. Perform the operations and write your answer in standard form.

$$(30 - i) - (18 + 6i) + 3i^2$$

Solution

$$(30 - i) - (18 + 6i) + 3i^2 = 30 - i - 18 - 6i - 3$$
$$= 9 - 7i$$

69. Perform the operations and write your answer in standard form.

$$(-2i)(-10i)$$

Solution

$$(2i)(-10i) = 20i^2 = -20$$

71. Perform the operations and write your answer in standard form.

$$(-6i)(-i)(6i)$$

Solution

$$(-6i)(-i)(6i) = 36i^3 = -36i$$

73. Perform the operation and write your answer in standard form.

$$(-3i)^3$$

Solution

$$(-3i)^3 = -27i^3 = 27i$$

75. Perform the operations and write your answer in standard form.

$$-5(13 + 2i)$$

Solution

$$-5(13 + 2i) = -65 - 10i$$

77. Perform the operations and write your answer in standard form.

$$4i(-3 - 5i)$$

Solution

$$4i(-3 - 5i) = -12i - 20i^2 = 20 - 12i$$

79. Perform the operations and write your answer in standard form.

$$(-7 + 7i)(4 - 2i)$$

Solution

$$\begin{aligned}(-7 + 7i)(4 - 2i) &= -28 + 14i + 28i - 14i^2 \\ &= -28 + 42i + 14 \\ &= -14 + 42i\end{aligned}$$

81. Perform the operations and write your answer in standard form.

$$(3 - 4i)^2$$

Solution

$$\begin{aligned}(3 - 4i)^2 &= 3^2 - 2(3)(4i) + (4i)^2 \\ &= 9 - 24i + 16i^2 \\ &= 9 - 16 - 24i \\ &= -7 - 24i\end{aligned}$$

83. Perform the operations and write your answer in standard form.

$$\left(-2 + \sqrt{-5}\right)\left(-2 - \sqrt{-5}\right)$$

Solution

$$\begin{aligned}\left(-2 + \sqrt{-5}\right)\left(-2 - \sqrt{-5}\right) &= \left(-2 + i\sqrt{5}\right)\left(-2 - i\sqrt{5}\right) \\ &= 4 + 2i\sqrt{5} - 2i\sqrt{5} - 5i^2 \\ &= 4 + 5 \\ &= 9\end{aligned}$$

85. Find the conjugate of $2 + i$. Then find the product of the number and its conjugate.

Solution

$$2 + i, \text{conjugate} = 2 - i$$

$$\begin{aligned}\text{product} &= (2 + i)(2 - i) \\ &= 2^2 - i^2 \\ &= 4 + 1 \\ &= 5\end{aligned}$$

87. Find the conjugate of $5 - \sqrt{6}i$. Then find the product of the number and its conjugate.

Solution

$$5 - \sqrt{6}i, \text{conjugate} = 5 + \sqrt{6}i$$

$$\begin{aligned}\text{product} &= \left(5 - \sqrt{6}i\right)\left(5 + \sqrt{6}i\right) \\ &= 5^2 - \left(\sqrt{6}i\right)^2 \\ &= 25 - 6i^2 = 25 + 6 = 31\end{aligned}$$

89. Find the conjugate of $10i$. Then find the product of the number and its conjugate.

Solution

$10i$, conjugate $= -10i$

$$\text{product} = (10i)(-10i)$$
$$= -(10i)^2$$
$$= -100i^2$$
$$= 100$$

91. Perform the division and write your answer in standard form.

$$\frac{4}{1-i}$$

Solution

$$\frac{4}{1-i} = \frac{4}{1-i} \cdot \frac{1+i}{1+i} = \frac{4(1+i)}{1+1}$$
$$= \frac{4(1+i)}{2}$$
$$= 2(1+i)$$
$$= 2+2i$$

93. Perform the division and write your answer in standard form.

$$\frac{4i}{1-3i}$$

Solution

$$\frac{4i}{1-3i} = \frac{4i}{1-3i} \cdot \frac{1+3i}{1+3i} = \frac{4i(1+3i)}{1+9}$$
$$= \frac{4i+12i^2}{10}$$
$$= \frac{-12+4i}{10} = \frac{4(-3+i)}{10}$$
$$= \frac{2(-3+i)}{5} = \frac{-6+2i}{5}$$
$$= -\frac{6}{5} + \frac{2}{5}i$$

95. Perform the division and write your answer in standard form.

$$\frac{2+3i}{1+2i}$$

Solution

$$\frac{2+3i}{1+2i} = \frac{2+3i}{1+2i} \cdot \frac{1-2i}{1-2i} = \frac{2-4i+3i-6i^2}{1+4}$$
$$= \frac{2+6-i}{5}$$
$$= \frac{8-i}{5}$$
$$= \frac{8}{5} - \frac{1}{5}i$$

97. Find the sum and write your answer in standard form.

$$\frac{1}{1-2i} + \frac{4}{1+2i}$$

Solution

$$\frac{1}{1-2i} + \frac{4}{1+2i} = \frac{1}{1-2i} \cdot \frac{1+2i}{1+2i} + \frac{4}{1+2i} \cdot \frac{1-2i}{1-2i}$$
$$= \frac{1+2i}{1+4} + \frac{4-8i}{1+4} = \frac{1+2i}{5} + \frac{4-8i}{5}$$
$$= \frac{(1+4)+(2-8)i}{5} = \frac{5-6i}{5}$$
$$= 1 - \frac{6}{5}i$$

99. Find the difference and write your answer in standard form.

$$\frac{i}{4-3i} - \frac{5}{2+i}$$

Solution

$$\frac{i}{4-3i} - \frac{5}{2+i} = \frac{i}{4-3i} \cdot \frac{4+3i}{4+3i} - \frac{5}{2+i} \cdot \frac{2-i}{2-i}$$

$$= \frac{4i + 3i^2}{16+9} - \frac{10-5i}{4+1} = \frac{-3+4i}{25} - \frac{10-5i}{5} \cdot \frac{5}{5}$$

$$= \frac{-3+4i}{25} - \frac{50-25i}{25} = \frac{(-3-50)+(4+25)i}{25}$$

$$= \frac{-53+29i}{25} = -\frac{53}{25} + \frac{29}{25}i$$

101. Decide whether each number is a solution of the equation $x^2 + 2x + 5 = 0$.

Solution

(a) $x = -1 + 2i$

$$(-1+2i)^2 + 2(-1+2i) + 5 \overset{?}{=} 0$$

$$1 - 4i + 4i^2 - 2 + 4i + 5 = 0$$

$$1 - 4 - 2 + 5 = 0$$

$$0 = 0$$

(b) $x = -1 - 2i$

$$(-1-2i)^2 + 2(-1-2i) + 5 \overset{?}{=} 0$$

$$1 + 4i + 4i^2 - 2 - 4i + 5 = 0$$

$$1 - 4 - 2 + 5 = 0$$

$$0 = 0$$

103. Decide whether each number is a solution of the equation $x^3 + 4x^2 + 9x + 36 = 0$.

Solution

(a) $x = -4$

$$(-4)^3 + 4(-4)^2 + 9(-4) + 36 \overset{?}{=} 0$$

$$-64 + 64 - 36 + 36 = 0$$

$$0 = 0$$

(b) $x = -3i$

$$(-3i)^3 + 4(-3i)^2 + 9(-3i) + 36 \overset{?}{=} 0$$

$$-27i^3 + 36i^2 - 27i + 36 = 0$$

$$27i - 36 - 27i + 36 = 0$$

$$0 = 0$$

105. Perform the operation.

$$(a+bi) + (a-bi)$$

Solution

$$(a+bi) + (a-bi) = (a+a) + (b-b)i$$

$$= 2a + 0i$$

107. Perform the operation.

$$(a+bi) - (a-bi)$$

Solution

$$(a+bi) - (a-bi) = (a-a) + (b+b)i$$

$$= 0 + 2bi$$

Review Exercises for Chapter 5

1. Evaluate $(2^3 \cdot 3^2)^{-1}$.

Solution

$$(2^3 \cdot 3^2)^{-1} = (8 \cdot 9)^{-1}$$
$$= 72^{-1}$$
$$= \frac{1}{72}$$

3. Evaluate $\left(\frac{2}{5}\right)^{-3}$.

Solution

$$\left(\frac{2}{5}\right)^{-3} = \left(\frac{5}{2}\right)^3$$
$$= \frac{125}{8}$$

5. Evaluate $(6 \times 10^3)^2$.

Solution

$$(6 \times 10^3)^2 = 6^2 \times 10^6$$
$$= 36 \times 10^6$$
$$= 36{,}000{,}000$$

7. Evaluate $\dfrac{3.5 \times 10^7}{7 \times 10^4}$.

Solution

$$\frac{3.5 \times 10^7}{7 \times 10^4} = \frac{3.5}{7} \times 10^{7-4}$$
$$= .5 \times 10^3$$
$$= 5 \times 10^2$$
$$= 500$$

9. Simplify $\dfrac{4x^2}{2x}$.

Solution

$$\frac{4x^2}{2x} = 2x$$

11. Simplify $(x^3 y^{-4})^2$.

Solution

$$(x^3 y^{-4})^2 = x^6 y^{-8}$$
$$= \frac{x^6}{y^8}$$

13. Simplify $\dfrac{t^{-5}}{t^{-2}}$.

Solution

$$\frac{t^{-5}}{t^{-2}} = t^{(-5)-(-2)}$$
$$= t^{-5+2}$$
$$= t^{-3} = \frac{1}{t^3}$$

15. Simplify $\left(\dfrac{y}{3}\right)^{-3}$.

Solution

$$\left(\frac{y}{3}\right)^{-3} = \left(\frac{3}{y}\right)^3$$
$$= \frac{27}{y^3}$$

17. Evaluate $\sqrt{1.44}$.

Solution

$$\sqrt{1.44} = \sqrt{144 \times 10^{-2}}$$
$$= \sqrt{144} \times \sqrt{10^{-2}}$$
$$= 12 \times 10^{-1} = 1.2$$

19. Evaluate $\sqrt{\dfrac{25}{36}}$.

Solution

$$\sqrt{\frac{25}{36}} = \frac{5}{6}$$

21. Evaluate $\sqrt{169 - 25}$.

Solution

$\sqrt{169 - 25} = \sqrt{144} = 12$

25. Evaluate $\sqrt{13^2 - 4(2)(7)}$. Round the result to two decimal places.

Solution

$\sqrt{13^2 - 4(2)(7)} = 10.630146 \approx 10.63$

29. Fill in the missing description.

Radical Form   Rational Exponent Form
$$216^{1/3} = 6$$

Solution

$\sqrt[3]{216} = 6$

33. Evaluate $25^{3/2}$.

Solution

$25^{3/2} = \left(\sqrt{25}\right)^3 = 5^3 = 125$

37. Simplify $x^{3/4} \cdot x^{-1/6}$.

Solution

$$\begin{aligned}
x^{3/4} \cdot x^{-1/6} &= x^{3/4 + (-1/6)} \\
&= x^{9/12 + (-2)/12} \\
&= x^{7/12}
\end{aligned}$$

41. Simplify $\sqrt{360}$.

Solution

$$\begin{aligned}
\sqrt{360} &= \sqrt{36 \cdot 10} \\
&= 6\sqrt{10}
\end{aligned}$$

23. Evaluate $1800(1 + 0.08)^{24}$. Round the result to two decimal places.

Solution

$1800(1 + 0.08)^{24} = 11414.13$

27. Fill in the missing description.

Radical Form   Rational Exponent Form
$\sqrt{49} = 7$

Solution

$49^{1/2} = 7$

31. Evaluate $27^{4/3}$.

Solution

$27^{4/3} = \left(\sqrt[3]{27}\right)^4 = 3^4 = 81$

35. Evaluate $75^{-3/4}$. Round the result to two decimal places.

Solution

$75^{-3/4} = 0.0392377 \approx 0.04$

39. Simplify $\dfrac{15x^{1/4}y^{3/5}}{5x^{1/2}y}$.

Solution

$$\begin{aligned}
\frac{15x^{1/4}y^{3/5}}{5x^{1/2}y} &= 3x^{1/4 - 1/2}y^{3/5 - 1} \\
&= 3x^{1/4 - 2/4}y^{3/5 - 5/5} \\
&= 3x^{-1/4}y^{-2/5} \\
&= \frac{3}{x^{1/4}y^{2/5}}
\end{aligned}$$

43. Simplify $\sqrt{0.25x^4 y}$.

Solution

$$\begin{aligned}
\sqrt{0.25x^4 y} &= \sqrt{25 \times 10^{-2}x^4 y} \\
&= 5 \times 10^{-1}x^2\sqrt{y} \\
&= 0.5x^2\sqrt{y}
\end{aligned}$$

45. Simplify $\sqrt[3]{48a^3b^4}$.

Solution

$$\sqrt[3]{48a^3b^4} = \sqrt[3]{8 \cdot 6a^3b^3b}$$

$$= 2ab\sqrt[3]{6b}$$

47. Rationalize the denominator of $\sqrt{\dfrac{5}{6}}$ and simplify further when possible.

Solution

$$\sqrt{\frac{5}{6}} = \sqrt{\frac{5}{6}} \cdot \frac{\sqrt{6}}{\sqrt{6}} = \frac{\sqrt{30}}{6}$$

49. Rationalize the denominator of $\dfrac{3}{\sqrt{12x}}$ and simplify further when possible.

Solution

$$\frac{3}{\sqrt{12x}} = \frac{3}{\sqrt{4 \cdot 3x}} = \frac{3}{2\sqrt{3x}} \cdot \frac{\sqrt{3x}}{\sqrt{3x}} = \frac{3\sqrt{3x}}{6x} = \frac{\sqrt{3x}}{2x}$$

51. Rationalize the denominator of $\dfrac{2}{\sqrt[3]{2x}}$ and simplify further when possible.

Solution

$$\frac{2}{\sqrt[3]{2x}} = \frac{2}{\sqrt[3]{2x}} \cdot \frac{\sqrt[3]{2^2x^2}}{\sqrt[3]{2^2x^2}} = \frac{2\sqrt[3]{4x^2}}{\sqrt[3]{8x^3}} = \frac{2\sqrt[3]{4x^2}}{2x} = \frac{\sqrt[3]{4x^2}}{x}$$

53. Rationalize the denominator of $\dfrac{6}{7 - \sqrt{7}}$ and simplify further when possible.

Solution

$$\frac{6}{7 - \sqrt{7}} = \frac{6}{7 - \sqrt{7}} \cdot \frac{7 + \sqrt{7}}{7 + \sqrt{7}} = \frac{6\left(7 + \sqrt{7}\right)}{49 - 7} = \frac{6\left(7 + \sqrt{7}\right)}{42} = \frac{7 + \sqrt{7}}{7}$$

55. Perform the operations and simplify.

$$3\sqrt{40} - 10\sqrt{90}$$

Solution

$$3\sqrt{40} - 10\sqrt{90} = 3\sqrt{4 \cdot 10} - 10\sqrt{9 \cdot 10}$$

$$= 6\sqrt{10} - 30\sqrt{10}$$

$$= -24\sqrt{10}$$

57. Perform the operations and simplify.

$$\sqrt{25x} + \sqrt{49x} - \sqrt{x}$$

Solution

$$\sqrt{25x} + \sqrt{49x} - \sqrt{x} = 5\sqrt{x} + 7\sqrt{x} - \sqrt{x} = 11\sqrt{x}$$

59. Perform the operations and simplify.

$$\left(3 - \sqrt{x}\right)\left(3 + \sqrt{x}\right)$$

Solution

$$\left(3 - \sqrt{x}\right)\left(3 + \sqrt{x}\right) = 9 - 3\sqrt{x} + 3\sqrt{x} - x = 9 - x$$

61. Use a graphing utility to graph the function.

$$y = 3\sqrt[3]{2x}$$

Solution

Keystrokes:

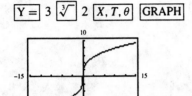

63. Use a graphing utility to graph the function.

$$g(x) = 4x^{3/4}$$

Solution

Keystrokes:

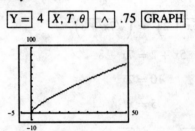

65. Solve $\sqrt{y} = 15$.

Solution

$$\sqrt{y} = 15 \qquad \text{Check:} \quad \sqrt{225} \stackrel{?}{=} 15$$
$$\left(\sqrt{y}\right)^2 = (15)^2 \qquad\qquad 15 = 15$$
$$y = 225$$

67. Solve $\sqrt{2(a - 7)} = 14$.

Solution

$$\sqrt{2(a - 7)} = 14 \qquad \text{Check:} \quad \sqrt{2(105 - 7)} \stackrel{?}{=} 14$$
$$\left(\sqrt{2(a - 7)}\right)^2 = (14)^2 \qquad\qquad \sqrt{196} = 14$$
$$2(a - 7) = 196 \qquad\qquad 14 = 14$$
$$2a - 14 = 196$$
$$2a = 210$$
$$a = 105$$

69. Solve $\sqrt{2(x + 5)} = x + 5$.

Solution

$$\sqrt{2(x + 5)} = x + 5 \qquad \text{Check:} \quad \sqrt{2(-5 + 5)} \stackrel{?}{=} -5 + 5$$
$$\left(\sqrt{2(x + 5)}\right)^2 = (x + 5)^2 \qquad\qquad \sqrt{0} = 0$$
$$2(x + 5) = x^2 + 10x + 25 \qquad\qquad 0 = 0$$
$$2x + 10 = x^2 + 10x + 25$$
$$0 = x^2 + 8x + 15 \qquad \sqrt{2(-3 + 5)} \stackrel{?}{=} -3 + 5$$
$$0 = (x + 5)(x + 3) \qquad\qquad \sqrt{4} = 2$$
$$-5 = x \qquad x = -3 \qquad\qquad 2 = 2$$

71. Solve $\sqrt[3]{5x+2} - \sqrt[3]{7x-8} = 0$.

Solution

$$\sqrt[3]{5x+2} - \sqrt[3]{7x-8} = 0$$

$$\sqrt[3]{5x+2} = \sqrt[3]{7x-8}$$

$$\left(\sqrt[3]{5x+2}\right)^3 = \left(\sqrt[3]{7x-8}\right)^3$$

$$5x+2 = 7x-8$$

$$10 = 2x$$

$$5 = x$$

Check: $\sqrt[3]{5(5)+2} - \sqrt[3]{7(5)-8} \overset{?}{=} 0$

$$\sqrt[3]{27} - \sqrt[3]{27} = 0$$

$$0 = 0$$

73. Solve $\sqrt{1+6x} = 2 - \sqrt{6x}$.

Solution

$$\sqrt{1+6x} = 2 - \sqrt{6x}$$

$$\left(\sqrt{1+6x}\right)^2 = \left(2 - \sqrt{6x}\right)^2$$

$$1+6x = 4 - 4\sqrt{6x} + 6x$$

$$1 = 4 - 4\sqrt{6x}$$

$$-3 = -4\sqrt{6x}$$

$$(3)^2 = \left(4\sqrt{6x}\right)^2$$

$$9 = 16(6x)$$

$$\frac{9}{96} = x$$

$$\frac{3}{32} = x$$

Check: $\sqrt{1+6\left(\dfrac{3}{32}\right)} \overset{?}{=} 2 - \sqrt{6\left(\dfrac{3}{32}\right)}$

$$\sqrt{\frac{32}{32} + \frac{18}{32}} = 2 - \sqrt{\frac{18}{32}}$$

$$\sqrt{\frac{50}{32}} = 2 - \sqrt{\frac{9 \cdot 2}{16 \cdot 2}}$$

$$\sqrt{\frac{25 \cdot 2}{16 \cdot 2}} = 2 - \sqrt{\frac{9 \cdot 2}{16 \cdot 2}}$$

$$\sqrt{\frac{25}{16}} = 2 - \sqrt{\frac{9}{16}}$$

$$\frac{5}{4} = 2 - \frac{3}{4}$$

$$\frac{5}{4} = \frac{8}{4} - \frac{3}{4}$$

$$\frac{5}{4} = \frac{5}{4}$$

75. Use a graphing utility to approximate the solution of the equation $4\sqrt[3]{x} = 7 - x$.

Solution

Keystrokes:

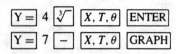

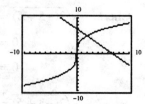

Approximate solution is 2

Exact solution is 1.978

77. Write $\sqrt{-48}$ in standard form.

Solution

$$\sqrt{-48} = \sqrt{16 \cdot 3 \cdot -1} = 4i\sqrt{3}$$

79. Write $10 - 3\sqrt{-27}$ in standard form.

Solution

$$10 - 3\sqrt{-27} = 10 - 3\sqrt{-1 \cdot 9 \cdot 3}$$
$$= 10 - 3\sqrt{-1} \cdot \sqrt{9} \cdot \sqrt{3}$$
$$= 10 - 9i\sqrt{3}$$

81. Write $\dfrac{3}{4} - 5\sqrt{-\dfrac{3}{25}}$ in standard form.

Solution

$$\frac{3}{4} - 5\sqrt{-\frac{3}{25}} = \frac{3}{4} - 5\sqrt{\frac{3}{25} \cdot -1}$$
$$= \frac{3}{4} - \frac{5}{5}i\sqrt{3}$$
$$= \frac{3}{4} - i\sqrt{3}$$

83. Perform the operations and write the answer in standard form.

$$(-4 + 5i) - (-12 + 8i)$$

Solution

$$(-4 + 5i) - (-12 + 8i) = (-4 + 12) + (5 - 8)i$$
$$= 8 - 3i$$

85. Perform the operations and write the answer in standard form.

$$(-2)(15i)(-3i)$$

Solution

$$(-2)(15i)(-3i) = 90i^2 = -90$$

87. Perform the operations and write the answer in standard form.

$$(4 - 3i)(4 + 3i)$$

Solution

$$(4 - 3i)(4 + 3i) = 4^2 - (3i)^2$$
$$= 16 + 9$$
$$= 25$$

89. Perform the operations and write the answer in standard form.

$$(12 - 5i)(2 + 7i)$$

Solution

$$(12 - 5i)(2 + 7i) = 24 + 84i - 10i - 35i^2$$
$$= 24 + 74i + 35$$
$$= 59 + 74i$$

91. Perform the operation and write the answer in standard form.

$$\frac{5i}{2 + 9i}$$

Solution

$$\frac{5i}{2 + 9i} = \frac{5i}{2 + 9i} \cdot \frac{2 - 9i}{2 - 9i} = \frac{10i - 45i^2}{4 + 81}$$
$$= \frac{45 + 10i}{85}$$
$$= \frac{45}{85} + \frac{10}{85}i$$
$$= \frac{9}{17} + \frac{2}{17}i$$

93. *Geometry* The four corners are cut from an $8\frac{1}{2}$-inch-by-11 inch sheet of paper (see figure). Find the perimeter of the remaining piece of paper.

Solution

$$c = \sqrt{2^2 + 2^2} = \sqrt{4+4} = \sqrt{8}$$

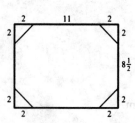

Equation: $P = 2(7) + 2\left(4\frac{1}{2}\right) + 4\left(\sqrt{8}\right)$

$$= 14 + 9 + 8\sqrt{2}$$

$$= 23 + 8\sqrt{2}$$

95. *Length of a Pendulum* The time t in seconds for a pendulum of length L in feet to go through one complete cycle (its period) is given by

$$t = 2\pi\sqrt{\frac{L}{32}}.$$

How long is the pendulum of a grandfather clock with a period of 1.3 seconds?

Solution

$$1.3 = 2\pi\sqrt{\frac{L}{32}}$$

$$\frac{1.3}{2\pi} = \sqrt{\frac{L}{32}}$$

$$\left(\frac{1.3}{2\pi}\right)^2 = \left(\sqrt{\frac{L}{32}}\right)^2$$

$$\frac{1.69}{4\pi^2} = \frac{L}{32}$$

$$\frac{1.69}{4\pi^2}(32) = L$$

$$1.3698624 = L$$

Test for Chapter 5

1. (a) Evaluate $2^{-2} + 2^{-3}$, without using a calculator.

Solution

$$2^{-2} + 2^{-3} = \frac{1}{2^2} + \frac{1}{2^3}$$

$$= \frac{1}{4} + \frac{1}{8}$$

$$= \frac{2}{8} + \frac{1}{8} = \frac{3}{8}$$

(b) Evaluate $\dfrac{6.3 \times 10^{-3}}{2.1 \times 10^2}$, without using a calculator.

Solution

$$\frac{6.3 \times 10^{-3}}{2.1 \times 10^2} = 3 \times 10^{-3-2}$$

$$= 3 \times 10^{-5}$$

$$= 0.00003$$

2. (a) Evaluate $27^{-2/3}$, without using a calculator.

Solution

$$27^{-2/3} = \frac{1}{27^{2/3}}$$

$$= \frac{1}{9}$$

(b) Evaluate $\sqrt{2}\sqrt{18}$, without using a calculator.

Solution

$$\sqrt{2}\sqrt{18} = \sqrt{2 \cdot 18} = \sqrt{36} = 6$$

3. Write the real number 0.000032 in scientific notation.

Solution

$$0.000032 = 3.2 \times 10^{-5}$$

4. Write the real number 3.04×10^7 in decimal notation.

Solution

$$3.04 \times 10^7 = 30,400,000$$

5. (a) Simplify $\dfrac{12t^{-2}}{20t^{-1}}$.

Solution

$$\frac{12t^{-2}}{20t^{-1}} = \frac{3t^{(-2)-(-1)}}{5}$$

$$= \frac{3}{5}t^{-1}$$

$$= \frac{3}{5t}$$

(b) Simplify $\left(x + y^{-2}\right)^{-1}$.

Solution

$$\left(x + y^{-2}\right)^{-1} = \frac{1}{x + y^{-2}}$$

$$= \frac{1}{x + \dfrac{1}{y^2}} \cdot \frac{y^2}{y^2}$$

$$= \frac{y^2}{xy^2 + 1}$$

6. (a) Simplify $\left(\dfrac{x^{1/2}}{x^{1/3}}\right)^2$.

Solution

$$\left(\frac{x^{1/2}}{x^{1/3}}\right)^2 = \frac{x}{x^{2/3}}$$

$$= x^{1-2/3} = x^{1/3}$$

(b) Simplify $5^{1/4} \cdot 5^{7/4}$.

Solution

$$5^{1/4} \cdot 5^{7/4} = 5^{1/4+7/4}$$

$$= 5^{8/4} = 5^2 = 25$$

7. (a) Simplify $\sqrt{\dfrac{32}{9}}$.

Solution

$$\sqrt{\frac{32}{9}} = \sqrt{\frac{16 \cdot 2}{9}} = \frac{4}{3}\sqrt{2}$$

(b) Simplify $\sqrt[3]{24}$.

Solution

$$\sqrt[3]{24} = \sqrt[3]{8 \cdot 3} = 2\sqrt[3]{3}$$

8. Rationalize the denominator of $\dfrac{3}{\sqrt{6}}$.

Solution

$$\frac{3}{\sqrt{6}} = \frac{3}{\sqrt{6}} \cdot \frac{\sqrt{6}}{\sqrt{6}} = \frac{3\sqrt{6}}{6} = \frac{\sqrt{6}}{2}$$

9. Combine $5\sqrt{3x} - 3\sqrt{75x}$.

Solution

$$5\sqrt{3x} - 3\sqrt{75x} = 5\sqrt{3x} - 3\sqrt{25 \cdot 3x}$$

$$= 5\sqrt{3x} - 15\sqrt{3x}$$

$$= -10\sqrt{3x}$$

10. Multiply and simplify $\sqrt{5}\left(\sqrt{15x}+3\right)$.

Solution

$$\sqrt{5}\left(\sqrt{15x}+3\right) = \sqrt{75x}+3\sqrt{5}$$
$$= \sqrt{25\cdot 3x}+3\sqrt{5}$$
$$= 5\sqrt{3x}+3\sqrt{5}$$

11. Expand $\left(4-\sqrt{2x}\right)^2$.

Solution

$$\left(4-\sqrt{2x}\right)^2 = 16-8\sqrt{2x}+2x$$

12. Factor $7\sqrt{27}+14y\sqrt{12} = 7\sqrt{3}(\quad)$.

Solution

$$7\sqrt{27}+14y\sqrt{12} = 7\sqrt{9\cdot 3}+14y\sqrt{4\cdot 3}$$
$$= 21\sqrt{3}+28y\sqrt{3}$$
$$= 7\sqrt{3}(3+4y)$$

13. Solve $\sqrt{x^2-1}=x-2$.

Solution

$$\sqrt{x^2-1}=x-2$$
$$\left(\sqrt{x^2-1}\right)^2 = (x-2)^2$$
$$x^2-1 = x^2-4x+4$$
$$4x = 5$$
$$x = \frac{5}{4}$$
$$\emptyset$$

Check: $\sqrt{\left(\frac{5}{4}\right)^2-1} \overset{?}{=} \frac{5}{4}-2$

$$\sqrt{\frac{25}{16}-\frac{16}{16}} = \frac{5}{4}-\frac{8}{4}$$
$$\sqrt{\frac{9}{16}} = -\frac{3}{4}$$
$$\frac{3}{4} \neq -\frac{3}{4}$$

14. Solve $\sqrt{x}-x+6=0$.

Solution

$$\sqrt{x}-x+6=0$$
$$\left(\sqrt{x}\right)^2 = (x-6)^2$$
$$x = x^2-12x+36$$
$$0 = x^2-13x+36$$
$$0 = (x-9)(x-4)$$
$$0 = x-9 \qquad 0 = x-4$$
$$9 = x \qquad\quad 4 = x$$

Check: $\sqrt{9}-9+6 \overset{?}{=} 0$

$$3-9+6 = 0$$
$$0 = 0$$

$\sqrt{4}-4+6 \overset{?}{=} 0$

$$2-4+6 = 0$$
$$4 \neq 0$$

15. Subtract $(2 + 3i) - \sqrt{-25}$ and simplify.

Solution

$$(2 + 3i) - \sqrt{-25} = 2 + 3i - 5i = 2 - 2i$$

16. Multiply $(2 - 3i)^2$ and simplify.

Solution

$$(2 - 3i)^2 = (2 - 3i)(2 - 3i)$$
$$= 4 - 6i - 6i + 9i^2$$
$$= 4 - 12i - 9$$
$$= -5 - 12i$$

17. Multiply $\sqrt{-16}\left(1 + \sqrt{4}\right)$ and simplify.

Solution

$$\sqrt{-16}\left(1 + \sqrt{4}\right) = 4i(1 + 2i)$$
$$= 4i + 8i^2$$
$$= -8 + 4i$$

18. Multiply $(3 - 2i)(1 + 5i)$ and simplify.

Solution

$$(3 - 2i)(1 + 5i) = 3 + 13i - 10i^2$$
$$= 3 + 13i + 10$$
$$= 13 + 13i$$

19. Divide $5 - 2i$ by i. Write the result in standard form.

Solution

$$\frac{5 - 2i}{i} = \frac{5 - 2i}{i} \cdot \frac{-i}{-i} = \frac{-5i + 2i^2}{-i^2} = -2 - 5i$$

20. The velocity v (in feet per second) of an object is given by $v = \sqrt{2gh}$, where $g = 32$ feet per second per second and h is the distance (in feet) the object has fallen. Find the height from which the rock has been dropped if it strikes the ground with a velocity of 80 feet per second.

Solution

$$v = \sqrt{2gh}$$
$$80 = \sqrt{2(32)h}$$
$$80 = \sqrt{64h}$$
$$80^2 = \left(\sqrt{64h}\right)^2$$
$$6400 = 64h$$
$$100 \text{ feet} = h$$

CHAPTER 6
Quadratic Equations and Inequalities

6.1 Factoring and Square Root

7. Solve $4x^2 - 12x = 0$ by factoring.

Solution

$$4x^2 - 12x = 0$$
$$4x(x - 3) = 0$$
$$4x = 0 \qquad x - 3 = 0$$
$$x = 0 \qquad x = 3$$

9. Solve $x^2 - 12x + 36 = 0$ by factoring.

Solution

$$x^2 - 12x + 36 = 0$$
$$(x - 6)(x - 6) = 0$$
$$x - 6 = 0 \qquad x - 6 = 0$$
$$x = 6 \qquad = 6$$

11. Solve $(y - 4)(y - 3) = 6$ by factoring.

Solution

$$(y - 4)(y - 3) = 6$$
$$y^2 - 7y + 12 - 6 = 0$$
$$y^2 - 7y + 6 = 0$$
$$(y - 6)(y - 1) = 0$$
$$y - 6 = 0 \qquad y - 1 = 0$$
$$y = 6 \qquad y = 1$$

13. Solve $3x(x - 6) - 5(x - 6) = 0$ by factoring.

Solution

$$3x(x - 6) - 5(x - 6) = 0$$
$$(x - 6)(3x - 5) = 0$$
$$x - 6 = 0 \qquad 3x - 5 = 0$$
$$x = 6 \qquad x = \frac{5}{3}$$

15. Solve $6x^2 = 54$ by factoring.

Solution

$$6x^2 = 54$$
$$6x^2 - 54 = 0$$
$$6\left(x^2 - 9\right) = 0$$
$$6(x - 3)(x + 3) = 0$$
$$x - 3 = 0 \qquad x + 3 = 0$$
$$x = 3 \qquad x = -3$$

17. Solve $x^2 = 64$ by extracting square roots.

Solution

$$x^2 = 64$$
$$x = \pm\sqrt{64}$$
$$x = \pm 8$$

19. Solve $(x + 4)^2 = 169$ by extracting square roots.

Solution

$$(x + 4)^2 = 169$$
$$x + 4 = \pm\sqrt{169}$$
$$x = -4 \pm 13$$
$$x = 9, -17$$

21. Solve $(2x + 1)^2 = 50$ by extracting square roots.

Solution

$$(2x + 1)^2 = 50$$
$$2x + 1 = \pm\sqrt{50}$$
$$2x = -1 \pm 5\sqrt{2}$$
$$x = \frac{-1 \pm 5\sqrt{2}}{2}$$

23. Solve $x^2 + 4 = 0$ by extracting complex square roots.

Solution

$$x^2 + 4 = 0$$
$$x^2 = -4$$
$$x = \pm\sqrt{-4}$$
$$x = \pm 2i$$

25. Solve $9(x + 6)^2 = -121$ by extracting complex square roots.

Solution

$$9(x + 6)^2 = -121$$
$$(x + 6)^2 = \frac{-121}{9}$$
$$x + 6 = \pm\sqrt{\frac{-121}{9}}$$
$$x = -6 \pm \frac{11}{3}i$$

27. Solve $(x - 1)^2 = -27$ by extracting complex square roots.

Solution

$$(x - 1)^2 = -27$$
$$x - 1 = \pm\sqrt{-27}$$
$$x = 1 \pm 3i\sqrt{3}$$

29. Find all real and complex solutions.

$$2x^2 - 5x = 0$$

Solution

$$2x^2 - 5x = 0$$
$$x(2x - 5) = 0$$
$$x = 0 \qquad 2x - 5 = 0$$
$$x = \frac{5}{2}$$

31. Find all real and complex solutions.

$$x^2 - 100 = 0$$

Solution

$$x^2 - 100 = 0$$
$$(x - 10)(x + 10) = 0$$
$$x - 10 = 0 \qquad x + 10 = 0$$
$$x = 10 \qquad x = -10$$

33. Find all real and complex solutions.

$$(x - 5)^2 - 100 = 0$$

Solution

$$(x - 5)^2 - 100 = 0$$
$$\text{let } u = (x - 5)$$
$$u^2 - 100 = 0$$
$$(u - 10)(u + 10) = 0$$
$$u - 10 = 0 \qquad u + 10 = 0$$
$$u = 10 \qquad u = -10$$
$$x - 5 = 10 \qquad x - 5 = -10$$
$$x = 15 \qquad x = -5$$

35. Solve $x^4 + 7x^2 - 8 = 0$.

List all real and complex solutions.

Solution

$$x^4 + 7x^2 - 8 = 0$$

$$\text{let } u = x^2$$

$$\left(x^2\right)^2 + 7\left(x^2\right) - 8 = 0$$

$$u^2 + 7u - 8 = 0$$

$$(u + 8)(u - 1) = 0$$

$u + 8 = 0$	$u - 1 = 0$
$u = -8$	$u = 1$
$x^2 = -8$	$x^2 = 1$
$x = \pm 2\sqrt{2}i$	$x = \pm 1$

37. Solve $2x - 9\sqrt{x} + 10 = 0$.

List all real and complex solutions.

Solution

$$2x - 9\sqrt{x} + 10 = 0$$

$$\text{let } u = \sqrt{x}$$

$$2\left(\sqrt{x}\right)^2 - 9\left(\sqrt{x}\right) + 10 = 0$$

$$2u^2 - 9u + 10 = 0$$

$$(2u - 5)(u - 2) = 0$$

$2u - 5 = 0$	$u - 2 = 0$
$u = \dfrac{5}{2}$	$u = 2$
$\sqrt{x} = \dfrac{5}{2}$	$\sqrt{x} = 2$
$\left(\sqrt{x}\right)^2 = \left(\dfrac{5}{2}\right)^2$	$\left(\sqrt{x}\right)^2 = 2^2$
$x = \dfrac{25}{4}$	$x = 4$

39. Solve $\left(x^2 - 2\right)^2 - 36 = 0$.

List all real and complex solutions.

Solution

$$\left(x^2 - 2\right)^2 - 36 = 0$$

$$\text{let } u = \left(x^2 - 2\right)$$

$$u^2 - 36 = 0$$

$$(u - 6)(u + 6) = 0$$

$u - 6 = 0$	$u + 6 = 0$
$u = 6$	$u = -6$
$x^2 - 2 = 6$	$x^2 - 2 = -6$
$x^2 = 8$	$x^2 = -4$
$x = \pm\sqrt{8}$	$x = \pm\sqrt{-4}$
$x = \pm 2\sqrt{2}$	$x = \pm 2i$
	not real

41. List all real and complex solutions.

Solve $\left(x^2 - 5\right)^2 - 100 = 0$.

Solution

$$\left(x^2 - 5\right)^2 - 100 = 0$$

$$\text{let } u = \left(x^2 - 5\right)$$

$$u^2 - 100 = 0$$

$$(u - 10)(u + 10) = 0$$

$u - 10 = 0$	$u + 10 = 0$
$u = 10$	$u = -10$
$x^2 - 5 = 10$	$x^2 - 5 = -10$
$x^2 = 15$	$x^2 = -5$
$x = \pm\sqrt{15}$	$x = \pm i\sqrt{5}$

43. Use a graphing utility to graph $y = x^2 - 9$. Use the graph to approximate any x-intercepts of the graph. Set $y = 0$ and solve the resulting equation. Compare the results with the x-intercepts of the graph.

Solution

Keystrokes:

x-intercepts are -3 and 3.

$$0 = x^2 - 9$$
$$= (x - 3)(x + 3)$$

$x - 3 = 0 \qquad x + 3 = 0$

$\qquad x = 3 \qquad\qquad x = -3$

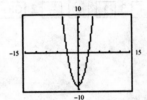

45. Use a graphing utility to graph $y = 4 - (x - 3)^2$. Use the graph to approximate any x-intercepts of the graph. Set $y = 0$ and solve the resulting equation. Compare the results with the x-intercepts of the graph.

Solution

Keystrokes:

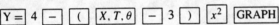

x-intercepts are 1 and 5.

$$0 = 4 - (x - 3)^2$$
$$(x - 3)^2 = 4$$
$$x - 3 = \pm 2$$
$$x = 5, 1$$

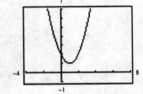

47. Use a graphing utility to graph $y = (x - 1)^2 + 1$ and observe that the graph has no x-intercepts. Set $y = 0$ and solve the resulting equation.

Solution

Keystrokes:

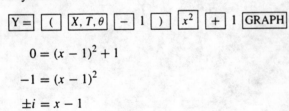

$$0 = (x - 1)^2 + 1$$
$$-1 = (x - 1)^2$$
$$\pm i = x - 1$$
$$1 \pm i = x$$

49. Use a graphing utility to graph $y = -(x + 3)^2 - 2$ and observe that the graph has no x-intercepts. Set $y = 0$ and solve the resulting equation.

Solution

Keystrokes:

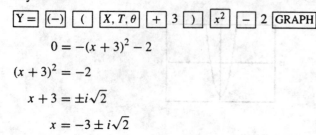

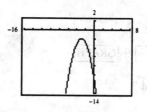

$$0 = -(x + 3)^2 - 2$$

$$(x + 3)^2 = -2$$

$$x + 3 = \pm i\sqrt{2}$$

$$x = -3 \pm i\sqrt{2}$$

51. *Graphical Reasoning* Use the model,

$$y = (44.17 + 2.82t)^2, \ 0 \le t \le 22$$

which approximates the amount of fire loss to private property in the United States from 1970 to 1992. In this model, y represents the value of private property lost to fire (in millions of dollars) and t represents the year, with $t = 0$ corresponding to 1970 (see figure in textbook). (*Source:* Insurance Information Institute.)

Graphically estimate the year in which private property loss was approximately 4500 million. Verify algebraically.

Solution

Estimate year 1978

$$4500 = (44.17 + 2.82t)^2$$

$$\pm\sqrt{4500} = 44.17 + 2.82t$$

$$t = \frac{-44.17 \pm \sqrt{4500}}{2.82}$$

$$t \approx 8$$

53. Factor $16x^2 - 121$ completely.

Solution

$$16x^2 - 121 = (4x + 11)(4x - 11)$$

55. Factor $x(x - 10) - 4(x - 10)$ completely.

Solution

$$x(x - 10) - 4(x - 10) = (x - 10)(x - 4)$$

57. *Speed* A boat's still-water speed is 18 miles per hour. It travels 35 miles upstream, and then returns to its starting point, in a total of 4 hours. Find the speed of the current.

Solution

Verbal model: $\boxed{\begin{array}{c}\text{Time}\\\text{Upstream}\end{array}} + \boxed{\begin{array}{c}\text{Time}\\\text{Downstream}\end{array}} = \boxed{\begin{array}{c}\text{Total}\\\text{Time}\end{array}}$

Equation:
$$\frac{35}{18-x} + \frac{35}{18+x} = 4$$

$$35(18+x) + 35(18-x) = 4(18+x)(18-x)$$

$$630 + 35x + 630 - 35x = 4\left(324 - x^2\right)$$

$$1260 = 4\left(324 - x^2\right)$$

$$315 = 324 - x^2$$

$$x^2 = 9$$

$$x = \pm 3$$

$$x = 3 \text{ mph}$$

59. Solve $u(u-9) - 12(u-9) = 0$ by factoring.

Solution

$$u(u-9) - 12(u-9) = 0$$

$$(u-9)(u-12) = 0$$

$$u - 9 = 0 \qquad u - 12 = 0$$

$$u = 9 \qquad u = 12$$

61. Solve $4x^2 - 25 = 0$ by factoring.

Solution

$$4x^2 - 25 = 0$$

$$(2x - 5)(2x + 5) = 0$$

$$2x - 5 = 0 \qquad 2x + 5 = 0$$

$$x = \frac{5}{2} \qquad x = -\frac{5}{2}$$

63. Solve $x^2 + 10x + 600 = 0$ by factoring.

Solution

$$x^2 + 10x + 600 = 0$$

Prime

65. Solve $2x(3x + 2) = 5 - 6x^2$ by factoring.

Solution

$$2x(3x + 2) = 5 - 6x^2$$

$$6x^2 + 4x = 5 - 6x^2$$

$$12x^2 + 4x - 5 = 0$$

$$(6x + 5)(2x - 1) = 0$$

$$6x + 5 = 0 \qquad 2x - 1 = 0$$

$$x = -\frac{5}{6} \qquad x = \frac{1}{2}$$

67. Solve $\frac{1}{2}y^2 = 32$ by factoring.

Solution

$$\frac{1}{2}y^2 = 32$$
$$y^2 = 64$$
$$y^2 - 64 = 0$$
$$(y - 8)(y + 8) = 0$$
$$y - 8 = 0 \qquad y + 8 = 0$$
$$y = 8 \qquad y = -8$$

69. Solve $25x^2 = 16$ by extracting square roots.

Solution

$$25x^2 = 16$$
$$x^2 = \frac{16}{25}$$
$$x = \pm\sqrt{\frac{16}{25}}$$
$$x = \pm\frac{4}{5}$$

71. Solve $4u^2 - 225 = 0$ by extracting square roots.

Solution

$$4u^2 - 225 = 0$$
$$u^2 = \frac{225}{4}$$
$$u = \pm\sqrt{\frac{225}{4}}$$
$$u = \pm\frac{15}{2}$$

73. Solve $(x - 3)^2 = 0.25$ by extracting square roots.

Solution

$$(x - 3)^2 = 0.25$$
$$x - 3 = \pm\sqrt{0.25}$$
$$x = 3 \pm 0.5$$
$$x = 3.5, 2.5$$

75. Solve $(x - 2)^2 = 7$ by extracting square roots.

Solution

$$(x - 2)^2 = 7$$
$$x - 2 = \pm\sqrt{7}$$
$$x = 2 \pm \sqrt{7}$$

77. Solve $(4x - 3)^2 - 98 = 0$ by extracting square roots.

Solution

$$(4x - 3)^2 - 98 = 0$$
$$(4x - 3)^2 = 98$$
$$4x - 3 = \pm\sqrt{98}$$
$$4x = 3 \pm 7\sqrt{2}$$
$$x = \frac{3 \pm 7\sqrt{2}}{4}$$

79. Solve $u^2 + 17 = 0$ by extracting complex square roots.

Solution

$$u^2 + 17 = 0$$
$$u^2 = -17$$
$$u = \pm\sqrt{-17}$$
$$= \pm i\sqrt{17}$$

81. Solve $(t - 3)^2 = -25$ by extracting complex square roots.

Solution

$$(t - 3)^2 = -25$$
$$t - 3 = \pm\sqrt{-25}$$
$$t = 3 \pm 5i$$

83. Solve $(2y - 3)^2 + 25 = 0$ by extracting complex square roots.

Solution

$$(2y - 3)^2 + 25 = 0$$

$$(2y - 3)^2 = -25$$

$$2y - 3 = \pm\sqrt{-25}$$

$$2y = 3 \pm 5i$$

$$y = \frac{3}{2} \pm \frac{5}{2}i$$

85. Solve $\left(c - \frac{2}{3}\right)^2 + \frac{1}{9} = 0$ by extracting square roots.

Solution

$$\left(c - \frac{2}{3}\right)^2 + \frac{1}{9} = 0$$

$$\left(c - \frac{2}{3}\right)^2 = -\frac{1}{9}$$

$$c - \frac{2}{3} = \pm\sqrt{-\frac{1}{9}}$$

$$c = \frac{2}{3} \pm \frac{1}{3}i$$

87. Solve $\left(x + \frac{7}{3}\right)^2 = -\frac{38}{9}$ by extracting square roots.

Solution

$$\left(x + \frac{7}{3}\right)^2 = -\frac{38}{9}$$

$$x + \frac{7}{3} = \pm\sqrt{-\frac{38}{9}}$$

$$x = -\frac{7}{3} \pm \frac{i}{3}\sqrt{38}$$

89. Find all the real and complex solutions.

$$x^2 - 900 = 0$$

Solution

$$x^2 - 900 = 0$$

$$x^2 = 900$$

$$x = \pm 30$$

91. Find all the real and complex solutions.

$$x^2 + 900 = 0$$

Solution

$$x^2 + 900 = 0$$

$$x^2 = -900$$

$$x = \pm\sqrt{-900}$$

$$x = \pm 30i$$

93. Find all the real and complex solutions.

$$\frac{2}{3}x^2 = 6$$

Solution

$$\frac{2}{3}x^2 = 6$$

$$\frac{3}{2} \cdot \frac{2}{3}x^2 = 6 \cdot \frac{3}{2}$$

$$x^2 = 9$$

$$x = \pm 3$$

95. Find all the real and complex solutions.

$$(x - 5)^2 - 100 = 0$$

Solution

$$(x - 5)^2 - 100 = 0$$

$$(x - 5)^2 = 100$$

$$x - 5 = \pm 10$$

$$x = 15, -5$$

97. Find all the real and complex solutions.

$$(x - 5)^2 + 100 = 0$$

Solution

$$(x - 5)^2 + 100 = 0$$

$$\text{let } u = (x - 5)$$

$$u^2 + 100 = 0$$

$$(u + 10i)(u - 10i) = 0$$

$$u + 10i = 0 \qquad u - 10i = 0$$

$$u = -10i \qquad u = 10i$$

$$x - 5 = -10i \qquad x - 5 = 10i$$

$$x = 5 - 10i \qquad x = 5 + 10i$$

99. Solve $x^4 - 5x^2 + 4 = 0$. List all the real and complex solutions.

Solution

$$x^4 - 5x^2 + 4 = 0$$

$$\left(x^2 - 4\right)\left(x^2 - 1\right) = 0$$

$$(x - 2)(x + 2)(x - 1)(x + 1) = 0$$

$$x - 2 = 0 \qquad x + 2 = 0 \qquad x - 1 = 0 \qquad x + 1 = 0$$

$$x = 2 \qquad x = -2 \qquad x = 1 \qquad x = -1$$

101. Solve $x^4 - 5x^2 + 6 = 0$.
List all real and complex solutions.

Solution

$$x^4 - 5x^2 + 6 = 0$$

$$\left(x^2 - 3\right)\left(x^2 - 2\right) = 0$$

$$x^2 - 3 = 0 \qquad x^2 - 2 = 0$$

$$x^2 = 3 \qquad x^2 = 2$$

$$x = \pm\sqrt{3} \qquad x = \pm\sqrt{2}$$

103. Solve $x^4 - 3x^2 - 4 = 0$.
List all real and complex solutions.

Solution

$$x^4 - 3x^2 - 4 = 0$$

$$\left(x^2 - 4\right)\left(x^2 + 1\right) = 0$$

$$x^2 - 4 = 0 \qquad x^2 + 1 = 0$$

$$x^2 = 4 \qquad x^2 = -1$$

$$x = \pm 2 \qquad x = \pm i$$

105. Solve $\left(x^2 - 4\right)^2 + 2\left(x^2 - 4\right) - 3 = 0$.
List all real and complex solutions.

Solution

$$\left(x^2 - 4\right)^2 + 2\left(x^2 - 4\right) - 3 = 0$$

$$\left[\left(x^2 - 4\right) + 3\right]\left[\left(x^2 - 4\right) - 1\right] = 0$$

$$\left(x^2 - 1\right)\left(x^2 - 5\right) = 0$$

$$x^2 - 1 = 0 \qquad x^2 - 5 = 0$$

$$x^2 = 1 \qquad x^2 = 5$$

$$x = \pm 1 \qquad x = \pm\sqrt{5}$$

107. Solve $\left(\sqrt{x} - 1\right)^2 + 3\left(\sqrt{x} - 1\right) - 4 = 0$.
List all real and complex solutions.

Solution

$$\left(\sqrt{x} - 1\right)^2 + 3\left(\sqrt{x} - 1\right) - 4 = 0$$

$$\left[\left(\sqrt{x} - 1\right) + 4\right]\left[\left(\sqrt{x} - 1\right) - 1\right] = 0$$

$$\left(\sqrt{x} + 3\right)\left(\sqrt{x} - 2\right) = 0$$

$$\sqrt{x} + 3 = 0 \qquad \sqrt{x} - 2 = 0$$

$$\sqrt{x} = -3 \qquad \sqrt{x} = 2$$

$$\text{no solution} \qquad x = 4$$

109. Solve $\dfrac{1}{x^2} - \dfrac{3}{x} + 2 = 0$. List all real and complex solutions.

Solution

$$\frac{1}{x^2} - \frac{3}{x} + 2 = 0$$

$$1 - 3x + 2x^2 = 0$$

$$2x^2 - 3x + 1 = 0$$

$$(2x - 1)(x - 1) = 0$$

$$2x - 1 = 0 \qquad x - 1 = 0$$

$$x = \frac{1}{2} \qquad x = 1$$

111. Solve $3\left(\dfrac{x}{x+1}\right)^2 + 7\left(\dfrac{x}{x+1}\right) - 6 = 0$. List all real and complex solutions.

Solution

$$3\left(\frac{x}{x+1}\right)^2 + 7\left(\frac{x}{x+1}\right) - 6 = 0$$

$$\left[3\left(\frac{x}{x+1}\right) - 2\right]\left[\left(\frac{x}{x+1}\right) + 3\right] = 0$$

$$\frac{3x}{x+1} - 2 = 0 \qquad \frac{x}{x+1} + 3 = 0$$

$$\frac{3x}{x+1} = 2 \qquad \frac{x}{x+1} = -3$$

$$3x = 2x + 2 \qquad x = -3x - 3$$

$$x = 2 \qquad 4x = -3$$

$$x = -\frac{3}{4}$$

113. Solve $x^{2/3} - x^{1/3} - 6 = 0$. List all real and complex solutions.

Solution

$$x^{2/3} - x^{1/3} - 6 = 0$$

$$\left(x^{1/3} - 3\right)\left(x^{1/3} + 2\right) = 0$$

$$x^{1/3} - 3 = 0 \qquad x^{1/3} + 2 = 0$$

$$x^{1/3} = 3 \qquad\qquad x^{1/3} = -2$$

$$x = 27 \qquad\qquad x = -8$$

115. Use a graphing utility to graph the function $y = 5x - x^2$. Use the graph to approximate any x-intercepts of the graph. Set $y = 0$ and solve the resulting equation. Compare the result with the x-intercepts of the graph.

Solution

Keystrokes:

$\boxed{Y=}\ 5\ \boxed{X, T, \theta}\ \boxed{-}\ \boxed{X, T, \theta}\ \boxed{x^2}\ \boxed{\text{GRAPH}}$

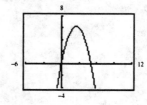

$$0 = 5x - x^2$$

$$0 = x(5 - x)$$

$$0 = x \qquad 5 - x = 0$$

$$5 = x$$

x-intercepts are 0 and 5.

117. Use a graphing utility to graph the function $y = (x - 2)^2 + 3$ and observe that the graph has no x-intercepts. Set $y = 0$ and solve the resulting equation. What type of roots does the equation have?

Solution

Keystrokes:

$\boxed{Y=}\ \boxed{(}\ \boxed{X, T, \theta}\ \boxed{-}\ 2\ \boxed{)}\ \boxed{x^2}\ \boxed{+}\ 3\ \boxed{\text{GRAPH}}$

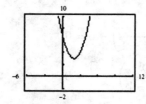

$$0 = (x - 2)^2 + 3$$

$$-3 = (x - 2)^2$$

$$\pm\sqrt{-3} = x - 2$$

$$2 \pm \sqrt{3}i = x$$

not real, therefore, there are no x-intercepts

119. Find a quadratic equation having the solutions 5, −2.

Solution

$$(x - 5)(x - (-2)) = 0$$
$$(x - 5)(x + 2) = 0$$
$$x^2 - 5x + 2x - 10 = 0$$
$$x^2 - 3x - 10 = 0$$

121. Find a quadratic equation having the solutions $1 + \sqrt{2}$, $1 - \sqrt{2}$.

Solution

$$\left[x - \left(1 + \sqrt{2}\right)\right]\left[x - \left(1 - \sqrt{2}\right)\right] = 0$$
$$\left[(x - 1) - \sqrt{2}\right]\left[(x - 1) + \sqrt{2}\right] = 0$$
$$(x - 1)^2 - \left(\sqrt{2}\right)^2 = 0$$
$$x^2 - 2x + 1 - 2 = 0$$
$$x^2 - 2x - 1 = 0$$

123. *Free-Falling Object* Find the time required for an object to reach the ground when it is dropped from a height of s_0 feet. The height h (in feet) is given by $h = -16t^2 + s_0$, where t measures time in seconds from the time the object is released.

$$s_0 = 256$$

Solution

$$0 = -16t^2 + 256$$
$$16t^2 = 256$$
$$t^2 = 16$$
$$t = 4 \text{ sec}$$

125. *Free-Falling Object* The height h (in feet) of an object thrown upward from a tower 144 feet high (see figure in textbook) is given by $h = 144 + 128t - 16t^2$ where t measures the time in seconds from the time the object is released. How long does it take for the object to reach the ground?

Solution

$$0 = 144 + 128 - 16t^2$$
$$0 = -16t^2 + 128t + 144$$
$$0 = -16\left(t^2 - 8t - 9\right)$$
$$0 = -16(t - 9)(t + 1)$$
$$t - 9 = 0 \qquad t + 1 = 0$$
$$t = 9 \qquad \cancel{t = 1}$$

127. *Reading a Graph* Use the following model which gives the federal funding for health research in the United States from 1985 to 1992.

$$y = (52.9 + 3.9t)^2, \; 5 \le t \le 12$$

In this model, y represents funding (in millions of dollars) and $t = 0$ represents 1980 (see figure in textbook). (*Source:* U.S. National Science Foundation)

Use the graph to find the year in which federal funding for health was approximately $6500 million. Verify algebraically.

Solution

From the graph, $t = 7$, so the year is 1987. $6500 \overset{?}{=} [52.9 + 3.9(7)]^2$

$$6500 = [52.9 + 27.3]^2$$

$$6500 = [80.2]^2$$

$$6500 \approx 6432.04$$

129. *Compound Interest* A principal of $1500 is deposited in an account at an annual interest rate r compounded annually. If the amount after 2 years is $1685.40, the annual interest rate is the solution to the equation

$$1685.40 = 1500(1 + r)^2. \; \text{Find } r.$$

Solution

$$1685.40 = 1500(1 + r)^2$$

$$1.1236 = (1 + r)^2$$

$$1.06 = 1 + r$$

$$.06 = r$$

$$6\% = r$$

131. *Free-Falling Object* Find the time required for an object to reach the ground when it is dropped from a height of s_0 feet. The height h (in feet) is given by $h = -16t^2 + s_0$ where t measures time in seconds from the time the object is released.

$$s_0 = 500$$

Solution

$$0 = -16t^2 + 500$$

$$16t^2 = 500$$

$$t^2 = 31.25$$

$$t = \sqrt{31.25} \approx 5.59 \text{ sec}$$

6.2 Completing the Square

7. Find the term that must be added to the expression so that it becomes a perfect square trinomial.

$$x^2 + 8x+$$

Solution

$$x^2 + 8x + 16$$

9. Find the term that must be added to the expression so that it becomes a perfect square trinomial.

$$t^2 + 5t+$$

Solution

$$t^2 + 5t + \frac{25}{4}$$

11. Solve $x^2 - 6x = 0$ (a) by completing the square and (b) by factoring.

Solution

(a) $x^2 - 6x + 9 = 0 + 9$

$\qquad (x-3)^2 = 9$

$\qquad x - 3 = \pm 3$

$\qquad x = 3 \pm 3$

$\qquad x = 6, 0$

(b) $\quad x^2 - 6x = 0$

$\qquad x(x-6) = 0$

$\qquad x = 0 \qquad x - 6 = 0$

$\qquad\qquad\qquad\quad x = 6$

13. Solve $x^2 + 7x + 12 = 0$ (a) by completing the square and (b) by factoring.

Solution

(a) $x^2 + 7x + \dfrac{49}{4} = -12 + \dfrac{49}{4}$

$\qquad \left(x + \dfrac{7}{2}\right)^2 = \dfrac{1}{4}$

$\qquad x + \dfrac{7}{2} = \pm\dfrac{1}{2}$

$\qquad x = \dfrac{-7}{2} \pm \dfrac{1}{2}$

$\qquad x = \dfrac{-6}{2}, \dfrac{-8}{2}$

$\qquad x = -3, -4$

(b) $\quad x^2 + 7x + 12 = 0$

$\qquad (x+4)(x+3) = 0$

$\qquad x = -4 \qquad x = -3$

15. Solve $x^2 - 4x - 3 = 0$ by completing the square. Give the solutions in exact form and in decimal form rounded to two decimal places. (The solutions may be complex numbers.)

Solution

$$x^2 - 4x - 3 = 0$$

$$x^2 - 4x + 4 = 3 + 4$$

$$(x-2)^2 = 7$$

$$x - 2 = \pm\sqrt{7}$$

$$x = 2 \pm \sqrt{7}$$

$$x = 4.65, -0.65$$

17. Solve $x^2 + 2x + 3 = 0$ by completing the square. Give the solutions in exact form and in decimal form rounded to two decimal places. (The solutions may be complex numbers.)

Solution

$$x^2 + 2x + 3 = 0$$

$$x^2 + 2x + 1 = -3 + 1$$

$$(x+1)^2 = -2$$

$$x + 1 = \pm\sqrt{-2}$$

$$x = -1 \pm i\sqrt{2}$$

19. Solve $x^2 - \dfrac{2}{3}x - 3 = 0$ by completing the square. Give the solutions in exact form and in decimal form rounded to two decimal places. (The solutions may be complex numbers.)

Solution

$$x^2 - \frac{2}{3}x - 3 = 0$$

$$x^2 - \frac{2}{3}x + \frac{1}{9} = 3 + \frac{1}{9}$$

$$\left(x - \frac{1}{3}\right)^2 = \frac{28}{9}$$

$$x - \frac{1}{3} = \pm\sqrt{\frac{28}{9}}$$

$$x = \frac{1}{3} \pm \frac{2}{3}\sqrt{7}$$

$$x = \frac{1 \pm 2\sqrt{7}}{3}$$

$$x = 2.10, -1.43$$

21. Solve $t^2 + 5t + 3 = 0$ by completing the square. Give the solutions in exact form and in decimal form rounded to two decimal places. (The solutions may be complex numbers.)

Solution

$$t^2 + 5t + 3 = 0$$

$$t^2 + 5t + \frac{25}{4} = -3 + \frac{25}{4}$$

$$\left(t + \frac{5}{2}\right)^2 = \frac{13}{4}$$

$$t + \frac{5}{2} = \pm\sqrt{\frac{13}{4}}$$

$$t = -\frac{5}{2} \pm \frac{\sqrt{13}}{2}$$

$$t = \frac{-5 \pm \sqrt{13}}{2}$$

$$t = -0.70, -4.30$$

23. Solve $2x^2 + 8x + 3 = 0$ by completing the square. Give the solutions in exact form and in decimal form rounded to two decimal places. (The solutions may be complex numbers.)

Solution

$$2x^2 + 8x + 3 = 0$$

$$x^2 + 4x + 4 = -\frac{3}{2} + 4$$

$$(x + 2)^2 = \frac{5}{2}$$

$$x + 2 = \pm\sqrt{\frac{5}{2}} \cdot \frac{\sqrt{2}}{\sqrt{2}}$$

$$x = -2 \pm \frac{\sqrt{10}}{2}$$

$$x = -0.42, -3.58$$

25. Solve $0.1x^2 + 0.5x + 0.2 = 0$ by completing the square. Give the solutions in exact form and in decimal form rounded to two decimal places. (The solutions may be complex numbers.)

Solution

$$0.1x^2 + 0.5x + 0.2 = 0$$

$$x^2 + 5x + 2 = 0$$

$$x^2 + 5x + \frac{25}{4} = -2 + \frac{25}{4}$$

$$\left(x + \frac{5}{2}\right)^2 = \frac{-8 + 25}{4}$$

$$\left(x + \frac{5}{2}\right)^2 = \frac{17}{4}$$

$$x + \frac{5}{2} = \pm\sqrt{\frac{17}{4}}$$

$$x = -\frac{5}{2} \pm \frac{\sqrt{17}}{2}$$

$$x = \frac{-5 \pm \sqrt{17}}{2}$$

$$x = -0.44, -4.56$$

27. Find the real solutions to $\dfrac{x}{2} - \dfrac{1}{x} = 1$.

Solution

$$\frac{x}{2} - \frac{1}{x} = 1$$

$$2x\left[\frac{x}{2} - \frac{1}{x} = 1\right]2x$$

$$x^2 - 2 = 2x$$

$$x^2 - 2x + 1 = 2 + 1$$

$$(x - 1)^2 = 3$$

$$x - 1 = \pm\sqrt{3}$$

$$x = 1 \pm \sqrt{3}$$

29. Find the real solutions to $\sqrt{2x + 1} = x - 3$.

Solution

$$\sqrt{2x + 1} = x - 3$$

$$\left(\sqrt{2x + 1}\right)^2 = (x - 3)^2$$

$$2x + 1 = x^2 - 6x + 9$$

$$0 = x^2 - 8x + 8$$

$$+16 - 8 = x^2 - 8x + 16$$

$$8 = (x - 4)^2$$

$$\pm\sqrt{8} = x - 4$$

$$4 \pm \sqrt{8} = x$$

31. Use a graphing utility to approximate any x-intercepts of the graph of $y = x^2 + 4x - 1$. Set $y = 0$ and solve the resulting equation. Compare the result with the x-intercepts of the graph.

Solution

Keystrokes:

$\boxed{Y=}$ $\boxed{X,T,\theta}$ $\boxed{x^2}$ $\boxed{+}$ 4 $\boxed{X,T,\theta}$ $\boxed{-}$ 1 $\boxed{\text{GRAPH}}$

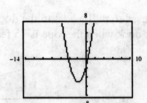

$$0 = x^2 + 4x - 1$$

$$1 = x^2 + 4x$$

$$1 + 4 = x^2 + 4x + 4$$

$$5 = (x + 2)^2$$

$$\pm\sqrt{5} = x + 2$$

$$-2 \pm \sqrt{5} = x$$

$$x = .236$$

$$x = -4.236$$

33. Use a graphing utility to approximate any x-intercepts of the graph of $y = x^2 - 2x - 5$. Set $y = 0$ and solve the resulting equation. Compare the result with the x-intercepts of the graph.

Solution

Keystrokes:

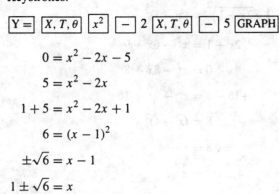

$$0 = x^2 - 2x - 5$$

$$5 = x^2 - 2x$$

$$1 + 5 = x^2 - 2x + 1$$

$$6 = (x - 1)^2$$

$$\pm\sqrt{6} = x - 1$$

$$1 \pm \sqrt{6} = x$$

$$x = 3.449$$

$$x = -1.449$$

35. Consider a windlass that is used to pull a boat to the dock (see figure in textbook). Find the distance from the boat to the dock when the rope is 75 feet long.

Solution

$$x^2 + 15^2 = 75^2$$

$$x^2 + 225 = 5625$$

$$x^2 = 5400$$

$$x = \sqrt{5400} = 30\sqrt{6} \approx 73.48 \text{ feet}$$

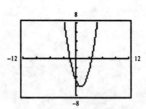

37. Simplify $\sqrt[3]{16x^4 y^3}$.

Solution

$$\sqrt[3]{16x^4 y^3} = \sqrt[3]{8 \cdot 2x^3 \cdot x \cdot y^3}$$

$$= 2xy\sqrt[3]{2x}$$

39. Simplify $\sqrt{3x^2} - 2\sqrt{12}$.

Solution

$$\sqrt{3x^2} - 2\sqrt{12} = |x|\sqrt{3} - 2\sqrt{4 \cdot 3}$$

$$= |x|\sqrt{3} - 4\sqrt{3}$$

$$= \sqrt{3}\,(|x| - 4)$$

41. *Time* A jogger starts running at 6 miles per hour. Five minutes later, another jogger starts on the same trail, running at 8 miles per hour. When will the second jogger overtake the first?

Solution

Verbal model: $\boxed{\text{Rate}} \cdot \boxed{\text{Time}} = \boxed{\text{Distance}}$

Equation: $\qquad 6(x + 5) = 8x$

$$6x + 30 = 8x$$

$$30 = 2x$$

$$15 \text{ minutes} = x$$

43. Find the term that must be added to the expression so that it becomes a perfect square trinomial.

$$y^2 - 20y+$$

Solution

$$y^2 - 20y + 100$$

45. Find the term that must be added to the expression so that it becomes a perfect square trinomial.

$$x^2 - \frac{6}{5}x+$$

Solution

$$x^2 - \frac{6}{5}x + \frac{9}{25}$$

47. Find the term that must be added to the expression so that it becomes a perfect square trinomial.

$$y^2 - \frac{3}{5}y+$$

Solution

$$y^2 - \frac{3}{5}y + \frac{9}{100}$$

49. Find the term that must be added to the expression so that it becomes a perfect square trinomial.

$$r^2 - 0.4r+$$

Solution

$$r^2 - 0.4r + 0.04$$

51. Solve $x^2 - 25x = 0$ (a) by completing the square and (b) by factoring.

Solution

(a) $x^2 - 25x + \dfrac{625}{4} = \dfrac{625}{4}$

$$\left(x - \frac{25}{2}\right)^2 = \frac{625}{4}$$

$$x - \frac{25}{2} = \pm\frac{25}{2}$$

$$x = \frac{25}{2} \pm \frac{25}{2}$$

$$x = 25, 0$$

(b) $x^2 - 25x = 0$

$$x(x - 25) = 0$$

$$x = 0 \qquad x = 25$$

53. Solve $t^2 - 8t + 7 = 0$ (a) by completing the square and (b) by factoring.

Solution

(a) $t^2 - 8t + 16 = -7 + 16$

$$(t - 4)^2 = 9$$

$$t - 4 = \pm 3$$

$$t = 4 \pm 3$$

$$t = 7, 1$$

(b) $t^2 - 8t + 7 = 0$

$$(t - 7)(t - 1) = 0$$

$$t = 7 \qquad t = 1$$

55. Solve $x^2 + 2x - 24 = 0$ (a) by completing the square and (b) by factoring.

Solution

(a) $x^2 + 2x + 1 = 24 + 1$

$\quad (x + 1)^2 = 25$

$\quad\quad x + 1 = \pm 5$

$\quad\quad\quad\quad x = -1 \pm 5$

$\quad\quad\quad\quad x = 4, -6$

(b) $\quad x^2 + 2x - 24 = 0$

$\quad\quad (x + 6)(x - 4) = 0$

$\quad\quad x = -6 \quad\quad x = 4$

57. Solve $x^2 - 3x - 18 = 0$ (a) by completing the square and (b) by factoring.

Solution

(a) $x^2 - 3x + \dfrac{9}{4} = 18 + \dfrac{9}{4}$

$\quad \left(x - \dfrac{3}{2}\right)^2 = \dfrac{81}{4}$

$\quad\quad x - \dfrac{3}{2} = \pm\dfrac{9}{2}$

$\quad\quad\quad x = \dfrac{3}{2} \pm \dfrac{9}{2}$

$\quad\quad\quad x = \dfrac{12}{2}, -\dfrac{6}{2}$

$\quad\quad\quad x = 6, -3$

(b) $\quad x^2 - 3x - 18 = 0$

$\quad\quad (x - 6)(x + 3) = 0$

$\quad\quad x = 6 \quad\quad x = -3$

59. Solve $2x^2 - 11x + 12 = 0$ (a) by completing the square and (b) by factoring.

Solution

(a) $\quad 2x^2 - 11x + 12 = 0$

$\quad\quad x^2 - \dfrac{11}{2}x = -6$

$\quad x^2 - \dfrac{11}{2}x + \dfrac{121}{16} = -6 + \dfrac{121}{16}$

$\quad\quad \left(x - \dfrac{11}{4}\right)^2 = -\dfrac{96}{16} + \dfrac{121}{16}$

$\quad\quad \left(x - \dfrac{11}{4}\right)^2 = \dfrac{25}{16}$

$\quad\quad\quad x - \dfrac{11}{4} = \pm\dfrac{5}{4}$

$\quad\quad\quad\quad x = \dfrac{11}{4} \pm \dfrac{5}{4}$

$\quad\quad\quad\quad x = \dfrac{16}{4}, \dfrac{6}{4}$

$\quad\quad\quad\quad x = 4, \dfrac{3}{2}$

(b) $2x^2 - 11x + 12 = 0$

$\quad (2x - 3)(x - 4) = 0$

$\quad 2x - 3 = 0 \quad\quad x - 4 = 0$

$\quad\quad x = \dfrac{3}{2} \quad\quad\quad x = 4$

61. Solve $x^2 + 4x - 3 = 0$ by completing the square. Give the solutions in exact form and in decimal form rounded to two decimal places. (The solutions may be complex numbers.)

Solution

$x^2 + 4x - 3 = 0$

$x^2 + 4x + 4 = 3 + 4$

$(x + 2)^2 = 7$

$x + 2 = \pm\sqrt{7}$

$x = -2 \pm \sqrt{7}$

$x = 0.65, -4.65$

63. Solve $u^2 - 4u + 1 = 0$ by completing the square. Give the solutions in exact form and in decimal form rounded to two decimal places. (The solutions may be complex numbers.)

Solution

$u^2 - 4u + 1 = 0$

$u^2 - 4u + 4 = -1 + 4$

$(u - 2)^2 = 3$

$u - 2 = \pm\sqrt{3}$

$u = 2 \pm \sqrt{3}$

$u = 3.73, 0.27$

65. Solve $x^2 - 10x - 2 = 0$ by completing the square. Give the solutions in exact form and in decimal form rounded to two decimal places. (The solutions may be complex numbers.)

Solution

$x^2 - 10x - 2 = 0$

$x^2 - 10x + 25 = 2 + 25$

$(x - 5)^2 = 27$

$x - 5 = \pm\sqrt{27}$

$x = 5 \pm 3\sqrt{3}$

$x = 10.20, 0.20$

67. Solve $y^2 + 20y + 10 = 0$ by completing the square. Give the solutions in exact form and in decimal form rounded to two decimal places. (The solutions may be complex numbers.)

Solution

$y^2 + 20y + 10 = 0$

$y^2 + 20y + 100 = -10 + 100$

$(y + 10)^2 = 90$

$y + 10 = \pm\sqrt{90}$

$y = -10 \pm 3\sqrt{10}$

$y = -0.51, -19.49$

69. Solve $v^2 + 3v - 2 = 0$ by completing the square. Give the solutions in exact form and in decimal form rounded to two decimal places. (The solutions may be complex numbers.)

Solution

$v^2 + 3v - 2 = 0$

$v^2 + 3v + \dfrac{9}{4} = 2 + \dfrac{9}{4}$

$\left(v + \dfrac{3}{2}\right)^2 = \dfrac{17}{4}$

$v + \dfrac{3}{2} = \pm\sqrt{\dfrac{17}{4}}$

$v = -\dfrac{3}{2} \pm \dfrac{\sqrt{17}}{2}$

$v = \dfrac{-3 \pm \sqrt{17}}{2}$

$v = 0.56, -3.56$

71. Solve $-x^2 + x - 1 = 0$ by completing the square. Give the solutions in exact form and in decimal form rounded to two decimal places. (The solutions may be complex numbers.)

Solution

$-x^2 + x - 1 = 0$

$x^2 - x + 1 = 0$

$x^2 - x + \dfrac{1}{4} = -1 + \dfrac{1}{4}$

$\left(x - \dfrac{1}{2}\right)^2 = -\dfrac{3}{4}$

$x - \dfrac{1}{2} = \pm\sqrt{-\dfrac{3}{4}}$

$x = \dfrac{1}{2} \pm \dfrac{i\sqrt{3}}{2}$

$x = \dfrac{1 \pm i\sqrt{3}}{2}$

73. Solve $3x^2 + 9x + 5 = 0$ by completing the square. Give the solutions in exact form and in decimal form rounded to two decimal places. (The solutions may be complex numbers.)

Solution

$3x^2 + 9x + 5 = 0$

$x^2 + 3x + \dfrac{9}{4} = -\dfrac{5}{3} + \dfrac{9}{4}$

$\left(x + \dfrac{3}{2}\right)^2 = \dfrac{-20 + 27}{12}$

$\left(x + \dfrac{3}{2}\right)^2 = \dfrac{7}{12}$

$x + \dfrac{3}{2} = \pm\sqrt{\dfrac{7}{12}} \cdot \dfrac{\sqrt{3}}{\sqrt{3}}$

$x = -\dfrac{3}{2} \pm \dfrac{\sqrt{21}}{6}$

$x = \dfrac{-9 \pm \sqrt{21}}{6}$

$x = -0.74, -2.26$

75. Solve $4y^2 + 4y - 9 = 0$ by completing the square. Give the solutions in exact form and in decimal form rounded to two decimal places. (The solutions may be complex numbers.)

Solution

$4y^2 + 4y - 9 = 0$

$y^2 + y + \dfrac{1}{4} = \dfrac{9}{4} + \dfrac{1}{4}$

$\left(y + \dfrac{1}{2}\right)^2 = \dfrac{10}{4}$

$y + \dfrac{1}{2} = \pm\sqrt{\dfrac{10}{4}}$

$y = -\dfrac{1}{2} \pm \dfrac{\sqrt{10}}{2}$

$y = \dfrac{-1 \pm \sqrt{10}}{2}$

$y = 1.08, -2.08$

77. Solve $0.1x^2 + 0.2x + 0.5 = 0$ by completing the square. Give the solutions in exact form and in decimal form rounded to two decimal places. (The solutions may be complex numbers.)

Solution

$0.1x^2 + 0.2x + 0.5 = 0$

$x^2 + 2x + 5 = 0$

$x^2 + 2x + 1 = -5 + 1$

$(x + 1)^2 = -4$

$x + 1 = \pm\sqrt{-4}$

$x = -1 \pm 2i$

79. Solve $x(x - 7) = 2$ by completing the square. Give the solutions in exact form and in decimal form rounded to two decimal places. (The solutions may be complex numbers.)

Solution

$x(x - 7) = 2$

$x^2 - 7x + \dfrac{49}{4} = 2 + \dfrac{49}{4}$

$\left(x - \dfrac{7}{2}\right)^2 = \dfrac{8 + 49}{4}$

$\left(x - \dfrac{7}{2}\right)^2 = \dfrac{57}{4}$

$x - \dfrac{7}{2} = \pm\sqrt{\dfrac{57}{4}}$

$x = \dfrac{7}{2} \pm \dfrac{\sqrt{57}}{2}$

$x = \dfrac{7 \pm \sqrt{57}}{2}$

$x = 7.27, -0.27$

81. Use a graphing utility to approximate any x-intercepts of the graph. Set $y = 0$ and solve the resulting equation. Compare the result with the x-intercepts of the graph.

$$\frac{1}{3}x^2 + 2x - 6 = 0$$

Solution

Keystrokes:

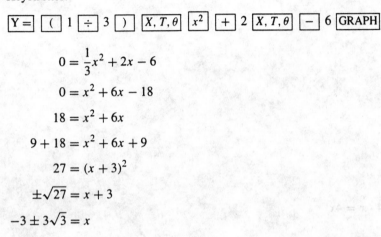

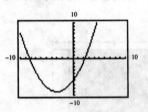

$$0 = \frac{1}{3}x^2 + 2x - 6$$

$$0 = x^2 + 6x - 18$$

$$18 = x^2 + 6x$$

$$9 + 18 = x^2 + 6x + 9$$

$$27 = (x + 3)^2$$

$$\pm\sqrt{27} = x + 3$$

$$-3 \pm 3\sqrt{3} = x$$

$$x = 2.20$$

$$x = -8.20$$

83. Use a graphing utility to approximate any x-intercepts of the graph. Set $y = 0$ and solve the resulting equation. Compare the result with the x-intercepts of the graph.

$$\frac{3}{x} - x - 1 = 0$$

Solution

Keystrokes:

$\boxed{Y=}$ 3 $\boxed{\div}$ $\boxed{X, T, \theta}$ $\boxed{-}$ $\boxed{X, T, \theta}$ $\boxed{-}$ 1 $\boxed{GRAPH}$

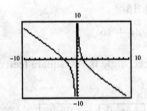

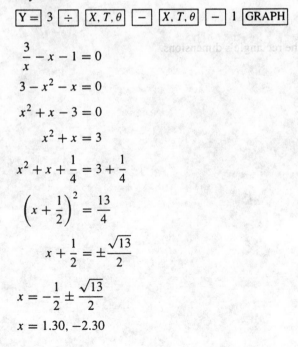

$$\frac{3}{x} - x - 1 = 0$$

$$3 - x^2 - x = 0$$

$$x^2 + x - 3 = 0$$

$$x^2 + x = 3$$

$$x^2 + x + \frac{1}{4} = 3 + \frac{1}{4}$$

$$\left(x + \frac{1}{2}\right)^2 = \frac{13}{4}$$

$$x + \frac{1}{2} = \pm\frac{\sqrt{13}}{2}$$

$$x = -\frac{1}{2} \pm \frac{\sqrt{13}}{2}$$

$$x = 1.30, -2.30$$

85. *Geometrical Modeling*

(a) Find the area of the two adjoining rectangles and the square (see figure in textbook).

(b) Find the area of the small square in the lower right-hand corner of the figure and add it to the area found in part (a).

(c) Find the dimensions and the area of the entire figure after adjoining the small square in the lower right-hand corner. Note that you have shown completing the square geometrically.

Solution

(a)

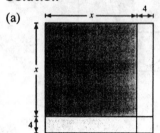

Area of square $= x \cdot x = x^2$

Area of 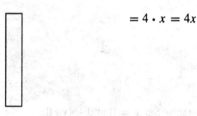 $= 4 \cdot x = 4x$

Area of [ ] $= 4 \cdot x = 4x$

Total area $= x^2 + 4x + 4x = x^2 + 8x$

(b) Area of small square $= 4 \cdot 4 = 16$

Total area $= x^2 + 8x + 16$

(c) $(x + 4)(x + 4) = x^2 + 8x + 16$

87. *Geometry* The area of the rectangle is 160 square feet. Find the rectangle's dimensions.

Solution

Verbal model: [Area] $=$ [Length] $\cdot$ [Width]

Equation:

$$160 = x\left(\frac{1}{4}x + 3\right)$$

$$160 = \frac{1}{4}x^2 + 3x$$

$$0 = x^2 + 12x - 640$$

$$0 = (x + 32)(x - 20)$$

$$x + 32 = 0 \qquad x - 20 = 0$$

$$\cancel{x = -32} \qquad\qquad x = 20 \text{ length}$$

$$\frac{1}{4}x + 3 = 8 \text{ width}$$

89. *Revenue* The revenue R from selling x units of a certain product is given by

$$R = x\left(50 - \frac{1}{2}x\right).$$

Find the number of units that must be sold to produce a revenue of $1218.

Solution

$$1218 = x\left(50 - \frac{1}{2}x\right)$$

$$1218 = 50x - \frac{1}{2}x^2$$

$$\frac{1}{2}x^2 - 50x = -1218$$

$$x^2 - 100x + 2500 = -2436 + 2500$$

$$(x - 50)^2 = 64$$

$$x - 50 = \pm 8$$

$$x = 50 \pm 8$$

$$x = 58, 42$$

91. *Fencing in a Corral* You have 200 feet of fencing to enclose two adjacent rectangular corrals (see figure in textbook). The total area of the enclosed region is 1400 square feet. What are the dimensions of each corral? (The corrals are the same size.)

Solution

Verbal model: | Area | = | Length | · | Width |

Equation:

$$1400 = 2\left[x \cdot \left(\frac{200 - 4x}{3}\right)\right]$$

$$1400 = 2\left[\frac{200}{3}x - \frac{4x^2}{3}\right]$$

$$1400 = \frac{400x}{3} - \frac{8x^2}{3}$$

$$4200 = 400x - 8x^2$$

$$8x^2 - 400x + 4200 = 0$$

$$x^2 - 50x + 525 = 0$$

$$(x - 35)(x - 15) = 0$$

$$x - 35 = 0 \qquad\qquad x - 15 = 0$$

$$x = 35 \text{ ft.} \qquad\qquad x = 15 \text{ ft.}$$

$$\frac{200 - 4x}{3} = 20 \text{ ft.} \qquad \frac{200 - 4x}{3} = 46\frac{2}{3} \text{ ft}$$

93. *Geometry* A closed box has a square base with edge x and a height of 3 feet (see figure in textbook). The material for constructing the base costs \$1.50 per square foot and the material for the top and sides costs \$1 per square foot. Find the dimensions of the box if its volume is 128 cubic feet.

Solution

Verbal model: $\boxed{\text{Volume}} = \boxed{\text{Length}} \cdot \boxed{\text{Width}} \cdot \boxed{\text{Height}}$

Equation: $128 = x \cdot x \cdot 3$

$$\frac{128}{3} = x^2$$

$$\sqrt{\frac{128}{3}} = x$$

$$6.53 \text{ ft.} = x$$

6.3 The Quadratic Formula and the Discriminant

5. Write $2x^2 = 7 - 2x$ in standard form.

Solution

$$2x^2 = 7 - 2x$$

$$2x^2 + 2x - 7 = 0$$

7. Write $x(10 - x) = 5$ in standard form.

Solution

$$x(10 - x) = 5$$

$$10x - x^2 = 5$$

$$-x^2 + 10x - 5 = 0$$

$$x^2 - 10x + 5 = 0$$

9. Solve $x^2 - 11x + 28 = 0$ (a) by the Quadratic Formula and (b) by factoring.

Solution

(a) $x = \dfrac{11 \pm \sqrt{11^2 - 4(1)(28)}}{2(1)}$

$x = \dfrac{11 \pm \sqrt{121 - 112}}{2}$

$x = \dfrac{11 \pm \sqrt{9}}{2}$

$x = \dfrac{11 \pm 3}{2}$

$x = 7, 4$

(b) $(x - 7)(x - 4) = 0$

$x - 7 = 0 \qquad x - 4 = 0$

$x = 7 \qquad\quad x = 4$

11. Solve $4x^2 + 12x + 9 = 0$ (a) by the Quadratic Formula and (b) by factoring.

Solution

(a) $x = \dfrac{-12 \pm \sqrt{12^2 - 4(4)(9)}}{2(4)}$

$x = \dfrac{-12 \pm \sqrt{144 - 144}}{8}$

$x = \dfrac{12 \pm 0}{8}$

$x = -\dfrac{12}{8} = -\dfrac{3}{2}$

(b) $(2x + 3)(2x + 3) = 0$

$2x + 3 = 0 \qquad 2x + 3 = 0$

$x = -\dfrac{3}{2} \qquad\quad x = -\dfrac{3}{2}$

13. Use the discriminant to determine the type of solutions of $2x^2 - 5x - 4 = 0$.

Solution

$b^2 - 4ac = (-5)^2 - 4(2)(-4)$

$= 25 + 32$

$= 57$

2 irrational solutions

15. Use the discriminant to determine the type of solutions of $3x^2 - x + 2 = 0$.

Solution

$b^2 - 4ac = (-1)^2 - 4(3)(2)$

$= 1 - 24$

$= -23$

2 complex solutions

17. Use the Quadratic Formula to find all real or complex solutions to

$$x^2 - 2x - 4 = 0.$$

Solution

$x = \dfrac{-(-2) \pm \sqrt{(-2)^2 - 4(1)(-4)}}{2(1)}$

$x = \dfrac{2 \pm \sqrt{4 + 16}}{2}$

$x = \dfrac{2 \pm \sqrt{20}}{2}$

$x = \dfrac{2 \pm 2\sqrt{5}}{2}$

$x = \dfrac{2\left(1 \pm \sqrt{5}\right)}{2}$

$x = 1 \pm \sqrt{5}$

19. Use the Quadratic Formula to find all real or complex solutions to

$$t^2 + 4t + 1 = 0.$$

Solution

$t = \dfrac{-4 \pm \sqrt{4^2 - 4(1)(1)}}{2(1)}$

$t = \dfrac{-4 \pm \sqrt{16 - 4}}{2}$

$t = \dfrac{4 \pm \sqrt{12}}{2}$

$t = \dfrac{-4 \pm 2\sqrt{3}}{2}$

$t = \dfrac{2\left(-2 \pm \sqrt{3}\right)}{2}$

$t = -2 \pm \sqrt{3}$

21. Use the Quadratic Formula to find all real or complex solutions to

$$x^2 + 3x + 3 = 0.$$

Solution

$$x = \frac{-3 \pm \sqrt{3^2 - 4(1)(3)}}{2(1)}$$

$$x = \frac{-3 \pm \sqrt{9 - 12}}{2}$$

$$x = \frac{-3 \pm \sqrt{-3}}{2}$$

$$x = \frac{-3 \pm i\sqrt{3}}{2}$$

23. Use the Quadratic Formula to find all real or complex solutions to

$$9z^2 + 6z - 4 = 0.$$

Solution

$$z = \frac{-6 \pm \sqrt{6^2 - 4(9)(-4)}}{2(9)}$$

$$z = \frac{-6 \pm \sqrt{36 + 144}}{18}$$

$$z = \frac{-6 \pm \sqrt{180}}{18}$$

$$z = \frac{-6 \pm 6\sqrt{5}}{18}$$

$$z = \frac{6\left(-1 \pm \sqrt{5}\right)}{18}$$

$$z = \frac{-1 \pm \sqrt{5}}{3}$$

25. Use the Quadratic Formula to find all real or complex solutions to

$$2.5x^2 + x - 0.9 = 0.$$

Solution

$$x = \frac{-1 \pm \sqrt{1^2 - 4(2.5)(-0.9)}}{2(2.5)}$$

$$x = \frac{-1 \pm \sqrt{1 + 9}}{5}$$

$$x = \frac{-1 \pm \sqrt{10}}{5}$$

27. Solve $y^2 + 15y = 0$ by the most convenient method. Find all real or complex solutions.

Solution

$$y^2 + 15y = 0$$

$$y(y + 15) = 0$$

$$y = 0 \qquad y + 15 = 0$$

$$y = 0 \qquad\qquad y = -15$$

29. Solve $x^2 + 8x + 25 = 0$ by the most convenient method. Find all real or complex solutions.

Solution

$$x^2 + 8x + 25 = 0$$

$$x^2 + 8x + 16 = -25 + 16$$

$$(x + 4)^2 = -9$$

$$x + 4 = \pm\sqrt{-9}$$

$$x = -4 \pm 3i$$

31. *Graphical Reasoning* Use a graphing utility to graph the function $y = 3x^2 - 6x + 1$. Use the graph to approximate any x-intercepts of the graph. Set $y = 0$ and solve the resulting equation. Compare the result with the x-intercepts of the graph.

Solution

Keystrokes:

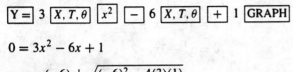

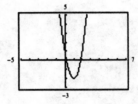

$$0 = 3x^2 - 6x + 1$$

$$x = \frac{-(-6) \pm \sqrt{(-6)^2 - 4(3)(1)}}{2(3)}$$

$$x = \frac{6 \pm \sqrt{36 - 12}}{6}$$

$$x = \frac{6 \pm \sqrt{24}}{6}$$

$$x = 1.82, 0.18$$

33. *Graphical Reasoning* Use a graphing utility to graph the function $y = -\left(4x^2 - 20x + 25\right)$. Use the graph to approximate any x-intercepts of the graph. Set $y = 0$ and solve the resulting equation. Compare the result with the x-intercepts of the graph.

Solution

Keystrokes:

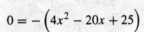

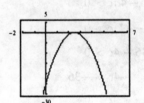

$$0 = -\left(4x^2 - 20x + 25\right)$$

$$= 4x^2 - 20x + 25$$

$$x = \frac{-(-20) \pm \sqrt{(-20)^2 - 4(4)(25)}}{2(4)}$$

$$x = \frac{20 \pm \sqrt{400 - 400}}{8}$$

$$x = \frac{20}{8} = \frac{5}{2}$$

35. Solve $\dfrac{2x^2}{5} - \dfrac{x}{2} = 1$.

Solution

$$\frac{2x^2}{5} - \frac{x}{2} = 1$$

$$10\left[\frac{2x^2}{5} - \frac{x}{2} = 1\right]10$$

$$4x^2 - 5x = 10$$

$$4x^2 - 5x - 10 = 0$$

$$x = \frac{-(-5) \pm \sqrt{(-5)^2 - 4(4)(-10)}}{2(4)}$$

$$x = \frac{5 \pm \sqrt{25 + 160}}{8}$$

$$x = \frac{5 \pm \sqrt{185}}{8}$$

37. Solve $\sqrt{x + 3} = x - 1$.

Solution

$$\sqrt{x + 3} = x - 1$$

$$\left(\sqrt{x + 3}\right)^2 = (x - 1)^2$$

$$x + 3 = x^2 - 2x + 1$$

$$0 = x^2 - 3x - 2$$

$$x = \frac{-(-3) \pm \sqrt{(-3)^2 - 4(1)(-2)}}{2(1)}$$

$$x = \frac{3 \pm \sqrt{9 + 8}}{2}$$

$$x = \frac{3 \pm \sqrt{17}}{2}$$

39. *Exploration* Determine the values of c such that the equation has (a) two real number solutions, (b) one real number solution, and (c) two imaginary number solutions.

$$x^2 - 6x + c = 0$$

(a) $\quad b^2 - 4ac > 0$

$(-6)^2 - 4(1)c > 0$

$36 - 4c > 0$

$-4c > -36$

$c < 9$

(b) $\quad b^2 - 4ac = 0$

$(-6)^2 - 4(1)c = 0$

$36 - 4c = 0$

$-4c = -36$

$c = 9$

(c) $\quad b^2 - 4ac < 0$

$(-6)^2 - 4(1)c < 0$

$36 - 4c < 0$

$-4c < -36$

$c > 9$

41. *Exploration* Determine the values of c such that the equation has (a) two real number solutions, (b) one real number solution, and (c) two imaginary number solutions.

$$x^2 + 8x + c = 0$$

(a) $\quad b^2 - 4ac > 0$

$8^2 - 4(1)c > 0$

$64 - 4c > 0$

$-4c > -64$

$c < 16$

(b) $\quad b^2 - 4ac = 0$

$8^2 - 4(1)c = 0$

$64 - 4c = 0$

$-4c = -64$

$c = 16$

(c) $\quad b^2 - 4ac < 0$

$8^2 - 4(1)c < 0$

$64 - 4c < 0$

$-4c < -64$

$c > 16$

43. *Geometry* A rectangle has a width of x inches, a length of $x + 6.3$ inches, and an area of 58.14 square inches (see figure in textbook). Find its dimensions.

Solution

Verbal model: $\boxed{\text{Area}} = \boxed{\text{Length}} \cdot \boxed{\text{Width}}$

Equation: $58.14 = (x + 6.3) \cdot x$

$$58.14 = x^2 + 6.3x$$

$$0 = x^2 + 6.3x - 58.14$$

$$x = \frac{-6.3 \pm \sqrt{6.3^2 - 4(1)(-58.14)}}{2(1)}$$

$$x = \frac{-6.3 \pm \sqrt{39.69 + 232.56}}{2}$$

$$x = \frac{-6.3 \pm \sqrt{272.25}}{2}$$

$$x = 5.1 \text{ inches}$$

$$x + 6.3 = 11.4 \text{ inches}$$

45. Perform the operation and simplify.

$$\left(\sqrt{x} + 3\right)\left(\sqrt{x} - 3\right)$$

Solution

$$\left(\sqrt{x} + 3\right)\left(\sqrt{x} - 3\right) = x - 9$$

47. Perform the operation and simplify.

$$\left(2\sqrt{t} + 3\right)^2$$

Solution

$$\left(2\sqrt{t} + 3\right)^2 = \left(2\sqrt{t} + 3\right)\left(2\sqrt{t} + 3\right)$$

$$= 4t + 6\sqrt{t} + 6\sqrt{t} + 9$$

$$= 4t + 12\sqrt{t} + 9$$

49. *Mixture* Determine the number of gallons of a 30% solution that must be mixed with a 60% solution to obtain 20 gallons of a 40% solution.

Solution

Verbal model: $\boxed{\begin{array}{c}\text{Amount}\\\text{Solution 1}\end{array}} + \boxed{\begin{array}{c}\text{Amount}\\\text{Solution 2}\end{array}} = \boxed{\begin{array}{c}\text{Amount}\\\text{Final Solution}\end{array}}$

Equation: $.30x + .60(20 - x) = .40(20)$

$$.30x + 12 - .60x = 8$$

$$-.30x = -4$$

$$x = 13\tfrac{1}{3} \text{ gal}$$

$$20 - x = 6\tfrac{2}{3} \text{ gal}$$

51. Solve $x^2 + 6x + 8 = 0$ (a) by the Quadratic Formula and (b) by factoring.

Solution

(a) $x = \dfrac{-6 \pm \sqrt{6^2 - 4(1)(8)}}{2(1)}$

$x = \dfrac{6 \pm \sqrt{36 - 32}}{2}$

$x = \dfrac{-6 \pm \sqrt{4}}{2}$

$x = \dfrac{-6 \pm 2}{2}$

$x = -2, -4$

(b) $(x + 4)(x + 2) = 04$

$x + 4 = 0 \qquad x + 2 = 0$

$x = -4 \qquad x = -2$

53. Solve $x^2 - \dfrac{4}{3}x + \dfrac{4}{9} = 0$ (a) by the Quadratic Formula and (b) by factoring.

Solution

(a) $9x^2 - 12x + 4 = 0$

$x = \dfrac{12 \pm \sqrt{(-12)^2 - 4(9)(4)}}{2(9)}$

$x = \dfrac{12 \pm \sqrt{144 - 144}}{18}$

$x = \dfrac{12 \pm 0}{18}$

$x = \dfrac{12}{18} = \dfrac{2}{3}$

(b) $\left(x - \dfrac{2}{3}\right)\left(x - \dfrac{2}{3}\right) = 0$

$x - \dfrac{2}{3} = 0 \qquad x - \dfrac{2}{3} = 0$

$x = \dfrac{2}{3} \qquad x = \dfrac{2}{3}$

55. Solve $6x^2 - x - 2 = 0$ (a) by the Quadratic Formula and (b) by factoring.

Solution

(a) $x = \dfrac{1 \pm \sqrt{(-1)^2 - 4(6)(-2)}}{2(6)}$

$x = \dfrac{1 \pm \sqrt{1 + 48}}{12}$

$x = \dfrac{1 \pm \sqrt{49}}{12}$

$x = \dfrac{1 \pm 7}{12}$

$x = \dfrac{8}{12}, -\dfrac{6}{12} = \dfrac{2}{3}, -\dfrac{1}{2}$

(b) $(3x - 2)(2x + 1) = 0$

$3x - 2 = 0 \qquad 2x + 1 = 0$

$x = \dfrac{2}{3} \qquad x = -\dfrac{1}{2}$

57. Use the discriminant to determine the type of solutions of $x^2 + 7x + 15 = 0$.

Solution

$$b^2 - 4ac = 7^2 - 4(1)(15)$$

$$= 49 - 60$$

$$= -11$$

2 complex solutions

59. Use the discriminant to determine the type of solutions of $4x^2 - 12x + 9 = 0$.

Solution

$$b^2 - 4ac = (-12)^2 - 4(4)(9)$$

$$= 144 - 144$$

$$= 0$$

1 rational repeated solution

61. Use the Quadratic Formula to find all real or complex solutions of

$$x^2 + 6x - 3 = 0.$$

Solution

$$x = \frac{-6 \pm \sqrt{6^2 - 4(1)(-3)}}{2(1)}$$

$$x = \frac{-6 \pm \sqrt{36 + 12}}{2}$$

$$x = \frac{-6 \pm \sqrt{48}}{2}$$

$$x = \frac{-6 \pm 4\sqrt{3}}{2}$$

$$x = \frac{2\left(-3 \pm 2\sqrt{3}\right)}{2}$$

$$x = -3 \pm 2\sqrt{3}$$

63. Use the Quadratic Formula to find all real or complex solutions to

$$x^2 - 10x + 23 = 0.$$

Solution

$$x = \frac{-(-10) \pm \sqrt{(-10)^2 - 4(1)(23)}}{2(1)}$$

$$x = \frac{10 \pm \sqrt{100 - 92}}{2}$$

$$x = \frac{10 \pm \sqrt{8}}{2}$$

$$x = \frac{10 \pm 2\sqrt{2}}{2}$$

$$x = \frac{2\left(5 \pm \sqrt{2}\right)}{2}$$

$$x = 5 \pm \sqrt{2}$$

65. Use the Quadratic Formula to find all real or complex solutions to

$$2v^2 - 2v - 1 = 0.$$

Solution

$$v = \frac{-(-2) \pm \sqrt{(-2)^2 - 4(2)(-1)}}{2(2)}$$

$$v = \frac{2 \pm \sqrt{4 + 8}}{4}$$

$$v = \frac{2 \pm \sqrt{12}}{4}$$

$$v = \frac{2 \pm 2\sqrt{3}}{4}$$

$$v = \frac{2\left(1 \pm \sqrt{3}\right)}{4}$$

$$v = \frac{1 \pm \sqrt{3}}{2}$$

67. Use the Quadratic Formula to find all real or complex solutions to

$$2x^2 + 4x - 3 = 0.$$

Solution

$$x = \frac{-4 \pm \sqrt{4^2 - 4(2)(-3)}}{2(2)}$$

$$x = \frac{-4 \pm \sqrt{16 + 24}}{4}$$

$$x = \frac{-4 \pm \sqrt{40}}{4}$$

$$x = \frac{-4 \pm 2\sqrt{10}}{4}$$

$$x = \frac{2\left(-2 \pm \sqrt{10}\right)}{4}$$

$$x = \frac{-2 \pm \sqrt{10}}{2}$$

69. Use the Quadratic Formula to find all real or complex solutions to

$$x^2 - 0.4x - 0.16 = 0.$$

Solution

$$x = \frac{-(-0.4) \pm \sqrt{(-0.4)^2 - 4(1)(-0.16)}}{2(1)}$$

$$x = \frac{0.4 \pm \sqrt{0.16 + 0.64}}{2}$$

$$x = \frac{0.4 \pm \sqrt{0.80}}{2}$$

$$x = \frac{0.4 \pm 2\sqrt{0.2}}{2}$$

$$x = 0.2 \pm \sqrt{0.2} \text{ or } \frac{1 \pm \sqrt{5}}{5}$$

71. Use the Quadratic Formula to find all real or complex solutions to

$$4x^2 - 6x + 3 = 0.$$

Solution

$$x = \frac{-(-6) \pm \sqrt{(-6)^2 - 4(4)(3)}}{2(4)}$$

$$x = \frac{6 \pm \sqrt{36 - 48}}{8}$$

$$x = \frac{6 \pm \sqrt{-12}}{8}$$

$$x = \frac{6}{8} \pm \frac{2i\sqrt{3}}{8}$$

$$x = \frac{3}{4} \pm \frac{\sqrt{3}}{4}i$$

73. Use the Quadratic Formula to find all real or complex solutions to

$$9x^2 = 1 + 9x.$$

Solution

$$9x^2 = 1 + 9x$$

$$9x^2 - 9x - 1 = 0$$

$$x = \frac{-(-9) \pm \sqrt{(-9)^2 - 4(9)(-1)}}{2(9)}$$

$$x = \frac{9 \pm \sqrt{81 + 36}}{18}$$

$$x = \frac{9 \pm \sqrt{117}}{18}$$

$$x = \frac{9}{18} \pm \frac{3\sqrt{13}}{6}$$

$$x = \frac{1}{2} \pm \frac{\sqrt{13}}{6} \text{ or } \frac{3 \pm \sqrt{13}}{6}$$

75. Solve $z^2 - 169 = 0$ by the most convenient method. Find all real or complex solutions.

Solution

$$z^2 - 169 = 0$$

$$z^2 = 169$$

$$z = \pm 13$$

77. Solve $25(x-3)^2 - 36 = 0$ by the most convenient method. Find all real or complex solutions.

Solution

$$25(x-3)^2 - 36 = 0$$

$$(x-3)^2 = \frac{36}{25}$$

$$x - 3 = \pm\sqrt{\frac{36}{25}}$$

$$x = 3 \pm \frac{6}{5}$$

$$x = \frac{15}{5} \pm \frac{6}{5}$$

$$x = \frac{21}{5}, \frac{9}{5}$$

79. Solve $x^2 - 24x + 128 = 0$ by the most convenient method. Find all real or complex solutions.

Solution

$$x^2 - 24x + 128 = 0$$

$$x^2 - 24x + 144 = -128 + 144$$

$$(x - 12)^2 = 16$$

$$x - 12 = \pm\sqrt{16}$$

$$x = 12 \pm 4$$

$$x = 16, 8$$

81. Use a calculator to solve $5x^2 - 18x + 6 = 0$. Round to three decimal places.

Solution

$$x = \frac{-(-18) \pm \sqrt{(-18)^2 - 4(5)(6)}}{2(5)}$$

$$x = \frac{18 \pm \sqrt{324 - 120}}{10}$$

$$x = \frac{18 \pm \sqrt{+204}}{10}$$

$$x = \frac{18 \pm 2\sqrt{51}}{10}$$

$$x = \frac{2\left(9 \pm \sqrt{51}\right)}{10}$$

$$x = \frac{9 \pm \sqrt{51}}{5}$$

$$x = 3.228, 0.372$$

83. Use a calculator to solve $-0.04x^2 + 4x - 0.8 = 0$. Round to three decimal places.

Solution

$$x = \frac{-4 \pm \sqrt{4^2 - 4(-0.04)(-0.8)}}{2(-0.04)}$$

$$x = \frac{-4 \pm \sqrt{16 - 0.128}}{-0.08}$$

$$x = \frac{-4 \pm \sqrt{15.872}}{-0.08}$$

$$x = \frac{-4 \pm 16\sqrt{0.062}}{-0.08}$$

$$x = \frac{4\left(-1 \pm 4\sqrt{0.062}\right)}{-0.08}$$

$$x = \frac{-1 \pm 4\sqrt{0.062}}{0.02}$$

$$x = -0.200, -99.800$$

85. *Free-Falling Object* A ball is thrown upward at a velocity of 40 feet per second from a bridge that is 50 feet above the level of the water (see figure in textbook). The height h (in feet) of the ball above the water at time t seconds after it is thrown is given by

$$h = -16t^2 + 40t + 50.$$

(a) Find the time when the ball is again at the 50-foot level above the water.

(b) Find the time when the ball stikes the water.

Solution

(a) $50 = -16t^2 + 40t + 50$

$0 = -16t^2 + 40t$

$0 = -8\left(2t^2 - 5t\right)$

$0 = 2t^2 - 5t$

$0 = t(2t - 5)$

$0 = t \qquad 2t - 5 = 0$

$$t = \frac{+5}{2}$$

(b) $0 = -16t^2 + 40t + 50$

$0 = -2\left(8t^2 - 20t - 25\right)$

$$t = \frac{-(-20) \pm \sqrt{(-20)^2 - 4(8)(-25)}}{2(8)}$$

$$t = \frac{20 \pm \sqrt{400 + 800}}{16}$$

$$t = \frac{20 \pm \sqrt{1200}}{16}$$

$$t = \frac{20 \pm 20\sqrt{3}}{16}$$

$$t = \frac{4\left(5 \pm 5\sqrt{3}\right)}{16}$$

$$t = \frac{5 + 5\sqrt{3}}{4}, \frac{5 - 5\sqrt{3}}{4}$$

$t \approx 3.415 \text{ sec}$

reject

87. Use the following model, which approximates the number of people employed in the aerospace industry in the United States from 1987 to 1992.

$$y = 803.49 - 33.23t - 10.77t^2, \qquad -3 \le t \le 2.$$

In this model, y represents the number employed in the aerospace industry (in thousands) and t represents the year, with $t = 0$ corresponding to 1990. (*Source:* U.S. Department of Commerce)

Use a graphing utility to graph the model.

Solution

Keystrokes:

 $\boxed{Y =}$ 803.49 $\boxed{-}$ 33.23 $\boxed{X, T, \theta}$ $\boxed{-}$ 10.77 $\boxed{X, T, \theta}$ $\boxed{x^2}$ $\boxed{\text{GRAPH}}$

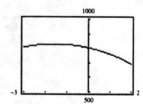

89. Use the model in problem #87 to estimate the number employed in the aerospace industry in 1993.

Solution

$$y = 803.49 - 33.23(3) - 10.77(3)^2$$

$$y = 803.49 - 99.69 - 96.93$$

$$y = 606.87 \text{ thousand}$$

$$y = 606,870$$

Mid-Chapter Quiz for Chapter 6

1. Solve $2x^2 - 72 = 0$ by factoring.

Solution

$$2x^2 - 72 = 0$$

$$2\left(x^2 - 36\right) = 0$$

$$2(x - 6)(x + 6) = 0$$

$$x - 6 = 0 \qquad x + 6 = 0$$

$$x = 6 \qquad\quad x = -6$$

2. Solve $2x^2 + 3x - 20 = 0$ by factoring.

Solution

$$2x^2 + 3x - 20 = 0$$

$$(2x - 5)(x + 4) = 0$$

$$2x - 5 = 0 \qquad x + 4 = 0$$

$$x = \frac{5}{2} \qquad\qquad x = -4$$

3. Solve $t^2 = 12$ by extracting square roots.

Solution

$$t^2 = 12$$

$$t = \pm\sqrt{12}$$

$$t = \pm 2\sqrt{3}$$

4. Solve $(u - 3)^2 - 16 = 0$ by extracting square roots.

Solution

$$(u - 3)^2 - 16 = 0$$

$$(u - 3)^2 = 16$$

$$u - 3 = \pm 4$$

$$u = 3 \pm 4 = 7, -1$$

5. Solve $s^2 + 10s + 1 = 0$ by completing the square.

Solution

$$s^2 + 10s + 1 = 0$$

$$s^2 + 10s = -1$$

$$s^2 + 10s + 25 = -1 + 25$$

$$(s + 5)^2 = 24$$

$$s + 5 = \pm\sqrt{24}$$

$$s = -5 \pm 2\sqrt{6}$$

6. Solve $2y^2 + 6y - 5 = 0$ by completing the square.

Solution

$$2y^2 + 6y - 5 = 0$$

$$y^2 + 3y = \frac{5}{2}$$

$$y^2 + 3y + \frac{9}{4} = \frac{5}{2} + \frac{9}{4}$$

$$\left(y + \frac{3}{2}\right)^2 = \frac{10}{4} + \frac{9}{4}$$

$$\left(y + \frac{3}{2}\right)^2 = \frac{19}{4}$$

$$y + \frac{3}{2} = \pm\frac{\sqrt{19}}{2}$$

$$y = -\frac{3}{2} \pm \frac{\sqrt{19}}{2}$$

7. Solve $x^2 + 4x - 6 = 0$ by the Quadratic Formula.

Solution

$$x = \frac{-4 \pm \sqrt{4^2 - 4(1)(-6)}}{2(1)}$$

$$x = \frac{-4 \pm \sqrt{16 + 24}}{2}$$

$$x = \frac{-4 \pm \sqrt{40}}{2}$$

$$x = \frac{-4 \pm 2\sqrt{10}}{2} = -2 \pm \sqrt{10}$$

8. Solve $6v^2 - 3v - 4 = 0$ by the Quadratic Formula.

Solution

$$v = \frac{-(-3) \pm \sqrt{(-3)^2 - 4(6)(-4)}}{2(6)}$$

$$v = \frac{3 \pm \sqrt{9 + 96}}{12}$$

$$v = \frac{3 \pm \sqrt{105}}{12}$$

9. Solve $x^2 + 5x + 7 = 0$ by the most convenient method. (Find all real *and* complex solutions.)

Solution

$$x = \frac{-5 \pm \sqrt{5^2 - 4(1)(7)}}{2(1)}$$

$$x = \frac{-5 \pm \sqrt{25 - 28}}{2}$$

$$x = \frac{-5 \pm \sqrt{-3}}{2}$$

$$x = \frac{-5 \pm i\sqrt{3}}{2} = -\frac{5}{2} \pm \frac{\sqrt{3}}{2}i$$

10. Solve $36 - (t - 4)^2 = 0$ by the most convenient method. (Find all real *and* complex solutions.)

Solution

$$36 = (t - 4)^2$$

$$\pm 6 = t - 4$$

$$4 \pm 6 = t$$

$$10, -2 = t$$

11. Solve $x(x - 10) + 3(x - 10) = 0$ by the most convenient method. (Find all real *and* complex solutions.)

Solution

$$(x - 10)(x + 3) = 0$$

$$(x - 10) = 0 \qquad x + 3 = 0$$

$$x = 10 \qquad x = -3$$

12. Solve $x(x - 3) = 10$ by the most convenient method. (Find all real *and* complex solutions.)

Solution

$$x^2 - 3x - 10 = 0$$

$$(x - 5)(x + 2) = 0$$

$$x - 5 = 0 \qquad x + 2 = 0$$

$$x = 5 \qquad x = -2$$

13. Solve $4b^2 - 12b + 9 = 0$ by the most convenient method. (Find all real *and* complex solutions.)

Solution

$$(2b - 3)(2b - 3) = 0$$

$$2b - 3 = 0 \qquad 2b - 3 = 0$$

$$b = \frac{3}{2} \qquad b = \frac{3}{2}$$

14. Solve $3m^2 + 10m + 5 = 0$ by the most convenient method. (Find all real *and* complex solutions.)

Solution

$$m = \frac{-10 \pm \sqrt{10^2 - 4(3)(5)}}{2(3)}$$

$$m = \frac{-10 \pm \sqrt{100 - 60}}{6}$$

$$m = \frac{-10 \pm \sqrt{40}}{6}$$

$$m = \frac{-10 \pm 2\sqrt{10}}{6}$$

$$m = \frac{-5 \pm \sqrt{10}}{3}$$

15. Solve $\dfrac{4}{u} - u = 3$ by the most convenient method. (Find all real *and* complex solutions.)

Solution

$$4 - u^2 = 3u$$

$$0 = u^2 + 3u - 4$$

$$0 = (u + 4)(u - 1)$$

$$u + 4 = 0 \qquad u - 1 = 0$$

$$u = -4 \qquad u = 1$$

16. Solve $\sqrt{2x + 5} = x + 1$ by the most convenient method. (Find all real *and* complex solutions.)

Solution

$$\left(\sqrt{2x + 5}\right)^2 = (x + 1)^2 \qquad \textbf{Check:} \qquad \sqrt{2(2) + 5} \stackrel{?}{=} 2 + 1$$

$$2x + 5 = x^2 + 2x + 1 \qquad\qquad\qquad \sqrt{9} = 3$$

$$0 = x^2 - 4 \qquad\qquad\qquad \sqrt{2(-2) + 5} \stackrel{?}{=} -2 + 1$$

$$4 = x^2 \qquad\qquad\qquad\qquad \sqrt{1} \neq -1$$

$$\pm 2 = x$$

17. Use a graphing utility to graph the function $y = \frac{1}{2}x^2 - 3x - 1$. Use the graph to approximate any x-intercepts of the graph. Set $y = 0$ and solve the resulting equation.

Solution

Keystrokes:

$\boxed{Y=}$ .5 $\boxed{X, T, \theta}$ $\boxed{x^2}$ $\boxed{-}$ 3 $\boxed{X, T, \theta}$ $\boxed{-}$ 1 $\boxed{\text{GRAPH}}$

$$0 = .5x^2 - 3x - 1$$

$$0 = x^2 - 6x - 2$$

$$x = \frac{-(-6) \pm \sqrt{(-6)^2 - 4(1)(-2)}}{2(1)}$$

$$x = \frac{6 \pm \sqrt{36 + 8}}{2}$$

$$x = \frac{6 \pm \sqrt{44}}{2}$$

$$x = \frac{6 \pm 2\sqrt{11}}{2}$$

$$x = 3 \pm \sqrt{11}$$

$$x = 6.32 \text{ and } -0.32$$

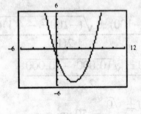

18. Use a graphing utility to graph the function $y = x^2 + 0.45x - 4$. Use the graph to approximate any x-intercepts of the graph. Set $y = 0$ and solve the resulting equation.

Solution

Keystrokes:

$\boxed{Y=}$ $\boxed{X,T,\theta}$ $\boxed{x^2}$ $\boxed{+}$ 0.45 $\boxed{X,T,\theta}$ $\boxed{-}$ 4 $\boxed{GRAPH}$

$0 = x^2 + 0.45x - 4$

$x = \dfrac{-0.45 \pm \sqrt{(0.45)^2 - 4(1)(-4)}}{2(1)}$

$x = \dfrac{-0.45 \pm \sqrt{0.2025 + 16}}{2}$

$x = \dfrac{-0.45 \pm \sqrt{16.2025}}{2}$

$x = 1.79$ and -2.24

19. The revenue R from selling x units of a certain product is given by $R = x(20 - 0.2x)$. Find the number of units that must be sold to produce a revenue of $500.

Solution

$$500 = x(20 - 0.2x)$$
$$= 20x - 0.2x^2$$
$$0.2x^2 - 20x + 500 = 0$$
$$x^2 - 100x + 2500 = 0$$

$$x = \frac{-(-100) \pm \sqrt{(-100)^2 - 4(1)(2500)}}{2(1)}$$

$$x = \frac{100 \pm \sqrt{10,000 - 10,000}}{2}$$

$$x = \frac{100 \pm \sqrt{0}}{2} = 50 \text{ units}$$

20. The perimeter of a rectangle with sides x and $100 - x$ is 200 meters. Its area A is given by $A = x(100 - x)$. Determine the dimensions of the rectangle if its area is 2275 square meters.

Solution

Verbal model: $\boxed{\text{Area}} = \boxed{\text{Length}} \cdot \boxed{\text{Width}}$

Equation: $2275 = x \cdot (100 - x)$

$$2275 = 100x - x^2$$
$$0 = x^2 - 100x + 2275$$
$$0 = (x - 35)(x - 65)$$
$$x - 35 = 0 \qquad\qquad x - 65 = 0$$
$$x = 35 \text{ meters} \qquad\qquad x = 65 \text{ meters}$$

6.4 | Applications of Quadratic Equations

7. Find two positive integers that satisfy the requirement.

The product of two consecutive integers is 240.

Solution

Verbal model: $\boxed{\text{Integer}} \cdot \boxed{\text{Integer}} = \boxed{\text{Product}}$

Equation:

$$n \cdot (n + 1) = 240$$

$$n^2 + n + \frac{1}{4} = 240 + \frac{1}{4}$$

$$\left(n + \frac{1}{2}\right)^2 = \frac{960 + 1}{4}$$

$$n + \frac{1}{2} = \pm\sqrt{\frac{961}{4}}$$

$$n = -\frac{1}{2} \pm \frac{\sqrt{961}}{2}$$

$$n = \frac{-1 \pm 31}{2}$$

$$n = 15, -16$$

$$n + 1 = 16, -15$$

9. Find two positive integers that satisfy the requirement.

The product of two consecutive *even* integers is 224.

Solution

Verbal model: $\boxed{\text{Even integer}} \cdot \boxed{\text{Even integer}} = \boxed{\text{Product}}$

Equation:

$$2n \cdot (2n + 2) = 224$$

$$4n^2 + 4n = 224$$

$$n^2 + n = 56$$

$$n^2 + n - 56 = 0$$

$$(n + 8)(n - 7) = 0$$

$n + 8 = 0$	$n - 7 = 0$
$n = -8$	$n = 7$
$2n = -16$	$2n = 14$
$2n + 2 = -14$	$2n + 2 = 16$

11. Find the area of the rectangle.

Width: $0.75l$ Length: l Perimeter: 42 in.   Area:

Solution

Verbal model: $2\boxed{\text{Length}} + 2\boxed{\text{Width}} = \boxed{\text{Perimeter}}$

Equation: $2l + 2(0.75l) = 42$

$2l + 1.5l = 42$

$3.5l = 42$

$l = 12$

$w = 0.75 \quad l = 9$

Verbal model: $\boxed{\text{Length}} \cdot \boxed{\text{Width}} = \boxed{\text{Area}}$

Equation: $12 \cdot 9 = A$

$108 = A$

13. Find the perimeter of the rectangle.

Width: $l - 20$ Length: l Perimeter: Area: $12{,}000\text{m}^2$

Solution

Verbal model: $\boxed{\text{Length}} \cdot \boxed{\text{Width}} = \boxed{\text{Area}}$

Equation: $l \cdot (l - 20) = 12{,}000$

$l^2 - 20l = 12{,}000$

$l^2 - 20l + 100 = 12{,}000 + 100$

$(l - 10)^2 = 12{,}100$

$l - 10 = \pm\sqrt{12{,}100}$

$l = 10 + 110 = 120$

$w = l - 20 = 100$

Verbal model: $2\boxed{\text{Length}} + 2\boxed{\text{Width}} = \boxed{\text{Perimeter}}$

Equation: $2(120) + 2(100) = 440 = P$

15. *Compound Interest* Find the interest rate r. Use the formula $A = P(1 + r)^2$, where A is the amount after 2 years in an account earning r percent compounded annually and P is the original investment.

$P = \$3000, \quad A = \$3499.20$

Solution

$A = P(1 + r)^2$

$3499.20 = 3000(1 + r)^2$

$1.1664 = (1 + r)^2$

$1.08 = 1 + r$

$0.08 = r$ or 8%

17. *Compound Interest* Find the interest rate, r. Use the formula $A = P(1 + r)^2$, where A is the amount after 2 years in an account earning r percent compounded annually and P is the original investment.

$P = \$8000.00, \quad A = \$8420.20$

Solution

$A = P(1 + r)^2$

$8420.20 = 8000.00(1 + r)^2$

$1.052525 = (1 + r)^2$

$1.0259 = 1 + r$

$.0259 = r$ or 2.59%

19. *Geometry* A television station claims that it covers a circular region of approximately 25,000 square miles.

    (a) Assume that the station is located at the center of the circular region. How far is the station from its farthest listener?

    (b) Assume that the station is located on the edge of the circular region. How far is the station from its farthest listener?

Solution

    (a)   *Verbal model:* $\boxed{\text{Area}} = \boxed{\pi} \cdot \boxed{\text{Radius}^2}$

           Equation: $A = \pi r^2$

$$25,000 = \pi r^2$$

$$\frac{25,000}{\pi} = r^2$$

$$89.206 = r$$

    (b)   *Verbal model:* $\boxed{\text{Diameter}} = \boxed{2} \cdot \boxed{\text{Radius}}$

           Equation: $d = 2r$

$$d = 2(89.206)$$

$$d = 178.412 \text{ miles}$$

21. *Geometry* A retail lumber business plans to build a rectangular storage region adjoining the sales office (see figure in textbook). The region will be fenced on three sides, and the fourth side will be bounded by the existing building. Find the dimensions of the region if 350 feet of fencing is used and the area of the region is 12,500 square feet.

Solution

Verbal model: $\boxed{\text{Length}} \cdot \boxed{\text{Width}} = \boxed{\text{Area}}$

Equation: $(350 - 2x) \cdot x = 12,500$

$$350x - 2x^2 = 12,500$$

$$2x^2 - 350x + 12,500 = 0$$

$$x^2 - 175x + 6,250 = 0$$

$$x = \frac{175 \pm \sqrt{175^2 - 4(1)(6,250)}}{2(1)}$$

$$x = \frac{175 \pm \sqrt{5625}}{2} = \frac{175 \pm 75}{2}$$

$$x = 125, 50$$

$$350 - 2x = 100, 250$$

$$100 \text{ ft} \times 125 \text{ ft.}$$

23. *Selling Price* A store owner bought a case of grade A large eggs for $21.60. By the time all but 6 dozen of the eggs had been sold at a profit of $0.30 per dozen, the original investment of $21.60 had been regained. How many dozen eggs did the owner sell, and what was the selling price per dozen?

Solution

Verbal model:

$$\boxed{\begin{array}{c}\text{Selling price} \\ \text{per doz eggs}\end{array}} = \boxed{\begin{array}{c}\text{Cost per} \\ \text{doz eggs}\end{array}} + \boxed{\begin{array}{c}\text{Profit per} \\ \text{doz eggs}\end{array}}$$

Equation:

$$\frac{21.60}{x} = \frac{21.60}{x+6} + 0.30$$

$$21.60(x+6) = 21.60x + 0.30x(x+6)$$

$$21.6x + 129.6 = 21.6x + 0.3x^2 + 1.8x$$

$$0 = 0.3x^2 + 1.8x - 129.6$$

$$0 = 3x^2 + 18x - 1296$$

$$0 = x^2 + 6x - 432$$

$$0 = (x+24)(x-18)$$

$$x = -24 \quad x = 18$$

$$\text{Selling Price} = \frac{21.60}{18} = \$1.20$$

25. *Reduced Ticket Price* A service organization obtained a block of tickets to a ball game for $240. When eight more people decide to go to the game the price per ticket is decreased by $1. How many people are going to the game?

Solution

Verbal model:

$$\boxed{\begin{array}{c}\text{Cost per} \\ \text{member}\end{array}} \cdot \boxed{\begin{array}{c}\text{Number of} \\ \text{members}\end{array}} = \boxed{\$240}$$

Equation:

$$\left(\frac{240}{x} - 1\right) \cdot (x+8) = 240$$

$$\left(\frac{240 - x}{x}\right)(x+8) = 240$$

$$(240 - x)(x+8) = 240x$$

$$240x + 1920 - x^2 - 8x = 240x$$

$$-x^2 - 8x + 1920 = 0$$

$$x^2 + 8x - 1920 = 0$$

$$(x+48)(x-40) = 0$$

$$x = -48 \quad x = 40$$

$$x + 8 = 48$$

27. *Solving Graphically and Algebraically* An adjustable rectangular form has minimum dimensions of 3 meters by 4 meters. The length and width can be expanded by equal amounts x (see figure in textbook).

(a) Write the length d of the diagonal as a function of x. Use a graphing utility to graph the function. Use the graph to approximate the value of x when $d = 10$ meters.

(b) Find x algebraically when $d = 10$.

Solution

(a) $d = \sqrt{(3 + x)^2 + (4 + x)^2}$

Keystrokes:

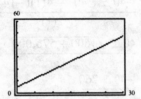

Approximate value of $x \approx 3.55$ when $d = 10$.

(b) $10 = \sqrt{(3 + x)^2 + (4 + x)^2}$

$100 = (3 + x)^2 + (4 + x)^2$

$ = 9 + 6x + x^2 + 16 + 8x + x^2$

$0 = 2x^2 + 14x - 75$

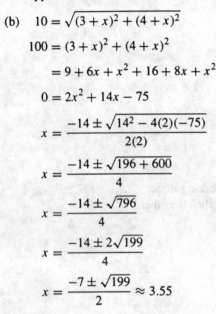

$x = \dfrac{-14 \pm \sqrt{14^2 - 4(2)(-75)}}{2(2)}$

$x = \dfrac{-14 \pm \sqrt{196 + 600}}{4}$

$x = \dfrac{-14 \pm \sqrt{796}}{4}$

$x = \dfrac{-14 \pm 2\sqrt{199}}{4}$

$x = \dfrac{-7 \pm \sqrt{199}}{2} \approx 3.55$

29. *Air Speed* An airline runs a commuter flight between two cities that are 720 miles apart. If the average speed of the planes could be increased by 40 miles per hour, the travel time would be decreased by 12 minutes. What air speed is required to obtain this decrease in travel time?

Solution

Verbal model: $\boxed{\text{Original time}} = \boxed{\text{New time}} + \boxed{\dfrac{1}{5}}$

Equation:

$$\frac{720}{x} = \frac{720}{x+40} + \frac{1}{5}$$

$$720(5)(x+40) = 720(5x) + x(x+40)$$

$$3600x + 144{,}000 = 3600x + x^2 + 40x$$

$$0 = x^2 + 40x - 144{,}000$$

$$x = \frac{-40 \pm \sqrt{40^2 - 4(1)(-144{,}000)}}{2(1)}$$

$$x = \frac{40 \pm \sqrt{1600 + 576{,}000}}{2}$$

$$x = \frac{-40 \pm 760}{2}$$

$$x = 360, -400$$

$$x + 40 = 400$$

31. *Work Rate* Working together, two people can complete a task in 5 hours. Working alone, how long would it take each to do the task if one person took 2 hours longer than the other?

Solution

Verbal model: $\boxed{\begin{array}{c}\text{Work done by}\\ \text{Person 1}\end{array}} + \boxed{\begin{array}{c}\text{Work done by}\\ \text{Person 2}\end{array}} = \boxed{\text{One complete job}}$

Equation:

$$\frac{1}{x}(5) + \frac{1}{x+2}(5) = 1$$

$$x(x+2)\left[(5)\left(\frac{1}{x} + \frac{1}{x+2}\right) = 1\right]x(x+2)$$

$$5(x+2) + 5x = x(x+2)$$

$$5x + 10 + 5x = x^2 + 2x$$

$$-x^2 + 8x + 10 = 0$$

$$x^2 - 8x - 10 = 0$$

$$x = \frac{8 \pm \sqrt{(-8)^2 - 4(1)(-10)}}{2(1)}$$

$$x = \frac{8 \pm \sqrt{64 + 40}}{2}$$

$$x = \frac{8 \pm \sqrt{104}}{2}$$

$$x = 9.0990195, -1.0990195$$

$$x + 2 = 11.0990195$$

33. *Fall-Falling Object* Find the time necessary for an object to fall to ground level from an initial height of $h_0 = 144$ feet if its height h at any time t (in seconds) is given by $h = h_0 - 16t^2$.

Solution

$$h = h_0 - 16t^2$$

$$0 = 144 - 16t^2$$

$$16t^2 = 144$$

$$t^2 = 9$$

$$t = 3 \text{ sec}$$

35. *Fall-Falling Object* Find the time necessary for an object to fall to ground level from an initial height of $h_0 = 1454$ feet (height of the Sears Tower) if its height h at any time t (in seconds) is given by $h = h_0 - 16t^2$.

Solution

$$h = h_0 - 16t^2$$

$$0 = 1454 - 16t^2$$

$$16t^2 = 1454$$

$$t^2 = 90.875$$

$$t = 9.532838 \text{ sec}$$

37. *Cost, Revenue, and Profit* You are given the cost C of producing x units, the revenue R from selling x units, and the profit P. Find the value of x that will produce the profit P.

$$C = 100 + 30x, \quad R = x(90 - x), \quad P = \$800$$

Solution

Verbal model: $\boxed{\text{Profit}} = \boxed{\text{Revenue}} - \boxed{\text{Cost}}$

Equation:

$$800 = x(90 - x) - (100 + 30x)$$

$$800 = 90x - x^2 - 100 - 30x$$

$$0 = -x^2 + 60x - 900$$

$$0 = x^2 - 60x + 900$$

$$0 = (x - 30)^2$$

$$0 = x - 30$$

$$30 \text{ units} = x$$

39. Solve for r in $A = P(1 + r)^2$.

Solution

$$A = P(1 + r)^2$$

$$\frac{A}{P} = (1 + r)^2$$

$$\sqrt{\frac{A}{P}} = 1 + r$$

$$\sqrt{\frac{A}{P}} - 1 = r$$

41. Find the product.

$$-2x^5 \left(5x^{-3}\right)$$

Solution

$$-2x^5 \left(5x^{-3}\right) = -10x^2$$

43. Find the product.

$$(2x - 15)^2$$

Solution

$$(2x - 15)^2 = (2x - 15)(2x - 15)$$

$$= 4x^2 - 30x - 30x + 225$$

$$= 4x^2 - 60x + 225$$

45. *List Price* A computer is discounted 15% from its list price to a sale price of $1955. Find the list price.

Solution

Verbal model: | List Price | · | Discount Rate | = | Sale Price |

Equation: $x \cdot 85\% = 1955$

$$x = \frac{1955}{.85}$$

$$x = \$2300$$

47. Find two positive integers that satisfy the requirement.

The product of two consecutive odd integers is 483.

Solution

Verbal model: | Odd integer | · | Odd integer | = | Product |

Equation: $(2n + 1) \cdot (2n + 3) = 483$

$$4n^2 + 8n + 3 = 483$$

$$4n^2 + 8n - 480 = 0$$

$$n^2 + 2n - 120 = 0$$

$$(n + 12)(n - 10) = 0$$

$$n + 12 = 0 \qquad n - 10 = 0$$

$$\cancel{n = -12} \qquad n = 10$$

$$\qquad\qquad 2n + 1 = 21$$

$$\qquad\qquad 2n + 3 = 23$$

49. Find two positive integers that satisfy the requirement.

The sum of the squares of two consecutive integers is 313.

Solution

Verbal model: | Integer squared | + | Integer squared | = | Sum |

Equation: $n^2 + (n + 1)^2 = 313$

$$n^2 + n^2 + 2n + 1 = 313$$

$$2n^2 + 2n - 312 = 0$$

$$n^2 + n - 156 = 0$$

$$(n - 12)(n + 13) = 0$$

$$n - 12 = 0 \qquad n + 13 = 0$$

$$n = 12 \qquad \cancel{n = -13}$$

$$n + 1 = 13 \qquad \cancel{n + 1 = -12}$$

51. Find the area of the rectangle.

Width: w Length: $w + 3$ Perimeter: 54 km   Area:

Solution

Verbal model: 2 $\boxed{\text{Length}}$ + 2 $\boxed{\text{Width}}$ = $\boxed{\text{Perimeter}}$

Equation: $2(w + 3) + 2w = 54$

$$2w + 6 + 2w = 54$$

$$4w = 48$$

$$w = 12$$

$$l = w + 3 = 15$$

Verbal model: $\boxed{\text{Length}}$ · $\boxed{\text{Width}}$ = $\boxed{\text{Area}}$

Equation: $15 \cdot 12 = 180 = A$

53. Find the perimeter of the rectangle.

Width: $\dfrac{3}{4}l$ Length: l Perimeter: Area: 2700 in^2

Solution

Verbal model: $\boxed{\text{Length}}$ · $\boxed{\text{Width}}$ = $\boxed{\text{Area}}$

Equation: $l \cdot \dfrac{3}{4}l = 2700$

$$\dfrac{3}{4}l^2 = 2700$$

$$l^2 = 3600$$

$$l = 60$$

$$w = \dfrac{3}{4}l = 45$$

Verbal model: 2 $\boxed{\text{Length}}$ + 2 $\boxed{\text{Width}}$ = $\boxed{\text{Perimeter}}$

Equation: $2(60) + 2(45) = P$

$$210 = P$$

55. *Geometry* An open-top rectangular conduit for carrying water in a manufacturing process is made by folding up the edges of a sheet of aluminum 48 inches wide (see figure in textbook). A cross section of the conduit must have an area of 288 square inches. Find the width and height of the conduit.

Solution

Verbal model: Height · Width = Area

Equation:
$$x \cdot (48 - 2x) = 288$$
$$2x^2 - 48x + 288 = 0$$
$$x^2 - 24x + 144 = 0$$
$$(x - 12)(x - 12) = 0$$
$$x = 12$$
$$\text{height} = 12 \text{ in}$$
$$\text{width} = 48 - 2(12)$$
$$= 48 - 24 = 24 \text{ in}$$

57. *Compound Interest* Find the interest rate r. Use the formula $A = P(1 + r)^2$, where A is the amount after 2 years in an account earning r percent compounded annually and P is the original investment.

$$P = \$250.00, \quad A = \$280.90$$

Solution

$$A = P(1 + r)^2$$
$$280.90 = 250.00(1 + r)^2$$
$$\frac{280.90}{250.00} = (1 + r)^2$$
$$1.1236 = (1 + r)^2$$
$$1.06 = 1 + r$$
$$.06 = r$$
$$6\% =$$

59. *Compound Interest* Find the interest rate r. Use the formula $A = P(1 + r)^2$, where A is the amount after 2 years in an account earning r percent compounded annually and P is the original investment.

$$P = \$10,000.00, \quad A = \$11,556.25$$

Solution

$$A = P(1 + r)^2$$
$$11,556.25 = 10,000.00(1 + r)^2$$
$$\frac{11,556.25}{10,000.00} = (1 + r)^2$$
$$1.155625 = (1 + r)^2$$
$$1.075 = 1 + r$$
$$.075 = r$$
$$7.5\% =$$

61. *Venture Capital* Eighty thousand dollars is needed to begin a small business. The cost will be divided equally among investors. Some have made a commitment to invest. If three more investors are found, the amount required from each would decrease by $6000. How many have made a commitment to invest in the business?

Solution

Verbal Model: $\boxed{\begin{array}{c}\text{Investment}\\\text{per person}\\\text{current group}\end{array}} - \boxed{\begin{array}{c}\text{Investment}\\\text{per person}\\\text{new group}\end{array}} = \boxed{6000}$

Equation:

$$\frac{80{,}000}{x} - \frac{80{,}000}{x+3} = 6000$$

$$x(x+3)\left(\frac{80{,}000}{x} - \frac{80{,}000}{x+3}\right) = (6000)x(x+3)$$

$$80{,}000(x+3) - 80{,}000x = 6000\left(x^2 + 3x\right)$$

$$80{,}000x + 240{,}000 - 80{,}000x = 6000x^2 + 18{,}000x$$

$$0 = 6000x^2 + 18{,}000x - 240{,}000$$

$$0 = x^2 + 3x - 40$$

$$0 = (x+8)(x-5)$$

$$x + 8 = 0 \qquad x - 5 = 0$$

$$\cancel{x = -8} \qquad x = 5 \text{ investors}$$

63. *Delivery Route* You are asked to deliver pizza to Offices B and C in your city (see figure in textbook), and you are required to keep a log of all the mileages between stops. You forget to look at the odometer at stop B, but after getting to stop C you record the total distance traveled from the pizza shop as 18 miles. The return distance from C to A is 16 miles. If the route approximates a right triangle, estimate the distance from A to B.

Solution

Common Formula: $a^2 + b^2 = c^2$

Equation:

$$x^2 + (18 - x)^2 = 16^2$$

$$x^2 + 324 - 36x + x^2 = 256$$

$$2x^2 - 36x + 68 = 0$$

$$x^2 - 18x + 34 = 0$$

$$x = \frac{18 \pm \sqrt{18^2 - 4(1)(34)}}{2(1)}$$

$$x = \frac{18 \pm \sqrt{324 - 136}}{2} = \frac{18 \pm \sqrt{188}}{2}$$

$$x = 15.855655, \quad \text{reject } 2.1443454$$

$$\approx 15.86 \text{ miles}$$

65. *Work Rate* A builder works with two plumbing companies. Company A is known to take 3 days longer than Company B to do the plumbing in a particular style of house. Using both companies it takes 4 days. How long would it take to do the plumbing using each company individually?

Solution

Verbal Model: $\boxed{\text{Rate Company A}} + \boxed{\text{Rate Company B}} = \boxed{\text{Rate Together}}$

Equation:

$$\frac{1}{x+3} + \frac{1}{x} = \frac{1}{4}$$

$$4x(x+3)\left(\frac{1}{x+3} + \frac{1}{x}\right) = \left(\frac{1}{4}\right)4x(x+3)$$

$$4x + 4(x+3) = x(x+3)$$

$$4x + 4x + 12 = x^2 + 3x$$

$$0 = x^2 - 5x - 12$$

$$x = \frac{-(-5) \pm \sqrt{(-5)^2 - 4(1)(-12)}}{2(1)}$$

$$x = \frac{5 \pm \sqrt{25 + 48}}{2}$$

$$x = \frac{5 \pm \sqrt{73}}{2}$$

$$x = 6.8 \text{ days} \qquad \cancel{x = -1.8}$$

$$x + 3 = 9.8 \text{ days}$$

67. *Height of a Baseball* The height h in feet of a baseball hit 3 feet above the ground is given by

$$h = 3 + 75t - 16t^2$$

where t is the time in seconds. Find the time when the ball hits the ground in the outfield.

Solution

$$h = 3 + 75t - 16t^2$$

$$0 = 3 + 75t - 16t^2$$

$$0 = 16t^2 - 75t - 3$$

$$t = \frac{75 \pm \sqrt{(-75)^2 - 4(16)(-3)}}{2(16)}$$

$$t = \frac{75 \pm \sqrt{5625 + 192}}{32}$$

$$t = \frac{75 \pm \sqrt{5817}}{32}$$

$$t = \frac{75 \pm 76.26926}{32}$$

$$t = 4.7271644, \quad \text{reject} \ -0.0396644$$

$$\approx 4.7 \text{ sec}$$

69. Solve for r in $S = 4\pi r^2$.

Solution

$$S = 4\pi r^2$$

$$\frac{S}{4\pi} = r^2$$

$$\sqrt{\frac{S}{4\pi}} = r$$

71. Solve for b in $I = \frac{1}{12}(M)(a^2 + b^2)$

Solution

$$I = \frac{1}{12}(M)\left(a^2 + b^2\right)$$

$$\frac{12I}{M} = a^2 + b^2$$

$$\frac{12I}{M} - a^2 = b^2$$

$$\sqrt{\frac{12I}{M} - a^2} = b = \sqrt{\frac{12I - a^2 M}{M}}$$

73. A small business uses a minivan to make deliveries. The cost per hour for fuel for the van is

$$C = \frac{v^2}{600},$$

where v is the speed in miles per hour. The driver is paid \$5 per hour. Find the speed if the cost for wages and fuel for a 110-mile trip is \$20.39.

Solution

Verbal Model: | Total Cost | = | Wage Cost | + | Fuel Cost |

Equation:

$$20.39 = 5x + x\left[\frac{\left(\frac{110}{x}\right)^2}{600}\right]$$

$$20.39 = 5x + \frac{121}{6x}$$

$$122.34x = 30x^2 + 121$$

$$0 = 30x^2 - 122.34x + 121$$

$$x = \frac{-(-122.34) \pm \sqrt{(-122.34)^2 - 4(30)(121)}}{2(30)}$$

$$x = \frac{122.34 \pm \sqrt{477.0756}}{60}$$

$$x \approx 2.39$$

$$v = \frac{110}{2.39} \approx 46 \text{ mi/hr}$$

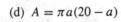 **75.** *Geometry* The area A of an ellipse is given by $A = \pi ab$ (see figure in textbook). For a certain ellipse it is required that $a + b = 20$.

(a) Show that $A = \pi a(20 - a)$.

(b) Complete the table.

a	4	7	10	13	16
A					

(c) Find two values of a such that $A = 300$.

(d) Use a graphing utility to graph the area function.

Solution

(a) $a + b = 20$ $\qquad$ $A = \pi ab$

 $b = 20 - a$ $\qquad$ $A = \pi a(20 - a)$

(b)

a	4	7	10	13	16
A	201.1	285.9	314.2	285.9	201.1

$A = \pi(4)(20 - 4)$ $\qquad$ $A = \pi(7)(20 - 7)$ $\qquad$ $A = \pi(10)(20 - 10)$

$ = \pi(4)(16)$ $\qquad\quad$ $= \pi(7)(13)$ $\qquad\quad$ $= \pi(10)(10)$

$ = 64\pi$ $\qquad\qquad\quad$ $= 91\pi$ $\qquad\qquad\quad$ $= 100\pi$

$ = 201.1$ $\qquad\qquad$ $= 285.9$ $\qquad\qquad$ $= 314.2$

$A = \pi(13)(20 - 13)$ $\qquad$ $A = \pi(16)(20 - 16)$

$ = \pi(13)(7)$ $\qquad\qquad$ $= \pi(16)(4)$

$ = 91\pi$ $\qquad\qquad\quad$ $= 64\pi$

$ = 285.9$ $\qquad\qquad$ $= 201.1$

(c) $300 = \pi a(20 - a)$

$ 0 = 20\pi a - \pi a^2 - 300$

$ 0 = \pi a^2 - 20\pi a + 300$

$ a = \dfrac{-(-20) \pm \sqrt{(-20\pi)^2 - 4(\pi)(300)}}{2(\pi)}$

$ a = \dfrac{20\pi \pm \sqrt{177.9305761}}{2\pi}$

$ a = 12.1, 7.9$

(d) $A = \pi a(20 - a)$

Keystrokes:

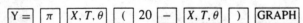

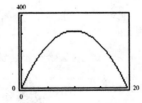

6.5 | Quadratic and Rational Inequalities

7. Find the critical numbers of $x(2x - 5)$.

Solution

Critical numbers $= 0, \dfrac{5}{2}$

9. Find the critical numbers of $x^2 - 4x + 3$.

Solution

$(x - 3)(x - 1)$

Critical numbers $= 3, 1$

11. Determine whether the x-values are solutions of $2x^2 - 7x - 4 > 0$.

(a) $x = 0$ (b) $x = 6$ (c) $x = -\dfrac{1}{2}$ (d) $x = -\dfrac{3}{2}$

Solution

(a) $2(0)^2 - 7(0) - 4 \overset{?}{>} 0$

$\qquad -4 > 0$ No

(b) $2(6)^2 - 7(6) - 4 \overset{?}{>} 0$

$\qquad 72 - 42 - 4 > 0$

$\qquad\qquad 26 > 0$ Yes

(c) $2\left(-\dfrac{1}{2}\right)^2 - 7\left(-\dfrac{1}{2}\right) - 4 \overset{?}{>} 0$

$\qquad 2\left(\dfrac{1}{4}\right) + \dfrac{7}{2} - 4 > 0$

$\qquad \dfrac{1}{2} + \dfrac{7}{2} - 4 > 0$

$\qquad\qquad 0 > 0$ No

(d) $2\left(-\dfrac{3}{2}\right)^2 - 7\left(-\dfrac{3}{2}\right) - 4 \overset{?}{>} 0$

$\qquad 2\left(\dfrac{9}{4}\right) + \dfrac{21}{2} - 4 > 0$

$\qquad \dfrac{9}{2} + \dfrac{21}{2} - \dfrac{8}{2} > 0$

$\qquad\qquad 11 > 0$ Yes

13. Determine whether the x-values are solutions of $\dfrac{2}{3 - x} \leq 0$.

(a) $x = 3$ (b) $x = 4$ (c) $x = -4$ (d) $x = -\dfrac{1}{3}$

Solution

(a) $\dfrac{2}{3 - 3} \leq 0$

$\qquad \dfrac{2}{0} \leq 0$

undefined no

(b) $\dfrac{2}{3 - 4} \leq 0$

$\qquad \dfrac{2}{-1} \leq 0$

$\qquad -2 \leq 0$

Yes

(c) $\dfrac{2}{3 - (-4)} \leq 0$

$\qquad \dfrac{2}{7} \leq 0$

No

(d) $\dfrac{2}{3 - \left(-\dfrac{1}{3}\right)} \leq 0$

$\qquad \dfrac{2}{\dfrac{9}{3} + \dfrac{1}{3}} \leq 0$

$\qquad \dfrac{2}{\dfrac{10}{3}} \leq 0$

$\qquad \dfrac{3}{5} \leq 0$ No

15. Determine the intervals for which $x - 4$ is entirely negative and entirely positive.

Solution

Negative: $(-\infty, 4)$

Positive: $(4, \infty)$

17. Determine the intervals for which $x^2 - 4x - 5$ is entirely negative and entirely positive.

Solution

$(x - 5)(x + 1)$

Positive: $(-\infty, -1)$

Negative: $(-1, 5)$

Positive: $(5, \infty)$

19. Solve the inequality $2x + 6 \geq 0$ and graph the solution on the real number line. (Some of the inequalities have no solution.)

Solution

$2(x + 3) \geq 0$

Critical number: $x = -3$

Test intervals:

Negative: $(-\infty, -3]$

Positive: $[-3, \infty)$

Solution: $[-3, \infty)$

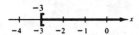

21. Solve the inequality $3x(2 - x) < 0$ and graph the solution on the real number line. (Some of the inequalities have no solution.)

Solution

Critical numbers: $x = 0, 2$

Test intervals:

Negative: $(-\infty, 0)$

Positive: $(0, 2)$

Negative: $(2, \infty)$

Solution: $(-\infty, 0)$ and $(2, \infty)$

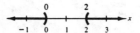

23. Solve the inequality $x^2 + 3x \leq 10$ and graph the solution on the real number line. (Some of the inequalities have no solution.)

Solution

$x^2 + 3x - 10 \leq 0$

$(x + 5)(x - 2) \leq 0$

Critical numbers: $x = -5, 2$

Test intervals:

Positive: $(-\infty, -5]$

Negative: $[-5, 2]$

Positive: $[2, \infty)$

Solution: $[-5, 2]$

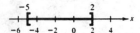

25. Solve the inequality $x^2 + 4x + 5 < 0$ and graph the solution on the real number line. (Some of the inequalities have no solution.)

Solution

$x = \dfrac{-4 \pm \sqrt{16 - 20}}{2}$

No critical numbers

$x^2 + 4x + 5$ is not less than zero for any value of x.

Solution: none

27. Solve the inequality $x^2 + 2x + 1 \geq 0$ and graph the solution on the real number line. (Some of the inequalities have no solution.)

Solution

$(x + 1)^2 \geq 0$

$(x + 1)^2 \geq 0$ for all real numbers

Solution: $(-\infty, \infty)$

29. Solve the inequality $x(x - 2)(x + 2) > 0$ and graph the solution on the real number line. (Some of the inequalities have no solution.)

Solution

Critical numbers: $x = 0, 2, -2$

Test intervals:

Negative: $(-\infty, -2)$

Positive: $(-2, 0)$

Negative: $(0, 2)$

Positive: $(2, \infty)$

Solution: $(-2, 0) \cup (2, \infty)$

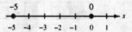

31. Determine the critical number(s) of $\dfrac{5}{x - 3}$ and plot them on the real number line.

Solution

Critical number: $x = 3$

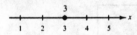

33. Determine the critical number(s) of $\dfrac{2x}{x + 5}$ and plot them on the real number line.

Solution

Critical number: $x = 0, -5$

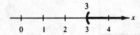

35. Solve the rational inequality $\dfrac{5}{x - 3} > 0$. As part of your solution, include a graph that shows the test intervals.

Solution

Critical number: $x = 3$

Test intervals:

Negative: $(-\infty, 3)$

Positive: $(3, \infty)$

Solution: $(3, \infty)$

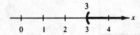

37. Solve the rational inequality $\dfrac{x}{x-3} \le 2$. As part of your solution, include a graph that shows the test intervals.

Solution

$$\frac{x}{x-3} \le 2$$

$$\frac{x}{x-3} - 2 \le 0$$

$$\frac{x - 2(x-3)}{x-3} \le 0$$

$$\frac{x - 2x + 6}{x-3} \le 0$$

$$\frac{-x+6}{x-3} \le 0$$

$$\frac{-1(x-6)}{x-3} \le 0$$

Critical numbers: $x = 3, 6$

Test intervals:

Negative: $(-\infty, 3)$

Positive: $(3, 6]$

Negative: $[6, \infty)$

Solution: $(-\infty, 3) \cup [6, \infty)$

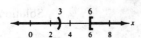

39. Use a graphing utility to solve $0.5x^2 + 1.25x - 3 > 0$. Use the strategies desribed on page 420 and on page 423 in the textbook.

Solution

Keystrokes:

$\boxed{Y=}$ 0.5 $\boxed{X,T,\theta}$ $\boxed{x^2}$ $\boxed{+}$ 1.25 $\boxed{X,T,\theta}$ $\boxed{-}$ 3 $\boxed{>}$ 0 $\boxed{\text{GRAPH}}$

$x < -4$ or $x > \dfrac{3}{2}$

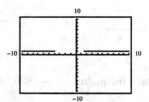

41. Use a graphing utility to solve $\dfrac{6x}{x+5} < 2$.

Solution

Keystrokes:

$\boxed{Y=}$ 6 $\boxed{X,T,\theta}$ $\boxed{\div}$ $\boxed{(}$ $\boxed{X,T,\theta}$ $\boxed{+}$ 5 $\boxed{)}$ $\boxed{<}$ 2 $\boxed{\text{GRAPH}}$

$-5 < x < 2.5$

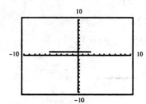

43. *Height of a Projectile* A projectile is fired straight up from ground level with an initial velocity of 128 feet per second, so that its height at any time t is given by

$$h = -16t^2 + 128t$$

where the height h is measured in feet and the time t is measured in seconds. During what interval of time will the height of the projectile exceed 240 feet?

Solution

$$\text{height} > 240$$

$$-16t^2 + 128t > 240$$

$$-16t^2 + 128t - 240 > 0$$

$$t^2 - 8t + 15 < 0$$

$$(t - 3)(t - 5) < 0$$

Critical numbers: $x = 3, 5$

Test intervals:

Positive: $(-\infty, 3)$

Negative: $(3, 5)$

Positive: $(5, \infty)$

Solution: $(3, 5)$

45. Solve $7 - 3x > 4 - x$.

Solution

$$7 - 3x > 4 - x$$

$$-2x > -3$$

$$x < \frac{3}{2}$$

47. Solve $|x - 3| < 2$.

Solution

$$|x - 3| < 2$$

$$x - 3 < 2 \quad \text{and} \quad x - 3 > -2$$

$$x < 5 \qquad\qquad x > 1$$

$$1 < x < 5$$

49. *Geometry* Determine the length and width of a rectangle with a perimeter of 50 inches and a diagonal of length $5\sqrt{13}$ inches.

Solution

$$P = 2l + 2w$$

$$50 = 2l + 2w$$

$$25 = l + w$$

$$= x + w$$

$$25 - x = w$$

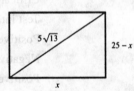

Formula: $c^2 = a^2 + b^2$

Equation: $\left(5\sqrt{13}\right)^2 = x^2 + (25 - x)^2$

$$325 = x^2 + 625 - 50x + x^2$$

$$0 = 2x^2 - 50x + 300$$

$$0 = x^2 - 25x + 150$$

$$0 = (x - 10)(x - 15)$$

$$x - 10 = 0 \qquad x - 15 = 0 \qquad\qquad 10 \text{ in.} \times 15 \text{ in.}$$

$$x = 10 \qquad\qquad x = 15$$

$$25 - x = 15 \qquad 25 - x = 10$$

51. Find the critical numbers.

$$4x^2 - 81$$

Solution

$$4x^2 - 81 = 0$$

$$x^2 = \frac{81}{4}$$

$$x = \pm\frac{9}{2}$$

Critical numbers: $\frac{9}{2}, -\frac{9}{2}$

53. Find the critical numbers.

$$4x^2 - 20x + 25$$

Solution

$$4x^2 - 20x + 25 = 0$$

$$(2x - 5)^2 = 0$$

$$2x - 5 = 0$$

$$x = \frac{5}{2}$$

Critical numbers: $\frac{5}{2}$

55. Determine the intervals for which $\frac{2}{3}x - 8$ is entirely negative and entirely positive.

Solution

Negative: $(-\infty, 12)$

Positive: $(12, \infty)$

57. Determine the intervals for which $2x(x - 4)$ is entirely negative and entirely positive.

Solution

Positive: $(-\infty, 0)$

Negative: $(0, 4)$

Positive: $(4, \infty)$

59. Solve $-\frac{3}{4}x + 6 < 0$ and graph the solution on the real number line. (Some of the inequalities have no solution.)

Solution

Critical number: $x = 8$

Test intervals:

Negative: $(-\infty, 8)$

Positive: $(8, \infty)$

Solution: $(8, \infty)$

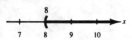

61. Solve $3x(x - 2) < 0$ and graph the solution on the real number line. (Some of the inequalities have no solution.)

Solution

Critical numbers: $x = 0, 2$

Test intervals:

Positive: $(-\infty, 0)$

Negative: $(0, 2)$

Positive: $(2, \infty)$

Solution: $(0, 2)$

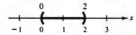

63. Solve $x^2 - 4 \geq 0$ and graph the solution on the real number line. (Some of the inequalities have no solution.)

Solution

$(x - 2)(x + 2) \geq 0$

Critical numbers: $x = 2, -2$

Test intervals:

Positive: $(-\infty, 2]$

Negative: $[-2, 2]$

Positive: $[2, \infty)$

Solution: $(-\infty, -2] \cup [2, \infty)$

67. Solve $x^2 - 4x + 2 > 0$ and graph the solution on the real number line. (Some of the inequalities have no solution.)

Solution

$x = \dfrac{4 \pm \sqrt{16 - 8}}{2}$

$= \dfrac{4 \pm \sqrt{8}}{2} = \dfrac{4 \pm 2\sqrt{2}}{2}$

$= 2 \pm \sqrt{2}$

Critical numbers: $x = 2 + \sqrt{2}, 2 - \sqrt{2}$

Test intervals:

Positive: $\left(-\infty, 2 - \sqrt{2}\right)$

Negative: $\left(2 - \sqrt{2}, 2 + \sqrt{2}\right)$

Positive: $\left(2 + \sqrt{2}, \infty\right)$

Solution: $\left(-\infty, 2 - \sqrt{2}\right) \cup \left(2 + \sqrt{2}, \infty\right)$

65. Solve $-2u^2 + 7u + 4 < 0$ and graph the solution on the real number line. (Some of the inequalities have no solution.)

Solution

$2u^2 - 7u - 4 > 0$

$(2u + 1)(u - 4) > 0$

Critical numbers: $u = -\dfrac{1}{2}, 4$

Test intervals:

Positive: $\left(-\infty, -\dfrac{1}{2}\right)$

Negative: $\left(-\dfrac{1}{2}, 4\right)$

Positive: $(4, \infty)$

Solution: $\left(-\infty, -\dfrac{1}{2}\right) \cup (4, \infty)$

69. Solve $(x - 5)^2 < 0$ and graph the solution on the real number line. (Some of the inequalities have no solution.)

Solution

$(x - 5)^2 > 0$ for all real numbers except 5.

Solution: none

71. Solve $6 - (x - 5)^2 < 0$ and graph the solution on the real number line. (Some of the inequalities have no solution.)

Solution

$$6 - \left(x^2 - 10x + 25\right) < 0$$

$$6 - x^2 + 10x - 25 < 0$$

$$x^2 - 10x + 19 > 0$$

$$x = \frac{10 \pm \sqrt{100 - 76}}{2}$$

$$= \frac{10 \pm \sqrt{24}}{2} = \frac{10 \pm 2\sqrt{6}}{2}$$

$$= 5 \pm \sqrt{6}$$

Critical numbers: $x = 5 + \sqrt{6}, 5 - \sqrt{6}$

Test intervals:

Positive: $\left(-\infty, 5 - \sqrt{6}\right)$

Negative: $\left(5 - \sqrt{6}, 5 + \sqrt{6}\right)$

Positive: $\left(5 + \sqrt{6}, \infty\right)$

Solution: $\left(-\infty, 5 - \sqrt{6}\right) \cup \left(5 + \sqrt{6}, \infty\right)$

73. Solve $x^2 - 6x + 9 \geq 0$ and graph the solution on the real number line. (Some of the inequalities have no solution.)

Solution

$$x^2 - 6x + 9 \geq 0$$

$$(x - 3)^2 \geq 0$$

Critical number: $x = 3$

Test intervals:

Positive: $(-\infty, 3]$

Positive: $[3, \infty)$

Solution: $(-\infty, \infty)$

75. Solve the rational inequality $\dfrac{4}{x - 3} < 0$. As part of your solution, include a graph that shows the test intervals.

Solution

Critical number: $x = 3$

Test intervals:

Negative: $(-\infty, 3)$

Positive: $(3, \infty)$

Solution: $(-\infty, 3)$

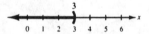

77. Solve the rational inequality $\dfrac{-5}{x - 3} > 0$. As part of your solution, include a graph that shows the test intervals.

Solution

Critical number: $x = 3$

Test intervals:

Positive: $(-\infty, 3)$

Negative: $(3, \infty)$

Solution: $(-\infty, 3)$

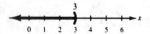

79. Solve the rational inequality $\dfrac{x}{x-4} \le 0$. As part of your solution, include a graph that shows the test intervals.

Solution

Critical numbers: $x = 0, 4$

Test intervals:

Positive: $(-\infty, 0]$

Negative: $[0, 4]$

Positive: $[4, \infty)$

Solution: $[0, 4)$

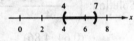

81. Solve the rational inequality $\dfrac{y-3}{2y-11} \ge 0$. As part of your solution, include a graph that shows the test intervals.

Solution

Critical numbers: $y = 3, \dfrac{11}{2}$

Test intervals:

Positive: $(-\infty, 3]$

Negative: $[3, \dfrac{11}{2}]$

Positive: $[\dfrac{11}{2}, \infty)$

Solution: $(-\infty, 3] \cup \left(\dfrac{11}{2}, \infty\right)$

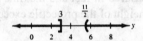

83. Solve the rational inequality $\dfrac{6}{x-4} > 2$. As part of your solution, include a graph that shows the test intervals.

Solution

$$\frac{6}{x-4} > 2$$

$$\frac{6}{x-4} - 2 > 0$$

$$\frac{6 - 2(x-4)}{x-4} > 0$$

$$\frac{6 - 2x + 8}{x-4} > 0$$

$$\frac{14 - 2x}{x-4} > 0$$

$$\frac{-2(-7+x)}{x-4} > 0$$

Critical numbers: $x = 7, 4$

Test intervals:

Negative: $(-\infty, 4)$

Positive: $(4, 7)$

Negative: $(7, \infty)$

Solution: $(4, 7)$

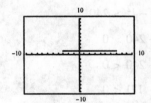

85. Use a graphing utility to solve $9 - 0.2(x-2)^2 > 4$. Use the strategies desribed on page 420 and on page 423 in the textbook.

Solution

Keystrokes:

$\boxed{Y=}$ 9 $\boxed{-}$ 0.2 $\boxed{(}$ $\boxed{X, T, \theta}$ $\boxed{-}$ 2 $\boxed{)}$ $\boxed{x^2}$ $\boxed{>}$ 4 $\boxed{\text{GRAPH}}$

$-3 < x < 7$

87. Use a graphing utility to solve $\dfrac{3x - 4}{x - 4} < -5$. Use the strategies desribed on page 420 and on page 423 in the textbook.

Solution

Keystrokes:

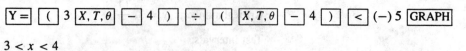

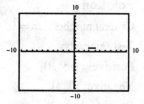

$3 < x < 4$

89. *Height of a Projectile* A projectile is fired straight up from ground level with an initial velocity of 88 feet per second, so that its height at any time t is given by

$$h = -16t^2 + 88t$$

where the height h is measured in feet and the time t is measured in seconds. During what interval of time will the height of the projectile exceed 50 feet?

Solution

$$\text{height} > 50$$

$$-16t^2 + 88t > 50$$

$$-16t^2 + 88t - 50 > 0$$

$$8t^2 - 44t + 25 < 0$$

Critical numbers: $t = \dfrac{11 + \sqrt{71}}{4}$ $t = \dfrac{11 - \sqrt{71}}{4}$

Test intervals:

Positive: $\left(-\infty, \dfrac{11 - \sqrt{71}}{4}\right)$

Negative: $\left(\dfrac{11 - \sqrt{71}}{4}, \dfrac{11 + \sqrt{71}}{4}\right)$

Positive: $\left(\dfrac{11 + \sqrt{71}}{4}, \infty\right)$

Solution: $\left(\dfrac{11 - \sqrt{71}}{4}, \dfrac{11 + \sqrt{71}}{4}\right)$

$(0.64, 4.86)$

91. *Geometry* You have 64 feet of fencing to enclose a rectangular region as shown in the figure in the textbook. Determine the interval for the length such that the area will exceed 240 square feet.

Solution

$$\text{Area} > 240$$

$$l(32 - l) > 240$$

$$32l - l^2 > 240$$

$$-l^2 + 32l - 240 > 0$$

$$l^2 - 32l + 240 < 0$$

$$(l - 20)(l - 12) < 0$$

Critical numbers: $l = 20, 12$

Test intervals:

Positive: $(-\infty, 12)$

Negative: $(12, 20)$

Positive: $(20, \infty)$

Solution: $(12, 20)$

93. *Geometry* Two circles are tangent to each other (see figure in textbook). The distance between their centers is 12 inches.

(a) If x is the radius of one of the circles, express the combined areas of the circles as a function of x. What is the domain of the function?

(b) Use a graphing utility to graph the function found in part (a).

(c) Determine the radii of the circles if the combined areas of the circles must be at least 300 square inches but no more than 400 square inches.

Solution

(a) *Verbal model:* $\boxed{\begin{array}{c}\text{Area}\\\text{Both}\\\text{Circles}\end{array}} = \boxed{\begin{array}{c}\text{Area}\\\text{Circle 1}\end{array}} + \boxed{\begin{array}{c}\text{Area}\\\text{Circle 2}\end{array}}$

Equation: $A = \pi x^2 + \pi (12 - x)^2$

Domain: $0 < x < 12$

(b) *Keystrokes:*

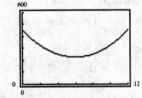

(c) $300 \le \pi x^2 + \pi(12 - x)^2 \le 400$

$$\frac{300}{\pi} < x^2 + 144 - 24x + x^2 < \frac{400}{\pi}$$

$$\frac{300}{\pi} < 2x^2 - 24x + 144 < \frac{400}{\pi}$$

$$\frac{150}{\pi} < x^2 - 12x + 72 < \frac{200}{\pi}$$

$0 < x^2 - 12x + 24.25$ and $x^2 - 12x + 8.34 < 0$

$x = \dfrac{12 \pm \sqrt{144 - 4(24.25)}}{2}$ $x = \dfrac{12 \pm \sqrt{144 - 4(8.34)}}{2}$

$x = 9.43, 2.57$ $x = 11.26, 0.74$

so $0.74 < x < 2.57$

Review Exercises for Chapter 6

1. Solve $x^2 + 12x = 0$ by factoring.

Solution

$$x^2 + 12x = 0$$
$$x(x + 12) = 0$$
$$x = 0 \qquad x + 12 = 0$$
$$x = 0 \qquad\qquad x = -12$$

3. Solve $3z(z + 10) - 8(z + 10) = 0$ by factoring.

Solution

$$3z(z + 10) - 8(z + 10) = 0$$
$$(z + 10)(3z - 8) = 0$$
$$z + 10 = 0 \qquad 3z - 8 = 0$$
$$z = -10 \qquad z = \frac{8}{3}$$

5. Solve $4y^2 - 1 = 0$ by factoring.

Solution

$$4y^2 - 1 = 0$$
$$(2y - 1)(2y + 1) = 0$$
$$2y - 1 = 0 \qquad 2y + 1 = 0$$
$$y = \frac{1}{2} \qquad y = -\frac{1}{2}$$

7. Solve $x^2 + \frac{8}{3}x + \frac{16}{9} = 0$ by factoring.

Solution

$$x^2 + \frac{8}{3}x + \frac{16}{9} = 0$$
$$9x^2 + 24x + 16 = 0$$
$$(3x + 4)(3x + 4) = 0$$
$$3x + 4 = 0 \qquad 3x + 4 = 0$$
$$x = -\frac{4}{3} \qquad x = -\frac{4}{3}$$

9. Solve $2x^2 - 2x = 180$ by factoring.

Solution

$$2x^2 - 2x - 180 = 0$$
$$2(x^2 - x - 90) = 0$$
$$2(x - 10)(x + 9) = 0$$
$$x - 10 = 0 \qquad x + 9 = 0$$
$$x = 10 \qquad x = -9$$

11. Solve $x^2 = 10,000$ by extracting square roots.

Solution

$$x^2 = 10,000$$
$$x = \pm\sqrt{10,000}$$
$$x = \pm 100$$

13. Solve $y^2 - 2.25 = 0$ by extracting square roots.

Solution

$$y^2 - 2.25 = 0$$
$$y^2 = 2.25$$
$$y = \pm\sqrt{2.25}$$
$$y = \pm 1.5$$

15. Solve $(x - 16)^2 = 400$ by extracting square roots.

Solution

$$(x - 16)^2 = 400$$
$$x - 16 = \pm\sqrt{400}$$
$$x = 16 \pm 20$$
$$x = 36, -4$$

17. Solve $x^2 - 6x - 3 = 0$ by completing the square. Find all real or complex solutions.

Solution

$$x^2 - 6x - 3 = 0$$
$$x^2 - 6x + 9 = 3 + 9$$
$$(x - 3)^2 = 12$$
$$x - 3 = \pm\sqrt{12}$$
$$x = 3 \pm 2\sqrt{3}$$

19. Solve $x^2 - 3x + 3 = 0$ by completing the square. Find all real or complex solutions.

Solution

$$x^2 - 3x + 3 = 0$$
$$x^2 - 3x + \frac{9}{4} = -3 + \frac{9}{4}$$
$$\left(x - \frac{3}{2}\right)^2 = \frac{-12 + 9}{4}$$
$$\left(x - \frac{3}{2}\right)^2 = -\frac{3}{4}$$
$$x - \frac{3}{2} = \pm\sqrt{-\frac{3}{4}}$$
$$x = \frac{3}{2} \pm \frac{i\sqrt{3}}{2}$$

21. Solve $2y^2 + 10y + 3 = 0$ by completing the square. Find all real or complex solutions.

Solution

$$2y^2 + 10y + 3 = 0$$
$$y^2 + 5y + \frac{25}{4} = -\frac{3}{2} + \frac{25}{4}$$
$$\left(y + \frac{5}{2}\right)^2 = \frac{-6 + 25}{4}$$
$$\left(y + \frac{5}{2}\right)^2 = \frac{19}{4}$$
$$y + \frac{5}{2} = \pm\sqrt{\frac{19}{4}}$$
$$y = -\frac{5}{2} \pm \frac{\sqrt{19}}{2}$$

23. Use the Quadratic Formula to solve $y^2 + y - 30 = 0$. Find all real or complex solutions.

Solution

$$y = \frac{-1 \pm \sqrt{1^2 - 4(1)(-30)}}{2(1)}$$
$$y = \frac{-1 \pm \sqrt{1 + 120}}{2}$$
$$y = \frac{-1 \pm \sqrt{121}}{2}$$
$$y = \frac{-1 \pm 11}{2}$$
$$y = 5, -6$$

25. Use the Quadratic Formula to solve $2y^2 + y - 21 = 0$. Find all real or complex solutions.

Solution

$$y = \frac{-1 \pm \sqrt{1^2 - 4(2)(-21)}}{2(2)}$$
$$y = \frac{-1 \pm \sqrt{1 + 168}}{4}$$
$$y = \frac{-1 \pm \sqrt{169}}{4}$$
$$y = \frac{-1 \pm 13}{4}$$
$$y = 3, -\frac{7}{2}$$

27. Use the Quadratic Formula to solve $0.3t^2 - 2t + 5 = 0$. Find all real or complex solutions.

Solution

$$t = \frac{-(-2) \pm \sqrt{(-2)^2 - 4(0.3)(5)}}{2(0.3)}$$
$$t = \frac{2 \pm \sqrt{4 - 6}}{0.6}$$
$$t = \frac{2 \pm \sqrt{-2}}{0.6}$$
$$t = \frac{2 \pm i\sqrt{2}}{0.6}$$

29. Solve $(v - 3)^2 = 250$ by the method of your choice. Find all real or complex solutions.

Solution

$$(v - 3)^2 = 250$$
$$v - 3 = \pm\sqrt{250}$$
$$v = 3 \pm 5\sqrt{10}$$

31. Solve $-x^2 + 5x + 84 = 0$ by the method of your choice. Find all real or complex solutions.

Solution

$$-x^2 + 5x + 84 = 0$$
$$x^2 - 5x - 84 = 0$$
$$(x - 12)(x + 7) = 0$$
$$x - 12 = 0 \qquad x + 7 = 0$$
$$x = 12 \qquad\qquad x = -7$$

33. Solve $(x - 9)^2 - 121 = 0$ by the method of your choice. Find all real or complex solutions.

Solution

$$(x - 9)^2 - 121 = 0$$
$$(x - 9)^2 = 121$$
$$x - 9 = \pm\sqrt{121}$$
$$x = 9 \pm 11$$
$$x = 20, -2$$

35. Solve $z^2 - 6z + 10 = 0$ by the method of your choice. Find all real or complex solutions.

Solution

$$z^2 - 6z + 10 = 0$$
$$z^2 - 6z + 9 = -10 + 9$$
$$(z - 3)^2 = -1$$
$$z - 3 = \pm\sqrt{-1}$$
$$z = 3 \pm i$$

37. Solve $2y^2 + 3y + 1 = 0$ by the method of your choice. Find all real or complex solutions.

Solution

$$y = \frac{-3 \pm \sqrt{3^2 - 4(2)(1)}}{2(2)}$$
$$y = \frac{-3 \pm \sqrt{9 - 8}}{4}$$
$$y = \frac{-3 \pm 1}{4}$$
$$y = -\frac{1}{2}, -1$$

39. Solve $\dfrac{1}{x} + \dfrac{1}{x + 1} = \dfrac{1}{2}$ by the method of your choice. Find all real or complex solutions.

Solution

$$\frac{1}{x} + \frac{1}{x + 1} = \frac{1}{2}$$
$$2x(x + 1)\left[\frac{1}{x} + \frac{1}{x + 1} = \frac{1}{2}\right]2x(x + 2)$$
$$2(x + 1) + 2x = x(x + 1)$$
$$2x + 2 + 2x = x^2 + x$$
$$0 = x^2 - 3x - 2$$
$$x = \frac{-(-3) \pm \sqrt{(-3)^2 - 4(1)(-2)}}{2(1)}$$
$$x = \frac{3 \pm \sqrt{9 + 8}}{2}$$
$$x = \frac{3 \pm \sqrt{17}}{2}$$

41. Solve the equation of quadratic form.

$$\left(x^2 - 2x\right)^2 - 4\left(x^2 - 2x\right) - 5 = 0$$

Solution

$$\left(x^2 - 2x\right)^2 - 4\left(x^2 - 2x\right) - 5 = 0$$

$$\left[\left(x^2 - 2x\right) - 5\right]\left[\left(x^2 - 2x\right) + 1\right] = 0$$

$$\left(x^2 - 2x - 5\right)\left(x^2 - 2x + 1\right) = 0$$

$$x = \frac{-(-2) \pm \sqrt{(-2)^2 - 4(1)(-5)}}{2(1)}$$

$$x = \frac{2 \pm \sqrt{4 + 20}}{2}$$

$$x = \frac{2 \pm \sqrt{24}}{2}$$

$$x = \frac{2 \pm 2\sqrt{6}}{2}$$

$$x = 1 \pm \sqrt{6}$$

$$(x - 1)^2 = 0$$

$$x = 1$$

43. Solve the equation of quadratic form.

$$6\left(\frac{1}{x}\right)^2 + 7\left(\frac{1}{x}\right) - 3 = 0$$

Solution

$$6\left(\frac{1}{x}\right)^2 + 7\left(\frac{1}{x}\right) - 3 = 0$$

$$\left[3\left(\frac{1}{x}\right) - 1\right]\left[2\left(\frac{1}{x}\right) + 3\right] = 0$$

$$3\left(\frac{1}{x}\right) - 1 = 0 \qquad 2\left(\frac{1}{x}\right) + 3 = 0$$

$$\frac{3}{x} = 1 \qquad\qquad \frac{2}{x} = -3$$

$$3 = x \qquad\qquad 2 = -3x$$

$$-\frac{2}{3} = x$$

45. Use a graphing utility to graph the function. Use the graph to approximate any x-intercepts of the graph. Set $y = 0$ and solve the resulting equation. Compare the result with the x-intercepts of the graph.

$$y = x^2 - 7x$$

Solution

Keystrokes:

$\boxed{Y=}$ $\boxed{X, T, \theta}$ $\boxed{x^2}$ $\boxed{-}$ 7 $\boxed{X, T, \theta}$ $\boxed{\text{GRAPH}}$

$$0 = x^2 - 7x$$

$$0 = x(x - 7)$$

$$x = 0 \qquad x - 7 = 0$$

$$x = 7$$

x-intercepts are 0 and 7

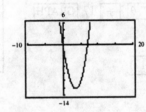

47. Use a graphing utility to graph the function. Use the graph to approximate any x-intercepts of the graph. Set $y = 0$ and solve the resulting equation. Compare the result with the x-intercepts of the graph.

$$y = \frac{1}{16}x^4 - x^2 + 3$$

Solution

Keystrokes:

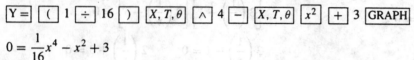

$$0 = \frac{1}{16}x^4 - x^2 + 3$$

$$0 = x^4 - 16x^2 + 48$$

$$0 = \left(x^2 - 4\right)\left(x^2 - 12\right)$$

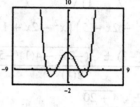

$$x^2 - 4 = 0 \qquad x^2 - 12 = 0$$

$$x^2 = 4 \qquad\qquad x^2 = 12$$

$$x = \pm 2 \qquad\qquad x = \pm\sqrt{12}$$

$$x = \pm 2\sqrt{3}$$

x-intercepts are $2, -2, 2\sqrt{3}, -2\sqrt{3}$

49. *Graphical Reasoning* Use a graphing utility to graph the function and observe that the graph has no x-intercepts. Set $y = 0$ and solve the resulting equation. Identify the type of roots of the equation.

$$y = x^2 - 8x + 17$$

Solution

Keystrokes:

Y= X,T,θ x² − 8 X,T,θ + 17 GRAPH

$$0 = x^2 - 8x + 17$$

$$x = \frac{-(-8) \pm \sqrt{(-8)^2 - 4(1)(17)}}{2(1)}$$

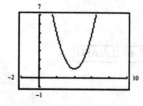

$$x = \frac{8 \pm \sqrt{64 - 68}}{2}$$

$$x = \frac{8 \pm \sqrt{-4}}{2}$$

$$x = \frac{8 \pm 2i}{2} = 4 \pm i \quad \text{(complex roots)}$$

51. Find two consecutive positive integers such that the sum of their squares is 265.

Solution

Verbal model: $\boxed{\begin{array}{c}\text{Positive integer}\\\text{squared}\end{array}} + \boxed{\begin{array}{c}\text{Positive integer}\\\text{squared}\end{array}} = \boxed{\text{Sum}}$

Equation:
$$n^2 + (n+1)^2 = 265$$
$$n^2 + n^2 + 2n + 1 = 265$$
$$2n^2 + 2n - 264 = 0$$
$$n^2 + n - 132 = 0$$
$$(n+12)(n-11) = 0$$

$$n + 12 = 0 \qquad n - 11 = 0$$

$$\cancel{n = -12} \qquad\qquad n = 11$$

$$\qquad\qquad\qquad n + 1 = 12$$

53. *Falling Time* The height h in feet of an object above the ground is given by

$$h = 200 - 16t^2, \quad t \geq 0$$

where t is the time in seconds.

(a) How high was the object when it was thrown?

(b) Was the object thrown upward or downward? Explain your reasoning.

(c) Find the time when the object strikes the ground.

Solution

(a) 200 feet   (b) downward, graph turns downward

(c)
$$h = 200 - 16t^2, t \geq 0$$
$$0 = 200 - 16t^2$$
$$16t^2 = 200$$
$$t^2 = \frac{200}{16}$$
$$t = +\sqrt{\frac{200}{16}}$$
$$t = \frac{10}{4}\sqrt{2} = \frac{5}{2}\sqrt{2} \approx 3.54 \text{ sec}$$

55. *Geometry* The perimeter of a rectangle of length l and width w is 48 feet (see figure in text book).

(a) Show that $w = 24 - l$.

(b) Show that the area is $A = lw = l(24 - l)$.

(c) Use the equation in part (b) to complete the table (see textbook).

Solution

(a)
$$P = 2l + 2w$$
$$48 = 2l + 2w$$
$$48 - 2l = 2w$$
$$\frac{48 - 2l}{2} = w$$
$$24 - l = w$$

(b)
$$A = lw$$
$$= l(24 - l)$$
$$= 24l - l^2$$

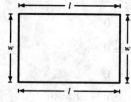

(c)

l	2	4	6	8	10	12	14	16	18
A	44	80	108	128	140	144	140	128	108

$A = 2(24 - 2)$

$\quad = 2(22)$

$\quad = 44$

$A = 4(24 - 4)$

$\quad = 4(20)$

$\quad = 80$

$A = 6(24 - 6)$

$\quad = 6(18)$

$\quad = 108$

$A = 8(24 - 8)$

$\quad = 8(16)$

$\quad = 128$

$A = 10(24 - 10)$

$\quad = 10(14)$

$\quad = 140$

$A = 12(24 - 12)$

$\quad = 12(12)$

$\quad = 144$

$A = 14(24 - 14)$

$\quad = 14(10)$

$\quad = 140$

$A = 16(24 - 16)$

$\quad = 16(8)$

$\quad = 128$

$A = 18(24 - 18)$

$\quad = 18(6)$

$\quad = 108$

57. *Decreased Price* A Little League baseball team obtains a block of tickets to a ball game for $96. When three more people decided to go to the game, the price per ticket is decreased by $1.60. How many people are going to the game?

Solution

Verbal model: $\boxed{\begin{array}{c}\text{Cost per}\\\text{ticket}\end{array}} \cdot \boxed{\begin{array}{c}\text{Number of}\\\text{tickets}\end{array}} = \boxed{\$96}$

Equation:

$$\left(\frac{96}{x} - 1.60\right)(x + 3) = 96$$

$$\left(\frac{96 - 1.60x}{x}\right)(x + 3) = 96$$

$$(96 - 1.6x)(x + 3) = 96x$$

$$96x - 1.6x^2 - 4.8x + 288 = 96x$$

$$1.6x^2 + 4.8x - 288 = 0$$

$$x^2 + 3x - 180 = 0$$

$$(x - 12)(x + 15) = 0$$

$$x - 12 = 0 \qquad x + 15 = 0$$

$$x = 12 \qquad\qquad x = -15 \text{ reject}$$

$$x + 3 = 15$$

59. *Work Rate* Working together, two people can complete a task in six hours. Working alone, how long would it take each to do the task if one person takes 3 hours longer than the other?

Solution

Verbal model: $\boxed{\begin{array}{c}\text{Work done}\\\text{by Person 1}\end{array}} + \boxed{\begin{array}{c}\text{Work done}\\\text{by Person 2}\end{array}} = \boxed{\text{One complete job}}$

Equation:

$$\frac{1}{x}(6) + \frac{1}{x + 3}(6) = 1$$

$$x(x + 3)\left[(6)\left(\frac{1}{x} + \frac{1}{x + 3}\right) = 1\right]x(x + 3)$$

$$6(x + 3) + 6x = x(x + 3)$$

$$6x + 18 + 6x = x^2 + 3x$$

$$-x^2 + 9x + 18 = 0$$

$$x^2 - 9x - 18 = 0$$

$$x = \frac{9 \pm \sqrt{(-9)^2 - 4(1)(-18)}}{2(1)}$$

$$x = \frac{9 \pm \sqrt{153}}{2}$$

$$x = \frac{9 \pm 12.369317}{2}$$

$$x = 10.684658, \quad \text{reject} -1.6846584$$

$$x + 3 = 13.684658$$

61. *Reduced Fare* The college wind ensemble charters a bus to attend a concert at a cost of $360. In an attempt to lower the bus fare per person, the ensemble invites nonmembers to go along. When eight nonmembers join the trip, the fare is decreased by $1.50. How many people are going on the excursion?

Solution

Verbal model: $\boxed{\begin{array}{c}\text{Cost per person}\\\text{Current Group}\end{array}} - \boxed{\begin{array}{c}\text{Cost per person}\\\text{New Group}\end{array}} = \boxed{\$1.50}$

Equation:

$$\frac{360}{x} - \frac{360}{x+8} = 1.50$$

$$[x(x+8)]\left(\frac{360}{x} - \frac{360}{x+8}\right) = (1.50)[x(x+8)]$$

$$360(x+8) - 360x = 1.50\left(x^2 + 8x\right)$$

$$360x + 2880 - 360x = 1.50x^2 + 12x$$

$$0 = 1.5x^2 + 12x - 2880$$

$$0 = x^2 + 8x - 1920$$

$$0 = (x+48)(x-40)$$

$$x + 48 = 0 \qquad x - 40 = 0$$

$$\cancel{x = -48} \qquad\quad x = 40$$

$$x + 8 = 48$$

63. *Work Rate* Working together, two people can complete a task in 10 hours. Working alone, how long would it take each to do the task if one person takes 2 hours longer than the other?

Solution

Verbal model: $\boxed{\begin{array}{c}\text{Work done}\\\text{by Person 1}\end{array}} + \boxed{\begin{array}{c}\text{Work done}\\\text{by Person 2}\end{array}} = \boxed{\text{One complete job}}$

Equation:

$$\frac{1}{x}(10) + \frac{1}{x+2}(10) = 1$$

$$x(x+2)\left[10\left(\frac{1}{x}+\frac{1}{x+2}\right)\right] = [1]x(x+2)$$

$$10(x+2) + 10x = x(x+2)$$

$$10x + 20 + 10x = x^2 + 2x$$

$$0 = x^2 - 18x - 20$$

$$x = \frac{-(-18) \pm \sqrt{(-18)^2 - 4(1)(-20)}}{2(1)}$$

$$x = \frac{18 \pm \sqrt{324 + 80}}{2}$$

$$x = \frac{18 \pm \sqrt{404}}{2}$$

$$x = \frac{18 \pm 2\sqrt{101}}{2}$$

$$x = 9 \pm \sqrt{101}$$

$$x = 19 \qquad \cancel{x < 1}$$

$$x + 2 = 21$$

65. Find a quadratic equation with the solutions $x = -3$ and $x = 3$.

Solution

$$x = -3 \qquad\qquad x = 3$$

$$x + 3 = 0 \qquad\quad x - 3 = 0$$

$$(x + 3)(x - 3) = 0$$

$$x^2 - 9 = 0$$

67. Find a quadratic equation with the solutions $x = -5$ and $x = -1$.

Solution

$$x = -5 \qquad\qquad x = -1$$

$$x + 5 = 0 \qquad\quad x + 1 = 0$$

$$(x + 5)(x + 1) = 0$$

$$x^2 + 6x + 5 = 0$$

69. Write a quadratic function that has the x-intercepts $(1, 0)$, $(3, 0)$.

Solution

$$y = (x - 1)(x - 3)$$

$$y = x^2 - 4x + 3$$

71. Write a quadratic function that has the x-intercepts $(-4, 0)$, $(-1, 0)$.

Solution

$$y = [x - (-4)][x - (-1)]$$

$$y = (x + 4)(x + 1)$$

$$y = x^2 + 5x + 4$$

73. *Think About It* Find a *different* quadratic equation that has the same solution as the equation you found in Exercise 65.

Solution

　　Answer will vary. Multiply both sides of equation $x^2 - 9 = 0$ by the same non-zero real number. For example, if you multiply by $\frac{1}{9}$ equation becomes $\frac{1}{9}x^2 - 1 = 0$.

75. Solve the inequality $4x - 12 < 0$ and graph its solution on the real number line. (Some of the inequalities have no solution.)

Solution

$4x - 12 < 0$

$4(x - 3) < 0$

Critical number: $x = 3$

Test intervals:

Negative: $(-\infty, 3)$

Positive: $(3, \infty)$

Solution: $(-\infty, 3)$

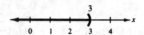

77. Solve the inequality $5x(7 - x) > 0$ and graph its solution on the real number line. (Some of the inequalities have no solution.)

Solution

Critical numbers: $x = 0, 7$

Test intervals:

Negative: $(-\infty, 0)$

Positive: $(0, 7)$

Negative: $(7, \infty)$

Solution: $(0, 7)$

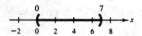

79. Solve the inequality $16 - (x - 2)^2 \leq 0$ and graph its solution on the real number line. (Some of the inequalities have no solution.)

Solution

$16 - (x - 2)^2 \leq 0$

$(4 - x + 2)(4 + x - 2) \leq 0$

$(6 - x)(2 + x) \leq 0$

Critical numbers: $x = -2, 6$

Test intervals:

Negative: $(-\infty, -2]$

Positive: $[-2, 6]$

Negative: $[6, \infty)$

Solution: $(-\infty, -2] \cup [6, \infty)$

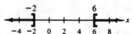

81. Solve the inequality $2x^2 + 3x - 20 < 0$ and graph its solution on the real number line. (Some of the inequalities have no solution.)

Solution

$2x^2 + 3x - 20 < 0$

$(2x - 5)(x + 4) < 0$

Critical numbers: $x = -4, \dfrac{5}{2}$

Test intervals:

Positive: $(-\infty, -4)$

Negative: $\left(-4, \dfrac{5}{2}\right)$

Positive: $\left(\dfrac{5}{2}, \infty\right)$

Solution: $\left(-4, \dfrac{5}{2}\right)$

83. Solve the inequality $\dfrac{x}{2x - 7} \geq 0$ and graph its solution on the real number line. (Some of the inequalities have no solution.)

Solution

Critical numbers: $x = 0, \dfrac{7}{2}$

Test intervals:

Positive: $(-\infty, 0]$

Negative: $\left[0, \dfrac{7}{2}\right)$

Positive: $\left(\dfrac{7}{2}, \infty\right)$

Solution: $(-\infty, 0] \cup \left(\dfrac{7}{2}, \infty\right)$

85. *College Completion* The percent p of the American population that graduated from college from 1960 to 1990 is approximated by the model

$p = (2.739 + 0.064t)^2, \quad 0 \leq t,$

where $t = 0$ represents 1960 (see figure in text book). According to this model, when will the percent of college graduates exceed 25% of the population? (*Source:* U.S. Bureau of Census)

Solution

$(2.739 + 0.064t)^2 \geq 25$

$2.739 + 0.064t \geq 5$

$0.064t \geq 2.261$

$t \geq 35.328125$

$t = 35$

$1960 + 35 = 1995$

1995

Test for Chapter 6

1. Solve $x(x + 5) - 10(x + 5) = 0$ by factoring.

Solution

$$x(x + 5) - 10(x + 5) = 0$$
$$(x + 5)(x - 10) = 0$$
$$x + 5 = 0 \qquad x - 10 = 0$$
$$x = -5 \qquad x = 10$$

2. Solve $8x^2 - 21x - 9 = 0$ by factoring.

Solution

$$8x^2 - 21x - 9 = 0$$
$$(8x + 3)(x - 3) = 0$$
$$8x + 3 = 0 \qquad x - 3 = 0$$
$$x = -\frac{3}{8} \qquad x = 3$$

3. Solve $(x - 2)^2 = 0.09$ by extracting square roots.

Solution

$$(x - 2)^2 = 0.09$$
$$x - 2 = \pm 0.3$$
$$x = 2 \pm 0.3$$
$$x = 2.3, 1.7$$

4. Solve $(x + 3)^2 + 81 = 0$ by extracting square roots.

Solution

$$(x + 3)^2 + 81 = 0$$
$$(x + 3)^2 = -81$$
$$x + 3 = \pm\sqrt{-81}$$
$$x = -3 \pm 9i$$

5. Find the real number c such that $x^2 - 3x + c$ is a perfect square trinomial.

Solution

$$c = \frac{9}{4}$$

$$\left[\frac{1}{2}(-3)\right]^2 = \left(-\frac{3}{2}\right)^2 = \frac{9}{4}$$

6. Solve $2x^2 - 6x + 3 = 0$ by completing the square.

Solution

$$2x^2 - 6x + 3 = 0$$
$$x^2 - 3x + \frac{9}{4} = -\frac{3}{2} + \frac{9}{4}$$
$$\left(x - \frac{3}{2}\right)^2 = \frac{-6 + 9}{4}$$
$$\left(x - \frac{3}{2}\right)^2 = \frac{3}{4}$$
$$x - \frac{3}{2} = \pm\sqrt{\frac{3}{4}}$$
$$x = \frac{3}{2} \pm \frac{\sqrt{3}}{2}$$

7. Find the discriminant and explain how it may be used to determine the type of solutions of the quadratic equation $5x^2 - 12x + 10 = 0$.

Solution

$$b^2 - 4ac = (-12)^2 - 4(5)(10)$$
$$= 144 - 200$$
$$= -56$$

2 complex solutions

8. Solve $3x^2 - 8x + 3 = 0$ by using the Quadratic Formula.

Solution

$$x = \frac{-(-8) \pm \sqrt{(-8)^2 - 4(3)(3)}}{2(3)}$$
$$= \frac{8 \pm \sqrt{64 - 36}}{6}$$
$$= \frac{8 \pm \sqrt{28}}{6} = \frac{8 \pm 2\sqrt{7}}{6} = \frac{4 \pm \sqrt{7}}{3}$$

9. Solve $2y(y-2) = 7$ by using the Quadratic Formula.

Solution

$$2y(y-2) = 7$$

$$2y^2 - 4y - 7 = 0$$

$$y = \frac{-(-4) \pm \sqrt{(-4)^2 - 4(2)(-7)}}{2(2)}$$

$$y = \frac{4 \pm \sqrt{16+56}}{4}$$

$$y = \frac{4 \pm \sqrt{72}}{4}$$

$$y = \frac{4 \pm 6\sqrt{2}}{4}$$

$$y = \frac{2 \pm 3\sqrt{2}}{2} \approx 7.41 \text{ and } -0.41$$

10. Solve $\sqrt{3x+7} - \sqrt{x+12} = 1$, and round the result to two decimal places.

Solution

$$\sqrt{3x+7} - \sqrt{x+12} = 1$$

$$\left(\sqrt{3x+7}\right)^2 = \left(1 + \sqrt{x+12}\right)^2$$

$$3x + 7 = 1 + 2\sqrt{x+12} + x + 12$$

$$2x - 6 = 2\sqrt{x+12}$$

$$(x-3)^2 = \left(\sqrt{x+12}\right)^2$$

$$x^2 - 6x + 9 = x + 12$$

$$x^2 - 7x - 3 = 0$$

$$x = \frac{-(-7) \pm \sqrt{(-7)^2 - 4(1)(-3)}}{2(1)}$$

$$x = \frac{7 \pm \sqrt{49+12}}{2}$$

$$x = \frac{7 \pm \sqrt{61}}{2} \approx 7.41 \text{ and } \cancel{-0.41}$$

$x \approx 7.41$ is only solution since $x = -0.41$ does not check in original equation.

11. Find a quadratic equation having the solutions -4 and 5.

Solution

$$(x - (-4))(x - 5) = 0$$

$$(x+4)(x-5) = 0$$

$$x^2 - x - 20 = 0$$

12. Find a quadratic equation that has $i - 1$ as a solution.

Solution

$$[x - (-1+i)][x - (-1-i)] = 0$$

$$(x+1-i)(x+1+i) = 0$$

$$[(x+1) - i][(x+1) + i] = 0$$

$$(x+1)^2 - i^2 = 0$$

$$(x+1)^2 + 1 = 0$$

$$x^2 + 2x + 1 + 1 = 0$$

$$x^2 + 2x + 2 = 0$$

13. Solve $2x(x - 3) < 0$ and sketch the solution.

Solution

Critical numbers: $x = 0, 3$

Test intervals:

Positive: $(-\infty, 0)$

Negative: $(0, 3)$

Positive: $(3, \infty)$

Solution: $(0, 3)$

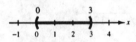

14. Solve $16 \leq (x - 2)^2$ and sketch the solution.

Solution

$$16 \leq (x - 2)^2$$

$$(x - 2)^2 \geq 16$$

$$x^2 - 4x + 4 \geq 16$$

$$x^2 - 4x - 12 \geq 0$$

$$x^2 - 4x - 12 \geq 0$$

$$(x - 6)(x + 2) \geq 0$$

Critical numbers: $x = -2, 6$

Test intervals:

Positive: $(-\infty, -2]$

Negative: $[-2, 6]$

Positive: $[6, \infty)$

Solution: $(-\infty, -2] \cup [6, \infty)$

15. Solve $\dfrac{3}{x - 2} > 4$ and sketch the solution.

Solution

$$\frac{3}{x - 2} > 4$$

$$\frac{3}{x - 2} - 4 > 0$$

$$\frac{3 - 4(x - 2)}{x - 2} > 0$$

$$\frac{3 - 4x + 8}{x - 2} > 0$$

$$\frac{11 - 4x}{x - 2} > 0$$

$$\frac{-1(4x - 11)}{x - 2} > 0$$

Critical numbers: $x = \dfrac{11}{4}, 2$

Test intervals:

Negative: $(-\infty, 2)$

Positive: $\left(2, \dfrac{11}{4}\right)$

Negative: $\left(\dfrac{11}{4}, \infty\right)$

Solution: $\left(2, \dfrac{11}{4}\right)$

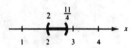

16. Solve $\dfrac{3u+2}{u-3} \le 2$ and sketch the solution.

Solution

$$\frac{3u+2}{u-3} \le 2$$

$$\frac{3u+2}{u-3} - \frac{2(u-3)}{u-3} \le 0$$

$$\frac{3u+2-2u+6}{u-3} \le 0$$

$$\frac{u+8}{u-3} \le 0$$

Critical numbers: $u = -8, 3$

Test intervals:

Positive: $(-\infty, -8]$

Negative: $[-8, +3)$

Positive: $(+3, \infty)$

Solution: $[-8, +3)$

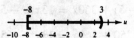

17. Find two consecutive positive integers whose product is 210.

Solution

Verbal model: | Positive integer | $\cdot$ | Consecutive positive integer | $=$ | Product |

Equation:
$$n \cdot (n+1) = 210$$

$$n^2 + n - 210 = 0$$

$$(n+15)(n-14) = 0$$

$$n + 15 = 0 \qquad n - 14 = 0$$

$$\text{reject } n = -15 \qquad n = 14$$

$$n + 1 = -14 \qquad n + 1 = 15$$

$$\text{reject}$$

18. The width of a rectangle is 8 feet less than the length. The area of the rectangle is 240 square feet. Find the dimensions of the rectangle.

Solution

Verbal model: | Area | $=$ | Length | $\cdot$ | Width |

Equation:
$$240 = l \cdot (l-8)$$

$$0 = l^2 - 8l - 240$$

$$0 = (l-20)(l+12)$$

$$0 = l - 20 \qquad 0 = l + 12$$

$$20 = l \qquad -12 = l$$

$$12 = l - 8 \qquad \text{reject}$$

19. A train traveled the first 125 miles of a trip at one speed and the last 180 miles at an average speed of 5 miles per hour less. If the total time for the trip was $6\frac{1}{2}$ hours, what was the average speed for the first part of the trip?

Solution

Verbal model: $\boxed{\text{Distance}} = \boxed{\text{Rate}} \cdot \boxed{\text{Time}}$

Equation:
$$\frac{125}{x} + \frac{180}{x-5} = \frac{13}{2}$$

$$2x(x-5)\left[\frac{125}{x} + \frac{180}{x-5} = \frac{13}{2}\right]2x(x-5)$$

$$250(x-5) + 360x = 13x(x-5)$$

$$250x - 1250 + 360x = 13x^2 - 65x$$

$$0 = 13x^2 - 675x + 1250$$

$$x = \frac{675 \pm \sqrt{675^2 - 4(13)(1,250)}}{26}$$

$$= \frac{675 \pm 625}{26}$$

$$x = 50, \quad \text{reject } \frac{50}{26}$$

$$x - 5 = 45$$

20. An object is dropped from a height of 75 feet. Its height (in feet) at any time is given by

$$h = -16t^2 + 75$$

where the time is measured in seconds. Find the time at which the object has fallen to a height of 35 feet.

Solution

$$35 = -16t^2 + 75$$

$$16t^2 = 40$$

$$t^2 = \frac{40}{16} = \frac{5}{2}$$

$$t = \sqrt{\frac{5}{2}}$$

$$t = \frac{\sqrt{10}}{2} = 1.5811388$$

$$t \approx 1.58 \text{ seconds}$$

Cumulative Test for Chapters 4–6

1. Perform the indicated operations and/or simplify $\dfrac{x^2 + 8x + 16}{18x^2} \cdot \dfrac{2x^4 + 4x^3}{x^2 - 16}$.

Solution

$$\frac{x^2 + 8x + 16}{18x^2} \cdot \frac{2x^4 + 4x^3}{x^2 - 16} = \frac{(x+4)^2}{18x^2} \cdot \frac{2x^3(x+2)}{(x-4)(x+4)} = \frac{x(x+4)(x+2)}{9(x-4)}$$

2. Perform the indicated operations and/or simplify $\dfrac{2}{x} - \dfrac{x}{x^3 + 3x^2} + \dfrac{1}{x+3}$.

Solution

$$\frac{2}{x} - \frac{x}{x^3 + 3x^2} + \frac{1}{x+3} = \frac{2}{x} - \frac{x}{x^2(x+3)} + \frac{1}{x+3}$$

$$= \frac{2}{x} - \frac{1}{x(x+3)} + \frac{1}{x+3}$$

$$= \frac{2}{x}\left(\frac{x+3}{x+3}\right) - \frac{1}{x(x+3)}\left(\frac{1}{1}\right) + \frac{1}{x+3}\left(\frac{x}{x}\right)$$

$$= \frac{2x+6}{x(x+3)} - \frac{1}{x(x+3)} + \frac{x}{x(x+3)}$$

$$= \frac{2x+6-1+x}{x(x+3)}$$

$$= \frac{3x+5}{x(x+3)}$$

3. Perform the indicated operation and/or simplify

$$\frac{\left(\dfrac{x}{y} - \dfrac{y}{x}\right)}{\left(\dfrac{x-y}{xy}\right)}.$$

Solution

$$\frac{\left(\dfrac{x}{y} - \dfrac{y}{x}\right)}{\left(\dfrac{x-y}{xy}\right)} = \frac{\left(\dfrac{x}{y} - \dfrac{y}{x}\right)}{\left(\dfrac{x-y}{xy}\right)} \cdot \frac{xy}{xy}$$

$$= \frac{x^2 - y^2}{x - y}$$

$$= \frac{(x-y)(x+y)}{x-y}$$

$$= x + y$$

4. Perform the indicated operations and/or simplify $\sqrt{-2}\left(\sqrt{-8} + 3\right)$.

Solution

$$\sqrt{-2}\left(\sqrt{-8} + 3\right) = i\sqrt{2}\left(2i\sqrt{2} + 3\right)$$

$$= 2i^2 \cdot 2 + 3i\sqrt{2}$$

$$= -4 + 3i\sqrt{2}$$

5. Perform the indicated operations and/or simplify $\dfrac{-4x^{-3}y^4}{6xy^{-2}}$.

Solution

$$\frac{-4x^{-3}y^4}{6xy^{-2}} = \frac{-2x^{-4}y^6}{3}$$

$$= -\frac{2y^6}{3x^4}$$

6. Perform the indicated operations and/or simplify $\left(\dfrac{t^{1/2}}{t^{1/4}}\right)^2$.

Solution

$$\left(\frac{t^{1/2}}{t^{1/4}}\right)^2 = \frac{t}{t^{1/2}}$$

$$= t^{1/2}$$

7. Perform the indicated operations and/or simplify $10\sqrt{20x} + 3\sqrt{125x}$.

Solution

$$10\sqrt{20x} + 3\sqrt{125x} = 10\sqrt{4 \cdot 5x} + 3\sqrt{25 \cdot 5x}$$

$$= 20\sqrt{5x} + 15\sqrt{5x}$$

$$= 35\sqrt{5x}$$

8. Perform the indicated operations and/or simplify $\left(\sqrt{2x} - 3\right)^2$.

Solution

$$\left(\sqrt{2x} - 3\right)^2 = 2x - 6\sqrt{2x} + 9.$$

9. Perform the indicated operations and/or simplify $\dfrac{6}{\sqrt{10} - 2}$.

Solution

$$\frac{6}{\sqrt{10} - 2} = \frac{6}{\sqrt{10} - 2} \cdot \frac{\sqrt{10} + 2}{\sqrt{10} + 2}$$

$$= \frac{6\left(\sqrt{10} + 2\right)}{10 - 4}$$

$$= \frac{6\left(\sqrt{10} + 2\right)}{6}$$

$$= \sqrt{10} + 2$$

10. Perform the indicated operations and/or simplify $\dfrac{x^3 + 27}{x + 3}$ (use synthetic division).

Solution

$$\frac{x^3 + 27}{x + 3}$$

$$
\begin{array}{r|rrrr}
-3 & 1 & 0 & 0 & 27 \\
   &   & -3 & 9 & -27 \\
\hline
   & 1 & -3 & 9 & 0
\end{array}
$$

$$\frac{x^3 + 27}{x + 3} = x^2 - 3x + 9$$

11. Graph the rational function $y = \dfrac{4}{x - 2}$.

Solution

y-intercept: $y = \dfrac{4}{0 - 2} = -2$

x-intercept: none, numerator is never zero

vertical asymptote: $x = 2$

horizontal asymptote: $y = 0$

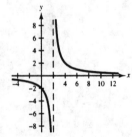

12. Graph the rational function $y = \dfrac{4x^2}{x^2 + 1}$.

Solution

y-intercept: $y = \dfrac{4(0)^2}{0^2 + 1} = 0$

x-intercept: $x = 0$

vertical
asymptote: none

horizontal
asymptote: $y = 4$

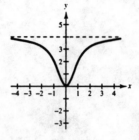

13. Solve $x + \dfrac{4}{x} = 4$.

Solution

$$x + \frac{4}{x} = 4$$

$$x\left[x + \frac{4}{x}\right] = [4]x$$

$$x^2 + 4 = 4x$$

$$x^2 - 4x + 4 = 0$$

$$(x - 2)^2 = 0$$

$$x - 2 = 0$$

$$x = 2$$

14. Solve $\sqrt{x + 10} = x - 2$.

Solution

$$\sqrt{x + 10} = x - 2$$

$$\left(\sqrt{x + 10}\right)^2 = (x - 2)^2$$

$$x + 10 = x^2 - 4x + 4$$

$$0 = x^2 - 5x - 6$$

$$0 = (x - 6)(x + 1)$$

$$0 = x - 6 \qquad 0 = x + 1$$

$$6 = x \qquad\qquad -1 = x$$

Check: $\sqrt{6 + 10} \overset{?}{=} 6 - 2$

$$\sqrt{16} = 4$$

$$4 = 4$$

$$\sqrt{-1 + 10} \overset{?}{=} -1 - 2$$

$$\sqrt{9} = -3$$

$$3 \neq -3$$

15. Solve $(x - 5)^2 + 50 = 0$.

Solution

$$(x - 5)^2 + 50 = 0$$

$$(x - 5)^2 = -50$$

$$x - 5 = \pm\sqrt{-50}$$

$$x = 5 \pm 5i\sqrt{2}$$

16. Solve $3x^2 + 6x + 2 = 0$.

Solution

$$3x^2 + 6x + 2 = 0$$

$$x^2 + 2x + 1 = -\frac{2}{3} + 1$$

$$(x + 1)^2 = \frac{1}{3}$$

$$x + 1 = \pm\sqrt{\frac{1}{3}}$$

$$x = -1 \pm \frac{\sqrt{3}}{3}$$

17. Use a graphing utility to graph the equation $y = x^2 - 6x - 8$. Use the graph to approximate any x-intercepts of the graph. Set $y = 0$ and solve the resulting equation. Compare the results with the x-intercepts of the graph.

Solution

Keystrokes:

$\boxed{Y=}$ $\boxed{X, T, \theta}$ $\boxed{x^2}$ $\boxed{-}$ 6 $\boxed{X, T, \theta}$ $\boxed{-}$ 8 $\boxed{\text{GRAPH}}$

$$0 = x^2 - 6x - 8$$

$$x = \frac{-(-6) \pm \sqrt{(-6)^2 - 4(1)(-8)}}{2(1)}$$

$$= \frac{6 \pm \sqrt{36 + 32}}{2}$$

$$= \frac{6 \pm \sqrt{68}}{2} \approx 7.12 \text{ and } -1.12$$

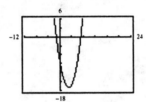

18. Find a quadratic equation having the solutions -2 and 6.

Solution

$$[x - (-2)][x - 6] = 0$$

$$(x + 2)(x - 6) = 0$$

$$x^2 - 4x - 12 = 0$$

19. Evaluate $\left(4 \times 10^3\right)^2$ without the aid of a calculator.

Solution

$$\left(4 \times 10^3\right)^2 = 16 \times 10^6$$

20. The volume V of a right circular cylinder is $V = \pi r^2 h$. The two cylinders in the figure in the text book have equal volumes. Write r_2 as a function of r_1.

Solution

$$\pi r_2{}^2(5) = \pi r_1{}^2(3) \qquad r_2 = \sqrt{\frac{3r_1}{5}}$$

$$r_2{}^2 = \frac{3}{5}r_1{}^2 \qquad r = \frac{\sqrt{15r_1}}{5}$$

CHAPTER 7
Lines, Conics, and Variation

7.1 Writing Equations of Lines

7. Match the equation $y = \frac{2}{3}x + 2$ with its graph.

Solution

(b)

11. From the equation $y - 3 = 6(x + 4)$, determine the slope of the line and the coordinates of one point through which it passes.

Solution

$m = 6 \quad (-4, 3)$

15. Find the point-slope form of the equation of the line passing through the point (0, 0) with the slope $m = -\frac{1}{2}$. Sketch the line.

Solution

$y - 0 = -\frac{1}{2}(x - 0)$

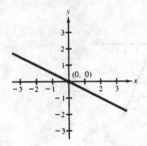

9. Match the equation $y = -\frac{3}{2}x + 2$ with its graph.

Solution

(a)

13. From the equation $3x - 2y = 0$, determine the slope of the line and the coordinates of one point through which it passes.

Solution

$3x - 2y = 0$

$-2y = -3x$

$y = \frac{3}{2}x$

$m = \frac{3}{2} \quad (0, 0)$

17. Find the point-slope form of the equation of the line passing through the point (0, −4) with the slope $m = 3$. Sketch the line.

Solution

$y + 4 = 3(x - 0)$

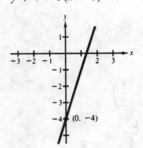

19. Find the point-slope form of the equation of the line passing through the point $\left(-2, \dfrac{7}{2}\right)$ with the slope $m = -4$. Sketch the line.

Solution

$$y - \frac{7}{2} = -4(x + 2)$$

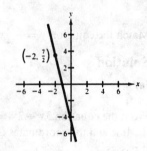

21. Find the general form of the equation of the line through the points $(0, 0)$ and $(2, 3)$. Sketch the line.

Solution

$$m = \frac{3 - 0}{2 - 0} = \frac{3}{2}$$

$$y - 0 = \frac{3}{2}(x - 0)$$

$$y = \frac{3}{2}x$$

$$2y = 3x$$

$$0 = 3x - 2y$$

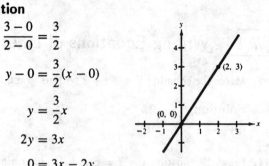

23. Find the general form of the equation of the line through the points $(-2, 3)$ and $(5, 0)$. Sketch the line.

Solution

$$m = \frac{0 - 3}{5 + 2} = -\frac{3}{7}$$

$$y - 0 = -\frac{3}{7}(x - 5)$$

$$y = -\frac{3}{7}x + \frac{15}{7}$$

$$7y = -3x + 15$$

$$3x + 7y - 15 = 0$$

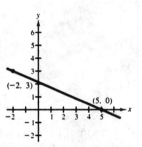

25. Find the general form of the equation of the line through the points $\left(\dfrac{3}{2}, 3\right)$ and $\left(\dfrac{9}{2}, -4\right)$. Sketch the line.

Solution

$$m = \frac{-4 - 3}{\dfrac{9}{2} - \dfrac{3}{2}} = -\frac{7}{\dfrac{6}{2}} = -\frac{7}{3}$$

$$y + 4 = -\frac{7}{3}\left(x - \frac{9}{2}\right)$$

$$y + 4 = -\frac{7}{3}x + \frac{63}{6}$$

$$6y + 24 = -14x + 63$$

$$14x + 6y - 39 = 0$$

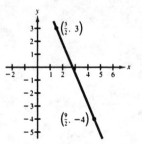

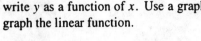 **27.** Write an equation of the line passing through the two points $(-2, 2)$ and $(4, 5)$. Use function notation to write y as a function of x. Use a graphing utility to graph the linear function.

Solution

$$m = \frac{5-2}{4+2} = \frac{3}{6} = \frac{1}{2}$$

$$y - 5 = \frac{1}{2}(x - 4)$$

$$y - 5 = \frac{1}{2}x - 2$$

$$y = \frac{1}{2}x + 3$$

$$f(x) = \frac{1}{2}x + 3$$

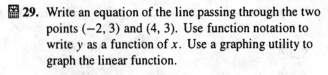 **29.** Write an equation of the line passing through the two points $(-2, 3)$ and $(4, 3)$. Use function notation to write y as a function of x. Use a graphing utility to graph the linear function.

Solution

$$m = \frac{3-3}{4+2} = \frac{0}{6} = 0$$

$$y - 3 = 0(x - 4)$$

$$y - 3 = 0$$

$$y = 3$$

$$f(x) = 3$$

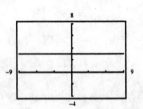

31. Write equations of the line through the point $(2, 1)$ that are (a) parallel and (b) perpendicular to the given line, $y = 3x$.

Solution

$y = 3x$ slope = 3

(a) $y - 1 = 3(x - 2)$

$\quad\ y - 1 = 3x - 6$

$\qquad\quad 0 = 3x - y - 5$

(b) $y - 1 = -\dfrac{1}{3}(x - 2)$

$\qquad y - 1 = -\dfrac{1}{3}x + \dfrac{2}{3}$

$\qquad\qquad 0 = -x - 3y + 5$

$\qquad\qquad\quad$ or

$\qquad\qquad 0 = x + 3y - 5$

33. Write equations of the line through the point $(-5, 4)$ that are (a) parallel and (b) perpendicular to the given line, $x + y = 1$.

Solution

$x + y = 1$ slope = -1

$\quad\ y = -x + 1$

(a) $y - 4 = -1(x - -5)$

$\quad\ y - 4 = -x - 5$

$\qquad\ 0 = x + y + 1$

(b) $y - 4 = 1(x - -5)$

$\quad\ y - 4 = x + 5$

$\qquad\ 0 = x - y + 9$

35. Write an equation of the line through the point $(-1, 2)$ (a) parallel and (b) perpendicular to $y + 5 = 0$.

Solution

$y + 5 = 0$

$y = -5$

(a) $y - 2 = 0(x + 1)$

$y - 2 = 0$

(b) $\quad x = -1$

$x + 1 = 0$

 37. Use a graphing utility on a square setting to graph both lines, $y = -0.6x + 1$ and $y = \frac{5}{3}x - 2$, on the same screen. Decide whether the lines are parallel, perpendicular, or neither.

Solution

Keystrokes:

y_1 [Y=] [(-)] .6 [X,T,θ] [+] 1 [ENTER]

y_2 [Y=] [(] 5 [÷] 3 [)] [X,T,θ] [–] 2 [GRAPH]

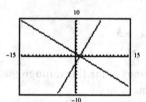

perpendicular

39. Use a graphing utility on a square setting to graph both lines, $y = 0.6x + 1$ and $y = \frac{1}{5}(3x + 22)$, on the same screen. Decide whether the lines are parallel, perpendicular, or neither.

Solution

Keystrokes:

y_1 [Y=] .6 [X,T,θ] [+] 1 [ENTER]

y_2 [Y=] [(] 1 [÷] 5 [)] [(] 3 [X,T,θ] [+] 22 [)] [GRAPH]

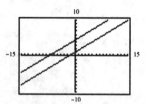

parallel

41. *Cost* The cost (in dollars) of producing x units of a certain product is given by $C = 20x + 5000$. Use this model to complete the table shown in the textbook.

Solution

x	0	50	100	500	1000
C	5000	6000	7000	15,000	25,000

(a) $C = 20(0) + 5000$

$= 5000$

(b) $6000 = 20x + 5000$

$1000 = 20x$

$50 = x$

(c) $C = 20(100) + 5000$

$= 2000 + 5000$

$= 7000$

(d) $15,000 = 20x + 5000$

$10,000 = 20x$

$500 = x$

(e) $C = 20(1000) + 5000$

$= 20,000 + 5000$

$= 25,000$

43. *Rental Occupancy* A real estate office handles an apartment complex with 50 units. When the rent per unit is $450 per month, all 50 units are occupied. However, when the rent is $525 per month, the average number of occupied units drops to 45. Assume the relationship between the monthly rent p and the demand x is linear.

(a) Write the equation of the line giving the demand x in terms of the rent p.

(b) *(Linear Extrapolation)* Use the equation from part (a) to predict the number of units occupied when the rent is $570.

(c) *(Linear Interpolation)* Use the equation from part (a) to predict the number of units occupied when the rent is $480.

Solution

(a) $m = \dfrac{450 - 525}{50 - 45} = -\dfrac{75}{5} = -15$

$p - 450 = -15(x - 50)$

$p - 450 = -15x + 750$

$p = -15x + 1200$

(b) $570 = -15x + 1200$

$-630 = -15x$

$42 = x$

(c) $480 = -15x + 1200$

$-720 = -15x$

$48 = x$

45. Factor $5x = 20x^2$ completely.

Solution

$5x - 20x^2 = 5x(1 - 4x)$

47. Factor $15x^2 - 16x - 15$ completely.

Solution

$15x^2 - 16x - 15 = (5x + 3)(3x - 5)$

49. *Graphical Reasoning* Consider the equation

$$2x^3 - 3x^2 - 18x + 27 = (2x - 3)(x + 3)(x - 3).$$

(a) Use a graphing utility to verify the equation by graphing both the left side and right side of the equation. Are the graphs the same?

(b) Verify the equation by multiplying the polynomials on the right side of the equation.

Solution

(a) Keystrokes:

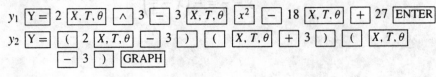

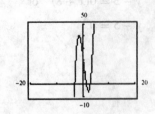

yes

(b) $2x^3 - 3x^2 - 18x + 27 = (2x - 3)(x + 3)(x - 3)$

$= (2x - 3)(x^2 - 9)$

$= 2x^3 - 3x^2 - 18x + 27$

51. From the equation $y + \dfrac{5}{8} = \dfrac{3}{4}(x + 2)$, find the slope of the line (if possible) and the coordinates of one point through which it passes.

Solution

$m = \dfrac{3}{4} \quad \left(-2, -\dfrac{5}{8}\right)$

53. From the equation $y = \dfrac{2}{3}x - 2$, find the slope of the line (if possible) and the coordinates of one point through which it passes.

Solution

$m = \dfrac{2}{3} \quad (0, -2)$

55. From the equation $5x - 2y + 24 = 0$, find the slope of the line (if possible) and the coordinates of one point through which it passes.

Solution

$5x - 2y + 24 = 0$

$\qquad -2y = -5x - 24$

$\qquad\qquad y = \dfrac{5}{2}x + 12$

$m = \dfrac{5}{2} \quad (0, 12)$

57. Find the point-slope form of the equation of the line passing through the point $(0, -6)$ with the slope $m = \dfrac{1}{2}$.

Solution

$y + 6 = \dfrac{1}{2}(x - 0)$

$y + 6 = \dfrac{1}{2}x$

59. Find the point-slope form of the equation of the line passing through the point $(-2, 8)$ with the slope $m = -2$.

Solution

$y - 8 = -2(x + 2)$

61. Find the point-slope form of the equation of the line passing through the point $(5, 0)$ with the slope $m = -\dfrac{2}{3}$.

Solution

$y - 0 = -\dfrac{2}{3}(x - 5)$

63. Find the point-slope form of the equation of the line passing through the point $(-8, 5)$ with the slope $m = 0$.

Solution

$y - 5 = 0(x + 8)$ or

$y - 5 = 0$

65. Find the point-slope form of the equation of the line passing through the point $\left(\dfrac{3}{4}, \dfrac{5}{2}\right)$ with the slope $m = \dfrac{4}{3}$.

Solution

$y - \dfrac{5}{2} = \dfrac{4}{3}\left(x - \dfrac{3}{4}\right)$

67. Find the general form of the equation of the line passing through the two points $(-5, 2)$ and $(5, -2)$.

Solution

$$m = \frac{-2-2}{5+5} = -\frac{4}{10} = -\frac{2}{5}$$

$$y - 2 = -\frac{2}{5}(x + 5)$$

$$y - 2 = -\frac{2}{5}x - 2$$

$$y = -\frac{2}{5}x$$

$$5y = -2x$$

$$2x + 5y = 0$$

71. Find the general form of the equation of the line passing through the two points $(7.5, 2)$ and $(7.5, 9)$.

Solution

$$m = \frac{9-2}{7.5-7.5} = \frac{7}{0} = \text{undefined}$$

$$x = 7.5$$

$$x - 7.5 = 0$$

69. Find the general form of the equation of the line passing through the two points $\left(10, \frac{1}{2}\right)$ and $\left(\frac{3}{2}, \frac{7}{4}\right)$.

Solution

$$m = \frac{7/4 - 1/2}{(3/2) - 10} \cdot \frac{4}{4} = \frac{7-2}{6-40} = \frac{5}{-34}$$

$$y - \frac{1}{2} = -\frac{5}{34}(x - 10)$$

$$y - \frac{1}{2} = -\frac{5}{34}x + \frac{50}{34}$$

$$34y - 17 = -5x + 50$$

$$5x + 34y - 67 = 0$$

73. Find the general form of the equation of the line passing through the two points $(2, 0.6)$ and $(8, -4.2)$.

Solution

$$m = \frac{-4.2 - 0.6}{8 - 2} = -\frac{4.8}{6} = -0.8$$

$$y - 0.6 = -0.8(x - 2)$$

$$y - 0.6 = -0.8x + 1.6$$

$$0.8x + y - 2.2 = 0$$

75. Write equations of the line through the given point $(3, 7)$ that are (a) parallel and (b) perpendicular to the given line, $4x - y - 3 = 0$.

Solution

$$4x - y - 3 = 0 \quad \text{slope} = 4$$

$$4x - 3 = y$$

(a) $\quad y - 7 = 4(x - 3)$

$$y - 7 = 4x - 12$$

$$0 = 4x - y - 5$$

(b) $\quad y - 7 = -\frac{1}{4}(x - 3)$

$$y - 7 = -\frac{1}{4}x + \frac{3}{4}$$

$$4y - 28 = -x + 3$$

$$0 = x + 4y - 31$$

77. Write equations of the line through the given point $(6, -4)$ that are (a) parallel and (b) perpendicular to the given line, $3x + 10y = 24$.

Solution

$$3x + 10y = 24 \qquad\qquad \text{slope} = -\frac{3}{10}$$

$$10y = -3x + 24$$

$$y = -\frac{3}{10}x + \frac{24}{10}$$

(a) $\quad y - -4 = -\dfrac{3}{10}(x - 6)$

$\qquad y + 4 = -\dfrac{3}{10}x + \dfrac{18}{10}$

$\qquad 10y + 40 = -3x + 18$

$\qquad\qquad 0 = 3x + 10y + 22$

(b) $\quad y - -4 = \dfrac{10}{3}(x - 6)$

$\qquad y + 4 = \dfrac{10}{3}x - 20$

$\qquad 3y + 12 = 10x - 60$

$\qquad\qquad 0 = 10x - 3y - 72$

79. Use a graphing utility on a square setting to decide whether the lines, $y = 3(x - 2)$ and $y = 3x + 2$, are parallel, perpendicular, or neither.

Solution

Keystrokes:

y_1 $\boxed{Y=}$ 3 $\boxed{(}$ $\boxed{X,T,\theta}$ $\boxed{-}$ 2 $\boxed{)}$ $\boxed{\text{ENTER}}$

y_2 $\boxed{Y=}$ 3 $\boxed{X,T,\theta}$ $\boxed{+}$ 2 $\boxed{\text{GRAPH}}$

Parallel

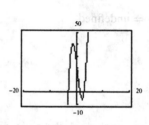

81. Use a graphing utility on a square setting to decide whether the lines, $y = 1 - 1.5x$ and $y = \frac{2}{3}x - 4$, are parallel, perpendicular, or neither.

Solution

Keystrokes:

y_1 $\boxed{Y=}$ 1 $\boxed{-}$ 1.5 $\boxed{X,T,\theta}$ $\boxed{\text{ENTER}}$

y_2 $\boxed{Y=}$ $\boxed{(}$ 2 $\boxed{\div}$ 3 $\boxed{)}$ $\boxed{X,T,\theta}$ $\boxed{-}$ 4 $\boxed{\text{GRAPH}}$

Perpendicular

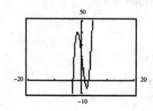

83. Use a graphing utility on a square setting to decide whether the lines $x + 2y - 3 = 0$ and $-2x - 4y + 1 = 0$ are parallel, perpendicular, or neither.

Solution

$x + 2y - 3 = 0$

$\qquad 2y = -x + 3$

$\qquad y = -\dfrac{1}{2}x + \dfrac{3}{2}$

$\qquad y = -.5x + 1.5$

$-2x - 4y + 1 = 0$

$\qquad -4y = 2x - 1$

$\qquad y = -\dfrac{1}{2}x + \dfrac{1}{4}$

$\qquad y = -.5x + .25$

Keystrokes:

y_1 [Y=] [(-)] .5 [X, T, θ] [+] 1.5

y_2 [Y=] [(-)] .5 [X, T, θ] [+] .25 [GRAPH]

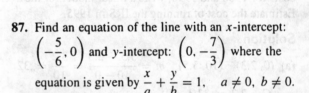

85. Find an equation of the line with an x-intercept: $(3, 0)$ and y-intercept: $(0, 2)$ where the equation is given by

$$\frac{x}{a} + \frac{y}{b} = 1, \quad a \neq 0, \ b \neq 0.$$

Solution

$$\frac{x}{3} + \frac{y}{2} = 1$$

87. Find an equation of the line with an x-intercept: $\left(-\dfrac{5}{6}, 0\right)$ and y-intercept: $\left(0, -\dfrac{7}{3}\right)$ where the equation is given by $\dfrac{x}{a} + \dfrac{y}{b} = 1, \quad a \neq 0, \ b \neq 0.$

Solution

$$\frac{x}{(-5)/6} + \frac{y}{(-7)/3} = 1$$

$$-\frac{6x}{5} - \frac{3y}{7} = 1$$

89. *Sales Commission* The salary for a sales representative is $1500 per month plus a 3% commission of total monthly sales. Write a linear function giving the salary S in terms of the monthly sales M.

Solution

$$S = 1500 + 0.03M$$

91. *Discount Price* A store is offering a 30% discount on all items in its inventory.

(a) Write a linear function giving the sale price S for an item in terms of its list price L.

(b) Use the function in part (a) to find the sale price of an item that has a list price of $135.

Solution

(a) $S = 0.70L$

(b) $S = 0.70(135)$

$\qquad S = \$94.50$

93. *College Enrollment* A small college had an enrollment of 1500 students in 1980. During the next 10 years, the enrollment increased by approximately 60 students per year.

(a) Write a linear function giving the enrollment N in terms of the year t. (Let $t = 0$ represent 1980.)

(b) *Linear Extrapolation* Use the function in part (a) to predict the enrollment in the year 2000.

(c) *Linear Interpolation* Use the function in part (a) to estimate the enrollment in 1985.

Solution

(a) $N = 1500 + 60t$

(b) $N = 1500 + 60(20)$
$= 1500 + 1200$
$= 2700$

(c) $N = 1500 + 60(5)$
$= 1500 + 300$
$= 1800$

95. Use the graph (see the textbook) which shows the cost C (in billions of dollars) of running the Internal Revenue Service from 1980 to 1990. (*Source:* Internal Revenue Service) Using the costs for 1980 and 1990, write a linear model for the average cost, letting $t = 0$ represent 1980. Estimate the cost of running the IRS in 1995.

Solution

(a) $(0, 2.3)$ $(10, 5.5)$ $m = \dfrac{5.5 - 2.3}{10 - 0} = \dfrac{3.2}{10} = .32$

$C - 2.3 = .32(t - 0)$

$C = .32t + 2.3$

(b) $C = .32(15) + 2.3$

$C = \$7.1$ billion

97. *Best-Fitting Line* An instructor gives 20-point quizzes and 100-point exams in a mathematics course. The average quiz and test scores for six students are given as ordered pairs (x, y) where x is the average quiz score and y is the average test score. The ordered pairs are (18, 87), (10, 55), (19, 96), (16, 79), (13, 76) and (15, 82).

(a) Plot the points.

(b) Use a ruler to sketch the best-fitting line through the points.

(c) Find an equation for the line sketched in part (b).

(d) Use the equation in part (c) to estimate the average test score for a person with an average quiz score of 17.

Solution

(a) & (b)

(c) Two points taken from the "best-fitting" line sketched in part (b) are (12, 67) and (20, 99).

$$m = \frac{99 - 67}{20 - 12} = \frac{32}{8} = 4$$

$$y - 67 = 4(x - 12)$$

$$y - 67 = 4x - 48$$

$$y = 4x + 19$$

(d) $y = 4(17) + 19 = 68 + 19 = 87$

7.2 Graphs of Linear Inequalities

7. Using the graphs in the textbook, match the inequality $y \geq -2$ with its graph. [The graphs are labeled (a), (b), (c), (d), (e), and (f).]

Solution

$y \geq -2$ (b)

9. Using the graphs in the textbook, match the inequality $3x - 2y < 0$ with its graph. [The graphs are labeled (a), (b), (c), (d), (e), and (f).]

Solution

$3x - 2x < 0$ (d)

11. Using the graphs in the textbook, match the inequality $x + y < 4$ with its graph. [The graphs are labeled (a), (b), (c), (d), (e), and (f).]

Solution

$x + y < 4$ (f)

13. Determine whether the points are solutions of the inequality $x - 2y < 4$.

(a) $(0, 0)$ (b) $(2, -1)$ (c) $(3, 4)$ (d) $(5, 1)$

Solution

(a) $0 - 2(0) \overset{?}{<} 4$

$0 < 4$

$(0, 0)$ is a solution.

(b) $2 - 2(-1) \overset{?}{<} 4$

$2 + 2 < 4$

$4 \not< 4$

$(2, -1)$ is not a solution.

(c) $3 - 2(4) \overset{?}{<} 4$

$3 - 8 < 4$

$-5 < 4$

$(3, 4)$ is a solution.

(d) $5 - 2(1) \overset{?}{<} 4$

$5 - 2 < 4$

$3 < 4$

$(5, 1)$ is a solution.

15. Determine whether the following points are solutions of the inequality $y > 0.2x - 1$.

(a) $(0, 2)$ (b) $(6, 0)$ (c) $(4, -1)$ (d) $(-2, 7)$

Solution

(a) $2 \overset{?}{>} 0.2(0) - 1$

$2 > -1$

$(0, 2)$ is a solution.

(b) $0 \overset{?}{>} 0.2(6) - 1$

$0 > 0.2$

$(6, 0)$ is a solution.

(c) $-1 \overset{?}{>} 0.2(4) - 1$

$-1 \not> -0.2$

$(4, -1)$ is not a solution.

(d) $7 \overset{?}{>} 0.2(-2) - 1$

$7 > -1.4$

$(-2, 7)$ is a solution.

17. Sketch the graph of the linear inequality $x \geq 2$.

Solution

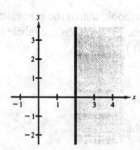

19. Sketch the graph of the linear inequality $y \leq x + 1$.

Solution

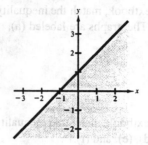

21. Sketch the graph of the linear inequality $x - 2y \geq 6$.

Solution

$x - 2y \geq 6$ or

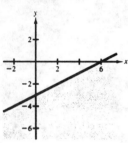

23. Use a graphing utility to graph the solution of the inequality $y \geq \frac{3}{4}x - 1$.

Solution

Keystrokes:

[DRAW] [7] .75 [X, T, θ] [−] 1 [,] 10 [)] [ENTER]

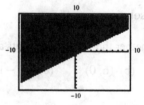

25. Use a graphing utility to graph the solution of the inequality $2x + 3y - 12 \leq 0$.

Solution

$$2x + 3y - 12 \leq 0$$
$$3y \leq -2x + 12$$
$$y \leq -\frac{2}{3}x + 4$$

Keystrokes:

[DRAW] [7] [(] −10, [(] (-)2 [÷] 3 [)] [X, T, θ] [+] 4 [)] [ENTER]

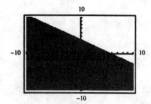

27. Represent all points above the line $y = x$ with an inequality.

Solution

$y > x$

29. Write an inequality for the shaded region shown in the figure (see textbook).

Solution

$m = \dfrac{2-5}{3+1} = -\dfrac{3}{4} \quad y - 2 > -\dfrac{3}{4}(x - 3)$

31. Write an inequality for the shaded region shown in the figure (see textbook).

Solution

$y < 2$

33. Write an inequality for the shaded region shown in the figure (see textbook).

Solution

$m = \dfrac{1-0}{2-0} = \dfrac{1}{2} \quad y > \dfrac{1}{2}x$

35. Sketch the region containing the solution points that satisfy both inequalities. Explain your reasoning.

$x \geq 2, \quad y \leq 4$

Solution

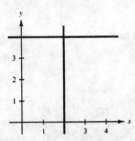

37. Sketch the region containing the solution points that satisfy both inequalities. Explain your reasoning.

$3x + 2y \geq 2, \quad y \geq -5$

Solution

$3x + 2y \geq 2$

$2y \geq -3x + 2$

$y \geq -\dfrac{3}{2}x + 1$

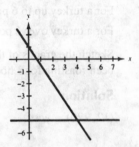

39. Use a graphing utility to graph (shade) the region containing the solution points that satisfy both inequalities.

$2x - 3y \leq 6, \quad y \leq 4$

Solution

$2x - 3y \leq 6$

$-3y \leq -2x + 6$

$y \geq \dfrac{2}{3}x + 2$

Keystrokes:

y_1 Y= (2 ÷ 3) X,T,θ + 2 ENTER

y_2 Y= 4

DRAW 7 Y-VARS 1 1 , Y-VARS 1 2) ENTER

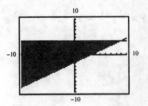

41. Use a graphing utility to graph (shade) the region containing the solution points that satisfy both inequalities.

$$2x + 2y \leq 2, \quad y \geq -4$$

Solution

$$2x + 2y \leq 2$$

$$y \leq -2x + 2$$

Keystrokes:

y_1 Y= (-) 2 X, T, θ + 2 ENTER

y_2 Y= (-) 4

DRAW 7 Y-VARS 1 2 , Y-VARS 1 1) ENTER

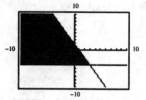

43. *Roasting a Turkey* The time t (in minutes) that it takes to roast a turkey weighing p pounds at 350° F is given by the following inequalities.

For a turkey up to 6 pounds: $t \geq 20p$

For a turkey over 6 pounds: $t \geq 15p + 30$

Sketch the graphs of these inequalities. What are the coordinates for a 12-pound turkey that has been roasting for 3 hours and 40 minutes? Is this turkey fully cooked?

Solution

$(12, 220)$ yes

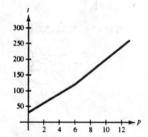

45. Simplify $(x^3 \cdot x^{-2})^{-3}$.

Solution

$$(x^3 \cdot x^{-2})^{-3} = x^{-9}x^6 = x^{-3} = \frac{1}{x^3}$$

47. Simplify $\left(\dfrac{2x}{3y}\right)^{-2}$.

Solution

$$\left(\frac{2x}{3y}\right)^{-2} = \left(\frac{3y}{2x}\right)^2 = \frac{9y^2}{4x^2}$$

49. *Geometry*   If a television set is advertised as having a 19-inch screen, it means that the diagonal measurement of the screen is 19 inches. If the screen is square, what are its height and width?

Solution

$$x^2 + x^2 = 19^2$$

$$2x^2 = 361$$

$$x^2 = 180.5$$

$$x = \sqrt{180.5} \approx 13.435 \text{ in.}$$

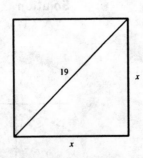

51. Determine whether the points are solutions of the inequality $3x + y \geq 10$.

(a) $(1, 3)$ (b) $(-3, 1)$ (c) $(3, 1)$ (d) $(2, 15)$

Solution

(a) $3(1) + 3 \overset{?}{\geq} 10$

$\qquad 9 \not\geq 10$

$(1, 3)$ is not a solution.

(c) $3(3) + 1 \overset{?}{\geq} 10$

$\qquad 10 \geq 10$

$(3, 1)$ is a solution.

(b) $3(-3) + 1 \overset{?}{\geq} 10$

$\qquad -8 \not\geq 10$

$(-3, 1)$ is not a solution.

(d) $3(2) + 15 \overset{?}{\geq} 10$

$\qquad 21 \geq 10$

$(2, 15)$ is a solution.

53. Determine whether the points are solutions of the inequality $y \leq 3 - |x|$.

(a) $(-1, 4)$ (b) $(2, -2)$ (c) $(6, 0)$ (d) $(5, -2)$

Solution

(a) $4 \overset{?}{\leq} 3 - |-1|$

$\quad 4 \not\leq 3 - 1$

$(-1, 4)$ is not a solution.

(c) $0 \overset{?}{\leq} 3 - |6|$

$\quad 0 \leq 3 - 6$

$\quad 0 \not\leq -3$

$(6, 0)$ is not a solution.

(b) $-2 \overset{?}{\leq} 3 - |2|$

$\quad -2 \leq 3 - 2$

$(2, -2)$ is a solution.

(d) $-2 \overset{?}{\leq} 3 - |5|$

$\quad -2 \leq 3 - 5$

$\quad -2 \leq -2$

$(5, -2)$ is a solution.

55. Sketch the graph of the solution of the linear inequality $y < 5$.

Solution

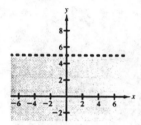

57. Sketch the graph of the solution of the linear inequality $y > \frac{1}{2}x$.

Solution

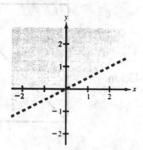

59. Sketch the graph of the solution of the linear inequality $y - 1 > -\frac{1}{2}(x - 2)$.

Solution

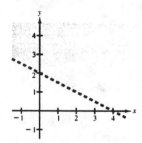

61. Sketch the graph of the solution of the linear inequality $\frac{x}{3} + \frac{y}{4} \le 1$.

Solution

$\frac{x}{3} + \frac{y}{4} \le 1$ or $y \le -\frac{4}{3}x + 4$

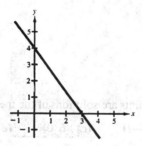

63. Sketch the graph of the solution of the linear inequality $3x - 2y \ge 4$.

Solution

$3x - 2y \ge 4$

$-2y \ge -3x + 4$

$y \le \frac{3}{2}x - 2$

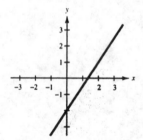

65. Sketch the graph of the solution of the linear inequality $0.2x + 0.3y < 2$.

Solution

$0.2x + 0.3y < 2$

or

$y < -\frac{2}{3}x + \frac{20}{3}$

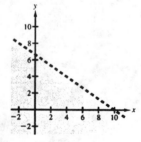

67. Use a graphing utility to graph (shade) the solution of the inequality $y \leq -\frac{2}{3}x + 6$.

Solution

Keystrokes:

[Y=] [(] [(-)] 2 [÷] 3 [)] [X, T, θ] [+] 6
[DRAW] [7] (-) 10 [,] [Y-VARS] [1] [1] [)] [ENTER]

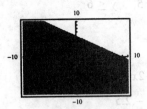

69. Use a graphing utility to graph (shade) the solution of the inequality $x - 2y - 4 \geq 0$.

Solution

$$x - 2y - 4 \geq 0$$
$$-2y \geq -x + 4$$
$$y \leq \frac{1}{2}x - 2$$

Keystrokes:

[Y=] .5 [X, T, θ] [−] 2
[DRAW] [7] [(-)] 10 [,] [Y-VARS] [1] [1] [)] [ENTER]

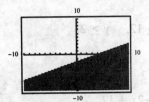

71. Sketch the region showing all points that satisfy both inequalities.

$$x \leq 5, \quad y \geq 2$$

Solution

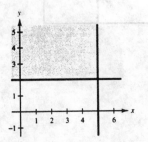

73. Sketch the region showing all points that satisfy both inequalities.

$$x + y \geq 4, \quad y \leq 4$$

Solution

$$x + y \geq 4$$
$$y \geq -x + 4$$

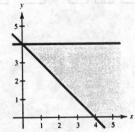

75. Use a graphing utility to graph (shade) the region containing the solution points satisfying both inequalities.

$$2x - 2y \le 5, \quad y \le 6$$

Solution

$$2x - 2y \le 5$$
$$-2y \le -2x + 5$$
$$y \ge x - 2.5$$

Keystrokes:

y_1 $\boxed{Y=}$ $\boxed{X, T, \theta}$ $\boxed{-}$ 2.5 $\boxed{\text{ENTER}}$

y_2 $\boxed{Y=}$ 6

$\boxed{\text{DRAW}}$ $\boxed{7}$ $\boxed{\text{Y-VARS}}$ $\boxed{1}$ $\boxed{1}$ $\boxed{,}$ $\boxed{\text{Y-VARS}}$ $\boxed{1}$ $\boxed{2}$ $\boxed{)}$ $\boxed{\text{ENTER}}$

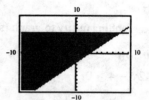

77. Use a graphing utility to graph (shade) the region containing the solution points satisfying both inequalities.

$$2x + 3y \ge 12, \quad y \ge 2$$

Solution

$$2x + 3y \ge 12$$
$$3y \ge -2x + 12$$
$$y \ge -\frac{2}{3}x + 4$$

Keystrokes:

y_1 $\boxed{Y=}$ $\boxed{(}$ $\boxed{(-)}$ 2 $\boxed{\div}$ 3 $\boxed{)}$ $\boxed{X, T, \theta}$ $\boxed{+}$ 4 $\boxed{\text{ENTER}}$

y_2 $\boxed{Y=}$ 2

$\boxed{\text{DRAW}}$ $\boxed{7}$ $\boxed{\text{Y-VARS}}$ $\boxed{1}$ $\boxed{1}$ $\boxed{,}$ 10 $\boxed{,}$ 3 $\boxed{)}$ $\boxed{\text{ENTER}}$

$\boxed{\text{DRAW}}$ $\boxed{7}$ $\boxed{\text{Y-VARS}}$ $\boxed{1}$ $\boxed{2}$ $\boxed{,}$ 10 $\boxed{,}$ 2 $\boxed{)}$ $\boxed{\text{ENTER}}$

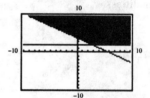

79. *Geometry*   The perimeter of a rectangle of length x and width y cannot exceed 500 feet. Write a linear inequality for this constraint and sketch its graph.

Solution

(a) $P = 2x + 2y$

$\quad 2x + 2y \le 500$

or

$\quad 0 \le x + y \le 250$

or

$\quad y \le -x + 250$

(Note: x and y cannot be negative.)

(b)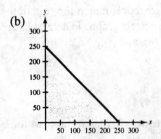

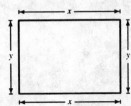

81. *Diet Supplement*   A dietician is asked to design a special diet supplement using two foods. Each ounce of food x contains 30 units of calcium and each ounce of food y contains 20 units of calcium. The minimum daily requirement in the diet is 300 units of calcium. Write an inequality that represents the different numbers of units of food x and food y required. Sketch the graph of the inequality. From the graph, find several ordered pairs with positive integer coordinates that are solutions of the inequality.

Solution

$30x + 20y \ge 300$

(x and y cannot be negative.)

$20y \ge -30x + 300$

$y \ge -\dfrac{3}{2}x + 15$

Solution

$(10, 0)$

$(0, 15)$

7.3 Graphs of Quadratic Functions

7. Using the graphs in the textbook, match the equation $f(x) = 4 - 2x$ with the correct graph. [The graphs are labeled (a), (b), (c), (d), (e), and (f).]

Solution

$f(x) = 4 - 2x$ (e)

9. Using the graphs in the textbook, match the equation $f(x) = x^2 - 3$ with the correct graph. [The graphs are labeled (a), (b), (c), (d), (e), and (f).]

Solution

$f(x) = x^2 - 3$ (b)

11. Using the graphs in the textbook, match the equation $f(x) = (x - 2)^2$ with the correct graph. [The graphs are labeled (a), (b), (c), (d), (e), and (f).]

Solution

$f(x) = (x - 2)^2$ (d)

13. State whether the graph of $y = 2(x - 0)^2 + 2$ opens up or down, and find the vertex.

Solution

$2 > 0$ opens up

vertex $= (0, 2)$

15. State whether the graph of $y = x^2 - 6$ opens up or down, and find the vertex.

Solution

$1 > 0$ opens up

vertex $= (0, -6)$

17. Find the x and y-intercepts of $y = 25 - x^2$.

Solution

$y = 25 - x^2$

$0 = 25 - x^2$

$x^2 = 25$

$x = \pm 5$

$(5, 0), (-5, 0)$

$y = 25 - x^2$

$y = 25 - 0^2$

$y = 25$

$(0, 25)$

19. Find the x and y-intercepts of $y = x^2 - 3x + 3$.

Solution

$y = x^2 - 3x + 3$

$0 = x^2 - 3x + 3$

$x = \dfrac{3 \pm \sqrt{9 - 12}}{2}$

$= \dfrac{3 \pm \sqrt{-3}}{2}$

no x-intercepts

$y = x^2 - 3x + 3$

$y = 0^2 - 3(0) + 3$

$y = 3$

$(0, 3)$

21. Write $y = x^2 - 4x + 7$ in standard form and find the vertex of its graph.

Solution

$y = x^2 - 4x + 7 = (x^2 - 4x + 4) + 7 - 4 = (x - 2)^2 + 3$

vertex $= (2, 8)$

23. Write $y = -x^2 + 2x - 7$ in standard form and find the vertex of its graph.

Solution

$y = -x^2 + 2x - 7 = -1(x^2 - 2x + 1) - 7 + 1 = -1(x-1)^2 - 6$

vertex $= (1, -6)$

25. Sketch the graph of $f(x) = x^2 - 4$. Identify the vertex and any x-intercepts. Use a graphing utility to verify your graph.

Solution

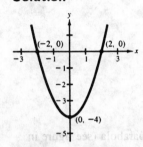

27. Sketch the graph of $f(x) = -x^2 + 3x$. Identify the vertex and any x-intercepts. Use a graphing utility to verify your graph.

Solution

$0 = -x^2 + 3x$

$0 = -x(x-3)$

$0 = x \quad x = 3$

$y = -1\left(x^2 - 3x + \dfrac{9}{4}\right) + \dfrac{9}{4}$

$= -1\left(x - \dfrac{3}{2}\right)^2 + \dfrac{9}{4}$

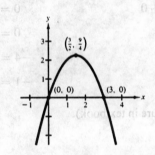

29. Sketch the graph of $f(x) = x^2 - 8x + 15$. Identify the vertex and any x-intercepts. Use a graphing utility to verify your graph.

Solution

$0 = x^2 - 8x + 15$

$0 = (x-5)(x-3)$

$5 = x \quad x = 3$

$y = \left(x^2 - 8x + 16\right) + 15 - 16$

$= (x-4)^2 - 1$

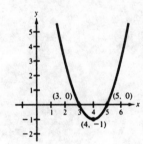

31. Sketch the graph of $f(x) = \dfrac{1}{5}\left(3x^2 - 24x + 38\right)$. Identify the vertex and any x-intercepts. Use a graphing utility to verify your graph.

Solution

$$y = \frac{3}{5}\left(x^2 - 8x + 16\right) + \frac{38}{5} - \frac{48}{5}$$

$$y = \frac{3}{5}(x - 4)^2 - 2$$

$$0 = 3x^2 - 24x + 38$$

$$x = \frac{24 \pm \sqrt{576 - 456}}{6}$$

$$x = \frac{24 \pm \sqrt{120}}{6}$$

$$\approx 5.83, 2.17$$

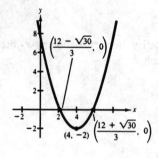

33. Write an equation of the parabola (see figure in textbook).

vertex $= (2, 0)$ point $= (0, 4)$

Solution

$4 = a(0 - 2)^2 + 0$ $y = 1(x - 2)^2 + 0$

$4 = a(4)$ $y = (x - 2)^2$

$1 = a$

35. Write an equation of the parabola (see figure in textbook).

vertex $= (-2, 4)$ point $= (0, 0)$

Solution

$0 = a(0 - -2)^2 + 4$ $y = -1(x - -2)^2 + 4$

$0 = a(4) + 4$ $y = -(x + 2)^2 + 4$

$-4 = a(4)$

$-1 = a$

37. Write an equation of the parabola (see figure in textbook).

vertex $= (-3, 3)$ point $= (-2, 1)$

Solution

$1 = a(-2 - -3)^2 + 3$ $y = -2(x - -3)^2 + 3$

$1 = a(1) + 3$ $y = -2(x + 3)^2 + 3$

$-2 = a$

39. *Graphical Interpretation* The height y (in feet) of a ball thrown by a child is given by

$$y = -\frac{1}{12}x^2 + 2x + 4$$

where x is the horizontal distance (in feet) from where the ball is thrown.

(a) Use a graphing utility to sketch the path of the ball.

(b) How high is the ball when it leaves the child's hand?

(c) How high is the ball when it is at its maximum height?

(d) How far from the child does the ball strike the ground?

Solution

(a)
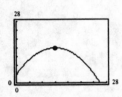

(b) $y = -\frac{1}{12}(0)^2 + 2(0) + 4$

$y = 4$ feet

(c) $y = -\frac{1}{12}x^2 + 2x + 4$

$y = -\frac{1}{12}\left(x^2 - 24x + 144\right) + 4 + 12$

$y = -\frac{1}{12}(x - 12)^2 + 16$

Maximum height $= 16$ feet

(d) $0 = -\frac{1}{12}x^2 + 2x + 4$

$0 = x^2 - 24x - 48$

$x = \frac{24 \pm \sqrt{576 + 192}}{2}$

≈ 25.86 feet

41. *Graphical Interpretation* A company manufactures radios that cost (the company) $60 each. For buyers who purchase 100 or fewer radios, the purchase price P is $90 per radio. To encourage large orders, the company will reduce the price *per radio* for orders over 100, as follows. If 101 radios are purchased, the price is $89.95 per unit. If 102 radios are purchased, the price is $89.70 per unit. If $(100 + x)$ radios are purchased, the price per unit is $P = 90 - x(0.15)$

(a) Show that the profit for orders over 100 is

$$P = (100 + x)[90 - x(0.15)] - (100 + x)60 = 3000 + 15x - \frac{3}{20}x^2$$

(b) Use a graphing utility to graph the profit function.

(c) Find the vertex of the profit curve and determine the order size for maximum profit.

(d) Would you recommend this pricing scheme? Explain your reasoning.

Solution

(a) $P = (100 + x)[90 - x(0.15)] - (100 + x)60$

$P = (100 + x)[90 - x(0.15) - 60]$

$P = (100 + x)(30 - 0.15x)$

$P = 3000 - 15x + 30x - 0.15x^2$

$P = 3000 + 15x - 0.15x^2$

$P = 3000 + 15x - \frac{3}{20}x^2$

(b) Keystrokes:

$\boxed{Y=}$ 3,000 $\boxed{+}$ 15 $\boxed{X, T, \theta}$ $\boxed{-}$ $\boxed{(}$ 3 $\boxed{\div}$ 20 $\boxed{)}$ $\boxed{X, T, \theta}$ $\boxed{x^2}$ $\boxed{\text{GRAPH}}$

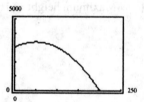

(c) $P = -\frac{3}{20}x^2 + 15x + 3000$

$P = -\frac{3}{20}\left(x^2 - 100x + 2500\right) + 3000 + 375$

$P = -\frac{3}{20}(x - 50)^2 + 3375$

vertex $= (50, 3375)$

order size for maximum profit

$P = 100 + 50 = 150$

(d) Recommend pricing scheme if price reductions are restricted to orders between 100 and 150 orders.

43. *Think About It* Write the equation of the parabola.

Points $(0, 0)$ and $(2, 0)$

Solution

Estimate the vertex $\left(1, -\dfrac{1}{2}\right)$

$$y = a(x - h)^2 + k$$

$$0 = a(0 - 1)^2 - \frac{1}{2}$$

$$\frac{1}{2} = a$$

$$y = \frac{1}{2}(x - 1)^2 - \frac{1}{2}$$

$$y = \frac{1}{2}\left(x^2 - 2x + 1\right) - \frac{1}{2}$$

$$y = \frac{1}{2}x^2 - x + \frac{1}{2} - \frac{1}{2}$$

$$y = \frac{1}{2}x^2 - x$$

$$y = \frac{1}{2}x(x - 2)$$

45. Solve the inequality $7 - 3x > 4 - x$ and sketch its graph on the real number line.

Solution

$$7 - 3x > 4 - x$$

$$-2x > -3$$

$$x < 3/2$$

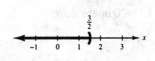

47. Solve the inequality $|x - 3| < 2$ and sketch its graph on the real number line.

Solution

$$|x - 3| < 2$$

$$-2 < x - 3 < 2$$

$$1 < x < 5$$

49. Solve $x^2 - 4x + 3 = 0$.

Solution

$$x^2 - 4x + 3 = 0$$

$$(x - 3)(x - 1) = 0$$

$$x - 3 = 0 \qquad x - 1 = 0$$

$$x = 3 \qquad\quad x = 1$$

51. *Free-Falling Object* Find the time for an object to reach the ground when it is dropped from a height of s_0 feet. The height h (in feet) is given by

$$h = -16t^2 + s_0,$$

where t is the time (in seconds).

Solution

$$h = -16t^2 + s_0$$

$$0 = -16t^2 + 80$$

$$16t^2 = 80$$

$$t^2 = \frac{80}{16}$$

$$t^2 = 5$$

$$t = \sqrt{5} \approx 2.236 \text{ sec}$$

53. State whether the graph of $y = 4 - (x - 10)^2$ opens up or down, and find the vertex.

Solution

$-1 < 0$ opens down

vertex $= (10, 4)$

55. State whether the graph of $y = x^2 - 6x$ opens up or down, and find the vertex.

Solution

$1 > 0$ opens down

$y = x^2 - 6x$

$y = x^2 - 6x + 9 - 9$

$y = (x - 3)^2 - 9$

  vertex $= (3, -9)$

57. Find the x and y-intercepts of $y = x^2 - 9x$.

Solution

$y = x^2 - 9x$

$0 = x^2 - 9x$

$0 = x(x - 9)$

$(0, 0), (9, 0)$

$y = x^2 - 9x$

$y = 0^2 - 9(0)$

$y = 0$

$(0, 0)$

59. Find the x and y-intercepts of $y = 4x^2 - 12x + 9$.

Solution

$y = 4x^2 - 12x + 9$

$0 = 4x^2 - 12x + 9$

$0 = (2x - 3)^2$

$0 = 2x - 3$

$\dfrac{3}{2} = x$

$\left(\dfrac{3}{2}, 0\right)$

$y = 4x^2 - 12x + 9$

$y = 4(0)^2 - 12(0) + 9$

$y = 9$

$(0, 9)$

61. Write $y = x^2 + 6x + 5$ in standard form and find the vertex of its graph.

Solution

$y = x^2 + 6x + 5$

$y = \left(x^2 + 6x + 9\right) + 5 - 9$

$y = (x + 3)^2 - 4$

  vertex $= (-3, -4)$

63. Write $y = -x^2 + 6x - 10$ in standard form and find the vertex of its graph.

Solution

$y = -x^2 + 6x - 10$

$y = -1\left(x^2 - 6x\right) - 10$

$y = -1\left(x^2 - 6x + 9\right) - 10 + 9$

$y = -1(x - 3)^2 - 1$

  vertex $= (3, -1)$

65. Write $y = 2x^2 + 6x + 2$ in standard form and find the vertex of its graph.

Solution

$$y = 2x^2 + 6x + 2 = 2\left(x^2 + 3x + \frac{9}{4}\right) + 2 - \frac{9}{2} = 2\left(x + \frac{3}{2}\right)^2 - \frac{5}{2}$$

$$\text{vertex} = \left(-\frac{3}{2}, -\frac{5}{2}\right)$$

67. Sketch the graph of $f(x) = -x^2 + 4$. Identify the vertex and any x-intercepts. Use a graphing utility to verify your graph.

Solution

$$0 = -x^2 + 4$$

$$x^2 = 4$$

$$x = \pm 2$$

$$f(x) = -\left(x^2 \quad \right) + 4$$

$$= -(x - 0)^2 + 4$$

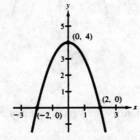

69. Sketch the graph of $f(x) = x^2 - 3x$. Identify the vertex and any x-intercepts. Use a graphing utility to verify your graph.

Solution

$$0 = x^2 - 3x$$

$$0 = x(x - 3)$$

$$0 = x \quad x = 3$$

$$f(x) = \left(x^2 - 3x + \frac{9}{4}\right) - \frac{9}{4}$$

$$f(x) = \left(x - \frac{3}{2}\right)^2 - \frac{9}{4}$$

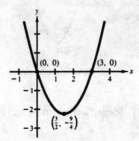

71. Sketch the graph of $f(x) = (x - 4)^2$. Identify the vertex and any x-intercepts. Use a graphing utility to verify your graph.

Solution

$$0 = (x - 4)^2$$

$$0 = x - 4$$

$$4 = x$$

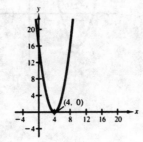

73. Sketch the graph of $f(x) = 5 - \dfrac{x^2}{3}$. Identify the vertex and any x-intercepts.
Use a graphing utility to verify your graph.

Solution

$f(x) = -\dfrac{1}{3}x^2 + 5$

$f(x) = -\dfrac{1}{3}(x - 0)^2 + 5$

$0 = -\dfrac{1}{3}x^2 + 5$

$\dfrac{1}{3}x^2 = 5$

$x^2 = 15$

$x = \pm\sqrt{15}$

$x \approx 3.87, -3.87$

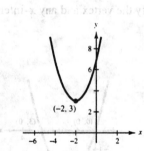

75. Sketch the graph of $f(x) = x^2 + 4x + 7$. Identify the vertex and any x-intercepts.
Use a graphing utility to verify your graph.

Solution

$f(x) = \left(x^2 + 4x + 4\right) + 7 - 4$

$f(x) = (x + 2)^2 + 3$

no x-intercepts

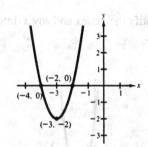

77. Sketch the graph of $f(x) = 2\left(x^2 + 6x + 8\right)$. Identify the vertex and any x-intercepts.
Use a graphing utility to verify your graph.

Solution

$f(x) = 2\left(x^2 + 6x + 9\right) + 16 - 18$

$f(x) = 2(x + 3)^2 - 2$

$0 = x^2 + 6x + 8$

$0 = (x + 4)(x + 2)$

$-4 = x \qquad x = -2$

79. Sketch the graph of $f(x) = -\left(x^2 + 6x + 5\right)$. Identify the vertex and any x-intercepts. Use a graphing utility to verify your graph.

Solution

$$0 = x^2 + 6x + 5$$

$$0 = (x + 5)(x + 1)$$

$$-5 = x \qquad x = -1$$

$$y = -\left(x^2 + 6x + 9\right) - 5 + 9$$

$$y = -(x + 3)^2 + 4$$

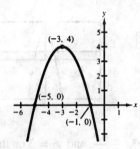

81. Sketch the graph of $f(x) = \dfrac{1}{2}\left(x^2 - 2x - 3\right)$. Identify the vertex and any x-intercepts. Use a graphing utility to verify your graph.

Solution

$$f(x) = \frac{1}{2}\left(x^2 - 2x + 1\right) - \frac{3}{2} - \frac{1}{2}$$

$$f(x) = \frac{1}{2}(x - 1)^2 - 2$$

$$0 = x^2 - 2x - 3$$

$$0 = (x - 3)(x + 1)$$

$$3 = x \qquad x = -1$$

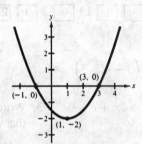

83. *Graphical and Algebraic Reasoning* Use a graphing utility to approximate the vertex of $y = \dfrac{1}{6}\left(2x^2 - 8x + 11\right)$. Then check your result algebraically.

Solution

Keystrokes:

$$\boxed{Y=} \;\boxed{(}\; 1 \;\boxed{\div}\; 6 \;\boxed{)}\; \boxed{(}\; 2 \;\boxed{X,T,\theta}\; \boxed{x^2}\; \boxed{-}\; 8 \;\boxed{X,T,\theta}\; \boxed{+}\; 11 \;\boxed{)}\; \boxed{\text{GRAPH}}$$

vertex $= (2, 0.5)$

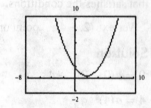

85. *Graphical and Algebraic Reasoning* Use a graphing utility to approximate the vertex of $y = -0.7x^2 - 2.7x + 2.3$. Then check your result algebraically.

Solution

Keystrokes:

$$\boxed{Y=} \;\boxed{(-)}\; .7 \;\boxed{X,T,\theta}\; \boxed{x^2}\; \boxed{-}\; 2.7 \;\boxed{X,T,\theta}\; \boxed{+}\; 2.3 \;\boxed{\text{GRAPH}}$$

vertex $= (-1.9, 4.9)$

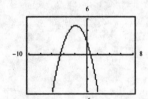

87. Use a graphing utility to graph $y_1 = -x^2 + 6$, and $y_2 = 2$ on the same screen. Do the graphs intersect? If so, approximate the point or points of intersection.

Solution

Keystrokes:

y_1 $\boxed{Y=}$ $\boxed{(-)}$ $\boxed{X, T, \theta}$ $\boxed{x^2}$ $\boxed{+}$ 6 $\boxed{\text{ENTER}}$

y_2 $\boxed{Y=}$ 2 $\boxed{\text{GRAPH}}$

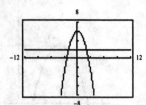

Points of intersection are $(2, 2)$ and $(-2, 2)$.

89. Use a graphing utility to graph $y_1 = \dfrac{1}{2}x^2 - 3x + \dfrac{13}{2}$, and $y_2 = 3$ on the same screen. Do the graphs intersect? If so, approximate the point or points of intersection.

Solution

Keystrokes:

y_1 $\boxed{Y=}$ .5 $\boxed{X, T, \theta}$ $\boxed{x^2}$ $\boxed{-}$ 3 $\boxed{X, T, \theta}$ $\boxed{+}$ $\boxed{(}$ 13 $\boxed{\div}$ 2 $\boxed{)}$ $\boxed{\text{ENTER}}$

y_2 $\boxed{Y=}$ 3 $\boxed{\text{GRAPH}}$

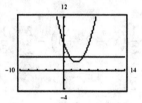

Points of intersection are $\left(3 + \sqrt{2}, 3\right)$ and $\left(3 - \sqrt{2}, 3\right)$.

91. Write an equation of the parabola $y = ax^2 + bx + c$ that satisfies the conditions.

Vertex: $(2, 1)$; $a = 1$

Solution

$y = 1(x - 2)^2 + 1 = x^2 - 4x + 5$

93. Write an equation of the parabola $y = ax^2 + bx + c$ that satisfies the conditions.

Vertex: $(-3, 4)$; $a = -1$

Solution

$y = -1(x + 3)^2 + 4 = -x^2 - 6x - 5$

95. Write an equation of the parabola $y = ax^2 + bx + c$ that satisfies the conditions.

Vertex: $(2, -4)$; point on graph: $(0, 0)$

Solution

$0 = a(0 - 2)^2 - 4$

$4 = a(4)$

$1 = a$

$y = 1(x - 2)^2 - 4 = x^2 - 4x$

97. Write an equation of the parabola $y = ax^2 + bx + c$ that satisfies the conditions.

Vertex: $(3, 2)$; point on graph: $(1, 4)$

Solution

$4 = a(1 - 3)^2 + 2$

$2 = a(4)$

$\dfrac{1}{2} = a$

$y = \dfrac{1}{2}(x - 3)^2 + 2 = \dfrac{1}{2}x^2 - 3x + \dfrac{13}{2}$

99. *Graphical Estimation* The profit (in thousands of dollars) for a company is given by

$P = 230 + 20s - \dfrac{1}{2}s^2$, where s is the amount (in hundreds of dollars) spent on advertising.

Use a graphing utility to graph the profit function and approximate the amount of advertising that yields a maximum profit. Verify the maximum algebraically.

Solution

Keystrokes:

Amount of advertising that
yields a maximum profit = $2000

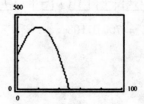

$$P = 230 + 20s - \frac{1}{2}s^2$$

$$P = -\frac{1}{2}s^2 + 20s + 230$$

$$P = -\frac{1}{2}\left(s^2 - 40s + 400\right) + 230 + 200$$

$$P = -\frac{1}{2}(s - 20)^2 + 430$$

101. *Highway Design* A highway department engineer must design a parabolic arc to create a turn in a freeway around a city. The vertex of the parabola is placed at the origin, and the parabola must connect with roads represented by the equations

$$y = -0.4x - 100, x < -500$$
$$y = 0.4x - 100, x > 500$$

(see figure in textbook). Find an equation of the parabolic arc.

Solution

$$100 = a(500 - 0)^2 + 0$$

$$100 = a(250{,}000)$$

$$\frac{100}{250{,}000} = a$$

$$\frac{1}{2500} = a$$

$$y = \frac{1}{2500}(x - 0)^2 + 0$$

$$y = \frac{1}{2500}x^2$$

103. *Graphical Estimation* The cost of producing x units of a product is given by $C = 800 - 10x + \frac{1}{4}x^2$, $0 < x < 40$. Use a graphing utility to sketch the graph of this equation and use the TRACE feature to approximate the value of x when C is minimum.

Solution

Keystrokes:

$x = 19.789474$ when $C = 700.01108$

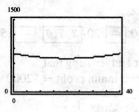

Mid-Chapter Quiz for Chapter 7

1. Write an equation of the line passing through $\left(0, -\frac{3}{2}\right)$ with slope 2.

Solution

$$y - -\frac{3}{2} = 2(x - 0)$$

$$y + \frac{3}{2} = 2x$$

$$0 = 4x - 2y - 3$$

2. Write an equation of the line passing through $(0, 6)$ with slope $-\frac{3}{2}$.

Solution

$$y - 6 = -\frac{3}{2}(x - 0)$$

$$y - 6 = -\frac{3}{2}x$$

$$0 = -3x - 2y + 12$$

or

$$0 = 3x + 2y - 12$$

3. Write an equation of the line passing through $\left(\frac{5}{2}, 6\right)$ with slope $-\frac{3}{4}$.

Solution

$$y - 6 = -\frac{3}{4}\left(x - \frac{5}{2}\right)$$

$$y - 6 = -\frac{3}{4}x + \frac{15}{8}$$

$$8y - 48 = -6x + 15$$

$$0 = 6x + 8y - 63$$

4. Write an equation of the line passing through $(-3.5, -1.8)$ with slope 3.

Solution

$$y - -1.8 = 3(x - -3.5)$$

$$y + 1.8 = 3x + 10.5$$

$$0 = 3x - y + 8.7$$

$$0 = 30x - 10y + 87$$

5. Write an equation of the line passing through the points $\left(\dfrac{1}{3}, 1\right)$ and $(4, 5)$.

Solution

$$m = \frac{5 - 1}{4 - \dfrac{1}{3}} = \frac{4}{\dfrac{11}{3}} = \frac{12}{11}$$

$$y - 5 = \frac{12}{11}(x - 4)$$

$$11y - 55 = 12x - 48$$

$$0 = 12x - 11y + 7$$

6. Write an equation of the line passing through the points $(0, 0.8)$ and $(3, -2.3)$.

Solution

$$m = \frac{-2.3 - 0.8}{3 - 0} = \frac{-3.1}{3} = -\frac{31}{30}$$

$$y - {-2.3} = -\frac{31}{30}(x - 3)$$

$$y + 2.3 = -\frac{31}{30}x + \frac{31}{10}$$

$$30y + 69 = -31x + 93$$

$$0 = 31x + 30y - 24$$

7. Write an equation of the line passing through the points $(3, -1)$ and $(10, -1)$.

Solution

$$m = \frac{-1 - {-1}}{10 - 3} = \frac{0}{7} = 0 \qquad y - {-1} = 0(x - 10)$$

$$y + 1 = 0$$

8. Write an equation of the line passing through the points $\left(4, \dfrac{5}{3}\right)$ and $(4, 8)$.

Solution

$$m = \frac{8 - \dfrac{5}{3}}{4 - 4} = \frac{\dfrac{19}{3}}{0} = \text{undefined} \qquad x = 4$$

$$x - 4 = 0$$

9. Write an equation of the line that passes through the point $(3, 5)$ and is (a) parallel to and (b) perpendicular to the line $2x - 3y = 1$.

Solution

$$2x - 3y = 1$$

$$-3y = -2x + 1$$

$$y = \frac{2}{3}x - \frac{1}{3} \quad \text{slope} = \frac{2}{3}$$

(a) $\quad y - 5 = \dfrac{2}{3}(x - 3)$

$$y - 5 = \frac{2}{3}x - \frac{6}{3}$$

$$3y - 15 = 2x - 6$$

$$0 = 2x - 3y + 9$$

(b) $\quad y - 5 = -\dfrac{3}{2}(x - 3)$

$$y - 5 = -\frac{3}{2}x + \frac{9}{2}$$

$$2y - 10 = -3x + 9$$

$$0 = 3x + 2y - 19$$

10. Decide whether the points are solutions of the inequality $2x - 3y \leq 4$.

(a) $(5, 2)$ (b) $(-2, 4)$ (c) $(2, -4)$ (d) $(3, 0)$

Solution

(a) $2(5) - 3(2) \overset{?}{\leq} 4$

$\qquad 10 - 6 \leq 4$

$\qquad\qquad 4 \leq 4$

$(5, 2)$ is a solution.

(c) $2(2) - 3(-4) \overset{?}{\leq} 4$

$\qquad 4 + 12 \leq 4$

$\qquad\quad 16 \leq 4$

$(2, -4)$ is not a solution.

(b) $2(-2) - 3(4) \overset{?}{\leq} 4$

$\qquad -4 - 12 \leq 4$

$\qquad\quad -16 \leq 4$

$(-2, 4)$ is a solution.

(d) $2(3) - 3(0) \overset{?}{\leq} 4$

$\qquad 6 - 0 \leq 4$

$\qquad\quad 6 \leq 4$

$(3, 0)$ is not a solution.

11. Write an inequality for the shaded region (see figure in textbook).

Solution

$m = \dfrac{3 - 5}{5 - 1} = -\dfrac{2}{4} = -\dfrac{1}{2}$

$y - 3 = -\dfrac{1}{2}(x - 5)$

$y - 3 = -\dfrac{1}{2}x + \dfrac{5}{2}$

$2y - 6 = -x + 5$

$x + 2y = 11$ line

Shaded region: $x + 2y \leq 11$

12. Write an inequality for the shaded region (see figure in textbook).

Solution

$m = \dfrac{3 - 1}{4 - -2} = \dfrac{2}{6} = \dfrac{1}{3}$

$y - 3 = \dfrac{1}{3}(x - 4)$

$y - 3 = \dfrac{1}{3}x - \dfrac{4}{3}$

$3y - 9 = x - 4$

$x - 3y = -5$ line

Shaded region: $x - 3y > -5$

13. Sketch the region that is determined by both inequalities, $2x + 3y \leq 9$ and $y \geq 1$.

Solution

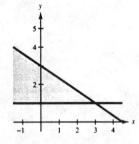

14. Sketch the region that is determined by both inequalities, $2x - y \leq 4$ and $y \leq 4$.

Solution

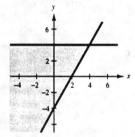

15. Write an equation for the parabola with vertex $(3, -1)$ and passing through the point $(5, 3)$.

Solution

$3 = a(5 - 3)^2 - 1$ $\quad y = 1(x - 3)^2 - 1$

$3 = a(4) - 1$

$4 = a(4)$

$1 = a$

16. Write an equation for the parabola with vertex $(5, 4)$ and passing through the point $(3, 3)$.

Solution

$3 = a(3 - 5)^2 + 4$ $\quad y = -\dfrac{1}{4}(x - 5)^2 + 4$

$3 = a(4) + 4$

$-1 = a(4)$

$-\dfrac{1}{4} = a$

17. Sketch the graph of the quadratic function $y = -\dfrac{1}{4}\left(x^2 + 6x + 1\right)$. Identify the vertex and the x-intercepts.

Solution

vertex $= (-3, 2)$

$y = -\dfrac{1}{4}\left(x^2 + 6x \quad\right) - \dfrac{1}{4}$

$y = -\dfrac{1}{4}\left(x^2 + 6x + 9\right) - \dfrac{1}{4} + \dfrac{9}{4}$

$y = -\dfrac{1}{4}(x + 3)^2 + 2$

x-intercepts

$0 = -\dfrac{1}{4}\left(x^2 + 6x + 1\right)$

$0 = x^2 + 6x + 1$

$x = \dfrac{-6 \pm \sqrt{6^2 - 4(4)(1)}}{2(1)}$

$x = \dfrac{-6 \pm \sqrt{36 - 16}}{2}$

$x = \dfrac{-6 \pm \sqrt{20}}{2}$

$x \approx -0.76$ and -5.24

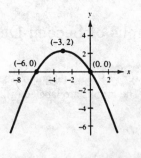

18. Sketch the graph of the quadratic function $y = 2x^2 - 4x - 7$. Identify the vertex and the x-intercepts.

Solution

vertex $= (1, -9)$

$y = 2\left(x^2 - 2x \quad\right) - 7$

$y = 2\left(x^2 - 2x + 1\right) - 7 - 2$

$y = 2(x - 1)^2 - 9$

x-intercepts

$0 = 2x^2 - 4x - 7$

$x = \dfrac{-(-4) \pm \sqrt{(-4)^2 - 4(2)(-7)}}{2(2)}$

$x = \dfrac{4 \pm \sqrt{16 + 56}}{4}$

$x = \dfrac{4 \pm \sqrt{72}}{4}$

$x = \dfrac{4 \pm 6\sqrt{2}}{4}$

$x = \dfrac{2 \pm 3\sqrt{2}}{2}$

$x \approx -1.12$ and 3.12

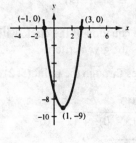

19. You purchase a used car for $12,400. It is estimated that after 4 years its depreciated value will be $5000. Assuming straight-line depreciation, write a linear function giving the value V of the car in terms of time t. State the domain of the function.

Solution

$(0, 12, 400) \quad (4, 5000)$

$$m = \frac{5000 - 12,400}{4 - 0} = \frac{-7400}{4} = -1850$$

$$y - 12,400 = -1850(x - 0)$$

$$y = -1850x + 12,400$$

$$V = -1850t + 12,400, \quad 0 \le t \le 4$$

20. The path of a ball is given by $y = -0.005x^2 + x + 5$. Determine the maximum height of the ball.

Solution

$$y = -0.005x^2 + x + 5$$

$$y = -0.005\left(x^2 - 200x \quad \right) + 5$$

$$y = -0.005\left(x^2 - 200x + 10,000\right) + 5 + 50$$

$$y = -0.005(x - 100)^2 + 55$$

maximum height $= 55$ feet

7.4 Graphs of Second-Degree Equations

7. Match $x^2 + y^2 = 9$ with its graph (see figures in textbook). [The graphs are labeled (a), (b), (c), (d), (e), and (f).]

Solution

$x^2 + y^2 = 9 \qquad$ (c)

9. Match $\dfrac{x^2}{4} + \dfrac{y^2}{9} = 1$ with the correct graph (see figures in textbook). [The graphs are labeled (a), (b), (c), (d), (e), and (f).]

Solution

$\dfrac{x^2}{4} + \dfrac{y^2}{9} = 1 \qquad$ (e)

11. Match $x^2 - y^2 = 4$ with the correct graph (see figures in textbook). [The graphs are labeled (a), (b), (c), (d), (e), and (f).]

Solution

$x^2 - y^2 = 4 \qquad$ (a)

13. Find an equation of the circle with center at $(0, 0)$ that satisfies the criteria of radius: 5.

Solution

$x^2 + y^2 = 25$

15. Find an equation of the circle with center at $(0, 0)$ that satisfies the given criteria.

Passes through the point $(5, 2)$

Solution

$$r = \sqrt{(5 - 0)^2 + (2 - 0)^2}$$

$$r = \sqrt{25 + 4}$$

$$r = \sqrt{29}$$

$$x^2 + y^2 = 29$$

17. Identify the center and radius of the circle $x^2 + y^2 = 16$ and sketch its graph.

Solution

$r = 4 \qquad$ center $= (0, 0)$

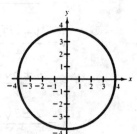

19. Identify the center and radius of the circle $25x^2 + 25y^2 - 144 = 0$ and sketch its graph.

Solution

$$25x^2 + 25y^2 - 144 = 0$$

$$x^2 + y^2 = \frac{144}{25}$$

$$r = \frac{12}{5} \qquad \text{center} = (0, 0)$$

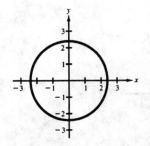

21. Write the standard form of the equation of the ellipse, centered at the origin, with vertices $(-4, 0)$, $(4, 0)$ and co-vertices $(0, -3)$, $(0, 3)$.

Solution

Equation: $\dfrac{x^2}{16} + \dfrac{y^2}{9} = 1$

23. Write the standard form of the equation of the ellipse, centered at the origin, with vertices $(0, -4)$, $(0, 4)$ and co-vertices $(-3, 0)$, $(3, 0)$.

Solution

Equation: $\dfrac{x^2}{9} + \dfrac{y^2}{16} = 1$

25. Write the standard form of the equation of the ellipse, centered at the origin, with a horizontal major axis of length 20 and a vertical minor axis of length 12.

Solution

Equation: $\dfrac{x^2}{100} + \dfrac{y^2}{36} = 1$

27. Sketch the ellipse $\dfrac{x^2}{16} + \dfrac{y^2}{4} = 1$. Identify its vertices and co-vertices.

Solution

Vertices: $(-4, 0)$, $(4, 0)$
Co-Vertices: $(0, 2)$, $(0, -2)$

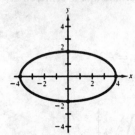

29. Sketch the ellipse $4x^2 + y^2 - 4 = 0$. Identify its vertices and co-vertices.

Solution

$$4x^2 + y^2 - 4 = 0$$

$$\frac{x^2}{1} + \frac{y^2}{4} = 1$$

Vertices: $(0, 2)$, $(0, -2)$
Co-vertices $(1, 0)$, $(-1, 0)$

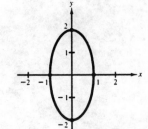

31. Sketch the hyberbola $\dfrac{x^2}{9} - \dfrac{y^2}{25} = 1$. Identify its vertices and asymptotes.

Solution

Vertices: $(3, 0)$, $(-3, 0)$

Asymptotes: $y = \dfrac{5}{3}x$

$y = -\dfrac{5}{3}x$

Equation: $\dfrac{x^2}{9} - \dfrac{y^2}{25} = 1$

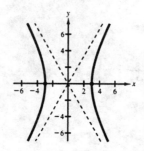

33. Sketch the hyberbola $\dfrac{y^2}{9} - \dfrac{x^2}{25} = 1$. Identify its vertices and asymptotes.

Solution

Vertices: $(0, 3)$, $(0, -3)$

Asymptotes: $y = \dfrac{3}{5}x$

$y = -\dfrac{3}{5}x$

Equation: $\dfrac{y^2}{9} - \dfrac{x^2}{25} = 1$

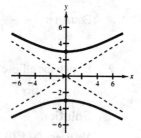

35. Find an equation of the hyperbola that is centered at the origin.

Vertices: $(-4, 0)$, $(4, 0)$

Asymptotes: $y = 2x$, $y = -2x$

Solution

$\dfrac{x^2}{16} - \dfrac{y^2}{64} = 1$

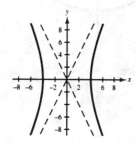

37. Find an equation of the hyperbola that is centered at the origin.

Vertices: $(0, -4)$, $(0, 4)$

Asymptotes: $y = \dfrac{1}{2}x$, $y = -\dfrac{1}{2}x$

Solution

$\dfrac{y^2}{16} - \dfrac{x^2}{64} = 1$

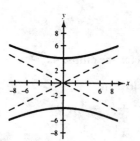

39. Sketch the circle $(x - 2)^2 + (y - 3)^2 = 4$ and identify its center and radius.

Solution

center $= (2, 3)$

radius $= 2$

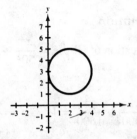

41. Write $x^2 + y^2 - 4x - 2y + 1 = 0$ in complete square form. Then sketch its graph.

Solution

$$x^2 + y^2 - 4x - 2y + 1 = 0$$

$$x^2 - 4x \quad + y^2 - 2y = -1$$

$$\left(x^2 - 4x + 4\right) + \left(y^2 - 2y + 1\right) = -1 + 4 + 1$$

$$(x - 2)^2 + (y - 1)^2 = 4$$

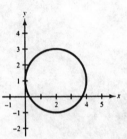

43. Use a graphing calculator to graph $x^2 + y^2 = 30$.

Solution

Keystrokes:

y_1 [Y=] [√] [( ] 30 [−] [X, T, θ] [x^2] [ )] [ENTER]

y_2 [Y=] [(−)] [√] [( ] 30 [−] [X, T, θ] [x^2] [ )] [GRAPH]

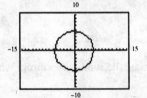

45. Use a graphing calculator to graph $6x^2 - y^2 = 40$.

Solution

Keystrokes:

y_1 [Y=] [√] [( ] 6 [X, T, θ] [x^2] [−] 40 [ )] [ENTER]

y_2 [Y=] [(−)] [√] [( ] 6 [X, T, θ] [x^2] [−] 40 [ )] [GRAPH]

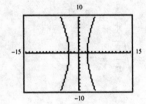

47. *Satellite Orbit* Find an equation of the circular orbit of a satellite 500 miles above the surface of the earth. Place the origin of the rectangular coordinate system at the center of the earth and assume the radius of the earth to be 4000 miles.

Solution

$x^2 + y^2 = 4500^2$

$x^2 + y^2 = 20{,}250{,}000$

49. *Height of an Arch* A *semi elliptical* arch for a tunnel under a river has a width of 100 feet and a height of 40 feet (see figure in textbook). Determine the height of the arch 5 feet from the edge of the tunnel.

Solution

Equation of ellipse $= \dfrac{x^2}{50^2} + \dfrac{y^2}{40^2} = 1$

or

$$\frac{x^2}{2500} + \frac{y^2}{1600} = 1$$

$$\frac{45^2}{2500} + \frac{y^2}{1600} = 1$$

$$\frac{y^2}{1600} = 0.19$$

$$y^2 = 304$$

$$y = 17.435596 \approx 17 \text{ feet}$$

51. Sketch the graph of $y = 2x - 3$.

Solution

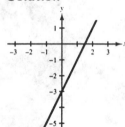

53. Sketch the graph of $y = x^2 - 4x + 4$.

Solution

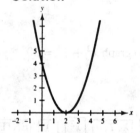

55. *Geometry* Approximate the radius of a circle whose area is 3 square feet.

Solution

Formula: $\quad A = \pi r^2$

Equation: $\qquad\qquad 3 = \pi r^2$

$$\frac{3}{\pi} = r^2$$

$$0.9722 \approx \sqrt{\frac{3}{\pi}} = r$$

57. Find an equation of the circle with center at $(0, 0)$ that satisfies the criteria of Radius: $\dfrac{2}{3}$.

Solution

$x^2 + y^2 = \dfrac{4}{9} \quad$ or $\quad 9x^2 + 9y^2 = 4$

59. Find an equation of the circle with center at $(0, 0)$ that satisfies the following criteria.

Passes through the point $(0, 8)$

Solution

$r = \sqrt{(0 - 0)^2 + (8 - 0)^2}$

$r = \sqrt{64}$

$r = 8$

$x^2 + y^2 = 64$

61. Identify the center and radius of the circle $x^2 + y^2 = 36$ and sketch its graph.

Solution

center $= (0, 0)$

radius $= 6$

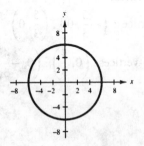

63. Identify the center and radius of the circle $4x^2 + 4y^2 = 1$ and sketch its graph.

Solution

$4x^2 + 4y^2 = 1$ center $= (0, 0)$

$x^2 + y^2 = \dfrac{1}{4}$ radius $= \dfrac{1}{2}$

$r = \dfrac{1}{2}$

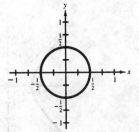

65. Write the standard form of the equation of the ellipse, centered at the origin, with vertices $(-2, 0)$, $(2, 0)$ and co-vertices $(0, -1)$, $(0, 1)$.

Solution

$\dfrac{x^2}{4} + \dfrac{y^2}{1} = 1$

67. Write the standard form of the equation of the ellipse, centered at the origin, with vertices $(0, -2)$, $(0, 2)$ and co-vertices $(-1, 0)$, $(1, 0)$.

Solution

$\dfrac{x^2}{1} + \dfrac{y^2}{4} = 1$

69. Write the standard form of the equation of the ellipse, centered at the origin, whose major axis is vertical with length 10 and whose minor axis has length 6.

Solution

$\dfrac{x^2}{9} + \dfrac{y^2}{25} = 1$

71. Sketch the ellipse $\dfrac{x^2}{4} + \dfrac{y^2}{16} = 1$. Identify its vertices and co-vertices.

Solution

Vertices: $(0, 4)$, $(0, -4)$

Co-vertices: $(2, 0)$, $(-2, 0)$

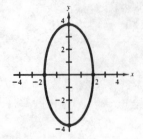

73. Sketch the ellipse $\dfrac{x^2}{25/9} + \dfrac{y^2}{16/9} = 1$. Identify its vertices and co-vertices.

Solution

Vertices: $\left(-\dfrac{5}{3}, 0\right), \left(\dfrac{5}{3}, 0\right)$

Co-vertices: $\left(0, \dfrac{4}{3}\right), \left(0, -\dfrac{4}{3}\right)$

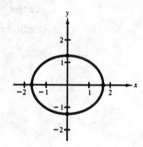

75. Sketch the hyperbola $x^2 - y^2 = 9$. Identify its vertices and asymptotes.

Solution

Vertices: $(3, 0), (-3, 0)$

Asymptotes: $y = \dfrac{9}{9}x \qquad y = -\dfrac{9}{9}x$

$\qquad\qquad\quad y = x \qquad\quad y = -x$

Equation: $x^2 - y^2 = 9$

$\qquad\qquad \dfrac{x^2}{9} - \dfrac{y^2}{9} = 1$

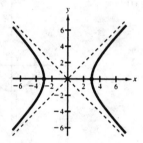

77. Sketch the hyperbola $y^2 - x^2 = 9$. Identify its vertices and asymptotes.

Solution

Vertices: $(0, 3), (0, -3)$

Asymptotes: $y = \dfrac{9}{9}x \qquad y = -\dfrac{9}{9}x$

$\qquad\qquad\quad y = x \qquad\quad y = -x$

Equation: $y^2 - x^2 = 9$

$\qquad\qquad \dfrac{y^2}{9} - \dfrac{x^2}{9} = 1$

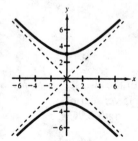

79. Sketch the hyperbola $4y^2 - x^2 + 16 = 0$. Identify its asymptotes.

Solution

Vertices: $(4, 0)$, $(-4, 0)$

Asymptotes: $y = \dfrac{2}{4}x = \dfrac{1}{2}x$

$\qquad\qquad y = -\dfrac{2}{4}x = -\dfrac{1}{2}x$

Equation: $4y^2 - x^2 + 16 = 0$

$$\frac{4y^2}{-16} - \frac{x^2}{-16} = \frac{-16}{-16}$$

$$\frac{-y^2}{4} + \frac{x^2}{16} = 1$$

$$\frac{x^2}{16} - \frac{y^2}{4} = 1$$

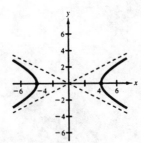

81. Find an equation of the hyberbola, centered at the orgin, with vertices $(-9, 0)$, $(9, 0)$ and asymptotes $y = \dfrac{2}{3}x$, $y = -\dfrac{2}{3}x$.

Solution

$$\frac{x^2}{81} - \frac{y^2}{36} = 1$$

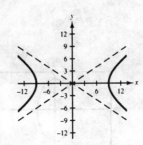

83. Write an equation of the hyberbola, centered at the origin, with vertices $(0, -1)$, $(0, 1)$ and asymptotes $y = 2x$, $y = -2x$.

Solution

$$\frac{y^2}{1} - \frac{x^2}{1/4} = 1$$

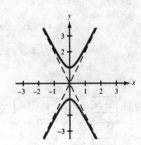

85. Identify the graph of $y = 2x^2 - 8x + 2$ as a line, circle, parabola, ellipse, or hyperbola.

Solution

Parabola

87. Identify the graph of $4x^2 + 9y^2 = 36$ as a line, circle, parabola, ellipse, or hyperbola.

Solution

Ellipse

89. Identify the graph of $4x^2 - 9y^2 = 36$ as a line, circle, parabola, ellipse, or hyperbola.

Solution

Hyperbola

91. Identify the graph of $x^2 + y^2 - 1 = 0$ as a line, circle, parabola, ellipse, or hyperbola.

Solution

Circle

93. Identify the graph of $3x + 2 = 0$ as a line, circle, parabola, ellipse, or hyperbola.

Solution

Line

95. Use a graphing utility to graph $x^2 - 2y^2 = 4$.

Solution

$$x^2 - 2y^2 = 4$$

$$x^2 - 4 = 2y^2$$

$$\frac{x^2 - 4}{2} = y^2$$

$$\pm\sqrt{\frac{x^2 - 4}{2}} = y$$

Keystrokes:

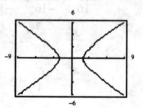

y_1 [Y=] [√] [(] [(] [X,T,θ] [x^2] [−] 4 [)] [÷] 2 [)] [ENTER]

y_2 [Y=] [(−)] [√] [(] [(] [X,T,θ] [x^2] [−] 4 [)] [÷] 2 [)] [GRAPH]

97. Use a graphing calculator to sketch the graph of $3x^2 + y^2 - 12 = 0$.

Solution

$$3x^2 + y^2 - 12 = 0$$

$$y^2 = 12 - 3x^2$$

$$y = \pm\sqrt{12 - 3x^2}$$

Keystrokes:

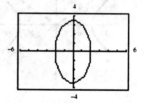

y_1 [Y=] [√] [(] 12 [−] 3 [X,T,θ] [x^2] [)] [ENTER]

y_2 [Y=] [(−)] [√] [(] 12 [−] 3 [X,T,θ] [x^2] [)] [GRAPH]

99. A rectangle centered at the origin with sides parallel to the coordinate axes is placed in a circle of radius 25 inches centered at the origin (see figure in textbook). The length of the rectangle is $2x$ inches.

(a) Show that the width and area of the rectangle are given by $2\sqrt{625 - x^2}$ and $4x\sqrt{625 - x^2}$, respectively.

(b) Use a graphing utility to graph the area function. Approximate the value of x for which the area is maximum.

Solution

(a) $x^2 + y^2 = 625$ (equation of circle)

(x, y) of the rectangle is also the point on the circle, so y-coordinate equals:

$$x^2 + y^2 = 625$$
$$y^2 = 625 - x^2$$
$$y = \sqrt{625 - x^2}$$
$$\text{width} = 2\left(\sqrt{625 - x^2}\right)$$
$$\text{area} = 2x \cdot 2\left(\sqrt{625 - x^2}\right)$$
$$\text{area} = 4x\sqrt{625 - x^2}$$

(b)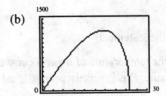

$$x \approx 17.68$$

101. *Area* The area A of the ellipse $\dfrac{x^2}{a^2} + \dfrac{y^2}{b^2} = 1$ is given by $A = \pi ab$. Find the equation of an ellipse with area 301.59 square units and $a + b = 20$.

Solution

$$A = \pi ab \qquad\qquad a + b = 20$$
$$301.59 = \pi ab \qquad\quad b = 20 - a$$
$$\frac{301.59}{\pi} = ab$$
$$96 \approx ab$$
$$96 = a(20 - a)$$
$$0 = -a^2 + 20a - 96$$
$$0 = a^2 - 20a + 96$$
$$0 = (a - 12)(a - 8)$$

$a = 12 \qquad a = 8$

$b = 8 \qquad\ \ b = 12$

$$\frac{x^2}{144} + \frac{y^2}{64} = 1$$

7.5   Variation

5. Write a model for the statement.

  I varies directly as V.

Solution

$I = kV$

7. Write a model for the statement.

  p varies inversely as d.

Solution

$p = \dfrac{k}{d}$

9. Write a model for the statement.

  Boyle's Law   If the temperature of a gas is constant, its absolute pressure P is inversely proportional to its volume V.

Solution

$P = \dfrac{k}{V}$

11. Write a sentence using variation terminology to describe the formula for the area of a triangle, $A = \dfrac{1}{2}bh$.

Solution

The area of a triangle is proportional to the product of the base and height.

13. Write a sentence using variation terminology to describe the formula for the volume of a right circular cylinder, $V = \pi r^2 h$.

Solution

The volume of a right circular cylinder is proportional to the product of the square of the radius and the height.

15. Find the constant of proportionality and give the equation relating variables.

  s varies directly as t, and $s = 20$ when $t = 4$.

Solution

$s = kt$

$20 = k(4)$

$5 = k$

$s = 5t$

17. Find the constant of proportionality and give the equation relating the variables.

  F is directly proportional to the square of x, and $F = 500$ when $x = 40$.

Solution

$F = kx^2$

$500 = k(40)^2$

$\dfrac{500}{1600} = k$

$\dfrac{5}{16} = k$

$F = \dfrac{5}{16}x^2$

19. Find the constant of proportionality and give the equation relating the variables.

  n varies inversely as m, and $n = 32$ when $m = 1.5$.

Solution

$n = \dfrac{k}{m}$

$32 = \dfrac{k}{1.5}$

$48 = k$

$n = \dfrac{48}{m}$

21. Find the constant of proportionality and give the equation relating the variables.

d varies directly as the square of x and inversely with r, and $d = 3000$ when $x = 10$ and $r = 4$.

Solution

$$d = k\left(\frac{x^2}{r}\right)$$

$$3000 = k\left(\frac{10^2}{4}\right)$$

$$3000 = k(25)$$

$$120 = k$$

$$d = \frac{120x^2}{r}$$

23. *Hooke's Law* A baby weighing $10\frac{1}{2}$ pounds compresses the spring of a baby scale 7 millimeters. What weight will compress the spring 12 millimeters?

Solution

$$d = kF$$

$$7 = k(10.5)$$

$$\frac{7}{10.5} = k$$

$$\frac{70}{105} = k$$

$$\frac{2}{3} = k$$

$$12 = \frac{2}{3}F$$

$$18 \text{ pounds} = F$$

25. *Stopping Distance* The stopping distance d of an automobile is directly proportional to the square of its speed s. On a certain road surface a car requires 75 feet to stop when its speed is 30 miles per hour. Estimate the stopping distance if the brakes are applied when the car is traveling at 50 miles per hour under similar road conditions.

Solution

$$d = ks^2$$

$$75 = k(30)^2$$

$$\frac{75}{900} = k$$

$$\frac{1}{12} = k$$

$$d = \frac{1}{12}(50)^2$$

$$d = 208.\overline{3} \text{ feet}$$

27. *Power Generation* The power P generated by a wind turbine varies directly as the cube of the wind speed w. The turbine generates 750 watts of power in a 25-mile-per-hour wind. Find the power it generates in a 40-mile-per-hour wind.

Solution

$$P = kw^2$$

$$750 = k(25)^3$$

$$\frac{750}{15,625} = k$$

$$0.048 = k$$

$$P = 0.048(40)^3$$

$$P = 3072 \text{ watts of power}$$

29. *Demand Function* A company has found that the daily demand x for its product is inversely proportional to the price p. When the price is $5, the demand is 800 units. Approximate the demand if the price is increased to $6.

Solution

$$x = \frac{k}{p}$$

$$800 = \frac{k}{5}$$

$$4000 = k$$

$$x = \frac{4000}{6}$$

$$x = 666.\overline{6} \approx 667 \text{ units}$$

31. *Amount of Illumination* The illumination I from a light source varies inversely as the square of the distance d from the light source. If you raise a lamp from 18 inches to 36 inches over your desk, the illumination will change by what factor?

Solution

$$I = \frac{k}{d^2}$$

$$I = \frac{k}{18^2} \qquad I = \frac{k}{36^2}$$

$$I = \frac{k}{324} \qquad I = \frac{k}{1296}$$

I will change by a factor of $\frac{324}{1296}$ or $\frac{1}{4}$.

33. *Load of a Beam* The load that can be safely supported by a horizontal beam varies jointly as the width of the beam and the square of its depth and inversely as the length of the beam. A beam of width 3 inches, depth 8 inches, and length 10 feet can safely support 2000 pounds. Determine the safe load of a beam made from the same material if its depth is increased to 10 inches.

Solution

$$L = \frac{kwd^2}{l}$$

$$2000 = \frac{k(3)8^2}{120}$$

$$2000 = \frac{k(192)}{120}$$

$$1250 = k$$

$$L = \frac{1250(3)10^2}{120}$$

$$L = 3125 \text{ pounds}$$

35. Solve for y in terms of x.

$$3x + 4y - 5 = 0$$

Solution

$$3x + 4y - 5 = 0$$

$$4y = -3x + 5$$

$$y = -\frac{3}{4}x + \frac{5}{4}$$

37. Solve for y in terms of x.

$$-2x^2 + 3y + 2 = 0$$

Solution

$$-2x^2 + 3y + 2 = 0$$

$$3y = 2x^2 - 2$$

$$y = \frac{2}{3}x^2 - \frac{2}{3}$$

39. Find two positive consecutive odd integers whose product is 255.

Solution

Verbal model: $\boxed{\begin{array}{c}\text{First Odd}\\\text{Integer}\end{array}} \cdot \boxed{\begin{array}{c}\text{Second Odd}\\\text{Integer}\end{array}} = 255$

Equation: $(2n + 1) \cdot (2n + 3) = 255$

$$4n^2 + 8n + 3 = 255$$

$$4n^2 + 8n - 252 = 0$$

$$n^2 + 2n - 63 = 0$$

$$(n + 9)(n - 7) = 0$$

$$n + 9 = 0 \qquad n - 7 = 0$$

$$\cancel{n = -9} \qquad n = 7$$

$$2n + 1 = 15$$

$$2n + 3 = 17$$

41. Write a model for the statement.

V is directly proportional to t.

Solution

$V = kt$

43. Write a model for the statement.

u is directly proportional to the square of v.

Solution

$u = kv^2$

45. Write a model for the statement.

P is inversely proportional to the square root of $1 + r$.

Solution

$P = \dfrac{k}{\sqrt{1 + r}}$

47. Write a model for the statement.

A varies jointly as l and w.

Solution

$A = klw$

49. Write a sentence using variation terminology to describe the formula for volume of a sphere,

$V = \dfrac{4}{3}\pi r^3$.

Solution

The volume of a sphere varies directly as the cube of the radius.

51. Write a sentence using variation terminology to describe the formula for area of an ellipse, $A = \pi ab$.

Solution

The area of an ellipse varies jointly as the semi-major axis and the semi-minor axis.

53. Find the constant of proportionality and give the equation relating the variables.

H is directly proportional to u, and $H = 100$ when $u = 40$.

Solution

$$H = ku \qquad H = \frac{5}{2}u$$

$$100 = k(40)$$

$$\frac{100}{40} = k$$

$$\frac{5}{2} = k$$

57. Find the constant of proportionality and give the equation relating the variables.

F varies jointly as x and y, and $F = 500$ when $x = 15$ and $y = 8$.

Solution

$$F = kxy$$

$$500 = k(15)(8)$$

$$\frac{500}{120} = k$$

$$\frac{25}{6} = k$$

$$F = \frac{25}{6}xy$$

61. *Free-Falling Object* The velocity v of a free-falling object is proportional to the time that it has fallen. The constant of proportionality is the acceleration due to gravity. Find the acceleration due to gravity if the velocity of a falling object is 96 feet per second after the object has fallen 3 seconds.

Solution

$$V = kt \qquad \text{acceleration} = 32 \text{ ft/sec}^2$$

$$96 = k(3)$$

$$32 = k$$

55. Find the constant of proportionality and give the equation relating the variables.

g varies inversely as the square root of z, and $g = \frac{4}{5}$ when $z = 25$.

Solution

$$g = \frac{k}{\sqrt{z}} \qquad g = \frac{4}{\sqrt{z}}$$

$$\frac{4}{5} = \frac{k}{\sqrt{25}}$$

$$4 = k$$

59. The total revenue R is directly proportional to the number of units sold x. When 500 units are sold, the revenue is \$3875. (a) Find the revenue when 635 units are sold. (b) Interpret the constant of proportionality.

Solution

(a) $R = kx$ \qquad\qquad (b) Price per unit

$$3875 = K(500)$$

$$7.75 = k$$

$$R = 7.75x$$

$$R = 7.75(635)$$

$$R = \$4921.25$$

63. *Travel Time* The travel time between two cities is inversely proportional to the average speed. If a train travels between two cities in 3 hours at an average speed of 65 miles per hour, how long would it take at an average speed of 80 miles per hour? What does the constant of proportionality measure in this problem?

Solution

$$t = \frac{k}{r} \qquad k = 195 \text{ miles}$$

$$3 = \frac{k}{65} \qquad t = \frac{k}{80}$$

$$t = \frac{195}{80} \approx 2.44 \text{ hr}$$

(b) Distance

65. *Weight* A person's weight on the moon varies directly with his or her weight on earth. Neil Armstrong, the first man on the moon, weighed 360 pounds on earth, including his equipment. On the moon he weighed only 60 pounds, with equipment. If the first woman in space, Valentina V. Tereshkova, had landed on the moon and weighed 54 pounds, with equipment, how much would she have weighed on earth?

Solution

$$W_e = k \cdot W_m \qquad W_e = k \cdot W_m$$
$$360 = k \cdot 60 \qquad x = 6 \cdot 54$$
$$6 = k \qquad x = 324 \text{ pounds}$$

67. *Reading a Graph* The graph (in the textbook) shows the percent p of oil that remained in Chedabucto Bay, Nova Scotia, after an oil spill. The cleaning of the spill was left primarily to natural actions such as wave motion, evaporation, photochemical decomposition, and bacterial decomposition. After about a year, the percent that remained varied inversely with time. Find a model that relates p and t, where t is the number of years since the spill. Then use the model to find the amount of oil that remained $6\frac{1}{2}$ years after the spill.

Solution

$$p = \frac{k}{t} \qquad 114 = k$$
$$38 = \frac{k}{3} \qquad \text{so } p = \frac{114}{t}$$

(b) $p = \dfrac{114}{6.5}$

$p = 17.5\%$

69. Use $k = 1$ to complete the table (see textbook) and plot the resulting points.

Solution

x	2	4	6	8	10
$y = kx^2$	4	16	36	64	100

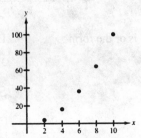

71. Use $k = \dfrac{1}{2}$ to complete the table (see textbook) and plot the resulting points.

Solution

x	2	4	6	8	10
$y = kx^2$	2	8	18	32	50

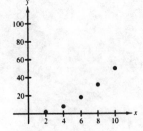

73. Use $k = 2$ to complete the table (see textbook) and plot the resulting points.

Solution

x	2	4	6	8	10
$y = \dfrac{k}{x^2}$	$\dfrac{1}{2}$	$\dfrac{1}{8}$	$\dfrac{1}{18}$	$\dfrac{1}{32}$	$\dfrac{1}{50}$

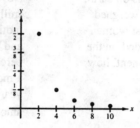

75. Use $k = 10$ to complete the table (see textbook) and plot the resulting points.

Solution

x	2	4	6	8	10
$y = \dfrac{k}{x^2}$	$\dfrac{5}{2}$	$\dfrac{5}{8}$	$\dfrac{5}{18}$	$\dfrac{5}{32}$	$\dfrac{1}{10}$

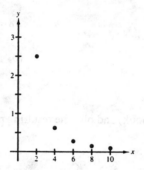

77. Determine whether the variation model is of the form $y = kx$ or $y = \dfrac{k}{x}$, and find k.

Solution

x	10	20	30	40	50
y	$\dfrac{2}{5}$	$\dfrac{1}{5}$	$\dfrac{2}{15}$	$\dfrac{1}{10}$	$\dfrac{2}{25}$

$y = \dfrac{4}{x}$

79. Determine whether the variation model is of the form $y = kx$ or $y = \dfrac{k}{x}$, and find k.

Solution

x	10	20	30	40	50
y	-3	-6	-9	-12	-15

$y = -\dfrac{3}{10}x$

Review Exercises for Chapter 7

1. Write an equation of the line passing through the point $(1, -4)$ with the slope of $m = 2$.

Solution

$$y + 4 = 2(x - 1)$$
$$y + 4 = 2x - 2$$
$$0 = 2x - y - 6$$

3. Write an equation of the line passing through the point $(-1, 4)$ with the slope of $m = -4$.

Solution

$$y - 4 = -4(x + 1)$$
$$y - 4 = -4x - 4$$
$$4x + y = 0$$

5. Write an equation of the line passing through the point $\left(\dfrac{5}{2}, 4\right)$ with the slope of $m = -\dfrac{2}{3}$.

Solution

$$y - 4 = -\frac{2}{3}\left(x - \frac{5}{2}\right)$$
$$y - 4 = -\frac{2}{3}x + \frac{5}{3}$$
$$3y - 12 = -2x + 5$$
$$2x + 3y - 17 = 0$$

7. Write an equation of the line passing through the point $(7, 8)$ with the slope of m as undefined.

Solution

$$x = 7$$
$$x - 7 = 0$$

9. Write an equation of the line passing through the points $(-6, 0)$ and $(0, -3)$.

Solution

$$m = \frac{0 + 3}{-6 - 0} = \frac{3}{-6} = -\frac{1}{2}$$
$$y - 0 = -\frac{1}{2}(x + 6)$$
$$y = -\frac{1}{2}x - 3$$
$$2y = -x - 6$$
$$x + 2y + 6 = 0$$

11. Write an equation of the line passing through the points $(0, 10)$ and $(6, 10)$.

Solution

$$m = \frac{10 - 10}{6 - 0} = \frac{0}{6} = 0$$
$$y = 10$$
$$y - 10 = 0$$

13. Write an equation of the line passing through the points $\left(\frac{4}{3}, \frac{1}{6}\right)$ and $\left(4, \frac{7}{6}\right)$.

Solution

$$m = \frac{\frac{7}{6} - \frac{1}{6}}{4 - \frac{4}{3}} \cdot \frac{6}{6} = \frac{7 - 1}{24 - 8} = \frac{6}{16} = \frac{3}{8}$$

$$y - \frac{7}{6} = \frac{3}{8}(x - 4)$$

$$y - \frac{7}{6} = \frac{3}{8}x - \frac{12}{8}$$

$$48y - 56 = 18x - 72$$

$$0 = 18x - 48y - 16 = 9x - 24y - 8$$

15. Find equations of the line passing through the point $\left(\frac{3}{5}, -\frac{4}{5}\right)$ that are (a) parallel and (b) perpendicular to the line $3x + y = 2$.

Solution

$$3x + y = 2$$

$$y = -3x + 2$$

(a) $y + \frac{4}{5} = -3\left(x - \frac{3}{5}\right)$ or $3x + y - 1 = 0$

(b) $y + \frac{4}{5} = \frac{1}{3}\left(x - \frac{3}{5}\right)$ or $x - 3y - 3 = 0$

17. Find equations of the line passing through the point $(12, 1)$ that are (a) parallel and (b) perpendicular to the line $5x = 3$.

Solution

$$5x = 3$$

$$x = \frac{3}{5} \quad m = \text{undefined}$$

(a) $x = 12$ or $x - 12 = 0$

(b) $y = 1$ or $y - 1 = 0$

19. Use a graphing utility on a square setting to sketch the lines, $y = 2x - 3$ and $y = \frac{1}{2}(4 - x)$. Decide whether the lines are parallel, perpendicular, or neither.

Solution

Keystrokes:

y_1 [Y=] 2 [X, T, θ] [−] 3 [ENTER]

y_2 [Y=] .5 [(] 4 [−] [X, T, θ] [)] [ZOOM 5] [GRAPH]

Perpendicular

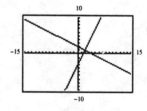

21. Use a graphing utility on a square setting to sketch the lines, $2x - 3y - 5 = 0$ and $x + 2y - 6 = 0$. Decide whether the lines are parallel, perpendicular, or neither.

Solution

$$2x - 3y - 5 = 0 \qquad\qquad x + 2y - 6 = 0$$

$$-3y = -2x + 5 \qquad\qquad 2y = -x - 6$$

$$y = \frac{2}{3}x - \frac{5}{3} \qquad\qquad y = -\frac{1}{2}x - 3$$

Keystrokes:

y_1 [Y=] [(] [2] [÷] [3] [)] [X, T, θ] [−] [(] [5] [÷] [3] [)] [ENTER]

y_2 [Y=] [(−)] [.5] [X, T, θ] [−] [3] [ZOOM 5] [GRAPH]

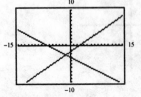

Neither

23. *Cost and Profit* A company produces a product for which the variable cost is $8.55 per unit and the fixed costs are $25,000. The product is sold for $12.60, and the company can sell all it produces.

(a) Write the cost C as a linear function, the number of units produced.

(b) Write the profit P as a linear function of x, the number of units sold.

Solution

(a) $C = 25{,}000 + 8.55x$

(b) $P = 12.60x - (25{,}000 + 8.55x)$

$$P = 12.60x - 25{,}000 - 8.55x$$

$$P = 4.05x - 25{,}000$$

25. *Rocker Arm Construction* Consider the rocker arm shown in the textbook. Find an equation of the line through the centers of the two small bolt holes in the rocker arm.

Solution

$$m = \frac{-0.88 - 4.75}{4.75 - -0.88} = \frac{-5.63}{5.63} = -1$$

$$y + 0.88 = -1(x - 4.75)$$

$$y = -x + 4.75 - 0.88$$

$$y = -x + 3.87$$

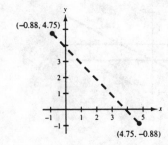

27. Sketch the graph of the inequality $x - 2 \geq 0$.

Solution

$x - 2 \geq 0$

$\quad x \geq 2$

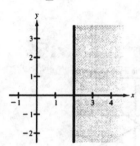

29. Sketch the graph of the inequality $2x + y < 1$.

Solution

$2x + y < 1$ or $y < -2x + 1$

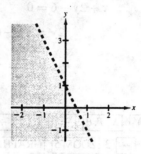

31. Sketch the graph of the inequality $x \leq 4y - 2$.

Solution

$x \leq 4y - 2$ or $y \geq \dfrac{1}{4}x + \dfrac{1}{2}$

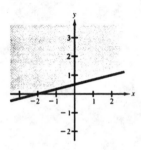

33. Sketch a graph of the solution points that satisfy both inequalities, $x \geq 5$ and $y \leq 2$.

Solution

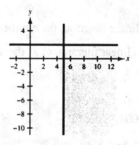

35. Sketch a graph of the solution points that satisfy both inequalities, $x - 2y \geq 2$ and $y \geq -1$

Solution

$x - 2y \geq 2$

$\quad -2y \geq -x + 2$

$\qquad y \leq \dfrac{1}{2}x - 1$

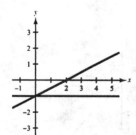

37. Use a graphing utility to graph the region containing the solution points that satisfy both inequalities, $x + y \geq 0$ and $y \leq 5$.

Solution

$x + y \geq 0$

$\quad y \geq -x$

Keystrokes:

y_1 [Y =] [(−)] [X, T, θ] [ENTER]

y_2 [Y =] 5

[DRAW] [7] [y-Vars] [1] [1] [,] [y-Vars] [1] [2] [)] [ENTER]

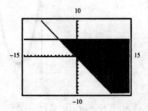

39. *Weekly Pay* You have two-part time jobs. One is at a grocery store, which pays $8 per hour, and the other is mowing lawns, which pays $10 per hour. Between the two jobs, you want to earn at least $200 a week. Write an inequality that shows the different numbers of hours you can work at each job, and sketch the graph of the inequality. From the graph, find several ordered pairs with positive integer coordinates that are solutions of the inequality.

Solution

$8x + 10y \geq 200$

$8x + 10y \geq 200$

$\quad 10y \geq -8x + 200$

$\quad\quad y \geq -.8x + 20$

Ordered pair solutions:

$(0, 20), (25, 0)$

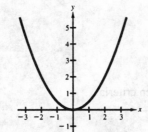

41. Identify and sketch the conic $x^2 - 2y = 0$.

Solution

$x^2 - 2y = 0$

$\quad x^2 = 2y$

$\quad \dfrac{x^2}{2} = y$

parabola

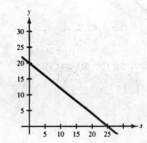

43. Identify and sketch the conic $x^2 - y^2 = 0$.

Solution

$\quad x^2 - y^2 = 64$

$\dfrac{x^2}{64} - \dfrac{y^2}{64} = 1$

hyperbola

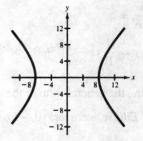

45. Identify and sketch the conic $y = x(x - 6)$.

Solution

$y = x(x - 6)$

$y = x^2 - 6x$

parabola

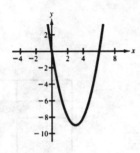

47. Identify and sketch the conic $\dfrac{x^2}{25} + \dfrac{y^2}{4} = 1$.

Solution

ellipse

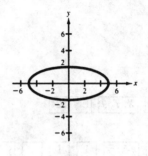

49. Identify and sketch the conic $4x^2 + 4y^2 - 9 = 0$.

Solution

$4x^2 + 4y^2 - 9 = 0$

$\dfrac{4x^2}{9} + \dfrac{4y^2}{9} = 1$

$\dfrac{x^2}{9/4} + \dfrac{y^2}{9/4} = 1$

circle

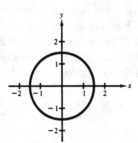

51. Find the general form of the equation for the conic meeting the given criteria.

Parabola: vertex: $(5, 0)$; passes through the point $(1, 1)$

Solution

$y = a(x - h)^2 + k$

$1 = a(1 - 5)^2 + 0$

$1 = a(16)$

$\dfrac{1}{16} = a$

$y = \dfrac{1}{16}(x - 5)^2 + 0$ or $y = \dfrac{1}{16}x^2 - \dfrac{5}{8}x + \dfrac{25}{16}$

53. Find the general form of the equation for the conic meeting the given criteria.

Ellipse: vertices: $(0, -5)$, $(0, 5)$; co-vertices: $(-2, 0)$, $(2, 0)$

Solution

$\dfrac{x^2}{4} + \dfrac{y^2}{25} = 1$

55. Find the general form of the equation for the conic meeting the given criteria.

Circle: center: $(0, 0)$; radius: 20

Solution

$x^2 + y^2 = 400$

57. *Graphical Estimation* The profit (in thousands of dollars) for a ceertain product is given by
$P = 320 + 10s - \dfrac{1}{2}s^2$, where s is the amount (in hundreds of dollars) spent on advertising.
Use a graphing utility to graph the profit function and approximate the amount of advertising
that yields a maximum profit. Verify the maximum algebraically.

Solution

Keystrokes:

$\boxed{Y=}$ 320 $\boxed{+}$ 10 $\boxed{X, T, \theta}$ $\boxed{-}$ .5 $\boxed{X, T, \theta}$ $\boxed{x^2}$ $\boxed{\text{GRAPH}}$

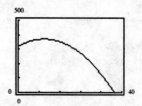

Maximum occurs at $s = 10$.

$$P = 320 + 10s - \frac{1}{2}s^2$$

$$P = -\frac{1}{2}\left(s^2 - 20s \qquad\right) + 320$$

$$P = -\frac{1}{2}\left(s^2 - 20s + 100\right) + 320 + 50$$

$$P = -\frac{1}{2}(s - 10)^2 + 370$$

59. *Graphical Estimation* The enrollment E (in millions) in public elementary schools
in the United States for the year 1970 through 1990 is approximated by the model
$E = 27.611 - 0.586t + 0.026t^2$, where t represents the calendar year, with $t = 0$
corresponding to 1970. (*Source:* U.S. National Center for Education Statistics)

(a) Use a graphing utility to graph the enrollment function.

(b) Use the graph of part (a) to approximate the year when the enrollment was minimum.

(c) Predict the enrollment in 1999 if this trend continues.

Solution

(a) Keystrokes:

$\boxed{Y=}$ 27.611 $\boxed{-}$ 0.586 $\boxed{X, T, \theta}$ $\boxed{+}$ 0.026 $\boxed{X, T, \theta}$ $\boxed{x^2}$ $\boxed{\text{GRAPH}}$

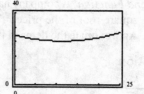

(b) Minimum at $t \approx 11$, so year is 1981

(c) $E = 27.611 - 0.586(29) + 0.026(29)^2$

$E = 32.483$ million

61. Match the equation $4x^2 + 4y^2 = 81$ with its graph (see textbook).

Solution

(c)

63. Match the equation $\dfrac{y^2}{4} - x^2 = 1$ with its graph (see textbook).

Solution

(a)

65. Match the equation $y = -x^2 + 6x - 5$ with its graph (see textbook).

Solution

(b)

67. Find the constant of proportionality and write an equation that relates the variables.

y varies directly as the cube root of x, and $y = 12$ when $x = 9$.

Solution

$$y = k\sqrt[3]{x}$$

$$12 = k\sqrt[3]{9}$$

$$\frac{12}{\sqrt[3]{9}} = k$$

$$y = 4\sqrt[3]{3}\sqrt[3]{x} = 4\sqrt[3]{3x}$$

69. Find the constant of proportionality and write an equation that relates the variables.

T varies jointly as r and the square of s, and $T = 5000$ when $r = 0.09$ and $s = 1000$.

Solution

$$T = krs^2$$

$$5000 = k(0.09)(1000)^2$$

$$\frac{5000}{90,000} = k$$

$$\frac{1}{18} = k$$

$$T = \frac{1}{18}rs^2$$

71. *Hooke's Law* A force of 100 pounds stretches a spring 4 inches. Find the force required to stretch the same spring 6 inches.

Solution

$$d = kF$$

$$4 = k(100)$$

$$\frac{4}{100} = k$$

$$\frac{1}{25} = k$$

$$6 = \frac{1}{25}F$$

$$150 \text{ pounds} = F$$

73. *Demand Function* A company has found that the daily demand x for its product varies inversely as the square root of the price p. When the price is $25, the demand is approximately 1000 units. Approximate the demand if the price is increased to $28.

Solution

$$x = \frac{k}{\sqrt{p}}$$

$$1000 = \frac{k}{\sqrt{25}}$$

$$5000 = k$$

$$x = \frac{5000}{\sqrt{28}}$$

$$x = 944.91118$$

$$x \approx 945 \text{ units}$$

Test for Chapter 7

1. Write an equation of the line which has a slope of -2 and passes through $(2, -4)$.

   Solution

   $$y - -4 = -2(x - 2)$$
   $$y + 4 = -2x + 4$$
   $$2x + y = 0$$

2. Write an equation of the line passing through the points $(25, -15)$ and $(75, 100)$.

   Solution

   $$m = \frac{10 + 15}{75 - 25} = \frac{25}{50} = \frac{1}{2}$$
   $$y - 10 = \frac{1}{2}(x - 75)$$
   $$y - 10 = \frac{1}{2}x - \frac{75}{2}$$
   $$2y - 20 = x - 75$$
   $$0 = x - 2y - 55$$

3. Write an equation of the line that is horizontal and passes through $(5, -1)$.

   Solution

   $$y = -1$$
   $$y + 1 = 0$$

4. Write an equation of the line that is vertical and passes through $(-2, 4)$.

   Solution

   $$x = -2$$
   $$x + 2 = 0$$

5. Find the slope of a line perpendicular to the line given by $5x + 3y - 9 = 0$.

   Solution

   $$5x + 3y - 9 = 0$$
   $$3y = -5x + 9$$
   $$y = -\frac{5}{3}x + 3$$
   $$m = \frac{3}{5}$$

6. After 4 years, a \$26,000 car will have depreciated to a value of \$10,000. Write a linear equation that gives the volume V in terms of t, the number of years. When will the car be worth \$16,000?

   Solution

   $(0, \$26,000), (4, \$10,000)$

   $$m = \frac{10,000 - 26,000}{4 - 0} = \frac{-16,000}{4} = -4000$$
   $$V - 26,000 = -4000(t - 0)$$
   $$V = -4000t + 26,000$$
   $$16,000 = -4000t + 26,000$$
   $$-10,000 = -4000t$$
   $$\frac{-10,000}{-4000} = t$$
   $$2.5 = \frac{5}{2} = t$$

7. Sketch the graph of the inequality $x + 2y \leq 4$.

Solution

$x + 2y \leq 4$ or $y \leq -\dfrac{1}{2}x + 2$

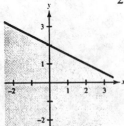

8. Sketch the graph of the inequality $3x - y \geq 6$.

Solution

$3x - y \geq 6$

$-y \geq -3x + 6$

$y \leq 3x - 6$

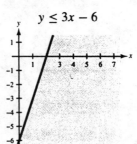

9. Sketch the graph of the parabola $y = -2(x - 2)^2 + 8$. Label its vertex and intercepts.

Solution

$y = -2(x - 2)^2 + 8$

$\text{vertex} = (2, 8)$

$y = -2(0 - 2)^2 + 8$

$y = -2(4) + 8$

$y = 0 \quad (0, 0); \ x \ \& \ y\text{-intercept}$

$0 = -2(x - 2)^2 + 8$

$-8 = -2(x - 2)^2$

$4 = (x - 2)^2$

$\pm 2 = x - 2$

$2 \pm 2 = x$

$0, 4 = x \quad (0, 0), (4, 0); \ x\text{-intercepts}$

10. Write an equation of the parabola with vertex $(3, -2)$ and passing through the point $(0, 4)$.

Solution

$4 = a(0 - 3)^2 - 2$

$6 = a(9)$

$\dfrac{6}{9} = a$

$\dfrac{2}{3} = a$

$y = \dfrac{2}{3}(x - 3)^2 - 2 \quad \text{or} \quad = \dfrac{2}{3}x^2 - 4x + 4$

11. The revenue R for a chartered bus trip is given by $R = -\dfrac{1}{20}(n^2 - 240n)$, when $80 \le n \le 160$ and n is the number of passengers. How many passengers will produce a maximum revenue?

Solution

$$R = -\frac{1}{20}\left(n^2 - 240n\right), 80 \le n \le 160$$

$$R = -\frac{1}{20}\left(n^2 - 240n + 14{,}400\right) + 720$$

$$R = -\frac{1}{20}(n - 120)^2 + 720$$

$n = 120$ passengers will produce a maximum revenue

12. Write an equation of the circle with center at the origin and passing through the point $(-3, 4)$.

Solution

$r = \sqrt{(-3 - 0)^2 + (4 - 0)^2} = \sqrt{9 + 16} = \sqrt{25} = 5$

$x^2 + y^2 = 25$

13. Write the standard form of the equation of the ellipse centered at the origin with vertices $(0, -10)$ and $(0, 10)$ and co-vertices $(-3, 0)$ and $(3, 0)$.

Solution

$$\frac{x^2}{9} + \frac{y^2}{100} = 1$$

14. Write the standard form of the equation of the hyperbola centered at the origin with vertices $(-3, 0)$ and $(3, 0)$ and asymptotes $y = \frac{1}{2}x$ and $y = -\frac{1}{2}x$.

Solution

$$\frac{x^2}{9} - \frac{y^2}{9/4} = 1$$

$$\frac{x^2}{9} - \frac{4y^2}{9} = 1$$

or

$$x^2 - 4y^2 = 9$$

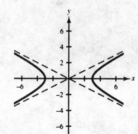

15. Sketch the graph of $x^2 + y^2 = 9$.

Solution

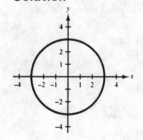

16. Sketch the graph of $\dfrac{x^2}{9} + \dfrac{y^2}{16} = 1$.

Solution

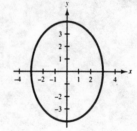

17. Sketch the graph of $\dfrac{x^2}{9} - \dfrac{y^2}{16} = 1$.

Solution

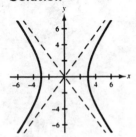

18. Sketch the graph of $\dfrac{x}{3} - \dfrac{y}{4} = 1$.

Solution

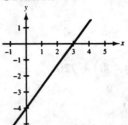

19. Write a mathematical model for the statement "S varies directly as the square of x and inversely as y."

Solution

$$S = \frac{kx^2}{y}$$

20. Find the constant of proportionality if v varies directly as the square root of u, and $v = \dfrac{3}{2}$ when $u = 36$.

Solution

$$v = k\sqrt{u}$$

$$\frac{3}{2} = k\sqrt{36}$$

$$\frac{1}{4} = k$$

$$v = \frac{1}{4}\sqrt{u}$$

CHAPTER 8
Introduction to Systems of Equations

8.1 | Introduction to Systems of Equations

7. Determine whether each ordered pair (a) $(1, 4)$ and (b) $(3, -1)$ is a solution of the following system of equations.

$$x + 2y = 9$$
$$-2x + 3y = 10$$

Solution

(a) $(1, 4)$

$$1 + 2(4) = 9$$
$$9 = 9$$
$$-2(1) + 3(4) = 10$$
$$-2 + 12 = 10$$
$$10 = 10$$

yes

(b) $(3, -1)$

$$3 + 2(-1) = 9$$
$$3 - 2 = 9$$
$$1 \neq 9$$

no

9. Determine whether each ordered pair (a) $(-3, 2)$ and (b) $(-2, 6)$ is a solution of the following system of equations.

$$-2x + 7y = 46$$
$$3x + y = 0$$

Solution

(a) $(-3, 2)$

$$-2(-3) + 7(2) = 46$$
$$6 + 14 = 46$$
$$20 \neq 46$$

no

(b) $(-2, 6)$

$$-2(-2) + 7(6) = 46$$
$$4 + 42 = 46$$
$$46 = 46$$
$$3(-2) + 6 = 0$$
$$-6 + 6 = 0$$
$$0 = 0$$

yes

11. Solve the system by substitution.

$$x - 2y = 0$$
$$3x + 2y = 8$$

Solution

$$x = 2y$$
$$3(2y) + 2y = 8$$
$$6y + 2y = 8$$
$$8y = 8$$
$$y = 1$$
$$x = 2(1)$$
$$= 2$$

$(2, 1)$

13. Solve the system by substitution.

$$x + y = 2$$
$$x^2 - y = 0$$

Solution

$$y = 2 - x$$
$$x^2 - (2 - x) = 0$$
$$x^2 + x - 2 = 0$$
$$(x + 2)(x - 1) = 0$$
$$x = -2 \qquad x = 1$$
$$y = 2 - (-2) \quad y = 2 - 1$$
$$= 4 \qquad\qquad = 1$$

$(-2, 4), (1, 1)$

15. Solve the system by substitution.

$$x + y = 3$$
$$2x - y = 0$$

Solution

$$y = 3 - x$$
$$2x - (3 - x) = 0$$
$$2x - 3 + x = 0$$
$$3x = 3$$
$$x = 1$$
$$y = 3 - 1$$
$$y = 2$$

$(1, 2)$

17. Solve the system by substitution.

$$4x - 14y = -15$$
$$18x - 12y = \quad 9$$

Solution

$$4x = -15 + 14y$$
$$x = \frac{-15 + 14y}{4}$$
$$18\left(\frac{-15 + 14y}{4}\right) - 12y = 9$$
$$18(-15 + 14y) - 48y = 36$$
$$-270 + 252y - 48y = 36$$
$$204y = 306$$
$$y = \frac{3}{2}$$

$$x = \frac{-15 + 14\left(\dfrac{3}{2}\right)}{4}$$
$$= \frac{-15 + 21}{4} = \frac{3}{2}$$

$$\left(\frac{3}{2}, \frac{3}{2}\right)$$

19. Solve the system by substitution.

$$x^2 + y^2 = 25$$
$$2x - y = -5$$

Solution

$$x^2 + y^2 = 25 \qquad\qquad x = 0$$
$$2x - y = 25 \qquad\qquad y = 5 + 2(0)$$
$$-y = -5 - 2x \qquad\qquad y = 5$$
$$y = 5 + 2x \qquad\qquad (0, 5)$$
$$x^2 + (5 + 2x)^2 = 25 \qquad\qquad x = -4$$
$$x^2 + 25 + 20x + 4x^2 - 25 = 0 \qquad\qquad y = 5 + 2(-4)$$
$$5x^2 + 20x = 0 \qquad\qquad y = -3$$
$$5x(x + 4) = 0 \qquad\qquad (-4, -3)$$

21. Use the graph in the textbook to determine whether the system has any solutions. Find any solutions that exist.

$$x + y = 4$$
$$x + y = -1$$

Solution

no solutions

23. Use the graph in the textbook to determine whether the system has any solutions. Find any solutions that exist.

$$5x - 3y = 4$$
$$2x + 3y = 3$$

Solution

one solution

$$3y = 3 - 2x$$
$$y = \frac{3 - 2x}{3}$$
$$5x - 3\left(\frac{3 - 2x}{3}\right) = 4$$
$$5x - 3 + 2x = 4$$
$$7x - 3 = 4$$
$$7x = 7$$
$$x = 1$$
$$y = \frac{3 - 2(1)}{3}$$
$$y = \frac{1}{3}$$

$$\left(1, \frac{1}{3}\right)$$

25. Use the graph in the textbook to determine whether the system has any solutions. Find any solutions that exist.

$$x - 2y = 4$$
$$x^2 - y = 0$$

Solution

no solutions

27. Use a graphing utility to graph the equations and approximate any solutions of the system of equations.

$$-2x + y = 1$$
$$x - 3y = 2$$

Solution

Solve each equation for y.

$$y = 2x + 1 \qquad -3y = -x + 2$$
$$y = \frac{1}{3}x - \frac{2}{3}$$

Keystrokes:

y_1 [Y=] [2] [X, T, θ] [+] [1] [ENTER]

y_2 [Y=] [(] [1] [÷] [3] [)] [X, T, θ] [−] [(] [2] [÷] [3] [)] [GRAPH]

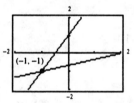

Point of intersection is $(-1, -1)$.

29. Use a graphing utility to graph the equations and approximate any solutions of the systems of equations.

$$x^2 - y^2 = 12$$
$$x - 2y = 0$$

Solution

Solve each equation for y.

$$x^2 - y^2 = 12 \qquad\qquad x - 2y = 0$$
$$y^2 = x^2 - 12 \qquad\qquad -2y = -x$$
$$y = \pm\sqrt{x^2 - 12} \qquad\qquad y = \frac{1}{2}x$$

Keystrokes:

y_1 [Y=] [√] [(] [X, T, θ] [x^2] [−] [12] [)] [ENTER]

y_2 [Y=] [(−)] [y_1] [ENTER]

y_3 [Y=] [.5] [X, T, θ] [GRAPH]

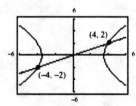

Points of intersection are $(4, 2)$ and $(-4, -2)$.

31. Use a graphing utility to graph the equations and approximate any solutions of the system of equations.

$$y = x^3$$
$$y = x^3 - 3x^2 + 3x$$

Solution

Keystrokes:

y_1 $\boxed{Y=}$ $\boxed{X, T, \theta}$ $\boxed{\wedge}$ 3 $\boxed{ENTER}$

y_2 $\boxed{Y=}$ $\boxed{X, T, \theta}$ $\boxed{\wedge}$ 3 $\boxed{-}$ 3 $\boxed{X, T, \theta}$ $\boxed{x^2}$ $\boxed{+}$ 3 $\boxed{X, T, \theta}$ $\boxed{GRAPH}$

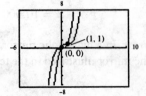

Points of intersection are $(0, 0)$ and $(1, 1)$.

33. Use a graphing utility to graph each equation in the system. The graphs appear parallel.

$$x - 100y = -200$$
$$3x - 275y = 198$$

Yet, from the slope-intercepts of the lines, you find that the slopes are not equal and thus the graphs intersect. Find the point of intersection of the two lines.

Solution

$$x - 100y = -200$$
$$3x - 275y = 198$$

$$x = -200 + 100y$$

$$3(-200 + 100y) - 275y = 198$$
$$-600 + 300y - 275y = 198$$
$$25y = 798$$
$$y = 31.92$$

$$x = -200 + 100(31.92) = -200 + 3192 = 2992$$
$$(2992, 31.92)$$

Solve each equation for y.

$$x - 100y = -200 \qquad\qquad 3x - 275y = 198$$
$$\qquad y = \frac{1}{100}x + 2 \qquad -275y = -3x + 198$$
$$\qquad y = .01x + 2 \qquad\qquad y = \frac{-3}{-275}x + \frac{198}{-275}$$
$$\qquad\qquad\qquad\qquad\qquad y = \frac{3}{275}x - .72$$

Keystrokes:

y_1 $\boxed{Y=}$ .01 $\boxed{X, T, \theta}$ $\boxed{+}$ 2 $\boxed{ENTER}$

y_2 $\boxed{Y=}$ $\boxed{(}$ 3 $\boxed{\div}$ 275 $\boxed{)}$ $\boxed{X, T, \theta}$ $\boxed{-}$ .72 $\boxed{GRAPH}$

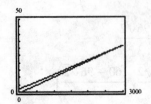

35. *Break-Even Analysis* A small business invests $8000 in equipment to produce a product. Each unit of the product costs $1.20 to produce and is sold for $2.00. How many items must be sold before the business breaks even?

Solution

$$C = 8000 + 1.20x \qquad 2.00x = 8000 + 1.20x$$

$$R = 2.00x \qquad\qquad 0.80x = 8000$$

$$R = C \qquad\qquad\quad x = 10{,}000 \text{ items}$$

37. *Hyperbolic Mirror* In a hyperbolic mirror, light rays directed to one focus are reflected to the other focus. The mirror illustrated in the text has the equation

$$\frac{x^2}{36} - \frac{y^2}{64} = 1$$

At which point on the mirror will light from the point $(0, 10)$ reflect to the focus?

Solution

Recall $c^2 = a^2 + b^2$. If c is the focus, then

$$\left[(10 - 0)^2 + (0 + 10)^2\right] + \left[(10 - y)^2 + (0 - x)^2\right] = \left[(-10 - x)^2 + (0 - y)^2\right]$$

$$100 + 100 + 100 - 20y + y^2 + x^2 = 100 + 20x + x^2 + y^2$$

$$200 - 20y - 20x = 0$$

$$10 = x + y$$

Now solve $10 = x + y$ and $\dfrac{x^2}{36} - \dfrac{y^2}{64} = 1$ by substitution

$$\frac{(10 - y)^2}{36} - \frac{x^2}{64} = 1 \qquad\qquad 10 = x + y$$

$$16(10 - y)^2 - 9y^2 = 576 \qquad 10 - y = x$$

$$1600 - 320y + 16y^2 - 9y^2 = 576$$

$$7y^2 - 320y + 1024 = 0$$

$$y = \frac{320 \pm \sqrt{102400 - 4(7)(1024)}}{14}$$

$$y = \frac{320 \pm \sqrt{73728}}{14}$$

$$y = \frac{160 \pm 96\sqrt{2}}{7} \qquad \text{reject} \quad \frac{160 + 96\sqrt{2}}{7}$$

$$x = 10 - \left(\frac{160 - 96\sqrt{2}}{7}\right)$$

$$x = \frac{-90 + 96\sqrt{2}}{7}$$

$$\left(\frac{-90 + 96\sqrt{2}}{7}, \frac{160 - 96\sqrt{2}}{7}\right)$$

39. Find the general form of the equation of the line through the two points $(-1, -2)$ and $(3, 6)$.

Solution

$$m = \frac{6 - -2}{3 - -1} = \frac{8}{4} = 2$$

$$y - 6 = 2(x - 3)$$

$$y - 6 = 2x - 6$$

$$0 = 2x - y$$

41. Find the general form of the equation of the line through the two points $\left(\frac{3}{2}, 8\right)$ and $\left(\frac{11}{2}, \frac{5}{2}\right)$.

Solution

$$m = \frac{\frac{5}{2} - 8}{\frac{11}{2} - \frac{3}{2}} = \frac{\frac{5}{2} - 8}{\frac{11}{2} - \frac{3}{2}} \cdot \frac{2}{2} = \frac{5 - 16}{11 - 3} = -\frac{11}{8}$$

$$y - 8 = -\frac{11}{8}\left(x - \frac{3}{2}\right)$$

$$y - 8 = -\frac{11}{8}x + \frac{33}{16}$$

$$16y - 128 = -22x + 33$$

$$22x + 16y - 161 = 0$$

43. *Work Rate* Machine A can complete a job in 4 hours and Machine B can complete it in 6 hours. How long will it take both machines working together to complete the job?

Solution

Verbal model: | Rate Machine 1 | + | Rate Machine 2 | = | Rate Together |

Equation:

$$\frac{1}{4} + \frac{1}{6} = \frac{1}{x}$$

$$3x + 2x = 12$$

$$5x = 12$$

$$x = \frac{12}{5} = 2.4 \text{ hours}$$

45. Solve the system by substitution.

$$x = 4$$
$$x - 2y = -2$$

Solution

$$x = 4$$

$$4 - 2y = -2$$

$$-2y = -6$$

$$y = 3$$

$$(4, 3)$$

47. Solve the system by substitution.

$$7x + 8y = 24$$
$$x - 8y = 8$$

Solution

$$x = 8 + 8y$$

$$7(8 + 8y) + 8y = 24$$

$$56 + 56y + 8y = 24$$

$$64y = -32$$

$$y = -\frac{1}{2}$$

$$x = 8 + 8\left(-\frac{1}{2}\right)$$

$$x = 4$$

$$\left(4, -\frac{1}{2}\right)$$

49. Solve the following system by the method of substitution.

$$y = \sqrt{8 - x}$$

$$y = -\frac{1}{5}(x - 14)$$

Solution

$$\sqrt{8 - x} = -\frac{1}{5}(x - 14)$$

$$\sqrt{8 - x} = -\frac{1}{5}x + \frac{14}{5}$$

$$8 - x = \frac{1}{25}x^2 - \frac{28}{25}x + \frac{196}{25}$$

$$200 - 25x = x^2 - 28x + 196$$

$$0 = x^2 - 3x - 4$$

$$0 = (x - 4)(x + 1)$$

$$x = 4 \qquad x = -1$$

$$y = 2 \qquad y = 3$$

$(4, 2)$ and $(-1, 3)$

51. Solve the system by substitution.

$$x + y = 2$$

$$x - 4y = 12$$

Solution

$$x = 2 - y$$

$$2 - y - 4y = 12$$

$$-5y = 10$$

$$y = -2$$

$$x = 2 - (-2)$$

$$x = 4$$

$(4, -2)$

53. Solve the system by substitution.

$$x + 6y = 19$$

$$x - 7y = -7$$

Solution

$$x = -7 + 7y$$

$$-7 + 7y + 6y = 19$$

$$13y = 26$$

$$y = 2$$

$$x = -7 + 7(2) = 7$$

$(7, 2)$

55. Solve the system by substitution.

$$2x + 5y = 29$$

$$5x + 2y = 13$$

Solution

$$2x + 5y = 29$$

$$5x + 2y = 13$$

$$2x = 29 - 5y$$

$$x = \frac{29 - 5y}{2}$$

$$5\left(\frac{29 - 5y}{2}\right) + 2y = 13$$

$$5(29 - 5y) + 4y = 26$$

$$145 - 25y + 4y = 26$$

$$-21y = -119$$

$$y = \frac{119}{21}$$

$$y = \frac{17}{3}$$

$$x = \frac{29 - 5\left(\frac{17}{3}\right)}{2} \cdot \frac{3}{3}$$

$$x = \frac{87 - 85}{6}$$

$$x = \frac{1}{3}$$

$$\left(\frac{1}{3}, \frac{17}{3}\right)$$

57. Solve the system by substitution.

$$y = 2x^2$$

$$y = -2x + 12$$

Solution

$$y = 2x^2$$

$$y = -2x + 12$$

$$2x^2 = -2x + 12$$

$$2x^2 + 2x - 12 = 0$$

$$x^2 + x - 6 = 0$$

$$(x + 3)(x - 2) = 0$$

$x = -3$	$x = 2$
$y = 2(-3)^2$	$y = 2(2)^2$
$= 2(9)$	$= 2(4)$
$= 18$	$= 8$

$(-3, 18)$ and $(2, 8)$

59. Solve the system by substitution.

$$x^2 + y = 9$$
$$x - y = -3$$

Solution

$$x^2 + y = 9$$

$$x - y = -3$$

$$x = -3 + y$$

$$4(-3 + y)^2 + y = 9$$

$$9 - 6y + y^2 + y - 9 = 0$$

$$y^2 - 5y = 0$$

$$y(y - 5) = 0$$

$y = 0$	$y = 5$
$x = -3 + 0$	$x = -3 + 5$
$x = -3$	$x = 2$

$(-3, 0)$ and $(2, 5)$

61. Solve the system by substitution.

$$3x + 2y = 90$$
$$xy = 300$$

Solution

$$3x + 2y = 90$$

$$xy = 300$$

$$x = \frac{300}{y}$$

$$3\left(\frac{300}{y}\right) + 2y = 90$$

$$900 + 2y^2 = 90y$$

$$2y^2 - 90y + 900 = 0$$

$$y^2 - 45y + 450 = 0$$

$$(y - 30)(y - 15) = 0$$

$y = 30$	$y = 15$
$x = \dfrac{300}{30}$	$x = \dfrac{300}{15}$
$x = 10$	$x = 20$

$(10, 30)$ and $(20, 15)$

63. Solve the system by substitution.

$$x^2 + y^2 = 100$$
$$x = 12$$

Solution

$$x^2 + y^2 = 100$$
$$x = 12$$
$$12^2 + y^2 = 100$$
$$144 + y^2 = 100$$
$$y^2 = -44$$

No real solution.

65. Solve the system by substitution.

$$y = \sqrt{4 - x}$$
$$x + 3y = 6$$

Solution

$$y = \sqrt{4 - x} \qquad\qquad y = 2 \qquad\qquad y = 1$$
$$x + 3y = 6 \qquad\qquad x = 6 - 3(2) \qquad x = 6 - 3(1)$$
$$x = 6 - 3y \qquad x = 0 \qquad\qquad x = 3$$
$$y = \sqrt{4 - (6 - 3y)} \qquad (0, 2) \text{ and } (3, 1)$$
$$y = \sqrt{4 - 6 + 3y}$$
$$y^2 = -2 + 3y$$
$$y^2 - 3y + 2 = 0$$
$$(y - 2)(y - 1) = 0$$

67. Use a graphing utility to graph the equations and approximate any solutions of the system.

$$2x - 5y = 20$$

$$4x - 5y = 40$$

Solution

Solve each equation for y.

$2x - 5y = 20$	$4x - 5y = 40$
$-5y = -2x + 20$	$-5y = -4x + 40$
$y = \dfrac{2}{5}x - 4$	$y = \dfrac{4}{5}x - 8$
$y = .4x - 4$	$y = .8x - 8$

Keystrokes:

y_1 [Y =] .4 [X, T, θ] [−] 4 [ENTER]

y_2 [Y =] .8 [X, T, θ] [−] 8 [GRAPH]

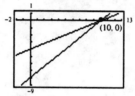

Solution: $(10, 0)$

69. Use a graphing utility to graph the equations and approximate any solutions of the system.

$$x - y = -3$$

$$2x - y = \;\;\; 6$$

Solution

Solve each equation for y.

$x - y = -3$	$2x - y = 6$
$-y = -x - 3$	$-y = -2x + 6$
$y = x + 3$	$y = 2x - 6$

Keystrokes:

y_1 [Y =] [X, T, θ] [+] 3 [ENTER]

y_2 [Y =] 2 [X, T, θ] [−] 6 [GRAPH]

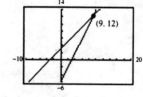

Solution: $(9, 12)$

71. Use a graphing utility to graph the equations and approximate any solutions of the system.

$$-5x + 3y = 15$$
$$x + y = 1$$

Solution

Solve each equation for y.

$$-5x + 3y = 15 \qquad\qquad x + y = 1$$
$$3y = 5x + 15 \qquad\qquad y = -x + 1$$
$$y = \frac{5}{3}x + 5$$

Keystrokes:

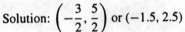

y_1 [Y=] [(] 5 [÷] 3 [)] [X,T,θ] [+] 5 [ENTER]

y_2 [Y=] [(−)] [X,T,θ] [+] 1 [GRAPH]

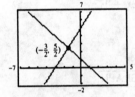

Solution: $\left(-\dfrac{3}{2}, \dfrac{5}{2}\right)$ or $(-1.5, 2.5)$

73. Use a graphing utility to graph the equations and approximate any solutions of the system.

$$y = x^2$$
$$y = 4x - x^2$$

Solution

Keystrokes:

y_1 [Y=] [X,T,θ] [x^2] [ENTER]

y_2 [Y=] [(−)] [X,T,θ] [x^2] [+] 4 [X,T,θ] [GRAPH]

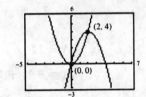

Solution: $(0, 0)$ and $(2, 4)$

75. Use a graphing utility to graph the equations and approximate any solutions of the system.

$$\sqrt{x} - y = 0$$
$$x - 5y = -6$$

Solution

Solve each equation for y.

$\sqrt{x} - y = 0$ $x - 5y = -6$

$\sqrt{x} = y$ $-5y = -x - 6$

$$y = \frac{1}{5}x + \frac{6}{5}$$

$$y = .2x + 1.2$$

Keystrokes:

y_1 | Y= | | √ | | X, T, θ | | ENTER |

y_2 | Y= | .2 | X, T, θ | | + | 1.2 | GRAPH |

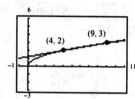

Solution: (4, 2) and (9, 3)

77. Use a graphing utility to graph the equations and approximate any solutions of the system.

$$16x^2 + 9y^2 = 144$$
$$4x + 3y = 12$$

Solution

Solve each equation for y.

$16x^2 + 9y^2 = 144$ $4x + 3y = 12$

$9y^2 = -16x^2 + 144$ $3y = -4x + 12$

$$y^2 = -\frac{16}{9}x^2 + 16 \qquad\qquad y = -\frac{4}{3}x + 4$$

$$y = \pm\sqrt{-\frac{16}{9}x^2 + 16}$$

Keystrokes:

y_1 | Y= | | √ | | (| | (| | (−) | 16 | ÷ | 9 |) | | X, T, θ | | x^2 | | + | 16 |) | | ENTER |

y_2 | Y= | | (−) | | y_1 | | ENTER |

y^3 | Y= | | (| | (−) | 4 | ÷ | 3 |) | | X, T, θ | | + | 4 | GRAPH |

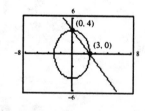

Solution: (0, 4) and (0, 3)

79. Find two positive integers that satisfy the given requirements.

The sum of two numbers is 80 and their difference is 18.

Solution

$x + y = 80$

$x - y = 18$

$x = 18 + y$

$18 + y + y = 80$

$2y = 62$

$y = 31$

$x = 18 + 31 = 49$

$(49, 31)$

81. Find two positive integers that satisfy the given requirements.

The sum of two numbers is 52 and the larger number is 8 less than twice the smaller number.

Solution

$x + y = 52$

$x = 2y - 8$

$2y - 8 + y = 52$

$3y = 60$

$y = 20$

$x = 2(20) - 8 = 32$

$(32, 20)$

83. *Geometry* Find the dimensions of the rectangle that meets the specified conditions.

Perimeter	*Condition*
50 feet	The length is 5 feet greater than the width.

Solution

$2x + 2y = 50$

$x = 5 + y$

$2(5 + y) + 2y = 50$

$10 + 2y + 2y = 50$

$4y = 40$

$y = 10$

$x = 5 + 10 = 15$

(length = 15, width = 10)

85. *Geometry* Find the dimensions of the rectangle that meets the specified conditions.

Perimeter	Condition
68 yards	The width is 7/10 of the length.

Solution

$$2x + 2y = 68$$

$$x = \frac{7}{10}y$$

$$2\left(\frac{7}{10}y\right) + 2y = 68$$

$$\frac{7}{5}y + 2y = 68$$

$$7y + 10y = 340$$

$$17y = 340$$

$$y = 20$$

$$x = \frac{7}{10}(20) = 7(2) = 14$$

(width = 14, length = 20)

87. *Simple Interest* A combined total of $20,000 is invested in two bonds that pay 8% and 9.5% simple interest. The annual interest is $1675. How much is invested in each bond?

Solution

$$x + y = 20,000$$
$$0.08x + 0.095y = 1675$$
$$x = 20,000 - y$$

$$0.08(20,000 - y) + 0.095y = 1675$$

$$1600 - 0.08y + 0.095y = 1675$$

$$0.015y = 75$$

$$y = \$5000$$

$$x = 20,000 - 5000 = \$15,000$$

($15,000, $5000)

89. *Busing Boundary*   To be eligible to ride the school bus to East High School, a student must live at least 1 mile from the school (see figure in textbook). Describe the portion of Clarke Street from which the residents are not eligible to ride the school bus. Use a coordinate system in which the school is at (0, 0) and each units represents 1 mile.

Solution

equation of circle: $x^2 + y^2 = 1$

equation of line: $m = \dfrac{0 - -1}{5 - -2} = \dfrac{1}{7}$ $y - 0 = \dfrac{1}{7}(x - 5)$

$$y = \dfrac{1}{7}x - \dfrac{5}{7}$$

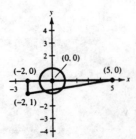

Solve system: $x^2 + y^2 = 1$

$$y = \dfrac{1}{7}x - \dfrac{5}{7}$$

Solution: $\left(-\dfrac{3}{5}, -\dfrac{4}{5}\right)$ and $\left(\dfrac{4}{5}, -\dfrac{3}{5}\right)$

91. *Comparing Population*   From 1982 to 1988, the northeastern part of the United States grew at a slower rate than the western part. Two models that represent the populations of the two regions are

$$P = 49{,}094.5 + 160.9t + 9.7t^2 \qquad \text{Northeast}$$
$$P = 43{,}331.9 + 907.8t \qquad \text{West}$$

where P is the population in thousands and t is the calendar year with $t = 3$ corresponding to 1983. Use a graphing utility to determine when the population of the West overtook the population of the Northeast. (*Source:* U.S. Bureau of Census)

Solution

Keystrokes:

y_1 [Y =] 49,094.5 [+] 160.9 [X, T, θ] [+] 9.7 [X, T, θ] [x^2] [ENTER]

y_2 [Y =] 43,331.9 [+] 907.8 [X, T, θ] [GRAPH]

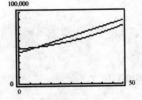

The point of intersection is (8.697, 51227.818), so $t \approx 9$.

1989

8.2   Systems of Linear Equations in Two Variables

7. Use a graphing utility to graph the equations in the system. Use the graphs to determine whether the system is consistent or inconsistent. If it is consistent, determine the number of solutions.

$$\frac{1}{3}x - \frac{1}{2}y = 1$$
$$-2x + 3y = 6$$

Solution

Solve each equation for y.

$$\frac{1}{3}x - \frac{1}{2}y = 1 \qquad\qquad -2x + 3y = 6$$
$$-\frac{1}{2}y = -\frac{1}{3}x + 1 \qquad\qquad 3y = 2x + 6$$
$$y = \frac{2}{3}x - 2 \qquad\qquad y = \frac{2}{3}x + 2$$

Keystrokes:

y_1 Y= (2 ÷ 3) X,T,θ − 2 ENTER
y_2 Y= (2 ÷ 3) X,T,θ + 2 GRAPH

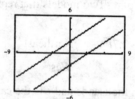

Inconsistent

9. Use a graphing utility to graph the equations in the system. Use the graphs to determine whether the system is consistent or inconsistent. If it is consistent, determine the number of solutions.

$$-2x + 3y = 6$$
$$x - y = -1$$

Solution

Solve each equation for y.

$$-2x + 3y = 6 \qquad\qquad x - y = -1$$
$$3y = 2x + 6 \qquad\qquad -y = -x - 1$$
$$y = \frac{2}{3}x + 2 \qquad\qquad y = x + 1$$

Keystrokes:

y_1 Y= (2 ÷ 3) X,T,θ + 2 ENTER
y_2 Y= X,T,θ + 1 GRAPH

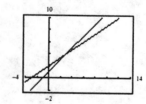

One solution.

11. Determine whether the system is consistent or inconsistent.

$$x + 2y = 6$$
$$x + 2y = 3$$

Solution

Solve each equation for y.

$$x + 2y = 6 \qquad\qquad x + 2y = 3$$
$$2y = -x + 6 \qquad\qquad 2y = -x + 3$$
$$y = -\frac{1}{2}x + 3 \qquad\qquad y = -\frac{1}{2}x + \frac{3}{2}$$

Slopes are equal; therefore the system is inconsistent.

13. Solve the system of linear equations by elimination. Identify and label each line with its equation and label the point of intersection (if any).

$$-x + 2y = 1$$
$$x - y = 2$$

Solution

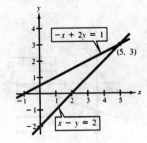

$$-x + 2y = 1$$
$$\underline{x - y = 2}$$
$$y = 3$$
$$x - 3 = 2$$
$$x = 5$$
$$(5, 3)$$

15. Solve the system of linear equations by elimination. Identify and label each line with its equation and label the point of intersection (if any).

$$x + y = 0$$
$$3x - 2y = 10$$

Solution

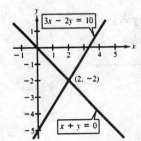

$$x + y = 0 \Rightarrow 2x + 2y = 0$$
$$3x - 2y = 10 \Rightarrow 3x - 2y = 10$$
$$\underline{}$$
$$5x = 10$$
$$x = 2$$
$$2 + y = 0$$
$$y = -2$$

$$(2, -2)$$

17. Solve the system of linear equations by elimination. Identify and label each line with its equation and label the point of intersection (if any).

$$x - y = 1$$
$$-3x + 3y = 8$$

Solution

$$x - y = 1 \Rightarrow 3x - 3y = 3$$
$$-3x + 3y = 8 \Rightarrow -3x + 3y = 8$$

$$\overline{ 0 \neq 11}$$

no solution

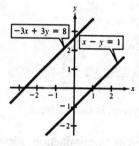

19. Solve the system of linear equations by elimination. Identify and label each line with its equation and label the point of intersection (if any).

$$x - 3y = 5$$
$$-2x + 6y = -10$$

Solution

$$x - 3y = 5 \Rightarrow 2x - 6y = 10$$
$$-2x + 6y = -10 \Rightarrow -2x + 6y = -10$$

$$\overline{ 0 = 0}$$

all solutions to $x - 3y = 5$

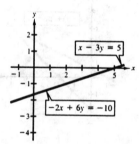

21. Solve the system of linear equations by elimination. Identify and label each line with its equation and label the point of intersection (if any).

$$2x - 8y = -11$$
$$5x + 3y = 7$$

Solution

$$2x - 8y = -11 \Rightarrow 6x - 24y = -33$$
$$5x + 3y = 7 \Rightarrow 40x + 24y = 56$$

$$\overline{}$$

$$46x = 23 \qquad 2\tfrac{1}{2} - 8y = -11$$
$$x = \frac{23}{46} \qquad -8y = -12$$
$$x = \frac{1}{2} \qquad y = \frac{-12}{-8}$$
$$y = \frac{3}{2}$$

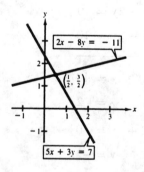

$\left(\frac{1}{2}, \frac{3}{2}\right)$

23. Solve the system of linear equations by the method of elimination.

$$3x - 2y = 5$$
$$x + 2y = 7$$

Solution

$$
\begin{array}{rcl}
3x - & 2y = & 5 \\
x + & 2y = & 7 \\
\hline
4x & = & 12 \\
x & = & 3 \\
3 + 2y = & 7 \\
2y = & 4 \\
y = & 2
\end{array}
$$

$(3, 2)$

25. Solve the system of linear equations by the method of elimination.

$$12x - 5y = 2$$
$$-24x + 10y = 6$$

Solution

$$
\begin{array}{rcl}
12x - 5y = 2 & \Rightarrow & 24x - 10y = 4 \\
-24x + 10y = 6 & \Rightarrow & -24x + 10y = 6 \\
\hline
& & 0 = 10
\end{array}
$$

no solution

27. Solve the system of linear equations by the method of elimination.

$$2x + 3y = 0$$
$$3x + 5y = -1000$$

Solution

$$
\begin{array}{rcl}
2x + 3y = & 0 & \Rightarrow & 6x + 9y = & 0 \\
3x + 5y = & -1000 & \Rightarrow & -6x - 10y = & 2000 \\
\hline
& & & -y = & 2000 \\
& & & y = & -2000
\end{array}
$$

$$
\begin{array}{rcl}
2x + 3(-2000) = & 0 \\
2x - 6000 = & 0 \\
2x = & 6000 \\
x = & 3000
\end{array}
$$

$(3000, -2000)$

29. Solve the system by any convenient method.

$$\frac{3}{2}x + 2y = 12$$

$$\frac{1}{4}x + y = 4$$

Solution

$$y = 4 - \frac{1}{4}x$$

$$\frac{3}{2}x + 2\left(4 - \frac{1}{4}x\right) = 12$$

$$\frac{3}{2}x + 8 - \frac{1}{2}x = 12$$

$$x = 4$$

$$\frac{1}{4}(4) + y = 4$$

$$y = 3$$

$$(4, 3)$$

33. Find a system of linear equations that has the given solution. (There are many correct answers.)

Solution

$$x + 2y = 0$$

$$x - 4y = 9$$

31. Solve the system by any convenient method.

$$y = 5x - 3$$

$$y = -2x + 11$$

Solution

$$y = 5x - 3 \Rightarrow y = 5x - 3$$

$$y = -2x + 11 \Rightarrow -y = 2x - 11$$

$$\begin{array}{ll} 0 = 7x - 14 & y = 5(2) - 3 \\ 14 = 7x & y = 10 - 3 \\ 2 = x & y = 7 \end{array}$$

$$(2, 7)$$

35. *Break-Even Analysis* You are planning to open a restaurant. You need an initial investment of $85,000. Each week your costs will be about $7400. If your weekly revenue is $8100, how many weeks will it take to break even?

Solution

Cost = initial investment + weekly cost

$$y = 85,000 + 7400x$$

Revenue = weekly revenue

$$y = 8100x$$

Break-Even occurs when Cost = Revenue.

$$y = 85,000 + 7400x$$

$$y = 8100x$$

$$0 = 85,000 - 700x$$

$$700x = 85,000$$

$$x = 121 \text{ weeks}$$

37. *Average Speed* A truck travels for 4 hours at an average speed of 42 miles per hour. How much longer must the truck travel at an average speed of 55 miles per hour so that the average speed for the entire trip will be 50 miles per hour?

Solution

Verbal model: | Average speed at 42 mph | $+$ | Average speed at 55 mph | $=$ | Average speed at 50 mph |

Equation:
$$42(4) + 55x = 50(4 + x)$$
$$168 + 55x = 200 + 50x$$
$$5x = 32$$
$$x = 6.4 \text{ hours}$$

39. *Rope Length* You must cut a rope that is 160 inches long into two pieces so that one piece is four times as long as the other. Find the length of each piece.

Solution

Verbal model: | Length Piece 1 | $+$ | Length Piece 2 | $=$ | 160 |

| Length Piece 1 | $= 4 \cdot$ | Length Piece 2 |

System of equations:
$$x + y = 160$$
$$x = 4y$$
$$4y + y = 160$$
$$5y = 160$$
$$y = 32 \text{ inches}$$
$$x = 4(32) = 128 \text{ inches}$$

41. *Vietnam Veterans Memorial* "The Wall" in Washington, D.C., designed by Maya Ling Lin when she was a student at Yale University, has two vertical triangular sections of black granite with a common side, (see figure in textbook). The top of each section is level with the ground. The bottoms of the two sections can be modeled by the equations $y = \frac{2}{25}x - 10$ and $y = -\frac{5}{61}x - 10$ when the x-axis is superimposed on the top of the wall. Each unit on the coordinate system represents 1 foot. How deep is the memorial at the point where the two sections meet? How long is each section?

Solution

(a) $y = \frac{2}{25}x - 10$ Solve by substitution.

$y = -\frac{5}{61}x - 10$

$\frac{2}{25}x - 10 = -\frac{5}{61}x - 10$ $y = \frac{2}{25}(0) - 10$

$\frac{2}{25}x = -\frac{5}{61}x$ $y = -10$

$x = 0$ $(0, -10)$

(b) $0 = \frac{2}{25}x - 10$ $0 = \frac{5}{61}x - 10$

$10 = \frac{2}{25}x$ $10 = \frac{5}{61}x$

$125 = x$ $122 = x$

122 feet and 125 feet

The memorial is 10 feet deep.

43. Find an equation of the line through the points $(-2, 7)$ and $(5, 5)$.

Solution

$$m = \frac{5 - 7}{5 - -2} = -\frac{2}{7} \qquad y - 5 = -\frac{2}{7}(x - 5)$$

$$y - 5 = -\frac{2}{7}x + \frac{10}{7}$$

$$7y - 35 = -2x + 10$$

$$2x + 7y - 45 = 0$$

45. Find an equation of the line through the points $(6, 3)$ and $(10, 3)$.

Solution

$$m = \frac{3 - 3}{10 - 6} = \frac{0}{4} = 0 \qquad y = 3 = 0(x - 10)$$

$$y - 3 = 0$$

47. Find the number of units x that produces a maximum revenue R. $R = 900x - 0.1x^2$. Confirm your equation graphically with a graphing utility.

Solution

Maximum occurs at $(4500, 2{,}025{,}000)$.

4500 units

Keystrokes:

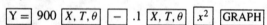

$\boxed{Y =}\ 900\ \boxed{X, T, \theta}\ \boxed{-}\ .1\ \boxed{X, T, \theta}\ \boxed{x^2}\ \boxed{\text{GRAPH}}$

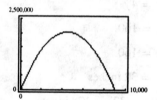

49. Solve the system of linear equations by the method of elimination.

$$2x + y = 9$$
$$3x - y = 16$$

Solution

$$
\begin{array}{ll}
2x + y = 9 & 2(5) + y = 9 \\
3x - y = 16 & 10 + y = 9 \\
\rule{3cm}{0.4pt} & y = -1 \\
5x = 25 & (5, -1) \\
x = 5 &
\end{array}
$$

51. Solve the system of linear equations by the method of elimination.

$$4x + y = -3$$
$$-4x + 3y = 23$$

Solution

$$4x + y = -3$$
$$-4x + 3y = 23$$

$$\overline{}$$

$$4y = 20$$
$$y = 5$$
$$4x + 5 = -3$$
$$4x = -8$$
$$x = -2$$

$$(-2, 5)$$

53. Solve the system of linear equations by the method of elimination.

$$x - 3y = 2$$
$$3x - 7y = 4$$

Solution

$$x - 3y = 2 \Rightarrow -3x + 9y = -6$$
$$3x - 7y = 4 \Rightarrow 3x - 7y = 4$$

$$\overline{}$$

$$2y = -2$$
$$y = -1$$
$$x - 3(-1) = 2$$
$$x = -1$$

$$(-1, -1)$$

55. Solve the system of linear equations by the method of elimination.

$$2u + 3v = 8$$
$$3u + 4v = 13$$

Solution

$$2u + 3v = 8 \Rightarrow -6u - 9v = -24$$
$$3u + 4v = 13 \Rightarrow 6u + 8v = 26$$

$$\overline{}$$

$$-v = 2 \qquad 2u + 3(-2) = 8$$
$$v = -2 \qquad 2u = 14$$
$$u = 7$$

$$(7, -2)$$

57. Solve the system of linear equations by the method of elimination.

$$\frac{2}{3}r - s = 0$$
$$10r + 4s = 19$$

Solution

$$\frac{2}{3}r - s = 0 \Rightarrow 2r - 3s = 0 \Rightarrow 8r - 12s = 0$$
$$10r + 4s = 19 \Rightarrow 10r + 4s = 19 \Rightarrow 30r + 12s = 57$$

$$
\begin{array}{ll}
38r = 57 & \frac{2}{3}\left(\frac{3}{2}\right) - s = 0 \\
r = \frac{57}{38} & -s = -1 \\
r = \frac{3}{2} & s = 1
\end{array}
$$

$\left(\frac{3}{2}, 1\right)$

59. Solve the system of linear equations by the method of elimination.

$$0.7u - v = -0.4$$
$$0.3u - 0.8v = 0.2$$

Solution

$$0.7u - v = -0.4 \Rightarrow 7u - 10v = -4 \Rightarrow 21u - 30v = -12$$
$$0.3u - 0.8v = 0.2 \Rightarrow 3u - 8v = 2 \Rightarrow -21u + 56v = -14$$

$$
\begin{aligned}
26v &= -26 \\
v &= -1 \\
7u - 10(1) &= -4 \\
7u &= -14 \\
u &= -2
\end{aligned}
$$

$(-2, -1)$

61. Solve the system of linear equations by the method of elimination.

$$5x + 7y = 25$$
$$x + 1.4y = 5$$

Solution

$$5x + 7y = 25 \Rightarrow 5x + 7y = 25$$
$$x + 1.4y = 5 \Rightarrow -5x - 7y = -25$$

$$0 = 0$$

All solutions of the form $x + 1.4y = 5$.

63. Solve the system of linear equations by the method of elimination.

$$\frac{3}{2}x - y = 4$$
$$-x + \frac{2}{3}y = -1$$

Solution

$$\frac{3}{2}x - y = 4 \Rightarrow 3x - 2y = 8$$
$$-x + \frac{2}{3}y = -1 \Rightarrow -3x + 2y = -3$$
$$\overline{0 = 5}$$

Inconsistent

65. Solve the system of linear equations by the method of elimination.

$$2x = 25$$
$$4x - 10y = 0.52$$

Solution

$$2x = 25 \Rightarrow -4x = -50$$
$$4x - 10y = 0.52 \Rightarrow 4x - 10y = 0.52$$
$$\overline{}$$

$$-10y = -49.48 \qquad 4x - 10(4.948) = 0.52$$
$$y = 4.948 \qquad 4x - 49.48 = 0.52$$
$$4x = 50$$
$$x = 12.5$$

(12.5, 4.948)

67. Solve the system by any convenient method.

$$2x - y = 20$$
$$-x + y = -5$$

Solution

$$2x - y = 20 \qquad -15 + y = -5$$
$$-x + y = -5 \qquad y = 10$$
$$\overline{}$$
$$x = 15$$

(15, 10)

69. Solve the system by any convenient method.

$$3y = 2x + 21$$
$$x = 50 - 4y$$

Solution

$$3y = 2x + 21$$
$$x = 50 - 4y$$

$$3y = 2(50 - 4y) + 21 \qquad x = 50 - 4(11)$$
$$3y = 100 - 8y + 21 \qquad x = 50 - 44$$
$$11y = 121 \qquad x = 6$$
$$y = 11$$

(6, 11)

71. Determine the value of k such that the system of linear equations is inconsistent.

$$5x - 10y = 40$$
$$-2x + ky = 30$$

Solution

$$5x - 10y = 40 \Rightarrow y = \frac{1}{2}x - 4$$

$$-2x + ky = 30 \Rightarrow y = \frac{2}{k}x + \frac{30}{16}$$

$$\text{so} \quad \frac{2}{k} = \frac{1}{2} \Rightarrow k = 4$$

$$5x - 10y = 40 \Rightarrow 10x - 20y = 80$$
$$-2x + 4y = 30 \Rightarrow -10x + 20y = 150$$

$$\overline{}$$

$$0 \neq 230$$

Inconsistent; no solution

73. Decide whether the system is consistent or inconsistent.

$$4x - 5y = 3$$
$$-8x + 10y = -6$$

Solution

$$4x - 5y = 3 \Rightarrow -5y = -4x + 3 \Rightarrow y = \frac{4}{5}x - \frac{3}{5}$$

$$-8x + 10y = -6 \Rightarrow 10y = 8x - 6 \Rightarrow y = \frac{4}{5}x - \frac{3}{5}$$

many solutions $\Rightarrow$ consistent

75. Decide whether the system is consistent or inconsistent.

$$-2x + 5y = 3$$
$$5x + 2y = 8$$

Solution

$$-2x + 5y = 3 \Rightarrow 5y = 2x + 3 \Rightarrow y = \frac{2}{5}x + \frac{3}{5}$$

$$5x + 2y = 8 \Rightarrow 2y = -5x + 8 \Rightarrow y = -\frac{5}{2}x + 4$$

one solution $\Rightarrow$ consistent

77. Decide whether the system is consistent or inconsistent.

$$-10x + 15y = 25$$
$$2x - 3y = -24$$

Solution

$$-10x + 5y = 25 \Rightarrow 15y = 10x + 25 \Rightarrow y = \frac{2}{3}x + \frac{5}{3}$$

$$2x - 3y = -24 \Rightarrow -3y = -2x - 24 \Rightarrow y = \frac{2}{3}x + 8$$

no solution $\Rightarrow$ inconsistent

79. Use a graphing utility to sketch the graphs of the two equations. Use the graphs to approximate the solution of the system.

$$5x + 4y = 35$$
$$-x + 3y = 12$$

Solution

Keystrokes:

y_1 Y= (35 − 5 X,T,θ) ÷ 4 ENTER
y_2 Y= 4 + (1 ÷ 3) X,T,θ GRAPH

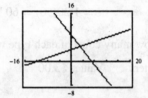

approximate solution $(3.05, 4.93) \approx (3, 5)$.

81. Use a graphing utility to sketch the graphs of the two equations. Use the graphs to approximate the solution of the system.

$$4x - y = 3$$
$$6x + 2y = 1$$

Solution

Keystrokes:

y_1 Y= 4 X,T,θ − 3 ENTER
y_2 Y= (−) 3 X,T,θ + .5 GRAPH

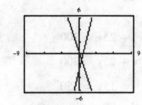

approximate solution $\approx \left(\dfrac{1}{2}, -1 \right)$.

83. *Dimensions of a Rectangle* Find the dimensions of the rectangle meeting the specified conditions.

Perimeter: 220 meters

Relationship Between the Length & Width:
Length is 120% of the width.

Solution

Verbal model:

2 [Length] + 2 [Width] = [Perimeter]

[Length] = 1.20 · [Width]

System of equations:

$$2x + 2y = 220$$
$$x = 1.20y$$
$$2(1.20y) + 2y = 220$$
$$2.4y + 2y = 220$$
$$4.4y = 220$$
$$y = 50 \text{ width}$$
$$x = 1.20(50) = 60 \text{ length}$$

85. *Coin Problem* Determine how many coins of each type will yield the given value.

Coins: 15 dimes and quarters   Value: $3.00

Solution

Verbal model:

[Number of dimes] + [Number of quarters] = [15]

[Value of dimes] + [Value of quarters] = [$3.00]

System of equations:

$$x + y = 15$$
$$0.10x + 0.25y = 3.00$$
$$x = 15 - y$$
$$10(15 - y) + 25y = 300$$
$$150 - 10y + 25y = 300$$
$$15y = 150$$
$$y = 10 \text{ quarters}$$
$$x = 15 - 10 = 5 \text{ dimes}$$

87. *Coin Problem* Determine how many coins of each type will yield the given value.

Coins: 25 nickels and dimes   Value: $1.95

Solution

Verbal model:

$$\boxed{\text{Number of nickels}} + \boxed{\text{Number of dimes}} = \boxed{25}$$

$$\boxed{\text{Value of nickels}} + \boxed{\text{Value of dimes}} = \boxed{\$1.95}$$

System of equations:

$$x + y = 25$$

$$0.05x + 0.10y = 1.95$$

$$x = 25 - y$$

$$5(25 - y) + 10y = 195$$

$$125 - 5y + 10y = 195$$

$$5y = 70$$

$$y = 14 \text{ dimes}$$

$$x = 25 - 14 = 11 \text{ nickels}$$

89. *Gasoline Mixture* Twelve gallons of regular unleaded gasoline plus 8 gallons of premium unleaded gasoline cost $23.08. The price of premium unleaded is 11 cents more per gallon than the price of regular unleaded. Find the price per gallon for each grade of gasoline.

Solution

Verbal model:

$$12 \left(\boxed{\text{Cost of regular gasoline}} \right) + 8 \left(\boxed{\text{Cost of premium gasoline}} \right) = \boxed{\$23.08}$$

$$\boxed{\text{Cost of premium gasoline}} = \boxed{\$0.11} + \boxed{\text{Cost of regular gasoline}}$$

System of equations:

$$12x + 8y = 23.08$$

$$y = 0.11 + x$$

$$12x + 8(0.11 + x) = 23.08$$

$$12x + 0.88 + 8x = 23.08$$

$$20x = 22.20$$

$$x = \$1.11 \text{ regular}$$

$$y = 0.11 + 1.11 = \$1.22 \text{ premium}$$

91. *Nut Mixture* Ten pounds of mixed nuts sell for $6.95 per pound. The mixture is obtained from two kinds of nuts, with one variety priced at $5.65 per pound and the other at $8.95 per pound. How many pounds of each variety of nuts were used in the mixture?

Solution

Verbal model:

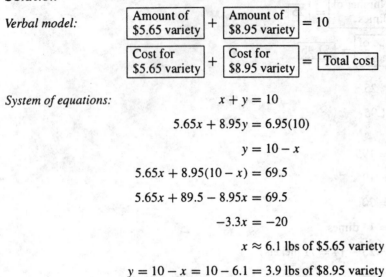

System of equations:

$$x + y = 10$$
$$5.65x + 8.95y = 6.95(10)$$
$$y = 10 - x$$
$$5.65x + 8.95(10 - x) = 69.5$$
$$5.65x + 89.5 - 8.95x = 69.5$$
$$-3.3x = -20$$
$$x \approx 6.1 \text{ lbs of \$5.65 variety}$$
$$y = 10 - x = 10 - 6.1 = 3.9 \text{ lbs of \$8.95 variety}$$

93. *Alcohol Mixture* How many liters of a 40% alcohol solution must be mixed with a 65% solution to obtain 20 liters of a 50% solution?

Solution

Verbal model:

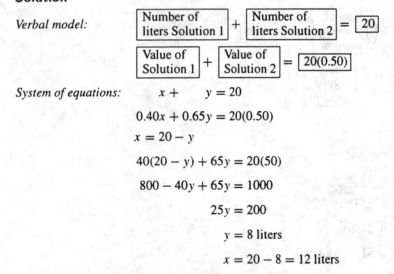

System of equations:

$$x + y = 20$$
$$0.40x + 0.65y = 20(0.50)$$
$$x = 20 - y$$
$$40(20 - y) + 65y = 20(50)$$
$$800 - 40y + 65y = 1000$$
$$25y = 200$$
$$y = 8 \text{ liters}$$
$$x = 20 - 8 = 12 \text{ liters}$$

95. *Best-Fitting Line* The slope and y-intercept of the line $y = mx + b$ that "best fits" the three non collinear points $(0, 0)$, $(1, 1)$, and $(2, 3)$ are given by the following system of linear equations.

$$3b + 6m = 7$$
$$3b + 3m = 4$$

(a) Solve the system and find the equation of the "best-fitting" line.

(b) Plot the three points and sketch the graph of the "best-fitting" line.

Solution

$$3b + 6m = 7$$
$$\underline{3b + 3m = 4}$$

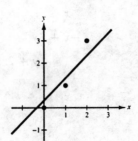

$$3m = 3$$

$$\boxed{b = 1}$$

$$3b + 3(1) = 4$$
$$3b = 4 - 3$$
$$3b = 1$$

$$\boxed{m = 1/3}$$

Therefore, $y = \dfrac{1}{3}x + 1$

8.3 Systems of Linear Equations in Three Variables

5. Use back-substitution to solve the system of linear equations.

$$x - 2y + 4z = 4$$
$$3y - z = 2$$
$$z = -5$$

Solution

$$3y - (-5) = 2$$
$$3y = -3$$
$$y = -1$$
$$x - 2(-1) + 4(-5) = 4$$
$$x + 2 - 20 = 4$$
$$x - 18 = 4$$
$$x = 22$$

$$(22, -1, -5)$$

7. Use back-substitution to solve the system of linear equations.

$$x - 2y + 4z = 4$$
$$y = 3$$
$$y + z = 2$$

Solution

$$3 + z = 2$$
$$z = -1$$
$$x - 2(3) + 4(-1) = 4$$
$$x - 6 - 4 = 4$$
$$x - 10 = 4$$
$$x = 14$$

$$(14, 3, -1)$$

9. Use Gaussian elimination to solve the system of linear equations.

$$x \quad + z = 4$$
$$y \quad = 2$$
$$4x \quad + z = 7$$

Solution

$$x \quad + \quad z = \quad 4$$
$$y \quad = \quad 2$$
$$4x \quad + \quad z = \quad 7$$

$$x \quad + \quad z = \quad 4$$
$$y \quad = \quad 2$$
$$-3z = -9$$

$$x \quad + \quad z = \quad 4$$
$$y \quad = \quad 2$$
$$z = \quad 3$$

$$x \quad = \quad 1$$
$$y \quad = \quad 2$$
$$z = \quad 3$$

$$(1, 2, 3)$$

11. Use Gaussian elimination to solve the system of linear equations.

$$x \quad = 3$$
$$-x + 3y \quad = 3$$
$$y + 2z = 4$$

Solution

$$x \quad = 3$$
$$-x + 3y \quad = 3$$
$$y + 2z = 4$$

$$x \quad = 3$$
$$3y \quad = 6$$
$$y + 2z = 4$$

$$x \quad = 3$$
$$y \quad = 2$$
$$y + 2z = 4$$

$$x \quad = 3$$
$$y \quad = 2$$
$$2z = 2$$

$$x \quad = 3$$
$$y \quad = 2$$
$$z = 1$$

$$(3, 2, 1)$$

13. Use Gaussian elimination to solve the system of linear equations.

$$x + 2y + 6z = 5$$
$$-x + y - 2z = 3$$
$$x - 4y - 2z = 1$$

Solution

$$x + 2y + 6z = 5$$
$$-x \quad y - 2z = 3$$
$$x - 4y - 2z = 1$$

$$x + 2y + 6z = 5$$
$$3y + 4z = 8$$
$$x - 4y - 2z = 1$$

$$x + 2y + 6z = 5$$
$$3y + 4z = 8$$
$$-6y - 8z = -4$$

$$x + 2y + 6z = 5$$
$$3y + 4z = 8$$
$$0 = 12$$

no solution
inconsistent

15. Use Gaussian elimination to solve the system of linear equations.

$$0.2x + 1.3y + 0.6z = 0.1$$
$$0.1x \quad + 0.3z = 0.7$$
$$2x + 10y + 8z = 8$$

Solution

$$0.2x + 1.3y + 0.6z = 0.1$$
$$0.1x \quad + 0.3z = 0.7$$
$$2x + 10y + 8z = 8$$

$$2x + 13y + 6z = 1$$
$$1x \quad + 3z = 7$$
$$2x + 10y + 8z = 8$$

$$1x \quad + 3z = 7$$
$$13y = -13$$
$$10y + 2z = -6$$

$$1x \quad + 3z = 7$$
$$y = -1$$
$$10y + 2z = -6$$

$$x \quad + 3z = 7$$
$$y = -1$$
$$2z = 4$$

$$x + \quad 3z = 7$$
$$y = -1$$
$$z = 2$$

$$x + 3(2) = 7$$
$$x = 1$$
$$(1, -1, 2)$$

17. Use Gaussian elimination to solve the system of linear equations.

$$x + 4y - 2z = 2$$
$$-3x + y + z = -2$$
$$5x + 7y - 5z = 6$$

Solution

$$x + 4y - 2z = 2$$
$$-3x + y + z = -2$$
$$5x + 7y - 5z = 6$$

$$x + 4y - 2z = 2$$
$$13y - 5z = 4$$
$$-13y + 5z = -4$$

$$x + 4y - 2z = 2$$
$$y - \frac{5}{13}z = \frac{4}{13}$$
$$-13y + 5z = -4$$

$$x + 4y - 2z = 2$$
$$y - \frac{5}{13}z = \frac{4}{13}$$
$$0 = 0$$

$$y = \frac{5}{13}z + \frac{4}{13}$$

$$x + 4\left(\frac{5}{13}x + \frac{4}{13}\right) - 2z = 2$$

$$x + \frac{20}{13}z + \frac{16}{13} - \frac{26}{13}z = \frac{26}{13}$$

$$x - \frac{6}{13}z = \frac{10}{13}$$

$$x = \frac{6}{13}z + \frac{10}{13}$$

let $z = a$

$$\left(\frac{6}{13}a + \frac{10}{13}, \frac{5}{13}a + \frac{4}{13}, a\right)$$

19. Find the equation of the parabola (see figure in textbook)

$$y = ax^2 + bx + c$$

that passes through the points $(0, 5)$, $(1, 6)$, and $(2, 5)$.

Solution

$$5 = a(0)^2 + b(0) + c$$
$$6 = a(1)^2 + b(1) + c$$
$$5 = a(2)^2 + b(2) + c$$

$$5 = c$$
$$6 = a + b + c$$
$$5 = 4a + 2b + c$$

$$5 = c$$
$$6 = a + b + c$$
$$-19 = -2b - 3c$$

$$-19 = -2b - 3(5)$$

$$-4 = -2b$$

$$2 = b$$

$$6 = a + 2 + 5$$

$$-1 = a$$

$$y = -x^2 + 2x + 5$$

21. *Curve Fitting* Find the equation of the circle

$$x^2 + y^2 + Dx + Ey + F = 0$$

that passes through the points $(0, 0)$, $(2, -2)$, and $(4, 0)$ (see figure in textbook).

Solution

$$0^2 + \quad 0^2 + D(0) + \quad E(0) + F = 0$$
$$2^2 + (-2)^2 + D(2) + E(-2) + F = 0$$
$$4^2 + \quad 0^2 + D(4) + \quad E(0) + F = 0$$

$$F = 0$$
$$2D - 2E + F = -8$$
$$4D + F = -16$$

$$4D + 0 = -16$$
$$4D = -16$$
$$D = -4$$

$$2(-4) - 2E + 0 = -8$$
$$-2E = 0$$
$$E = 0$$

$$x^2 + y^2 - 4x = 0$$

23. Find the position equation $s = \dfrac{1}{2}at^2 + v_0t + s_0$ for an object that has the indicated heights at the specified times.

$s = 128$ feet at $t = 1$ second

$s = 80$ feet at $t = 2$ seconds

$s = 0$ feet at $t = 3$ seconds

Solution

$$128 = \frac{1}{2}a(1)^2 + v_0(1) + \quad s_0$$
$$80 = \frac{1}{2}a(2)^2 + v_0(2) + \quad s_0$$
$$0 = \frac{1}{2}a(3)^2 + v_0(3) + \quad s_0$$

$$128 = \quad \frac{1}{2}a + \quad v_0 + \quad s_0$$
$$80 = \quad 2a + \quad 2v_0 + \quad s_0$$
$$0 = \quad \frac{9}{2}a + \quad 3v_0 + \quad s_0$$

$$256 = \quad a + \quad 2v_0 + \quad 2s_0$$
$$80 = \quad 2a + \quad 2v_0 + \quad s_0$$
$$0 = \quad \frac{9}{2}a + \quad 3v_0 + \quad s_0$$

$$256 = \quad a + \quad 2v_0 + \quad 2s_0$$
$$-432 = \quad - \quad 2v_0 - \quad 3s_0$$
$$-1152 = \quad - \quad 6v_0 - \quad 8s_0$$

$$256 = \quad a + \quad 2v_0 + \quad 2s_0$$
$$216 = \quad v_0 + \quad \frac{3}{2}s_0$$
$$1152 = \quad - \quad 6v_0 - \quad 8s_0$$

$$256 = \quad a + \quad 2v_0 + \quad 2s_0$$
$$216 = \quad v_0 + \quad \frac{3}{2}s_0$$
$$144 = \quad + \quad s_0$$

$$216 = \quad v_0 + \frac{3}{2}(144)$$
$$0 = \quad v_0$$
$$256 = \quad a + \quad 0 + \quad 288$$

$$-32 = \quad a$$
$$s = -16t^2 + 144$$

25. Find the position equation $s = \frac{1}{2}at^2 + v_0t + s_0$ for an object that has the indicated heights at the specified times.

$s = 32$ feet at $t = 1$ second

$s = 32$ feet at $t = 2$ seconds

$s = 0$ feet at $t = 3$ seconds

Solution

$$32 = \frac{1}{2}a(1)^2 + v_0(1) + s_0$$

$$32 = \frac{1}{2}a(2)^2 + v_0(2) + s_0$$

$$0 = \frac{1}{2}a(3)^2 + v_0(3) + s_0$$

$$64 = a + 2v_0 + 2s_0$$

$$32 = 2a + 2v_0 + s_0$$

$$0 = 9a + 6v_0 + 2s_0$$

$$64 = a + 2v_0 + 2s_0$$

$$-96 = - 2v_0 - 3s_0$$

$$-576 = - 12v_0 - 16s_0$$

$$64 = a + 2v_0 + 2s_0$$

$$48 = v_0 + \frac{3}{2}s_0$$

$$-576 = - 12v_0 - 16s_0$$

$$64 = a + 2v_0 + 2s_0$$

$$48 = v_0 + \frac{3}{2}s_0$$

$$0 = + 2s_0$$

$$0 = s_0$$

$$48 = v_0 + 0$$

$$64 = a + 2(48) + 0$$

$$-32 = a$$

$$s = -16t^2 + 48t$$

27. *Crop Spraying* A mixture of 12 gallons of chemical A, 16 gallons of chemical B, and 26 gallons of chemical C is required to kill a certain destructive crop insect. Commercial spray X contains 1, 2, and 2 parts, respectively, of these chemicals. Commercial spray Y contains only chemical C. Commercial spray Z contains only chemicals A and B in equal amounts. How much of each type of commercial spray is needed to get the desired mixture?

Solution

$$\begin{aligned}.20x &&+ .50z &= 12 \\ .40x &&+ .50z &= 16 \\ .40x &+ 1y && = 26\end{aligned}$$

$$\left\{\begin{aligned} x &&+ 2.5z &= 60 \\ .4x &&+ .5z &= 16 \\ .4x &+ 1y && = 26\end{aligned}\right\} \Rightarrow \left\{\begin{aligned} x &&+ 2.5z &= 60 \\ &&- .5z &= -8 \\ 1y &-& 1z &= 2\end{aligned}\right\} \Rightarrow \left\{\begin{aligned} x &&+ 2.5z &= 60 \\ y &-& z &= 2 \\ &&z &= 16\end{aligned}\right\} \Rightarrow \left\{\begin{aligned} x &&&= 20 \\ y &&&= 18 \\ &&z &= 16\end{aligned}\right\}$$

Spray X : 20 gal

Spray Y : 18 gal

Spray Z : 16 gal

29. *School Orchestra* The table in the textbook shows the percents of each section of the North High School Orchestra that were chosen to participate in the city orchestra, the county orchestra, and the state orchestra. Thirty members of the city orchestra, 17 members of the county orchestra, and 6 members of the state orchestra are from North. How many members are in each section of North High's orchestra?

Solution

$$\left\{\begin{aligned}.40x + .30y + .50z &= 30 \\ .20x + .25y + .25z &= 17 \\ .10x + .15y + .25z &= 10\end{aligned}\right\}$$

$$\left\{\begin{aligned} x + .75y + 1.25z &= 75 \\ .20x + .25y + .25z &= 17 \\ .10x + .15y + .25z &= 10\end{aligned}\right\} \Rightarrow \left\{\begin{aligned} x + .75y + 1.25z &= 75 \\ .1y &= 2 \\ .075y + .125z &= 2.5\end{aligned}\right\} \Rightarrow \left\{\begin{aligned} x + .75y + 1.25z &= 75 \\ y &= 20 \\ .075y + .125z &= 2.5\end{aligned}\right\} \Rightarrow$$

$$\left\{\begin{aligned} x &+ 1.25z &= 60 \\ y && = 20 \\ &+ .125z &= 1\end{aligned}\right\} \Rightarrow \left\{\begin{aligned} x &+ 1.25z &= 60 \\ y && = 20 \\ &z &= 8\end{aligned}\right\} \Rightarrow \left\{\begin{aligned} x && = 50 \\ y && = 20 \\ z && = 8\end{aligned}\right\}$$

string: 50

wind: 20

percussion: 8

31. Find the domain of $f(x) = x^3 - 2x$.

Solution

f is a polynomial; therefore the domain is $(-\infty, \infty)$.

33. Find the domain of $h(x) = \sqrt{16 - x^2}$.

Solution

$$16 - x^2 \geq 0$$

$$(4 - x)(4 + x) \geq 0$$

$4 - x$	$+$	$+$	$-$
$4 + x$	$-$	$+$	$+$
	-4		4
	$-$	$+$	$-$

Domain is $[-4, 4]$

35. *Predator-Prey* The number N of prey animals t months after a predator is introduced into a test area is inversely proportional to $t + 1$. When $t = 0$, $N = 300$. Find N when $t = 5$.

Solution

$$N = \frac{k}{t+1} \qquad N = \frac{300}{5+1}$$

$$300 = \frac{k}{0+1} \qquad N = \frac{300}{6}$$

$$300 = k \qquad N = 50$$

37. Perform the row operation and write the equivalent system.

Add Equation 1 to Equation 2.

$$\begin{array}{ll} x - 2y + 3z = 5 & \text{Equation 1} \\ -x + y + 5z = 4 & \text{Equation 2} \\ 2x \qquad - 3z = 0 & \text{Equation 3} \end{array}$$

Solution

$$\begin{array}{l} x - 2y + 3z = 5 \\ \underline{-x + y + 5z = 4} \\[4pt] -y + 8z = 9 \end{array}$$

39. Decide whether each ordered triple is a solution of the system.

$$\begin{array}{rl} x + 3y + 2z = & 1 \\ 5x - y + 3z = & 16 \\ -3x + 7y + z = & -14 \end{array}$$

(a) $(0, 3, -2)$ (b) $(12, 5, -13)$
(c) $(1, -2, 3)$ (d) $(-2, 5, -3)$

Solution

(a) $0 + 3(3) + 2(-2) \overset{?}{=} 1$

$9 - 4 \neq 1$

no

(b) $12 + 3(5) + 2(-13) \overset{?}{=} 1$

$12 + 15 - 26 = 1$

$1 = 1$

yes

(c) $1 + 3(-2) + 2(3) \overset{?}{=} 1$

$1 - 6 + 6 = 1$

$1 = 1$

yes

(d) $-2 + 3(5) + 2(-3) \overset{?}{=} 1$

$-2 + 15 - 6 = 1$

$7 \neq 1$

no

41. Use Gaussian elimination to solve the system of linear equations.

$$x + y + z = 6$$
$$2x - y + z = 3$$
$$3x \quad - z = 0$$

Solution

$$x + y + z = 6$$
$$2x - y + z = 3$$
$$3x \qquad -z = 0$$

$$x + y + z = 6$$
$$-3y - z = -9$$
$$-3y - 4z = -18$$

$$x + y + z = 6$$
$$y + \frac{1}{3}z = 3$$
$$-3y - 4z = -18$$

$$x + y + z = 6$$
$$y + \frac{1}{3}z = 3$$
$$-3x \qquad = -9$$

$$x + y + z = 6$$
$$y + \frac{1}{3}z = 3$$
$$z = 3$$

$$y + \frac{1}{3}(3) = 3$$
$$y = 2$$

$$x + 2 + 3 = 6$$
$$x = 1$$

$(1, 2, 3)$

43. Use Gaussian elimination to solve the system of linear equations.

$$x + y + 8z = 3$$
$$2x + y + 11z = 4$$
$$x \quad + 3z = 0$$

Solution

$$x + y + 8z = 3$$
$$2x + y + 11z = 4$$
$$x \qquad + 3z = 0$$

$$x + y + 8z = 3$$
$$-y - 5z = -2$$
$$-y - 5z = -3$$

$$x + y + 8z = 3$$
$$y + 5z = 2$$
$$y + 5z = \frac{3}{5}$$

inconsistent

45. Use Gaussian elimination to solve the system of linear equations.

$$\begin{aligned} 2x \qquad + 2z &= 2 \\ 5x + 3y \qquad &= 4 \\ 3y - 4z &= 4 \end{aligned}$$

Solution

$$\begin{aligned} 2x \qquad + \quad 2z &= \quad 2 \\ 5x + 3y \qquad &= \quad 4 \\ 3y - \quad 4z &= \quad 4 \end{aligned}$$

$$\begin{aligned} x \qquad + \quad z &= \quad 1 \\ 5x + 3y \qquad &= \quad 4 \\ 3y - \quad 4z &= \quad 4 \end{aligned}$$

$$\begin{aligned} x \qquad + \quad z &= \quad 1 \\ 3y - \quad 5z &= -1 \\ 3y - \quad 4z &= \quad 4 \end{aligned}$$

$$\begin{aligned} x \qquad + \quad z &= \quad 1 \\ y - \tfrac{5}{3}z &= -\tfrac{1}{3} \\ 3y - \quad 4z &= \quad 4 \end{aligned}$$

$$\begin{aligned} x \qquad + \quad z &= \quad 1 \\ y - \tfrac{5}{3}z &= -\tfrac{1}{3} \\ z &= \quad 5 \end{aligned}$$

$$\begin{aligned} y - \tfrac{5}{3}(5) &= -\tfrac{1}{3} \\ y &= \quad 8 \end{aligned}$$

$$\begin{aligned} x \qquad + \quad 5 &= 1 \\ x \qquad &= -4 \end{aligned}$$

$(-4, 8, 5)$

47. Use Gaussian elimination to solve the system of linear equations.

$$\begin{aligned} 3x - y - 2z &= 5 \\ 2x + y + 3z &= 6 \\ 6x - y - 4z &= 9 \end{aligned}$$

Solution

$$\begin{aligned} 3x - \quad y - \quad 2z &= \quad 5 \\ 2x + \quad y + \quad 3z &= \quad 6 \\ 6x - \quad y - \quad 4z &= \quad 9 \end{aligned}$$

$$\begin{aligned} x - \tfrac{1}{3}y - \tfrac{2}{3}z &= \tfrac{5}{3} \\ 2x + \quad y + \quad 3z &= \quad 6 \\ 6x - \quad y - \quad 4z &= \quad 9 \end{aligned}$$

$$\begin{aligned} x - \tfrac{1}{3}y - \tfrac{2}{3}z &= \tfrac{5}{3} \\ \tfrac{5}{3}y + \tfrac{13}{3}z &= \tfrac{8}{3} \\ y \qquad &= -1 \end{aligned}$$

$$\begin{aligned} x - \tfrac{1}{3}y - \tfrac{2}{3}z &= \tfrac{5}{3} \\ y \qquad &= -1 \\ 5y + 13z &= \quad 8 \end{aligned}$$

$$\begin{aligned} x - \tfrac{1}{3}y - \tfrac{2}{3}z &= \tfrac{5}{3} \\ y \qquad &= -1 \\ 13z &= 13 \end{aligned}$$

$$\begin{aligned} x - \tfrac{1}{3}y - \tfrac{2}{3}z &= \tfrac{5}{3} \\ y \qquad &= -1 \\ z &= \quad 1 \end{aligned}$$

$$\begin{aligned} x - \tfrac{1}{3}(-1) - \tfrac{2}{3}(1) &= \tfrac{5}{3} \\ x + \tfrac{1}{3} - \tfrac{2}{3} &= \tfrac{5}{3} \\ x &= 2 \end{aligned}$$

$(2, -1, 1)$

49. Use Gaussian elimination to solve the system of linear equations.

$$5x + 2y \quad = -8$$
$$z = 5$$
$$3x - y + z = 9$$

Solution

$$5x + 2y = -8$$
$$z = 5$$
$$3x - y + z = 9$$

$$x - \frac{1}{3}y + \frac{1}{3}z = 3$$
$$5x + 2y \phantom{+ \frac{1}{3}z} = -8$$
$$z = 5$$

$$x - \frac{1}{3}y + \frac{1}{3}z = 3$$
$$\frac{11}{3}y - \frac{5}{3}z = -23$$
$$z = 5$$

$$x - \frac{1}{3}y + \frac{1}{3}z = 3$$
$$y - \frac{5}{11}z = -\frac{69}{11}$$
$$z = 5$$

$$y - \frac{5}{11}(5) = -\frac{69}{11}$$
$$y = -\frac{44}{11} = -4$$

$$x - \frac{1}{3}(-4) + \frac{1}{3}(5) = 3$$
$$x + \frac{4}{3} + \frac{5}{3} = 3$$
$$x = 0$$

$$(0, -4, 5)$$

51. Use Gaussian elimination to solve the system of linear equations.

$$x + 2y - 2z = 4$$
$$2x + 5y - 7z = 5$$
$$3x + 7y - 9z = 10$$

Solution

$$x + 2y - 2z = 4$$
$$2x + 5y - 7z = 5$$
$$3x + 7y - 9z = 10$$

$$x + 2y - 2z = 4$$
$$y - 3z = -3$$
$$y - 3z = -2$$

inconsistent

53. Use Gaussian elimination to solve the system of linear equations.

$$2x + y - z = 4$$
$$y + 3z = 2$$
$$3x + 2y = 4$$

Solution

$$2x + y - z = 4$$
$$y + 3z = 2$$
$$3x + 2y = 4$$

$$x + \frac{1}{2}y - \frac{1}{2}z = 2$$
$$y + 3z = 2$$
$$3x + 2y = 4$$

$$x + \frac{1}{2}y - \frac{1}{2}z = 2$$
$$y + 3z = 2$$
$$\frac{1}{2}y + \frac{3}{2}z = -2$$

$$x + \frac{1}{2}y - \frac{1}{2}z = 2$$
$$y + 3z = 2$$
$$y + 3z = -4$$

inconsistent

55. Use Gaussian elimination to solve the system of linear equations.

$$3x + y + z = 2$$
$$4x + 2z = 1$$
$$5x - y + 3z = 0$$

Solution

$$3x + y + z = 2$$
$$4x + 2z = 1$$
$$5x - y + 3z = 0$$

$$x + \frac{1}{3}y + \frac{1}{3}z = \frac{2}{3}$$
$$4x + 2z = 1$$
$$5x - y + 3z = 0$$

$$x + \frac{1}{3}y + \frac{1}{3}z = \frac{2}{3}$$
$$-\frac{4}{3}y + \frac{2}{3}z = -\frac{5}{3}$$
$$-\frac{8}{3}y + \frac{4}{3}z = -\frac{10}{3}$$

$$x + \frac{1}{3}y + \frac{1}{3}z = \frac{2}{3}$$
$$y - \frac{1}{2}z = \frac{5}{4}$$
$$-8y + 4z = -10$$

$$x + \frac{1}{3}y + \frac{1}{3}z = \frac{2}{3}$$
$$y - \frac{1}{2}z = \frac{5}{4}$$
$$0 = 0$$

$$y = \frac{1}{2}z + \frac{5}{4}$$

$$x + \frac{1}{3}\left(\frac{1}{2}z + \frac{5}{4}\right) + \frac{1}{3}z = \frac{2}{3}$$

$$x + \frac{1}{6}z + \frac{5}{12} + \frac{1}{3}z = \frac{2}{3}$$

$$x + \frac{1}{2}z = \frac{1}{4} \quad x = \frac{1}{4} - \frac{1}{2}z$$

let $a = z$ $\quad \left(\frac{1}{4} - \frac{1}{2}a, \frac{1}{2}a + \frac{5}{4}, a\right)$

57. *Curve Fitting* Find the equation of the parabola (see figure in textbook) $y = ax^2 + bx + c$ that passes through the points $(1, 0)$, $(2, -1)$, and $(3, 0)$.

Solution

$$0 = a(1)^2 + b(1) + c \Rightarrow \quad 0 = a + b + c$$

$$-1 = a(2)^2 + b(2) + c \Rightarrow -1 = 4a + 2b + c$$

$$0 = a(3)^2 + b(3) + c \Rightarrow \quad 0 = 9a + 3b + c$$

$$
\begin{aligned}
a + \quad b + \quad c &= \quad 0 \\
-2b - \quad 3c &= -1 \\
-6b - \quad 8c &= \quad 0
\end{aligned}
$$

$$
\begin{aligned}
a + \quad b + \quad c &= \quad 0 \\
b + \tfrac{3}{2}c &= \tfrac{1}{2} \\
3b + \quad 4c &= \quad 0
\end{aligned}
$$

$$
\begin{aligned}
a + \quad\quad -\tfrac{1}{2}c &= -\tfrac{1}{2} \\
b + \tfrac{3}{2}c &= \tfrac{1}{2} \\
-\tfrac{1}{2}c &= -\tfrac{3}{2}
\end{aligned}
$$

$$
\begin{aligned}
a \quad\quad -\tfrac{1}{2}c &= -\tfrac{1}{2} \\
b + \tfrac{3}{2}c &= \tfrac{1}{2} \\
c &= \quad 3
\end{aligned}
$$

$$
\begin{aligned}
a \quad\quad\quad &= \quad 1 \\
b \quad\quad &= -4 \\
c &= \quad 3
\end{aligned}
$$

$$y = x^2 - 4x + 3$$

59. *Curve Fitting* Find the equation of the parabola (see figure in textbook) $y = ax^2 + bx + c$ that passes through the points $(-1, -3)$, $(1, 1)$, and $(2, 0)$.

Solution

$$
\begin{aligned}
-3 &= a(-1)^2 + b(-1) + \quad c \\
1 &= a(1)^2 + \quad b(1) + \quad c \\
0 &= a(2)^2 + \quad b(2) + \quad c
\end{aligned}
$$

$$
\begin{aligned}
-3 &= \quad a - \quad b + \quad c \\
1 &= \quad a + \quad b + \quad c \\
0 &= \quad 4 + \quad 2b + \quad c
\end{aligned}
$$

$$
\begin{aligned}
-3 &= \quad a - \quad b + \quad c \\
4 &= \quad\quad +2b \\
12 &= \quad\quad +6b - \quad 3c
\end{aligned}
$$

$$
\begin{aligned}
2 &= \quad b \\
12 &= 6(2) - \quad 3c \\
0 &= \quad\quad -3c
\end{aligned}
$$

$$
\begin{aligned}
0 &= \quad\quad c \\
-3 &= \quad a - \quad 2 + \quad 0 \\
-1 &= \quad a
\end{aligned}
$$

$$y = -1x^2 + 2x + 0$$

61. *Curve Fitting* Find the equation of the circle (see figure in textbook) $x^2 + y^2 + Dx + Ey + F = 0$ that passes through the points $(0, 0)$, $(0, 2)$, and $(3, 0)$.

Solution

$$0^2 + 0^2 + D(0) + E(0) + F = 0$$
$$0^2 + 2^2 + D(0) + E(2) + F = 0$$
$$3^2 + 0^2 + D(3) + E(0) + F = 0$$

$$F = 0$$
$$2E + F = 0$$
$$3D + F = 0$$

$$2E + 0 = -4$$
$$2E = -4$$
$$E = -2$$

$$3D + 0 = -9$$
$$3D = -9$$
$$D = -3$$
$$x^2 + y^2 - 3x - 2y = 0$$

63. *Curve Fitting* Find the equation of the circle (see figure in textbook) $x^2 + y^2 + Dx + Ey + F = 0$ that passes through the points $(-3, 5)$, $(4, 6)$, and $(5, 5)$.

Solution

$$(-3)^2 + 5^2 + D(-3) + E(5) + F = 0$$
$$4^2 + 6^2 + D(4) + E(6) + F = 0$$
$$5^2 + 5^2 + D(5) + E(5) + F = 0$$

$$-3D + 5E + F = -34$$
$$4D + 6E + F = -52$$
$$5D + 5E + F = -50$$

$$D + E + \frac{1}{5}F = -10$$
$$4D + 6E + F = -52$$
$$-3D + 5E + F = -34$$

$$D + E + \frac{1}{5}F = -10$$
$$2E + \frac{1}{5}F = -12$$
$$8E + \frac{8}{5}F = -64$$

$$D + E + \frac{1}{5}F = -10$$
$$E + \frac{1}{10}F = -6$$
$$8E + \frac{8}{5}F = -64$$

$$D + \frac{1}{10}F = -4$$
$$E + \frac{1}{10}F = -6$$
$$\frac{4}{5}F = -16$$

$$D + \frac{1}{10}F = -4$$
$$E + \frac{1}{10}F = -6$$
$$F = -20$$

$$D = -2$$
$$E = -4$$
$$F = -20$$
$$x^2 + y^2 - 2x - 4y - 20 = 0$$

65. *Graphical Estimateion* The table in the textbook gives the amounts y, in millions of short tons, of newsprint produced in the years 1990 through 1992 in the United States (Source: American Paper Institute).

(a) Find the equation of the parabola $y = at^2 + bt + c$ that passes through the three points, letting $t = 0$ correspond to 1990.

(b) Use a graphing utility to graph the model found in part (a).

(c) Use the model of part (a) to predict newsprint production in 1999 if the trend continues.

Solution

(a) Points $(0, 6.6)$, $(1, 6.8)$, $(2, 7.1)$

$$6.6 = a(0)^2 + b(0) + c \Rightarrow 6.6 = \qquad\qquad c$$
$$6.8 = a(1)^2 + b(1) + c \Rightarrow 6.8 = a + b + c$$
$$7.1 = a(2)^2 + b(2) + c \Rightarrow 7.1 = 4a + 2b + c$$

$$a + b + c = 6.8$$
$$4a + 2b + c = 7.1$$
$$c = 6.6$$

$$a + b + c = 6.8$$
$$-2b - 3c = -20.1$$
$$c = 6.6$$

$$a + b + c = 6.8$$
$$b + \frac{3}{2}c = 10.05$$
$$c = 6.6$$

$$a \qquad - \frac{1}{2}c = -3.25$$
$$b + \frac{3}{2}c = 10.05$$
$$c = 6.6$$

$$a \qquad = 0.05$$
$$b \qquad = 0.15$$
$$c = 6.6$$

$$y = 0.05t^2 + 0.15t + 6.6$$

(b) Keystrokes:

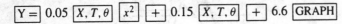

$$\boxed{Y=}\ 0.05\ \boxed{X, T, \theta}\ \boxed{x^2}\ \boxed{+}\ 0.15\ \boxed{X, T, \theta}\ \boxed{+}\ 6.6\ \boxed{\text{GRAPH}}$$

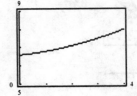

(c) $t = 9 \quad y = 0.05(9)^2 + 0.15(9) + 6.6$

$$= 12$$

67. Find a system of linear equations in three variables that has $(4, -3, 2)$ as a solution. (Note: There are many correct answers.)

Solution

$$x + y + z = 3$$

$$2x + y + 2z = 9$$

$$x \quad\ - 2z = 0$$

Only one example, many correct answers

Mid-Chapter Quiz for Chapter 8

1. Which is the solution of the system $5x - 12y = 2$ and $2x + 1.5y = 26$: $(1, -2)$ or $(10, 4)$? Explain your reasoning.

Solution

$(1, -2)$ $\qquad 5(1) - 12(-2) \stackrel{?}{=} 2$

$\qquad\qquad\qquad 5 + 24 \neq 2$

This is not the solution.

$(10, 4)$ $\qquad 5(10) - 12(4) \stackrel{?}{=} 2$

$\qquad\qquad\qquad 50 - 48 = 2$

$\qquad\qquad\qquad 2 = 2$

$\qquad\qquad 2(10) + 1.5(4) \stackrel{?}{=} 26$

$\qquad\qquad\qquad 20 + 6 = 26$

$\qquad\qquad\qquad 26 = 26$

This is the solution.

2. Graph the equations in the system. Use the graph to determine the number of solutions of the system.

$$-6x + 9y = 9$$

$$2x - 3y = 6$$

Solution

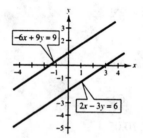

no solution

3. Graph the equations in the system. Use the graphs to determine the number of solutions of the system.

$$x - 2y = -4$$

$$3x - 2y = 4$$

Solution

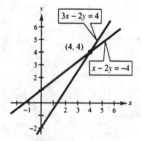

one solution

4. Graph the equations in the system. Use the graphs to determine the number of solutions of the system.

$$y = x - 1$$

$$y = 1 + 2x - x^2$$

Solution

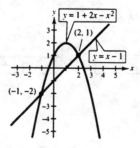

two solutions

5. Solve the system of equations graphically.

$$x = 4$$
$$2x - y = 6$$

Solution

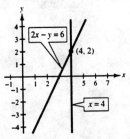

$(4, 2)$

6. Solve the system of equations graphically.

$$y = \frac{1}{3}(1 - 2x)$$
$$y = \frac{1}{3}(5x - 13)$$

Solution

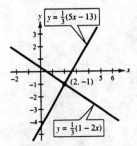

$(2, -1)$

7. Solve the system of equations graphically.

$$2x + 7y = 16$$
$$3x + 2y = 24$$

Solution

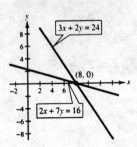

$(8, 0)$

8. Solve the system of equations graphically.

$$7x - 17y = -169$$
$$x^2 + y^2 = 169$$

Solution

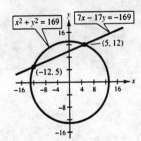

$(5, 12), (-12, 5)$

9. Use substitution to solve the system.

$$2x - 3y = 4$$
$$y = 2$$

Solution

$$2x - 3y = 4$$
$$y = 2$$

$$2x - 3(2) = 4$$
$$2x - 6 = 4$$
$$2x = 10$$
$$x = 5$$

$(5, 2)$

10. Use substitution to solve the system.

$$y = 5 - x^2$$
$$y = 2(x + 1)$$

Solution

$$y = 5 - x^2$$
$$y = 2(x + 1)$$

$$5 - x^2 = 2(x + 1)$$
$$5 - x^2 = 2x + 2$$
$$0 = x^2 + 2x - 3$$
$$0 = (x + 3)(x - 1)$$

$$x = -3 \qquad x = 1$$
$$y = -4 \qquad y = 4$$

$(-3, -4), (1, 4)$

11. Use substitution to solve the system.

$$5x - y = 32$$
$$6x - 9y = 18$$

Solution

$$5x - y = 32 \quad \Rightarrow \quad -y = -5x + 32$$
$$6x - 9y = 18 \qquad\qquad y = 5x - 32$$

$$6x - 9(5x - 32) = 18$$
$$6x - 45x + 288 = 18$$
$$-39x = -270$$
$$x = \frac{-270}{-39} = \frac{90}{13}$$

$$y = 5\left(\frac{90}{13}\right) - 32$$
$$= \frac{450}{13} - \frac{416}{13}$$
$$= \frac{34}{13}$$

$\left(\dfrac{90}{13}, \dfrac{34}{13}\right)$

12. Use substitution to solve the system.

$$0.2x + 0.7y = 8$$
$$-x + 2y = 15$$

Solution

$$0.2x + 0.7y = 8 \quad \Rightarrow \quad -x = -2y + 15$$
$$-x + 2y = 15 \qquad\qquad x = 2y - 15$$

$$0.2(2y - 15) + 0.7y = 8$$
$$0.4y - 3 + 0.7y = 8$$
$$1.1y = 11$$
$$y = 10$$

$$x = 2(10) - 15$$
$$= 20 - 15$$
$$= 5$$

$(5, 10)$

13. Use Gaussian elimination to solve the linear systems.

$$x + 10y = 18$$
$$5x + 2y = 42$$

Solution

$$x + 10y = 18$$
$$5x + 2y = 42$$

$$x + 10y = 18$$
$$-48y = -48$$

$$x + 10y = 18$$
$$y = 1$$

$$x \qquad = 8$$
$$y = 1$$

$$(8, 1)$$

14. Use Gaussian elimination to solve the linear systems.

$$3x + 11y = 38$$
$$7x - 5y = -34$$

Solution

$$3x + 11y = 38$$
$$7x - 5y = -34$$

$$x + \frac{11}{3}y = \frac{38}{3}$$
$$7x - 5y = -34$$

$$x + \frac{11}{3}y = \frac{38}{3}$$
$$-\frac{92}{3}y = -\frac{368}{3}$$

$$x + \frac{11}{3}y = \frac{38}{3}$$
$$y = 4$$

$$x \qquad = -2$$
$$y = 4$$

$$(-2, 4)$$

15. Use Gaussian elimination to solve the linear systems.

$$a + b + c = 1$$
$$4a + 2b + c = 2$$
$$9a + 3b + c = 4$$

Solution

$$a + b + c = 1$$
$$4a + 2b + c = 2$$
$$9a + 3b + c = 4$$

$$a + b + c = 1$$
$$-2b - 3c = -2$$
$$-6b - 8c = -5$$

$$a + b + c = 1$$
$$b + \frac{3}{2}c = 1$$
$$-6b - 8c = -5$$

$$a \quad - \frac{1}{2}c = 0$$
$$b + \frac{3}{2}c = 1$$
$$c = 1$$

$$a \quad = \frac{1}{2}$$
$$b \quad = -\frac{1}{2}$$
$$c = 1$$

$$\left(\frac{1}{2}, -\frac{1}{2}, 1\right)$$

16. Use Gaussian elimination to solve the linear systems.

$$x \quad + 4z = 17$$
$$-3x + 2y - z = -20$$
$$x - 5y + 3z = 19$$

Solution

$$x \quad + 4z = 17$$
$$-3x + 2y - z = -20$$
$$x - 5y + 3z = 19$$

$$x \quad + 4z = 17$$
$$2y + 11z = 31$$
$$-5y - z = 2$$

$$x \quad + 4z = 17$$
$$y + \frac{11}{2}z = \frac{31}{2}$$
$$\frac{53}{2}z = \frac{159}{2}$$

$$x \quad + 4z = 17$$
$$y + \frac{11}{2}z = \frac{31}{2}$$
$$z = 3$$

$$x \quad = 5$$
$$y \quad = -1$$
$$z = 3$$

$$(5, -1, 3)$$

17. Find a system of linear equations that has the unique solution $(10, -12)$. (There are many correct answers.)

Solution

$$x + y = -2$$
$$2x - y = 32$$

18. Find a system of linear equations that has the unique solution $(1, 3, -7)$. (There are many correct answers.)

Solution

$$x + y - z = 11$$
$$x + 2y - z = 14$$
$$-2x + y + z = -6$$

19. Twenty gallons of a 30% brine solution is obtained by mixing a 20% solution with a 50% solution. Let x represent the number of gallons of the 20% solution and let y represent the number of gallons of the 50% solution. Write a system of equations that models this problem and solve the system.

Solution

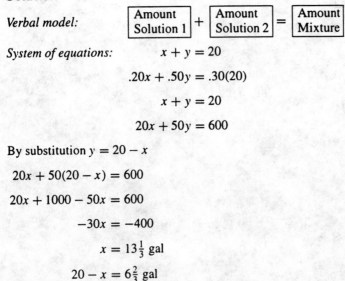

Verbal model: | Amount Solution 1 | + | Amount Solution 2 | = | Amount Mixture |

System of equations:

$$x + y = 20$$

$$.20x + .50y = .30(20)$$

$$x + y = 20$$

$$20x + 50y = 600$$

By substitution $y = 20 - x$

$$20x + 50(20 - x) = 600$$

$$20x + 1000 - 50x = 600$$

$$-30x = -400$$

$$x = 13\tfrac{1}{3} \text{ gal}$$

$$20 - x = 6\tfrac{2}{3} \text{ gal}$$

20. Find the equation of the parabola $y = ax^2 + bx + c$ that passes through the points $(1, 2)$, $(-1, -4)$, and $(2, 8)$.

Solution

$$2 = a(1)^2 + b(1) + c \Rightarrow a + b + c = 2$$
$$-4 = a(-1)^2 + b(-1) + c \Rightarrow a - b + c = -4$$
$$8 = a(2)^2 + b(2) + c \Rightarrow 4a + 2b + c = 8$$

$$
\begin{aligned}
a + b + c &= 2 \\
-2b &= -6 \\
-2b - 3c &= 0
\end{aligned}
$$

$$
\begin{aligned}
a + b + c &= 2 \\
b &= 3 \\
-2b - 3c &= 0
\end{aligned}
$$

$$
\begin{aligned}
a \quad + c &= -1 \\
b &= 3 \\
-3c &= 6
\end{aligned}
$$

$$
\begin{aligned}
a \quad + c &= -1 \\
b &= 3 \\
c &= -2
\end{aligned}
$$

$$
\begin{aligned}
a &= 1 \\
b &= 3 \\
c &= -2
\end{aligned}
$$

$$y = x^2 + 3x - 2$$

8.4 **Matrices and Systems of Linear Equations**

7. State the order of the matrix.

$$\begin{bmatrix} 3 & -2 \\ -4 & 0 \\ 2 & -7 \end{bmatrix}$$

Solution

3×2

9. State the order of the matrix.

$$\begin{bmatrix} 5 & -8 \\ 7 & 15 \end{bmatrix}$$

Solution

2×2

11. Write the augmented matrix for the system of linear equations.

$$4x - 5y = -2$$
$$-x + 8y = 10$$

Solution

$$\begin{bmatrix} 4 & -5 & \vdots & -2 \\ -1 & 8 & \vdots & 10 \end{bmatrix}$$

13. Write the augmented matrix for the system of linear equations.

$$x + 10y - 3z = 2$$
$$5x - 3y + 4z = 0$$
$$2x + 4y = 6$$

Solution

$$\begin{bmatrix} 1 & 10 & -3 & \vdots & 2 \\ 5 & -3 & 4 & \vdots & 0 \\ 2 & 4 & 0 & \vdots & 6 \end{bmatrix}$$

15. Write the system of linear equations represented by the augmented matrix. (Use variables x and y.)

$$\begin{bmatrix} 4 & 3 & \vdots & 8 \\ 1 & -2 & \vdots & 3 \end{bmatrix}$$

Solution

$$4x + 3y = 8$$
$$x - 2y = 3$$

17. Write the system of linear equations represented by the augmented matrix. (Use variables x, y, and z.)

$$\begin{bmatrix} 1 & 0 & 2 & \vdots & -10 \\ 0 & 3 & -1 & \vdots & 5 \\ 4 & 2 & 0 & \vdots & 3 \end{bmatrix}$$

Solution

$$x + 2z = -10$$
$$3y - z = 5$$
$$4x + 2y = 3$$

19. Fill in the blank by using elementary row operations to form a row-equivalent matrix.

$$\begin{bmatrix} 1 & 4 & 3 \\ 2 & 10 & 5 \end{bmatrix}$$

$$\begin{bmatrix} 1 & 4 & 3 \\ 0 & & -1 \end{bmatrix}$$

Solution

$$\begin{bmatrix} 1 & 4 & 3 \\ 2 & 10 & 5 \end{bmatrix}$$

$$-2R_1 + R_2 \begin{bmatrix} 1 & 4 & 3 \\ 0 & 2 & -1 \end{bmatrix}$$

21. Convert the matrix to row-echelon form. (Note: There is more than one correct answer.)

$$\begin{bmatrix} 1 & 2 & 3 \\ 2 & -1 & -4 \end{bmatrix}$$

Solution

$$\begin{bmatrix} 1 & 2 & 3 \\ 2 & -1 & -4 \end{bmatrix}$$

$$-2R_1 + R_2 \begin{bmatrix} 1 & 2 & 3 \\ 0 & -5 & -10 \end{bmatrix}$$

$$-\tfrac{1}{5} R_2 \begin{bmatrix} 1 & 2 & 3 \\ 0 & 1 & 2 \end{bmatrix}$$

23. Convert the matrix to row-echelon form. (Note: There is more than one correct answer.)

$$\begin{bmatrix} 1 & 1 & -1 & 3 \\ 2 & 1 & 2 & 5 \\ 3 & 2 & 1 & 8 \end{bmatrix}$$

Solution

$$\begin{bmatrix} 1 & 1 & -1 & 3 \\ 2 & 1 & 2 & 5 \\ 3 & 2 & 1 & 8 \end{bmatrix}$$

$$\begin{matrix} \\ -2R_1 + R_2 \\ -3R_1 + R_3 \end{matrix} \begin{bmatrix} 1 & 1 & -1 & 3 \\ 0 & -1 & 4 & -1 \\ 0 & 1 & 4 & -1 \end{bmatrix}$$

$$\begin{matrix} \\ \\ R_2 + R_3 \end{matrix} \begin{bmatrix} 1 & 1 & -1 & 3 \\ 0 & -1 & 4 & -1 \\ 0 & 0 & 8 & -2 \end{bmatrix}$$

$$-R_2 \begin{bmatrix} 1 & 1 & -1 & 3 \\ 0 & 1 & -4 & 1 \\ 0 & 0 & 8 & -2 \end{bmatrix}$$

25. Write the system of linear equations represented by the augmented matrix. Then use back-substitution to find the solution. (Use variables x and y.)

$$\begin{bmatrix} 1 & 5 & \vdots & 3 \\ 0 & 1 & \vdots & -2 \end{bmatrix}$$

Solution

$$x + 5y = 3$$
$$y = -2$$

$$x + 5(-2) = 3$$
$$x - 10 = 3$$
$$x = 13$$

$(13, -2)$

27. Use matrices to solve the system of linear equations.

$$6x - 4y = 2$$
$$5x + 2y = 7$$

Solution

$$\begin{bmatrix} 6 & -4 & \vdots & 2 \\ 5 & 2 & \vdots & 7 \end{bmatrix}$$

$$\tfrac{1}{6}R_1 \begin{bmatrix} 1 & -\tfrac{2}{3} & \vdots & \tfrac{1}{3} \\ 5 & 2 & \vdots & 7 \end{bmatrix}$$

$$-5R_1 + R_2 \begin{bmatrix} 1 & -\tfrac{2}{3} & \vdots & \tfrac{1}{3} \\ 0 & \tfrac{16}{3} & \vdots & \tfrac{16}{3} \end{bmatrix}$$

$$\tfrac{3}{16}R_2 \begin{bmatrix} 1 & -\tfrac{2}{3} & \vdots & \tfrac{1}{3} \\ 0 & 1 & \vdots & 1 \end{bmatrix}$$

$$\tfrac{2}{3}R_2 + R_1 \begin{bmatrix} 1 & 0 & \vdots & 1 \\ 0 & 1 & \vdots & 1 \end{bmatrix}$$

$(1, 1)$

29. Use matrices to solve the system of linear equations.

$$x - 2y - z = 6$$
$$y + 4z = 5$$
$$4x + 2y + 3z = 8$$

Solution

$$x - 2y - z = 6$$
$$y + 4z = 5$$
$$4x + 2y + 3z = 8$$

$$\begin{bmatrix} 1 & -2 & -1 & \vdots & 6 \\ 0 & 1 & 4 & \vdots & 5 \\ 4 & 2 & 3 & \vdots & 8 \end{bmatrix}$$

$$-4R_1 + R_3 \begin{bmatrix} 1 & -2 & -1 & \vdots & 6 \\ 0 & 1 & 4 & \vdots & 5 \\ 0 & 10 & 7 & \vdots & -16 \end{bmatrix}$$

$$-10R_2 + R_3 \begin{bmatrix} 1 & -2 & -1 & \vdots & 6 \\ 0 & 1 & 4 & \vdots & 5 \\ 0 & 0 & -33 & \vdots & -66 \end{bmatrix}$$

$$\tfrac{1}{-33}R_3 \begin{bmatrix} 1 & -2 & -1 & \vdots & 6 \\ 0 & 1 & 4 & \vdots & 5 \\ 0 & 0 & 1 & \vdots & 2 \end{bmatrix}$$

$$z = 2 \quad y + 4(2) = 5 \quad x - 2(-3) - (2) = 6$$
$$y = -3 \quad x + 6 - 2 = 6$$
$$x = 2$$

$(2, -3, 2)$

31. Use matrices to solve the system of linear equations.

$$x + y - 5z = 3$$
$$x \quad\quad - 2z = 1$$
$$2x - y - z = 0$$

Solution

$$\begin{bmatrix} 1 & 1 & -5 & \vdots & 3 \\ 1 & 0 & -2 & \vdots & 1 \\ 2 & -1 & -1 & \vdots & 0 \end{bmatrix}$$

$$\begin{matrix} \\ -R_1 + R_2 \\ -2R_1 + R_3 \end{matrix} \begin{bmatrix} 1 & 1 & -5 & \vdots & 3 \\ 0 & -1 & 3 & \vdots & -2 \\ 0 & -3 & 9 & \vdots & -6 \end{bmatrix}$$

$$\begin{matrix} \\ -R_2 \\ \\ \end{matrix} \begin{bmatrix} 1 & 1 & -5 & \vdots & 3 \\ 0 & 1 & -3 & \vdots & 2 \\ 0 & -3 & 9 & \vdots & -6 \end{bmatrix}$$

$$\begin{matrix} \\ \\ 3R_2 + R_3 \end{matrix} \begin{bmatrix} 1 & 1 & -5 & \vdots & 3 \\ 0 & 1 & -3 & \vdots & 2 \\ 0 & 0 & 0 & \vdots & 0 \end{bmatrix}$$

$$\begin{array}{lll} y - 3z = 2 & x + (2 + 3z) - 5z = 3 & \text{let } a = z \\ \quad y = 2 + 3z & \quad\quad x = 1 + 2z & (a \text{ is any real number}) \end{array}$$

$$(1 + 2a, \ 2 + 3a, \ a)$$

33. Use matrices to solve the following system of linear equations.

$$2x \quad\quad + 4z = 1$$
$$x + y + 3z = 0$$
$$x + 3y + 5z = 0$$

Solution

$$\begin{bmatrix} 2 & 0 & 4 & \vdots & 1 \\ 1 & 1 & 3 & \vdots & 0 \\ 1 & 3 & 5 & \vdots & 0 \end{bmatrix}$$

$$R_2 \leftrightarrow R_1 \begin{bmatrix} 1 & 1 & 3 & \vdots & 0 \\ 2 & 0 & 4 & \vdots & 1 \\ 1 & 3 & 5 & \vdots & 0 \end{bmatrix}$$

$$\begin{matrix} \\ -2R_1 + R_2 \\ -R_1 + R_3 \end{matrix} \begin{bmatrix} 1 & 1 & 3 & \vdots & 0 \\ 0 & -2 & -2 & \vdots & 1 \\ 0 & 2 & 2 & \vdots & 0 \end{bmatrix}$$

$$\begin{matrix} \\ -\frac{1}{2}R_2 \\ \\ \end{matrix} \begin{bmatrix} 1 & 1 & 3 & \vdots & 0 \\ 0 & 1 & 1 & \vdots & -\frac{1}{2} \\ 0 & 2 & 2 & \vdots & 0 \end{bmatrix}$$

$$\begin{matrix} \\ \\ -2R_2 + R_3 \end{matrix} \begin{bmatrix} 1 & 1 & 3 & \vdots & 0 \\ 0 & 1 & 1 & \vdots & -\frac{1}{2} \\ 0 & 0 & 0 & \vdots & 1 \end{bmatrix}$$

inconsistent; no solution

35. *Simple Interest*   A corporation borrowed $1,500,000 to expand its product line. Some of the money was borrowed at 8%, some at 9%, and the remainder at 12%. The total annual interest payment to the lenders was $133,000. If the amount borrowed at 8% was four times the amount borrowed at 12%, how much was borrowed at each rate?

Solution

Verbal model:

$$\boxed{\text{Money 1}} + \boxed{\text{Money 2}} + \boxed{\text{Money 3}} = \boxed{1,500,000}$$

$$\boxed{0.08 \cdot \text{Money 1}} + \boxed{0.09 \cdot \text{Money 2}} +$$

$$\boxed{0.12 \cdot \text{Money 3}} = \boxed{133,000}$$

$$\boxed{\text{Money 1}} = 4 \cdot \boxed{\text{Money 3}}$$

Systems of equations:

$$x + y + z = 1,500,000$$

$$0.08x + 0.09y + 0.12z = 133,000$$

$$x = 4z$$

$$\begin{bmatrix} 1 & 1 & 1 & \vdots & 1,500,000 \\ 8 & 9 & 12 & \vdots & 13,300,000 \\ 1 & 0 & -4 & \vdots & 0 \end{bmatrix}$$

$$\begin{matrix} \\ -8R_1 + R_2 \\ -R_1 + R_3 \end{matrix} \begin{bmatrix} 1 & 1 & 1 & \vdots & 1,500,000 \\ 0 & 1 & 4 & \vdots & 1,300,000 \\ 0 & -1 & -5 & \vdots & -1,500,000 \end{bmatrix}$$

$$\begin{matrix} \\ \\ R_2 + R_3 \end{matrix} \begin{bmatrix} 1 & 1 & 1 & \vdots & 1,500,000 \\ 0 & 1 & 4 & \vdots & 1,300,000 \\ 0 & 0 & -1 & \vdots & -200,000 \end{bmatrix}$$

$$\begin{matrix} \\ \\ -R_3 \end{matrix} \begin{bmatrix} 1 & 1 & 1 & \vdots & 1,500,000 \\ 0 & 1 & 4 & \vdots & 1,300,000 \\ 0 & 0 & 1 & \vdots & 200,000 \end{bmatrix}$$

$z = 200,000 \qquad y + 4(200,000) = 1,300,000 \qquad x + 500,000 + 200,000 = 1,500,000$

$$y = 500,000 \qquad\qquad x = 800,000$$

$(800,000, 500,000, 200,000)$

37. *Curve Fitting* Find the equation of the parabola (see figure in textbook)

$$y = ax^2 + bx + c$$

that passes through the given points $(1, 7)$, $(2, 12)$, and $(3, 19)$.

Solution

$$7 = a(1)^2 + b(1) + c \Rightarrow 7 = a + b + c$$

$$12 = a(2)^2 + b(2) + c \Rightarrow 12 = 4a + 2b + c$$

$$19 = a(3)^2 + b(3) + c \Rightarrow 19 = 9a + 3b + c$$

$$\begin{bmatrix} 1 & 1 & 1 & \vdots & 7 \\ 4 & 2 & 1 & \vdots & 12 \\ 9 & 3 & 1 & \vdots & 19 \end{bmatrix}$$

$$\begin{matrix} \\ -4R_1 + R_2 \\ -9R_1 + R_3 \end{matrix} \begin{bmatrix} 1 & 1 & 1 & \vdots & 7 \\ 0 & -2 & -3 & \vdots & -16 \\ 0 & -6 & -8 & \vdots & -44 \end{bmatrix}$$

$$\begin{matrix} \\ -\frac{1}{2}R_2 \\ -\frac{1}{2}R_3 \end{matrix} \begin{bmatrix} 1 & 1 & 1 & \vdots & 7 \\ 0 & 1 & \frac{3}{2} & \vdots & 8 \\ 0 & 3 & 4 & \vdots & 22 \end{bmatrix}$$

$$\begin{matrix} -R_2 + R_1 \\ \\ -3R_2 + R_3 \end{matrix} \begin{bmatrix} 1 & 0 & -\frac{1}{2} & \vdots & -1 \\ 0 & 1 & \frac{3}{2} & \vdots & 8 \\ 0 & 0 & -\frac{1}{2} & \vdots & -2 \end{bmatrix}$$

$$\begin{matrix} \\ \\ -2R_3 \end{matrix} \begin{bmatrix} 1 & 0 & -\frac{1}{2} & \vdots & -1 \\ 0 & 1 & \frac{3}{2} & \vdots & 8 \\ 0 & 0 & 1 & \vdots & 4 \end{bmatrix}$$

$$\begin{matrix} -\frac{3}{2}R_3 + R_2 \\ \frac{1}{2}R_3 + R_1 \end{matrix} \begin{bmatrix} 1 & 0 & 0 & \vdots & 1 \\ 0 & 1 & 0 & \vdots & 2 \\ 0 & 0 & 1 & \vdots & 4 \end{bmatrix}$$

$$a = 1, b = 2, c = 4$$

$$y = x^2 + 2x + 4$$

39. *Mathematical Modeling* A videotape of the path of a ball thrown by a baseball player was analyzed on a television set with a grid on the screen (see figure in textbook). The tape was paused three times and the position of the coordinates of the ball were measured each time. The coordinates were approximately (0, 6), (25, 18.5), and (50, 26).

(a) Find the equation of the parabola $y = ax^2 + bx + c$ that passes through the three points.

(b) Use a graphing utility to graph the parabola. Approximate the maximum height of the ball and the point at which the ball struck the ground.

(c) Find algebraically the maximum height of the ball and the point at which it struck the ground.

Solution

(a)
$$6 = a(0)^2 + b(0) + c \Rightarrow 6 = c$$
$$18.5 = a(25)^2 + b(25) + c \Rightarrow 18.5 = 625a + 25b + c$$
$$26 = a(50)^2 + b(50) + c \Rightarrow 26 = 2500a + 50b + c$$

$$\begin{bmatrix} 0 & 0 & 1 & \vdots & 6 \\ 625 & 25 & 1 & \vdots & 18.5 \\ 2500 & 50 & 1 & \vdots & 26 \end{bmatrix}$$

$$\begin{matrix} R_1 \leftrightarrow R_2 \\ R_2 \leftrightarrow R_3 \end{matrix} \begin{bmatrix} 625 & 25 & 1 & \vdots & 18.5 \\ 2500 & 50 & 1 & \vdots & 26 \\ 0 & 0 & 1 & \vdots & 6 \end{bmatrix}$$

$$\tfrac{1}{625} R_1 \begin{bmatrix} 1 & .04 & .0016 & \vdots & .0296 \\ 2500 & 50 & 1 & \vdots & 26 \\ 0 & 0 & 1 & \vdots & 6 \end{bmatrix}$$

$$-2500 R_1 + R_2 \begin{bmatrix} 1 & .04 & .0016 & \vdots & .0296 \\ 0 & -50 & -3 & \vdots & -48 \\ 0 & 0 & 1 & \vdots & 6 \end{bmatrix}$$

$$-\tfrac{1}{50} R_2 \begin{bmatrix} 1 & .04 & .0016 & \vdots & .0296 \\ 0 & 1 & .06 & \vdots & .96 \\ 0 & 0 & 1 & \vdots & 6 \end{bmatrix}$$

$$-.04 R_2 + R_1 \begin{bmatrix} 1 & 0 & -.0008 & \vdots & -.0088 \\ 0 & 1 & .06 & \vdots & .96 \\ 0 & 0 & 1 & \vdots & 6 \end{bmatrix}$$

$$\begin{matrix} -.06 R_3 + R_2 \\ .0008 R_3 + R_1 \end{matrix} \begin{bmatrix} 1 & 0 & 0 & \vdots & -.004 \\ 0 & 1 & 0 & \vdots & .6 \\ 0 & 0 & 1 & \vdots & 6 \end{bmatrix}$$

so $a = -.004$

$b = .6$

$c = 6$

$y = -.004x^2 + .6x + 6$

(b) Keystrokes:

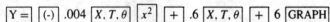

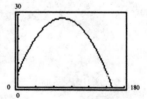

(c) maximum height = 28.5 feet

point at which the ball struck the ground = 159.4 feet

41. Evaluate the function.

$$f(x) = \frac{1}{3}x^2 \qquad \text{(a) } f(6) \quad \text{(b) } f\left(\frac{3}{4}\right)$$

Solution

(a) $f(6) = \frac{1}{3}(6)^2 = \frac{1}{3} \cdot 36 = 12$

(b) $f\left(\frac{3}{4}\right) = \frac{1}{3}\left(\frac{3}{4}\right)^2 = \frac{1}{3} \cdot \frac{9}{16} = \frac{3}{16}$

43. Evaluate the function.

$$g(x) = \frac{x}{x + 10} \qquad \text{(a) } g(5) \quad \text{(b) } g(c - 6)$$

Solution

(a) $g(5) = \dfrac{5}{5 + 10} = \dfrac{5}{15} = \dfrac{1}{3}$

(b) $g(c - 6) = \dfrac{c - 6}{c - 6 + 10} = \dfrac{c - 6}{c + 4}$

45. *Cost* The inventor of a new game believes that the variable cost for producing the game is $5.75 per unit and the fixed costs are $12,000. Let x represent the number of games produced. Express the total cost C as a function of x.

Solution

$C = 12,000 + 5.75x$

47. Fill in the blanks by using elementary row operations to form a row-equivalent matrix.

$$\begin{bmatrix} 9 & -18 & 6 \\ 2 & 8 & 15 \end{bmatrix}$$

$$\begin{bmatrix} 1 & & \\ 2 & 8 & 5 \end{bmatrix}$$

Solution

$$\begin{bmatrix} 9 & -18 & 6 \\ 2 & 8 & 15 \end{bmatrix}$$

$$\frac{1}{9}R_1 \begin{bmatrix} 1 & -2 & \frac{2}{3} \\ 2 & 8 & 15 \end{bmatrix}$$

49. Fill in the blanks by using elementary row operations to form a row-equivalent matrix.

$$\begin{bmatrix} 1 & 1 & 4 & -1 \\ 3 & 8 & 10 & 3 \\ -2 & 1 & 12 & 6 \end{bmatrix}$$

$$\begin{bmatrix} 1 & 1 & 4 & -1 \\ 0 & 5 & & \\ 0 & 3 & & \end{bmatrix}$$

$$\begin{bmatrix} 1 & 1 & 4 & -1 \\ 0 & 1 & & \\ 0 & 3 & 20 & 4 \end{bmatrix}$$

Solution

$$\begin{bmatrix} 1 & 1 & 4 & -1 \\ 3 & 8 & 10 & 3 \\ -2 & 1 & 12 & 6 \end{bmatrix}$$

$$\begin{matrix} \\ 2R_1 + R_3 \\ 2R_1 + R_3 \end{matrix} \begin{bmatrix} 1 & 1 & 4 & -1 \\ 0 & 5 & -2 & 6 \\ 0 & 3 & 20 & 4 \end{bmatrix}$$

$$\frac{1}{5}R_2 \begin{bmatrix} 1 & 1 & 4 & -1 \\ 0 & 1 & -\frac{2}{5} & \frac{6}{5} \\ 0 & 3 & 20 & 4 \end{bmatrix}$$

51. Convert the matrix to row-echelon form. (Note: There is more than one correct answer.)

$$\begin{bmatrix} 4 & 6 & 1 \\ -2 & 2 & 5 \end{bmatrix}$$

Solution

$$\begin{bmatrix} 4 & 6 & 1 \\ -2 & 2 & 5 \end{bmatrix}$$

$$\frac{1}{4}R_1 \begin{bmatrix} 1 & \frac{3}{2} & \frac{1}{4} \\ -2 & 2 & 5 \end{bmatrix}$$

$$2R_1 + R_2 \begin{bmatrix} 1 & \frac{3}{2} & \frac{1}{4} \\ 0 & 5 & \frac{11}{2} \end{bmatrix}$$

$$\frac{1}{5}R_2 \begin{bmatrix} 1 & \frac{3}{2} & \frac{1}{4} \\ 0 & 1 & \frac{11}{10} \end{bmatrix}$$

53. Convert the matrix to row-echelon form. (Note: There is more than one correct answer.)

$$\begin{bmatrix} 1 & 1 & 0 & 5 \\ -2 & -1 & 2 & -10 \\ 3 & 6 & 7 & 14 \end{bmatrix}$$

Solution

$$\begin{bmatrix} 1 & 1 & 0 & 5 \\ -2 & -1 & 2 & -10 \\ 3 & 6 & 7 & 14 \end{bmatrix}$$

$$\begin{matrix} \\ 2R_1 + R_2 \\ -3R_1 + R_3 \end{matrix} \begin{bmatrix} 1 & 1 & 0 & 5 \\ 0 & 1 & 2 & 0 \\ 0 & 3 & 7 & -1 \end{bmatrix}$$

$$-3R_2 + R_3 \begin{bmatrix} 1 & 1 & 0 & 5 \\ 0 & 1 & 2 & 0 \\ 0 & 0 & 1 & -1 \end{bmatrix}$$

55. Convert the matrix to row-echelon form. (Note: There is more than one correct answer.)

$$\begin{bmatrix} 1 & -1 & -1 & 1 \\ 4 & -4 & 1 & 8 \\ -6 & 8 & 18 & 0 \end{bmatrix}$$

Solution

$$\begin{bmatrix} 1 & -1 & -1 & 1 \\ 4 & -4 & 1 & 8 \\ -6 & 8 & 18 & 0 \end{bmatrix}$$

$$\begin{matrix} \\ -4R_1 + R_2 \\ 6R_1 + R_3 \end{matrix} \begin{bmatrix} 1 & -1 & -1 & 1 \\ 0 & 0 & 5 & 4 \\ 0 & 2 & 12 & 6 \end{bmatrix}$$

$$\begin{matrix} \\ R_3 \\ R_2 \end{matrix} \begin{bmatrix} 1 & -1 & -1 & 1 \\ 0 & 2 & 12 & 6 \\ 0 & 0 & 5 & 4 \end{bmatrix}$$

$$\begin{matrix} \\ \frac{1}{2}R_2 \\ \frac{1}{5}R_3 \end{matrix} \begin{bmatrix} 1 & -1 & -1 & 1 \\ 0 & 1 & 6 & 3 \\ 0 & 0 & 1 & \frac{4}{5} \end{bmatrix}$$

57. Write the system of linear equations represented by the augmented matrix. Then use back-substitution to find the solution.

$$\begin{bmatrix} 1 & -2 & \vdots & 4 \\ 0 & 1 & \vdots & -3 \end{bmatrix}$$

Solution

$$
\begin{array}{lll}
x - 2y = 4 & \qquad x - 2(-3) = 4 \\
\qquad y = -3 & \qquad x + 6 = 4 \\
& \qquad x = -2
\end{array}
$$

$(-2, -3)$

59. Write the system of linear equations represented by the augmented matrix. Then use back-substitution to find the solution.

$$\begin{bmatrix} 1 & -1 & 2 & \vdots & 4 \\ 0 & 1 & -1 & \vdots & 2 \\ 0 & 0 & 1 & \vdots & -2 \end{bmatrix}$$

Solution

$$
\begin{array}{lll}
x - y + 2z = 4 & \quad y - (-2) = 2 & \quad x - 0 + 2(-2) = 4 \\
\quad y - z = 2 & \quad y + 2 = 2 & \quad x \qquad - 4 = 4 \\
\qquad z = -2 & \qquad y = 0 & \qquad x = 8
\end{array}
$$

$(8, 0, -2)$

61. Use matrices to solve the system of linear equations.

$$
\begin{array}{rcl}
x + 2y &=& 7 \\
3x + y &=& 8
\end{array}
$$

Solution

$$\begin{bmatrix} 1 & 2 & \vdots & 7 \\ 3 & 1 & \vdots & 8 \end{bmatrix}$$

$-3R_1 + R_2 \begin{bmatrix} 1 & 2 & \vdots & 7 \\ 0 & -5 & \vdots & -13 \end{bmatrix}$

$-\frac{1}{5}R_2 \begin{bmatrix} 1 & 2 & \vdots & 7 \\ 0 & 1 & \vdots & \frac{13}{5} \end{bmatrix}$

$-2R_2 + R_1 \begin{bmatrix} 1 & 0 & \vdots & \frac{9}{5} \\ 0 & 1 & \vdots & \frac{13}{5} \end{bmatrix}$

$\left(\frac{9}{5}, \frac{13}{5}\right)$

63. Use matrices to solve the system of linear equations.

$$
\begin{array}{rcl}
-x + 2y &=& 1.5 \\
2x + 4y &=& 3
\end{array}
$$

Solution

$$\begin{bmatrix} -1 & 2 & \vdots & 1.5 \\ 2 & -4 & \vdots & 3 \end{bmatrix}$$

$-R_1 \begin{bmatrix} 1 & -2 & \vdots & -1.5 \\ 2 & -4 & \vdots & 3 \end{bmatrix}$

$-2R_1 + R_2 \begin{bmatrix} 1 & -2 & \vdots & -1.5 \\ 0 & 0 & \vdots & 6 \end{bmatrix}$

inconsistent; no solution

65. Use matrices to solve the system of linear equations.

$$x - 3y + 2z = 8$$
$$2y - z = -4$$
$$x + z = 3$$

Solution

$$\begin{bmatrix} 1 & -3 & 2 & \vdots & 8 \\ 0 & 2 & -1 & \vdots & -4 \\ 1 & 0 & 1 & \vdots & 3 \end{bmatrix}$$

$$\begin{matrix} \\ -R_1 + R_3 \\ \frac{1}{2}R_2 \end{matrix} \begin{bmatrix} 1 & -3 & 2 & \vdots & 8 \\ 0 & 1 & -\frac{1}{2} & \vdots & -2 \\ 0 & 3 & -1 & \vdots & -5 \end{bmatrix}$$

$$\begin{matrix} 3R_2 + R_1 \\ \\ -3R_2 + R_3 \end{matrix} \begin{bmatrix} 1 & 0 & \frac{1}{2} & \vdots & 2 \\ 0 & 1 & -\frac{1}{2} & \vdots & -2 \\ 0 & 0 & \frac{1}{2} & \vdots & 1 \end{bmatrix}$$

$$\begin{matrix} \\ \\ 2R_3 \end{matrix} \begin{bmatrix} 1 & 0 & \frac{1}{2} & \vdots & 2 \\ 0 & 1 & -\frac{1}{2} & \vdots & -2 \\ 0 & 0 & 1 & \vdots & 2 \end{bmatrix}$$

$$\begin{matrix} -\frac{1}{2}R_3 + R_1 \\ \frac{1}{2}R_3 + R_2 \end{matrix} \begin{bmatrix} 1 & 0 & 0 & \vdots & 1 \\ 0 & 1 & 0 & \vdots & -1 \\ 0 & 0 & 1 & \vdots & 2 \end{bmatrix}$$

$(1, -1, 2)$

67. Use matrices to solve the system of linear equations.

$$2x - y + 3z = 24$$
$$2y - z = 14$$
$$7x - 5y = 6$$

Solution

$$\begin{bmatrix} 2 & -1 & 3 & \vdots & 24 \\ 0 & 2 & -1 & \vdots & 14 \\ 7 & -5 & 0 & \vdots & 6 \end{bmatrix}$$

$$\begin{matrix} \frac{1}{2}R_1 \\ \\ \frac{1}{2}R_2 \end{matrix} \begin{bmatrix} 1 & -\frac{1}{2} & \frac{3}{2} & \vdots & 12 \\ 0 & 1 & -\frac{1}{2} & \vdots & 7 \\ 7 & -5 & 0 & \vdots & 6 \end{bmatrix}$$

$$\begin{matrix} \\ \\ -7R_1 + R_3 \end{matrix} \begin{bmatrix} 1 & -\frac{1}{2} & \frac{3}{2} & \vdots & 12 \\ 0 & 1 & -\frac{1}{2} & \vdots & 7 \\ 0 & -\frac{3}{2} & -\frac{21}{2} & \vdots & -78 \end{bmatrix}$$

$$\begin{matrix} \frac{1}{2}R_2 + R_1 \\ \\ \frac{3}{2}R_2 + R_3 \end{matrix} \begin{bmatrix} 1 & 0 & \frac{5}{4} & \vdots & \frac{31}{2} \\ 0 & 1 & -\frac{1}{2} & \vdots & 7 \\ 0 & 0 & -\frac{45}{4} & \vdots & -\frac{135}{2} \end{bmatrix}$$

$$\begin{matrix} \\ \frac{-4}{45}R_3 \\ \\ \end{matrix} \begin{bmatrix} 1 & 0 & \frac{5}{4} & \vdots & \frac{31}{2} \\ 0 & 1 & -\frac{1}{2} & \vdots & 7 \\ 0 & 0 & 1 & \vdots & 6 \end{bmatrix}$$

$$\begin{matrix} -\frac{5}{4}R_3 + R_1 \\ \frac{1}{2}R_3 + R_2 \end{matrix} \begin{bmatrix} 1 & 0 & 0 & \vdots & 8 \\ 0 & 1 & 0 & \vdots & 10 \\ 0 & 0 & 1 & \vdots & 6 \end{bmatrix}$$

$(8, 10, 6)$

69. Use matrices to solve the system of linear equations.

$$x + 3y \qquad = 2$$
$$2x + 6y \qquad = 4$$
$$2x + 5y + 4z = 3$$

Solution

$$\begin{bmatrix} 1 & 3 & 0 & \vdots & 2 \\ 2 & 6 & 0 & \vdots & 4 \\ 2 & 5 & 4 & \vdots & 3 \end{bmatrix}$$

$$\begin{matrix} \\ -2R_1 + R_2 \\ -2R_1 + R_3 \end{matrix} \begin{bmatrix} 1 & 3 & 0 & \vdots & 2 \\ 0 & 0 & 0 & \vdots & 0 \\ 0 & -1 & 4 & \vdots & -1 \end{bmatrix}$$

$$-R_3 \leftrightarrow R_2 \begin{bmatrix} 1 & 3 & 0 & \vdots & 2 \\ 0 & 1 & -4 & \vdots & 1 \\ 0 & 0 & 0 & \vdots & 0 \end{bmatrix}$$

let $z = a$

then $y = 1 + 4a$

$$x = 2 - 3(1 + 4a)$$
$$= 2 - 3 - 12a$$
$$= -1 - 12a$$

$(-12a - 1, \ 1 + 4a, \ a)$

71. Use matrices to solve the system of linear equations.

$$2x + 4y + 5z = 5$$
$$x + 3y + 3z = 2$$
$$2x + 4y + 4z = 2$$

Solution

$$\begin{bmatrix} 2 & 4 & 5 & \vdots & 5 \\ 1 & 3 & 3 & \vdots & 2 \\ 2 & 4 & 4 & \vdots & 2 \end{bmatrix}$$

$$-\tfrac{1}{2}R_3 \leftrightarrow R_1 \begin{bmatrix} 1 & 2 & 2 & \vdots & 1 \\ 1 & 3 & 3 & \vdots & 2 \\ 2 & 4 & 5 & \vdots & 5 \end{bmatrix}$$

$$\begin{matrix} -R_1 + R_2 \\ -2R_1 + R_3 \end{matrix} \begin{bmatrix} 1 & 2 & 2 & \vdots & 1 \\ 0 & 1 & 1 & \vdots & 1 \\ 0 & -2 & -1 & \vdots & 1 \end{bmatrix}$$

$$\begin{matrix} -2R_2 + R_1 \\ 2R_2 + R_3 \end{matrix} \begin{bmatrix} 1 & 0 & 0 & \vdots & -1 \\ 0 & 1 & 1 & \vdots & 1 \\ 0 & 0 & 1 & \vdots & 3 \end{bmatrix}$$

$$-R_3 + R_1 \begin{bmatrix} 1 & 0 & 0 & \vdots & -1 \\ 0 & 1 & 0 & \vdots & -2 \\ 0 & 0 & 1 & \vdots & 3 \end{bmatrix}$$

$(-1, \ -2, \ 3)$

73. Use matrices to solve the system of linear equations.

$$3x + 3y + z = 4$$
$$2x + 6y + z = 5$$
$$-x - 3y + 2z = -5$$

Solution

$$\begin{bmatrix} 3 & 3 & 1 & \vdots & 4 \\ 2 & 6 & 1 & \vdots & 5 \\ -1 & -3 & 2 & \vdots & -5 \end{bmatrix}$$

$$-R_3 \leftrightarrow R_1 \begin{bmatrix} 1 & 3 & -2 & \vdots & 5 \\ 2 & 6 & 1 & \vdots & 5 \\ 3 & 3 & 1 & \vdots & 4 \end{bmatrix}$$

$$\begin{matrix} -2R_1 + R_2 \\ -3R_1 + R_3 \end{matrix} \begin{bmatrix} 1 & 3 & -2 & \vdots & 5 \\ 0 & 0 & 5 & \vdots & -5 \\ 0 & -6 & 7 & \vdots & -11 \end{bmatrix}$$

$$\tfrac{1}{5}R_2 \leftrightarrow R_3 \begin{bmatrix} 1 & 3 & -2 & \vdots & 5 \\ 0 & -6 & 7 & \vdots & -11 \\ 0 & 0 & 1 & \vdots & -1 \end{bmatrix}$$

$$-\tfrac{1}{6}R_2 \begin{bmatrix} 1 & 3 & -2 & \vdots & 5 \\ 0 & 1 & -\tfrac{7}{6} & \vdots & \tfrac{11}{6} \\ 0 & 0 & 1 & \vdots & -1 \end{bmatrix}$$

$$-3R_2 + R_1 \begin{bmatrix} 1 & 0 & \tfrac{3}{2} & \vdots & -\tfrac{1}{2} \\ 0 & 1 & -\tfrac{7}{6} & \vdots & \tfrac{11}{6} \\ 0 & 0 & 1 & \vdots & -1 \end{bmatrix}$$

$$\begin{matrix} -\tfrac{3}{2}R_3 + R_1 \\ \tfrac{7}{6}R_3 + R_2 \end{matrix} \begin{bmatrix} 1 & 0 & 0 & \vdots & 1 \\ 0 & 1 & 0 & \vdots & \tfrac{2}{3} \\ 0 & 0 & 1 & \vdots & -1 \end{bmatrix}$$

$$\left(1, \tfrac{2}{3}, -1\right)$$

75. *Investment Portfolio* Consider an investor with a portfolio totaling $500,000 that is to be allocated among the following types of investments: certificates of deposit, municipal bonds, blue-chip stocks, and growth of speculative stocks.

The certificates of deposit pay 10% annually, and the municipal bonds pay 8% annually. Over a 5-year period, the investor expects the blue-chip stocks to return 12% annually, and expects the growth stocks to return 13% annually. The investor wants a combined annual return of 10% and also wants to have only one-fourth of the portfolio invested in stocks. How much should be allocated to each type of investment?

Solution

$x =$ certificates of deposit

$y =$ municipal bonds

$z =$ blue-chip stocks

$w =$ growth stocks

$$\begin{cases} .10x + .08y + .12z + .13w = 50{,}000 \\ z + w = 125{,}000 \\ x + y = 375{,}000 \end{cases}$$

$$\begin{bmatrix} 10 & 8 & 12 & 13 & \vdots & 5{,}000{,}000 \\ 0 & 0 & 1 & 1 & \vdots & 125{,}000 \\ 1 & 1 & 0 & 0 & \vdots & 375{,}000 \end{bmatrix}$$

$$R_1 \leftrightarrow R_3 \begin{bmatrix} 1 & 1 & 0 & 0 & \vdots & 375{,}000 \\ 0 & 0 & 1 & 1 & \vdots & 125{,}000 \\ 10 & 8 & 12 & 13 & \vdots & 5{,}000{,}000 \end{bmatrix}$$

$$-10R_1 + R_3 \begin{bmatrix} 1 & 1 & 0 & 0 & \vdots & 375{,}000 \\ 0 & 0 & 1 & 1 & \vdots & 125{,}000 \\ 0 & -2 & 12 & 13 & \vdots & 1{,}250{,}000 \end{bmatrix}$$

$$-\tfrac{1}{2}R_3 \begin{bmatrix} 1 & 1 & 0 & 0 & \vdots & 375{,}000 \\ 0 & 0 & 1 & 1 & \vdots & 125{,}000 \\ 0 & 1 & -6 & -\tfrac{13}{2} & \vdots & -625{,}000 \end{bmatrix}$$

$$-R_3 + R_1 \begin{bmatrix} 1 & 0 & 6 & \tfrac{13}{2} & \vdots & 1{,}000{,}000 \\ 0 & 0 & 1 & 1 & \vdots & 125{,}000 \\ 0 & 1 & -6 & -\tfrac{13}{2} & \vdots & -625{,}000 \end{bmatrix}$$

$$\begin{matrix} -6R_2 + R_1 \\ {} \\ 6R_2 + R_3 \end{matrix} \begin{bmatrix} 1 & 0 & 0 & .5 & \vdots & 250{,}000 \\ 0 & 0 & 1 & 1 & \vdots & 125{,}000 \\ 0 & 1 & 0 & -.5 & \vdots & 125{,}000 \end{bmatrix}$$

so let $w = s$

then $x + .5w = 250{,}000$

$\qquad x = -.5s + 250{,}000$

$\qquad z + w = 125{,}000$

$\qquad z = -s + 125{,}000$

$\qquad y - .5w = 125{,}000$

$\qquad y = .5s + 125{,}000$

Certificates of deposit: $250{,}000 - .5s$

Municipal bonds: $125{,}000 + .5s$

Blue-chip stocks: $125{,}000 - s$

Growth stocks: s

77. *Curve Fitting* Find the equation of the parabola

$$y = ax^2 + bx + c$$

that passes through the points $(1, 1)$, $(-3, 17)$ and $(2, -\frac{1}{2})$.

Solution

$$1 = a(1)^2 + b(1) + c \Rightarrow 1 = a + b + c$$

$$17 = a(-3)^2 + b(-3) + c \Rightarrow 17 = 9a - 3b + c$$

$$-\frac{1}{2} = a(2)^2 + b(2) + c \Rightarrow -1 = 8a + 4b + 2c$$

$$\begin{bmatrix} 1 & 1 & 1 & \vdots & 1 \\ 9 & -3 & 1 & \vdots & 17 \\ 8 & 4 & 2 & \vdots & -1 \end{bmatrix}$$

$$\begin{matrix} \\ -9R_1 + R_2 \\ -8R_1 + R_3 \end{matrix} \begin{bmatrix} 1 & 1 & 1 & \vdots & 1 \\ 0 & -12 & -8 & \vdots & 8 \\ 0 & -4 & -6 & \vdots & -9 \end{bmatrix}$$

$$-\frac{1}{12}R_2 \begin{bmatrix} 1 & 1 & 1 & \vdots & 1 \\ 0 & 1 & \frac{2}{3} & \vdots & -\frac{2}{3} \\ 0 & -4 & -6 & \vdots & -9 \end{bmatrix}$$

$$4R_2 + R_3 \begin{bmatrix} 1 & 1 & 1 & \vdots & 1 \\ 0 & 1 & \frac{2}{3} & \vdots & -\frac{2}{3} \\ 0 & 0 & -\frac{10}{3} & \vdots & -\frac{35}{3} \end{bmatrix}$$

$$-\frac{3}{10}R_3 \begin{bmatrix} 1 & 1 & 1 & \vdots & 1 \\ 0 & 1 & \frac{2}{3} & \vdots & -\frac{2}{3} \\ 0 & 0 & 1 & \vdots & \frac{7}{2} \end{bmatrix}$$

$$c = \frac{7}{2} \qquad b + \frac{2}{3}\left(\frac{7}{2}\right) = -\frac{2}{3} \qquad a + -3 + \frac{7}{2} = 1 \qquad y = \frac{1}{2}x^2 - 3x + \frac{7}{2}$$

$$b + \frac{7}{3} = -\frac{9}{3} \qquad a + -\frac{6}{2} + \frac{7}{2} = 1$$

$$b = -3 \qquad a = 1 - \frac{1}{2}$$

$$a = \frac{1}{2}$$

79. The sum of three positive numbers is 33. The second number is 3 greater than the first, and the third is four times the first. Find the three numbers.

Solution

Verbal model:

$$\boxed{\text{Number 1}} + \boxed{\text{Number 2}} + \boxed{\text{Number 3}} = \boxed{33}$$

$$\boxed{\text{Number 2}} = 3 + \boxed{\text{Number 1}}$$

$$\boxed{\text{Number 3}} = 4 \cdot \boxed{\text{Number 1}}$$

System of equations:

$$x + y + z = 33$$

$$y = 3 + x$$

$$z = 4x$$

$$\begin{bmatrix} 1 & 1 & 1 & \vdots & 33 \\ -1 & 1 & 0 & \vdots & 3 \\ -4 & 0 & 1 & \vdots & 0 \end{bmatrix}$$

$$\begin{array}{c} \\ R_1 + R_2 \\ 4R_1 + R_3 \end{array} \begin{bmatrix} 1 & 1 & 1 & \vdots & 33 \\ 0 & 2 & 1 & \vdots & 36 \\ 0 & 4 & 5 & \vdots & 132 \end{bmatrix}$$

$$\tfrac{1}{2}R_2 \begin{bmatrix} 1 & 1 & 1 & \vdots & 33 \\ 0 & 1 & \tfrac{1}{2} & \vdots & 18 \\ 0 & 4 & 5 & \vdots & 132 \end{bmatrix}$$

$$-4R_2 + R_3 \begin{bmatrix} 1 & 1 & 1 & \vdots & 33 \\ 0 & 1 & \tfrac{1}{2} & \vdots & 18 \\ 0 & 0 & 3 & \vdots & 60 \end{bmatrix}$$

$$\tfrac{1}{3}R_3 \begin{bmatrix} 1 & 1 & 1 & \vdots & 33 \\ 0 & 1 & \tfrac{1}{2} & \vdots & 18 \\ 0 & 0 & 1 & \vdots & 20 \end{bmatrix}$$

$$z = 20 \qquad y + \tfrac{1}{2}(20) = 18 \qquad x + 8 + 20 = 33$$

$$y = 8 \qquad\qquad x = 5$$

$(5, \ 8, \ 20)$

81. *Rewriting a Fraction* The fraction

$$\frac{2x^2 - 9x}{(x-2)^3}$$

can be written as a sum of three fractions as follows.

$$\frac{2x^2 - 9x}{(x-2)^3} = \frac{A}{x-2} + \frac{B}{(x-2)^2} + \frac{C}{(x-c)^3}$$

The numbers A, B, and C are the solutions of the system

$$4A - 2B + C = 0$$
$$-4A + B = -9$$
$$A = 2$$

Solve the system and verify that the sum of the three resulting fractions is the original fraction.

Solution

$$\begin{bmatrix} 4 & -2 & 1 & \vdots & 0 \\ -4 & 1 & 0 & \vdots & -9 \\ 1 & 0 & 0 & \vdots & 2 \end{bmatrix}$$

$$R_1 \leftrightarrow R_3 \begin{bmatrix} 1 & 0 & 0 & \vdots & 2 \\ -4 & 1 & 0 & \vdots & -9 \\ 4 & -2 & 1 & \vdots & 0 \end{bmatrix}$$

$$\begin{matrix} 4R_1 + R_2 \\ -4R_1 + R_3 \end{matrix} \begin{bmatrix} 1 & 0 & 0 & \vdots & 2 \\ 0 & 1 & 0 & \vdots & -1 \\ 0 & -2 & 1 & \vdots & -8 \end{bmatrix}$$

$$2R_2 + R_3 \begin{bmatrix} 1 & 0 & 0 & \vdots & 2 \\ 0 & 1 & 0 & \vdots & -1 \\ 0 & 0 & 1 & \vdots & -10 \end{bmatrix}$$

$$\begin{aligned} \frac{2x^2 - 9x}{(x-2)^3} &= \frac{2}{x-2} - \frac{1}{(x-2)^2} - \frac{10}{(x-2)^3} \\ &= \frac{2(x-2)^2 - (x-2) - 10}{(x-2)^3} \\ &= \frac{2\left(x^2 - 4x + 4\right) - x + 2 - 10}{(x-2)^3} \\ &= \frac{2x^2 - 8x + 8 - x - 8}{(x-2)^3} \\ &= \frac{2x^2 - 9x}{(x-2)^3} \end{aligned}$$

8.5 Linear Systems and Determinants

5. Find the determinant of the matrix.

$$\begin{bmatrix} 2 & 1 \\ 3 & 4 \end{bmatrix}$$

Solution

$$\det (A) = \begin{vmatrix} 2 & 1 \\ 3 & 4 \end{vmatrix} = 2(4) - 3(1) = 8 - 3 = 5$$

7. Find the determinant of the matrix.

$$\begin{bmatrix} -7 & 6 \\ \frac{1}{2} & 3 \end{bmatrix}$$

Solution

$$\det (A) = \begin{vmatrix} -7 & 3 \\ \frac{1}{2} & 6 \end{vmatrix} = (-7)(3) - \left(\frac{1}{2}\right)(6) = -21 - 3 = -24$$

9. Evaluate the determinant of the matrix six different ways by expanding by minors along each row and column.

$$\begin{bmatrix} 2 & 3 & -1 \\ 6 & 0 & 0 \\ 4 & 1 & 1 \end{bmatrix}$$

Solution

$$\det (A) = \begin{vmatrix} 2 & 3 & -1 \\ 6 & 0 & 0 \\ 4 & 1 & 1 \end{vmatrix}$$

(a)
$$\text{first row} = 2\begin{vmatrix} 0 & 0 \\ 1 & 1 \end{vmatrix} - (3)\begin{vmatrix} 6 & 0 \\ 4 & 1 \end{vmatrix} + (-1)\begin{vmatrix} 6 & 0 \\ 4 & 1 \end{vmatrix}$$
$$= (2)(0) - (3)(6) - (1)(6)$$
$$= -18 - 6$$
$$= -24$$

(b) $\text{second row} = -(6)\begin{vmatrix} 3 & -1 \\ 1 & 1 \end{vmatrix} + 0 + 0$
$$= (-6)(4)$$
$$= -24$$

(c) $\text{third row} = (4)\begin{vmatrix} 3 & -1 \\ 0 & 0 \end{vmatrix} - (1)\begin{vmatrix} 2 & -1 \\ 6 & 0 \end{vmatrix} + (1)\begin{vmatrix} 2 & 3 \\ 6 & 0 \end{vmatrix}$
$$= (4)(0) - (1)(6) + (1)(-18)$$
$$= -6 - 18$$
$$= -24$$

(d) $\text{first column} = (2)\begin{vmatrix} 0 & 0 \\ 1 & 1 \end{vmatrix} - (6)\begin{vmatrix} 3 & -1 \\ 1 & 1 \end{vmatrix} + (4)\begin{vmatrix} 3 & -1 \\ 0 & 0 \end{vmatrix}$
$$= (2)(0) - (6)(4) + 4(0)$$
$$= -24$$

(e) $\text{second column} = -(3)\begin{vmatrix} 6 & 0 \\ 4 & 1 \end{vmatrix} + (0)\begin{vmatrix} 2 & -1 \\ 4 & 1 \end{vmatrix} - (1)\begin{vmatrix} 2 & -1 \\ 6 & 0 \end{vmatrix}$
$$= (-3)(6) + 0 - (1)(6)$$
$$= -18 - 6$$
$$= -24$$

(f) $\text{third column} = +(-1)\begin{vmatrix} 6 & 0 \\ 4 & 1 \end{vmatrix} - (0)\begin{vmatrix} & \\ & \end{vmatrix} + (1)\begin{vmatrix} 2 & 3 \\ 6 & 0 \end{vmatrix}$
$$= (-1)(6) - 0 + (1)(-18)$$
$$= -6 - 18$$
$$= -24$$

11. Evaluate the determinant of the matrix. Expand by minors on the row or column that appears to make the computation easiest.

$$\begin{bmatrix} 2 & 4 & 6 \\ 0 & 3 & 1 \\ 0 & 0 & -5 \end{bmatrix}$$

Solution

$$\det(A) = \begin{vmatrix} 2 & 4 & 6 \\ 0 & 3 & 1 \\ 0 & 0 & -5 \end{vmatrix}$$

$$\text{first column} = (2)\begin{vmatrix} 3 & 1 \\ 0 & -5 \end{vmatrix} - 0 + 0$$

$$= (2)(-15) = -30$$

13. Evaluate the determinant of the matrix. Expand by minors on the row or column that appears to make the computation easiest.

$$\begin{bmatrix} 6 & 8 & -7 \\ 0 & 0 & 0 \\ 4 & -6 & 22 \end{bmatrix}$$

Solution

$$\begin{bmatrix} 6 & 8 & -7 \\ 0 & 0 & 0 \\ 4 & -6 & 22 \end{bmatrix} \quad \det(A) = \begin{vmatrix} 6 & 8 & -7 \\ 0 & 0 & 0 \\ 4 & -6 & 22 \end{vmatrix}$$

$$\text{second row} = -0 + 0 - 0$$

$$= 0$$

15. Use a graphing utility to evaluate the determinant of the matrix.

$$\begin{bmatrix} 5 & -3 & 2 \\ 7 & 5 & -7 \\ 0 & 6 & -1 \end{bmatrix}$$

Solution

Keystrokes:

[MATRIX] [EDIT 1] 5 [ENTER] [(−)] 3 [ENTER] 2 [ENTER] 7 [ENTER] 5 [ENTER] [(−)] 7 [ENTER]

0 [ENTER] 6 [ENTER] 1 [ENTER] [MATRIX] [MATH 1] [MATRIX 1] [ENTER]

Solution is 248.

17. Use a graphing utility to evaluate the determinant of the matrix.

$$\begin{bmatrix} 0.2 & 0.8 \\ -8 & -5 \end{bmatrix}$$

Solution

Keystrokes:

[MATRIX] [EDIT 2] .2 [ENTER] .8 [ENTER] [(−)] 8 [ENTER] [(−)] 5 [ENTER]

[MATRIX] [MATH 1] [MATRIX 2] [ENTER]

Solution is 5.4.

19. Use Cramer's Rule to solve the following system of equations. (If it is not possible, state the reason.)

$$3x + 4y = -2$$
$$5x + 3y = 4$$

Solution

$$\begin{bmatrix} 3 & 4 & \vdots & -2 \\ 5 & 3 & \vdots & 4 \end{bmatrix}$$

$$D = \begin{vmatrix} 3 & 4 \\ 5 & 3 \end{vmatrix} = 9 - 20 = -11$$

$$x = \frac{Dx}{D} = \frac{\begin{vmatrix} -2 & 4 \\ 4 & 3 \end{vmatrix}}{-11} = \frac{-6 - 16}{-11} = \frac{-22}{-11} = 2$$

$$y = \frac{Dy}{D} = \frac{\begin{vmatrix} 3 & -2 \\ 5 & 4 \end{vmatrix}}{-11} = \frac{12 - (-10)}{-11} = \frac{22}{-11} = -2$$

$(2, -2)$

21. Use Cramer's Rule to solve the following system of equations. (If it is not possible, state the reason.)

$$-0.4x + 0.8y = 1.6$$
$$2x - 4y = 5$$

Solution

$$\begin{bmatrix} -0.4 & 0.8 & \vdots & 1.6 \\ 2 & -4 & \vdots & 5 \end{bmatrix}$$

$$D = \begin{vmatrix} -0.4 & 0.8 \\ 2 & -4 \end{vmatrix} = 1.6 - 1.6 = 0$$

Cannot be solved by Carter's Rule because $D = 0$.

Solve by elimination.

$$-4x + 8y = 16 \implies -4x + 8y = 16$$
$$2x - 4y = 5 \implies \underline{4x - 8y = 10}$$
$$0 \neq 26$$

inconsistent; no solution

23. Use Cramer's Rule to solve the following system of equations. (If it is not possible, state the reason.)

$$4x - y + z = -5$$
$$2x + 2y + 3z = 10$$
$$5x - 2y + 6z = 1$$

Solution

$$\begin{bmatrix} 4 & -1 & 1 & \vdots & -5 \\ 2 & 2 & 3 & \vdots & 10 \\ 5 & -2 & 6 & \vdots & 1 \end{bmatrix}$$

$$D = \begin{vmatrix} 4 & -1 & 1 \\ 2 & 2 & 3 \\ 5 & -2 & 6 \end{vmatrix} = (1)\begin{vmatrix} 2 & 2 \\ 5 & -2 \end{vmatrix} - (3)\begin{vmatrix} 4 & -1 \\ 5 & -2 \end{vmatrix} + (6)\begin{vmatrix} 4 & -1 \\ 2 & 2 \end{vmatrix}$$

$$= (1)(-14) + (-3)(-3) + (6)(10)$$
$$= -14 + 9 + 60 = 55$$

$$x = \frac{\begin{vmatrix} -5 & -1 & 1 \\ 10 & 2 & 3 \\ 1 & -2 & 6 \end{vmatrix}}{55} = \frac{(1)\begin{vmatrix} 10 & 2 \\ 1 & -2 \end{vmatrix} - (3)\begin{vmatrix} -5 & -1 \\ 1 & -2 \end{vmatrix} + (6)\begin{vmatrix} -5 & -1 \\ 10 & 2 \end{vmatrix}}{55}$$

$$= \frac{(1)(-22) + (-3)(11) + (6)(0)}{55}$$

$$= \frac{-22 - 33}{55} = \frac{-55}{55} = -1$$

$$y = \frac{\begin{vmatrix} 4 & -5 & 1 \\ 2 & 10 & 3 \\ 5 & 1 & 6 \end{vmatrix}}{55} = \frac{(1)\begin{vmatrix} 2 & 10 \\ 5 & 1 \end{vmatrix} - (3)\begin{vmatrix} 4 & -5 \\ 5 & 1 \end{vmatrix} + (6)\begin{vmatrix} 4 & -5 \\ 2 & 10 \end{vmatrix}}{55}$$

$$= \frac{(1)(-48) + (-3)(29) + (6)(50)}{55}$$

$$= \frac{-48 - 87 + 300}{55} = \frac{165}{55} = 3$$

$$z = \frac{\begin{vmatrix} 4 & -1 & -5 \\ 2 & 2 & 10 \\ 5 & -2 & 1 \end{vmatrix}}{55} = \frac{(5)\begin{vmatrix} -1 & -5 \\ 2 & 10 \end{vmatrix} - (-2)\begin{vmatrix} 4 & -5 \\ 2 & 10 \end{vmatrix} + (1)\begin{vmatrix} 4 & -1 \\ 2 & 2 \end{vmatrix}}{55}$$

$$= \frac{(5)(0) + (2)(50) + (1)(10)}{55}$$

$$= \frac{0 + 100 + 10}{55} = \frac{110}{55} = 2$$

$(-1, 3, 2)$

25. Use Cramer's Rule to solve the following system of equations. (If it is not possible, state the reason.)

$$3x + 4y + 4z = 11$$
$$4x - 4y + 6z = 11$$
$$6x - 6y \qquad = 3$$

Solution

$$\begin{bmatrix} 3 & 4 & 4 & \vdots & 11 \\ 4 & -4 & 6 & \vdots & 11 \\ 6 & -6 & 0 & \vdots & 3 \end{bmatrix}$$

$$D = \begin{vmatrix} 3 & 4 & 4 \\ 4 & -4 & 3 \\ 6 & -6 & 0 \end{vmatrix} = (4)\begin{vmatrix} 4 & -4 \\ 6 & -6 \end{vmatrix} - (6)\begin{vmatrix} 3 & 4 \\ 6 & -6 \end{vmatrix} + 0$$

$$= (4)(0) - (6)(-42) + 0$$

$$= 252$$

$$x = \dfrac{\begin{vmatrix} 11 & 4 & 4 \\ 11 & -4 & 6 \\ 3 & -6 & 0 \end{vmatrix}}{252} = \dfrac{(4)\begin{vmatrix} 11 & -4 \\ 3 & -6 \end{vmatrix} - (6)\begin{vmatrix} 11 & 4 \\ 3 & -6 \end{vmatrix} + 0}{252}$$

$$= \dfrac{(4)(-54) - (6)(-78)}{252}$$

$$= \dfrac{-216 - 468}{252} = \dfrac{252}{252} = 1$$

$$y = \dfrac{\begin{vmatrix} 3 & 11 & 4 \\ 4 & 11 & 6 \\ 6 & 3 & 0 \end{vmatrix}}{252} = \dfrac{(4)\begin{vmatrix} 4 & 11 \\ 6 & 3 \end{vmatrix} - (6)\begin{vmatrix} 3 & 11 \\ 6 & 3 \end{vmatrix} + 0}{252}$$

$$= \dfrac{(4)(-54) - (6)(-57)}{252}$$

$$= \dfrac{-216 + 342}{252} = \dfrac{126}{252} = \dfrac{1}{2}$$

$$z = \dfrac{\begin{vmatrix} 3 & 4 & 11 \\ 4 & -4 & 11 \\ 6 & -6 & 3 \end{vmatrix}}{252} = \dfrac{(3)\begin{vmatrix} -4 & 11 \\ -6 & 3 \end{vmatrix} - (4)\begin{vmatrix} 4 & 11 \\ -6 & 3 \end{vmatrix} + (6)\begin{vmatrix} 4 & 11 \\ -4 & 11 \end{vmatrix}}{252}$$

$$= \dfrac{(3)(54) - (4)(78) + (6)(88)}{252}$$

$$= \dfrac{162 - 312 + 528}{252} = \dfrac{378}{252} = \dfrac{3}{2}$$

$$\left(1, \dfrac{1}{2}, \dfrac{3}{2}\right)$$

27. Use a graphing utility and Cramer's Rule to solve the system of equations.

$$-3x + 10y = 22$$
$$9x - 3y = 0$$

Solution

$$\begin{bmatrix} -3 & 10 & \vdots & 22 \\ 9 & -3 & \vdots & 0 \end{bmatrix}$$

$$D = \begin{vmatrix} -3 & 10 \\ 9 & -3 \end{vmatrix} = -81$$

$$x = \frac{Dx}{D} = \frac{\begin{vmatrix} 22 & 10 \\ 0 & -3 \end{vmatrix}}{-81} = \frac{-66}{-81} = \frac{22}{27}$$

$$y = \frac{Dy}{D} = \frac{\begin{vmatrix} -3 & 22 \\ 9 & 0 \end{vmatrix}}{-81} = \frac{-198}{-81} = \frac{22}{9}$$

Keystrokes:

det D

[MATRIX] [EDIT 2]
Enter each number in matrix followed by Enter.
[MATRIX] [MATH 1] [MATRIX 2] [ENTER]

det Dx

[MATRIX] [EDIT 2]
Enter each number in matrix followed by Enter.
[MATRIX] [MATH 1] [MATRIX 2] [ENTER]

det Dy

[MATRIX] [EDIT 2]
Enter each number in matrix followed by Enter.
[MATRIX] [MATH 1] [MATRIX 2] [ENTER]

29. Use a graphing utility and Cramer's Rule to solve the system of equations.

$$x + y - z = -2$$
$$6x + 4y + 3z = 4$$
$$3x + 6z = -3$$

Solution

$$\begin{bmatrix} 1 & 1 & -1 & \vdots & 2 \\ 6 & 4 & 3 & \vdots & 4 \\ 3 & 0 & 6 & \vdots & -3 \end{bmatrix}$$

$$D = \begin{vmatrix} 1 & 1 & -1 \\ 6 & 4 & 3 \\ 3 & 0 & 6 \end{vmatrix} = 9$$

$$x = \frac{Dx}{D} = \frac{\begin{vmatrix} 2 & 1 & -1 \\ 4 & 4 & 3 \\ -3 & 0 & 6 \end{vmatrix}}{9} = \frac{3}{9} = \frac{1}{3}$$

$$y = \frac{Dy}{D} = \frac{\begin{vmatrix} 1 & 2 & -1 \\ 6 & 4 & 3 \\ 3 & -3 & 6 \end{vmatrix}}{9} = \frac{9}{9} = 1$$

$$z = \frac{Dz}{D} = \frac{\begin{vmatrix} 1 & 1 & 2 \\ 6 & 4 & 4 \\ 3 & 0 & -3 \end{vmatrix}}{9} = \frac{-6}{9} = \frac{-2}{3}$$

Keystrokes:

det D

[MATRIX] [EDIT 3]
Enter each number in matrix followed by Enter.
[MATRIX] [MATH 1] [MATRIX 3] [ENTER]

det Dx

[MATRIX] [EDIT 3]
Enter each number in matrix followed by Enter.
[MATRIX] [MATH 1] [MATRIX 3] [ENTER]

det Dy

[MATRIX] [EDIT 3]
Enter each number in matrix followed by Enter.
[MATRIX] [MATH 1] [MATRIX 3] [ENTER]

det Dz

[MATRIX] [EDIT 3]
Enter each number in matrix followed by Enter.
[MATRIX] [MATH 1] [MATRIX 3] [ENTER]

31. Use a determinant to find the area of the triangle
with the given vertices.

$(0, 3)$, $(4, 0)$, $(8, 5)$

Solution

$(x_1, y_1) = (0, 3)$, $(x_2, y_2) = (4, 0)$, $(x_{3, y_3}) = (8, 5)$

$$\begin{vmatrix} x_1 & y_1 & 1 \\ x_2 & y_2 & 1 \\ x_3 & y_3 & 1 \end{vmatrix} = \begin{vmatrix} 0 & 3 & 1 \\ 4 & 0 & 1 \\ 8 & 5 & 1 \end{vmatrix} = 32$$

$\text{Area} = +\dfrac{1}{2}(32) = 16$

33. *Geometry*   A large region of forest has been
infested with gypsy moths. The region is roughly
triangular, as shown in the figure in the textbook.
Approximate the number of square miles in this
region.

Solution

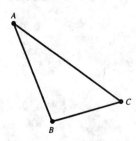

From diagram the coordinates of A, B, C are determined
to be $A(0, 20)$, $B(10, -5)$ and $C(28, 0)$.

$$\begin{vmatrix} x_1 & y_1 & 1 \\ x_2 & y_2 & 1 \\ x_3 & y_3 & 1 \end{vmatrix} = \begin{vmatrix} 0 & 20 & 1 \\ 10 & -5 & 1 \\ 28 & 0 & 1 \end{vmatrix} = -500$$

$\text{Area} = -\dfrac{1}{2}(-500) = 250 \text{ mi}^2$

35. Decide whether the points $(-1, -5)$, $(1, -1)$, and $(4, 5)$
are collinear.

Solution

$(x_1, y_1) = (-1, -5)$, $(x_2, y_2) = (1, -1)$, $(x_3, y_3) = (4, 5)$

$$\begin{vmatrix} x_1 & y_1 & 1 \\ x_2 & y_2 & 1 \\ x_3 & y_3 & 1 \end{vmatrix} = \begin{vmatrix} -1 & -5 & 1 \\ 1 & -1 & 1 \\ 4 & 5 & 1 \end{vmatrix} = 0$$

The three points are collinear.

37. Find an equation of the line through the points.

$$\left(-2, \frac{3}{2}\right), (3, -3)$$

Solution

$$(x_1, y_1) = \left(-2, \frac{3}{2}\right), \quad (x_2, y_2) = (3, -3)$$

$$\begin{vmatrix} x & y & 1 \\ -2 & \frac{3}{2} & 1 \\ 3 & -3 & 1 \end{vmatrix} = 0$$

$$x\begin{vmatrix} \frac{3}{2} & 1 \\ -3 & 1 \end{vmatrix} - y\begin{vmatrix} -2 & 1 \\ 3 & 1 \end{vmatrix} + 1\begin{vmatrix} -2 & \frac{3}{2} \\ 3 & -3 \end{vmatrix} = 0$$

$$\frac{9}{2}x + 5y + \frac{3}{2} = 0$$

$$9x + 10y + 3 = 0$$

39. Use Cramer's Rule to find the equation of the parabola that passes through the points.

$$(-2, 6), \quad (2, -2), \quad (4, 0)$$

Solution

$$6 = a(-2)^2 + b(-2) + c \implies 6 = 4a - 2b + c$$
$$-2 = a\ (2)^2 + b\ (2) + c \implies -2 = 4a + 2b + c$$
$$0 = a\ (4)^2 + b\ (4) + c \implies 6 = 16a + 4b + c$$

$$\begin{bmatrix} 4 & -2 & 1 & \vdots & 6 \\ 4 & 2 & 1 & \vdots & -2 \\ 16 & 4 & 1 & \vdots & 0 \end{bmatrix} \quad D = \begin{vmatrix} 4 & -2 & 1 \\ 4 & 2 & 1 \\ 16 & 4 & 1 \end{vmatrix} = -48$$

$$a = \frac{D_a}{D} = \frac{\begin{vmatrix} 6 & -2 & 1 \\ -2 & 2 & 1 \\ 0 & 4 & 1 \end{vmatrix}}{-48} = \frac{-24}{-48} = \frac{1}{2}$$

$$b = \frac{D_b}{D} = \frac{\begin{vmatrix} 4 & 6 & 1 \\ 4 & -2 & 1 \\ 16 & 0 & 1 \end{vmatrix}}{-48} = \frac{96}{-48} = -2$$

$$c = \frac{D_c}{D} = \frac{\begin{vmatrix} 4 & -2 & 6 \\ 4 & 2 & -2 \\ 16 & 4 & 0 \end{vmatrix}}{-48} = \frac{0}{-48} = 0$$

$$y = \frac{1}{2}x^2 - 2x$$

41. *Mathematical Modeling*   The table in the textbook gives the merchandise exports y_1 and the merchandise imports y_2 (in billions of dollars) for the years 1990 through 1992 in the United States. (Source: U.S. Bureau of Census)

(a) Find a quadratic model for exports. Let $t = 0$ represent 1990.

(b) Find a quadratic model for imports. Let $t = 0$ represent 1990.

(c) Use a graphing utility to graph the models found in parts (a) and (b).

(d) Find a model for the merchandise trade balance. (merchandise exports − merchandise imports)

Solution

(a) $(0, 393.6)$, $(1, 421.7)$, $(2, 448.2)$

$$393.6 = a(0)^2 + b(0) + c \implies 393.6 = \qquad c$$
$$421.7 = a(1)^2 + b(1) + c \implies 421.7 = a + b + c$$
$$448.2 = a(2)^2 + b(2) + c \implies 448.2 = 4a + 2b + c$$

$$\det(A) = \begin{vmatrix} 0 & 0 & 1 \\ 1 & 1 & 1 \\ 4 & 2 & 1 \end{vmatrix} = -2$$

$$a = \frac{\begin{vmatrix} 393.6 & 0 & 1 \\ 421.7 & 1 & 1 \\ 448.2 & 2 & 1 \end{vmatrix}}{-2} = \frac{1.6}{-2} = -.8 \qquad b = \frac{\begin{vmatrix} 0 & 393.7 & 1 \\ 1 & 421.7 & 1 \\ 4 & 448.2 & 1 \end{vmatrix}}{-2} = \frac{-57.8}{-2} = 28.9$$

$$c = 393.6 \qquad\qquad y_1 = -0.8t^2 + 28.9t + 393.6$$

(b) $(0, 495.3)$, $(1, 487.1)$, $(2, 532.5)$

$$495.3 = a(0)^2 + b(0) + c \implies 495.3 = \qquad c$$
$$487.1 = a(1)^2 + b(1) + c \implies 487.1 = a + b + c$$
$$532.5 = a(2)^2 + b(2) + c \implies 532.5 = 4a + 2b + c$$

$$\det(A) = \begin{vmatrix} 0 & 0 & 1 \\ 1 & 1 & 1 \\ 4 & 2 & 1 \end{vmatrix} = -2$$

$$a = \frac{\begin{vmatrix} 495.3 & 0 & 1 \\ 487.1 & 1 & 1 \\ 532.5 & 2 & 1 \end{vmatrix}}{-2} = \frac{-53.6}{-2} = 26.8 \qquad b = \frac{\begin{vmatrix} 0 & 495.3 & 1 \\ 1 & 487.1 & 1 \\ 4 & 532.5 & 1 \end{vmatrix}}{-2} = \frac{70}{-2} = -35$$

$$c = 495.3 \qquad\qquad y_2 = 26.8t^2 - 35t + 495.3$$

(c) Keystrokes:

y_1 $\boxed{Y=}$ $\boxed{(-)}$ .8 $\boxed{X, T, \theta}$ $\boxed{x^2}$ $\boxed{+}$ 28.9 $\boxed{X, T, \theta}$ $\boxed{+}$ 393.6 $\boxed{\text{ENTER}}$

y_2 $\boxed{Y=}$ 26.8 $\boxed{X, T, \theta}$ $\boxed{x^2}$ $\boxed{-}$ 35 $\boxed{X, T, \theta}$ $\boxed{+}$ 495.3 $\boxed{\text{GRAPH}}$

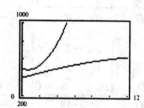

(d) Merchandise Trade Balance = Exports − Imports

$$= y_1 - y_2 = (-0.8t^2 + 28.8t + 393.6) - (26.8t - 35t + 495.3)$$
$$= -27.6t^2 + 63.9t - 101.7$$

43. Sketch a graph of the equation , $y = 3 - \dfrac{1}{2}x$.

Solution

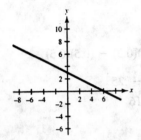

45. Sketch a graph of the equation, $\dfrac{y_2}{25} + x^2 = 1$.

Solution

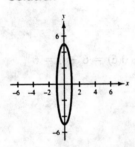

47. *Defective Units*   A quality control engineer for a certain buyer found two defective units in a sample of 75. At that rate, what is the expected number of defective units in a shipment of 10,000 units?

Solution

Verbal Model: $\boxed{\dfrac{\text{Defective Units}}{\text{Total Units}}} = \boxed{\dfrac{\text{Defective Units}}{\text{Total Units}}}$

Proportion:

$$\dfrac{2}{75} = \dfrac{x}{10,000}$$

$$75x = 20,000$$

$$x = 267 \text{ units}$$

49. Find the determinant of the matrix.

$$\begin{bmatrix} 5 & 2 \\ -6 & 3 \end{bmatrix}$$

Solution

$$\det (A) = \begin{vmatrix} 5 & 2 \\ -5 & 3 \end{vmatrix} = 5(3) - (-6)(2) = 15 + 12 = 27$$

51. Find the determinant of the matrix.

$$\begin{bmatrix} 5 & -4 \\ -10 & 8 \end{bmatrix}$$

Solution

$$\det (A) = \begin{vmatrix} 5 & -4 \\ -10 & 8 \end{vmatrix} = 5(8) - (-10)(-4) = 40 - 40 = 0$$

53. Find the determinant of the matrix.

$$\begin{bmatrix} 2 & 6 \\ 0 & 3 \end{bmatrix}$$

Solution

$$\det(A) = \begin{vmatrix} 2 & 6 \\ 0 & 3 \end{vmatrix} = 2(3) - 0(6) = 6 - 0 = 6$$

55. Find the determinant of the matrix.

$$\begin{bmatrix} 0.3 & 0.5 \\ 0.5 & 0.3 \end{bmatrix}$$

Solution

$$\det(A) = \begin{vmatrix} 0.3 & 0.5 \\ 0.5 & 0.3 \end{vmatrix} = (0.3)(0.3) - (0.5)(0.5)$$
$$= .09 - .25$$
$$= -0.16$$

57. Evaluate the determinant of the matrix six different ways by expanding by minors along each row and column.

$$\begin{bmatrix} 2 & -5 & 0 \\ 4 & 7 & 0 \\ -7 & 25 & 3 \end{bmatrix}$$

Solution

$$\det(A) = \begin{vmatrix} 2 & -5 & 0 \\ 4 & 7 & 0 \\ -7 & 25 & 3 \end{vmatrix}$$

(a) first row $= 2\begin{vmatrix} 7 & 0 \\ 25 & 3 \end{vmatrix} - (-5)\begin{vmatrix} 4 & 0 \\ -7 & 3 \end{vmatrix} + 0$

$\qquad\qquad = (2)(21) + (5)(12)$

$\qquad\qquad = 42 + 60$

$\qquad\qquad = 102$

(b) second row $= -(4)\begin{vmatrix} -5 & 0 \\ 25 & 3 \end{vmatrix} + (7)\begin{vmatrix} 2 & 0 \\ -7 & 3 \end{vmatrix} - 0$

$\qquad\qquad = (-4)(-15) + (7)(6)$

$\qquad\qquad = 60 + 42$

$\qquad\qquad = 102$

(c) third row $= (-7)\begin{vmatrix} -5 & 0 \\ 7 & 0 \end{vmatrix} - (25)\begin{vmatrix} 2 & 0 \\ 4 & 0 \end{vmatrix} + 3\begin{vmatrix} 2 & -5 \\ 4 & 7 \end{vmatrix}$

$\qquad\qquad = (-7)(0) - 25(0) + (3)(34)$

$\qquad\qquad = 102$

(d) first column $= 2\begin{vmatrix} 7 & 0 \\ 25 & 3 \end{vmatrix} - (4)\begin{vmatrix} -5 & 0 \\ 25 & 3 \end{vmatrix} + (-7)\begin{vmatrix} -5 & 0 \\ 7 & 0 \end{vmatrix}$

$\qquad\qquad = (2)(21) - (4)(-15) + (-7)(0)$

$\qquad\qquad = 42 + 60$

$\qquad\qquad = 102$

(e) second column $= -(-5)\begin{vmatrix} 4 & 0 \\ -7 & 3 \end{vmatrix} + 7\begin{vmatrix} 2 & 0 \\ -7 & 3 \end{vmatrix} - 25\begin{vmatrix} 2 & 0 \\ 4 & 0 \end{vmatrix}$

$\qquad\qquad = (5)(12) + (7)(6) - 25(0)$

$\qquad\qquad = 42 + 60$

$\qquad\qquad = 102$

(f) third column $= 0 - 0 + 3\begin{vmatrix} 2 & -5 \\ 4 & 7 \end{vmatrix}$

$\qquad\qquad = (3)(34)$

$\qquad\qquad = 102$

59. Evaluate the determinant of the matrix six different ways by expanding by minors along each row and column.

$$\begin{bmatrix} 1 & 1 & 2 \\ 3 & 1 & 0 \\ -2 & 0 & 3 \end{bmatrix}$$

Solution

$$\det(A) = \begin{vmatrix} 1 & 1 & 2 \\ 3 & 1 & 0 \\ -2 & 0 & 3 \end{vmatrix}$$

(a)   first row $= (1)\begin{vmatrix} 1 & 0 \\ 0 & 3 \end{vmatrix} - (1)\begin{vmatrix} 3 & 0 \\ -2 & 3 \end{vmatrix} + (2)\begin{vmatrix} 3 & 1 \\ -2 & 0 \end{vmatrix}$

$= (1)(3) - (1)(9) + (2)(2)$

$= 3 - 9 + 4$

$= -2$

(b)   second row $= -(3)\begin{vmatrix} 1 & 2 \\ 0 & 3 \end{vmatrix} + (1)\begin{vmatrix} 1 & 2 \\ -2 & 3 \end{vmatrix} - (0)\begin{vmatrix} 1 & 1 \\ -2 & 0 \end{vmatrix}$

$= (-3)(3) + (1)(7) - 0$

$= -9 + 7$

$= -2$

(c)   third row $= (-2)\begin{vmatrix} 1 & 2 \\ 1 & 0 \end{vmatrix} - (0)\begin{vmatrix} 1 & 2 \\ 3 & 0 \end{vmatrix} + 3\begin{vmatrix} 1 & 1 \\ 3 & 1 \end{vmatrix}$

$= (-2)(-2) - 0 + (3)(-2)$

$= 4 - 6$

$= -2$

(d)   first column $= (1)\begin{vmatrix} 1 & 0 \\ 0 & 3 \end{vmatrix} - (3)\begin{vmatrix} 1 & 2 \\ 0 & 3 \end{vmatrix} + (-2)\begin{vmatrix} 1 & 2 \\ 1 & 0 \end{vmatrix}$

$= (1)(3) - (3)(3) - 2(-2)$

$= 3 - 9 + 4$

$= -2$

(e)   second column $= -(1)\begin{vmatrix} 3 & 0 \\ -2 & 3 \end{vmatrix} + (1)\begin{vmatrix} 1 & 2 \\ -2 & 3 \end{vmatrix} - (0)\begin{vmatrix} 1 & 2 \\ 3 & 0 \end{vmatrix}$

$= (-1)(9) + (1)(7) - 0$

$= -9 + 7$

$= -2$

(f)   third column $= (2)\begin{vmatrix} 3 & 1 \\ -2 & 0 \end{vmatrix} - (0)\begin{vmatrix} 1 & 1 \\ -2 & 0 \end{vmatrix} + (3)\begin{vmatrix} 1 & 1 \\ 3 & 1 \end{vmatrix}$

$= (2)(2) - 0 + (3)(-2)I$

$= 4 - 6 = -2$

61. Evaluate the determinant of the matrix. Expand by minors on the row or column that appears to make the computation easiest.

$$\begin{bmatrix} -2 & 2 & 3 \\ 1 & -1 & 0 \\ 0 & 1 & 4 \end{bmatrix}$$

Solution

$$\det(A) = \begin{vmatrix} -2 & 2 & 3 \\ 1 & -1 & 0 \\ 0 & 1 & 4 \end{vmatrix}$$

$$\text{second row} = -(1)\begin{vmatrix} 2 & 3 \\ 1 & 4 \end{vmatrix} + (-1)\begin{vmatrix} -2 & 3 \\ 0 & 4 \end{vmatrix} - 0$$

$$= (-1)(5) + (-1)(-8)$$

$$= -5 + 8 = 3$$

63. Evaluate the determinant of the matrix. Expand by minors on the row or column that appears to make the computation easiest.

$$\begin{bmatrix} 1 & 4 & -2 \\ 3 & 6 & -6 \\ -2 & 1 & 4 \end{bmatrix}$$

Solution

$$\det(A) = \begin{vmatrix} 1 & 4 & -2 \\ 3 & 6 & -6 \\ -2 & 1 & 4 \end{vmatrix}$$

$$\text{first row} = (1)\begin{vmatrix} 6 & -6 \\ 1 & 4 \end{vmatrix} - (4)\begin{vmatrix} 3 & -6 \\ -2 & 4 \end{vmatrix} + (-2)\begin{vmatrix} 3 & 6 \\ -2 & 1 \end{vmatrix}$$

$$= (1)(30) - (4)(0) + (-2)(15)$$

$$= 30 - 0 - 30 = 0$$

65. Evaluate the determinant of the matrix. Expand by minors on the row or column that appears to make the computation easiest.

$$\begin{bmatrix} -3 & 2 & 1 \\ 4 & 5 & 6 \\ 2 & -3 & 1 \end{bmatrix}$$

Solution

$$\det (A) = \begin{vmatrix} -3 & 2 & 1 \\ 4 & 5 & 6 \\ 2 & 3 & 1 \end{vmatrix}$$

$$\text{third column} = (1)\begin{vmatrix} 4 & 5 \\ 2 & -3 \end{vmatrix} - (6)\begin{vmatrix} -3 & 2 \\ 2 & -3 \end{vmatrix} + (1)\begin{vmatrix} -3 & 2 \\ 4 & 5 \end{vmatrix}$$

$$= (1)(-22) - (6)(5) + (1)(-23)$$

$$= -22 - 30 - 23 = -75$$

67. Evaluate the determinant of the matrix. Expand by minors on the row or column that appears to make the computation easiest.

$$\begin{bmatrix} 1 & 4 & -2 \\ 3 & 2 & 0 \\ -1 & 4 & 3 \end{bmatrix}$$

Solution

$$\det (A) = \begin{vmatrix} 1 & 4 & -2 \\ 3 & 2 & 0 \\ -1 & 4 & 3 \end{vmatrix}$$

$$\text{second row} = -(3)\begin{vmatrix} 4 & -2 \\ 4 & 3 \end{vmatrix} + (2)\begin{vmatrix} 1 & -2 \\ -1 & 3 \end{vmatrix} - 0$$

$$= (-3)(20) + (2)(1)$$

$$= -60 + 2$$

$$= -58$$

69. Evaluate the determinant of the matrix. Expand by minors on the row or column that appears to make the computation easiest.

$$\begin{bmatrix} 0.4 & 0.3 & 0.3 \\ -0.2 & 0.6 & 0.6 \\ 3 & 1 & 1 \end{bmatrix}$$

Solution

$$\det (A) = \begin{vmatrix} 0.4 & 0.3 & 0.3 \\ -0.2 & 0.6 & 0.6 \\ 3 & 1 & 1 \end{vmatrix}$$

$$\text{third row} = +(3)\begin{vmatrix} 0.3 & 0.3 \\ 0.6 & 0.6 \end{vmatrix} - (1)\begin{vmatrix} 0.4 & 0.3 \\ -0.2 & 0.6 \end{vmatrix} + (1)\begin{vmatrix} 0.4 & 0.3 \\ -0.2 & 0.6 \end{vmatrix}$$

$$= (3)(0) - (1)(.30) + (1)(.30)$$

$$= 0$$

71. Evaluate the determinant of the matrix. Expand by minors on the row or column that appears to make the computation easiest.

$$\begin{bmatrix} x & y & 1 \\ 3 & 1 & 1 \\ -2 & 0 & 1 \end{bmatrix}$$

Solution

$$\det (A) = \begin{vmatrix} x & y & 1 \\ 3 & 1 & 1 \\ -2 & 0 & 1 \end{vmatrix}$$

$$\text{third row} = (-2)\begin{vmatrix} y & 1 \\ 1 & 1 \end{vmatrix} - 0 + (1)\begin{vmatrix} x & y \\ 3 & 1 \end{vmatrix}$$

$$= (-2)(y - 1) + (1)(x - 3y)$$

$$= -2y + 2 + x - 3y$$

$$= x - 5y + 2$$

73. Use a graphing utility to evaluate the determinant of the matrix.

$$\begin{bmatrix} 35 & 15 & 70 \\ -8 & 20 & 3 \\ -5 & 6 & 20 \end{bmatrix}$$

Solution

Keystrokes:

MATRIX   EDIT 3

Enter each number in matrix followed by Enter.

MATRIX   MATH 1

MATRIX 3   ENTER

Solution: det (A) = 19,185

75. Use a graphing utility to evaluate the determinant of the matrix.

$$\begin{bmatrix} 0.3 & -0.2 & 0.5 \\ 0.6 & 0.4 & -0.3 \\ 1.2 & 0 & 0.7 \end{bmatrix}$$

Solution

Keystrokes:

MATRIX   EDIT 3

Enter each number in matrix followed by Enter.

MATRIX   MATH 1

MATRIX 3   ENTER

Solution: det $(A) = 0$

77. Use Cramer's Rule to solve the system. (If it is not possible, state the reason.)

$$\begin{aligned} x + 2y &= 5 \\ -x + y &= 1 \end{aligned}$$

Solution

$$\begin{bmatrix} 1 & 2 & \vdots & 5 \\ -1 & 1 & \vdots & 1 \end{bmatrix}$$

$$D = \begin{vmatrix} 1 & 2 \\ -1 & 1 \end{vmatrix} = 1 - (-2) = 3$$

$$x = \frac{Dx}{D} = \frac{\begin{vmatrix} 5 & 2 \\ 1 & 1 \end{vmatrix}}{3} = \frac{5 - 2}{3} = \frac{3}{3} = 1$$

$$y = \frac{Dy}{D} = \frac{\begin{vmatrix} 1 & 5 \\ -1 & 1 \end{vmatrix}}{3} = \frac{1 - (-5)}{3} = \frac{6}{3} = 2$$

$(1, 2)$

79. Use Cramer's Rule to solve the system. (If it is not possible, state the reason.)

$$\begin{aligned} 20x + 8y &= 11 \\ 12x - 24y &= 21 \end{aligned}$$

Solution

$$\begin{bmatrix} 20 & 8 & \vdots & 11 \\ 12 & -24 & \vdots & 21 \end{bmatrix}$$

$$D = \begin{vmatrix} 20 & 8 \\ 12 & -24 \end{vmatrix} = -480 - 96 = -576$$

$$x = \frac{Dx}{D} = \frac{\begin{vmatrix} 11 & 8 \\ 21 & -24 \end{vmatrix}}{-576} = \frac{-264 - 168}{-576} = \frac{-432}{-576} = \frac{3}{4}$$

$$y = \frac{Dy}{D} = \frac{\begin{vmatrix} 20 & 11 \\ 12 & 21 \end{vmatrix}}{-576} = \frac{420 - 132}{-576} = \frac{288}{-576} = -\frac{1}{2}$$

$\left(\dfrac{3}{4}, -\dfrac{1}{2}\right)$

81. Use Cramer's Rule to solve the following system. (If it is not possible, state the reason.)

$3u + 6v = 5$

$6u + 14v = 11$

Solution

$$\begin{bmatrix} 3 & 6 & \vdots & 5 \\ 6 & 14 & \vdots & 11 \end{bmatrix}$$

$$D = \begin{vmatrix} 3 & 6 \\ 6 & 14 \end{vmatrix} = 42 - 36 = 6$$

$$x = \frac{Dx}{D} = \frac{\begin{vmatrix} 5 & 6 \\ 11 & 14 \end{vmatrix}}{6} = \frac{70 - 66}{6} = \frac{4}{6} = \frac{2}{3}$$

$$y = \frac{Dy}{D} = \frac{\begin{vmatrix} 3 & 5 \\ 6 & 11 \end{vmatrix}}{6} = \frac{33 - 30}{6} = \frac{3}{6} = \frac{1}{2}$$

$$\left(\frac{2}{3}, \frac{1}{2} \right)$$

83. Use Cramer's Rule to solve the following system.
(If it is not possible, state the reason.)

$$3a + 3b + 4c = 1$$
$$3a + 5b + 9c = 2$$
$$5a + 9b + 17c = 4$$

Solution

$$\begin{bmatrix} 3 & 3 & 4 & \vdots & 1 \\ 3 & 5 & 9 & \vdots & 2 \\ 5 & 9 & 14 & \vdots & 4 \end{bmatrix}$$

$$D = \begin{vmatrix} 3 & 3 & 4 \\ 3 & 5 & 9 \\ 5 & 9 & 17 \end{vmatrix} = (3)\begin{vmatrix} 5 & 9 \\ 9 & 17 \end{vmatrix} - (3)\begin{vmatrix} 3 & 4 \\ 9 & 17 \end{vmatrix} + (5)\begin{vmatrix} 3 & 4 \\ 5 & 9 \end{vmatrix}$$

$$= (3)(4) - (3)(15) + (5)(7)$$
$$= 12 - 45 + 35$$
$$= 2$$

$$a = \dfrac{\begin{vmatrix} 1 & 3 & 4 \\ 2 & 5 & 9 \\ 4 & 9 & 17 \end{vmatrix}}{2} = \dfrac{(1)\begin{vmatrix} 5 & 9 \\ 9 & 17 \end{vmatrix} - (2)\begin{vmatrix} 3 & 4 \\ 9 & 17 \end{vmatrix} + (4)\begin{vmatrix} 3 & 4 \\ 5 & 9 \end{vmatrix}}{2}$$

$$= \dfrac{(1)(4) - (2)(15) + (4)(7)}{2}$$

$$= \dfrac{4 - 30 + 28}{2} = \dfrac{2}{2} = 1$$

$$b = \dfrac{\begin{vmatrix} 3 & 1 & 4 \\ 3 & 2 & 9 \\ 5 & 4 & 17 \end{vmatrix}}{2} = \dfrac{(3)\begin{vmatrix} 2 & 9 \\ 4 & 17 \end{vmatrix} - (2)\begin{vmatrix} 3 & 9 \\ 5 & 17 \end{vmatrix} + (4)\begin{vmatrix} 3 & 2 \\ 5 & 4 \end{vmatrix}}{2}$$

$$= \dfrac{(3)(-2) - (1)(6) + (4)(2)}{2}$$

$$= \dfrac{-6 - 6 + 8}{2} = \dfrac{-4}{2} = -2$$

$$c = \dfrac{\begin{vmatrix} 3 & 3 & 1 \\ 3 & 5 & 2 \\ 5 & 9 & 4 \end{vmatrix}}{2} = \dfrac{(1)\begin{vmatrix} 3 & 5 \\ 5 & 9 \end{vmatrix} - (2)\begin{vmatrix} 3 & 3 \\ 5 & 9 \end{vmatrix} + (4)\begin{vmatrix} 3 & 3 \\ 3 & 5 \end{vmatrix}}{2}$$

$$= \dfrac{(1)(2) - (2)(12) + (4)(6)}{2}$$

$$= \dfrac{2 - 24 + 24}{2} = \dfrac{2}{2} = 1$$

$$(1, -2, 1)$$

85. Use Cramer's Rule to solve the system. (If it is not possible, state the reason.)

$$5x - 3y + 2z = 2$$
$$2x + 2y - 3z = 3$$
$$x - 7y + 8z = -4$$

Solution

$$\begin{bmatrix} 5 & -3 & 2 & \vdots & 2 \\ 2 & 2 & -3 & \vdots & 3 \\ 1 & -7 & 8 & \vdots & -4 \end{bmatrix}$$

$$D = \begin{vmatrix} 5 & -3 & 2 \\ 2 & 2 & -3 \\ 1 & 7 & 8 \end{vmatrix} = (5)\begin{vmatrix} 2 & -3 \\ -7 & 8 \end{vmatrix} - (2)\begin{vmatrix} -3 & 2 \\ -7 & 8 \end{vmatrix} + (1)\begin{vmatrix} -3 & 2 \\ 2 & -3 \end{vmatrix}$$

$$= (5)(-5) - (2)(-10) + (1)(5)$$

$$= -25 + 20 + 5$$

$$= 0$$

Cannot be solved by Cramer's Rule because $D = 0$.

87. Use a graphing utility and Cramer's Rule to solve the following system of equations.

$$4x - y = -2$$
$$-2x + y = 3$$

Solution

$$\begin{bmatrix} 4 & -1 & \vdots & -2 \\ -2 & 1 & \vdots & 3 \end{bmatrix}$$

$$D = \begin{vmatrix} 4 & -1 \\ -2 & 1 \end{vmatrix} = 2$$

$$x = \frac{Dx}{D} = \frac{\begin{vmatrix} -2 & -1 \\ 3 & 1 \end{vmatrix}}{2} = \frac{1}{2}$$

$$y = \frac{Dy}{D} = \frac{\begin{vmatrix} 4 & -2 \\ -2 & 3 \end{vmatrix}}{2} = \frac{8}{2} = 4$$

$$\left(\frac{1}{2}, 4\right)$$

Keystrokes:

det D

| MATRIX | | EDIT 3 |

Enter each number in matrix followed by Enter.

| MATRIX | | MATH 1 | | MATRIX 3 | | ENTER |

det Dy

| MATRIX | | EDIT 3 |

Enter each number in matrix followed by Enter.

| MATRIX | | MATH 1 | | MATRIX 3 | | ENTER |

det Dx

| MATRIX | | EDIT 3 |

Enter each number in matrix followed by Enter.

| MATRIX | | MATH 1 | | MATRIX 3 | | ENTER |

89. Use Cramer's Rule to solve the following system. (If it is not possible, state the reason.)

$$3x + y + z = 6$$
$$x - 4y + 2z = -1$$
$$x - 3y + z = 0$$

Solution

$$\begin{bmatrix} 3 & 1 & 1 & \vdots & 6 \\ 1 & -4 & 2 & \vdots & -1 \\ 1 & -3 & 1 & \vdots & 0 \end{bmatrix}$$

$$D = \begin{vmatrix} 3 & 1 & 1 \\ 1 & -4 & 2 \\ 1 & -3 & 1 \end{vmatrix} = 8$$

$$x = \frac{Dx}{D} = \frac{\begin{vmatrix} 6 & 1 & 1 \\ -1 & -4 & 2 \\ 0 & -3 & 1 \end{vmatrix}}{8} = \frac{6}{8} = 2$$

$$y = \frac{Dy}{D} = \frac{\begin{vmatrix} 3 & 6 & 1 \\ 1 & -1 & 2 \\ 1 & 0 & 1 \end{vmatrix}}{8} = \frac{4}{8} = \frac{1}{2}$$

$$z = \frac{Dz}{D} = \frac{\begin{vmatrix} 3 & 1 & 6 \\ 1 & -4 & -1 \\ 1 & -3 & 0 \end{vmatrix}}{8} = \frac{-4}{8} = \frac{-1}{2}$$

$$\left(2, \frac{1}{2}, \frac{-1}{2}\right)$$

91. *Geometry*   Use a determinant to find the area of the triangle with the following vertices.

$$(0, 0),\ (3, 1),\ (1, 5)$$

Solution

$$(x_1, y_1) = (0, 0),\ (x_2, y_2) = (3, 1),\ (x_3, y_3) = (1, 5)$$

$$\begin{vmatrix} x_1 & y_1 & 1 \\ x_2 & y_2 & 1 \\ x_3 & y_3 & 1 \end{vmatrix} = \begin{vmatrix} 0 & 0 & 1 \\ 3 & 1 & 1 \\ 1 & 5 & 1 \end{vmatrix} = (1)\begin{vmatrix} 3 & 1 \\ 1 & 5 \end{vmatrix}$$

$$= (1)(14) = 14$$

$$\text{Area} = +\frac{1}{2}(14) = 7$$

93. *Geometry*   Use a determinant to find the area of the triangle with the following vertices.

$$(-2, 1),\ (3, -1),\ (1, 6)$$

Solution

$$(x_1, y_1) = (-2, 1),\ (x_2, y_2) = (3, -1),\ (x_3, y_3) = (1, 6)$$

$$\begin{vmatrix} x_1 & y_1 & 1 \\ x_2 & y_2 & 1 \\ x_3 & y_3 & 1 \end{vmatrix} = \begin{vmatrix} -2 & 1 & 1 \\ 3 & -1 & 1 \\ 1 & 6 & 1 \end{vmatrix} = (1)\begin{vmatrix} 3 & -1 \\ 1 & 6 \end{vmatrix} - (1)\begin{vmatrix} -2 & 1 \\ 1 & 6 \end{vmatrix} + (1)\begin{vmatrix} -2 & 1 \\ 3 & -1 \end{vmatrix}$$

$$= (1)(19) - (1)(-13) + (1)(-1)$$

$$= 19 + 13 - 1$$

$$= 31$$

$$\text{Area} = +\frac{1}{2}(31) = \frac{31}{2} \text{ or } 15\frac{1}{2}$$

95. *Geometry*   Find the area of the shaded region of the figure (see textbook).

Solution

Verbal Model: | Area of Shaded Region | = | Area of Rectangle | − | Area of Triangle |

Equation: $A = (9)(4) - 9.5$

$= 36 - 9.5$

$= 26.5$

Let $(x_1, y_1) = (-3, -1)$, $(x_2, y_2) = (2, -2)$, $(x_{3, y_3}) = (1, 2)$

$$\begin{vmatrix} x_1 & y_1 & 1 \\ x_2 & y_2 & 1 \\ x_3 & y_3 & 1 \end{vmatrix} = \begin{vmatrix} -3 & -1 & 1 \\ 2 & -2 & 1 \\ 1 & 2 & 1 \end{vmatrix} = 19$$

Area $= \dfrac{1}{2}(19) = 9.5$

97. *Collinear*   Decide whether the points are collinear.

$(-1, 11)$, $(0, 8)$, $(2, 2)$

Solution

Let $(x_1, y_1) = (-1, 11)$, $(x_2, y_2) = (0, 8)$, $(x_3, y_3) = (2, 2)$

$$\begin{vmatrix} x_1 & y_1 & 1 \\ x_2 & y_2 & 1 \\ x_3 & y_3 & 1 \end{vmatrix} = \begin{vmatrix} -1 & 11 & 1 \\ 0 & 8 & 1 \\ 2 & 2 & 1 \end{vmatrix} = (-1)\begin{vmatrix} 8 & 1 \\ 2 & 1 \end{vmatrix} + 0 + (2)\begin{vmatrix} 11 & 1 \\ 8 & 1 \end{vmatrix}$$

$$= (-1)(6) + (2)(3)$$

$$= -6 + 6$$

$$= 0$$

The three points are collinear.

99. *Collinear* Decide whether the points are collinear.

$$\left(-2, \frac{1}{3}\right), \ (2, 1), \ \left(3, \frac{1}{5}\right)$$

Solution

Let $(x_1, y_1) = \left(-2, \frac{1}{3}\right), \ (x_2, y_2) = (2, 1), \ (x_{3, y_3}) = \left(3, \frac{1}{5}\right)$

$$\begin{vmatrix} x_1 & y_1 & 1 \\ x_2 & y_2 & 1 \\ x_3 & y_3 & 1 \end{vmatrix} = \begin{vmatrix} -2 & \frac{1}{3} & 1 \\ 2 & 1 & 1 \\ 3 & \frac{1}{5} & 1 \end{vmatrix} = (1)\begin{vmatrix} 2 & 1 \\ 3 & \frac{1}{5} \end{vmatrix} - (1)\begin{vmatrix} -2 & \frac{1}{3} \\ 3 & \frac{1}{5} \end{vmatrix} + (1)\begin{vmatrix} -2 & \frac{1}{3} \\ 2 & 1 \end{vmatrix}$$

$$= (1)\left(-\frac{13}{5}\right) - (1)\left(-\frac{7}{5}\right) + (1)\left(-\frac{8}{3}\right)$$

$$= -\frac{13}{5} + \frac{7}{5} - \frac{8}{3}$$

$$= -\frac{18}{15} - \frac{40}{15}$$

$$= -\frac{58}{15}$$

The three points are not collinear.

101. *Equation of a Line* Find an equation of the line through the points $(0, 0)$ and $(5, 3)$.

Solution

$(x_1, y_1) = (0, 0), \ (x_2, y_2) = (5, 3)$

$$\begin{vmatrix} x & y & 1 \\ 0 & 0 & 1 \\ 5 & 3 & 1 \end{vmatrix} = 0$$

$$(1)\begin{vmatrix} x & y \\ 5 & 3 \end{vmatrix} = 0$$

$$(1)(3x - 5y) = 0$$

$$3x - 5y = 0$$

103. *Equation of a Line* Find an equation of the line through the points $(10, 7)$ and $(-2. -7)$.

Solution

$(x_1, y_1) = (10, 7), \ (x_2, y_2) = (-2. -7)$

$$\begin{vmatrix} x & y & 1 \\ 10 & 7 & 1 \\ -2 & -7 & 1 \end{vmatrix} = 0$$

$$(1)\begin{vmatrix} 10 & 7 \\ -2 & -7 \end{vmatrix} - (1)\begin{vmatrix} x & y \\ -2 & -7 \end{vmatrix} + (1)\begin{vmatrix} x & y \\ 10 & 7 \end{vmatrix} = 0$$

$$(1)(-56) - (-7x + 2y) + (1)(7x - 10y) = 0$$

$$-56 + 7x - 2y + 7x - 10y = 0$$

$$14x - 12y - 56 = 0$$

105. *Curve Fitting*   Use Cramer's Rule to find the equation of the parabola that passes through the points $(0, 1)$, $(1, -3)$, and $(-2, 21)$.

Solution

$$1 = a(0)^2 + b(0) + c \implies 1 = + c$$
$$-3 = a(1)^2 + b(1) + c \implies -3 = a + b + c$$
$$21 = a(-2)^2 + b(-2) + c \implies 21 = 4a + 2b + c$$

$$\begin{bmatrix} 0 & 0 & 1 & \vdots & 1 \\ 1 & 1 & 1 & \vdots & -3 \\ 4 & -2 & 1 & \vdots & 21 \end{bmatrix}$$

$$D = \begin{vmatrix} 0 & 0 & 1 \\ 1 & 1 & 1 \\ 4 & -2 & 1 \end{vmatrix} = (1)\begin{vmatrix} 1 & 1 \\ 4 & -2 \end{vmatrix} = (1)(-6) = -6$$

$$a = \frac{\begin{vmatrix} 1 & 0 & 1 \\ -3 & 1 & 1 \\ 21 & -2 & 1 \end{vmatrix}}{-6} = \frac{(1)\begin{vmatrix} 1 & 1 \\ -2 & 1 \end{vmatrix} - 0 + (1)\begin{vmatrix} -3 & 1 \\ 21 & -2 \end{vmatrix}}{-6}$$

$$= \frac{(1)(3) + (1)(-15)}{-6} = \frac{-12}{-6} = 2$$

$$b = \frac{\begin{vmatrix} 0 & 1 & 1 \\ 1 & -3 & 1 \\ 4 & 21 & 1 \end{vmatrix}}{-6} = \frac{-(1)\begin{vmatrix} 1 & 1 \\ 4 & 1 \end{vmatrix} + (1)\begin{vmatrix} 1 & -3 \\ 4 & 21 \end{vmatrix}}{-6}$$

$$= \frac{(-1)(3) + (1)(33)}{-6} = \frac{36}{-6} = -6$$

$$c = \frac{\begin{vmatrix} 0 & 0 & 1 \\ 1 & 1 & -3 \\ 4 & -2 & 21 \end{vmatrix}}{-6} = \frac{(1)\begin{vmatrix} 1 & 1 \\ 4 & -2 \end{vmatrix}}{-6} = \frac{(1)(-6)}{-6} = 1$$

$$y = 2x^2 - 6x + 1$$

107. *Curve Fitting*   Use Cramer's Rule to find the equation of the parabola that passes through the points $(1, -1)$, $(-1, -5)$, and $\left(\frac{1}{2}, \frac{1}{4}\right)$.

Solution

$$-1 = a(1)^2 + b(1) + c \implies -1 = a + b + c$$

$$-5 = a(-1)^2 + b(-1) + c \implies -5 = a - b + c$$

$$\frac{1}{4} = a\left(\frac{1}{2}\right)^2 + b\left(\frac{1}{2}\right) + c \implies \frac{1}{4} = \frac{1}{4}a + \frac{1}{2}b + c \quad \text{or} \quad 1 = a + 2b + 4c$$

$$\begin{bmatrix} 1 & 1 & 1 & \vdots & -1 \\ 1 & -1 & 1 & \vdots & -5 \\ 1 & 2 & 4 & \vdots & 1 \end{bmatrix}$$

$$D = \begin{vmatrix} 1 & 1 & 1 \\ 1 & -1 & 1 \\ 1 & 2 & 4 \end{vmatrix} = (1)\begin{vmatrix} -1 & 1 \\ 2 & 4 \end{vmatrix} - (1)\begin{vmatrix} 1 & 1 \\ 2 & 4 \end{vmatrix} + (1)\begin{vmatrix} 1 & 1 \\ -1 & 1 \end{vmatrix}$$

$$= (1)(-6) - (1)(2) + (1)(2)$$

$$= -6 - 2 + 2$$

$$= -6$$

$$a = \dfrac{\begin{vmatrix} -1 & 1 & 1 \\ -5 & -1 & 1 \\ 1 & 2 & 4 \end{vmatrix}}{-6} = \dfrac{(-1)\begin{vmatrix} -1 & 1 \\ 2 & 4 \end{vmatrix} - (1)\begin{vmatrix} -5 & 1 \\ 1 & 4 \end{vmatrix} + (1)\begin{vmatrix} -5 & -1 \\ 1 & 2 \end{vmatrix}}{-6}$$

$$= \dfrac{(-1)(-6) - (1)(-21) + (1)(-9)}{-6}$$

$$= \dfrac{6 + 21 - 9}{-6} = \dfrac{18}{-6} = -3$$

$$b = \dfrac{\begin{vmatrix} 1 & -1 & 1 \\ 1 & -5 & 1 \\ 1 & 1 & 4 \end{vmatrix}}{-6} = \dfrac{(1)\begin{vmatrix} -5 & 1 \\ 1 & 4 \end{vmatrix} - (1)\begin{vmatrix} -1 & 1 \\ 1 & 4 \end{vmatrix} + (1)\begin{vmatrix} -1 & 1 \\ -5 & 1 \end{vmatrix}}{-6}$$

$$= \dfrac{(1)(-21) - (1)(-5) + (1)(4)}{-6}$$

$$= \dfrac{-21 + 5 + 4}{-6} = \dfrac{-12}{-6} = 2$$

$$c = \dfrac{\begin{vmatrix} 1 & 1 & -5 \\ 1 & -1 & -5 \\ 1 & 2 & 1 \end{vmatrix}}{-6} = \dfrac{(1)\begin{vmatrix} -1 & -5 \\ 2 & 1 \end{vmatrix} - (1)\begin{vmatrix} 1 & -1 \\ 2 & 1 \end{vmatrix} + (1)\begin{vmatrix} 1 & -1 \\ -1 & -5 \end{vmatrix}}{-6}$$

$$= \dfrac{(1)(9) - (1)(3) + (1)(-5)}{-6} = \dfrac{9 - 3 - 6}{-6} = 0$$

$$y = -3x^2 + 2x$$

109. Solve the equation.

$$\begin{vmatrix} 5 - x & 4 \\ 1 & 2 - x \end{vmatrix} = 0$$

Solution

$$(5 - x)(2 - x) - 4 = 0$$
$$10 - 7x + x^2 - 4 = 0$$
$$x^2 - 7x + 6 = 0$$
$$(x - 6)(x - 1) = 0$$
$$x = 6 \qquad x = 1$$

Review Exercises for Chapter 8

1. Use substitution to solve the system.

$$2x + 3y = 1$$
$$x + 4y = -2$$

Solution

$$x = -2 - 4y$$
$$2(-2 - 4y) + 3y = 1$$
$$-4 - 8y + 3y = 1$$
$$-5y = 5$$
$$y = -1$$
$$x = -2 - 4(-1)$$
$$x = 2$$

$$(2, -1)$$

3. Use substitution to solve the system.

$$-5x + 2y = 4$$
$$10x - 4y = 7$$

Solution

$$2y = 5x + 4$$
$$y = \frac{5x + 4}{2}$$
$$10x - 4\left(\frac{5x + 4}{2}\right) = 7$$
$$10x - 2(5x + 4) = 7$$
$$10x - 10x - 8 = 7$$
$$-8 \neq 7$$

no solution

5. Use substitution to solve the system.

$3x - 7y = 5$
$5x - 9y = -5$

Solution

$3x = 7y + 5$

$x = \dfrac{7y + 5}{3}$

$5\left(\dfrac{7y + 5}{3}\right) - 9y = -5$

$5(7y + 5) - 27y = -15$

$35y + 25 - 27y = -15$

$8y = -40$

$y = -5$

$x = \dfrac{7(-5) + 5}{3}$

$x = -10$

$(-10, -5)$

7. Use substitution to solve the system.

$y = 5x^2$
$y = -15x - 10$

Solution

$y = 5x^2$ $5x^2 = -15x - 10$
$y = -15x - 10$ $5x^2 + 15x + 10 = 0$
                    $x^2 + 3x + 2 = 0$
                    $(x + 2)(x + 1) = 0$
$x = -2$ $x = -1$
$y = 5(-2)^2$ $y = 5(-1)^2$
$= 20$ $= 5$

$(-2, 20), (-1, 5)$

9. Use substitution to solve the system.

$x^2 + y^2 = 1$
$x + y = -1$

Solution

$x^2 + y^2 = 1$ $y = -1 - x$
$x + y = -1$ $x^2 + (-1 - x)^2 = 1$
                         $x^2 + 1 + 2x + x^2 = 1$
                         $2x^2 + 2x = 0$
                         $2x(x + 1) = 0$
$x = 0$ $x = -1$
$y = -1$ $y = 0$

$(0, -1), (-1, 0)$

11. Solve the system graphically.

$2x - y = 0$
$-x + y = 4$

Solution

$2x - y = 0$
$-x + y = 4$
$\overline{x = 4}$
$-4 + y = 4$
$y = 8$

$(4, 8)$

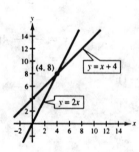

13. Solve the system graphically.

$\dfrac{x^2}{16} + \dfrac{y^2}{4} = 1$

$y = x + 2$

Solution

$(0, 2), \left(-\dfrac{16}{5}, \dfrac{-6}{5}\right)$

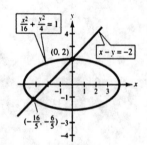

15. Solve the system graphically.

$\dfrac{x^2}{25} + \dfrac{y^2}{9} = 1$

$\dfrac{x^2}{25} - \dfrac{y^2}{9} = 1$

Solution

$(5, 0), (-5, 0)$

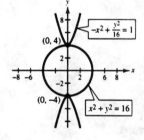

17. Use a graphing utility to solve the system.

$$5x - 3y = 3$$
$$2x + 2y = 14$$

Solution

Solve each equation for y.

$5x - 3y = 3$ $2x + 2y = 14$

$\quad -3y = -5x + 3$ $\quad 2y = -2x + 14$

$\quad\quad y = \dfrac{5}{3}x - 1$ $\quad\quad y = -1x + 7$

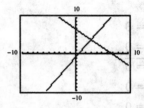

Keystrokes:

y_1 [Y =] [(] [5] [÷] [3] [)] [X, T, θ] [−] 1 [ENTER]

y_2 [Y =] [(−)] [X, T, θ] [+] 7 [GRAPH]

Solution is $(3, 4)$

19. Use a graphing utility to solve the system.

$$x^2 + y^2 = 25$$
$$y^2 - x^2 = 7$$

Solution

Solve each equation for y.

$x^2 + y^2 = 25$ $y^2 - x^2 = 7$

$\quad y^2 = 25 - x^2$ $\quad y^2 = 7 + x^2$

$\quad y = \pm\sqrt{25 - x^2}$ $\quad y = \pm\sqrt{7 + x^2}$

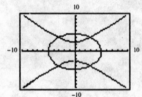

Keystrokes:

y_1 [Y =] [√‾] [(] 25 [−] [X, T, θ] [x^2] [)] [ENTER]

y_2 [Y =] [(−)] y_1 [ENTER]

y_3 [Y =] [√‾] [(] 7 [+] [X, T, θ] [x^2] [)] [ENTER]

y_4 [Y =] [(−)] y_3 [GRAPH]

Solutions: $(3, 4)$, $(3, -4)$, $(-3, 4)$, $(-3, -4)$

21. Use elimination to solve the system.

$$x + y = 0$$
$$2x + y = 0$$

Solution

$$x + y = 0 \implies \quad x + y = 0$$
$$2x + y = 0 \implies \underline{-2x - y = 0}$$
$$ \quad -x = 0$$
$$ \quad x = 0$$
$$ \quad 0 + y = 0$$
$$ \quad y = 0$$

$(0, 0)$

23. Use elimination to solve the system.

$$-x + y + 2z = 1$$
$$2x + 3y + z = -2$$
$$5x + 4y + 2z = 4$$

Solution

$$x - y - 2z = -1$$
$$2x + 3y + z = -2$$
$$5x + 4y + 2z = 4$$

$$x - y - 2z = -1$$
$$5y + 5z = 0$$
$$9y + 12z = 9$$

$$x - y - 2z = -1$$
$$y + z = 0$$
$$9y + 12z = 9$$

$$x - y - 2z = -1$$
$$y + z = 0$$
$$3z = 9$$

$$x - y - 2z = -1$$
$$y + z = 0$$
$$z = 3$$
$$y + 3 = 0$$
$$y = -3$$
$$x - (-3) - 2(3) = -1$$
$$x + 3 - 6 = -1$$
$$x = 2$$

$(2, -3, 3)$

25. Use matrices and elementary row operations to solve the system.

$$5x + 4y = 2$$
$$-x + y = -22$$

Solution

$$\begin{bmatrix} 5 & 4 & \vdots & 2 \\ -1 & 1 & \vdots & -22 \end{bmatrix}$$

$$R_1 \leftrightarrow R_2 \begin{bmatrix} -1 & 1 & \vdots & -22 \\ 5 & 4 & \vdots & 2 \end{bmatrix}$$

$$-R_1 \begin{bmatrix} 1 & -1 & \vdots & 22 \\ 5 & 4 & \vdots & 2 \end{bmatrix}$$

$$-5R_1 + R_2 \begin{bmatrix} 1 & -1 & \vdots & 22 \\ 0 & 9 & \vdots & -108 \end{bmatrix}$$

$$\tfrac{1}{9}R_2 \begin{bmatrix} 1 & -1 & \vdots & 22 \\ 0 & 1 & \vdots & -12 \end{bmatrix}$$

$$y = -12 \quad x - (-12) = 22 \quad (10, -12)$$
$$x = 10$$

27. Use matrices and elementary row operations to solve the system.

$$\begin{aligned} x + 2y + 6z &= 4 \\ -3x + 2y - z &= -4 \\ 4x + 2z &= 16 \end{aligned}$$

Solution

$$\begin{bmatrix} 1 & 2 & 6 & \vdots & 4 \\ -3 & 2 & -1 & \vdots & -4 \\ 4 & 0 & 2 & \vdots & 16 \end{bmatrix}$$

$$\begin{matrix} \\ 3R_1 + R_2 \\ -4R_1 + R_3 \end{matrix} \begin{bmatrix} 1 & 2 & 6 & \vdots & 4 \\ 0 & 8 & 17 & \vdots & 8 \\ 0 & -8 & -22 & \vdots & 0 \end{bmatrix}$$

$$\tfrac{1}{8}R_2 \begin{bmatrix} 1 & 2 & 6 & \vdots & 4 \\ 0 & 1 & \tfrac{17}{8} & \vdots & 1 \\ 0 & -8 & -22 & \vdots & 0 \end{bmatrix}$$

$$8R_2 + R_3 \begin{bmatrix} 1 & 2 & 6 & \vdots & 4 \\ 0 & 1 & \tfrac{17}{8} & \vdots & 1 \\ 0 & 0 & -5 & \vdots & 8 \end{bmatrix}$$

$$-\tfrac{1}{5}R_3 \begin{bmatrix} 1 & 2 & 6 & \vdots & 4 \\ 0 & 1 & \tfrac{17}{8} & \vdots & 1 \\ 0 & 0 & 1 & \vdots & -\tfrac{8}{5} \end{bmatrix}$$

$$z = -\frac{8}{5} \qquad y + \frac{17}{8}\left(-\frac{8}{5}\right) = 1 \qquad x + 2\left(\frac{22}{5}\right) + 6\left(-\frac{8}{5}\right) = 4$$

$$y = \frac{22}{5} \qquad\qquad x = \frac{24}{5}$$

$$\left(\frac{24}{5}, \frac{22}{5}, -\frac{8}{5}\right)$$

29. Find the determinant of the matrix.

$$\begin{bmatrix} 7 & 10 \\ 10 & 15 \end{bmatrix}$$

Solution

$$\det(A) = \begin{vmatrix} 7 & 10 \\ 10 & 15 \end{vmatrix} = (7)(15) - (10)(10) = 105 - 100 = 5$$

31. Evaluate the determinant of the matrix six different ways by expanding by minors along each row and column.

$$\begin{bmatrix} 8 & 6 & 3 \\ 6 & 3 & 0 \\ 3 & 0 & 2 \end{bmatrix}$$

Solution

$$\det (A) = \begin{vmatrix} 8 & 6 & 3 \\ 6 & 3 & 0 \\ 3 & 0 & 2 \end{vmatrix}$$

(a) first row $= 8\begin{vmatrix} 3 & 0 \\ 0 & 2 \end{vmatrix} - 6\begin{vmatrix} 6 & 0 \\ 3 & 2 \end{vmatrix} + 3\begin{vmatrix} 6 & 3 \\ 3 & 0 \end{vmatrix}$

$= (8)(6) - (6)(12) + (3)(-9)$

$= 48 - 72 - 27$

$= -51$

(b) second row $= -(6)\begin{vmatrix} 6 & 3 \\ 0 & 2 \end{vmatrix} + 3\begin{vmatrix} 8 & 3 \\ 3 & 2 \end{vmatrix} - 0\begin{vmatrix} 8 & 6 \\ 3 & 0 \end{vmatrix}$

$= -(6)(12) + 3(7) - 0$

$= -72 + 21$

$= -51$

(c) third row $= (3)\begin{vmatrix} 6 & 3 \\ 3 & 0 \end{vmatrix} - 0\begin{vmatrix} 8 & 3 \\ 6 & 0 \end{vmatrix} + 2\begin{vmatrix} 8 & 6 \\ 6 & 3 \end{vmatrix}$

$= (3)(-9) - 0 + (2)(-12)$

$= -27 - 24$

$= -51$

(d) first column $= 8\begin{vmatrix} 3 & 0 \\ 0 & 2 \end{vmatrix} - 6\begin{vmatrix} 6 & 3 \\ 0 & 2 \end{vmatrix} + 3\begin{vmatrix} 6 & 3 \\ 3 & 0 \end{vmatrix}$

$= (8)(6) - (6)(12) + 3(-9)$

$= 48 - 72 - 27$

$= -51$

(e) Second column = same as second row

(f) third column = same as third row

33. Find the determinant of the matrix.

$$\begin{bmatrix} 8 & 3 & 2 \\ 1 & -2 & 4 \\ 6 & 0 & 5 \end{bmatrix}$$

Solution

$$\det (A) = \begin{bmatrix} 8 & 3 & 2 \\ 1 & -2 & 4 \\ 6 & 0 & 5 \end{bmatrix}$$

$$\text{third row} = 6\begin{vmatrix} 3 & 2 \\ -2 & 4 \end{vmatrix} - 0 + 5\begin{vmatrix} 8 & 3 \\ 1 & -2 \end{vmatrix}$$

$$= (6)(16) + (5)(-19)$$

$$= 1$$

35. Solve the system of linear equations by using Cramer's Rule. (If it is not possible, state the reason.)

$$7x + 12y = 63$$
$$2x + 3y = 15$$

Solution

$$\begin{bmatrix} 7 & 12 & \vdots & 63 \\ 2 & 3 & \vdots & 15 \end{bmatrix}$$

$$D = \begin{vmatrix} 7 & 12 \\ 2 & 3 \end{vmatrix} = 21 - 24 = -3$$

$$x = \frac{Dx}{D} = \frac{\begin{vmatrix} 63 & 12 \\ 15 & 3 \end{vmatrix}}{-3} = \frac{189 - 180}{-3} = \frac{9}{-3} = -3$$

$$y = \frac{Dy}{D} = \frac{\begin{vmatrix} 7 & 63 \\ 2 & 15 \end{vmatrix}}{-3} = \frac{105 - 126}{-3} = \frac{-21}{-3} = 7$$

$$(-3, 7)$$

37. Solve the system of linear equations by using Cramer's Rule. (If it is not possible, state the reason.)

$$3x - 2y = 16$$
$$12x - 8y = -5$$

Solution

$$\begin{bmatrix} 3 & -2 & \vdots & 16 \\ 12 & -8 & \vdots & -5 \end{bmatrix}$$

$$D = \begin{vmatrix} 3 & -2 \\ 12 & -8 \end{vmatrix} = -24 + 24 = 0$$

cannot be solved by Cramer's Rule because $D = 0$
by elimination

$$-12x + 8y = -64$$
$$\underline{12x - 8y = -5}$$
$$0 \neq -69$$

inconsistent; no solution

39. Solve the system of linear equations by using Cramer's Rule. (If it is not possible, state the reason.)

$$-x + y + 2z = 1$$
$$2x + 3y + z = -2$$
$$5x + 4y + 2z = 4$$

Solution

$$\begin{bmatrix} -1 & 1 & 2 & \vdots & 1 \\ 2 & 3 & 1 & \vdots & -2 \\ 5 & 4 & 2 & \vdots & 4 \end{bmatrix}$$

$$D = \begin{vmatrix} -1 & 1 & 2 \\ 2 & 3 & 1 \\ 5 & 4 & 2 \end{vmatrix} = (-1)\begin{vmatrix} 3 & 1 \\ 4 & 2 \end{vmatrix} - (1)\begin{vmatrix} 2 & 1 \\ 5 & 2 \end{vmatrix} + (2)\begin{vmatrix} 2 & 3 \\ 5 & 4 \end{vmatrix}$$

$$= (-1)(2) - (1)(-1) + (2)(-7)$$
$$= -2 + 1 - 14 = -15$$

$$x = \frac{\begin{vmatrix} 1 & 1 & 2 \\ -2 & 3 & 1 \\ 4 & 4 & 2 \end{vmatrix}}{-15} = \frac{(1)\begin{vmatrix} 3 & 1 \\ 4 & 2 \end{vmatrix} - (1)\begin{vmatrix} -2 & 1 \\ 4 & 2 \end{vmatrix} + (2)\begin{vmatrix} -2 & 3 \\ 4 & 4 \end{vmatrix}}{-15}$$

$$= \frac{(1)(2) - (1)(-8) + (2)(-20)}{-15}$$

$$= \frac{2 + 8 - 40}{-15} = \frac{-30}{-15} = 2$$

$$y = \frac{\begin{vmatrix} -1 & 1 & 2 \\ 2 & -2 & 1 \\ 5 & 4 & 2 \end{vmatrix}}{-15} = \frac{(-1)\begin{vmatrix} -2 & 1 \\ 4 & 2 \end{vmatrix} - (1)\begin{vmatrix} 2 & 1 \\ 5 & 2 \end{vmatrix} + (2)\begin{vmatrix} 2 & -2 \\ 5 & 4 \end{vmatrix}}{-15}$$

$$= \frac{(-1)(-8) - (1)(-1) + (2)(18)}{-15}$$

$$= \frac{8 + 1 + 36}{-15} = \frac{45}{-15} = -3$$

$$z = \frac{\begin{vmatrix} -1 & 1 & 1 \\ 2 & 2 & -2 \\ 5 & 4 & 4 \end{vmatrix}}{-15} = \frac{(-1)\begin{vmatrix} 3 & -2 \\ 4 & 4 \end{vmatrix} - (1)\begin{vmatrix} 2 & -2 \\ 5 & 4 \end{vmatrix} + (1)\begin{vmatrix} 2 & 3 \\ 5 & 4 \end{vmatrix}}{-15}$$

$$= \frac{(-1)(20) - (1)(18) + (1)(-7)}{-15}$$

$$= \frac{-20 - 18 - 7}{-15} = \frac{-45}{-15} = 3 \qquad (2, -3, 3)$$

41. Use a determinant to find the equation of the line through the points $(-4, 0)$ and $(4, 4)$.

Solution

$$\begin{vmatrix} x & y & 1 \\ -4 & 0 & 1 \\ 4 & 4 & 1 \end{vmatrix} = 0$$

$$-(-4)\begin{vmatrix} y & 1 \\ 4 & 1 \end{vmatrix} + 0 - (1)\begin{vmatrix} x & y \\ 4 & 4 \end{vmatrix} = 0$$

$$(4)(y - 4) - (1)(4x - 4y) = 0$$

$$4y - 16 - 4x + 4y = 0$$

$$-4x + 8y - 16 = 0$$

43. Use a determinant to find the equation of the line through the points $\left(-\dfrac{5}{2}, 3\right)$ and $\left(\dfrac{7}{2}, 1\right)$.

Solution

$$\begin{vmatrix} x & y & 1 \\ -\frac{5}{2} & 3 & 1 \\ \frac{7}{2} & 1 & 1 \end{vmatrix} = 0$$

$$(1)\begin{vmatrix} -\frac{5}{2} & 3 \\ \frac{7}{2} & 1 \end{vmatrix} - (1)\begin{vmatrix} x & y \\ \frac{7}{2} & 1 \end{vmatrix} + (1)\begin{vmatrix} x & y \\ -\frac{5}{2} & 3 \end{vmatrix} = 0$$

$$(1)(-13) - (1)\left(x - \frac{7}{2}y\right) + (1)\left(3x + \frac{5}{2}y\right) = 0$$

$$-13 - x + \frac{7}{2}y + 3x + \frac{5}{2}y = 0$$

$$2x + 6y - 13 = 0$$

45. Use a determinant to find the area of the triangle with the vertices $(1, 0)$, $(5, 0)$, and $(5, 8)$.

Solution

$(x_1, y_1) = (1, 0)$, $(x_2, y_2) = (5, 0)$, $(x_3, y_3) = (5, 8)$

$$\begin{vmatrix} x_1 & y_1 & 1 \\ x_2 & y_2 & 1 \\ x_3 & y_3 & 1 \end{vmatrix} = \begin{vmatrix} 1 & 0 & 1 \\ 5 & 0 & 1 \\ 5 & 8 & 1 \end{vmatrix} = -0 + 0 - (8)\begin{vmatrix} 1 & 1 \\ 5 & 1 \end{vmatrix} = (-8)(-4) = 32$$

$$\text{Area} = +\frac{1}{2}(32) = 16$$

47. Use a determinant to find the area of the triangle with the vertices $(1, 2)$, $(4, -5)$, and $(3, 2)$.

Solution

$(x_1, y_1) = (1, 2)$, $(x_2, y_2) = (4, -5)$, $(x_3, y_3) = (3, 2)$

$$\begin{vmatrix} x_1 & y_1 & 1 \\ x_2 & y_2 & 1 \\ x_3 & y_3 & 1 \end{vmatrix} = \begin{vmatrix} 1 & 2 & 1 \\ 4 & -5 & 1 \\ 3 & 2 & 1 \end{vmatrix} = (1)\begin{vmatrix} 4 & -5 \\ 3 & 2 \end{vmatrix} - (1)\begin{vmatrix} 1 & 2 \\ 3 & 2 \end{vmatrix} + (1)\begin{vmatrix} 1 & 2 \\ 4 & -5 \end{vmatrix}$$

$$= (1)(23) - (1)(-4) + (1)(-13)$$

$$= 23 + 4 - 13$$

$$= 14$$

$$\text{Area} = +\frac{1}{2}(14) = 7$$

49. Create a system of equations having the solution $\left(\dfrac{2}{3}, -4\right)$.

Solution

$$3x - y = 6$$
$$-3x + 2y = -10$$

There are many other correct solutions.

51. *Break-Even Analysis* A small business invests \$25,000 in equipment to produce a product. Each unit of the product costs \$3.75 to produce and is sold for \$5.25. How many items must be sold before the business breaks even?

Solution

$$C = 25,000 + 3.75x$$
$$R = 5.25x$$
$$R = C$$
$$5.25x = 25,000 + 3.75x$$
$$1.50x = 25,000$$
$$x = 16,666.\overline{6} \approx 16,667 \text{ items}$$

53. *Acid Mixture* One hundred gallons of a 60% acid solution is obtained by mixing a 75% solution with a 50% solution. How many gallons of each solution must be used to obtain the desired mixture?

Solution

Verbal model:

$$\boxed{\begin{array}{c}\text{Gallons}\\\text{Solution 1}\end{array}} + \boxed{\begin{array}{c}\text{Gallons}\\\text{Solution 2}\end{array}} = \boxed{100}$$

$$\boxed{\begin{array}{c}\text{Value}\\\text{Solution 1}\end{array}} + \boxed{\begin{array}{c}\text{Value}\\\text{Solution 2}\end{array}} = \boxed{0.60(100)}$$

System of equations:

$$x + y = 100$$
$$0.75x + 0.50y = 0.60(100)$$
$$x = 100 - y$$
$$75(100 - y) + 50y = 60(100)$$
$$7500 - 75y + 50y = 6000$$
$$-25y = -1500$$
$$y = 60 \text{ gallons}$$
$$x = 100 - 60 = 40 \text{ gallons}$$

55. *Flying Speeds*   Two planes leave Pittsburgh and Philadelphia at the same time, each going
to the other city. Because of the wind, one plane flies 25 miles per hour faster than the other.
Find the ground speed of each plane if the cities are 275 miles apart and the planes pass one
another after 40 minutes.

Solution

Verbal model: $\boxed{\text{Rate}} \cdot \boxed{\text{Time}} = \boxed{\text{Distance}}$

Equation: $\dfrac{2}{3}x + \dfrac{2}{3}(x + 25) = 275$

$$\frac{2}{3}(2x + 25) = 275$$

$$2x + 25 = 412.5$$

$$2x = 387.5$$

$$x = 193.5 \text{ mph}$$

$$x + 25 = 218.5 \text{ mph}$$

57. Find an equation of the parabola passing through the points $(0, -6)$, $(1, -3)$, and $(2, 4)$.

Solution

$-6 = a(0)^2 + b(0) + c \implies -6 = \qquad c$
$-3 = a(1)^2 + b(1) + c \qquad -3 = a + b + c$
$4 = a(2)^2 + b(2) + c \qquad 4 = 4a + 2b + c$

$$\begin{bmatrix} 0 & 0 & 1 & \vdots & -6 \\ 1 & 1 & 1 & \vdots & -3 \\ 4 & 2 & 1 & \vdots & 4 \end{bmatrix}$$

$$D = \begin{vmatrix} 0 & 0 & 1 \\ 1 & 1 & 1 \\ 4 & 2 & 1 \end{vmatrix} = -2$$

$$a = \frac{D_a}{D} = \frac{\begin{vmatrix} -6 & 0 & 1 \\ -3 & 1 & 1 \\ 4 & 2 & 1 \end{vmatrix}}{-2} = \frac{-4}{-2} = 2$$

$$b = \frac{D_b}{D} = \frac{\begin{vmatrix} 0 & -6 & 1 \\ 1 & -3 & 1 \\ 4 & 4 & 1 \end{vmatrix}}{-2} = \frac{-2}{-2} = 1$$

$$y = 2x^2 + 1x - 6$$

59. Find an equation of the circle $x^2 + y^2 + Dx + Ey + F = 0$ passing through the points $(2, 2)$, $(5, -1)$, and $(-1, -1)$.

Solution

$$2^2 + 2^2 + D(2) + E(2) + F = 0 \implies 2D + 2E + F = -8$$
$$5^2 + (-1)^2 + D(5) + E(-1) + F = 0 \implies 5D - E + F = -26$$
$$(-1)^2 + (-1)^2 + D(-1) + E(-1) + F = 0 \implies -D - E + F = -2$$

$$\begin{bmatrix} 2 & 2 & 1 & \vdots & -8 \\ 5 & -1 & 1 & \vdots & -26 \\ -1 & -1 & 1 & \vdots & -2 \end{bmatrix}$$

$$D = \begin{vmatrix} 2 & 2 & 1 \\ 5 & -1 & 1 \\ -1 & -1 & 1 \end{vmatrix} = -18$$

$$D = \frac{D_D}{D} = \frac{\begin{vmatrix} -8 & 2 & 1 \\ -26 & -1 & 1 \\ -2 & -1 & 1 \end{vmatrix}}{-18} = \frac{72}{-18} = -4$$

$$E = \frac{D_E}{D} = \frac{\begin{vmatrix} 2 & -8 & 1 \\ 5 & -26 & 1 \\ -1 & -2 & 1 \end{vmatrix}}{-18} = \frac{-36}{-18} = 2$$

$$F = \frac{D_F}{D} = \frac{\begin{vmatrix} 2 & 2 & -8 \\ 5 & -1 & -26 \\ -1 & -1 & -2 \end{vmatrix}}{-18} = \frac{72}{-18} = -4$$

$$x^2 + y^2 - 4x + 2y - 4 = 0$$

61. *Mathematical Modeling* A child throws a softball over a garage. The location of the eaves and the peak of the roof are given by (0, 10), (15, 15), and (30, 10).

(a) Find the equation of the parabola for the path of the ball if the ball follows a path 1 foot over the eaves and the peak of the roof.

(b) Use a graphing utility to graph the path.

(c) How far from the edge of the garage is the child standing if the ball is at a height of 5 feet when it leaves the child's hand?

Solution

(a)

Points: (0, 11), (15, 16), (30, 11)

$$11 = a(0)^2 + b(0) + c \implies 11 = c$$
$$16 = a(15)^2 + b(15) + c \implies 16 = 225a + 15b + c$$
$$11 = a(30)^2 + b(30) + c \implies 11 = 900a + 30b + c$$

$$\det(A) = \begin{vmatrix} 0 & 0 & 1 \\ 225 & 15 & 1 \\ 900 & 30 & 1 \end{vmatrix} = -6750$$

$$a = \frac{\begin{vmatrix} 11 & 0 & 1 \\ 16 & 15 & 1 \\ 11 & 30 & 1 \end{vmatrix}}{-6750} = \frac{150}{-6750} = -\frac{1}{45} \qquad c = 11$$

$$b = \frac{\begin{vmatrix} 0 & 11 & 1 \\ 225 & 16 & 1 \\ 900 & 11 & 1 \end{vmatrix}}{-6750} = \frac{-4500}{-6750} = \frac{2}{3}$$

$$y = -\frac{1}{45}x^2 + \frac{2}{3}x + 11$$

(b) Keystrokes:

$\boxed{Y =}$ $\boxed{(}$ $\boxed{(-)}$ 1 $\boxed{\div}$ 45 $\boxed{)}$ $\boxed{X, T, \theta}$ $\boxed{x^2}$ $\boxed{+}$ $\boxed{(}$ 2 $\boxed{\div}$ 3 $\boxed{)}$ $\boxed{X, T, \theta}$ $\boxed{+}$ 11 $\boxed{\text{GRAPH}}$

(c) Trace until the y-coordinate is approximately 5 feet. The x-coordinate is approximately −7.25 so the child is $7\frac{1}{4}$ feet from the edge of the garage.

Test for Chapter 8

1. Determine whether the ordered pair (a) $(3, -4)$ or (b) $\left(1, \dfrac{1}{2}\right)$ is a solution of the following system of equation.

$$2x - 2y = 1$$
$$-x + 2y = 0$$

Solution

(a) $2(3) - 2(-4) \stackrel{?}{=} 1$
$$6 + 8 \neq 1$$

   no

(b) $2(1) - 2\left(\dfrac{1}{2}\right) \stackrel{?}{=} 1$ $-1 + 2\left(\dfrac{1}{2}\right) \stackrel{?}{=} 0$

$$2 - 1 = 1 \qquad\qquad\qquad -1 + 1 = 0$$
$$1 = 1 \qquad\qquad\qquad\qquad 0 = 0$$

                                yes

2. Solve the system by the method of substitution.

$$5x - y = 6$$
$$4x - 3y = -4$$

Solution

$5x - y = 6$ $4x - 3(5x - 6) = -4$ $y = 5(2) - 6$

$4x - 3y = -4$ $4x - 15x + 18 = -4$ $y = 4$

$\quad -y = -5x + 6$ $-11x = -22$

$\qquad y = 5x - 6$ $x = 2$

$(2, 4)$

3. Solve the system by the method of substitution.

$$x + y = 8$$
$$xy = 12$$

Solution

$x + y = 8$ $\dfrac{12}{y} + y = 8$ $x = \dfrac{12}{6} = 2$

$\quad xy = 12$ $12 + y^2 = 8y$ $x = \dfrac{12}{2} = 6$

$\qquad x = \dfrac{12}{y}$ $y^2 - 8y + 12 = 0$

                 $(y - 6)(y - 2) = 0$

                        $y = 6 \quad y = 2$

$(2, 6), \quad (6, 2)$

4. Solve the system graphically.

$$x - 2y = -1$$
$$2x + 3y = 12$$

Solution

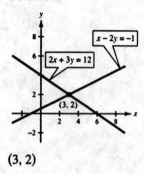

(3, 2)

5. Solve the system by the method of elimination.

$$3x - 4y = -14$$
$$-3x + y = 8$$

Solution

$$
\begin{array}{r}
3x - 4y = -14 \\
-3x + y = 8 \\
\hline
-3y = -6 \\
y = 2
\end{array}
$$

$$3x - 4(2) = -14$$
$$3x = -6$$
$$x = -2$$

$(-2, 2)$

6. Solve the system of equations by the method of elimination.

$$8x + 3y = 3$$
$$4x - 6y = -1$$

Solution

$$
\begin{array}{rcl}
8x + 3y = 3 & \Rightarrow & 16x + 6y = 6 \\
4x - 6y = -1 & \Rightarrow & 4x - 6y = -1 \\
\hline
& & 20x = 5 \\
& & x = \dfrac{1}{4}
\end{array}
$$

$$8\left(\frac{1}{4}\right) + 3y = 3$$
$$3y = 1$$
$$y = \frac{1}{3}$$

$\left(\dfrac{1}{4}, \dfrac{1}{3}\right)$

7. Use Gaussian elimination to solve the system.

$$x + 2y - 4z = 0$$
$$3x + y - 2z = 5$$
$$3x - y + 2z = 7$$

Solution

$$x + 2y - 4z = 0$$
$$3x + y - 2z = 5$$
$$3x - y + 2z = 7$$

$$x + 2y - 4z = 0$$
$$-5y + 10z = 5$$
$$-7y + 14z = 7$$

$$x + 2y - 4z = 0$$
$$y - 2z = -1$$
$$-7y + 14z = 7$$

$$x + 2y - 4z = 0$$
$$y - 2z = -1$$
$$0 = 0$$

let $a = z$ (a is any real number)

$$y = 2z - 1$$
$$x = -2y + 4z$$
$$= -2(2z - 1) + 4z$$
$$= -4z + 2 + 4z$$
$$x = 2$$

$(2, 2a - 1, a)$

8. Use matrices to solve the system.

$$x \qquad - 3z = -10$$
$$-2y + 2z = 0$$
$$x - 2y \qquad = -7$$

Solution

$$\begin{bmatrix} 1 & 0 & -3 & \vdots & -10 \\ 0 & -2 & 2 & \vdots & 0 \\ 1 & -2 & 0 & \vdots & -7 \end{bmatrix}$$

$$\begin{matrix} \\ -\frac{1}{2}R_2 \\ -R_1 + R_3 \end{matrix} \begin{bmatrix} 1 & 0 & -3 & \vdots & -10 \\ 0 & 1 & -1 & \vdots & 0 \\ 0 & -2 & 3 & \vdots & 3 \end{bmatrix}$$

$$\begin{matrix} \\ \\ 2R_2 + R_3 \end{matrix} \begin{bmatrix} 1 & 0 & -3 & \vdots & -10 \\ 0 & 1 & -1 & \vdots & 0 \\ 0 & 0 & 1 & \vdots & 3 \end{bmatrix}$$

$z = 3 \qquad y - 3 = 0 \qquad x - 3(3) = -10$
$\qquad\qquad\quad y = 3 \qquad\qquad x = -1$

$(-1, 3, 3)$

9. Use matrices to solve the system.

$$x - 3y + z = -3$$
$$3x + 2y - 5z = 18$$
$$y + z = -1$$

Solution

$$\begin{bmatrix} 1 & -3 & 1 & \vdots & -3 \\ 3 & 2 & -5 & \vdots & 18 \\ 0 & 1 & 1 & \vdots & -1 \end{bmatrix}$$

$$-3R_1 + R_2 \begin{bmatrix} 1 & -3 & 1 & \vdots & -3 \\ 0 & 11 & -8 & \vdots & 27 \\ 0 & 1 & 1 & \vdots & -1 \end{bmatrix}$$

$$R_2 \leftrightarrow R_3 \begin{bmatrix} 1 & -3 & 1 & \vdots & -3 \\ 0 & 1 & 1 & \vdots & -1 \\ 0 & 11 & -8 & \vdots & 27 \end{bmatrix}$$

$$\begin{matrix} -11R_2 + R_3 \\ 3R_2 + R_1 \end{matrix} \begin{bmatrix} 1 & 0 & 4 & \vdots & -6 \\ 0 & 1 & 1 & \vdots & -1 \\ 0 & 0 & -19 & \vdots & 38 \end{bmatrix}$$

$$-\frac{1}{19}R_3 \begin{bmatrix} 1 & 0 & 4 & \vdots & -6 \\ 0 & 1 & 1 & \vdots & -1 \\ 0 & 0 & 1 & \vdots & -2 \end{bmatrix}$$

$$z = -2 \qquad y + (-2) = -1 \qquad x + 4(-2) = -6$$
$$y = 1 \qquad x = 2$$

$$(2, 1, -2)$$

11. Use the graphical method to solve the system.

$$x - 2y = -3$$
$$2x + 3y = 22$$

Solution

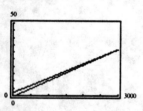

Solution: (5, 4)

10. Use Cramer's Rule to solve the system.

$$2x - 7y = 7$$
$$3x + 7y = 13$$

Solution

$$\begin{bmatrix} 2 & -7 & \vdots & 7 \\ 3 & 7 & \vdots & 13 \end{bmatrix}$$

$$D = \begin{vmatrix} 2 & -7 \\ 3 & 7 \end{vmatrix} = 14 + 21 = 35$$

$$x = \frac{Dx}{D} = \frac{\begin{vmatrix} 7 & -7 \\ 13 & 7 \end{vmatrix}}{35} = \frac{49 + 91}{35} = \frac{140}{35} = 4$$

$$y = \frac{Dy}{D} = \frac{\begin{vmatrix} 2 & 7 \\ 3 & 13 \end{vmatrix}}{35} = \frac{26 - 21}{35} = \frac{5}{35} = \frac{1}{7}$$

$$\left(4, \frac{1}{7}\right)$$

12. Use any method to solve the system.

$$3x - 2y + z = 12$$
$$x - 3y \qquad = 2$$
$$-3x \qquad - 9z = -6$$

Solution

$$\begin{bmatrix} 3 & -2 & 1 & \vdots & 12 \\ 1 & -3 & 0 & \vdots & 2 \\ -3 & 0 & -9 & \vdots & -6 \end{bmatrix}$$

$$R_1 \leftrightarrow \tfrac{-1}{3}R_3 \begin{bmatrix} 1 & 0 & 3 & \vdots & 2 \\ 1 & -3 & 0 & \vdots & 2 \\ 3 & -2 & 1 & \vdots & 12 \end{bmatrix}$$

$$\begin{matrix} -1R_1 + R_2 \\ -3R_1 + R_3 \end{matrix} \begin{bmatrix} 1 & 0 & 3 & \vdots & 2 \\ 0 & -3 & -3 & \vdots & 0 \\ 0 & -2 & -8 & \vdots & 6 \end{bmatrix}$$

$$\begin{matrix} -\tfrac{1}{3}R_2 \\ -\tfrac{1}{2}R_3 \end{matrix} \begin{bmatrix} 1 & 0 & 3 & \vdots & 2 \\ 0 & 1 & 1 & \vdots & 0 \\ 0 & 1 & 4 & \vdots & -3 \end{bmatrix}$$

$$-R_2 + R_3 \begin{bmatrix} 1 & 0 & 3 & \vdots & 2 \\ 0 & 1 & 1 & \vdots & 0 \\ 0 & 0 & 3 & \vdots & -3 \end{bmatrix}$$

$$\tfrac{1}{3}R_3 \begin{bmatrix} 1 & 0 & 3 & \vdots & 2 \\ 0 & 1 & 1 & \vdots & 0 \\ 0 & 0 & 1 & \vdots & -1 \end{bmatrix}$$

$$\begin{matrix} -3R_3 + R_1 \\ -R_3 + R_2 \end{matrix} \begin{bmatrix} 1 & 0 & 0 & \vdots & 5 \\ 0 & 1 & 0 & \vdots & 1 \\ 0 & 0 & 1 & \vdots & -1 \end{bmatrix}$$

$$(5, 1, -1)$$

13. Use any method to solve the system.

$$4x + y + 2z = -4$$
$$3y + z = 8$$
$$-3x + y - 3z = 5$$

Solution

$$\begin{bmatrix} 4 & 1 & 2 & \vdots & -4 \\ 0 & 3 & 1 & \vdots & 8 \\ -3 & 1 & -3 & \vdots & 5 \end{bmatrix}$$

$$\tfrac{1}{4}R_1 \begin{bmatrix} 1 & \tfrac{1}{4} & \tfrac{1}{2} & \vdots & -1 \\ 0 & 3 & 1 & \vdots & 8 \\ -3 & 1 & -3 & \vdots & 5 \end{bmatrix}$$

$$\begin{matrix} 3R_1 + R_3 \\ \tfrac{1}{3}R_2 \end{matrix} \begin{bmatrix} 1 & \tfrac{1}{4} & \tfrac{1}{2} & \vdots & -1 \\ 0 & 1 & \tfrac{1}{3} & \vdots & \tfrac{8}{3} \\ 0 & \tfrac{7}{4} & -\tfrac{3}{2} & \vdots & 2 \end{bmatrix}$$

$$\begin{matrix} -\tfrac{1}{4}R_2 + R_1 \\ -\tfrac{7}{4}R_2 + R_3 \end{matrix} \begin{bmatrix} 1 & 0 & \tfrac{5}{12} & \vdots & -\tfrac{5}{3} \\ 0 & 1 & \tfrac{1}{3} & \vdots & \tfrac{8}{3} \\ 0 & 0 & -\tfrac{25}{12} & \vdots & -\tfrac{8}{3} \end{bmatrix}$$

$$-\tfrac{12}{25}R_3 \begin{bmatrix} 1 & 0 & \tfrac{5}{12} & \vdots & -\tfrac{5}{3} \\ 0 & 1 & \tfrac{1}{3} & \vdots & \tfrac{8}{3} \\ 0 & 0 & 1 & \vdots & \tfrac{32}{25} \end{bmatrix}$$

$$\begin{matrix} -\tfrac{5}{12}R_3 + R_1 \\ -\tfrac{1}{3}R_3 + R_2 \end{matrix} \begin{bmatrix} 1 & 0 & 0 & \vdots & -\tfrac{11}{5} \\ 0 & 1 & 0 & \vdots & \tfrac{56}{25} \\ 0 & 0 & 1 & \vdots & \tfrac{32}{25} \end{bmatrix}$$

$$\left(-\frac{11}{5}, \frac{56}{25}, \frac{32}{25} \right)$$

14. Describe the number of possible solutions of a system of linear equations.

Solution

Inconsistent: no solutions

Consistent: one solution or infinitely many solutions

15. Evaluate the determinant of the matrix.

$$\begin{bmatrix} 3 & -2 & 0 \\ -1 & 5 & 3 \\ 2 & 7 & 1 \end{bmatrix}$$

Solution

$$\begin{vmatrix} 3 & -2 & 0 \\ -1 & 5 & 3 \\ 2 & 7 & 1 \end{vmatrix} = 0 - (3)\begin{vmatrix} 3 & -2 \\ 2 & 7 \end{vmatrix} + (1)\begin{vmatrix} 3 & -2 \\ -1 & 5 \end{vmatrix}$$

$$= (-3)(25) + (1)(13)$$

$$= -75 + 13$$

$$= -62$$

16. Find the value of a such that the system is inconsistent.

$$5x - 8y = 3$$
$$3x + ay = 0$$

Solution

$$\begin{vmatrix} 5 & -8 \\ 3 & a \end{vmatrix} = 0$$

$$5a + 24 = 0$$

$$5a = -24$$

$$a = -\frac{24}{5}$$

17. Find a system of linear equations, with integer coefficients that has the solution $(5, -3)$. (The problem has many correct answers.)

Solution

$$x + y = 2$$
$$2x - y = 13$$

There are many correct answers.

18. Two people share the driving on a 200-mile trip. One person drives four times as far as the other. Write a system of linear equations that models the problem. Find the distance each person drives.

Solution

$$x + y = 200$$
$$x = 4y$$

by substitution

$$4y + y = 200$$
$$5y = 200$$
$$y = 40 \text{ miles}$$
$$x = 4(40) = 160 \text{ miles}$$

19. Find the equation of the parabola $y = ax^2 + bx + c$ that passes through the points $(0, 4)$, $(1, 3)$, and $(2, 6)$.

Solution

$4 = a(0)^2 + b(0) + c \implies 4 = \qquad + c$

$3 = a(1)^2 + b(1) + c \implies 3 = a + b + c$

$6 = a(2)^2 + b(2) + c \implies 6 = 4a + 2b + c$

$$\begin{bmatrix} 0 & 0 & 1 & \vdots & 4 \\ 1 & 1 & 1 & \vdots & 3 \\ 4 & 2 & 1 & \vdots & 6 \end{bmatrix}$$

$$\begin{bmatrix} 1 & 1 & 1 & \vdots & 3 \\ 4 & 2 & 1 & \vdots & 6 \\ 0 & 0 & 1 & \vdots & 4 \end{bmatrix} = \begin{bmatrix} 1 & 1 & 1 & \vdots & 3 \\ 0 & -2 & -3 & \vdots & -6 \\ 0 & 0 & 1 & \vdots & 4 \end{bmatrix} = \begin{bmatrix} 1 & 1 & 1 & \vdots & 3 \\ 0 & 1 & \frac{3}{2} & \vdots & 3 \\ 0 & 0 & 1 & \vdots & 4 \end{bmatrix}$$

$c = 4 \qquad b + \dfrac{3}{2}(4) = 3 \qquad a + (-3) + 4 = 3$

$\qquad\qquad\qquad\qquad b = -3 \qquad\qquad\quad a = 2$

$y = 2x^2 - 3x + 4$

20. Find the area of the triangle with vertices $(0, 0)$, $(5, 4)$, and $(6, 0)$.

Solution

$(x_1, y_1) = (0, 0), \ (x_2, y_2) = (5, 4), \ (x_3, y_3) = (6, 0)$

$$\begin{vmatrix} x_1 & y_1 & 1 \\ x_2 & y_2 & 1 \\ x_3 & y_3 & 1 \end{vmatrix} = \begin{vmatrix} 0 & 0 & 1 \\ 5 & 4 & 1 \\ 6 & 0 & 1 \end{vmatrix} = (1)\begin{vmatrix} 5 & 4 \\ 6 & 0 \end{vmatrix} = (1)(-24) = -24$$

$\text{Area} = -\dfrac{1}{2}(-24) = 12$

CHAPTER 9
Exponential and Logarithmic Functions

9.1 Exponential Functions

7. Simplify $2^x \cdot 2^{x-1}$.

Solution

$2^x \cdot 2^{x-1} = 2^{x+(x-1)} = 2^{2x-1}$

9. Simplify $(2e^x)^3$.

Solution

$(2e^x)^3 = 8e^{3x}$

11. Evaluate the function $f(x) = 3^x$ at the following values.

(a) $x = -2$ (b) $x = 0$ (c) $x = 1$

Solution

(a) $f(-2) = 3^{-2} = \dfrac{1}{9}$

(b) $f(0) = 3^0 = 1$

(c) $f(1) = 3^1 = 3$

13. Evaluate the function $g(x) = 10e^{-0.5x}$ at the following values.

(a) $x = -4$ (b) $x = 4$ (c) $x = 8$

Solution

(a) $g(-4) = 10e^{-0.5(-4)} = 10e^2 \approx 73.89$

(b) $g(4) = 10e^{-0.5(4)} = 10e^{-2} \approx 1.35$

(c) $g(8) = 10e^{-0.5(8)} = 10e^{-4} \approx 0.18$

15. Using the graphs in the textbook, match the function $f(x) = 2^x$ with its graph. [The graphs are labeled (a), (b), (c), (d), (e), and (f).]

Solution

$f(x) = 2^x$ (b)

17. Using the graphs in the textbook, match the function $f(x) = 2^{-x}$ with its graph. [The graphs are labeled (a), (b), (c), (d), (e), and (f).]

Solution

$f(x) = 2^{-x}$ (e)

19. Using the graphs in the textbook, match the function $f(x) = 2^{x-1}$ with its graph. [The graphs are labeled (a), (b), (c), (d), (e), and (f).]

Solution

$f(x) = 2^{x-1}$ (f)

21. Sketch the graph of $f(x) = 3^x$.

Solution

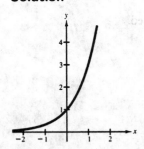

23. Sketch the graph of $g(x) = 3^x - 2$.

Solution

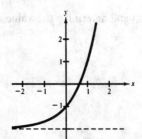

25. Sketch the graph of $f(t) = 2^{-t^2}$.

Solution

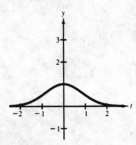

27. Sketch the graph of $f(x) = -2^{0.5x}$.

Solution

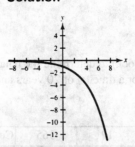

29. Use a graphing utility to graph $y = 5^{x/3}$.

Solution

Keystrokes:

| Y = | 5 | ∧ | (| X, T, θ | ÷ | 3 |) | GRAPH |

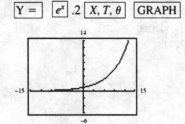

31. Use a graphing utility to graph $f(x) = e^{0.2x}$.

Solution

Keystrokes:

| Y = | e^x | .2 | X, T, θ | GRAPH |

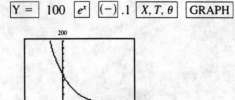

33. Using a graphing utility to graph $P(t) = 100e^{-0.1t}$.

Solution

Keystrokes:

| Y = | 100 | e^x | (−) | .1 | X, T, θ | GRAPH |

35. *Depreciation*   After t years, the value of a car that costs \$16,000 is given by

$$V(t) = 16,000\left(\frac{3}{4}\right)^t.$$

Sketch a graph of the function and determine the value of the car 2 years after it was purchased.

Solution

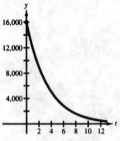

$$V(2) = 16,000\left(\frac{3}{4}\right)^2 = 9000$$

37. Complete the table (see textbook) to determine the balance A when the principal (P) of \$100 is invested at a rate (r) of 8% for a time (t) of 20 years and compounded n times per year.

Solution

n	1	4	12	365	Continuous
A	\$466.10	\$487.54	\$492.68	\$495.22	\$495.30

(a) $A = 100\left(1 + \dfrac{0.08}{1}\right)^{1(20)}$

   $= \$466.10$

(b) $A = 100\left(1 + \dfrac{0.08}{4}\right)^{4(20)}$

   $= \$487.54$

(c) $A = 100\left(1 + \dfrac{0.08}{12}\right)^{12(20)}$

   $= \$492.68$

(d) $A = 100\left(1 + \dfrac{0.08}{365}\right)^{365(20)}$

   $= \$495.22$

(e) $A = Pe^{rt}$

   $= 100e^{0.08(20)}$

   $= 495.30$

39. Complete the table (see textbook) to determine the principal P that yields a balance of (A) $5000 when invested at a rate (r) of 7% for a time (t) of 10 years and compounded n times per year.

Solution

n	1	4	12	365	Continuous
P	$2541.75	$2498.00	$2487.98	$2483.09	$2482.93

(a) $5000 = P\left(1 + \dfrac{0.07}{1}\right)^{1(10)}$

$\dfrac{5000}{(1.07)^{10}} = P$

$\$2541.75 = P$

(b) $5000 = \left(1 + \dfrac{0.07}{4}\right)^{4(10)}$

$\dfrac{5000}{(1.0175)^{40}} = P$

$\$2498.00 = P$

(c) $5000 = P\left(1 + \dfrac{0.07}{12}\right)^{12(10)}$

$\dfrac{5000}{(1.0058\overline{3})^{120}} = P$

$\$2487.98 = P$

(d) $5000 = P\left(1 + \dfrac{0.07}{365}\right)^{365(10)}$

$\dfrac{5000}{(1.0001918)^{3.650}} = P$

$\$2483.09 = P$

(e) $5000 = Pe^{0.07(10)}$

$\dfrac{5000}{e^{0.7}} = P$

$\$2482.93 = P$

41. *Graphical Interpretation*   An investment of $500 in two different accounts with respective interest rates of 6% and 8% is compounded continuously. The balances in the accounts after t years are modeled by $A_1 = 500e^{0.06t}$ and $A_2 = 500e^{0.08t}$.

(a) Use a graphing utility to graph each of the models.

(b) Use a graphing utility to graph the function $A_2 - A_1$ in the same window as the graphs of part (a).

(c) Use the graphs to discuss the rates of increase of the balances in the two accounts.

Solution

(a) Keystrokes:

y_1 [Y =] 500 [e^x] 0.06 [X, T, θ] [ENTER]

y_2 [Y =] 500 [e^x] 0.08 [X, T, θ] [GRAPH]

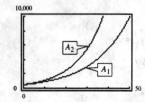

(b) $A_2 - A_1 = 500e^{0.08t} - 500e^{0.06t}$

$\qquad\qquad = 500(e^{0.08t} - e^{0.06t})$

Keystrokes:

[Y =] 500 [(] [e^x] 0.08 [X, T, θ] [−] [e^x] 0.06 [X, T, θ] [GRAPH]

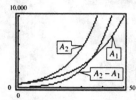

(c) The difference between the functions increases at an increasing rate.

43. Simplify $(4x + 3y) - 3(5x + y)$.

Solution

$(4x + 3y) - 3(5x + y) = 4x + 3y - 15x - 3y$

$\qquad\qquad\qquad\qquad = -11x$

45. Simplify $2x^2 + (2x - 3)^2 + 12x$.

Solution

$2x^2 + (2x - 3)^2 + 12x = 2x^2 + 4x^2 - 12x + 9 + 12x$

$\qquad\qquad\qquad\qquad = 6x^2 + 9$

47. *Geometry* A circle has a circumference of 1 inch.

(a) Find the radius of the circle.

(b) Find the area of the circle.

Solution

(a) $C = 2\pi r$

$1 = 2\pi r$

$\dfrac{1}{2\pi} = r$

(b) $A = \pi r^2$

$A = \pi\left(\dfrac{1}{2\pi}\right)^2$

$= \pi\left(\dfrac{1}{4\pi^2}\right)$

$= \dfrac{1}{4\pi}$

49. Evaluate $4^{\sqrt{3}}$

Solution

$4^{\sqrt{3}} \approx 11.036$

51. Evaluate $e^{1/3}$

Solution

$e^{1/3} \approx 1.396$

53. Evaluate the function $g(x) = 5^x$ at the following values.

(a) $x = -1$ (b) $x = 1$ (c) $x = 3$

(a) $g(-1) = 5^{-1} = \dfrac{1}{5}$

(b) $g(1) = 5^1 = 5$

(c) $g(3) = 5^3 = 125$

55. Evaluate the function $f(t) = 500\left(\dfrac{1}{2}\right)^t$ at the following values.

(a) $t = 0$ (b) $t = 1$ (c) $t = \pi$

(a) $f(0) = 500\left(\dfrac{1}{2}\right)^0 = 500$

(b) $f(1) = 500\left(\dfrac{1}{2}\right)^1 = 250$

(c) $f(\pi) = 500\left(\dfrac{1}{2}\right)^\pi = 56.657366$

57. Evaluate the function $f(x) = 1000(1.05)^{2x}$ at the following values.

(a) $x = 0$ (b) $x = 5$ (c) $x = 10$

Solution

(a) $f(0) = 1000(1.05)^{(2)(0)} = 1000$

(b) $f(5) = 1000(1.05)^{2(5)} = 1628.8946$

(c) $f(10) = 1000(1.05)^{2(10)} = 2653.2977$

59. Evaluate the function $f(x) = e^x$ at the following values.

(a) $x = -1$ (b) $x = 0$ (c) $x = \dfrac{1}{2}$

Solution

(a) $f(-1) = e^{-1} = 0.3678794$

(b) $f(0) = e^0 = 1$

(c) $f\left(\dfrac{1}{2}\right) = e^{1/2} = 1.6487213$

61. Using the graphs in the textbook, match the function $f(x) = \left(\dfrac{1}{2}\right)^x - 2$ with its graph. [The graphs are labeled (a), (b), (c), (d), (e), and (f).]

Solution

(d)

63. Using the graphs in the textbook, match the function $f(x) = 4^x - 1$ with its graph. [The graphs are labeled (a), (b), (c), (d), (e), and (f).]

Solution

(a)

65. Using the graphs in the textbook, match the function $f(x) = e^{-x^2}$ with its graph. [The graphs are labeled (a), (b), (c), (d), (e), and (f).]

Solution

(f)

67. Sketch the graph of $h(x) = \dfrac{1}{2}(3^x)$.

Solution

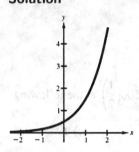

69. Sketch the graph of $h(x) = 2^{0.5x}$.

Solution

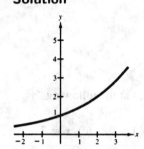

71. Sketch the graph of $f(x) = 4^{x-5}$.

Solution

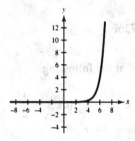

73. Sketch the graph of $f(x) = 4^x - 5$.

Solution

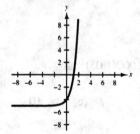

75. Sketch the graph of $f(x) = -\left(\dfrac{1}{3}\right)^x$.

Solution

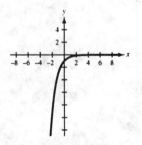

77. Sketch the graph of $g(t) = 200\left(\dfrac{1}{2}\right)^t$.

Solution

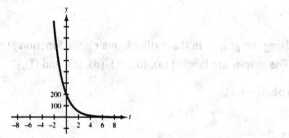

79. Use a graphing utility to sketch the graph of $y = 3^{-x/2}$.

Solution

Keystrokes:

| Y = | 3 | ∧ | (| (−) | X, T, θ | ÷ | 2 |) | GRAPH |

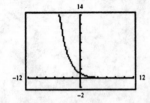

81. Use a graphing utility to sketch the graph of $y = 5^{-x/3}$.

Solution

Keystrokes:

| Y = | 5 | ∧ | (| (−) | X, T, θ | ÷ | 3 |) | GRAPH |

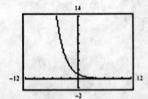

83. Use a graphing utility to sketch the graph of $y = 500(1.06)^t$.

Solution

Keystrokes:

| Y = | 500 | (| 1.06 |) | ∧ | X, T, θ | GRAPH |

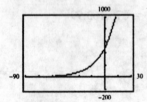

85. Use a graphing utility to sketch the graph of $y = 3e^{0.2x}$.

Solution

Keystrokes:

| Y = | 3 | e^x | 0.2 | X, T, θ | GRAPH |

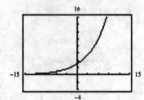

87. Use a graphing utility to sketch the graph of $y = 6e^{-x^2/3}$.

Solution

Keystrokes:

| Y = | 6 | e^x | (−) | (| X, T, θ | x^2 | ÷ | 3 |) | GRAPH |

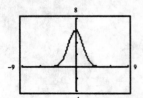

89. *Population Growth*   The population of the United States (in recent years) can be approximated by the exponential function.

$$P(t) = 203(1.0118)^{t-1970}$$

where t is the year and P is the population in millions. Use the model to approximate the population for the years (a) 1995 and (b) 2000.

Solution

(a) $P(1995) = 203(1.0118)^{1995-1970} = 272.1842$ million

(b) $P(2000) = 203(1.0118)^{2000-1970} = 288.62656$ million

91. *Inflation Rate*   Suppose the annual rate of inflation averages 5% over the next 10 years. With this rate of inflation, the approximate cost C of goods or services during any year in that decade will be given by $C(t) = P(1.05)^t$, $0 \le t \le 10$ where t is the time in years and P is the present cost. If the price of an oil change for your car is presently $19.95, estimate the price 10 years from now.

Solution

$C(10) = 19.95(1.05)^{10} = 32.496448 \approx \$32.50$

93. Complete the table (see textbook) for principal (P) of $2000 invested at a rate (r) of 9% for a time (t) of 40 years and compounded n times per year.

Solution

Compounded 1 time: $A = 2000\left(1 + \dfrac{0.09}{1}\right)^{1(40)}$

$\qquad\qquad = \$62,818.84$

Compounded 4 times: $A = 2000\left(1 + \dfrac{0.09}{4}\right)^{4(40)}$

$\qquad\qquad = \$70,333.25$

Compounded 12 times: $A = 2000\left(1 + \dfrac{0.09}{12}\right)^{12(40)}$

$\qquad\qquad = \$72,219.80$

Compounded 365 times: $A = 2000\left(1 + \dfrac{0.09}{365}\right)^{365(40)}$

$\qquad\qquad = \$73,163.99$

Compounded continuously: $A = 2000e^{0.09(40)}$

$\qquad\qquad = \$73,196.47$

n	1	4	12	365	Continuous
A	$62,818.84	$70,333.25	$72,219.80	$73,163.99	$73,196.47

95. Complete the table (see textbook) for the principal (P) of $5000 invested at a rate (r) of 10% for a time (t) of 40 years and compounded n times per year.

Solution

Compounded 1 time: $\quad A = 5000\left(1 + \dfrac{0.10}{1}\right)^{1(40)}$

$\qquad\qquad\qquad\qquad = \$226{,}296.28$

Compounded 4 times: $\quad A = 5000\left(1 + \dfrac{0.10}{4}\right)^{4(40)}$

$\qquad\qquad\qquad\qquad = \$259{,}889.34$

Compounded 12 times: $\quad A = 5000\left(1 + \dfrac{0.10}{12}\right)^{12(40)}$

$\qquad\qquad\qquad\qquad = \$268{,}503.31$

Compounded 365 times: $\quad A = 5000\left(1 + \dfrac{0.10}{365}\right)^{365(40)}$

$\qquad\qquad\qquad\qquad = \$272{,}841.22$

Compounded continuously: $\quad A = 5000^{0.10(40)}$

$\qquad\qquad\qquad\qquad = \$272{,}990.75$

n	1	4	12	365	Continuous
A	\$226,296.28	\$259,889.34	\$268,503.31	\$272,841.22	\$272,990.75

97. Complete the table (see textbook) to determine the principal (P) that will yield a balance of (A) \$1,000,000 when invested at a rate (r) of 10.5% for a time (t) of 40 years and compounded n times per year.

Solution

Compounded 1 time: $1,000,000 = P\left(1 + \dfrac{0.105}{1}\right)^{1(40)}$

$\dfrac{1,000,000}{(1.105)^{40}} = P$

$\$18,429.30 = P$

Compounded 4 times: $1,000,000 = P\left(1 + \dfrac{0.105}{4}\right)^{4(40)}$

$\dfrac{1,000,000}{(1.02625)^{160}} = P$

$\$15,830.43 = P$

Compounded 12 times: $1,000,000 = P\left(1 + \dfrac{0.105}{12}\right)^{12(40)}$

$\dfrac{1,000,000}{(1.00875)^{480}} = P$

$\$15,272.04 = P$

Compounded 365 times: $1,000,000 = P\left(1 + \dfrac{0.105}{365}\right)^{365(40)}$

$\dfrac{1,000,000}{(1.002877)^{14,600}} = P$

$\$15,004.64 = P$

Compounded continuously: $1,000,000 = Pe^{0.105(40)}$

$\dfrac{1,000,000}{e^{4.2}} = P$

$\$14,995.58 = P$

n	1	4	12	365	Continuous
P	\$18,429.30	\$15,830.43	\$15,272.04	\$15,004.64	\$14,995.58

99. On the same set of coordinate axes, sketch the graph of the following functions. Which functions are exponential?

(a) $f(x) = 2x$ (b) $f(x) = 2x^2$

(c) $f(x) = 2^x$ (d) $f(x) = 2^{-x}$

Solution

(c) and (d) are exponential

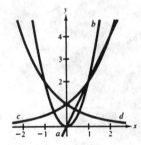

101. *Graphical Estimation*   A parachutist jumps from a plane and opens the parachute at a height of 2000 feet (see figure in textbook). The height of the parachutist is then given by

$$h = 1950 + 50e^{-1.6t} - 20t$$

where h is the height if feet and t is the time in seconds. (The time $t = 0$ corresponds to the time when the parachute is opened.)

(a) Use a graphing utility to graph the height function.

(b) Find the height of the parachutist when $t = 0, 25, 50,$ and 75.

(c) Approximate the time when the parachutist will reach the ground.

Solution

(a) Keystroke:

$\boxed{Y =}$ 1950 $\boxed{+}$ 50 $\boxed{e^x}$ $\boxed{(-)}$ 1.6 $\boxed{X, T, \theta}$ $\boxed{-}$ 20 $\boxed{X, T, \theta}$ $\boxed{GRAPH}$

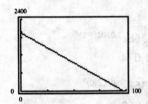

(b) $h = 1950 + 50e^{-1.6(0)} - 20(0) = 2000$ feet

$h = 1950 + 50e^{-1.6(25)} - 20(25) = 1450$ feet

$h = 1950 + 50e^{-1.6(50)} - 20(50) = 950$ feet

$h = 1950 + 50e^{-1.6(75)} - 20(75) = 450$ feet

(c) The parachutist will reach the ground at 97.5 seconds.

103. *Creating a Table* The median price of a home in the United States for the years 1987 through 1992 is given in the table in the textbook. A model for this data is given by

$$y = 131{,}368e^{0.0102t^3}$$

where t is time in years, with $t = 0$ representing 1990.

(a) Use the model to complete the table and compare the results with the actual data.

Year	1987	1988	1989	1990	1991	1992
Price						

(b) Use a graphing utility to graph the model.

(c) If the model were used to predict home prices in the years ahead, would the predictions be increasing at a higher rate of a lower rate? Do you think the model would be reliable for predicting the future prices of homes? Explain.

Solution

1987: $t = -3$ $y = 131{,}368e^{0.0102(-3)^3}$
                 $= \$99{,}744$

1988: $t = -2$ $y = 131{,}368e^{0.0102(-2)^3}$
                 $= \$121{,}074$

1989: $t = -1$ $y = 131{,}368e^{0.0102(-1)^3}$
                 $= \$130{,}035$

1990: $t = 0$ $y = 131{,}368e^{0.0102(0)^3}$
                 $= \$131{,}368$

1991: $t = 1$ $y = 131{,}368e^{0.0102(1)^3}$
                 $= \$132{,}715$

1992: $t = 2$ $y = 131{,}368e^{0.0102(2)^3}$
                 $= \$142{,}537$

(b) Keystrokes:

Y= 131368 x e^x (.0102 X, T, θ MATH 3) GRAPH

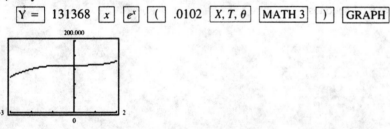

(c) Increasing at a higher rate. Probably home prices will not increase at a higher rate indefinitely.

9.2 | Composite and Inverse Functions

7. Find the indicated composites.

(a) $(f \circ g)(4)$ (b) $(g \circ f)(7)$

(c) $(f \circ g)(x)$ (d) $(g \circ f)(x)$

when $f(x) = x - 3$ and $g(x) = x^2$.

Solution

(a) $(f \circ g)(4) = 4^2 - 3 = 13$

(b) $(g \circ f)(7) = (7 - 3)^2 = 16$

(c) $(f \circ g)(x) = x^2 - 3$

(d) $(g \circ f)(x) = (x - 3)^2$

11. Use the functions f and g to find the indicated values.

$f = \{(-2, 3), (-1, 1), (0, 0), (1, -1), (2, -3)\}$ and

$g = \{(-3, 1), (-1, -2), (0, 2), (2, 2), (3, 1)\}$.

(a) $f(1)$ (b) $g(-1)$ (c) $(g \circ f)(1)$

Solution

(a) $f(1) = -1$ (b) $g(-1) = -2$ (c) $(g \circ f)(1) = g[f(1)]$

$= g[-1]$

$= -2$

13. Use the functions f and g to find the indicated values.

$f = \{(-2, 3), (-1, 1), (0, 0), (1, -1), (2, -3)\}$ and

$g = \{(-3, 1), (-1, -2), (0, 2), (2, 2), (3, 1)\}$.

(a) $(f \circ g)(-3)$ (b) $(g \circ f)(-2)$

Solution

(a) $(f \circ g)(-3) = f[g(-3)] = f[1] = -1$

(b) $(g \circ f)(-2) = g[f(-2)] = g[3] = 1$

15. Find the domain of the composition (a) $f \circ g$ and (b) $g \circ f$ when $f(x) = \sqrt{x}$ and $g(x) = x - 2$.

Solution

(a) $f \circ g = \sqrt{x - 2}$ Domain: Real numbers ≥ 2

(b) $g \circ f = \sqrt{x} - 2$ Domain: Nonnegative real numbers

9. Find the indicated composites.

(a) $(f \circ g)(4)$ (b) $(g \circ f)(9)$

(c) $(f \circ g)(x)$ (d) $(g \circ f)(x)$

when $f(x) = \sqrt{x}$ and $g(x) = x + 5$.

Solution

(a) $(f \circ g)(4) = \sqrt{4 + 5} = 3$

(b) $(g \circ f)(9) = \sqrt{9} + 5 = 8$

(c) $(f \circ g)(x) = \sqrt{x + 5}$

(d) $(g \circ f)(x) = \sqrt{x} + 5$

17. *Sales Bonus* You are a sales representative for a clothing manufacturer. You are paid an annual salary plus a bonus of 2% of your sales over $200,000. Consider the two functions

$$f(x) = x - 200,000 \text{ and } g(x) = 0.02x.$$

If x is greater than $200,000, which of the following represents your bonus? Explain.

(a) $f(g(x))$ (b) $g(f(x))$

Solution

$g(f(x)) = 0.02(x - 200,000), \quad x > 200,000$

---

19. Find the inverse of $f(x) = 5x$ informally.

Solution

$f^{-1}(x) = \dfrac{x}{5}$

21. Find the inverse of $f(x) = x^7$ informally.

Solution

$f^{-1}(x) = \sqrt[7]{x}$

---

23. Verifying algebraically that the functions $f(x) = x + 15$ and $g(x) = x - 15$ are inverses of each other.

$f(g(x)) = (x - 15) + 15 = x$

$g(f(x)) = (x + 15) - 15 = x$

25. Verifying algebraically that the functions $f(x) = \sqrt[3]{x + 1}$ and $g(x) = x^3 - 1$ are inverses of each other.

$f(g(x)) = \sqrt[3]{x^3 - 1 + 1} = \sqrt[3]{x^3} = x$

$g(f(x)) = \left(\sqrt[3]{x + 1}\right)^3 - 1 = x + 1 - 1 = x$

---

27. Using the graphs in the textbook, match the graph with the graph of its inverse. [The graphs of the inverse functions are labeled (a), (b), (c), and (d).]

Solution

(a)

29. Using the graphs in the textbook, match the graph with the graph of its inverse. [The graphs of the inverse functions are labeled (a), (b), (c), and (d).]

Solution

(b)

---

31. Use a graphing utility to verify that the functions are inverses of each other.

$f(x) = \dfrac{1}{3}x$

$g(x) = 3x$

Solution

Keystrokes:

y_1 [Y =] [( ] [1] [÷] [3] [ )] [X, T, θ] [ENTER]

y_2 [Y =] [3] [X, T, θ] [GRAPH]

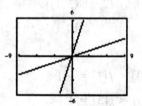

33. Use a graphing utility to verify that the functions are inverses of each other.

$$f(x) = \sqrt[3]{x} + 2$$

$$g(x) = x^3 - 2$$

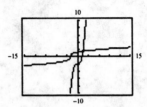

Solution

Keystrokes:

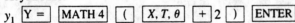

y_1 [Y =] [MATH 4] [(] [X, T, θ] [+] [2] [)] [ENTER]

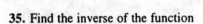

y_2 [Y =] [X, T, θ] [MATH 3] [−] [2] [GRAPH]

35. Find the inverse of the function

$$g(x) = 3 - 4x.$$

Solution

$$g(x) = 3 - 4x$$
$$y = 3 - 4x$$
$$x = 3 - 4y$$
$$x - 3 = -4y$$
$$\frac{x - 3}{-4} = y$$

$$\frac{3 - x}{4} \text{ or } \frac{x - 3}{-4} = g^{-1}(x)$$

37. Find the inverse of the function

$$f(t) = t^3 - 1.$$

Solution

$$f(t) = t^3 - 1$$
$$y = t^3 - 1$$
$$t = y^3 - 1$$
$$t + 1 = y^3$$
$$\sqrt[3]{t + 1} = y$$

$$\sqrt[3]{t + 1} = f^{-1}(t)$$

39. Using the graph in the textbook, decide whether the function $f(x) = x^2 - 2$ has an inverse.

Solution

$$f(x) = x^2 - 2$$

No, it does not have an inverse.

41. Using the graph in the textbook, decide whether the function $g(x) = \sqrt{25 - x^2}$ has an inverse.

Solution

No, it does not have an inverse.

43. Place a restriction on the domain of the function $f(x) = x^4$ so that the graph (see textbook) of the restricted function satisfies the Horizontal Line Test. Then find the inverse of the restricted function.
(*Note:* There is more than one correct answer.)

Solution

$$f(x) = x^4, \ x \geq 0$$
$$y = x^4$$
$$x = y^4$$
$$\sqrt[4]{x} = y$$
$$\sqrt[4]{x} = f^{-1}(x), \ x \geq 0$$

45. Place a restriction on the domain of the function $f(x) = (x - 2)^2$ so that the graph (see textbook) of the restricted function satisfies the Horizontal Line Test. Then find the inverse of the restricted function.
(*Note:* There is more than one correct answer.)

Solution

$$f(x) = (x - 2)^2, \ x \geq 2$$
$$y = (x - 2)^2$$
$$x = (y - 2)^2$$
$$\sqrt{x} = y - 2$$
$$\sqrt{x} + 2 = y$$
$$\sqrt{x} + 2 = f^{-1}(x), \ x \geq 2$$

47. Use the graph of f in the textbook to sketch the graph of f^{-1}.

Solution

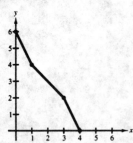

49. Decide whether the statement is true of false. If the inverse of f exists, the y-intercept of f is an x-intercept of f^{-1}.

Solution

True

51. Solve $3x^2 + 9x - 12 = 0$ by factoring.

Solution

$$3x^2 + 9x - 12 = 0$$
$$x^2 + 3x - 4 = 0$$
$$(x + 4)(x - 1) = 0$$
$$x = -4 \quad x = 1$$

53. Solve $4x^2 - 7 = -6x$ by the quadratic formula.

Solution

$$4x^2 - 7 = -6x$$
$$4x^2 + 6x - 7 = 0$$
$$x = \frac{-6 \pm \sqrt{6^2 - 4(4)(-7)}}{2(4)}$$
$$x = \frac{-6 \pm \sqrt{36 + 112}}{8}$$
$$x = \frac{-6 \pm \sqrt{148}}{8} = \frac{-6 \pm 2\sqrt{37}}{8} = \frac{-3 \pm \sqrt{37}}{4}$$

55. Use $y = -x^2 + 4x$.

(a) Does the graph open up or down? Explain.

(b) Find the x-intercepts algebraically.

(c) Find the coordinates of the vertex of the parabola.

Solution

(a) It opens down because the coefficient of x^2 is negative.

(b) $0 = -x^2 + 4x$
$$0 = -x(x - 4)$$
$$0 = -x \quad 0 = x - 4$$
$$0 = \quad x \quad 4 = x$$
$(0, 0)$ and $(4, 0)$

(c) $y = -x^2 + 4x$
$$y = -1(x^2 - 4x + 4) + 4$$
$$= -1(x - 2) + 4$$
vertex $= (2, 4)$

57. Find the following composite functions
  (a) $(f \circ g)(3)$ (b) $(g \circ f)(3)$
  (c) $(f \circ g)(x)$ (d) $(g \circ f)(x)$
  when $f(x) = x + 1$ and $g(x) = 2x$.
  Solution
  (a) $(f \circ g)(3) = f[g(3)] = f[6] = 7$
  (b) $(g \circ f)(3) = g[f(3)] = g[4] = 8$
  (c) $(f \circ g)(x) = f[g(x)] = 2x + 1$
  (d) $(g \circ f)(x) = g[f(x)] = 2(x + 1) = 2x + 2$

59. Find the following composite functions
  (a) $(f \circ g)(1)$ (b) $(g \circ f)(2)$
  (c) $(f \circ g)(x)$ (d) $(g \circ f)(x)$
  when $f(x) = |x - 3|$ and $g(x) = 3x$.
  Solution
  (a) $(f \circ g)(1) = |3 - 3| = 0$
  (b) $(g \circ f)(2) = 3|2 - 3| = 3$
  (c) $(f \circ g)(x) = |3x - 3|$
  (d) $(g \circ f)(x) = 3|x - 3|$

61. Find the following composite functions
  (a) $(f \circ g)(49)$ (b) $(g \circ f)(12)$ (c) $(f \circ g)(x)$ (d) $(g \circ f)(x)$
  when $f(x) = \dfrac{1}{x - 3}$ and $g(x) = \sqrt{x}$.
  Solution
  (a) $(f \circ g)(49) = \dfrac{1}{\sqrt{49} - 3} = \dfrac{1}{4}$

  (b) $(g \circ f)(12) = \sqrt{\dfrac{1}{12 - 3}} = \dfrac{1}{3}$

  (c) $(f \circ g)(x) = \dfrac{1}{\sqrt{x} - 3}$

  (d) $(g \circ f)(x) = \sqrt{\dfrac{1}{x - 3}}$ or $\dfrac{1}{\sqrt{x - 3}}$

63. Use the functions f and g to find the indicated values.
  $f = \{(0, 1), (1, 2), (2, 5), (3, 10), (4, 17)\}$
  $g = \{(5, 4), (10, 1), (2, 3), (17, 0), (1, 2)\}$
  (a) $f(3)$ (b) $(g \circ f)(3)$

  Solution
  (a) $f(3) = 10$
  (b) $(g \circ f)(3) = g[f(3)] = g[10] = 1$

65. Use the functions f and g to find the indicated values.
  $f = \{(0, 1), (1, 2), (2, 5), (3, 10), (4, 17)\}$
  $g = \{(5, 4), (10, 1), (2, 3), (17, 0), (1, 2)\}$
  (a) $(g \circ f)(4)$ (b) $(f \circ g)(2)$

  Solution
  (a) $(g \circ f)(4) = g[f(4)] = g[17] = 0$
  (b) $(f \circ g)(2) = f[g(2)] = f[3] = 10$

67. Find the domain of the composition (a) $f \circ g$ and (b) $g \circ f$ when $f(x) = x^2 + 1$
  and $g(x) = 2x$.

  Solution
  (a) $f \circ g = (2x)^2 + 1 = 4x^2 + 1$ Domain: Real numbers
  (b) $g \circ f = 2(x^2 + 1) = 2x^2 + 2$ Domain: Real numbers

69. Find the domain of the composition (a) $f \circ g$ and (b) $g \circ f$ when $f(x) = \dfrac{9}{x + 9}$ and $g(x) = x^2$.

Solution

$f(x) = \dfrac{9}{x + 9}$, $g(x) = x^2$

(a) $f \circ g = \dfrac{9}{x^2 + 9}$ Domain: Real numbers

(b) $g \circ f = \left(\dfrac{9}{x + 9}\right)^2$

$\quad\;\; = \dfrac{81}{x^2 + 18x + 81}$ Domain: Real numbers $\neq -9$

71. *Production Cost*   The daily cost of producing x units of a product is $C(x) = 8.5x + 300$. The number of units produced in t hours during a day is given by $x(t) = 12t$, $0 \leq t \leq 8$. Find, simplify, and interpret $(C \circ x)(t)$.

Solution

$(C \circ x)(t) = C[x(t)] = 8.5(12t) + 300$

$\qquad\qquad\qquad = 102t + 300$

Total cost after t hours of production.

73. Find the inverse of $f(x) = 6x$ informally.

Solution

$f^{-1}(x) = \dfrac{x}{6}$

75. Find the inverse of $f(x) = x + 10$ informally.

Solution

$f^{-1}(x) = x - 10$

77. Find the inverse of $f(x) = \sqrt[3]{x}$ informally.

Solution

$f^{-1}(x) = x^3$

79. Find the inverse of $f(x) = 2x - 1$ informally.

Solution

$f^{-1}(x) = \dfrac{1}{2}(x + 1)$

81. Verify algebraically that the functions $f(x) = 10x$ and $g(x) = \dfrac{1}{10}x$ are inverses of each other.

Solution

$f(g(x)) = 10\left(\dfrac{1}{10}x\right) = x$ $\qquad\qquad$ $g(f(x)) = \dfrac{10x}{10} = x$

83. Verify algebraically that the functions $f(x) = 1 - 2x$ and $g(x) = \frac{1}{2}(1 - x)$ are inverses of each other.

Solution

$$f(g(x)) = 1 - 2\left[\frac{1}{2}(1 - x)\right] = 1 - (1 - x) = 1 - 1 + x = x$$

$$g(f(x)) = \frac{1}{2}[1 - (1 - 2x)] = \frac{1}{2}[1 - 1 + 2x] = \frac{1}{2}[2x] = x$$

85. Verify algebraically that the functions $f(x) = 2 - 3x$ and $g(x) = \frac{1}{3}(2 - x)$ are inverses of each other.

Solution

$$f(g(x)) = 2 - 3\left[\frac{1}{3}(2 - x)\right] = 2 - (2 - x) = x$$

$$g(f(x)) = \frac{1}{3}[2 - (2 - 3x)] = \frac{1}{3}[3x] = x$$

87. Verify algebraically that the functions $f(x) = \frac{1}{x}$ and $g(x) = \frac{1}{x}$ are inverses of each other.

Solution

$$f(g(x)) = \frac{1}{\frac{1}{x}} = x \qquad g(f(x)) = \frac{1}{\frac{1}{x}} = x$$

89. Use a graphing utility to verify that $f(x) = 3x + 4$ and $g(x) = \frac{1}{3}(x - 4)$ are inverses of each other.

Solution

Keystrokes:

y_1 [Y =] 3 [X, T, θ] [+] 4 [ENTER]

y_2 [Y =] [(] [X, T, θ] [−] 4 [)] [÷] 3 [GRAPH]

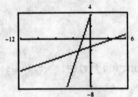

91. Use a graphing utility to verify that the functions $f(x) = \frac{1}{8}x^3$ and $g(x) = 2\sqrt[3]{x}$ are inverses of each other.

Solution

Keystrokes:

y_1 [Y =] [(] 1 [÷] 8 [)] [X, T, θ] [MATH 3] [ENTER]

y_2 [Y =] 2 [MATH 4] [X, T, θ] [GRAPH]

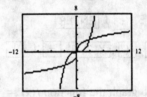

93. Find the inverse of the function $f(x) = 8x$.

Solution

$$f^{-1}(x) = \frac{x}{8}$$

95. Find the inverse of the function $g(x) = x + 25$.

Solution

$$g^{-1}(x) = x - 25$$

97. Find the inverse of the function $g(t) = -\frac{1}{4}t + 2$.

Solution

$g^{-1}(t) = -4(t - 2)$

99. Find the inverse of the function $h(x) = \sqrt{x}$.

Solution

$$h(x) = \sqrt{x}$$
$$y = \sqrt{x}$$
$$x = \sqrt{y}$$
$$x^2 = y$$
$$x^2 = h^{-1}(x), \ x \geq 0$$

101. Find the inverse of the function $g(s) = \frac{5}{s}$.

Solution

$$g(s) = \frac{5}{s}$$
$$y = \frac{5}{s}$$
$$s = \frac{5}{y}$$
$$y = \frac{5}{s}$$
$$g^{-1}(s) = \frac{5}{s}, \ s \neq 0$$

103. Find the inverse of the function $f(x) = x^3 + 1$. Use a graphing utility to sketch the graphs of f and f^{-1}.

Solution

$f^{-1}(x) = \sqrt[3]{x - 1}$

Keystrokes:

y_1 Y = $\boxed{X, T, \theta}$ $\boxed{\wedge}$ 3 $\boxed{+}$ 1 ENTER

y_2 Y = MATH $\boxed{4}$ $\boxed{(}$ $\boxed{X, T, \theta}$ $\boxed{-}$ 1 $\boxed{)}$ GRAPH

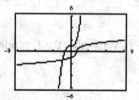

105. Use a graphing utility to graph the function $f(x) = \frac{1}{4}x^3$ and determine whether $f(x)$ is one-to-one.

Solution

Keystrokes:

$\boxed{Y =}$ .25 $\boxed{X, T, \theta}$ MATH 3 GRAPH

one-to-one

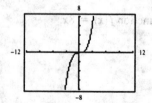

107. Use a graphing utility to graph the function $f(t) = \sqrt[3]{5 - t}$ and determine whether $f(t)$ is one-to-one.

Solution

Keystrokes:

$\boxed{Y =}$ $\boxed{\text{MATH 4}}$ $\boxed{(}$ $\boxed{5}$ $\boxed{-}$ $\boxed{X, T, \theta}$ $\boxed{)}$ $\boxed{\text{GRAPH}}$

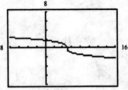

one-to-one

109. Use a graphing utility to graph the function $g(x) = x^4$ and determine whether $g(x)$ is one-to-one.

Solution

Keystrokes:

$\boxed{Y =}$ $\boxed{X, T, \theta}$ $\boxed{\wedge}$ $\boxed{4}$ $\boxed{\text{GRAPH}}$

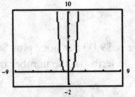

not one-to-one

111. Use a graphing utility to graph the function $h(t) = \dfrac{5}{t}$ and determine whether $h(t)$ is one-to-one.

Solution

Keystrokes:

$\boxed{Y =}$ $\boxed{5}$ $\boxed{\div}$ $\boxed{X, T, \theta}$ $\boxed{\text{GRAPH}}$

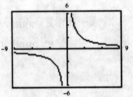

not one-to-one

113. Use a graphing utility to graph the function $f(s) = \dfrac{4}{s^2 + 1}$ and determine whether $f(s)$ is one-to-one.

Solution

Keystrokes:

$\boxed{Y =}$ $\boxed{4}$ $\boxed{\div}$ $\boxed{(}$ $\boxed{X, T, \theta}$ $\boxed{x^2}$ $\boxed{+}$ $\boxed{1}$ $\boxed{)}$ $\boxed{\text{GRAPH}}$

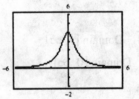

not one-to-one

115. Use the graph of f to sketch the graph of f^{-1}.

Solution

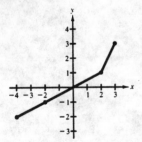

117. Consider the function $f(x) = 3 - 2x$.

(a) Find $f^{-1}(x)$.   (b) Find $(f^{-1})^{-1}(x)$.

Solution

(a) $y = 3 - 2x$

$x = 3 - 2y$

$2y = 3 - x$

$y = \dfrac{3 - x}{2}$

$f^{-1}(x) = \dfrac{3 - x}{2}$

(b) $y = \dfrac{3 - x}{2}$

$x = \dfrac{3 - y}{2}$

$2x = 3 - y$

$y = 3 - 2x$

$(f^{-1})^{-1}(x) = 3 - 2x$

119. *Hourly Wage*   Your wage is $9.00 per hour plus $0.65 for each unit produced per hour. Thus, your hourly wage y in terms of the number of units produced is given by $y = 9 + 0.65x$.

(a) Determine the inverse of the function.

(b) What does each variable represent in the inverse function?

(c) Determine the number of units produced when your hourly wage averages $14.20.

Solution

(a) $y = 9 + 0.65x$

$x = 9 + 0.65y$

$\dfrac{x - 9}{0.65} = y$

(b) x: hourly wage

y: number of units produced

(c) $14.20 = 9 + 0.65x$

$5.20 = 0.65x$

$8 = x$

121. Decide whether the statement is true or false.

If the inverse of f exists, the domains of f and f^{-1} are the same. If false, give an example to show why.

Solution

False: $f(x) = \sqrt{x - 1}$ Domain $[1, \infty)$

$f^{-1}(x) = x^2 + 1$ Domain $[0, \infty)$

9.3 Logarithmic Functions

7. Write $\log_5 25 = 2$ in exponential form.

Solution

$$\log_5 25 = 2$$
$$5^2 = 25$$

9. Write $\log_{36} 6 = \frac{1}{2}$ in exponential form.

Solution

$$\log_{36} 6 = \frac{1}{2}$$
$$36^{1/2} = 6$$

11. Write $3^{-2} = \frac{1}{9}$ in logarithmic form.

Solution

$$3^{-2} = \frac{1}{9}$$
$$\log_3 \frac{1}{9} = -2$$

13. Write $8^{2/3} = 4$ in logarithmic form.

Solution

$$8^{2/3} = 4$$
$$\log_8 4 = \frac{2}{3}$$

15. Evaluate $\log_2 8$ without the aid of a calculator.
(If it is not possible, state the reason.)

Solution

$$\log_2 8 = 3$$

17. Evaluate $\log_2 \frac{1}{4}$ without the aid of a calculator.
(If it is not possible, state the reason.)

Solution

$$\log_2 \frac{1}{4} = -2$$

19. Evaluate $\log_2 (-3)$ without the aid of a calculator.
(If it is not possible, state the reason.)

Solution

$\log_2 (-3)$ is not possible, because the domain is
is positive real numbers.

21. Evaluate $\log_9 3$ without the aid of a calculator.
(If it is not possible, state the reason.)

Solution

$$\log_9 3 = \frac{1}{2}$$

23. Use the properties of natural logarithms to
evaluate the expression $\ln e^2$.

Solution

$\ln e^2 = 2 \ln e = 2(1) = 2$

25. Use the properties of natural logarithms to
evaluate the expression $\ln e$.

Solution

$\ln e = 1$

27. Evaluate $\log_{10} 31$ with the aid of a calculator.
Round your answer to four decimal places.

Solution

$\log_{10} 31 = 1.4914$

29. Evaluate $\ln 0.75$ with the aid of a calculator.
Round your answer to four decimal places.

Solution

$\ln 0.75 = -0.2877$

31. Use the graph of $y = \log_3 x$ to match
$f(x) = 4 + \log_3 x$ to its graph (see textbook).
[The graphs are labeled (a), (b), (c), (d), (e),
and (f).]

Solution

$f(x) = 4 + \log_3 x$ (e)

33. Use the graph of $y = \log_3 x$ to match
$f(x) = -\log_3 x$ to its graph (see textbook).
[The graphs are labeled (a), (b), (c), (d), (e),
and (f).]

Solution

$f(x) = -\log_3 x$ (c)

35. Use the graph of $y = \log_3 x$ to match $f(x) = \log_3 (x - 4)$ to its graph (see textbook).
[The graphs are labeled (a), (b), (c), (d), (e), and (f).]

Solution

$f(x) = \log_3 (x - 4)$ (a)

37. Sketch the graph of $y = \log_3 x$ and $g(x) = 3^x$ on the same set of coordinates axes. Describe
the relationship between the graphs of f and g.

Solution

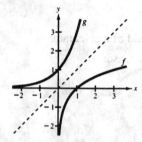

f and g are inverse functions.

39. Sketch the graph of $f(x) = \log_5 x$.

Solution

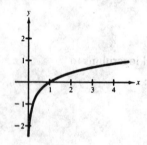

41. Sketch the graph of $f(x) = 3 + \log_2 x$.

Solution

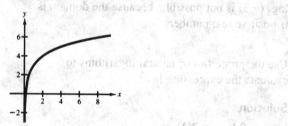

43. Use a graphing utility to sketch the graph of
$y = 5 \log_{10} x$.

Solution

Keystrokes:

$\boxed{Y =}$ 5 $\boxed{\log}$ $\boxed{X, T, \theta}$ $\boxed{\text{GRAPH}}$

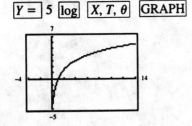

45. Use a graphing utility to sketch the graph of
$f(x) = 3 \ln x$.

Solution

Keystrokes:

$\boxed{Y =}$ 3 $\boxed{\ln}$ $\boxed{X, T, \theta}$ $\boxed{\text{GRAPH}}$

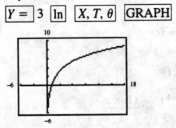

47. Use a calculator to evaluate $\log_8 132$ by means of the change-of-base formula.

Solution

$$\log_8 132 = \frac{\log 132}{\log 8} = 2.3481$$

$$= \frac{\ln 132}{\ln 8} = 2.3481$$

49. Use a calculator to evaluate $\log_2 0.72$ by means of the change-of-base formula.

Solution

$$\log_2 0.72 = \frac{\log 0.72}{\log 2} = -0.4739$$

$$= \frac{\ln 0.72}{\ln 2} = -0.4739$$

51. *Creating a Table* The time t in years for an investment to double in value when compounded continuously at annual interest rate r is given by

$$t = \frac{\ln 2}{r}.$$

Complete the table (see textbook), which shows the "doubling time" for several annual interest rates.

Solution

r of 0.07: $t = \dfrac{\ln 2}{0.07} = 0.9021$

r of 0.09: $t = \dfrac{\ln 2}{0.09} = 7.7016$

r of 0.11: $t = \dfrac{\ln 2}{0.11} = 6.3013$

r of 0.08: $t = \dfrac{\ln 2}{0.08} = 8.6643$

r of 0.10: $t = \dfrac{\ln 2}{0.10} = 6.9315$

r of 0.12: $t = \dfrac{\ln 2}{0.12} = 5.7762$

r	0.07	0.08	0.09	0.10	0.11	0.12
t	9.9021	8.6643	7.7016	6.9315	6.3013	5.7762

53. Use a graphing utility to graph each equation and use the graph to approximate any points of intersection. Find the solution of the system of equations algebraically.

$$x + 3y = 11$$
$$x^2 - 3y = -5$$

Solution

Solve each equation for y.

$$x + 3y = 11$$
$$3y = -x + 11$$
$$y = -\frac{1}{3}x + \frac{11}{3} = \frac{11 - x}{3}$$

$$x^2 - 3y = -5$$
$$-3y = -x^2 - 5$$
$$y = \frac{x^2 + 5}{3}$$

Keystrokes:

y_1 [Y =] [(] 11 [−] [X, T, θ] [)] [÷] 3 [ENTER]

y_2 [Y =] [(] [X, T, θ] [x²] [+] 5 [)] [÷] 3 [GRAPH]

$$x + 3y = 11$$
$$x^2 - 3y = -5$$

$$x^2 + x = 6$$
$$x^2 + x - 6 = 0$$
$$(x + 3)(x - 2) = 0$$
$$x = -3 \quad x = 2$$
$$y = \frac{14}{3} \quad y = 3$$

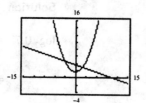

$$\left\{ \left(-3, \frac{14}{3} \right), (2, 3) \right\}$$

55. *Geometry* Find the area of the shaded region (see figure in textbook).

Solution

Verbal Model:

| Area of Shaded Area | = | Area of Outside Rectangle | − | Area of Inside Rectangle |

Equation: Area $= (x + 2)(2x - 3) - x(x + 1)$
$$A = 2x^2 + x - 6 - x^2 - x$$
$$A = x^2 - 6$$

57. From 1986 through 1990, the number of compact discs C (in millions) and the number of record albums R (in millions) sold in the United States can be approximated by the models.

$$C = 4.3t^2 - 9.8t - 42.7 \quad \text{and} \quad R = -28.8t + 301$$

where $t = 6$ represents 1986. How many compact discs and how many records were sold in 1986? (Source: Recording Industry Association of America)

Solution

$$C(6) = 4.3(6)^2 - 9.8(6) - 42.7 = 53.3 \text{ million}$$
$$R(6) = -28.8(6) + 301 = 128.2 \text{ million}$$

59. Write $\log_6 36 = 2$ in exponential form.

Solution

$\log_6 36 = 2$

$6^2 = 36$

61. Write $\log_4 \dfrac{1}{16} = -2$ in exponential form.

Solution

$\log_4 \dfrac{1}{16} = -2$

$4^{-2} = \dfrac{1}{16}$

63. Write $\log_8 4 = \dfrac{2}{3}$ in exponential form.

Solution

$\log_8 4 = \dfrac{2}{3}$

$8^{2/3} = 4$

65. Write $7^2 = 49$ in logarithmic form.

Solution

$7^2 = 49$

$\log_7 49 = 2$

67. Write $25^{-1/2} = \dfrac{1}{5}$ in logarithmic form.

Solution

$25^{-1/2} = \dfrac{1}{5}$

$\log_{25} \dfrac{1}{5} = -\dfrac{1}{2}$

69. Write $4^0 = 1$ in logarithmic form.

Solution

$4^0 = 1$

$\log_4 1 = 0$

71. Evaluate $\log_4 1$ without the aid of a calculator. (If it is not possible, state the reason.)

Solution

$\log_4 1 = 0$

73. Evaluate $\log_{10} 10$ without the aid of a calculator. (If it is not possible, state the reason.)

Solution

$\log_{10} 10 = 1$

75. Evaluate $\log_{10} 1000$ without the aid of a calculator. (If it is not possible, state the reason.)

Solution

$\log_{10} 1000 = 3$

77. Evaluate $\log_4 (-4)$ without the aid of a calculator. (If it is not possible, state the reason.)

Solution

$\log_4 (-4) = $ not possible

There is no power to which 4 can be raised to obtain -4.

79. Evaluate $\log_6 8$ without the aid of a calculator. (It it is not possible, state the reason.)

Solution

$\log_{16} 8 = \dfrac{3}{4}$

81. Evaluate $\log_7 7^4$ without the aid of a calculator. (It it is not possible, state the reason.)

Solution

$\log_7 7^4 = 4$

83. Use the properties of natural logarithms to evaluate the expression ln 1.

Solution

ln 1 = 0

85. Use the properties of natural logarithms to evaluate the expression $\ln \dfrac{e^3}{e^2}$.

Solution

$\ln \dfrac{e^3}{e^2} = \ln e = 1$

87. Evaluate $\log_{10}\left(\sqrt{2} + 4\right)$ with the aid of a calculator. Round your answer to four decimal places.

Solution

$\log_{10}\left(\sqrt{2} + 4\right) = 0.7335$

89. Evaluate $\log_{10} 0.85$ with the aid of a calculator. Round your answer to four decimal places.

Solution

$\log_{10} 0.85 = -0.0706$

91. Evaluate ln 25 with the aid of a calculator. Round your answer to four decimal places.

Solution

ln 25 = 3.2189

93. Describe the relationship between the graphs of f and g.

$f(x) = \log_6 x$

$g(x) = 6^x$

Solution

f and g are inverse functions.

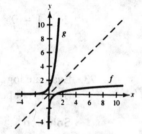

95. Sketch the graph of $g(t) = -\log_2 t$.

Solution

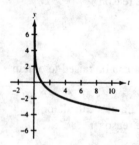

97. Sketch the graph of $g(x) = \log_2 (x - 3)$.

Solution

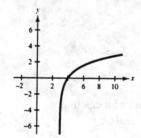

99. Sketch the graph of $f(x) = \log_{10}(10x)$.

Solution

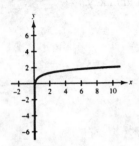

101. Use a graphing utility to sketch the graph of $y = -3 + 5 \log_{10} x$.

Solution

Keystrokes:

$\boxed{Y =}$ $\boxed{(-)}$ 3 $\boxed{+}$ 5 $\boxed{\log}$ $\boxed{X, T, \theta}$ $\boxed{\text{GRAPH}}$

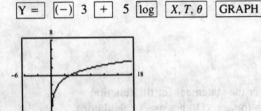

103. Use a graphing utility to sketch the graph of $h(t) = -2 \ln t$.

Solution

Keystrokes:

$\boxed{Y =}$ $\boxed{(-)}$ 2 $\boxed{\ln}$ $\boxed{X, T, \theta}$ $\boxed{\text{GRAPH}}$

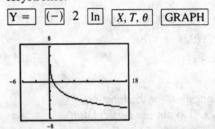

105. Use a graphing utility to sketch the graph of $g(x) = \ln(x + 6)$.

Solution

Keystrokes:

$\boxed{Y =}$ $\boxed{\ln}$ $\boxed{(}$ $\boxed{X, T, \theta}$ $\boxed{+}$ 6 $\boxed{)}$ $\boxed{\text{GRAPH}}$

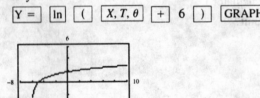

107. Use a graphing utility to sketch the graph of $f(x) = \ln(-x)$.

Solution

Keystrokes:

$\boxed{Y =}$ $\boxed{\ln}$ $\boxed{(}$ $\boxed{(-)}$ $\boxed{X, T, \theta}$ $\boxed{)}$ $\boxed{\text{GRAPH}}$

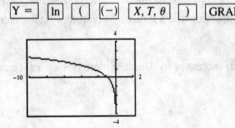

109. Evaluate $\log_3 7$ by means of the change-of-base formula. Use (a) the common logarithm key and (b) the natural logarithm key.

Solution

$$\log_3 7 = \frac{\log 7}{\log 3} = 1.7712$$

$$= \frac{\ln 7}{\ln 3} = 1.7712$$

111. Evaluate $\log_{15} 1250$ by means of the change-of-base formula. Use (a) the common logarithm key and (b) the natural logarithm key.

Solution

$$\log_{15} 1250 = \frac{\log 1250}{\log 15} = 2.6332$$

$$= \frac{\ln 1250}{\ln 15} = 2.6332$$

113. Evaluate $\log_4 \sqrt{42}$ by means of the change-of-base formula. Use (a) the common logarithm key and (b) the natural logarithm key.

Solution

$$\log_4 \sqrt{42} = \frac{\log \sqrt{42}}{\log 4} = 1.3481$$

$$= \frac{\ln \sqrt{42}}{\ln 4} = 1.3481$$

115. Evaluate $\log_{(1/2)} 4$ by means of the change-of-base formula. Use (a) the common logarithm key and (b) the natural logarithm key.

Solution

$$\log_{(1/2)} 4 = \frac{\log 4}{\log .5} = -2$$

$$= \frac{\ln 4}{\ln .5} = -2$$

117. *American Elk*   The antler spread a (in inches) and shoulder height h (in inches) of an adult male American elk are related by the model

$$h = 116 \log_{10} (a + 40) - 176.$$

Approximate the shoulder height of a male American elk with an antler spread of 55 inches.

Solution

$$h = 116 \log_{10} (55 + 40) - 176$$
$$= 116 \log_{10} (95) - 176$$
$$= 53.4 \text{ inches}$$

119. Answer the question for the function $f(x) = \log_{10} x$. (Do not use a calculator.)

What is the domain of f?

Solution

Domain = positive real numbers $(x > 0)$

121. Answer the statement for the function $f(x) = \log_{10} x$. (Do not use a calculator.)

Describe the values of $f(x)$ for $1000 \le x \le 10,000$.

Solution

If $1000 \le x \le 10,000$, then $f(x) = \log_{10} x$ lies $3 \le f(x) \le 4$.

123. Answer the question for the function $f(x) = \log_{10} x$. (Do not use a calculator.)

By what amount will x increase, given that $f(x)$ is increased by one unit?

Solution

When $f(x)$ increases by 1 unit, x increases by a factor of 10.

Mid-Chapter Quiz

1. Given $f(x) = \left(\dfrac{4}{3}\right)^x$, find (a) $f(2)$, (b) $f(0)$, (c) $f(-1)$, and (d) $f(1.5)$.

Solution

(a) $f(2) = \left(\dfrac{4}{3}\right)^2 = \dfrac{16}{9}$

(b) $f(0) = \left(\dfrac{4}{3}\right)^0 = 1$

(c) $f(-1) = \left(\dfrac{4}{3}\right)^{-1} = \dfrac{3}{4}$

(d) $f(1.5) = \left(\dfrac{4}{3}\right)^{1.5} = 1.54$ or $\dfrac{8\sqrt{3}}{9}$

2. Find the domain and range of $g(x) = 2^{-0.5x}$.

Solution

$g(x) = 2^{-0.5x}$

Domain: $(-\infty, \infty)$ Range: $(0, \infty)$

3. Sketch the graph of $y = \dfrac{1}{2}(4^x)$.

Solution

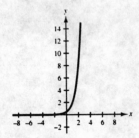

4. Sketch the graph of $y = 5(2^{-x})$.

Solution

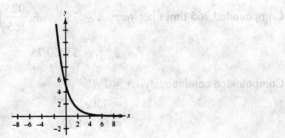

5. Sketch the graph of $f(t) = 12e^{-0.4t}$.

Solution

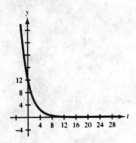

6. Sketch the graph of $g(x) = 100(1.08)^x$.

Solution

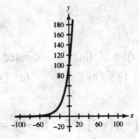

7. You deposit \$750 at $7\frac{1}{2}\%$ interest, compounded n times per year or continuously. Find the balance A after 20 years.

n	1	4	12	365	Continuous compounding
A					

Solution

Compounded 1 time per year: $A = 750\left(1 + \dfrac{.075}{1}\right)^{1(20)}$

$$= \$3185.89$$

Compounded 4 times per year: $A = 750\left(1 + \dfrac{.075}{4}\right)^{4(20)}$

$$= \$3314.90$$

Compounded 12 times per year: $A = 750\left(1 + \dfrac{.075}{12}\right)^{12(20)}$

$$= \$3345.61$$

Compounded 365 times per year: $A = 750\left(1 + \dfrac{.075}{365}\right)^{365(20)}$

$$= \$3360.75$$

Compounded continuously: $A = Pe^{rt}$

$$= 750e^{.075(20)}$$

$$= \$3361.27$$

8. A gallon of milk costs \$2.23 now. If the price increases by 4% each year, what will the price be after 5 years?

Solution

$A = 2.23e^{(.04)(5)} = \$2.72$

9. Given $f(x) = 2x - 3$ and $g(x) = x^3$, find the indicated composition.

    (a) $(f \circ g)(-2)$ (b) $(g \circ f)(4)$ (c) $(f \circ g)(x)$ (d) $(g \circ f)(x)$

Solution

(a) $(f \circ g)(-2) = f[g(-2)] = f[-8] = 2(-8) - 3 = -19$

(b) $(g \circ f)(4) = g[f(4)] = g[5] = 5^3 = 125$

(c) $(f \circ g)(x) = f[g(x)] = 2x^3 - 3$

(d) $(g \circ f)(x) = g[f(x)] = (2x - 3)^3$

10. Verify algebraically and graphically that $f(x) = 3 - 5x$ and $g(x) = \frac{1}{5}(3 - x)$ are inverses of each other.

Solution

$$f[g(x)] = 3 - 5\left[\frac{1}{5}(3 - x)\right] = 3 - 1(3 - x) = 3 - 3 + x = x$$

$$g[f(x)] = \frac{1}{5}[3 - (3 - 5x)] = \frac{1}{5}[3 - 3 + 5x] = \frac{1}{5}[5x] = x$$

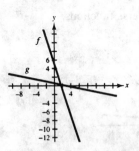

11. Find the inverse of $h(x) = 10x + 3$.

Solution

$$h(x) = 10x + 3$$
$$y = 10x + 3$$
$$x = 10y + 3$$
$$x - 3 = 10y$$
$$\frac{x - 3}{10} = y$$
$$\frac{x - 3}{10} = h^{-1}(x)$$

12. Find the inverse of $g(t) = \frac{1}{2}t^3 + 2$.

Solution

$$g(t) = \frac{1}{2}t^3 + 2$$
$$y = \frac{1}{2}t^3 + 2$$
$$t = \frac{1}{2}y^3 + 2$$
$$t - 2 = \frac{1}{2}y^3$$
$$2t - 4 = y^3$$
$$\sqrt[3]{2t - 4} = y$$
$$\sqrt[3]{2t - 4} = g^{-1}(t)$$

13. Write the logarithmic equation $\log_4\left(\frac{1}{16}\right) = -2$ in exponential form.

Solution

$$\log_4\left(\frac{1}{16}\right) = -2$$
$$4^{-2} = \frac{1}{16}$$

14. Write the exponential equation $3^4 = 81$ in logarithmic form.

Solution

$$3^4 = 81$$
$$\log_3 81 = 4$$

15. Evaluate $\log_5 125$ without the aid of a calculator.

Solution

$$\log_5 125 = 3$$

16. Write a paragraph comparing the graph of $f(x) = \log_5 x$ and $g(x) = 5^x$.

Solution

f and g are inverse functions.

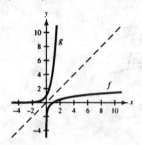

17. Use a graphing utility to sketch the graph of the function $f(t) = \dfrac{1}{2} \ln t$.

Solution

Keystrokes:

$\boxed{Y =}$ .5 $\boxed{\ln}$ $\boxed{X, T, \theta}$ $\boxed{\text{GRAPH}}$

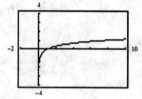

18. Use a graphing utility to sketch the graph of the function $h(x) = 3 - \ln x$.

Solution

Keystrokes:

$\boxed{Y =}$ 3 $\boxed{-}$ $\boxed{\ln}$ $\boxed{X, T, \theta}$ $\boxed{\text{GRAPH}}$

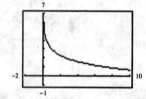

19. Use the graph of f in the textbook to determine h and k if $f(x) = \log_5(x - h) + k$.

Solution

$f(x) = \log_5(x - 3) + 1$

The graph of $f(x) = \log_5 x$ has been shifted 3 units right and 1 unit up.

20. Use a calculator and the change-of-base formula to evaluate $\log_6 450$.

Solution

$$\log_6 450 = \frac{\log 450}{\log 6} = 3.4096$$

9.4 | Properties of Logarithms

5. Evaluate $\log_5 5^2$ without using a calculator. (If it is not possible, state the reason.)

Solution

$\log_5 5^2 = 2$

7. Evaluate $\log_2 \dfrac{1}{8}$ without using a calculator. (If it is not possible, state the reason.)

Solution

$\log_2 \dfrac{1}{8} = -3$

9. Use $\log_{10} 3 \approx 0.477$ and $\log_{10} 12 \approx 1.079$ to approximate the value of the expression $\log_{10} 9$.

Solution

$\log_{10} 9 = \log_{10} 3^2 = 2 \log_{10} 3 \approx 2(0.477) \approx 0.954$

11. Use $\log_{10} 3 \approx 0.477$ and $\log_{10} 12 \approx 1.079$ to approximate the value of the expression $\log_{10} 36$.

Solution

$\log_{10} 36 = \log_{10} (3 \cdot 12) = \log_{10} 3 + \log_{10} 12$
$\approx 0.477 + 1.079$
≈ 1.556

13. Use $\log_{10} 3 \approx 0.477$ and $\log_{10} 12 \approx 1.079$ to approximate the value of the expression $\log_{10} \sqrt{36}$.

Solution

$\log_{10} \sqrt{36} = \log_{10} 36^{1/2} = \dfrac{1}{2} \log_{10} 36$

$\approx \dfrac{1}{2}(1.556)$

≈ 0.778

15. Use the properties of logarithms to expand $\log_2 3x$ as a sum, difference, or multiple of logarithms.

Solution

$\log_2 3x = \log_2 3 + \log_2 x$

17. Use the properties of logarithms to expand $\log_5 x^{-2}$ as a sum, difference, or multiple of logarithms.

Solution

$\log_5 x^{-2} = -2 \log_5 x$

19. Use the properties of logarithms to expand $\ln \dfrac{5}{x-2}$ as a sum, difference, or multiple of logarithms.

Solution

$\ln \dfrac{5}{x-2} = \ln 5 - \ln(x-2)$

21. Use the properties of logarithms to expand $\ln \sqrt{x(x+2)}$ as a sum, difference, or multiple of logarithms.

Solution

$\ln \sqrt{x(x+2)} = \dfrac{1}{2}[\ln x + \ln(x+2)]$

23. Use the properties of logarithms to condense $\log_{12} x - \log_{12} 3$.

Solution

$$\log_{12} x - \log_{12} 3 = \log_{12} \frac{x}{3}$$

25. Use the properties of logarithms to condense $-2 \log_5 2x$.

Solution

$$-2 \log_5 2x = \log_5 (2x)^{-2} = \log_5 \frac{1}{4x^2}$$

27. Use the properties of logarithms to condense $5 \ln 2 - \ln x + 3 \ln y$.

Solution

$$5 \ln 2 - \ln x + 3 \ln y = \ln 2^5 - \ln x + \ln y^3$$
$$= \ln 32 - \ln x + \ln y^3$$
$$= \ln \frac{32 y^3}{x}$$

29. Use the properties of logarithms to condense $\frac{1}{2}(\ln 8 + \ln 2x)$.

Solution

$$\frac{1}{2}(\ln 8 + \ln 2x) = \frac{1}{2}[\ln(8 \cdot 2x)]$$
$$= \frac{1}{2} \ln(16x)$$
$$= \ln \sqrt{16x}$$
$$= \ln 4\sqrt{x}$$

31. Simplify $\log_4 \frac{4}{x}$.

Solution

$$\log_4 \frac{4}{x} = \log_4 4 - \log_4 x = 1 - \log_4 x$$

33. Simplify $\log_5 \sqrt{50}$.

Solution

$$\log_5 \sqrt{50} = \frac{1}{2}[\log_5(5^2 \cdot 2)] = \frac{1}{2}[2 \log_5 5 + \log_5 2] = \frac{1}{2}[2 + \log_5 2] = 1 + \frac{1}{2}\log_5 2$$

35. Simplify $\ln 3e^2$.

Solution

$$\ln 3e^2 = 2 \ln 3e = 2[\ln 3 + \ln e] = 2[\ln 3 + 1] = 2 \ln 3 + 2$$

37. Use a graphing utility to verify that the expressions $\ln[10/(x^2 + 1)]^2$ and $2[\ln 10 - \ln(x^2 + 1)]$ are equivalent.

Solution

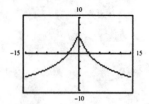

Keystrokes:

y_1 | Y = | | ln | | (| | 10 | | ÷ | | (| | X, T, θ | | x^2 | | + | 1 |) | |) | | x^2 | | ENTER |

y_2 | Y = | 2 | (| | ln | 10 | | − | | ln | | (| | X, T, θ | | x^2 | | + | 1 |) | |) | | GRAPH |

39. *Think About It*   Explain how you could show that

$$\frac{(\ln x)}{(\ln y)} \neq \ln\left(\frac{x}{y}\right).$$

Solution

Choose 2 values for x and y, such as $x = 3$ and $y = 5$, and show the 2 expressions are not equal.

$$\frac{\ln 3}{\ln 5} \neq \ln\frac{3}{5} = \ln 3 - \ln 5$$

$$0.6826062 \neq -0.5108256 = -0.5108256$$

41. *Biology*   The energy E (in kilocalories per gram molecule) required to transport a substance from the outside to the inside of a living cell is

$$E = 1.4(\log_{10} C_2 - \log_{10} C_1)$$

where C_1 is the concentration of the substance outside the cell and C_2 is the concentration inside. Condense the expression.

Solution

$$E = 1.4(\log_{10} C_2 - \log_{10} C_1) = 1.4\left(\log_{10}\frac{C_2}{C_1}\right)$$

43. Use the properties of exponents to simplify $(x^2 \cdot x^3)^4$.

Solution

$$(x^2 \cdot x^3)^4 = (x^5)^4 = x^{20}$$

45. Use the properties of exponents to simplify $\dfrac{15y^{-3}}{10y^2}$.

Solution

$$\frac{15y^{-3}}{10y^2} = \frac{3y^{-5}}{2} = \frac{3}{2y^5}$$

47. Use the properties of exponents to simplify $\dfrac{3x^2y^3}{18x^{-1}y^2}$.

Solution

$$\frac{3x^2y^3}{18x^{-1}y^2} = \frac{x^3y}{6}$$

49. *Ticket Prices* A service organization paid $288 for a block of tickets to a game. The block contained three more tickets than the organization needed for its members. By inviting three more people to share in the cost, the organization lowered the price per ticket by $8. How many people will attend?

Solution

Verbal Model: $\boxed{\begin{array}{c}\text{Cost per}\\\text{Person,}\\\text{Current}\\\text{Group}\end{array}} - \boxed{\begin{array}{c}\text{Cost per}\\\text{Person,}\\\text{New}\\\text{Group}\end{array}} = 8$

Equation: $\dfrac{288}{x} - \dfrac{288}{x+3} = 8$

$288(x+3) - 288x = 8x(x+3)$

$288x + 864 - 288x = 8x^2 + 24x$

$0 = 8x^2 + 24x - 864$

$0 = x^2 + 3x - 108$

$0 = (x+12)(x-9)$

$x = 9$

$x + 3 = 12$

51. Evaluate $\log_4 2 + \log_4 8$ without using a calculator. (If it is not possible, state the reason.)

Solution

$\log_4 2 + \log_4 8 = \log_4 16 = 2$ because $4^2 = 16$

53. Evaluate $\log_3 9$ without using a calculator. (If it is not possible, state the reason.)

Solution

$\log_3 9 = 2$ because $3^2 = 9$

55. Evaluate $\ln e^5 - \ln e^2$ without using a calculator. (If it is not possible, state the reason.)

Solution

$\ln e^5 - \ln e^2 = \ln \dfrac{e^5}{e^2} = \ln e^3 = 3 \ln e = 3$

57. Evaluate $\ln 1$ without using a calculator. (If it is not possible, state the reason.)

Solution

$\ln 1 = 0$

59. Approximate $\log_4 4$ given that $\log_4 2 = 0.5000$ and $\log_4 3 \approx 0.7925$. Do not use a calculator.

Solution

$\log_4 4 = \log_4 2 + \log_4 2 = 0.5000 + 0.5000 = 1$

61. Approximate $\log_4 6$ given that $\log_4 2 = 0.5000$ and $\log_4 3 \approx 0.7925$. Do not use a calculator.

Solution

$\log_4 6 = \log_4 2 + \log_4 3 = 0.5000 + 0.7925 \approx 1.2925$

63. Approximate $\log_4 \dfrac{3}{2}$ given that $\log_4 2 = 0.5000$ and $\log_4 3 \approx 0.7925$. Do not use a calculator.

Solution

$\log_4 \dfrac{3}{2} = \log_4 3 - \log_4 2 = 0.7925 - 0.5.000 \approx 0.2925$

65. Approximate $\log_4 \sqrt{2}$ given that $\log_4 2 = 0.5000$ and $\log_4 3 \approx 0.7925$. Do not use a calculator.

Solution

$\log_4 \sqrt{2} = \frac{1}{2} \log_4 2 = \frac{1}{2}(0.5000) = 0.25$

67. Approximate $\log_4 3 \cdot 2^4$ given that $\log_4 2 = 0.5000$ and $\log_4 3 \approx 0.7925$. Do not use a calculator.

Solution

$\log_4 3 \cdot 2^4 = \log_4 3 + 4 \log_4 2 = 0.7925 + 4(0.5000) \approx 2.7925$

69. Approximate $\log_4 3^0$ given that $\log_4 2 \approx 0.5000$ and $\log_4 3 \approx 0.7925$. Do not use a calculator.

Solution

$\log_4 3^0 = \log_4 1 = 0$

71. Use the properties of logarithms to expand $\log_3 11x$ as a sum, difference, or multiple of logarithms.

Solution

$\log_3 11x = \log_3 11 + \log_3 x$

73. Use the properties of logarithms to expand $\ln y^3$ as a sum, difference, or multiple of logarithms.

Solution

$\ln y^3 = 3 \ln y$

75. Use the properties of logarithms to expand $\log_2 \frac{z}{17}$ as a sum, difference, or multiple of logarithms.

Solution

$\log_2 \frac{z}{17} = \log_2 z - \log_2 17$

77. Use the properties of logarithms to expand $\log_3 \sqrt[3]{x+1}$ as a sum, difference, or multiple of logarithms.

Solution

$\log_3 \sqrt[3]{x+1} = \frac{1}{3} \log_3(x+1)$

79. Use the properties of logarithms to expand $\ln 3x^2 y$ as a sum, difference, or multiple of logarithms.

Solution

$\ln 3x^2 y = \ln 3 + 2 \ln x + \ln y$

81. Use the properties of logarithms to expand $\log_2 \frac{x^2}{x-3}$ as a sum, difference, or multiple of logarithms.

Solution

$\log_2 \frac{x^2}{x-3} = 2 \log_2 x - \log_2(x-3)$

83. Use the properties of logarithms to expand $\ln \sqrt[3]{x(x+5)}$ as a sum, difference, or multiple of logarithms.

Solution

$\ln \sqrt[3]{x(x+5)} = \frac{1}{3}[\ln x + \ln(x+5)] \quad$ or $\quad \frac{1}{3}\ln x + \frac{1}{3}\ln(x+5)$

85. Use the properties of logarithms to condense $\log_2 3 + \log_2 x$.

Solution

$$\log_2 3 + \log_2 x = \log_2 3x$$

87. Use the properties of logarithms to condense $\log_{10} 4 - \log_{10} x$.

Solution

$$\log_{10} 4 - \log_{10} x = \log_{10} \frac{4}{x}$$

89. Use the properties of logarithms to condense $4 \ln b$.

Solution

$$4 \ln b = \ln b^4$$

91. Use the properties of logarithms to condense $\frac{1}{3} \ln(2x + 1)$.

Solution

$$\frac{1}{3} \ln(2x + 1) = \ln \sqrt[3]{2x + 1}$$

93. Use the properties of logarithms to condense $\log_3 2 + \frac{1}{2} \log_3 y$.

Solution

$$\log_3 2 + \frac{1}{2} \log_3 y = \log_3 2 + \log_3 \sqrt{y}$$
$$= \log_3 2\sqrt{y}$$

95. Use the properties of logarithms to condense $2 \ln x + 3 \ln y - \ln z$.

Solution

$$2 \ln x + 3 \ln y - \ln z = \ln \frac{x^2 y^3}{z}$$

97. Use the properties of logarithms to condense $4(\ln x + \ln y)$.

Solution

$$4(\ln x + \ln y) = \ln(xy)^4 \quad \text{or} \quad \ln x^4 y^4$$

99. Determine whether the equation is true or false.
$\ln e^{2-x} = 2 - x$

Solution

true, $\ln e^{2-x} = (2 - x) \ln e = (2 - x)(1)$
$$= 2 - x$$

101. Determine whether the equation is true or false.
$\log_8 4 + \log_8 16 = 2$

Solution

true, $\log_8 4 + \log_8 16 = \log_8 4 \cdot 16$
$$= \log_8 64$$
$$= 2$$

103. Determine whether the equation is true or false.
$\log_3(u + v) = \log_3 u \cdot \log_3 v$

Solution

false, $\log_3 (u \cdot v) = \log_3 u + \log_3 v$

105. Use a graphing utility to graph the two equations in the same viewing rectangle. Use the graphs to verify that the expressions are equivalent.

$$y_1 = \ln[x^2(x + 2)], x > 0$$
$$y_2 = 2 \ln x + \ln(x + 2)$$

Solution

Keystrokes:

y_1 [Y =] [ln] [(] [X, T, θ] [x²] [(] [X, T, θ] [+] [2] [)] [)] [ENTER]

y_2 [Y =] [2] [ln] [X, T, θ] [+] [ln] [(] [X, T, θ] [+] [2] [)] [GRAPH]

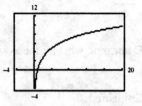

107. *Intensity of Sound*   The relationship between the number of decibels B and the intensity of a sound I in watts per meter squared is given by

$$B = 10 \log_{10}\left(\frac{I}{10^{-16}}\right).$$

Use properties of logarithms to write the formula in simpler form, and determine the number of decibels of a sound with intensity of 10^{-10} watts per meter squared.

Solution

(a) $B = 10 \log_{10}\left(\dfrac{I}{10^{-16}}\right)$

$\qquad = 10[\log_{10} I - \log_{10} 10^{-16}]$

$\qquad = 10[\log_{10} I - (-16)]$

$\qquad = 10[\log_{10} I + 16]$

(b) $B = 10[\log_{10} 10^{-10} + 16]$

$\qquad = 10[-10 + 16]$

$\qquad = 60$

109. Determine if the statement is true or false given that $f(x) = \ln x$. If false, state why or give an example to show that it is false.

$$f(0) = 0$$

Solution

False; 0 is not in the domain of f.

111. Determine if the statement is true or false given that $f(x) = \ln x$. If false, state why or give an example to show that it is false.

$$f(x - 3) = \ln x - \ln 3, x > 3$$

Solution

False; $f(x - 3) = \ln(x - 3) \neq \ln x - \ln 3$.

113. Determine if the statement is true or false given that $f(x) = \ln x$. If false, state why or give an example to show that it is false.

If $f(u) = 2f(v)$, then $v = u^2$.

Solution

False; if $v = u^2$, then $f(v) = \ln u^2 = 2 \ln u = 2 f(u)$.

9.5 Solving Exponential and Logarithmic Equations

5. Determine whether the x-values are solutions of the equation $3^{2x-5} = 27$.

   (a) $x = 1$ (b) $x = 4$

Solution

(a) $3^{2(1)-5} \overset{?}{=} 27$ (b) $3^{2(4)-5} \overset{?}{=} 27$

      $3^{-3} \neq 27$ $3^3 = 27$

    no   yes

7. Determine whether the x-values are solutions of the equation $e^{x+5} = 45$.

   (a) $x = -5 + \ln 45$ (b) $x = -5 + e^{45}$

Solution

(a) $e^{-5 + \ln 45 + 5} \overset{?}{=} 45$ (b) $e^{-5 + e^{45} + 5} \overset{?}{=} 45$

        $e^{\ln 45} \overset{?}{=} 45$ $e^{e^{45}} \overset{?}{\neq} 45$

             $45 = 45$ no

    yes

9. Determine whether the x-values are solutions of the equation $\log_9 (6x) = \dfrac{3}{2}$.

   (a) $x = 27$ (b) $x = \dfrac{9}{2}$

Solution

(a) $\log_9 (6 \cdot 27) \overset{?}{=} \dfrac{3}{2}$ (b) $\log_9 \left(6 \cdot \dfrac{9}{2} \right) \overset{?}{=} \dfrac{3}{2}$

       $\log_9 162 \overset{?}{\neq} \dfrac{3}{2}$ $\log_9 27 = \dfrac{3}{2}$

    no   yes

11. Solve the equation $3^{x-1} = 3^7$.

Solution

     $3^{x-1} = 3^7$

so $x - 1 = 7$

        $x = 8$

13. Solve the equation $2^{x+2} = \dfrac{1}{16}$.

Solution

     $2^{x+2} = \dfrac{1}{16}$

     $2^{x+2} = 2^{-4}$

so $x + 2 = -4$

        $x = -6$

15. Solve the equation $\log_3(2 - x) = 2$.

Solution

$$\log_3(2 - x) = 2$$
$$2 - x = 3^2$$
$$-x = 7$$
$$x = -7$$

17. Solve the equation $\ln(2x - 3) = \ln 15$.

Solution

$$\ln(2x - 3) = \ln 15$$
$$2x - 3 = 15$$
$$2x = 18$$
$$x = 9$$

19. Simplify $\ln e^{2x-1}$.

Solution

$$\ln e^{2x-1} = 2x - 1$$

21. Simplify $10^{\log_{10} 2x}$.

Solution]

$$10^{\log_{10} 2x} = 2x$$

23. Solve $4^x = 8$.

Solution

$$4^x = 8$$
$$\log_4 4^x = \log_4 8$$
$$x = \log_4 8$$
$$x = 1.5 \text{ or } \frac{3}{2}$$

25. Solve $\frac{1}{2} e^{3x} = 20$.

Solution

$$\frac{1}{2} e^{3x} = 20$$
$$e^{3x} = 40$$
$$\ln e^{3x} = \ln 40$$
$$3x = \ln 40$$
$$x = \frac{\ln 40}{3} \approx 1.23$$

27. Solve $5(2)^{3x} - 4 = 13$.

Solution

$$5(2)^{3x} - 4 = 13$$
$$5(2)^{3x} = 17$$
$$2^{3x} = \frac{17}{5}$$
$$\log_2 2^{3x} = \log_2 \frac{17}{5}$$
$$3x = \log_2 \frac{17}{5}$$
$$x = \frac{\log_2 \dfrac{17}{5}}{3}$$
$$x \approx 0.59$$

29. Solve $8 - 12 e^{-x} = 7$.

Solution

$$8 - 12e^{-x} = 7$$
$$-12 e^{-x} = -1$$
$$e^{-x} = \frac{1}{12}$$
$$\ln e^{-x} = \ln \frac{1}{12}$$
$$-x = \ln \frac{1}{12}$$
$$x = -\ln \frac{1}{12} \approx 2.48$$

31. Solve $\dfrac{8000}{(1.03)^t} = 6000$.

Solution

$$\frac{8000}{(1.03)^t} = 6000$$

$$\frac{8000}{6000} = (1.03)^t$$

$$\frac{4}{3} = (1.03)^t$$

$$\log_{1.03} \frac{4}{3} = \log_{1.03} 1.03^t$$

$$\log_{1.03} \frac{4}{3} = t$$

$$9.73 \approx t$$

33. Solve $\log_2 x = 4.5$.
(Round the solution to two decimal places.)

Solution

$$\log_2 x = 4.5$$

$$x = 2^{4.5}$$

$$x = 22.63$$

35. Solve $4 \log_3 x = 28$.
(Round the solution to two decimal places.)

Solution

$$4 \log_3 x = 28$$

$$\log_3 x = 7$$

$$x = 3^7$$

$$x = 2187$$

37. Solve $16 \ln x = 30$.
(Round the solution to two decimal places.)

Solution

$$16 \ln x = 30$$

$$\ln x = \frac{30}{16}$$

$$x = e^{15/8}$$

$$x \approx 6.52$$

39. Solve $1 - 2 \ln x = -4$. (Round the solution to two decimal places.)

Solution

$$1 - 2 \ln x = -4$$

$$-2 \ln x = -5$$

$$\ln x = \frac{5}{2}$$

$$x = e^{2.5}$$

$$x \approx 12.18$$

41. Solve $\log_2 (x - 1) + \log_2 (x + 3) = 3$. (Round the solution to two decimal places.)

Solution

$$\log_2 (x - 1) + \log_2 (x + 3) = 3$$
$$\log_2 (x - 1)(x + 3) = 3$$
$$x^2 + 2x - 3 = 2^3$$
$$x^2 + 2x - 11 = 0$$
$$x = \frac{-2 \pm \sqrt{4 - 4(1)(-11)}}{2(1)} = \frac{-2 \pm \sqrt{4 + 44}}{2} = \frac{-2 \pm \sqrt{48}}{2}$$

$x \approx 2.46$ and -4.46 (which is extraneous)

43. Use a graphing utility to approximate the x-intercept of the graph of $y = 10^{x/2} - 5$.

Solution

Keystrokes:

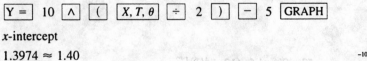

x-intercept

$1.3974 \approx 1.40$

$(1.40, 0)$

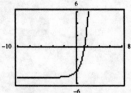

45. Use a graphing utility to approximate the x-intercept of the graph of $y = 6 \ln (0.4x) - 13$.

Solution

Keystrokes:

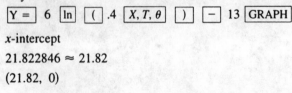

x-intercept

$21.822846 \approx 21.82$

$(21.82, 0)$

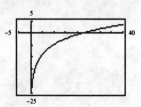

47. *Doubling Time* At 9% interest, compounded continuously, how long does it take an investment to double?

Solution

$$A = Pe^{rt}$$
$$2P = Pe^{.09t}$$
$$2 = e^{.09t}$$
$$\ln 2 = \ln e^{.09t}$$
$$\ln 2 = .09t$$
$$\frac{\ln 2}{.09} = t$$

7.70 years $\approx t$

49. *Friction* In order to restrain an untrained horse, a person partially wraps the rope around a cylindrical post in the corral (see figure in textbook). If the horse is pulling on the rope with a force of 200 pounds, the force F in pounds required by the person is

$$F = 200e^{-0.2\pi\theta/180}$$

where θ is the angle of the wrap in degrees. Find the smallest value of θ if F cannot exceed 80 pounds.

Solution

$$F = 200e^{-0.2\pi\theta/180}$$
$$80 = 200e^{-0.2\pi\theta/180}$$
$$0.4 = e^{-0.2\pi\theta/180}$$
$$\ln 0.4 = \ln e^{-0.2\pi\theta/180}$$
$$\ln 0.4 = \frac{-0.2\pi\theta}{180}$$
$$(\ln 0.4)\left(\frac{180}{-0.2\pi}\right) = \theta$$
$$262.50^0 = \theta$$

51. Simplify $\sqrt{24x^2y^3}$.

Solution

$$\sqrt{24x^2y^3} = \sqrt{4 \cdot 6 \cdot x^2 \cdot y^2 \cdot y} = 2xy\sqrt{6y}$$

53. Simplify $(12a^{-4}b^6)^{1/2}$.

Solution

$$(12a^{-4}b^6)^{1/2} = 4^{1/2} \cdot 3^{1/2} \cdot a^{-2}b^3 = \frac{2\sqrt{3}\,b^3}{a^2}$$

55. *Balance* $4000 is deposited at 6%, compounded monthly. Find the account balance after 5 years.

Solution

$$A = 4000\left(1 + \frac{.06}{12}\right)^{12(5)}$$
$$= \$5395.40$$

57. Solve $2^x = 2^5$. (Do not use a calculator.)

Solution

$$2^x = 2^5$$
$$x = 5$$

59. Solve $3^{x+4} = 3^{12}$. (Do not use a calculator.)

Solution

$$3^{x+4} = 3^{12}$$
$$x + 4 = 12$$
$$x = 8$$

61. Solve $4^{x-1} = 16$. (Do not use a calculator.)

Solution

$$4^{x-1} = 16$$
$$4^{x-1} = 4^2$$
$$x - 1 = 2$$
$$x = 3$$

63. Solve $5^x = \frac{1}{125}$. (Do not use a calculator.)

Solution

$$5^x = \frac{1}{125}$$
$$5^x = 5^{-3}$$
$$x = -3$$

65. Solve $\log_5 2x = \log_5 36$. (Do not use a calculator.)

Solution

$$\log_5 2x = \log_5 36$$
$$2x = 36$$
$$x = 18$$

67. Solve $\ln 5x = \ln 22$. (Do not use a calculator.)

Solution

$$\ln 5x = \ln 22$$
$$5x = 22$$
$$x = \frac{22}{5}$$

69. Solve $\log_3 x = 4$. (Do not use a calculator.)

Solution

$$\log_3 x = 4$$
$$3^{\log_3 x} = 3^4$$
$$x = 3^4$$
$$x = 81$$

71. Solve $\log_{10} 2x = 6$. (Do not use a calculator.)

Solution

$$\log_{10} 2x = 6$$
$$10^{\log_{10} 2x} = 10^6$$
$$2x = 10^6$$
$$x = \frac{1,000,000}{2}$$
$$x = 500,000$$

73. Solve $2^x = 45$.
(Round the solution to two decimal places.)

Solution

$$2x = 45$$
$$\log_2 2x = \log_2 45$$
$$x = \frac{\log 45}{\log 2}$$
$$x \approx 5.49$$

75. Solve $10^{2y} = 52$.
(Round the solution to two decimal places.)

Solution

$$10^{2y} = 52$$
$$\log 10^{2y} = \log 52$$
$$2y = \log 52$$
$$y = \frac{\log 52}{2}$$
$$y \approx 0.86$$

77. Solve $\frac{1}{5} 4^{x+2} = 300$.
(Round the solution to two decimal places.)

Solution

$$\frac{1}{5} 4^{x+2} = 300$$
$$4^{x+2} = 1500$$
$$\log_4 4^{x+2} = \log_4 1500$$
$$x + 2 = \frac{\log 1500}{\log 4}$$
$$x = \frac{\log 1500}{\log 4} - 2$$
$$x \approx 3.28$$

79. Solve $4 + e^{2x} = 150$.
(Round the solution to two decimal places.)

Solution

$$4 + e^{2x} = 150$$
$$e^{2x} = 146$$
$$\ln e^{2x} = \ln 146$$
$$2x = \ln 146$$
$$x = \frac{\ln 146}{2}$$
$$x \approx 2.49$$

81. Solve $23 - 5e^{x+1} = 3$.
(Round the solution to two decimal places.)

Solution

$$23 - 5e^{x+1} = 3$$
$$-5e^{x+1} = -20$$
$$e^{x+1} = 4$$
$$\ln e^{x+1} = \ln 4$$
$$x + 1 = \ln 4$$
$$x = \ln 4 - 1$$
$$x \approx 0.39$$

83. Solve $300e^{x/2} = 9000$.
(Round the solution to two decimal places.)

Solution

$$300e^{x/2} = 9000$$
$$e^{x/2} = 30$$
$$\ln e^{x/2} = \ln 30$$
$$\frac{x}{2} = \ln 30$$
$$x = 2 \ln 30$$
$$x \approx 6.80$$

85. Solve $6000e^{-2t} = 1200$.
(Round the solution to two decimal places.)

Solution

$$6000e^{-2t} = 1200$$
$$e^{-2t} = 0.2$$
$$\ln e^{-2t} = \ln 0.2$$
$$-2t = \ln 0.2$$
$$t = \frac{\ln 0.2}{-2}$$
$$t \approx 0.80$$

87. Solve $32(1.5)^x = 640$.
(Round the solution to two decimal places.)

Solution

$$32(1.5)^x = 640$$
$$1.5^x = 20$$
$$\log_{1.5} 1.5^x = \log_{1.5} 20$$
$$x = \frac{\log 20}{\log 1.5}$$
$$x \approx 7.39$$

89. Solve $\dfrac{1600}{(1.1)^x} = 200$.
(Round the solution to two decimal places.)

Solution

$$\frac{1600}{(1.1)^x} = 200$$
$$\frac{1600}{200} = (1.1)^x$$
$$8 = 1.1^x$$
$$\log_{1.1} 8 = \log_{1.1} 1.1^x$$
$$\frac{\log 8}{\log 1.1} = x$$
$$21.82 \approx x$$

91. Solve $4(1 + e^{x/3}) = 84$.
(Round the solution to two decimal places.)

Solution

$$4(1 + e^{x/3}) = 84$$
$$1 + e^{x/3} = 21$$
$$e^{x/3} = 20$$
$$\ln e^{x/3} = \ln 20$$
$$\frac{x}{3} = \ln 20$$
$$x = 3 \ln 20$$
$$x \approx 8.99$$

93. Solve $\log_{10} x = 0$.
(Round the solution to two decimal places.)

Solution

$$\log_{10} x = 0$$
$$10^{\log_{10} x} = 10^0$$
$$x = 1$$

95. Solve $\log_{10} x = 3$.
(Round the solution to two decimal places.)

Solution

$$\log_{10} x = 3$$
$$10^{\log_{10} x} = 10^3$$
$$x = 1000$$

97. Solve $\log_{10} 4x = \dfrac{3}{2}$.

(Round the solution to two decimal places.)

Solution

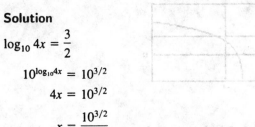

$$\log_{10} 4x = \frac{3}{2}$$

$$10^{\log_{10} 4x} = 10^{3/2}$$

$$4x = 10^{3/2}$$

$$x = \frac{10^{3/2}}{4}$$

$$x \approx 7.91$$

99. Solve $\ln x = 2.1$.

(Round the solution to two decimal places.)

Solution

$$\ln x = 2.1$$

$$e^{\ln x} = e^{2.1}$$

$$x = e^{2.1}$$

$$x \approx 8.17$$

101. Solve $\dfrac{2}{3} \ln(x + 1) = -1$.

(Round the solution to two decimal places.)

Solution

$$\frac{2}{3} \ln(x + 1) = -1$$

$$\ln(x + 1) = -1.5$$

$$e^{\ln(x + 1)} = e^{-1.5}$$

$$x + 1 = e^{-1.5} - 1$$

$$x \approx 0.78$$

103. Solve $2 \log_{10}(x + 5) = 15$.

(Round the solution to two decimal places.)

Solution

$$2 \log_{10}(x + 5) = 15$$

$$\log_{10}(x + 5) = 7.5$$

$$10^{\log_{10}(x + 5)} = 10^{7.5}$$

$$x + 5 = 10^{7.5}$$

$$x = 10^{7.5} - 5$$

$$x \approx 31{,}622{,}772$$

105. Solve $\ln x^2 = 6$.

(Round the solution to two decimal places.)

Solution

$$\ln x^2 = 6$$

$$e^{\ln x^2} = e^6$$

$$x^2 = e^6$$

$$x = \pm\sqrt{e^6}$$

$$x \approx \pm 20.09$$

107. Solve $\log_{10} x + \log_{10}(x - 3) = 1$.

(Round the solution to two decimal places.)

Solution

$$\log_{10} x + \log_{10}(x - 3) = 1$$

$$\log_{10} x(x - 3) = 1$$

$$10^{\log_{10} x(x - 3)} = 10^1$$

$$x(x - 3) = 10$$

$$x^2 - 3x - 10 = 0$$

$$(x - 5)(x + 2) = 0$$

$$x = 5, x = -2 \text{ (which is extraneous)}$$

109. Use a graphing utility to approximate the point of intersection of the graphs

$$y_1 = 2 \text{ and } y_2 = e^x.$$

Solution

Keystrokes:

y_1 [Y =] 2 [ENTER]

y_2 [Y =] [e] [X, T, θ] [GRAPH]

Point of intersection $= (0.69, 2)$

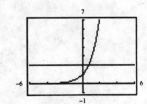

111. Use a graphing utility to approximate the point of intersection of the graph
$y_1 = 3$ and $y_2 = 2 \ln (x + 3)$.

Solution

Keystroke:

y_1 [Y =] 3 [ENTER]

y_2 [Y =] 2 [ln] [(] [X, T, θ] [+] 3 [)] [GRAPH]

Point of intersection = (1.48, 3)

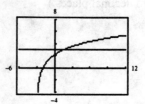

113. *Intensity of Sound*   The relationship between the number of decibels B and the intensity of a sound I in watts per meter squared is given by

$$B = 10 \log_{10}\left(\frac{I}{10^{-16}}\right).$$

Determine the intensity of a sound I if it registers 75 decibels on an intensity meter.

Solution

$$B = 10 \log_{10}\left(\frac{I}{10^{-16}}\right)$$

$$75 = 10 \log_{10}\left(\frac{I}{10^{-16}}\right)$$

$$7.5 = \log_{10}\left(\frac{I}{10^{-16}}\right)$$

$$10^{7.5} = 10^{\log_{10}}\left(\frac{I}{10^{-16}}\right)$$

$$10^{7.5} = \frac{I}{10^{-16}}$$

$$(10^{7.5})(10^{-16}) = I$$

$$10^{-8.5} = I$$

$$3.1623 \times 10^{-9} = I$$

115. *Muon Decay*   A muon is an elementary particle that is similar to an electron, but much heavier. Muons are unstable—they quickly decay to form electrons and other particles. In an experiment conducted in 1943, the number of muon decays m (of original 5000 muons) was related to the time t (in microseconds) by the model

$T = 15.7 - 2.48 \ln m$.

How many decays were recorded when $T = 2.5$?

Solution

$$2.5 = 15.7 - 2.48 \ln m$$

$$-13.2 = -2.48 \ln m$$

$$5.322580645 = \ln m$$

$$e^{5.322580645} = m$$

$$205 \approx m$$

9.6   Applications

7. *Annual Interest Rate*   Find the annual interest rate when the principal P is \$1000, the balance A is \$36,581.00, the time t is 40 years, and compounding is daily.

Solution

$$36{,}581.00 = 1000\left(1 + \frac{r}{365}\right)^{365(40)}$$

$$36.581 = \left(1 + \frac{r}{365}\right)^{14{,}600}$$

$$(36.581)^{1/14{,}600} = 1 + \frac{r}{365}$$

$$1.0002466 = 1 + \frac{r}{365}$$

$$0.0002466 = \frac{r}{365}$$

$$0.0899981 = r$$

$$8.9\% = r$$

9. *Annual Interest Rate*   Find the annual interest rate when the principal P is \$750, the balance A is \$8267.38, the time t is 30 years, and compounding is continuously.

Solution

$$8267.38 = 750e^{r(30)}$$

$$11.023173 = e^{r(30)}$$

$$\ln 11.023173 = \ln e$$

$$\ln 11.023173 = 30r$$

$$\frac{\ln 11.023173}{30} = r$$

$$0.08 = r$$

$$8\% = r$$

11. *Doubling Time*   Find the time for an investment to double in value when the principal P is \$6000, the annual interest rate r is 8%, and compounding is quarterly.

Solution

$$12{,}000 = 6000\left(1 + \frac{0.08}{4}\right)^{4t}$$

$$2 = (1.02)^{4t}$$

$$\log_{1.02} 2 = \log_{1.02} 1.02^{4t}$$

$$\frac{\log 2}{\log 1.02} = 4t$$

$$\frac{\log 2}{\log 1.02} \div 4 = t$$

$$8.75 \text{ years} = t$$

13. *Doubling Time*   Find the time for an investment to double in value when the principal P is \$2000, the annual interest rate r is 10.5%, and compounding is daily.

Solution

$$4000 = 2000\left(1 + \frac{0.105}{365}\right)^{365t}$$

$$2 = (1.0002877)^{365t}$$

$$\log_{1.0002877} 2 = \log_{1.0002877} 1.0002877^{365t}$$

$$\frac{\log 2}{\log 1.0002877} = 365t$$

$$\frac{\log 2}{\log 1.0002877} \div 365 = t$$

$$6.60 \text{ years} = t$$

15. *Type of Compounding* Determine the type of compounding when the principal P is \$750, the balance B is \$1587.75, the time t is 10 years, and the interest rate is r is 7.5%.

Solution

$1587.75 = 750\left(1 + \dfrac{0.075}{n}\right)^{n(10)}$

$1587.75 = 750e^{0.075(10)}$

$1587.75 = 1587.75$

continuous compounding

17. *Type of Compounding* Determine the type of compounding when the principal P is \$100, the balance B is \$141.48, the time t is 5 years, and the interest rate r is 7%.

Solution

$141.48 = 100\left(1 + \dfrac{0.07}{n}\right)^{n(5)}$

$141.48 = 100\left(1 + \dfrac{0.07}{4}\right)^{4(5)}$

$141.48 = 141.48$

quarterly compounding

19. *Effective Yield* Find the effective yield when the annual interest rate is 8% and the compounding is continuously.

Solution

$A = Pe^{rt}$

$A = 1000e^{0.08(1)}$

$A = \$1083.29$

effective yield $= \dfrac{83.29}{1000} = 0.08329 = 8.329\%$

21. *Effective Yield* Find the effective yield when the annual interest rate is 7% and the compounding is monthly.

Solution

$A = 1000\left(1 + \dfrac{0.07}{12}\right)^{12(1)}$

$A = \$1072.29$

effective yield $= \dfrac{72.29}{1000} = 0.07229 = 7.229\%$

23. *Principal* Find the principal P that must be deposited in an account to obtain the balance of \$10,000 at an interest rate r of 9%, for a time t of 20 years and compounded continuously.

Solution

$10,000 = Pe^{0.09(20)}$

$\dfrac{10,000}{e^{1.8}} = P$

$\$1652.99 = P$

25. *Principal* Find the principal P that must be deposited in an account to obtain the balance of \$750 at an interest rate r of 6%, for a time t of 3 years and compounded daily.

Solution

$750 = P\left(1 + \dfrac{0.06}{365}\right)^{365(3)}$

$\dfrac{750}{(1.0001644)^{1095}} = P$

$\$626.46 = P$

27. *Exponential Growth and Decay*   Using the graph in the textbook, find the constant k such that the graph of $y = Ce^{kt}$ passes through the given points.

Solution

(a) $y = Ce^{kt}$

$3 = Ce^{k(0)}$

$3 = C$

(b) $8 = 3e^{k(2)}$

$\dfrac{8}{3} = e^{2k}$

$\ln \dfrac{8}{3} = \ln e^{2k}$

$\ln \dfrac{8}{3} = 2k$

$\dfrac{\ln \dfrac{8}{3}}{2} = k = 0.4904146$

29. *Exponential Growth and Decay*   Using the graph in the textbook, find the constant k such that the graph of $y = Ce^{kt}$ passes through the given points.

Solution

(a) $y = Ce^{kt}$

$400 = Ce^{k(0)}$

$400 = C$

(b) $200 = 400e^{k(3)}$

$\dfrac{1}{2} = e^{3k}$

$\ln \dfrac{1}{2} = \ln e^{3k}$

$\ln \dfrac{1}{2} = 3k$

$\dfrac{\ln \dfrac{1}{2}}{3} = k = -0.2310491 = k$

31. *Population of a City*   The population of Los Angeles is 10.1 million for the year 1992, and the predicted population for the year 2000 is 10.7 million. Find the constants C and k to obtain exponential growth model $y = Ce^{kt}$ for the population growth. (Let $t = 0$ correspond to the year 1992.) Use the model to predict the population of Los Angeles in the year 2005.

Solution

(a) $y = Ce^{kt}$

$10.1 = Ce^{k(0)}$

$10.1 = C$

(b) $10.7 = 10.1e^{k(8)}$

$\dfrac{10.7}{10.1} = e^{8k}$

$\ln \dfrac{107}{101} = \ln e^{8k}$

$\ln \dfrac{107}{101} = 8k$

$\dfrac{1}{8} \ln \dfrac{107}{110} = k$

$0.0072 \approx k$

(c) $y = 10.1e^{0.0072(13)}$

$y \approx 11.1$ million

33. *Population of a City* The population of Dhaka, Bangladesh, is 4.6 million for the year 1992, and the predicted for the year 2000 is 6.5 million. Find the constants C and k to obtain the exponential growth model $y = Ce^{kt}$ for the population growth. (Let $t = 0$ correspond to the year 1992.) Use the model to predict the population of Dhaka, Bangladesh in the year 2005.

Solution

(a) $y = Ce^{kt}$

$\quad 4.6 = Ce^{k(0)}$

$\quad 4.6 = C$

(b) $6.5 = 4.6e^{k(8)}$

$\quad \dfrac{6.5}{4.6} = e^{8k}$

$\quad \ln \dfrac{65}{46} = \ln e^{8k}$

$\quad \ln \dfrac{65}{46} = 8k$

$\quad \dfrac{1}{8} \ln \dfrac{65}{46} = k$

$\quad 0.0432 \approx k$

(c) $y = 4.6^{0.0432(13)}$

$\quad y \approx 8.1$ million

35. *Rate of Growth*

(a) Compare the values of k in Exercises 31 and 33. Which is larger? Explain.

(b) What variable in the continuous compounding interest formula is equivalent to k in the model for population growth? Use your answer to give an interpretation of k.

Solution

(a) k is larger in Exercise 33, since the population of Dhaka is increasing faster then the population of Los Angeles.

(b) k corresponds to r; k gives the annual percentage rate of growth.

37. *Radioactive Decay* Radioactive radium (Ra^{226}) has a half-life of 1620 years. If you start with 5 grams of the isotope, how much remains after 1000 years?

Solution

(a) $y = Ce^{kt}$

$\quad 5 = Ce^{k(0)}$

$\quad 5 = C$

(b) $2.5 = 5e^{k(1620)}$

$\quad 0.5 = e^{1620k}$

$\quad \ln 0.5 = \ln e^{1620k}$

$\quad \dfrac{\ln 0.5}{1620} = k$

$\quad -0.0004279 = k$

(c) $y = 5e^{-0.0004279(1000)}$

$\quad y = 3.2594852$ grams

39. *Earthquake Intensity* On March 27, 1964, Alaska experienced and earthquake that measured 8.4 on the Richter scale. On February 9, 1971, an earthquake in the San Fernando Valley measured 6.6 on the Richter scale. Compare the intensities of these earthquakes.

Solution

$R = \log_{10} I$

(a) $8.4 = \log_{10} I$

$10^{8.4} = 10^{\log_{10} I}$

$10^{8.4} = I$

(b) $6.6 = \log_{10} I$

$10^{6.6} = 10^{\log_{10} I}$

$10^{6.6} = I$

(c) $\dfrac{I \text{ for Alaska}}{I \text{ for San Fernando Valley}} = \dfrac{10^{8.8}}{10^{6.6}} = 10^{8.8-6.6} = 10^{2.2} \approx 158.49$

41. *Acidity Model* Use the acidity model pH $= -\log_{10}[H^+]$, where acidity (pH) is a measure of the hydrogen ion concentration $[H^+]$ (measured in moles of hydrogen per liter) of solution. Find the pH of a solution that has a hydrogen ion concentration of 9.2×10^{-8}.

Solution

pH $= -\log_{10}[H^+]$

pH $= -\log_{10}(9.2 \times 10^{-8}) = 7.0362$

43. Simplify $\dfrac{\left(\dfrac{9}{x}\right)}{\left(\dfrac{6}{x}+2\right)}$.

Solution

$\dfrac{\left(\dfrac{9}{x}\right)}{\left(\dfrac{6}{x}+2\right)} \cdot \dfrac{x}{x} = \dfrac{9}{6+2x}$

45. Simplify $\dfrac{\left(\dfrac{4}{x^2-9}+\dfrac{2}{x-2}\right)}{\left(\dfrac{1}{x+3}+\dfrac{1}{x-3}\right)}$.

Solution

$\dfrac{\left(\dfrac{4}{x^2-9}+\dfrac{2}{x-2}\right)}{\left(\dfrac{1}{x+3}+\dfrac{1}{x-3}\right)} \cdot \dfrac{(x-3)(x+3)(x-2)}{(x-3)(x+3)(x-2)} =$

$\dfrac{4(x-2)+2(x^2-9)}{(x-3)(x-2)+(x+3)(x+2)} =$

$\dfrac{4x-8+2x^2-18}{x^2-5x+6+x^2+x-6} =$

$\dfrac{2x^2+4x-26}{2x^2-4x} =$

$\dfrac{2(x^2+2x-13)}{2(x^2-2x)}$

$\dfrac{x^2+2x-13}{x(x-2)}$

47. *Equal Parts* Find two real numbers that divide the real number line between $\frac{x}{2}$ and $\frac{4x}{3}$ into three equal parts.

Solution

Distance between $\frac{x}{2}$ and $\frac{4x}{3} = \frac{4x}{3} - \frac{x}{2} = \frac{8x - 3x}{6} = \frac{5x}{6}$

Divide $\frac{5x}{6}$ by $3 = \frac{5x}{6} \div 3 = \frac{5x}{6} \cdot \frac{1}{3} = \frac{5x}{18}$

First real number $= \frac{x}{2} + \frac{5x}{18} = \frac{9x + 5x}{18} = \frac{14x}{18} = \frac{7x}{9}$

Second real number $= \frac{7x}{9} + \frac{5x}{18} = \frac{14x + 5x}{18} = \frac{19x}{18}$

49. *Annual Interest Rate* Find the annual interest rate when the principal P is \$500, the balance A is \$1004.83, the time t is 10 years, and compounding is monthly.

Solution

$$1004.83 = 500\left(1 + \frac{r}{12}\right)^{12(10)}$$

$$2.00966 = \left(1 + \frac{r}{12}\right)^{120}$$

$$(2.00966)^{1/120} = 1 + \frac{r}{12}$$

$$1.0058333 = 1 + \frac{r}{12}$$

$$0.0058333 = \frac{r}{12}$$

$$0.07 = r$$

$$7\% = r$$

51. *Annual Interest Rate* Find the annual interest rate when the principal P is \$5000, the balance A is \$22,405.68, the time t is 25 years, and compounding is daily.

Solution

$$22,405.68 = 5000\left(1 + \frac{r}{365}\right)^{365(25)}$$

$$4.481136 = \left(1 + \frac{r}{365}\right)^{9125}$$

$$(4.481136)^{1/9125} = 1 + \frac{r}{365}$$

$$1.000164384 = 1 + \frac{r}{365}$$

$$0.00164384 = \frac{r}{365}$$

$$0.059 = r$$

$$6\% = r$$

53. *Annual Interest Rate* Find the annual interest rate when the principal P is \$1500, the balance A is \$24,666.97, the time t is 40 years, and compounding is continuous.

Solution

$$24,666.97 = 1500e^{r(40)}$$

$$16.44464667 = e^{40r}$$

$$\ln 16.44464667 = 40r \ln e$$

$$\frac{\ln 16.44464667}{40} = r$$

$$0.069 = r = 7\%$$

55. *Doubling Time* Find the time for an investment to double when the principal P is $300, the annual interest rate r is 5%, and compounding is yearly.

Solution

$$600 = 300\left(1 + \frac{.05}{1}\right)^{1(t)}$$
$$2 = 1.05^t$$
$$\log_{1.05} 2 = \log_{1.05} 1.05^t$$
$$\log_{1.05} 2 = t$$
$$14.21 \text{ years} \approx t$$

57. *Doubling Time* Find the time for an investment to double when the principal P is $6000, the annual interest rate r is 7%, and compounding is quarterly.

Solution

$$12000 = 6000\left(1 + \frac{.07}{4}\right)^{4t}$$
$$2 = 1.0175^{4t}$$
$$\log_{1.0175} 2 = \log_{1.0175} 1.0175^{4t}$$
$$\log_{1.0175} 2 = 4t$$
$$9.99 \text{ years} \approx t$$

59. *Doubling Time* Find the time for an investment to double when the principal P is $1500, the annual interest rate r is 7.5%, and compounding is continuous.

Solution

$$3000 = 1500e^{0.075t}$$
$$2 = e^{0.075t}$$
$$\ln 2 = \ln e^{0.075t}$$
$$\ln 2 = 0.075t$$
$$\frac{\ln 2}{0.075} = t$$
$$9.24 \text{ years} = t$$

61. Find the effective yield when the annual interest rate is 6% and the compounding is quarterly.

Solution

$$A = 1000\left(1 + \frac{0.06}{4}\right)^{4(1)}$$
$$A = \$1061.36$$

$$\text{effective yield} = \frac{61.36}{1000} = 0.06136 = 6.136\%$$

63. Find the effective yield when the annual interest rate is 8% and the compounding is monthly.

Solution

$$A = 1000\left(1 + \frac{0.08}{12}\right)^{12(1)}$$
$$A = \$1083.00$$

$$\text{effective yield} = \frac{83.00}{1000} = 0.083 = 8.300\%$$

65. Find the effective yield when the annual interest rate is 7.5% and the compounding is continuous.

Solution

$$A = 1000 e^{0.075(1)}$$
$$= \$1077.88$$

$$\text{effective yield} = \frac{77.88}{1000} = 0.07788 = 7.788\%$$

67. *Doubling Time* Is it necessary to know the principal P to find the doubling time in Exercises 11–14 and Exercises 55–60? Explain.

Solution

No. Each time the amount is divided by the principal, the result is always 2.

69. Find the principal P that must be deposited in an account to obtain the balance of $25,000 at an interest rate r of 7%, for a time t of 30 years, and compounded monthly.

Solution

$$25,000 = P\left(1 + \frac{0.07}{12}\right)^{12(30)}$$

$$\frac{25,000}{(1.005833)^{360}} = P$$

$$\$3080.15 = P$$

71. Find the principal P that must be deposited in an account to obtain the balance of $1000 at an interest rate r of 5%, for a time t of 1 year, and compounded daily.

Solution

$$1000 = P\left(1 + \frac{0.05}{365}\right)^{365(1)}$$

$$\frac{1000}{(1.000136986)^{365}} = P$$

$$\$951.23 = P$$

73. Find the principal P that must be deposited in an account to obtain the balance of $500,000 at an interest rate r of 8%, for a time t of 25 years and compounded continuously.

Solution

$$500,000 = Pe^{0.08(25)}$$

$$\frac{500,000}{e^2} = P$$

$$\$67,667.64 = P$$

75. *Balance After Monthly Deposits* Suppose you make monthly deposits of P of $30 into a savings account at an annual interest rate r of 8%, compounded continuously. Find the balance A after 10 years (time t) given that

$$A = \frac{P(e^{rt} - 1)}{e^{r/12} - 1}.$$

Solution

$$A = \frac{P(e^{rt} - 1)}{e^{r/12} - 1}$$

$$A = \frac{30(e^{0.08(10)} - 1)}{e^{0.08/12} - 1}$$

$$A = \$5496.57$$

77. *Balance After Monthly Deposits* Suppose you make monthly deposits of P of $50 into a savings account at an annual interest rate r of 10%, compounded continuously. Find the balance A after 40 years (time t) given that

$$A = \frac{P(e^{rt} - 1)}{e^{r/12} - 1}.$$

Solution

$$A = \frac{P(e^{rt} - 1)}{e^{r/12} - 1}$$

$$A = \frac{50(e^{0.10(40)} - 1)}{e^{0.10/12} - 1}$$

$$A = \$320,250.81$$

79. *Balance After Monthly Deposits*   Suppose you make monthly deposits of $30 into a savings at an annual interest rate r of 8%, compounded continuously (see figure in textbook). Find the total amount that you deposited in the account in the first 20 years and the total interest earned.

Solution

$$A = \frac{P(e^{rt} - 1)}{e^{r/12} - 1}$$

$$A = \frac{30(e^{0.08(20)} - 1)}{e^{0.08(20)} - 1}$$

$$A = \$17,729.42$$

Total interest = $17,729.42 - 7200 = $10,529.42

81. *Radioactive Decay*   The isotope Pu^{230} has a half-life of 24,360 years. If you start with 2 grams of this isotope, how much remains after 30,000 years?

Solution

(a) $y = Ce^{kt}$

$2 = Ce^{k(0)}$

$2 = C$

(b) $1 = 2e^{k(24,360)}$

$0.5 = e^{24,360k}$

$\ln 0.5 = \ln e^{24,360k}$

$\dfrac{\ln 0.5}{24,360} = k$

$-0.0000284543178 = k$

(c) $y = 2e^{-0.0000284543178(30,000)}$

$y = 0.8517$ grams

83. *Population of a City*   The population of Chicago is 6.5 million for the year 1992, and the predicted population for the year 2000 is 6.6 million. Find the constants C and k to obtain the exponential growth model $y = Ce^{kt}$ for the population growth. (Let $t = 0$ correspond to the year 1992.) Use the model to predict the population of Chicago in the year 2005. (Source: U.S. Bureau of the Census)

Solution

(a) $y = Ce^{kt}$

$6.5 = Ce^{k(0)}$

$6.5 = C$

(b) $6.6 = 6.5e^{k(8)}$

$\dfrac{6.6}{6.5} = e^{8k}$

$\ln \dfrac{66}{65} = \ln e^{8k}$

$\dfrac{\ln\left(\dfrac{66}{65}\right)}{8} = k$

$0.001908434 = k$

(c) $y = 6.5e^{0.001908434(13)}$

$y = 6.7$ million

85. *Population of a City*   The population of Mexico City, Mexico, is 21.6 million for the year 1992, and the predicted population for the year 2000 is 27.9 million. Find the constants C and k to obtain the exponential growth model $y = Ce^{kt}$ for the population growth. (Let $t = 0$ correspond to the year 1992.) Use the model to predict the population of Mexico City, Mexico, in the year 2005.

Solution

(a) $y = Ce^{kt}$

$21.6 = Ce^{k(0)}$

$21.6 = C$

(b) $27.9 = 21.6e^{k(8)}$

$\dfrac{27.9}{21.6} = e^{8k}$

$\ln\left(\dfrac{279}{216}\right) = \ln e^{8k}$

$\dfrac{\ln\left(\dfrac{279}{216}\right)}{8} = k$

$0.0319916718 = k$

(c) $y = 21.6e^{0.0320(13)}$

$y = 32.7$ million

87. *Graphical Interpretation*   The population p of a certain species t years after it is introduced into a new habitat is given by

$$p(t) = \frac{5000}{1 + 4e^{-t/6}}.$$

(a) Use a graphing utility to graph the population function.

(b) Determine th population size that was introduced into the habitat.

(c) Determine the population size after 9 years.

(d) After how many years will the population be 2000?

Solution

(a) Keystrokes:

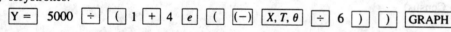

$\boxed{Y =}$ 5000 $\boxed{\div}$ $\boxed{(}$ $\boxed{1}$ $\boxed{+}$ $\boxed{4}$ $\boxed{e}$ $\boxed{(}$ $\boxed{(-)}$ $\boxed{X, T, \theta}$ $\boxed{\div}$ $\boxed{6}$ $\boxed{)}$ $\boxed{)}$ $\boxed{\text{GRAPH}}$

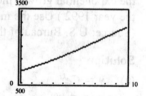

(b) $p(0) = \dfrac{5000}{1 + 4e^{-0/6}} = \dfrac{5000}{5} = 1000$

(c) $p(9) = \dfrac{5000}{1 + 4e^{-9/6}} \approx 2642$

(d) $2000 = \dfrac{5000}{1 + 4e^{-t/6}}$

$1 + 4e^{-t/6} = 2.5$

$4e^{-t/6} = 1.5$

$e^{-t/6} = .375$

$\ln e^{-t/6} = \ln .375$

$-\dfrac{t}{6} = \ln .375$

$+t = (\ln .375)(-6)$

$t \approx 5.88$ years

89. *Depreciation*   A car that costs $22,000 new has a depreciated value of $16,500 after 1 year. Find the value of the car when it is 3 years old by using the exponential model $y = Ce^{kt}$.

Solution

$$16,500 = 22,000e^{k(1)}$$

$$\frac{16,500}{22,000} = e^k$$

$$\ln \frac{16,500}{22,000} = \ln e^k$$

$$\ln \frac{16,500}{22,000} = k$$

$$-0.2876821 = k$$

$$y = 22,000e^{-0.2876821(3)}$$

$$= \$9281.25$$

91. *Graphical Estimation*   Annual sales y of a product x years after it is introduced are approximated by

$$y = \frac{2000}{1 + 4e^{-x/2}}$$

(a) Use a graphing utility to graph the equation.

(b) Graphically approximate sales when $x = 4$.

(c) Graphically approximate the time when annual sales will be 1100 units.

(d) Graphically estimate the maximum level annual sales will approach.

Solution

(a) Keystrokes:

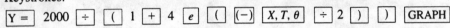

(b) 1300 units

(c) 3 years

(d) 2000 units

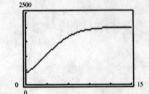

93. *Earthquake Intensity*   Compare the earthquake intensities. On August 16, 1906, Chile had an earthquake that measured 8.6 on the Richter scale. On December 7, 1988, an earthquake in Armenia, USSR, measured 6.8 on the Richter scale.

Solution

$R = \log_{10} I$

(a) Chile: $8.6 = \log_{10} I$ (b) Armenia: $6.8 = \log_{10} I$

       $10^{8.6} = I$ $10^{6.8} = I$

Ratio of 2 intensities:

$$\frac{I \text{ for Chile}}{I \text{ for Armenia}} = \frac{10^{8.6}}{10^{6.8}} = 10^{8.6 - 6.8} = 10^{1.8} \approx 63$$

Earthquake in Chile is 63 times as great.

95. *Acidity Model* Use the acidity model $pH = -\log_{10}[H^+]$, where acidity (pH) is a measure of the hydrogen ion concentration $[H^+]$ (measured in moles of hydrogen per liter) of solution.

A certain fruit has a pH of 2.5, and an antacid tablet has a pH of 9.5. The hydrogen ion concentation of the fruit is how many times the concentration of the tablet?

Solution

$pH = -\log_{10}[H^+]$

(a)
$$2.5 = -\log_{10}[H^+]$$
$$-2.5 = \log_{10}[H^+]$$
$$10^{-2.5} = 10^{\log_{10}[H^+]}$$
$$0.0031623 = H^+$$

(b)
$$9.5 = -\log_{10}[H^+]$$
$$-9.5 = \log_{10}[H^+]$$
$$10^{-9.5} = 10^{\log_{10}[H^+]}$$
$$3.1623 \times 10^{-10} = H^+$$

(c) $\dfrac{H^+ \text{ of fruit}}{H^+ \text{ of tablet}} = \dfrac{0.0031623}{3.1623 \times 10^{-10}} = 10{,}000{,}071$

97. *Comparing Models*   The figure in the textbook gives the earnings per share of common stock for the years 1981 through 1993 for Automatic Data Processing, Inc. A list of models ($t = 1$ is 1981) for the data is also given. For each of the models, (a) use a graphing utility to obtain its graph, and (b) find the sum of the squares of the differences between the actual data and the approximations given by the model. Use this sum to determine which model "best fits" the data. (Source: Automatic Data Processing, Inc.)

Linear: $E = 0.146t + 0.007$

Quadratic: $E = 0.009t^2 + 0.018t + 0.325$

Exponential: $E = 0.301(1.165)^t$

Exponential: $E = 0.301e^{0.153t}$

Logarithmic: $E = 0.201 + 0.228t - 0.444 \ln t$

Solution

(a) Keystrokes:

y_1 [Y =] .146 [X, T, θ] + .007 [GRAPH]

y_2 [Y =] .009 [X, T, θ] [x^2] [+] .018 [X, T, θ] + [.325] [GRAPH]

y_3 [Y =] .301 [×] 1.165 [∧] [X, T, θ] [GRAPH]

y_4 [Y =] .301 [e] .153 [X, T, θ] [GRAPH]

y_5 [Y =] .201 [+] .228 [X, T, θ] [−] .444 [ln] [X, T, θ] [GRAPH]

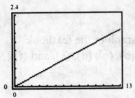
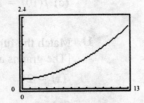
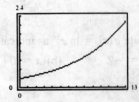

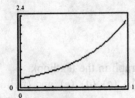
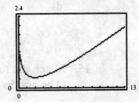

(b) Linear: Sum of Square 0.1725

    Quadratic: Sum of Squares 0.0076

    Exponential: Sum of Squares 0.0315

    Exponential: Sum of Squares 0.0313

    Logarithmic: Sum of Squares 0.0209

(c) Quadratic is the "best fits" model.

Review Exercises for Chapter 9

1. Evaluate $f(x) = 2^x$ as indicated.

(a) $x = -3$ (b) $x = 1$ (c) $x = 2$

Solution

(a) $f(-3) = 2^{-3} = \dfrac{1}{8}$

(b) $f(1) = 2^1 = 2$

(c) $f(2) = 2^2 = 4$

3. Evaluate $g(t) = e^{-t/3}$ as indicated.

(a) $t = -3$ (b) $t = \pi$ (c) $t = 6$

Solution

(a) $g(-3) = e^{-(-3)/3} = e^1 \approx 2.72$

(b) $g(\pi) = e^{-\pi/3} \approx 0.35$

(c) $g(6) = e^{-6/3} = e^{-2} \approx 0.14$

5. Evaluate $f(x) = \log_3 x$ as indicated.

(a) $x = 1$ (b) $x = 27$ (c) $x = 0.5$

Solution

(a) $f(1) = \log 31 = 0$

(b) $f(27) = \log_3 27 = 3$

(c) $f(0.5) = \log_3 0.5 = \dfrac{\log 0.5}{\log 3} = -0.63$

7. Evaluate $f(x) = \ln x$ as indicated.

(a) $x = e$ (b) $x = \dfrac{1}{3}$ (c) $x = 10$

Solution

(a) $f(e) = \ln 3 = 1$

(b) $f\left(\dfrac{1}{3}\right) = \ln \dfrac{1}{3} \approx -1.10$

(c) $f(10) = \ln 10 \approx 2.30$

9. Evaluate $g(x) = \ln e^{3x}$ as indicated.

(a) $x = -2$ (b) $x = 0$ (c) $x = 7.5$

Solution

(a) $g(-2) = \ln e^{3(-2)} = -6$

(b) $g(0) = \ln e^{3}(0) = 0$

(c) $g(7.5) = \ln e^{3(75)} = \ln e^{22.5} = 22.5$

11. Match the function with its graph in the textbook. [The graphs are labeled (a), (b), (c), (d), (e), and (f).]

$f(x) = 2^x$

Solution

(d)

13. Match the function with its graph in the textbook. [The graphs are labeled (a), (b), (c), (d), (e), and (f).]

$f(x) = -2^x$

Solution

(a)

15. Match the function with its graph in the textbook. [The graphs are labeled (a), (b), (c), (d), (e), and (f).]

$f(x) = \log_2 x$

Solution

(c)

17. Sketch the graph of $y = 3^{x/2}$.

Solution

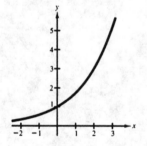

19. Sketch the graph of $f(x) = 3^{-x/2}$.

Solution

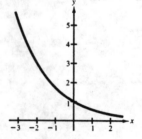

21. Sketch the graph of $f(x) = 3^{-x^2}$.

Solution

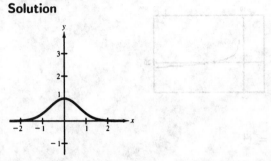

23. Sketch the graph of $f(x) = -2 + \log_3 x$.

Solution

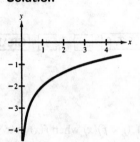

25. Sketch the graph of $y = \log_2(x - 4)$.

Solution

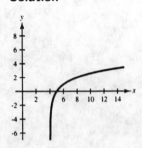

27. Use a graphing utility to graph $y = 5e^{-x/4}$.

Solution

Keystrokes:

$\boxed{Y =}$ 5 $\boxed{e}$ $\boxed{(}$ $\boxed{(-)}$ $\boxed{X, T, \theta}$ $\boxed{\div}$ 4 $\boxed{)}$ $\boxed{GRAPH}$

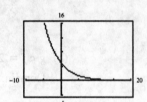

29. Use a graphing utility to graph $f(x) = e^{x+2}$.

Solution

Keystrokes:

$\boxed{Y =}$ $\boxed{e}$ $\boxed{(}$ $\boxed{X, T, \theta}$ $\boxed{+}$ 2 $\boxed{)}$ $\boxed{GRAPH}$

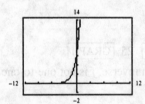

31. Use a graphing utility to graph $g(x) = \ln 2x$.

Solution

Keystrokes:

$\boxed{Y =}$ $\boxed{\ln}$ 2 $\boxed{X, T, \theta}$ $\boxed{GRAPH}$

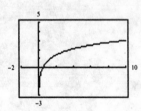

33. Use a graphing utility to graph $g(t) = 2 - \ln(t - 1)$.

Solution

Keystrokes:

$\boxed{Y =}$ 2 $\boxed{-}$ $\boxed{\ln}$ $\boxed{(}$ $\boxed{X, T, \theta}$ $\boxed{-}$ 1 $\boxed{)}$ $\boxed{\text{GRAPH}}$

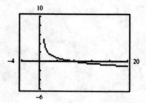

35. Find (a) $(f \circ g)(x)$ and (b) $(g \circ f)(x)$ when $f(x) = x + 2$ and $g(x) = x^2$.

Solution

(a) $(f \circ g)(x) = x^2 + 2$

(b) $(g \circ f)(x) = (x + 2)^2$
$$= x^2 + 4x + 4$$

37. Find (a) $(f \circ g)(x)$ and (b) $(g \circ f)(x)$ when $f(x) = \sqrt{x + 1}$ and $g(x) = x^2 - 1$.

Solution

(a) $(f \circ g)(x) = \sqrt{x^2 - 1 + 1}$
$$= \sqrt{x^2}$$
$$= |x|$$

(b) $(g \circ f)(x) = \left(\sqrt{x + 1}\right)^2 - 1$
$$= x + 1 - 1$$
$$= x$$

39. Find the domain of the composition (a) $(f \circ g)$ and (b) $(g \circ f)$ where $f(x) = \sqrt{x - 4}$ and $g(x) = 2x$.

Solution

(a) $f \circ g = \sqrt{2x - 4}$
    Domain: $x \geq 2$

(b) $g \circ f = 2\sqrt{x - 4}$
    Domain: $x \geq 4$

41. Use a graphing utility to decide if $f(x) = x^2 - 25$ has an inverse.

Solution

Keystrokes:

$\boxed{Y =}$ $\boxed{X, T, \theta}$ $\boxed{x^2}$ $\boxed{-}$ 25 $\boxed{\text{GRAPH}}$

No, $f(x)$ does not have an inverse. f is not one-to-one.

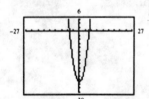

43. Use a graphing utility to decide if $h(x) = 4\sqrt[3]{x}$ has an inverse.

Solution

Keystrokes:

$\boxed{Y =}$ 4 $\boxed{\sqrt[3]{}}$ $\boxed{X, T, \theta}$ $\boxed{\text{GRAPH}}$

Yes, $h(x)$ does have an inverse.

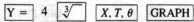
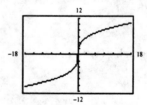

45. Find the inverse of $f(x) = \dfrac{1}{4}x$.

(If it is not possible, state the reason.)

Solution

$$y = \frac{1}{4}x$$

$$x = \frac{1}{4}y$$

$$4x = y$$

$$4x = f^{-1}(x)$$

47. Find the inverse of $h(x) = \sqrt{x}$.

(If it is not possible, state the reason.)

Solution

$$y = \sqrt{x}$$

$$x = \sqrt{y}$$

$$x^2 = y$$

$$x^2 = f^{-1}(x)$$

$$(x \geq 0)$$

49. Find the inverse of $f(t) = |t + 3|$. (If it is not possible, state the reason.)

Solution

$f(t)$ does not have an inverse, because the graph of $f(t)$ does not pass the horizontal line test.

51. Write $4^3 = 64$ in logarithmic form.

Solution

$\log_4 64 = 3$

53. Write $\ln e = 1$ in exponential form.

Solution

$e^1 = e$

55. Evaluate $\log_{10} 1000$.

Solution

$\log_{10} 1000 = 3$ because $10^3 = 1000$.

57. Evaluate $\log_3 \dfrac{1}{9}$.

Solution

$\log_3 \dfrac{1}{9} = -2$ because $3^{-2} = \dfrac{1}{9}$.

59. Evaluate $\ln e^7$.

Solution

$\ln e^7 = 7 \ln e = 7$

61. Evaluate $\ln 1$.

Solution

$\ln 1 = 0$

63. Use the properties of logarithms to expand $\log_4 6x^4$.

Solution

$\log_4 6x^4 = \log_4 6 + 4 \log_4 x$

65. Use the properties of logarithms to expand $\log_5 \sqrt{x + 2}$.

Solution

$\log_5 \sqrt{x + 2} = \dfrac{1}{2} \log_5 (x + 2)$

67. Use the properties of logarithms to expand $\ln \dfrac{x + 2}{x - 2}$.

Solution

$\ln \dfrac{x + 2}{x - 2} = \ln (x + 2) - \ln (x - 2)$

69. Use the properties of logarithms to condense $\log_4 x - \log_4 10$.

Solution

$\log_4 x - \log_4 10 = \log_4 \left(\dfrac{x}{10} \right)$

71. Use the properties of logarithms to condense $4(1 + \ln x + \ln x)$.

Solution

$$4(1 + \ln x + \ln x) = 4 + 4 \ln x + 4 \ln x$$
$$= 4 + \ln x^4 + \ln x^4$$
$$= 4 + \ln x^8$$

73. Use the properties of logarithms to condense $-2(\ln 2x - \ln 3)$.

Solution

$$-2(\ln 2x - \ln 3) = \ln \left(\frac{2x}{3}\right)^{-2}$$

75. Use a graphing utility to graphically verify that the expressions are equivalent.

$$y_1 = \ln \left(\frac{x + 1}{x - 1}\right)^2, \quad x > 1$$

$$y_2 = 2[\ln (x + 1) - \ln (x - 1)]$$

Solution

Keystrokes:

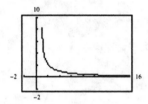

77. Decide whether $\log_2 4x = 2 \log_2 x$ is true or false.

Solution

false

79. Decide whether $\log_{10} 10^{2x} = 2x$ is true or false.

Solution

true

81. Decide whether $\log_2 \dfrac{16}{x} = 2 - \log_4 x$ is true or false.

Solution

true

83. Approximate $\log_5 18$ given that $\log_5 2 \approx 0.43068$ and $\log_5 3 \approx 0.68260$.

Solution

$$\log_5 18 = \log_5 3^2 + \log_5 2$$
$$= 2 \log_5 3 + \log 2$$
$$\approx 2(0.6826) + 0.43068$$
$$\approx 1.79628$$

85. Approximate $\log_5 \dfrac{1}{2}$ given that $\log_5 2 \approx 0.43068$ and $\log_5 3 \approx 0.68260$.

Solution

$$\log_5 \frac{1}{2} = \log_5 1 - \log_5 2 \approx 0 - (0.43068) \approx -0.43068$$

87. Approximate $\log_5 (12)^{2/3}$ given that $\log_5 2 \approx 0.43068$ and $\log_5 3 \approx 0.68260$.

Solution

$$\log_5 (12)^{2/3} = \frac{2}{3}[2 \log_5 2 + \log_5 3]$$

$$\approx \frac{2}{3}[2(0.43068) + 0.6826]$$

$$\approx 1.0293067$$

89. Evaluate $\log_4 9$ using the change-of-base formula. Round each logarithm to three decimal places.

Solution

$$\log_4 9 = \frac{\log 9}{\log 4} \approx 1.585$$

91. Evaluate $\log_{12} 200$ using the change-of-base formula. Round each logarithm to three decimal places.

Solution

$$\log_{12} 200 = \frac{\log 200}{\log 12} \approx 2.132$$

93. Solve $2^x = 64$.

Solution

$2^x = 64$

$2^x = 2^6$

$x = 6$

95. Solve $4^{x-3} = \frac{1}{16}$.

Solution

$$4^{x-3} = \frac{1}{16}$$

$$4^{x-3} = 4^{-2}$$

$$x - 3 = -2$$

$$x = 1$$

97. Solve $\log_3 x = 5$.

Solution

$\log_3 x = 5$

$3^{\log_3 x} = 3^5$

$x = 243$

99. Solve $3x = 500$.

(Round your answer to two decimal places.)

Solution

$3^x = 500$

$\log_3 3^x = \log_3 500$

$$x = \frac{\log 500}{\log 3}$$

$$x \approx 5.66$$

101. Solve $2e^{x/2} = 45$.

(Round your answer to two decimal places.)

Solution

$2e^{x/2} = 45$

$e^{x/2} = 22.5$

$\ln e^{x/2} = \ln 22.5$

$$\frac{x}{2} = \ln 22.5$$

$$x = 2 \ln 22.5$$

$$x \approx 6.23$$

103. Solve $\dfrac{500}{(1.05)^x} = 100$.

Solution

$$\frac{500}{(1.05)^x} = 100$$

$$\frac{500}{100} = (1.05)^x$$

$$5 = 1.05^x$$

$$\log_{1.05} 5 = \log_{1.05} 1.05^x$$

$$\frac{\log 5}{\log 1.05} = x$$

$$32.99 \approx x$$

105. Solve $\log_{10} 2x = 1.5$.
(Round your answer to two decimal places.)

Solution

$$\log_{10} 2x = 1.5$$

$$2x = 10^{1.5}$$

$$x = \frac{10^{1.5}}{2} \approx 15.81$$

107. Solve $\ln x = 7.25$.
(Round your answer to two decimal places.)

Solution

$$\ln x = 7.25$$

$$e^{\ln x} = e^{7.25}$$

$$x = e^{7.25}$$

$$x \approx 1408.10$$

109. Solve $\log_2 2x = -0.65$.
(Round your answer to two decimal places.)

Solution

$$\log_2 2x = -0.65$$

$$2 \log_2 2x = 2^{-0.65}$$

$$2x = 2^{-0.65}$$

$$x = \frac{2^{-0.65}}{2}$$

$$x \approx 0.32$$

111. *Creating a Table* Complete the table (see textbook) to determine the balance A when the principal P of $500 is invested at a rate r of 7% for time t of 30 years, and compounded n times per year.

Solution

(a) 1 time per year: $A = P\left(1 + \dfrac{r}{n}\right)^{nt}$

$$= 500\left(1 + \frac{0.07}{1}\right)^{1(30)}$$

$$= \$3806.13$$

(b) 4 times per year $A = 500\left(1 + \dfrac{0.07}{4}\right)^{4(30)}$

$$= \$4009.59$$

(c) 12 times per year: $A = 500\left(1 + \dfrac{0.07}{12}\right)^{12(30)}$

$$= \$4058.25$$

(d) 365 times per year $A = 500\left(1 + \dfrac{0.07}{365}\right)^{365(30)}$

$$= \$4082.26$$

(e) Continuously: $A = 500e^{0.07(30)}$

$$= \$4083.09$$

n	1	4	12	365	Continuous
A	$3806.13	$4009.59	$4058.25	$4082.26	$4083.09

113. *Creating a Table*   Complete the table (see textbook) to determine the balance A when the principal P of \$10,000 is invested at a rate r of 10% for time t of 20 years, and compounded n times per year.

Solution

(a) 1 time per year: $A = 10{,}000\left(1 + \dfrac{0.10}{1}\right)^{1(20)}$

$\qquad\qquad\qquad\qquad = \$67{,}275.00$

(b) 4 times per year $A = 10{,}000\left(1 + \dfrac{0.10}{4}\right)^{4(20)}$

$\qquad\qquad\qquad\qquad = \$72{,}095.68$

(c) 12 times per year: $A = 10{,}000\left(1 + \dfrac{0.10}{12}\right)^{12(20)}$

$\qquad\qquad\qquad\qquad = \$73{,}280.74$

(d) 365 times per year $A = 10{,}000\left(1 + \dfrac{0.10}{365}\right)^{365(20)}$

$\qquad\qquad\qquad\qquad = \$73{,}870.32$

(e) Continuously: $A = 10{,}000e^{0.10(20)}$

$\qquad\qquad\qquad = \$73{,}890.56$

n	1	4	12	365	Continuous
A	\$67,275.00	\$72,095.68	\$73,280.74	\$73,870.32	\$73,890.56

115. *Creating a Table* Complete the table (see textbook) to determine the principal P that will yield a balance A of $50,000 when invested at a rate r of 8% for time t of 40 years, and compounded n times per year.

Solution

(a) 1 time per year:
$$50,000 = P\left(1 + \frac{0.08}{1}\right)^{1(40)}$$

$$\frac{50,000}{(1.08)^{40}} = P$$

$$\$2301.55 = P$$

(b) 4 times per year
$$50,000 = P\left(1 + \frac{0.08}{4}\right)^{4(40)}$$

$$\frac{50,000}{(1.02)^{160}} = P$$

$$\$2103.50 = P$$

(c) 12 times per year:
$$50,000 = P\left(1 + \frac{0.08}{12}\right)^{12(40)}$$

$$\frac{50,000}{(1.0066667)^{480}} = P$$

$$\$2059.87 = P$$

(d) 365 times per year
$$50,000 = P\left(1 + \frac{0.08}{365}\right)^{365(40)}$$

$$\frac{50,000}{(1.0002192)^{14,600}} = P$$

$$\$2038.82 = P$$

(e) Continuously:
$$50,000 = Pe^{0.08(40)}$$

$$\frac{50,000}{e^{3.2}} = P$$

$$\$2038.11 = P$$

n	1	4	12	365	Continuous
P	$2301.55	$2103.50	$2059.87	$2038.82	$2038.11

117. *Inflation Rate*   if the annual rate of inflation averages 5% over the next 10 years, the approximate cost C of goods or services during any year in that decade will be given by

$$C(t) = P(1.05)^t, \quad 0 \le t \le 10$$

where t is the time in years and P is the present cost. If the price of an oil change for your car is presently $19.95, estimate the number of year until it will cost $25.00.

Solution

$$25.00 = 19.95(1.05)^t$$

$$\frac{25.00}{19.95} = 1.05^t$$

$$\log_{1.05} \frac{25.00}{19.95} = \log_{1.05} 1.05^t$$

$$\frac{\log \dfrac{2,500}{1,995}}{\log 1.05} = t$$

$$4.62 \text{ years} \approx t$$

119. *Produce Demand*   The demand x and price p for a certain product are related by

$$p = 25 - 0.4e^{0.02x}.$$

Approximate the demand when the price is $16.97.

Solution

$$p = 25 - 0.4e^{0.02x}$$

$$16.97 = 25 - 0.4e^{0.02x}$$

$$-8.03 = -0.43^{0.02x}$$

$$20.075 = e^{0.02x}$$

$$\ln 20.075 = \ln e^{0.02x}$$

$$\ln 20.075 = 0.02x$$

$$\frac{\ln 20.075}{0.02} = x$$

$$150 \text{ units} \approx x$$

121. *Graphical Interpretation*   The population p of a certain species t years after it is introduced into a new habitat is given by

$$p(t) = \frac{600}{1 + 2e^{-0.2t}}$$

Use a graphing utility to graph the function. Use the graph to determine the limiting size of the population in this habitat.

Solution

Keystrokes:

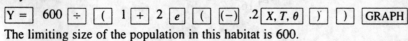

The limiting size of the population in this habitat is 600.

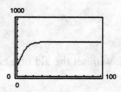

Test for Chapter 9

1. Evaluate $f(t) = 54 \left(\dfrac{2}{3}\right)^t$ when $t = -1, 0, \dfrac{1}{2},$ and 2.

Solution

(a) $f(-1) = 54 \left(\dfrac{2}{3}\right)^{-1}$

$= 54 \left(\dfrac{3}{2}\right) = 81$

(b) $f(0) = 54 \left(\dfrac{2}{3}\right)^{0}$

$= 54$

(c) $f\left(\dfrac{1}{2}\right) = 54 \left(\dfrac{2}{3}\right)^{1/2}$

$= 44.090815$

(d) $f(2) = 54 \left(\dfrac{2}{3}\right)^{2}$

$= 54 \left(\dfrac{4}{7}\right) = 24$

2. Sketch a graph of the function $f(x) = 2^{x/3}$.

Solution

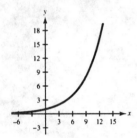

3. Write the logarithmic equation $\log_5 125 = 3$ in exponential form.

Solution

$\log_5 125 = 3$

$\qquad 5^3 = 125$

4. Write the logarithmic equation $4^{-2} = \dfrac{1}{16}$ in logarithmic form.

Solution

$4^{-2} = \dfrac{1}{16}$

$\log_4 \dfrac{1}{16} = -2$

5. Evaluate $\log_8 2$ without the aid of a calculator.

Soltuion

$\log_8 2 = \dfrac{1}{3}$

6. Describe the relationship between the graphs of $f(x) = \log_5$ and $g(x) = 5^x$.

Solution

f and g are inverse functions.

7. Use the properties of logarithms to expand $\log_4 \left(5x^2 \sqrt{y}\,\right)$.

Solution

$\log_4 \dfrac{5x^2}{\sqrt{y}} = \log_4 5 + 2 \log_4 x - \dfrac{1}{2} \log_4 y$

8. Use the properties of logarithms to condense
$\ln x - 4 \ln y$.

Solution

$$\ln x - \ln y = \ln \frac{x}{y^4}$$

9. Simplify $\log_5 (5^3 \cdot 6)$.

Solution

$$\log_5(5^3 \cdot 6) = 3 \log_5 5 + \log_5 6 = 3 + \log_5 6$$

10. Solve $\log_4 x = 3$.

Solution

$$\log_4 x = 3$$
$$4^{\log_4 x} = 4^3$$
$$x = 64$$

11. Solve $10^{3y} = 832$.

Solution

$$10^{3y} = 832$$
$$\log 10^{3y} = \log 832$$
$$3y = \log 832$$
$$y = \frac{\log 832}{3}$$
$$y \approx 0.97$$

12. Solve $400e^{0.08t} = 1200$.

Solution

$$400e^{0.08t} = 1200$$
$$e^{0.08t} = 3$$
$$\ln e^{0.08t} = \ln 3$$
$$0.08t = \ln 3$$
$$t = \frac{\ln 3}{0.08}$$
$$t \approx 13.73$$

13. Solve $3 \ln (2x - 3) = 10$.

Solution

$$3 \ln (2x - 3) = 10$$
$$\ln(2x - 3) = \frac{10}{3}$$
$$e^{\ln(2x-3)} = e^{10/3}$$
$$2x - 3 = e^{10/3}$$
$$x = \frac{e^{10/3} + 3}{2}$$
$$x \approx 15.52$$

14. Determine the balance after 20 years if $2000
is invested at 7% compounded (a) quarterly and
(b) continuously.

Solution

(a) $A = 2000\left(1 + \dfrac{0.07}{4}\right)^{4(20)}$

$\qquad = \$8012.78$

(b) $A = 2000e^{0.07(20)}$

$\qquad = \$8110.40$

15. Determine the principal that will yield $100,000
when invested at 9% compounded quarterly for
25 years.

Solution

$$100,000 = P\left(1 + \frac{0.09}{4}\right)^{4(25)}$$
$$\frac{100,000}{(1.0225)^{100}} = P$$
$$\$10,806.08 = P$$

16. A principal of $500 yields a balance of $1006.88 in 10 years when the interest is compounded continuously. What is the annual interest rate?

Solution

$$1006.88 = 500e^{r(10)}$$

$$2.01376 = e^{10r}$$

$$\ln 2.01376 = \ln e^{10r}$$

$$\ln 2.01376 = 10r$$

$$\frac{\ln 2.01376}{10} = r$$

$$0.07 = r$$

$$7\% = r$$

17. A new car that cost $18,000 has a depreciated value of $14,000 after 1 year. Find the value of the car when it is three years old by using the exponential model $y = Ce^{kt}$.

Solution

(a) $y = Ce^{kt}$

$$18,000 = Ce^{k(0)}$$

$$18,000 = C$$

(b) $$14,000 = 18,000e^{k(1)}$$

$$\frac{14,000}{18,000} = e^{k}$$

$$\ln\frac{14}{18} = \ln e^{k}$$

$$\ln\frac{14}{18} = k$$

$$-0.2513144 = k$$

(c) $y = 18,000^{-0.2513144(3)} = \$8469.14$

18–20. The population of a certain species t years after it is introduced into a new habitat is given by

$$p(t) = \frac{2400}{1 + 3e^{-t/4}}.$$

18. Determine the population size that was introduced into the habitat.

19. Determine the population after 4 years.

20. After how many years will the population be 1200?

Solutions

18. $p(0) = \dfrac{2400}{1 + 3e^{-0/4}} = 600$

19. $p(4) = \dfrac{2400}{1 + 3e^{-4/4}} = 1141$

20.
$$1200 = \frac{2400}{1 + 3e^{-t/4}}$$

$$1 + 3e^{-t/4} = \frac{2400}{1200}$$

$$3e^{-t/4} = 1$$

$$e^{-t/4} = \frac{1}{3}$$

$$\ln e^{-t/4} = \ln \frac{1}{3}$$

$$-\frac{t}{4} = \ln \frac{1}{3}$$

$$t = -4 \ln \frac{1}{3} \approx 4.39 \text{ years}$$

Cumulative Test: Chapters 7–9

1. Find an equation of the line through $(-4, 0)$ and $(4, 6)$.

Solution

$$m = \frac{6 - 0}{4 - -4} = \frac{6}{8} = \frac{3}{4}$$

$$y - 0 = \frac{3}{4}(x - -4)$$

$$y = \frac{3}{4}x + 3$$

$$4y = 3x + 12$$

$$0 = 3x - 4y + 12$$

2. Sketch the graph of $x^2 + y^2 = 8$.

Solution

$x^2 + y^2 = 8$

$\quad$ center $= (0, 0) \quad r = \sqrt{8} \approx 2.8$

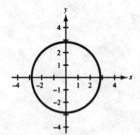

3. Sketch the graph of $\dfrac{x^2}{1} + \dfrac{y^2}{1} = 1$.

Solution

$\dfrac{x^2}{1} + \dfrac{y^2}{4} = 1$

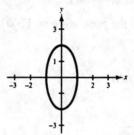

4. Sketch the graph of $\dfrac{x^2}{1} - \dfrac{y^2}{4} = 1$

Solution

$\dfrac{x^2}{1} + \dfrac{y^2}{4} = 1$

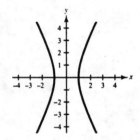

5. Graph the inequality $5x + 2y > 10$.

Solution

$5x + 2y > 10 \quad$ or $\quad y > -\dfrac{5}{2}x + 5$

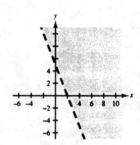

6. Find an equation of the parabola in the textbook with vertex $(3, -2)$ and y-intercept $(0, 4)$.

Solution $\quad y = a(x - h)^2 + k$

$4 = a(0 - 3)^2 - 2$

$4 = a(9) - 2$

$6 = 9a$

$\dfrac{6}{9} = a$

$\dfrac{2}{3} = a$

$y = \dfrac{2}{3}(x - 3)^2 - 2$

7. A *semicircular* arch is positioned over the roadway onto the grounds of an estate. The roadway is 10 feet wide and the arch is sitting on pillars that are 8 feet tall. Find the maximum height of a truck that can be driven onto the estate if its width is 8 feet.

Solution

$x^2 + y^2 = 25$ equation at circular arch

$y = \sqrt{25 - x^2}$

$y = \sqrt{25 - 16}$

$\quad = \sqrt{9} = 3$

Maximum height of truck: $8 + 3 = 11$ feet

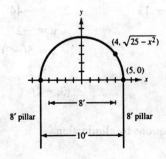

8. The stopping distance d of a car is directly proportional to the square of its speed s. On a certain type of pavement, a car requires 50 feet to stop when its speed is 25 miles per hour. Estimate the stopping distance when the speed of the car is 40 miles per hour.

Solution

$d = ks^2$

$50 = k(25)^2$

$\dfrac{50}{625} = k$

$\dfrac{2}{25} = k$

$d = \dfrac{2}{25}(40)^2$

$d = 128$ feet

9. Solve the system of equations graphically.

$x - y = 1$

$2x + y = 5$

Solution

$x - y = 1 \qquad\qquad 2x + y = 5$

$\quad -y = -x + 1 \qquad\qquad y = -2x + 5$

$\quad\quad y = x - 1$

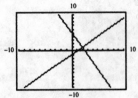

Keystrokes:

y_1 [Y =] [X, T, θ] [−] 1 [ENTER]

y_2 [Y =] [(−)] 2 [X, T, θ] [+] 5 [GRAPH]

10. Solve the system of equations by substitution.

$$4x + 2y = 8$$
$$x - 5y = 13$$

Solution

$x = 5y + 13$	$22y = -44$
$4(5y + 13) + 2y = 8$	$y = -2$
$20y + 52 + 2y = 8$	$x = 5(-2) + 13$
	$= 3$
	$(3, -2)$

11. Solve the system of equations by elimination.

$$4x - 3y = 8$$
$$-2x + y = -6$$

Solution

$$4x - 3y = 8 \implies 4x - 3y = 8$$
$$-2x + y = -6 \implies -6x + 3y = -18$$

$$-2x = -10 \qquad 4(5) - 3y = 8$$
$$x = 5 \qquad -3y = -12$$
$$y = 4$$

$(5, 4)$

12. Solve the system of equations by Cramer's Rule.

$$2x - y = 4$$
$$3x + y = -5$$

Solution

$$D = \begin{vmatrix} 2 & -1 \\ 3 & 1 \end{vmatrix} = (2) - (-3) = 5$$

$$x = \frac{\begin{vmatrix} 4 & -1 \\ -5 & 1 \end{vmatrix}}{5} = -\frac{1}{5}$$

$$y = \frac{\begin{vmatrix} 2 & 4 \\ 3 & -5 \end{vmatrix}}{5} = -\frac{22}{5}$$

$$\left(-\frac{1}{5}, -\frac{22}{5}\right)$$

13. The sum of three positive numbers is 44. The second number is 4 greater than the first, and the third number is three times the magnitude of the first. Find the numbers.

Solution

Verbal model:

$$\boxed{\text{Number 1}} + \boxed{\text{Number 2}} + \boxed{\text{Number 3}} = \boxed{44}$$

$$\boxed{\text{Number 2}} = 4 + \boxed{\text{Number 1}}$$

$$\boxed{\text{Number 3}} = 3 \cdot \boxed{\text{Number 1}}$$

System of equations: $x + y + z = 44$

$$y = 4 + x$$

$$z = 3x$$

$$\begin{bmatrix} 1 & 1 & 1 & \vdots & 44 \\ -1 & 1 & 0 & \vdots & 4 \\ -3 & 0 & 1 & \vdots & 0 \end{bmatrix}$$

$$\begin{matrix} \\ R_1 + R_2 \\ 3R_1 + R_3 \end{matrix} \begin{bmatrix} 1 & 1 & 1 & \vdots & 44 \\ 0 & 2 & 1 & \vdots & 48 \\ 0 & 3 & 4 & \vdots & 132 \end{bmatrix}$$

$$\tfrac{1}{2}R_2 \begin{bmatrix} 1 & 1 & 1 & \vdots & 44 \\ 0 & 1 & \tfrac{1}{2} & \vdots & 24 \\ 0 & 3 & 4 & \vdots & 132 \end{bmatrix}$$

$$-3R_2 + R_3 \begin{bmatrix} 1 & 1 & 1 & \vdots & 48 \\ 0 & 1 & \tfrac{1}{2} & \vdots & 24 \\ 0 & 0 & \tfrac{5}{2} & \vdots & 60 \end{bmatrix}$$

$$\tfrac{2}{5}R_3 \begin{bmatrix} 1 & 1 & 1 & \vdots & 48 \\ 0 & 1 & \tfrac{1}{2} & \vdots & 24 \\ 0 & 0 & 1 & \vdots & 24 \end{bmatrix}$$

$z = 24$ $y + \tfrac{1}{2}(24) = 24$ $x + 12 + 24 = 48$

$y = 12$ $x = 12$

$(12, 12, 24)$

14. Graph $y = 4e^{-x^2/4}$.

Solution

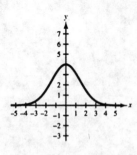

15. Evaluate $\log_4 \dfrac{1}{16}$ without using a calculator.

Solution

$$\log_4 \frac{1}{16} = -2 \quad \text{because} \quad 4^{-2} = \frac{1}{16}$$

16. Describe the relationship between the graphs of $f(x) = e^x$ and $g(x) = \ln x$.

Solution

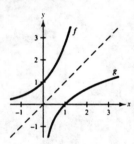

f and g are inverse functions.

17. Use the properties of logarithms to condense $3(\log_2 x + \log_2 + y) - \log_2 z$.

Solution

$$3(\log_2 x + \log_2 y) - \log_2 z = \log_2(xy)^3 - \log_2 z = \log_2 \frac{(xy)^3}{z}$$

18. Solve each equation.

(a) $\log_x\left(\dfrac{1}{9}\right) = -2$ (b) $4\ln x = 10$ (c) $500(1.08)^t = 2000$ (d) $3(1 + e^{2x}) = 20$

Solution

(a) $\log_x\left(\dfrac{1}{9}\right) = -2$

$x^{-2} = \dfrac{1}{9}$

$x^{-2} = 3^{-2}$

$x = 3$

(b) $4\ln x = 10$

$\ln x = \dfrac{5}{2}$

$e^{\ln x} = e^{5/2}$

$x = e^{5/2}$

(c) $500(1.08)^t = 2000$

$(1.08)^t = \dfrac{2000}{500}$

$(1.08)^t = 4$

$\log_{1.08} 1.08^t = \log_{1.08} 4$

$t = \dfrac{\log 4}{\log 1.08}$

$t \approx 18.01$

(d) $3(1 + e^{2x}) = 20$

$1 + e^{2x} = \dfrac{20}{3}$

$e^{2x} = \dfrac{17}{3}$

$\ln e^{2x} = \ln \dfrac{17}{3}$

$2x = \ln \dfrac{17}{3}$

$x = \dfrac{\ln \dfrac{17}{3}}{2}$

$x \approx 0.87$

19. Determine the effective yield of an 8% interest rate compounded continuously.

Solution

$A = Pe^{rt}$

$A = 1000e^{0.08(1)}$

$A = \$1,083.29$

$\textit{effective yield} = \dfrac{83.29}{1000} = 0.8329 = 8.329\%$

20. Determine the length of time for an investment of $1000 to quadruple in value if the investment earns 9% compounded continuously.

Solution

$A = Pe^{rt}$

$4000 = 1000e^{0.09t}$

$4 = e^{0.09t}$

$\ln 4 = \ln e^{0.09t}$

$\ln 4 = 0.09t$

$\dfrac{\ln 4}{0.09} = t$

$15.40 \text{ years} \approx t$

CHAPTER 10
Additional Topics in Algebra

10.1 Sequences

7. Write the first five terms of the sequence $a_n = 2n$. (Begin with $n = 1$.)

Solution

$a_1 = 2(1) = 2$

$a_2 = 2(2) = 4$

$a_3 = 2(3) = 6$

$a_4 = 2(4) = 8$

$a_5 = 2(5) = 10$

$2, 4, 6, 8, 10, \ldots, 2n, \ldots$

9. Write the first five terms of the sequence $a_n = \left(-\dfrac{1}{2}\right)^n$. (Begin with $n = 1$.)

Solution

$a_1 = \left(-\dfrac{1}{2}\right)^1 = -\dfrac{1}{2}$

$a_2 = \left(-\dfrac{1}{2}\right)^2 = \dfrac{1}{4}$

$a_3 = \left(-\dfrac{1}{2}\right)^3 = -\dfrac{1}{8}$

$a_4 = \left(-\dfrac{1}{2}\right)^4 = \dfrac{1}{16}$

$a_5 = \left(-\dfrac{1}{2}\right)^5 = -\dfrac{1}{32}$

$-\dfrac{1}{2}, \dfrac{1}{4}, -\dfrac{1}{8}, \dfrac{1}{16}, -\dfrac{1}{32}, \ldots, \left(-\dfrac{1}{2}\right)^n, \ldots$

11. Write the first five terms of the sequence $a_n = \dfrac{2n}{3n + 2}$. (Begin with $n = 1$.)

Solution

$a_1 = \dfrac{2(1)}{3(1) + 2} = \dfrac{2}{5}$

$a_2 = \dfrac{2(2)}{3(2) + 2} = \dfrac{4}{8} = \dfrac{1}{2}$

$a_3 = \dfrac{2(3)}{3(3) + 2} = \dfrac{6}{11}$

$a_4 = \dfrac{2(4)}{3(4) + 2} = \dfrac{8}{14} = \dfrac{4}{7}$

$a_5 = \dfrac{2(5)}{3(5) + 2} = \dfrac{10}{17}$

$\dfrac{2}{5}, \dfrac{1}{2}, \dfrac{6}{11}, \dfrac{4}{7}, \dfrac{10}{17}, \ldots, \dfrac{2n}{3n + 2}, \ldots$

13. Write the first five terms of the sequence $a_n = \dfrac{2n}{n!}$. (Begin with $n = 1$.)

Solution

$a_1 = \dfrac{2^1}{1!} = \dfrac{2}{1} = 2$

$a_2 = \dfrac{2^2}{2!} = \dfrac{4}{2} = 2$

$a_3 = \dfrac{2^3}{3!} = \dfrac{8}{6} = \dfrac{4}{3}$

$a_4 = \dfrac{2^4}{4!} = \dfrac{16}{24} = \dfrac{2}{3}$

$a_5 = \dfrac{2^5}{5!} = \dfrac{32}{120} = \dfrac{4}{15}$

$2, 2, \dfrac{4}{3}, \dfrac{2}{3}, \dfrac{4}{15}, \ldots, \dfrac{2^n}{n!}, \ldots$

15. Using the graphs in the textbook, match the sequence with the graph of its first 10 terms. [The graphs are labeled (a), (b), (c), and (d).]

$$a_n = \frac{6}{n+1}$$

Solution

(c)

17. Using the graphs in the textbook, match the sequence with the graph of its first 10 terms. [The graphs are labeled (a), (b), (c), and (d).]

$$a_n = (0.6)^{n-1}$$

Solution

(b)

19. Use a graphing utility to graph the first 10 terms of the sequence.

$$a_n = (-0.8)^{n-1}$$

Solution

Keystrokes (calculator in sequence and dot mode):

$$\boxed{Y=} \quad \boxed{(} \quad \boxed{(-)} \quad .8 \quad \boxed{)} \quad \boxed{\wedge} \quad \boxed{(} \quad \boxed{n} \quad \boxed{-} \quad 1 \quad \boxed{)} \quad \boxed{\text{TRACE}}$$

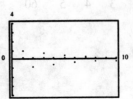

21. Find the indicated term of the sequence.

$$a_{15} = (-1)^n(5n - 3) \quad a_{15} = ?$$

Solution

$$a_{15} = (-1)^{15}[5(15) - 3]$$

$$= -1[72]$$

$$= -72$$

23. Simplify $\dfrac{25!}{27!}$.

Solution

$$\frac{25!}{27!} = \frac{25!}{27 \cdot 26 \cdot 25!} = \frac{1}{27 \cdot 26} = \frac{1}{702}$$

25. Simplify $\dfrac{(n+1)!}{(n-1)!}$.

Solution

$$\frac{(n+1)!}{(n-1)!} = \frac{(n+1)n(n-1)!}{(n-1)!} = (n+1)n$$

27. Find the sum $\displaystyle\sum_{k=1}^{6} 3k$.

Solution

$$\sum_{k=1}^{6} 3k = 3(1) + 3(2) + 3(3) + 3(4) + 3(5) + 3(6) = 3 + 6 + 9 + 12 + 15 + 18 = 63$$

29. Find the sum $\displaystyle\sum_{i=0}^{6} (2i + 5)$.

Solution

$$\sum_{i=0}^{6} (2i + 5) = [2(0) + 5] + [2(1) + 5] + [2(2) + 5] + [2(3) + 5] + [2(4) + 5] +$$

$$[2(5) + 5] + [2(6) + 5]$$

$$= 5 + 7 + 9 + 11 + 13 + 15 + 17 = 77$$

31. Find the sum $\displaystyle\sum_{m=2}^{6} \frac{2m}{2(m-1)}$.

Solution

$$\sum_{m=2}^{6} \frac{2m}{2(m-1)} = \frac{2(2)}{2(2-1)} + \frac{2(3)}{2(3-1)} + \frac{2(4)}{2(4-1)} + \frac{2(5)}{2(5-1)} + \frac{2(6)}{2(6-1)}$$

$$= \frac{4}{2} + \frac{6}{4} + \frac{8}{6} + \frac{10}{8} + \frac{12}{10}$$

$$= 2 + \frac{3}{2} + \frac{4}{3} + \frac{5}{4} + \frac{6}{5} = \frac{437}{60} = 7.283$$

33. Find the sum $\displaystyle\sum_{i=0}^{4} (i! + 4)$.

Solution

$$\sum_{i=0}^{4} (i! + 4) = [0! + 4] + [1! + 4] + [2! + 4] + [3! + 4] + [4! + 4]$$

$$= 5 + 5 + 6 + 10 + 28 = 54$$

35. Write the sum $2 + 4 + 6 + 8 + 10$ using sigma notation.

Solution

$$\sum_{k=1}^{5} 2k$$

37. Write the sum of $\dfrac{4}{1+3} + \dfrac{4}{2+3} + \dfrac{4}{3+3} + \cdots + \dfrac{4}{20+3}$ using sigma notation.

Solution

$$\sum_{k=1}^{20} \frac{4}{k+3}$$

39. Write the sum of $\dfrac{2}{4} + \dfrac{4}{5} + \dfrac{6}{6} + \dfrac{8}{7} + \cdots + \dfrac{40}{23}$ using sigma notation.

Solution

$$\sum_{k=1}^{20} \frac{2k}{k+3}$$

41. *Compound Interest* A deposit of $500 is made in an account that earns 7% interest compounded monthly. The balance in the account after N months is given by

$$A_N = 500(1 + 0.07)^N, \quad N = 1, 2, 3, \ldots$$

(a) Compute the first eight terms of this sequence.

(b) Find the balance in this account after 40 years by computing A_{40}.

(c) Use a graphing utility to graph the first 40 terms of the sequence.

(d) The terms of the sequence are increasing. Is the rate of growth of the terms increasing? Explain.

Solution

(a) $A_1 = 500(1 + 0.07)^1 = \$535.00$ $A_5 = 500(1 + 0.07)^5 = \$701.28$

$A_2 = 500(1 + 0.07)^2 = \$572.45$ $A_6 = 500(1 + 0.07)^6 = \$750.37$

$A_3 = 500(1 + 0.07)^3 = \$612.52$ $A_7 = 500(1 + 0.07)^7 = \$802.89$

$A_4 = 500(1 + 0.07)^4 = \$655.40$ $A_8 = 500(1 + 0.07)^8 = \$859.09$

(b) $A_{40} = 500(1 + 0.07)^{40} = \$7487.23$

(c) Keystrokes (calculator in sequence and dot mode):

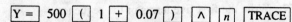

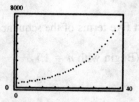

(d) Yes. Investment earning compound interest increases at an increasing rate.

43. Simplify $(x + 10)^2$, $x \neq -10$.

Solution

$(x + 10)^2 = x^2 + 20x + 100$

45. Simplify $(a^2)^{-4}$, $a \neq 0$.

Solution

$(a^2)^{-4} = a^{-8} = \dfrac{1}{a^8}$

47. Find an equation of the line through $(2, 3)$ and $(5, 6)$.

Solution

$m = \dfrac{6 - 3}{5 - 2} = \dfrac{3}{3} = 1$ $y - 6 = 1(x - 5)$

$y = x + 1$

$0 = x - y + 1$

49. Write the first five terms of the sequence $a_n = (-1)^n 2n$. (Begin with $n = 1$.)

Solution

$a_1 = (-1)^1 \cdot 2(1) = -2$

$a_2 = (-1)^2 \cdot 2(2) = 4$

$a_3 = (-1)^3 \cdot 2(3) = -6$

$a_4 = (-1)^4 \cdot 2(4) = 8$

$a_5 = (-1)^5 \cdot 2(5) = -10$

$-2, 4, -6, 8, -10, \ldots, (-1)^n 2n, \ldots$

51. Write the first five terms of the sequence

$a_n = \left(\dfrac{1}{2}\right)^n$. (Begin with $n = 1$.)

Solution

$a_1 = \left(\dfrac{1}{2}\right)^1 = \dfrac{1}{2}$

$a_2 = \left(\dfrac{1}{2}\right)^2 = \dfrac{1}{4}$

$a_3 = \left(\dfrac{1}{2}\right)^3 = \dfrac{1}{8}$

$a_4 = \left(\dfrac{1}{2}\right)^4 = \dfrac{1}{16}$

$a_5 = \left(\dfrac{1}{2}\right)^5 = \dfrac{1}{32}$

$\dfrac{1}{2}, \dfrac{1}{4}, \dfrac{1}{8}, \dfrac{1}{16}, \dfrac{1}{32}, \ldots, \left(\dfrac{1}{2}\right)^n, \ldots$

53. Write the first five terms of the sequence

$a_n = \dfrac{1}{n+1}$. (Begin with $n = 1$.)

Solution

$a_1 = \dfrac{1}{1+1} = \dfrac{1}{2}$

$a_2 = \dfrac{1}{2+1} = \dfrac{1}{3}$

$a_3 = \dfrac{1}{3+1} = \dfrac{1}{4}$

$a_4 = \dfrac{1}{4+1} = \dfrac{1}{5}$

$a_5 = \dfrac{1}{5+1} = \dfrac{1}{6}$

$\dfrac{1}{2}, \dfrac{1}{3}, \dfrac{1}{4}, \dfrac{1}{5}, \dfrac{1}{6}, \ldots, \dfrac{1}{n+1}, \ldots$

55. Write the first five terms of the sequence

$a_n = \dfrac{(-1)^n}{n^2}$. (Begin with $n = 1$.)

Solution

$a_1 = \dfrac{(-1)^1}{1^2} = -1$

$a_2 = \dfrac{(-1)^2}{2^2} = \dfrac{1}{4}$

$a_3 = \dfrac{(-1)^3}{3^2} = -\dfrac{1}{9}$

$a_4 = \dfrac{(-1)^4}{4^2} = \dfrac{1}{16}$

$a_5 = \dfrac{(-1)^5}{5^2} = -\dfrac{1}{25}$

$-1, \dfrac{1}{4}, -\dfrac{1}{9}, \dfrac{1}{16}, -\dfrac{1}{25}, \ldots, \dfrac{(-1)^n}{n^2}, \ldots$

57. Write the first five terms of the sequence

$a_n = 5 - \dfrac{1}{2^n}$. (Begin with $n = 1$.)

Solution

$a_1 = 5 - \dfrac{1}{2^1} = \dfrac{9}{2}$

$a_2 = 5 - \dfrac{1}{2^2} = \dfrac{19}{4}$

$a_3 = 5 - \dfrac{1}{2^3} = \dfrac{39}{8}$

$a_4 = 5 - \dfrac{1}{2^4} = \dfrac{79}{16}$

$a_5 = 5 - \dfrac{1}{2^5} = \dfrac{159}{32}$

$\dfrac{9}{2}, \dfrac{19}{4}, \dfrac{39}{8}, \dfrac{79}{16}, \dfrac{159}{32}, \ldots, 5 - \dfrac{1}{2^n}, \ldots$

59. Write the first five terms of the sequence $a_n = 2 + (-2)^n$. (Begin with $n = 1$.)

Solution

$a_1 = 2 + (-2)^1 = \quad 0$

$a_2 = 2 + (-2)^2 = \quad 6$

$a_3 = 2 + (-2)^3 = \quad -6$

$a_4 = 2 + (-2)^4 = \quad 18$

$a_5 = 2 + (-2)^5 = -30$

$0, 6, -6, 18, -30, \ldots, 2 + (-2)^n, \ldots$

61. Using a graphing utility to graph the first 10 terms of the sequence $a_n = \dfrac{1}{2}n$.

Solution

Keystrokes (calculator in sequence and dot mode):

$\boxed{Y =}$ .5 $\boxed{n}$ $\boxed{TRACE}$

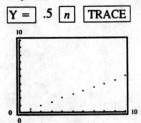

63. Using a graphing utility to graph the first 10 terms of the sequence $a_n = 3 - \dfrac{4}{n}$.

Solution

Keystrokes (calculator in sequence and dot mode):

$\boxed{Y =}$ 3 $\boxed{-}$ $\boxed{(}$ 4 $\boxed{\div}$ $\boxed{n}$ $\boxed{)}$ $\boxed{TRACE}$

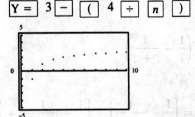

65. Simplify $\dfrac{5!}{4!}$.

Solution

$$\frac{5!}{4!} = \frac{5 \cdot 4 \cdot 3 \cdot 2 \cdot 1}{4 \cdot 3 \cdot 2 \cdot 1} = 5$$

67. Simplify $\dfrac{10!}{12!}$.

Solution

$$\frac{10!}{12!} = \frac{10!}{12 \cdot 11 \cdot 10!} = \frac{1}{132}$$

69. Simplify $\dfrac{n!}{(n+1)!}$.

Solution

$$\frac{n!}{(n+1)!} = \frac{1}{n+1}$$

71. Simplify $\dfrac{(2n)!}{(2n-1)!}$.

Solution

$$\frac{(2n)!}{(2n-1)!} = \frac{(2n)(2n-1)!}{(2n-1)!} = 2n$$

73. Find the sum $\displaystyle\sum_{i=0}^{4} (2i+3)$.

Solution

$$\sum_{i=0}^{4} (2i+3) = [2(0)+3] + [2(1)+3] + [2(2)+3] + [2(3)+3] + [2(4)+3]$$

$$= 3 + 5 + 7 + 9 + 11$$

$$= 35$$

75. Find the sum $\displaystyle\sum_{j=1}^{5} \frac{(-1)^{j+1}}{j}$.

Solution

$$\sum_{j=1}^{5} \frac{(-1)^{j+1}}{j} = \frac{(-1)^{1+1}}{1} + \frac{(-1)^{2+1}}{2} + \frac{(-1)^{3+1}}{3} + \frac{(-1)^{4+1}}{4} + \frac{(-1)^{5+1}}{5}$$

$$= 1 + -\frac{1}{2} + \frac{1}{3} + -\frac{1}{4} + \frac{1}{5}$$

$$= \frac{47}{60}$$

77. Find the sum $\displaystyle\sum_{k=1}^{6} (-8)$.

Solution

$$\sum_{k=1}^{6} (-8) = (-8) + (-8) + (-8) + (-8) + (-8) + (-8) = -48$$

79. Find the sum $\displaystyle\sum_{i=1}^{8} \left(\frac{1}{i} - \frac{1}{i+1}\right)$.

Solution

$$\sum_{i=1}^{8} \left(\frac{1}{i} - \frac{1}{i+1}\right) = \left[\frac{1}{1} - \frac{1}{1+1}\right] + \left[\frac{1}{2} - \frac{1}{2+1}\right] + \left[\frac{1}{3} - \frac{1}{3+1}\right] + \left[\frac{1}{4} - \frac{1}{4+1}\right] +$$

$$\left[\frac{1}{5} - \frac{1}{5+1}\right] + \left[\frac{1}{6} - \frac{1}{6+1}\right\} + \left[\frac{1}{7} - \frac{1}{7+1}\right] + \left[\frac{1}{8} - \frac{1}{8+1}\right]$$

$$= \frac{1}{2} + \frac{1}{6} + \frac{1}{12} + \frac{1}{20} + \frac{1}{30} + \frac{1}{42} + \frac{1}{56} + \frac{1}{72}$$

$$= \frac{1260 + 420 + 210 + 126 + 84 + 60 + 45 + 35}{2520}$$

$$= \frac{2240}{2520} = \frac{8}{9}$$

81. Find the sum $\displaystyle\sum_{n=0}^{5} \left(-\frac{1}{3}\right)^{n}$.

Solution

$$\sum_{n=0}^{5} \left(-\frac{1}{3}\right)^{n} = \left(-\frac{1}{3}\right)^{0} + \left(-\frac{1}{3}\right)^{1} + \left(-\frac{1}{3}\right)^{2} + \left(-\frac{1}{3}\right)^{3} + \left(-\frac{1}{3}\right)^{4} + \left(-\frac{1}{3}\right)^{5}$$

$$= 1 + \left(-\frac{1}{3}\right) + \frac{1}{9} + \left(-\frac{1}{27}\right) + \frac{1}{81} + \left(-\frac{1}{243}\right)$$

$$= \frac{243 - 81 + 27 - 9 + 3 - 1}{243}$$

$$= \frac{182}{243}$$

83. Find the sum $\displaystyle\sum_{n=1}^{6} n(n+1)$.

Solution

$$\sum_{n=1}^{6} n(n+1) = [1(1+1)] + [2(2+1)] + [3(3+1)] + [4(4+1)] + [5(5+1)] + [6(6+1)]$$

$$= 2 + 6 + 12 + 20 + 30 + 42$$

$$= 112$$

85. Find the sum $\displaystyle\sum_{j=2}^{6} (j! - j)$.

Solution

$$\sum_{j=2}^{6} (j! - j) = (2! - 2) + (3! - 3) + (4! - 4) + (5! - 5) + (6! - 6)$$
$$= 0 + 3 + 20 + 115 + 714$$
$$= 852$$

87. Find the sum $\displaystyle\sum_{k=1}^{6} \ln k$.

Solution

$$\sum_{k=1}^{6} \ln k = \ln 1 + \ln 2 + \ln 3 + \ln 4 + \ln 5 + \ln 6 = 6.5792512$$

89. Write the sum of $1 + 2 + 3 + 4 + 5$ using sigma notation. (Begin with $k = 0$ or $k = 1$.)

Solution

$$\sum_{k=1}^{5} k$$

91. Write the sum of $\dfrac{1}{2(1)} + \dfrac{1}{2(2)} + \dfrac{1}{2(3)} + \dfrac{1}{2(4)} + \cdots + \dfrac{1}{2(10)}$ using sigma notation.
(Begin with $k = 0$ or $k = 1$.)

Solution

$$\sum_{k=1}^{10} \frac{1}{2k}$$

93. Write the sum of $\dfrac{1}{1^2} + \dfrac{1}{2^2} + \dfrac{1}{3^2} + \dfrac{1}{4^2} + \cdots + \dfrac{1}{20^2}$
using sigma notation. (Begin with $k = 0$ or $k = 1$.)

Solution

$$\sum_{k=1}^{20} \frac{1}{k^2}$$

95. Write the sum of $\dfrac{1}{3^0} - \dfrac{1}{3^1} + \dfrac{1}{3^2} - \dfrac{1}{3^3} + \cdots - \dfrac{1}{3^9}$
using sigma notation. (Begin with $k = 0$ or $k = 1$.)

Solution

$$\sum_{k=0}^{9} \frac{1}{(-3)^k}$$

97. Write the sum of $\dfrac{1}{2} + \dfrac{2}{3} + \dfrac{3}{4} + \dfrac{4}{5} + \dfrac{5}{6} + \cdots + \dfrac{11}{12}$
using sigma notation. (Begin with $k = 0$ or $k = 1$.)

Solution

$$\sum_{k=1}^{11} \frac{k}{k+1}$$

99. Write the sum of $1 + 1 + 2 + 6 + 24 + 120 + 720$
using sigma notation. (Begin with $k = 0$ or $k = 1$.)

Solution

$$\sum_{k=0}^{6} k!$$

101. *Arithmetic Mean*   Find the arithmetic mean of 1, 2, 3, 4, 5. The *arithmetic mean* $\bar{x}$ of a set of n measurements $x_1, x_2, x_3, \ldots, x_n$, is

$$\bar{x} = \frac{1}{n} \sum_{i=1}^{n} x_i.$$

Solution

$$\bar{x} = \frac{1 + 2 + 3 + 4 + 5}{5} = 3$$

103. *Arithmetic Mean*   Find the arithmetic mean of 5, 8, 11, 14, 17, 20. The *arithmetic mean x* of a set of n measurements $x_1, x_2, x_3, \ldots, x_n$, is

$$\bar{x} = \frac{1}{n} \sum_{i=1}^{n} x_i.$$

Solution

$$\bar{x} = \frac{5 + 8 + 11 + 14 + 17 + 20}{6} = 12.5$$

105. *Soccer Ball*   The number of degrees a_n in each angle of a regular n-sided polygon is

$$a_n = \frac{180(n - 2)}{n}, \quad n \geq 3.$$

The surface of a soccer ball is made of regular hexagons and pentagons. If a soccer ball is taken apart and flattened (see figure in textbook), the sides of the hexagons do not meet each other. Use the terms a_5 and a_6 to explain why there are gaps between adjacent hexagons.

Solution

$$a_5 = \frac{180(5 - 2)}{5} = 108°$$

$$a_6 = \frac{180(6 - 2)}{6} = 120°$$

$$a_5 + 2a_6 = 108° + 240° = 348° < 360°$$

107. *Stars*   The stars in Exercise 106 (see figure in textbook) are formed by placing n equally spaced points on a circle and connecting each point with the second point from it on the circle. The stars in the figure for this exercise (see figure in textbook) are formed in a similar way except each point is connected with the third point from it. For these stars, the number of degrees in a tip is

$$d_n = \frac{180(n - 6)}{n}, \quad n \geq 7.$$

Write the first five terms of this sequence.

Solution

$$d_7 = \frac{180(7 - 6)}{7} = 25.7°$$

$$d_8 = \frac{180(8 - 6)}{8} = 45°$$

$$d_9 = \frac{180(9 - 6)}{9} = 60°$$

$$d_{10} = \frac{180(10 - 6)}{10} = 72°$$

$$d_{11} = \frac{180(11 - 6)}{11} = 81.8°$$

10.2 Arithmetic Sequences

7. Find the common difference of

10, 22, 34, 46, 58, ...

Solution

$d = 12$

9. Find the common difference of

$\dfrac{7}{2}, \dfrac{9}{4}, 1, -\dfrac{1}{4}, -\dfrac{3}{2}, \ldots .$

Solution

$d = -\dfrac{5}{4}$

11. Determine whether 10, 8, 6, 4, 2, . . . is arithmetic. If it is, find the common difference.

Solution

arithmetic; $d = -2$

13. Determine whether $3, \dfrac{5}{2}, 2, \dfrac{3}{2}, 1, \ldots$ is arithmetic. If it is, find the common difference.

Solution

arithmetic; $d = -\dfrac{1}{2}$

15. Determine whether $-12, -8, -4, 0, 4, \ldots$ is arithmetic. If it is, find the common difference.

Solution

arithmetic; $d = 4$

17. Determine whether ln 4, ln 8, ln 12, ln 16, . . . is arithmetic. If it is, find the common difference.

Solution

The sequence is not arithmetic.

19. Write the first five terms of $a_n = -5n + 45$. (Begin with $n = 1$.)

Solution

$a_1 = -5(1) + 45 = 40$

$a_2 = -5(2) + 45 = 35$

$a_3 = -5(3) + 45 = 30$

$a_4 = -5(4) + 45 = 25$

$a_5 = -5(5) + 45 = 20$

21. Write the first five terms of $a_n = \dfrac{3}{5}n + 1$. (Begin with $n = 1$.)

Solution

$a_1 = \dfrac{3}{5}(1) + 1 = \dfrac{8}{5}$

$a_2 = \dfrac{3}{5}(2) + 1 = \dfrac{11}{5}$

$a_3 = \dfrac{3}{5}(3) + 1 = \dfrac{14}{5}$

$a_4 = \dfrac{3}{5}(4) + 1 = \dfrac{17}{5}$

$a_5 = \dfrac{3}{5}(5) + 1 = \dfrac{20}{5} = 4$

23. Write the first five terms of $a_1 = 25$, $a_{k+1} = a_k + 3$.

Solution

$a_n = dn + c$

$25 = (a_k - 22)(1) + c$

$47 - a_k = c$

$a_n = (a_k - 22)n + 47 - a_k$

$a_1 = (a_k - 22)(1) + 47 - a_k$

$\quad = a_k - 22 + 47 - a_k$

$\quad = 25$

$a_2 = (a_k - 22)(2) + 47 - a_k$

$\quad = 2a_k - 44 + 47 - a_k$

$\quad = a_k + 3$

$a_3 = (a_k - 22)(3) + 47 - a_k$

$\quad = 3a_k - 66 + 47 - a_k$

$\quad = 2a_k - 19$

$a_4 = (a_k - 22)(4) + 47 - a_k$

$\quad = 4a_k - 88 + 47 - a_k$

$\quad = 3a_k - 41$

$a_5 = (a_k - 22)(5) + 47 - a_k$

$\quad = 5a_k - 110 + 47 - a_k$

$\quad = 4a_k - 63$

25. Using the graphs in the textbook, match the sequence $a_n = -\frac{1}{3}n + 2$ with its graph. [The graphs are labeled (a), (b), (c), and (d).]

Solution

(d)

27. Using the graphs in the textbook, match the sequence $a_n = 2n - 3$ with its graph. [The graphs are labeled (a), (b), (c), and (d).]

Solution

(c)

29. Find a formula for the nth term of the term of the arithmetic sequence $a_1 = 3$ $d = \frac{3}{2}$.

Solution

$a_n = a_1 + (n - 1)d$

$a_n = 3 + (n - 1)\frac{3}{2}$

$a_n = 3 + \frac{3}{2}n - \frac{3}{2}$

$a_n = \frac{3}{2}n + \frac{3}{2}$

31. Find a formula for the nth term of the arithmetic sequence $a_3 = 20$, $d = -4$.

Solution

$a_n = a_1 + (n - 1)d$

$20 = a_1 + (3 - 1)(-4)$ so $a_n = 28 + (n - 1)(-4)$

$20 = a_1 - 8$ $\qquad\qquad\qquad = 28 - 4n + 4$

$28 = a_1$ $\qquad\qquad\qquad\quad = -4n + 32$

33. Find a formula for the nth term of the arithmetic sequence $a_1 = 5$, $a_5 = 13$.

Solution

$a_n = a_1 + (n - 1)d$

$13 = 5 + (5 - 1)d$ $\qquad a_n = 5 + (n - 1)2$

$13 = 5 + 4d$

$8 = 4d$

$2 = \frac{8}{4} = d$

35. Find the sum $\displaystyle\sum_{k=1}^{10} 5k$.

Solution

$$\sum_{k=1}^{10} 5k = 10\left(\frac{5 + 50}{2}\right) = 275$$

37. Find the sum $\displaystyle\sum_{j=1}^{25} (750 - 30j)$.

Solution

$$\sum_{j=1}^{25} (750 - 30j) = 25\left(\frac{720 + 0}{2}\right) = 9000$$

39. Find the nth partial sum of the arithmetic sequence 2, 8, 14, 20, . . . , $n = 25$.

Solution

$$\sum_{n=1}^{25} (6n - 4) = 25\left(\frac{2 + 146}{2}\right) = 1850$$

41. Find the nth partial sum of the arithmetic sequence 0.5, 0.9, 1.3, 1.7, . . . , $n = 10$.

Solution

$$\sum_{n=1}^{10} (.4n + .1) = 10\left(\frac{.5 + 4.1}{2}\right) = 23$$

43. Find the sum of the multiples of 6 from 12 to 240.

Solution

Sequence = 12, 18, 24, 30, 36, . . .

$$a_n = a_1 + (n - 1)d$$
$$240 = 12 + (n - 1)6$$
$$228 = 6(n - 1)$$
$$228 = 6n - 6$$
$$234 = 6n$$
$$39 = n$$

$$a_n = 12 + (n - 1)6$$
$$= 12 + 6n - 6$$
$$= 6n + 6$$

$$\sum_{n=1}^{39} (6n + 6) = 39\left(\frac{12 + 240}{2}\right) = 4914$$

45. *Free-Falling Object*   A free-falling object will fall 16 feet during the first second, 48 more feet during the second second, 80 more feet during the third, and so on. What is the total distance the object will fall in 8 seconds if this pattern continues?

Solution

Sequence = 16, 48, 80, . . . $n = 8$ $d = 32$

$a_n = a_1 + (n - 1)d$ $a_n = 16 + (n - 1)32$

$a_n = 16 + (8 - 1)32$ $= 16 + 32n - 32$

$a_n = 16 + 224$ $a_n = 32n - 16$

$a_n = 240$

$$\sum_{n=1}^{8} (32n - 16) = 8\left(\frac{16 + 240}{2}\right) = 1024 \text{ feet}$$

47. Solve the system of equations.

$$y = x^2$$
$$-3x + 2y = 2$$

Solution

$$-3x + 2(x^2) = 2$$
$$2x^2 - 3x - 2 = 0$$
$$(2x + 1)(x - 2) = 0$$

$$x = -\frac{1}{2} \quad x = 2 \qquad \left(-\frac{1}{2}, \frac{1}{4}\right), \ (2, 4)$$

$$y = \frac{1}{4} \quad y = 4$$

49. Solve the system of equations.

$$-x + y \quad = 1$$
$$x + 2y - 2z = 3$$
$$3x - y + 2z = 3$$

Solution

$$\begin{bmatrix} -1 & 1 & 0 & \vdots & 1 \\ 1 & 2 & -2 & \vdots & 3 \\ 3 & -1 & 2 & \vdots & 3 \end{bmatrix} R_1 \leftarrow R_2 \Rightarrow \begin{bmatrix} 1 & 2 & -2 & \vdots & 3 \\ -1 & 1 & 0 & \vdots & 1 \\ 3 & -1 & 2 & \vdots & 3 \end{bmatrix} \begin{matrix} R_1 + R_2 \\ \\ -3R_1 + R_3 \end{matrix} \Rightarrow$$

$$\begin{bmatrix} 1 & 2 & -2 & \vdots & 3 \\ 0 & 3 & -2 & \vdots & 4 \\ 0 & -7 & 8 & \vdots & -6 \end{bmatrix} \frac{1}{3}R_2 \Rightarrow \begin{bmatrix} 1 & 2 & -2 & \vdots & 3 \\ 0 & 1 & -\frac{2}{3} & \vdots & \frac{4}{3} \\ 0 & -7 & 8 & \vdots & -6 \end{bmatrix} \begin{matrix} -2R_2 + R_1 \\ \\ 7R_2 + R_3 \end{matrix} \Rightarrow$$

$$\begin{bmatrix} 1 & 0 & -\frac{2}{3} & \vdots & \frac{1}{3} \\ 0 & 1 & -\frac{2}{3} & \vdots & \frac{4}{3} \\ 0 & 0 & \frac{10}{3} & \vdots & \frac{10}{3} \end{bmatrix} \frac{3}{10}R_3 \Rightarrow \begin{bmatrix} 1 & 0 & -\frac{2}{3} & \vdots & \frac{1}{3} \\ 0 & 1 & -\frac{2}{3} & \vdots & \frac{4}{3} \\ 0 & 0 & 1 & \vdots & 1 \end{bmatrix} \begin{matrix} \frac{2}{3}R_3 + R_1 \\ \\ \frac{2}{3}R_3 + R_2 \end{matrix} \Rightarrow$$

$$\begin{bmatrix} 1 & 0 & 0 & \vdots & 1 \\ 0 & 1 & 0 & \vdots & 2 \\ 0 & 0 & 1 & \vdots & 1 \end{bmatrix} \ (1, 2, 1)$$

51. *Ticket Sales* Twelve hundred tickets are sold for a total of $21,120. Adult tickets cost $20 and children's tickets cost $12.50. How many of each type of tickets were sold?

Solution

Verbal model: $\boxed{\begin{array}{c}\text{Cost} \\ \text{Children} \\ \text{Tickets}\end{array}} + \boxed{\begin{array}{c}\text{Cost} \\ \text{Adult} \\ \text{Tickets}\end{array}} = \boxed{\begin{array}{c}\text{Total} \\ \text{Cost}\end{array}}$

System of equations:

$$x + y = 1200$$
$$12.50x + 20y = 21,120$$

Solve by substitution:

$$y = 1200 - x$$
$$12.50x + 20(1200 - x) = 21,120$$
$$12.50x + 24,000 - 20x = 21,120$$
$$-7.5x = -2880$$
$$x = 384 \text{ children's tickets}$$
$$1200 - x = 814 \text{ adult tickets}$$

53. Find the common difference of 2, 5, 8, 11, ...

Solution
$d = 3$

55. Find the common difference of 100, 94, 88, 82, ...

Solution
$d = -6$

57. Find the common difference of $1, \dfrac{5}{3}, \dfrac{7}{3}, 3, \ldots$.

Solution
$d = \dfrac{2}{3}$

59. Determine whether 2, 4, 6, 8, . . . is arithmetic. If it is, find the common difference.

Solution
The sequence is arithmetic.
$d = 2$

61. Determine whether $2, \dfrac{7}{2}, 5, \dfrac{13}{2}, \ldots$ is arithmetic. If it is, find the common difference.

Solution
The sequence is arithmetic.
$d = \dfrac{3}{2}$

63. Find whether 32, 16, 8, 4, . . . is arithmetic. If it is, find the common difference.

Solution
The sequence is not arithmetic.

65. Determine whether $\dfrac{1}{3}, \dfrac{1}{2}, \dfrac{2}{3}, \dfrac{5}{6}, 1, \ldots$ is arithmetic. If it is, find the common difference.

Solution
The sequence is arithmetic.
$d = \dfrac{1}{6}$

67. Determine whether 3.2, 4, 4.8, 5.6, . . . is arithmetic. If it is, find the common difference.

Solution
The sequence is arithmetic.
$d = 0.8$

69. Determine whether $1, \sqrt{2}, \sqrt{3}, 2, \sqrt{5}, \ldots$ is arithmetic. If it is, find the common difference.

Solution

The sequence is not arithmetic.

71. Write the first five terms of $a_n = 3n + 4$. (Begin with $n = 1$.)

Solution

$a_1 = 3(1) + 4 = 7$
$a_2 = 3(2) + 4 = 10$
$a_3 = 3(3) + 4 = 13$
$a_4 = 3(4) + 4 = 16$
$a_5 = 3(5) + 4 = 19$

73. Write the first five terms of $a_n = -2n + 8$. (Begin with $n = 1$.)

Solution

$a_1 = -2(1) + 8 = 6$
$a_2 = -2(2) + 8 = 4$
$a_3 = -2(3) + 8 = 2$
$a_4 = -2(4) + 8 = 0$
$a_5 = -2(5) + 8 = -2$

75. Write the first five terms of $a_n = \dfrac{5}{2}n - 1$. (Begin with $n = 1$.)

Solution

$a_1 = \dfrac{5}{2}(1) - 1 = \dfrac{3}{2}$

$a_2 = \dfrac{5}{2}(2) - 1 = 4$

$a_3 = \dfrac{5}{2}(3) - 1 = \dfrac{13}{2}$

$a_4 = \dfrac{5}{2}(4) - 1 = 9$

$a_5 = \dfrac{5}{2}(5) - 1 = \dfrac{23}{2}$

77. Write the first five terms of $a_n = -\dfrac{1}{4}(n - 1) + 4$. (Begin with $n = 1$.)

Solution

$a_1 = -\dfrac{1}{4}(1 - 1) + 4 = 4$

$a_2 = -\dfrac{1}{4}(2 - 1) + 4 = \dfrac{15}{4}$

$a_3 = -\dfrac{1}{4}(3 - 1) + 4 = \dfrac{7}{2}$

$a_4 = -\dfrac{1}{4}(4 - 1) + 4 = \dfrac{13}{4}$

$a_5 = -\dfrac{1}{4}(5 - 1) + 4 = 3$

79. Use a graphing utility to graph the first 10 terms of the sequence $a_n = -2n + 21$.

Solution

Keystrokes (calculator in sequence and dot mode):

$\boxed{Y =}$ $\boxed{(-)}$ 2 $\boxed{n}$ $\boxed{+}$ 21 $\boxed{\text{TRACE}}$

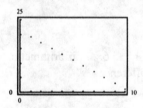

81. Use a graphing utility to graph the first 10 terms of the sequence $a_n = \dfrac{3}{5}n + \dfrac{3}{2}$.

Solution

Keystrokes (calculator in sequence and dot mode):

$\boxed{Y =}$ .6 $\boxed{n}$ $\boxed{+}$ 1.5 $\boxed{\text{TRACE}}$

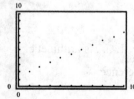

83. Find a formula for the *n*th term of $a_1 = 3$, $d = \dfrac{1}{2}$.

Solution

$$a_n = a_1 + (n - 1)d$$
$$a_n = 3 + (n - 1)\dfrac{1}{2}$$
$$a_n = 3 + \dfrac{1}{2}n - \dfrac{1}{2}$$
$$a_n = \dfrac{1}{2}n + \dfrac{5}{2}$$

85. Find a formula for the *n*th term of $a_1 = 64$, $d = -8$.

Solution

$$a_n = a_1 + (n - 1)d$$
$$a_n = 64 + (n - 1)(-8)$$
$$a_n = 64 - 8n + 8$$
$$a_n = -8n + 72$$

87. Find a formula for the *n*th term of $a_3 = 14$, $a_4 = 20$.

Solution

$$a_n = a_1 + (n - 1)d$$
$$16 = a_1 + (3 - 1)4$$
$$8 = a_1$$
$$a_n = 8 + (n - 1)(4)$$
$$a_n = 4n + 4$$

89. Find a formula for the *n*th term of $a_1 = 50$, $a_3 = 30$.

Solution

$$a_n = a_1 + (n - 1)d$$
$$30 = 50 + (3 - 1)d$$
$$-20 = 2d$$
$$-10 = d$$
$$a_n = 50 + (n - 1)(-10)$$
$$a_n = -10n + 60$$

91. Write the first five terms of the arithmetic sequence defined recursively:
$$a_1 = 9 \quad a_{k+1} = a_k - 3.$$

Solution

$$a_n = a_1 + (n - 1)d$$
$$a_1 = 9$$
$$a_2 = a_{1+1} = a_1 - 3 = 9 - 3 = 6$$
$$a_3 = a_{2+1} = a_2 - 3 = 6 - 3 = 3$$
$$a_4 = a_{3+1} = a_3 - 3 = 3 - 3 = 0$$
$$a_5 = a_{4+1} = a_4 - 3 = 0 - 3 = -3$$

93. Write the first five terms of the arithmetic sequence defined recursively.

First term: -10 $(k + 1)$st term: $a_k + 6$

Solution

$$a_n = a_1 + (n - 1)d$$
$$a_1 = -10$$
$$a_2 = a_{1+1} = a_1 + 6 = -10 + 6 = -4$$
$$a_3 = a_{2+1} = a_2 + 6 = -4 + 6 = 2$$
$$a_4 = a_{3+1} = a_3 + 6 = 2 + 6 = 8$$
$$a_5 = a_{4+1} = a_4 + 6 = 8 + 6 = 14$$

95. Using the graphs in the textbook, match the sequence with its graph. [The graphs are labeled (a), (b), (c), (d), and (e).]

$$a_n = \dfrac{1}{2}n + 1$$

Solution

(a)

97. Using the graphs in the textbook, match the sequence with its graph.

[The graphs are labeled (a), (b), (c), (d),and (e).]

$a_n = -2n + 10$

Solution

(e)

99. Using the graphs in the textbook, match the sequence with its graph.

[The graphs are labeled (a), (b), (c), (d),and (e).]

$a_1 = 12 \qquad a_{k+1} = a_k - 2$

Solution

(c)

101. Find the sum $\displaystyle\sum_{n=1}^{20} n$.

Solution

$$\sum_{n=1}^{20} n = 20\left(\frac{1+20}{2}\right)$$
$$= 210$$

103. Find the sum $\displaystyle\sum_{n=1}^{50} (2n + 3)$.

Solution

$$\sum_{n=1}^{50} (2n + 3) = 50\left(\frac{5+103}{2}\right)$$
$$= 2700$$

105. Find the sum $\displaystyle\sum_{n=1}^{500} \frac{n}{2}$.

Solution

$$\sum_{n=1}^{500} \frac{n}{2} = 500\left(\frac{\frac{1}{2}+250}{2}\right)$$
$$= 62{,}625$$

107. Use a graphing utility to find the sum $\displaystyle\sum_{n=1}^{40} (1000 - 25n)$.

Solution

Keystrokes:

| LIST | | MATH 5 | | LIST | | OPS 5 | 1000 | − | 25 | X, T, θ | , | X, T, θ | , | 1 | , | 40 | , | 1 |) | | ENTER |

$$\sum_{n=1}^{40} (1000 - 25n) = 19{,}500$$

109. Find the *n*th partial sum of
5, 12, 19, 26, 33, . . . , *n* = 12.

Solution

$$\sum_{n=1}^{12} (7n - 2) = 12\left(\frac{5+82}{2}\right) = 522$$

111. Find the *n*th partial sum of
200, 175, 150, 125, 100, . . . , *n* = 8.

Solution

$$\sum_{n=1}^{8} (225 - 25n) = 8\left(\frac{200+0}{2}\right) = 800$$

113. Find the *n*th partial sum of
−50, −38, −26, −14, −2, . . . , *n* = 50.

Solution

$$\sum_{n=1}^{50} (12n - 62) = 50\left(\frac{-50+538}{2}\right) = 12{,}200$$

115. Find the *n*th partial sum of
1, 4.5, 8, 11.5, 15, . . ., *n* = 12.

Solution

$$\sum_{n=1}^{12} (3.5n - 2.5) = 12\left(\frac{1+39.5}{2}\right) = 243$$

117. Find the sum of the first 75 positive integers.

Solution

$$\sum_{n=1}^{75} = 75\left(\frac{1+75}{2}\right) = 2850$$

119. Find the sum of the first 50 even positive integers.

Solution

$$\sum_{n=1}^{50} 2n = 50\left(\frac{2+100}{2}\right) = 2500$$

121. *Salary Increases* In your new job you are told that your starting salary will be $36,000 with and increase of $2000 at the end of each of the first five years. How much will you be paid through the end of your first six years of employment with the company?

Solution

36,000, 38,000, 40,000, 42,000, 44,000, 46,000

$$\text{Total salary} = 6\left(\frac{36{,}000 + 46{,}000}{2}\right) = \$246{,}000$$

123. *Ticket Prices* There are 20 rows of seats on the main floor of a concert hall—20 seats in the first row, 21 seats in the second row, 22 seats in the third row, and so on. How much should you charge per ticket in order to obtain $15,000 for the sale of all of the seats on the main floor?

Solution

Sequence = 20, 21, 22, . . . $n = 20$ $d = 1$

$a_n = a_1 + (n - 1)d$

$a_n = 22 + (n - 1)1$

$a_n = 19 + n$

$$\sum_{n=1}^{20} (19 + n) = 20\left(\frac{20 + 39}{2}\right)$$

$$= 590 \text{ seats}$$

$$\frac{\text{Total cost}}{\text{Total seats}} = \text{Cost per ticket}$$

$$\frac{\$15{,}000}{590} = \$25.42$$

125. *Pattern Recognition*

(a) Compute the sums of positive odd integers.

$1 + 3 =$

$1 + 3 + 5 =$

$1 + 3 + 5 + 7 =$

$1 + 3 + 5 + 7 + 9 =$

$1 + 3 + 5 + 7 + 9 + 11 =$

(b) Use the sums of part (a) to make a conjecture about the sums of positive odd integers. Check you conjecture for the sum

$1 + 3 + 5 + 7 + 9 + 11 + 13 =$

(c) Verify your conjecture algebraically.

Solution

(a) $1 + 3 = 4$

$1 + 3 + 5 = 9$

$1 + 3 + 5 + 7 = 16$

$1 + 3 + 5 + 7 + 9 = 25$

$1 + 3 + 5 + 7 + 9 + 11 = 36$

(b) The sums of positive odd integers yield perfect squares.

$1 + 3 + 5 + 7 + 9 + 11 + 13 = 49$

(c) $\displaystyle\sum_{k=1}^{n} [1 + (k - 1)2] = n\left(\frac{1 + (2n - 1)}{2}\right) = n\left(\frac{2n}{2}\right) = n^2$

10.3 Geometric Sequences

7. Find the common ratio for $1, -3, 9, -27, \ldots$.

Solution

$r = -3$

9. Find the common ratio for $1, \pi, \pi^2, \pi^3, \ldots$.

Solution

$r = \pi$

11. Determine whether the sequence $5, 10, 20, 40, \ldots$ is geometric. It it is, find the common ratio.

Solution

The sequence is geometric; $r = 2$.

13. Determine whether the sequence $1, 8, 27, 64, 125, \ldots$ is geometric. It it is, find the common ratio.

Solution

The sequence is not geometric.

15. Write the first five terms of $a_1 = 4, r = -\dfrac{1}{2}$.

Solution

$a_n = a_1 r^{n-1}$

$a_1 = 4\left(-\dfrac{1}{2}\right)^{1-1} = 4$

$a_2 = 4\left(-\dfrac{1}{2}\right)^{2-1} = -2$

$a_3 = 4\left(-\dfrac{1}{2}\right)^{3-1} = 1$

$a_4 = 4\left(-\dfrac{1}{2}\right)^{4-1} = -\dfrac{1}{2}$

$a_5 = 4\left(-\dfrac{1}{2}\right)^{5-1} = \dfrac{1}{4}$

17. Write the first five terms of $a_1 = 20, r = 1.07$.

Solution

$a_n = a_1 r^{n-1}$

$a_1 = 20(1.07)^{1-1} = 20$

$a_2 = 20(1.07)^{2-1} = 21.40$

$a_3 = 20(1.07)^{3-1} = 22.90$

$a_4 = 20(1.07)^{4-1} = 24.50$

$a_5 = 20(1.07)^{5-1} = 26.22$

19. Find the nth term of $a_1 = 120, r = -\dfrac{1}{3}, a_{10} = \boxed{}$.

Solution

$a_n = a_1 r^{n-1}$

$a_{10} = 120\left(-\dfrac{1}{3}\right)^{10-1}$

$a_{10} = -0.0061$

21. Find the nth term of $a_1 = 200, r = 1.2, a_{12} = \boxed{}$.

Solution

$a_n = a_1 r^{n-1}$

$a_{12} = 200(1.2)^{12-1}$

$a_{12} = 1486.02$

23. Find the formula for the nth term of $a_1 = 25, r = 4$. (Begin with n = 1.)

Solution

$a_n = a_1 r^{n-1}$

$a_n = 25(4)^{n-1}$

25. Find the formula for the nth term of $1, \dfrac{3}{2}, \dfrac{9}{4}, \dfrac{27}{8}, \ldots$. (Begin with n = 1.)

Solution

$a_n = a_1 r^{n-1}$

$a_n = 1\left(\dfrac{3}{2}\right)^{n-1}$

27. Using the graphs in the textbook, match the sequence $a_n = 12\left(\dfrac{3}{4}\right)^{n-1}$ with its graph.

[The graphs are labeled (a), (b), (c), and (d).]

Solution

(b)

29. Using the graphs in the textbook, match the sequence $a_n = 2\left(\dfrac{4}{3}\right)^{n-1}$ with its graph.

[The graphs are labeled (a), (b), (c), and (d).]

Solution

(a)

 31. Use a graphing utility to graph the first 10 terms of the sequence $a_n = 20(-0.6)^{n-1}$.

Solution

Keystrokes (calculator in sequence and dot mode):

$\boxed{Y=}$ 20 $\boxed{(}$ $\boxed{(-)}$ .6 $\boxed{)}$ $\boxed{\wedge}$ $\boxed{(}$ $\boxed{n}$ $\boxed{-}$ 1 $\boxed{)}$ $\boxed{\text{TRACE}}$

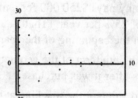

33. Find the sum $\displaystyle\sum_{i=1}^{12} 4(-2)^{i-1}$.

Solution

$$\sum_{i=1}^{12} 4(-2)^{i-1} = 4\left(\frac{(-2)^{12} - 1}{-2 - 1}\right)$$

$$= 4\left(\frac{4095}{-3}\right)$$

$$= -5460$$

35. Find the sum $\displaystyle\sum_{i=1}^{8} 6(0.1)^{i-1}$.

Solution

$$\sum_{i=1}^{8} 6(0.1)^{i-1} = 6\left(\frac{(0.1)^8 - 1}{0.1 - 1}\right)$$

$$= 6\left(\frac{-.99\overline{9}}{-.9}\right)$$

$$= 6(1.\overline{1})$$

$$\approx 6.67$$

 37. Use a graphing utility to find the sum.

$$\sum_{i=1}^{30} 100(0.75)^{i-1}$$

Solution

Keystrokes:

$\boxed{\text{LIST}}$ $\boxed{\text{MATH 5}}$ $\boxed{\text{LIST}}$ $\boxed{\text{OPS 5}}$ 100 $\boxed{(}$ 0.75 $\boxed{)}$ $\boxed{\wedge}$ $\boxed{(}$ $\boxed{X, T, \theta}$ $\boxed{-}$ 1 $\boxed{)}$ $\boxed{,}$

$\boxed{X, T, \theta}$ $\boxed{,}$ 1 $\boxed{,}$ 30 $\boxed{,}$ 1 $\boxed{)}$ $\boxed{\text{ENTER}}$

$$\sum_{i=1}^{30} 100(0.75)^{i-1} \approx 399.93$$

39. Find the nth partial sum of 4, 12, 36, 108, . . . , $n = 8$.

Solution

$$\sum_{i=1}^{8} 4(3)^{i-1} = 4\left(\frac{3^8 - 1}{3 - 1}\right) = 4\left(\frac{6560}{2}\right) = 13{,}120$$

41. Find the nth sum of $60, -15, \dfrac{15}{4}, \dfrac{-15}{16}, \ldots, n = 12$.

Solution

$$\sum_{i=1}^{12} 60\left(-\frac{1}{4}\right)^{i-1} = 60\left(\frac{\left(-\frac{1}{4}\right)^{12} - 1}{\left(-\frac{1}{4}\right) - 1}\right) = 60\left(\frac{-1.000\overline{0}}{-1.25}\right) = 48.00$$

43. *Depreciation*   A company pays \$250,000 for a machine. During the next 5 years, the machine depreciates at the rate of 25% per year. (That is, at the end of each year, the depreciated value is 75% of what it was at the beginning of the year.)

(a) Find a formula for the nth term of the geometric sequence that gives the value of the machine n full years after it was purchased.

(b) Find the depreciated value of the machine at the end of 5 full years.

(c) During which year did the machine depreciate the most?

Solution

$a_0 = 250{,}000$

$a_1 = 250{,}000(0.75)$

$a_2 = 250{,}000(0.75)^2$

$a_3 = 250{,}000(0.75)^3$

$a_4 = 250{,}000(0.75)^4$

(a) $a_n = 250{,}000(0.75)^n$

(b) $a_5 = 250{,}000(0.75)^5 = \$59{,}326.17$

(c) the first year

45. *Increasing Annuity*   Find the balance in an increasing annuity when P of \$75 is invested each month for 30 (t) years, compounded monthly at rate $r = 6\%$.

Solution

$$A = P\left(1 + \frac{r}{n}\right)^{nt}$$

$$a_{360} = 75\left(1 + \frac{0.06}{12}\right)^{12(30)} = 75(1.005)^{360}$$

$$a_1 = 75(1.005)^1$$

$$\text{balance} = [75(1.005)]\left[\frac{1.005^{360} - 1}{1.005 - 1}\right] = \$75{,}715.32$$

47. *Geometry* A square has 12-inch sides. A new square is formed by connecting the midpoints of the sides of the square, and two of the triangles are shaded (see figure in textbook). This process is repeated five more times. What is the total area of the shaded region?

Solution

$a_1 = 6^2 = 36$

$a_2 = \left(3\sqrt{2}\right)^2 = 18$

$r = \dfrac{a_2}{a_1} = \dfrac{18}{36} = \dfrac{1}{2}$

$a_n = 36\left(\dfrac{1}{2}\right)^{n-1}$

Total area $= \displaystyle\sum_{n=1}^{6} 36\left(\dfrac{1}{2}\right)^{n-1} = 36\left(\dfrac{\left(\dfrac{1}{2}\right)^6 - 1}{\dfrac{1}{2} - 1}\right) = 70.875$ square inches.

49. Evaluate the determinant.

$\begin{bmatrix} 10 & 25 \\ 6 & -5 \end{bmatrix}$

Solution

$\det(A) = \begin{vmatrix} 10 & 25 \\ 6 & -5 \end{vmatrix} = (10)(-5) - (6)(25) = -200$

51. Evaluate the determinant.

$\begin{bmatrix} 3 & -2 & 1 \\ 0 & 5 & 3 \\ 6 & 1 & 1 \end{bmatrix}$

Solution

$\det(A) = \begin{vmatrix} 3 & -2 & 1 \\ 0 & 5 & 3 \\ 6 & 1 & 1 \end{vmatrix} = -60$

53. *Geometry* Use a determinant to find the area of the triangle with vertices $(-5, 8)$, $(10,0)$ and $(3, -4)$.

Solution

let $(x_1, y_1) = (-5, 8)$, $(x_2, y_2) = (10, 0)$, $(x_3, y_3) = (3, -4)$.

$\begin{vmatrix} x_1 & y_1 & 1 \\ x_2 & y_2 & 1 \\ x_3 & y_3 & 1 \end{vmatrix} = \begin{vmatrix} -5 & 8 & 1 \\ 10 & 0 & 1 \\ 3 & -4 & 1 \end{vmatrix} = -116$

Area $= -\dfrac{1}{2}(-116) = 58$

55. Find the common ratio of 2, 6, 18, 54, ...

Solution

$r = 3$

57. Find the common ratio of 54, 18, 6, 2, ...

Solution

$r = \dfrac{1}{3}$

59. Find the common ratio of $1, -\dfrac{3}{2}, \dfrac{9}{4}, -\dfrac{27}{8}, \ldots$.

Solution

$r = -\dfrac{3}{2}$

61. Find the common ratio of $e, e^2, e^3, e^4, \ldots$.

Solution

$r = e$

63. Determine whether $10, 15, 20, 25, \ldots$ is geometric. If it is, find the common ratio.

Solution

The sequence is not geometric.

65. Determine whether $64, 32, 16, 8, \ldots$ is geometric. If it is, find the common ratio.

Solution

The sequence is geometric. $r = \dfrac{1}{2}$

67. Determine whether $1, -\dfrac{2}{3}, \dfrac{4}{9}, -\dfrac{8}{27}, \ldots$ is geometric. If it is, find the common ratio.

Solution

The sequence is geometric. $r = -\dfrac{2}{3}$

69. Determine whether $10(1 + 0.02), 10(1 + 0.02)^2,$ $10(1 + 0.02)^3, 10(1 + 0.02)^4, \ldots$ is geometric. If it is, find the common ratio.

Solution

The sequence is geometric. $r = (1 + 0.02)$

71. Write the first five terms of $a_1 = 4, r = 2$.

Solution

$a_n = a_1 r^{n-1}$

$a_n = 4(2)^{n-1}$

$a_1 = 4(2)^{1-1} = 4$

$a_2 = 4(2)^{2-1} = 8$

$a_3 = 4(2)^{3-1} = 16$

$a_4 = 4(2)^{4-1} = 32$

$a_5 = 4(2)^{5-1} = 64$

73. Write the first five terms of $a_1 = 6, r = \dfrac{1}{3}$.

Solution

$a_n = a_1 r^{n-1}$

$a_n = 6\left(\dfrac{1}{3}\right)^{n-1}$

$a_1 = 6\left(\dfrac{1}{3}\right)^{1-1} = 6$

$a_2 = 6\left(\dfrac{1}{3}\right)^{2-1} = 2$

$a_3 = 6\left(\dfrac{1}{3}\right)^{3-1} = \dfrac{2}{3}$

$a_4 = 6\left(\dfrac{1}{3}\right)^{4-1} = \dfrac{2}{9}$

$a_5 = 6\left(\dfrac{1}{3}\right)^{5-1} = \dfrac{2}{27}$

75. Write the first five terms of $a_1 = 1, r = -\dfrac{1}{2}$.

Solution

$a_n = a_1 r^{n-1}$

$a_n = 1\left(-\dfrac{1}{2}\right)^{n-1}$

$a_1 = 1\left(-\dfrac{1}{2}\right)^{1-1} = 1$

$a_2 = 1\left(-\dfrac{1}{2}\right)^{2-1} = -\dfrac{1}{2}$

$a_3 = 1\left(-\dfrac{1}{2}\right)^{3-1} = \dfrac{1}{4}$

$a_4 = 1\left(-\dfrac{1}{2}\right)^{4-1} = -\dfrac{1}{8}$

$a_5 = 1\left(-\dfrac{1}{2}\right)^{5-1} = \dfrac{1}{16}$

77. Write the first five terms of $a_1 = 1000$, $r = 1.01$.

Solution

$a_n = a_1 r^{n-1}$

$a_n = 1000(1.01)^{n-1}$

$a_1 = 1000(1.01)^{1-1} = 1000$

$a_2 = 1000(1.01)^{2-1} = 1010$

$a_3 = 1000(1.01)^{3-1} = 1020.1$

$a_4 = 1000(1.01)^{4-1} = 1030.301$

$a_5 = 1000(1.01)^{5-1} = 1040.604$

79. Find the nth term of $a_1 = 6$, $r = \dfrac{1}{2}$, $a_{10} = \boxed{}$.

Solution

$a_n = a_1 r^{n-1}$

$a_{10} = 6\left(\dfrac{1}{2}\right)^{10-1} = \dfrac{3}{256}$

81. Find the nth term of $a_1 = 3$, $r = \sqrt{2}$, $a_{10} = \boxed{}$.

Solution

$a_n = a_1 r^{n-1}$

$a_{10} = 3\left(\sqrt{2}\right)^{10-1} = 48\sqrt{2}$

83. Find the nth term of $a_1 = 4$, $a_2 = 3$, $a_5 = \boxed{}$.

Solution

$a_n = a_1 r^{n-1}$

$a_5 = 4\left(\dfrac{3}{4}\right)^{5-1} = \dfrac{81}{64}$

85. Find the nth term of $a_1 = 1$, $a_3 = \dfrac{9}{4}$, $a_6 = \boxed{}$.

Solution

$a_n = a_1 r^{n-1}$

$a_6 = 1\left(\pm\dfrac{3}{2}\right)^{6-1} = \pm\dfrac{243}{32}$

87. Find the formula for the nth term of $a_1 = 2$, $r = 3$. (Begin with $n = 1$.)

Solution

$a_n = a_1 r^{n-1}$

$a_n = 2(3)^{n-1}$

89. Find the formula for the nth term of $a_1 = 1$, $r = 2$. (Begin with $n = 1$.)

Solution

$a_n = a_1 r^{n-1}$

$a_n = 1(2)^{n-1}$

91. Find the formula for the nth term of $a_1 = 4$, $r = -\dfrac{1}{2}$. (Begin with $n = 1$.)

Solution

$a_n = a_1 r^{n-1}$

$a_n = 4\left(-\dfrac{1}{2}\right)^{n-1}$

93. Find the formula for the nth term of $a_1 = 8$, $a_2 = 2$. (Begin with $n = 1$.)

Solution

$a_n = a_1 r^{n-1}$

$a_n = 8\left(\dfrac{1}{4}\right)^{n-1}$

95. Find the sum $\displaystyle\sum_{i=1}^{10} 2^{i-1}$.

Solution

$$\sum_{i=1}^{10} 2^{i-1} = 1\left(\dfrac{2^{10} - 1}{2 - 1}\right) = \dfrac{1024 - 1}{1} = 1023$$

97. Find the sum $\displaystyle\sum_{i=1}^{12} 3\left(\frac{3}{2}\right)^{i-1}$.

Solution

$$\sum_{i=1}^{12} 3\left(\frac{3}{2}\right)^{i-1} = 3\left(\frac{\left(\frac{3}{2}\right)^{12} - 1}{\frac{3}{2} - 1}\right) = 3\left(\frac{128.74634}{0.5}\right) = 772.47803$$

99. Find the sum $\displaystyle\sum_{i=1}^{15} 3\left(-\frac{1}{3}\right)^{i-1}$.

Solution

$$\sum_{i=1}^{15} 3\left(-\frac{1}{3}\right)^{i-1} = 3\left(\frac{\left(-\frac{1}{3}\right)^{15} - 1}{-\frac{1}{3} - 1}\right) = 3\left(\frac{-1.0000001}{-1.3333333}\right) = 2.2500002$$

101. Use a graphing utility to find the sum.

$$\sum_{i=1}^{20} 100(1.1)^i$$

Solution

Keystrokes:

| LIST | | MATH 5 | | LIST | | OPS 5 | 100 | (| 1.1 |) | ∧ | X, T, θ | , | X, T, θ | | 1 | ,

20 | , | 1 |) | | ENTER |

$$\sum_{i=1}^{20} 100(1.1)^i = 6300.250$$

103. Find the nth partial sum of
$1, -3, 9, -27, 81, \ldots, n = 10$.

Solution

$$\sum_{i=1}^{10} 1(-3)^{i-1} = \left(\frac{(-3)^{10} - 1}{-3 - 1}\right) = -14,762$$

105. Find the nth partial sum of $8, 4, 2, 1, \frac{1}{2}, \ldots, n = 15$.

Solution

$$\sum_{i=1}^{15} 8\left(\frac{1}{2}\right)^{i-1} = 8\left(\frac{\left(\frac{1}{2}\right)^{15} - 1}{\frac{1}{2} - 1}\right) \approx 15.999512$$

107. Find the nth partial sum of
$1, \sqrt{2}, 2, 2\sqrt{2}, 4, \ldots, n = 12$.

Solution

$$\sum_{i=1}^{12} 1\left(\sqrt{2}\right)^{i-1} = \left(\frac{\sqrt{2}^{12} - 1}{\sqrt{2} - 1}\right) \approx 152.09545$$

109. Find the nth partial sum of $3, 3(1.06), 3(1.06)^2, 3(1.06)^3,$
$3(1.06)^4, \ldots, n = 20$.

Solution

$$\sum_{i=1}^{20} 3(1.06)^{i-1} = 3\left(\frac{1.06^{20} - 1}{1.06 - 1}\right) \approx 110.357$$

111. *Power Supply*   The electrical power for an implanted medical device decreases by 0.1% each day.

  (a) Find a formula for the nth term of the geometric sequence that gives the power n days after the device is implanted.

  (b) What percent of the initial power is still available 1 year after the device is implanted?

  (c) The power supply needs to be changed when half the power is depleted. Use a graphing utility to graph the first 750 terms and estimate when the power source should be changed.

Solution

  (a) $P = (0.999)^n$ (b) $P = (0.999)^{365}$

                                                 $= .694069887$

                                                 $\approx 69.4\%$

  (c) Keystrokes (calculator in sequence and dot mode):

  $\boxed{Y =}$ .999 $\boxed{\wedge}$ $\boxed{n}$ $\boxed{\text{TRACE}}$

  700 days

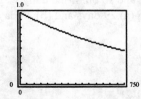

113. *Salary Increase*   You accept a job that pays a salary of $30,000 the first year. During the next 39 years, you receive a 5% raise each year. What is your *total* salary over the 40-year period?

Solution

$$\text{Total salary} = \sum_{n=1}^{40} 30{,}000(1.05)^n$$

$$= 30{,}000\left(\frac{1.05^{40} - 1}{1.05 - 1}\right)$$

$$= \$3{,}623{,}993.23$$

115. *Increasing Annuity*   Find the balance in an increasing annuity when a principal P of $100 is invested each month for 10 years, compounded monthly at a rate r of 9%.

Solution

$$A = P\left(1 + \frac{r}{n}\right)^{nt}$$

$$a_{120} = 100\left(1 + \frac{0.09}{12}\right)^{12(10)} = 100(1.0075)^{120}$$

$$a_1 = 100(1.0075)^1$$

$$\text{balance} = [100(1.0075)]\left[\frac{1.0075^{120} - 1}{1.0075 - 1}\right] = \$19{,}496.56$$

117. *Increasing Annuity* Find the balance in an increasing annuity when a principal P of $30 is invested each month for 40 years (t), compounded monthly at a rate r of 8%.

Solution

$$A = P\left(1 + \frac{r}{n}\right)^{nt}$$

$$a_{480} = 30\left(1 + \frac{0.08}{112}\right)^{12(40)} = 30\left(\frac{151}{150}\right)^{480}$$

$$a_1 = 30\left(\frac{151}{150}\right)^{1}$$

$$\text{balance} = \left[30\left(\frac{151}{150}\right)\right]\left[\frac{\left(\frac{151}{150}\right)^{480} - 1}{\left(\frac{151}{150}\right) - 1}\right] = \$105,428.44$$

119. *Salary* You start work at a company that pays $0.01 for the first day, $0.02 for the second day, $0.04 for the third day, and so on. If the daily wage keeps doubling, what would your total income be for working (a) 29 days and (b) 30 days?

Solution

$$a_n = 0.01(2)^{n-1}$$

(a) Total income $= \displaystyle\sum_{n=1}^{29} 0.01(2)^{n-1}$

$$= 0.01\left[\frac{2^{29} - 1}{2 - 1}\right]$$

$$= \$5,368,709.1$$

(b) Total income $= \displaystyle\sum_{n=1}^{30} 0.01(2)^{n-1}$

$$= 0.01\left[\frac{2^{30} - 1}{2 - 1}\right]$$

$$= \$10,737,418$$

121. *Geometry* A square has 12-inch sides. The square is divided into nine smaller squares and the center square is shaded (see figure in textbook). Each of the eight unshaded squares is then divided into nine smaller squares and each center square is shaded. This process is repeated four more times. What is the total area of the shaded region?

Solution

The square is partitioned a total of 6 times.

$$\sum_{i=1}^{6} \frac{1}{9}\left(\frac{8}{9}\right)^{i-1} = \frac{1}{9}\left(\frac{\left(\frac{8}{9}\right)^{6} - 1}{\frac{8}{9} - 1}\right) = \frac{1}{9} \cdot \frac{\dfrac{269,297}{531,441}}{\dfrac{1}{9}} = \frac{269,297}{531,441}$$

The original area of the square was 12^2 or 144 square inches. Therefore, the total area is

$$144 \cdot \frac{269,297}{531,441} \approx 72.969 \text{ square inches.}$$

Mid-Chapter Quiz

1. Write the first five terms of the sequence.

$a_n = 32\left(\dfrac{1}{4}\right)^{n-1}$ (Begin with $n = 1$.)

Solution

$a_1 = 32\left(\dfrac{1}{4}\right)^{1-1} = 32$

$a_2 = 32\left(\dfrac{1}{4}\right)^{2-1} = 8$

$a_3 = 32\left(\dfrac{1}{4}\right)^{3-1} = 2$

$a_4 = 32\left(\dfrac{1}{4}\right)^{4-1} = \dfrac{1}{2}$

$a_5 = 32\left(\dfrac{1}{4}\right)^{5-1} = \dfrac{1}{8}$

2. Write the first five terms of the sequence.

$a_n = \dfrac{(-3)^n n}{n + 4}$ (Begin with $n = 1$.)

Solution

$a_1 = \dfrac{(-3)^1 \cdot 1}{1 + 4} = -\dfrac{3}{5}$

$a_2 = \dfrac{(-3)^2 \cdot 2}{2 + 4} = 3$

$a_3 = \dfrac{(-3)^3 \cdot 3}{3 + 4} = -\dfrac{81}{7}$

$a_4 = \dfrac{(-3)^4 \cdot 4}{4 + 4} = \dfrac{81}{2}$

$a_5 = \dfrac{(-3)^5 \cdot 5}{5 + 4} = -135$

3. Find the sum $\displaystyle\sum_{k=1}^{4} 10k$.

Solution

$\displaystyle\sum_{k=1}^{4} 10k = 4\left(\dfrac{10 + 40}{2}\right) = 100$

4. Find the sum $\displaystyle\sum_{i=1}^{10} 4$.

Solution

$\displaystyle\sum_{i=1}^{10} 4 = 10\left(\dfrac{4 + 4}{2}\right) = 40$

5. Find the sum $\displaystyle\sum_{j=1}^{5} \dfrac{60}{j + 1}$.

Solution

$\displaystyle\sum_{j=1}^{5} \dfrac{60}{j + 1} = \dfrac{60}{2} + \dfrac{60}{3} + \dfrac{60}{4} + \dfrac{60}{5} + \dfrac{60}{6}$

$= 30 + 20 + 15 + 12 + 10$

$= 87$

6. Find the sum $\displaystyle\sum_{n=1}^{8} 8\left(-\dfrac{1}{2}\right)$.

Solution

$\displaystyle\sum_{n=1}^{8} 8\left(-\dfrac{1}{2}\right) = 8(-4) = -32$

7. Write the sum using sigma notation.

$\dfrac{2}{3(1)} + \dfrac{2}{3(2)} + \dfrac{2}{3(3)} + \cdots + \dfrac{2}{3(20)}$

Solution

$\displaystyle\sum_{k=1}^{20} \dfrac{2}{3k}$

8. Write the sum using sigma notation.

$\dfrac{1}{1^3} - \dfrac{1}{2^3} + \dfrac{1}{3^3} - \cdots + \dfrac{1}{25^3}$

Solution

$\displaystyle\sum_{k=1}^{25} \dfrac{(-1)^{k-1}}{k^3}$

9. Find the common difference of the arithmetic sequence. $1, \frac{3}{2}, 2, \frac{5}{2}, 3, \ldots$

Solution

$d = \frac{1}{2}$

10. Find the common difference of the arithmetic sequence. $100, 94, 88, 82, 76, \ldots$

Solution

$d = -6$

11. Find a formula for a_n.
Arithmetic: $a_1 = 20, a_4 = 11$

Solution

$$a_n = a_1 + (n - 1)d \qquad a_n = 20 + (n - 1)(-3)$$
$$11 = 20 + (4 - 1)d \qquad a_n = 20 - 3n + 3$$
$$-9 = 3d \qquad\qquad a_n = -3n + 23$$
$$-3 = d$$

12. Find a formula for a_n.
Geometric: $a_1 = 32, r = -\frac{1}{4}$

Solution

$$a_n = a_1 r^{n-1}$$
$$a_n = 32\left(-\frac{1}{4}\right)^{n-1}$$

13. Find the sum $\sum_{n=1}^{50}(3n + 5)$.

Solution

$$\sum_{n=1}^{50}(3n + 5) = 50\left(\frac{8 + 155}{2}\right) = 4075$$

14. Find the sum $\sum_{n=1}^{300}\frac{n}{5}$.

Solution

$$\sum_{n=1}^{300}\frac{n}{5} = 300\left(\frac{\frac{1}{5} + 60}{2}\right) = 9030$$

15. Find the sum $\sum_{i=1}^{8}9\left(\frac{2}{3}\right)^{i-1}$.

Solution

$$\sum_{i=1}^{8}9\left(\frac{2}{3}\right)^{i-1} = 9\left(\frac{\left(\frac{2}{3}\right)^8 - 1}{\frac{2}{3} - 1}\right)$$

$$= 9\left(\frac{\frac{256}{6561} - 1}{-\frac{1}{3}}\right)$$

$$= 9\left(\frac{-.96098}{-.33\overline{3}}\right)$$

$$= 25.947$$

16. Find the sum $\sum_{j=1}^{20}500(1.06)^{j-1}$.

Solution

$$\sum_{j=1}^{20}500(1.06)^{j-1} = 500\left(\frac{1.06^{20} - 1}{1.06 - 1}\right)$$

$$= 500\left(\frac{2.2071}{.06}\right)$$

$$= 18,392.796$$

17. Find the 12th term of $625, -250, 100, -40, \ldots$.

Solution

Geometric sequence with $a_1 = 625$ and $r = -.4$.

$$a_n = a_1 r^{n-1}$$
$$a_n = 625(-.4)^{n-1}$$
$$a_{12} = 625(-.4)^{12-1}$$
$$= -0.026$$

18. Match $a_n = 10\left(\dfrac{1}{2}\right)^{n-1}$ and $b_n = 10\left(-\dfrac{1}{2}\right)^{n-1}$ with the graphs shown in the textbook.

Solution

$$a_n = 10\left(\frac{1}{2}\right)^{n-1} \implies \text{upper graph}$$

$$b_n = 10\left(-\frac{1}{2}\right)^{n-1} \implies \text{lower graph}$$

19. The temperature of a coolant decreases by 25.75° F the first hour. For each subsequent hour, the temperature decreases by 2.25° F less than it decreased the previous hour. How much does the temperature decrease during the 10th hour?

Solution

Sequence $= 25.75, 23.5, 21.25, 19, \ldots$

arithmetic with $a_1 = 25.75, d = -2.25$

$$a_n = 25.75 + (n-1)(-2.25)$$
$$a_n = 25.75 + (10-1)(-2.25)$$
$$a_{10} = 5.5°$$

20. The sequence given by $a_n = 2^{n-1}$ is geometric. Describe the sequence given by $b_n = \ln a_n$.

Solution

$b_n = \ln a_n$ is arithmetic.

10.4 The Binomial Theorem

7. Evaluate $_6C_4$.

Solution

$$_6C_4 = {_6C_2} = \frac{6 \cdot 5}{2 \cdot 1} = 15$$

9. Evaluate $_{20}C_{20}$.

Solution

$$_{20}C_{20} = 1$$

11. Use a graphing utility to evaluate $_{30}C_6$.

Solution

Keystrokes:

30 [MATH] [PRB 3] 6 [ENTER] $_{30}C_6 = 593{,}775$

13. Use a graphing utility to evaluate $_{52}C_5$.

Solution

Keystrokes:

52 [MATH] [PRB 3] 5 [ENTER] $_{52}C_5 = 2{,}598{,}960$

15. Evaluate the binomial coefficient $_{15}C_3$. Also, evaluate its symmetric coefficient $_{15}C_{15-3}$ to demonstrate that it is equal to $_{15}C_3$.

Solution

$$_{15}C_3 = \frac{15 \cdot 14 \cdot 13}{3 \cdot 2 \cdot 1} = 455$$

$$_{15}C_{12} = \frac{15 \cdot 14 \cdot 13 \cdot 12 \cdot 11 \cdot 10 \cdot 9 \cdot 8 \cdot 7 \cdot 6 \cdot 5 \cdot 4}{12 \cdot 11 \cdot 10 \cdot 9 \cdot 8 \cdot 7 \cdot 6 \cdot 5 \cdot 4 \cdot 3 \cdot 2 \cdot 1} = 455$$

17. Evaluate the binomial coefficient $_{25}C_5$. Also, evaluate its symmetric coefficient $_{25}C_{25-5}$ to demonstrate that it is equal to $_{25}C_5$.

Solution

$$_{25}C_5 = \frac{25 \cdot 24 \cdot 23 \cdot 22 \cdot 21}{5 \cdot 4 \cdot 3 \cdot 2 \cdot 1} = 53{,}130$$

$$_{25}C_{20} = \frac{25 \cdot 24 \cdot 23 \cdot 22 \cdot 21 \cdot 20 \cdot 19 \cdot 18 \cdot 17 \cdot 16 \cdot 15 \cdot 14 \cdot 13 \cdot 12 \cdot 11 \cdot 10 \cdot 9 \cdot 8 \cdot 7 \cdot 6}{20 \cdot 19 \cdot 18 \cdot 17 \cdot 16 \cdot 15 \cdot 14 \cdot 13 \cdot 12 \cdot 11 \cdot 10 \cdot 9 \cdot 8 \cdot 7 \cdot 6 \cdot 5 \cdot 4 \cdot 3 \cdot 2 \cdot 1}$$

$$= 53{,}130$$

19. Find the ninth row of Pascal's Triangle.

Solution

1 8 28 56 70 56 28 8 1
1 9 36 84 126 126 84 36 9 1

21. Use the Binomial Theorem to expand $(x + 3)^6$.

Solution

$$(x + 3)^6 = 1x^6 + 6x^5(3) + 15x^4(3)^2 + 20x^3(3)^3 + 15x^2(3)^4 + 6x(3)^5 + 1(3)^6$$
$$= x^6 + 18x^5 + 135x^4 + 540x^3 + 1125x^2 + 1458x + 729$$

23. Use the Binomial Theorem to expand $(u - 2v)^3$.

Solution

$$(u - 2v)^3 = 1u^3 - 3u^2(2v) + 3u(2v)^2 - 1(2v)^3$$
$$= u^3 - 6u^2v + 12uv - 8v^3$$

25. Use Pascal's Triangle to expand $(x + y)^8$.

Solution

$$(x + y)^8 = 1x^8 + 8x^7y + 28x^6y^2 + 56x^5y^3 + 70x^4y^4 + 56x^3y^5 + 28x^2y^6 + 8xy^7 + 1y^8$$

27. Use Pascal's Triangle to expand $(x - 2)^6$.

Solution

$$(x - 2)^6 = 1x^6 - 6x^5(2) + 15x^4(2)^2 + 20x^3(2)^3 + 15x^2(2)^4 - 6x(2)^5 + 1(2)^6$$
$$= x^6 - 12x^5 + 60x^4 - 160x^3 + 240x^2 - 192x + 64$$

29. Find the coefficient of the term x^7 of the expression $(x + 1)^{10}$.

Solution

$$_{10}C_3 = \frac{10 \cdot 9 \cdot 8}{3 \cdot 2 \cdot 1} = 120$$

31. Find the coefficient of the term x^4y^{11} of the expression $(x - y)^{15}$.

Solution

$$_{15}C_{11} = {}_{15}C_4 = \frac{15 \cdot 14 \cdot 13 \cdot 12}{4 \cdot 3 \cdot 2 \cdot} = 1365$$

33. Use the Binomial Theorem to expand $\left(\dfrac{1}{2} + \dfrac{1}{2}\right)^5$. In the study of probability, it is sometimes necessary to use the expansion $(p + q)^n$, where $p + q = 1$.

Solution

$$\left(\frac{1}{2} + \frac{1}{2}\right)^5 = 1\left(\frac{1}{2}\right)^5 + 5\left(\frac{1}{2}\right)^4\left(\frac{1}{2}\right) + 10\left(\frac{1}{2}\right)^3\left(\frac{1}{2}\right)^2 + 10\left(\frac{1}{2}\right)^2\left(\frac{1}{2}\right)^3 + 5\left(\frac{1}{2}\right)\left(\frac{1}{2}\right)^4 + 1\left(\frac{1}{2}\right)^5$$

35. Use the Binomial Theorem to expand $\left(\dfrac{1}{4} + \dfrac{3}{4}\right)^4$. In the study of probability, it is sometimes necessary to use the expansion $(p + q)^n$, where $p + q = 1$.

Solution

$$\left(\frac{1}{4} + \frac{3}{4}\right)^4 = 1\left(\frac{1}{4}\right)^4 + 4\left(\frac{1}{4}\right)^3\left(\frac{3}{4}\right) + 6\left(\frac{1}{4}\right)^2\left(\frac{3}{4}\right)^2 + 4\left(\frac{1}{4}\right)\left(\frac{3}{4}\right)^3 + 1\left(\frac{3}{4}\right)^4$$

37. Use the Binomial Theorem to approximate the quantity $(1.02)^8$ accurate to three decimal places.

For example,

$(1.02)^{10} \approx 1 + 10(0.02) + 45(0.02)^2 \approx 1.22$.

Solution

$(1.02)^8 = (1 + 0.02)^8$

$= (1)^8 + 8(1)^7(0.02) + 28(1)^6(0.02)^2 + 56(1)^5(0.02)^3 + \ldots$

$\approx 1 + 0.16 + 0.0112 + 0.000448$

≈ 1.172

39. Use the Binomial Theorem to approximate the quantity $(2.99)^{12}$ accurate to three decimal places.

For example,

$(1.02)^{10} \approx 1 + 10(0.02) + 45(0.02)^2 \approx 1.22$.

Solution

$(2.99)^{12} = (3 - 0.01)^{12}$

$= 1(3)^{12} - 12(3)^{11}(0.01) + 66(3)^{10}(0.01)^2 - 220(3)^9(0.01)^3 + 495(3)^8(0.01)^4 - 792(3)^7(0.01)^5 + \ldots$

$\approx 531{,}441 - 21{,}257.64 + 389.7234 - 4.33026 + 0.03247695 - 0.0001732104$

$\approx 510{,}568.785$

41. Rewrite $\log_4 64 = 3$ in exponential form.

Solution

$\log_4 64 = 3$ is $4^3 = 64$

43. Rewrite $\ln 1 = 0$ in exponential form.

Solution

$\ln 1 = 0$ is $e^0 = 1$

45. After t years, the value of a car is given by $V(t) = 22{,}000(0.8)^t$. Graphically determine when the value of the car will be \$15,000.

Solution

Keystrokes:

$\boxed{Y =}$ 22,000 $\boxed{\times}$ .8 $\boxed{\wedge}$ $\boxed{X, T, \theta}$ $\boxed{GRAPH}$

$t = 1.72$

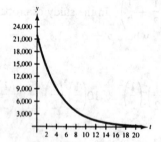

47. Evaluate $_{10}C_5$.

Solution

$_{10}C_5 = \dfrac{10 \cdot 9 \cdot 8 \cdot 7 \cdot 6}{5 \cdot 4 \cdot 3 \cdot 2 \cdot 1} = 252$

49. Evaluate $_{18}C_{18}$.

Solution

$_{18}C_{18} = 1$

51. Evaluate $_{50}C_{48}$.

Solution

$$_{50}C_{48} = {_{50}C_2} = \frac{50 \cdot 49}{2 \cdot 1} = 1225$$

53. Evaluate $_{25}C_4$.

Solution

$$_{25}C_4 = \frac{25 \cdot 24 \cdot 23 \cdot 22}{4 \cdot 3 \cdot 2 \cdot 1} = 12,650$$

55. Using a graphing utility to evaluate $_{12}C_7$.

Solution

Keystrokes:

12 [MATH] [PRB 3] 7 [ENTER] $_{12}C_7 = 792$

57. Using a graphing utility to evaluate $_{200}C_{10}$.

Solution

Keystrokes:

200 [MATH] [PRB 3] 10 [ENTER]

$_{200}C_{10} = 22,451,004,309,013,280$

59. Use a graphing utility to evaluate $_{25}C_{12}$.

Solution

Keystrokes:

25 [MATH] [PRB 3] 12 [ENTER] $_{25}C_{12} = 5,200,300$

61. Evaluate the binomial coefficient $_5C_2$. Also, evaluate its symmetric coefficient $_5C_3$.

Solution

$$_5C_2 = \frac{5!}{3! \, 2!} = \frac{5 \cdot 4 \cdot 3!}{2 \cdot 3!} = 5 \cdot 2 = 10$$

$$_5C_3 = \frac{5!}{2! \, 3!} = 10$$

63. Evaluate the binomial coefficient $_{12}C_5$. Also, evaluate its symmetric coefficient $_{12}C_7$.

Solution

$$_{12}C_5 = \frac{12!}{5! \, 7!} = \frac{12 \cdot 11 \cdot 10 \cdot 9 \cdot 8 \cdot 7!}{5 \cdot 4 \cdot 3 \cdot 2 \cdot 7!} = 792$$

$$_{12}C_7 = \frac{12!}{7! \, 5!} = \frac{12 \cdot 11 \cdot 10 \cdot 9 \cdot 8 \cdot 7!}{5 \cdot 4 \cdot 3 \cdot 2 \cdot 7!} = 792$$

65. Evaluate the binomial coefficient $_{10}C_0$. Also, evaluate its symmetric coefficient $_{10}C_{10}$.

Solution

$$_{10}C_0 = 1 \qquad\qquad _{10}C_{10} = 1$$

67. Use Pascal's Triangle to evaluate $_7C_3$.

Solution

$$_7C_3 = 35$$

69. Use Pascal's Triangle to evaluate $_8C_4$.

Solution

$$_8C_4 = 70$$

71. Use the Binomial Theorem to expand $(x + 1)^5$.

Solution

$$(x + 1)^5 = 1x^5 + 5x^4(1) + 10x^3(1)^2 + 10x^2(1)^3 + 5x(1)^4 + (1)^5$$
$$= x^5 + 5x^4 + 10x^3 + 10x^2 + 5x + 1$$

73. Use the Binomial Theorem to expand $(x - 4)^6$.

Solution

$(x - 4)^6 = (1)x^6 - (6)x^5(4) + (15)x^4(4^2) - (20)x^3(4^3) + (15)x^2(4^4) - (6)x(4^5) + (1)4^6$

$\quad = x^6 - 24x^5 + 240x^4 - 1280x^3 + 3840x^2 - 6144x + 4096$

75. Use the Binomial Theorem to expand $(x + y)^4$.

Solution

$(x + y)^4 = 1x^4 + 4x^3y + 6x^2y^2 + 4xy^3 + 1y^4$

77. Use the Binomial Theorem to expand $(x - y)^5$.

Solution

$(x - y)^5 = 1x^5 - 5x^4y + 10x^3y^2 - 10x^2y^3 + 5xy^4 - 1y^5$

79. Use Pascal's Triangle to expand $(a + 2)^3$.

Solution

$(a + 2)^3 = (1)a^3 + (3)a^2(2) + (3)\, a\, (2^2) + 1(2^3)$

$\quad = a^3 + 6a^2 + 12a + 8$

81. Use Pascal's Triangle to expand $(2x - 1)^4$.

Solution

$(2x - 1)^4 = (1)(2x)^4 - (4)(2x)^3(1) + (6)(2x)^2(1)^2$

$\quad\quad\quad - (4)(2x)^1(1)^3 + (1)(2x)^0(1)^4$

$\quad = 16x^4 - 32x^3 + 24x^2 - 8x + 1$

83. Use Pascal's Triangle to expand $(2y + z)^6$.

Solution

$(2y + z)^6 = (1)(2y)^6 + 6(2y)^5z + 15(2y)^4z^2 + 20(2y)^3z^3 + 15(2y)^2z^4 + 6(2y)z^5 + 1z^6$

$\quad = 64y^6 + 192y^5z + 240y^4z^2 + 160y^3z^3 + 60y^2z^4 + 12yz^5 + z^6$

85. Find the coefficient of the term x^7 of the expression $(x - 1)^{10}$.

Solution

$_{10}C_7 = -120$

87. Find the coefficient of the term x^3y^9 of the expression $(2x + y)^{12}$.

Solution

$_{12}C_9 = {}_{12}C_3 = 12 \cdot 11 \cdot 10 \cdot 3 \cdot 2 \cdot 1 = 220$

89. Find the coefficient of the term x^4 of the expression $(x^2 - 3)^4$.

Solution

$_4C_2 = 6 \cdot (-3)^2 = 54$

91. *Patterns in Pascal's Triangle* Use each encircled group of numbers (see the textbook) to form a 2×2 matrix. Find the determinant of each matrix. Describe the pattern.

Solution

Circle 1: $\begin{vmatrix} 1 & 1 \\ 1 & 2 \end{vmatrix} = 2 - 1 = 1$

Circle 4: $\begin{vmatrix} 4 & 1 \\ 10 & 5 \end{vmatrix} = 20 - 10 = 10$

Circle 2: $\begin{vmatrix} 2 & 1 \\ 3 & 3 \end{vmatrix} = 6 - 3 = 3$

Circle 5: $\begin{vmatrix} 5 & 1 \\ 15 & 6 \end{vmatrix} = 30 - 15 = 15$

Circle 3: $\begin{vmatrix} 3 & 1 \\ 6 & 4 \end{vmatrix} = 12 - 6 = 6$

The difference between each determinant increases by 1.

10.5 Counting Principles

5. *Random Selection* Find the number of ways the specified event can occur when one or more marbles are selected from a bowl containing 10 marbles numbered 0 through 9.
One number is drawn and it is even.

Solution

{0, 2, 4, 6, 8}   5 ways

7. *Random Selection* Find the number of ways the specified event can occur when one or more marbles are selected from a bowl containing 10 marbles numbered 0 through 9.
Two marbles are drawn one after the other. The first is replaced before the second is drawn. The sum of the numbers is 10.

Solution

First number	Second number
1	9
2	8
3	7
4	6
5	5
6	4
7	3
8	2
9	1

9 ways

9. *Random Selection* Find the number of ways the specified event can occur when one or more marbles are selected from a bowl containing 10 marbles numbered 0 through 9.

Two marbles are drawn without replacement. The sum of the numbers is 10.

Solution

First number	Second number
1	9
2	8
3	7
4	6
6	4
7	3
8	2
9	1

8 ways

11. *Staffing Choices* A small grocery store needs to open another checkout line. Three people who can run the cash register are available, and two people are available to bag groceries. How many different ways can the additional checkout line be staffed?

Solution

$3 \cdot 2 = 6$ ways

13. *Taking a Trip* Five people are taking a long trip in a car. Two sit in the front seat and three in the back seat. Three of the people agree to share the driving. In how many different arrangements can the five people sit?

Solution

$3 \cdot 4 \cdot 3 \cdot 2 \cdot 1 = 72$ ways

15. *Permutations* List all the permutations of the letters X, Y, and Z.

Solution

X, Y, Z; X, Z, Y; Y, X, Z; Y, Z, X; Z, X, Y; Z, Y, X

17. *Seating Arrangement* In how many ways can five children be seated in a single row of chairs.?

Solution

$5! = 5 \cdot 4 \cdot 3 \cdot 2 \cdot 1 = 120$ ways

19. *Work Assignments* Eight workers are assigned to eight different tasks. In how many ways can this be done assuming there are no restriction in making the assignments?

Solution

$8! = 40,320$ ways

21. *Number of Subsets* List all the subsets with two elements that can be formed from the set of letters {A, B, C, D, E, F}.

Solution

$$_6C_2 = \frac{6!}{4!\,2!} = \frac{6 \cdot 5}{2 \cdot 1} = 15 \text{ subsets}$$

{A, B}, {A, C}, {A, D}, {A, E}, {A, F}, {B, C}, {B, D}, {B, E}, {B, F}, {C, D}, {C, E}, {C, F}, {D, E}, {D, F}, {E, F}

23. *Relationships* As the size of a group increases, the number of relationships increases dramatically (see figure in textbook). Determine the number of two-person relationships in a group that has the following numbers.

(a) 3   (b) 4   (c) 6   (d) 8   (e) 10   (f) 12

Solution

(a) $_3C_2 = \dfrac{3!}{1!\,2!} = \dfrac{3 \cdot 2}{2 \cdot 1} = 3$

(b) $_4C_2 = \dfrac{4!}{2!\,2!} = \dfrac{4 \cdot 3}{2 \cdot 1} = 6$

(c) $_6C_2 = \dfrac{6!}{4!\,2!} = \dfrac{6 \cdot 5}{2 \cdot 1} = 15$

(d) $_8C_2 = \dfrac{8!}{6!\,2!} = \dfrac{8 \cdot 7}{2 \cdot 1} = 28$

(e) $_{10}C_2 = \dfrac{10!}{8!\,2!} = \dfrac{10 \cdot 9}{2 \cdot 1} = 45$

(f) $_{12}C_2 = \dfrac{12!}{10!\,2!} = \dfrac{12 \cdot 11}{2 \cdot 1} = 66$

25. *Number of Triangles* Eight points are located in the coordinate plane such that no three are collinear. How many different triangles can be formed having their vertices as three of the eight points?

Solution

$$_8C_3 = \frac{8!}{5!\,3!} = \frac{8 \cdot 7 \cdot 6}{3 \cdot 2 \cdot 1} = 56 \text{ triangles}$$

27. *Defective Units*   A shipment of 10 microwave ovens contains two defective units. In how many ways can a vending company purchase three of these units and receive (a) all good units, (b) two good units, and (c) one good unit?

Solution

(a) $_8C_3 = 56$ (b) $_8C_2 \cdot _2C_1 = 56$ (c) $_8C_1 \cdot _2C_2 = 8$

29. Solve $\sqrt{x - 5} = 6$. (Round your answer to two decimal places.)

Solution

$$\sqrt{x - 5} = 6$$
$$x - 5 = 36$$
$$x = 41$$

31. Solve $\log_2 (x - 5) = 6$. (Round your answer to two decimal places.)

Solution

$$\log_2 (x - 5) = 6$$
$$x - 5 = 2^6$$
$$x = 69$$

33. Write the equation of the line that passes through (3, 5) and (6, 7).

Solution

$$m = \frac{7 - 5}{6 - 3} = \frac{2}{3} \qquad y - 7 = \frac{2}{3}(x - 6)$$
$$3y - 21 = 2x - 12$$
$$0 = 2x - 3y + 9$$

35. *Random Selection*   Determine the number of ways the specified event can occur when one or more marbles are selected from a bowl containing 20 marbles numbered 1 through 20.

One number is drawn, and it is odd.

Solution

{1, 3, 5, 7, 9, 11, 13, 15, 17, 19}   10 ways

37. *Random Selection*   Determine the number of ways the specified event can occur when one or more marbles are selected from a bowl containing 20 marbles numbered 1 through 20.

One number is drawn, and it is prime.

Solution

{2, 3, 5, 7, 11, 13, 17, 19}   8 ways

39. *Random Selection*   Determine the number of ways the specified event can occur when one or more marbles are selected from a bowl containing 20 marbles numbered 1 through 20.

One number is drawn, and it is divisible by 3.

Solution

{3, 6, 9, 12, 15, 18}   6 ways

41. *Random Selection* Determine the number of ways the specified event can occur when one or more marbles are selected from a bowl containing 20 marbles numbered 1 through 20.

Two marbles are drawn one after the other. The first is replaced before the second is drawn. The sum of the numbers is 8.

Solution

First number	Second number
1	7
2	6
3	5
4	4
5	3
6	2
7	1

7 ways

43. *Random Selection* Determine the number of ways the specified event can occur when one or more marbles are selected from a bowl containing 20 marbles numbered 1 through 20.

Two marbles are drawn without replacement. The sum of the numbers is 8.

Solution

First number	Second number
1	7
2	6
3	5
4	4
5	3
6	2
7	1

6 ways

45. *License Plates* How many distinct automobile license plates can be formed by using a four-digit number followed by two letters?

Solution

plate = digit   digit   digit   digit   letter   letter

$\quad\quad$ 10 $\cdot$ 10 $\cdot$ 10 $\cdot$ 10 $\cdot$ 26 $\cdot$ 26 = 6,760,000 plates

47. *Permutations* List all the permutations of two letters selected from the letters A, B, C, and D.

Solution

AB	BA
AC	CA
AD	DA
BC	CB
BD	DB
CD	DC

49. *Posing for a Photograph* In how many ways can four children line up in one row to have their picture taken.

Solution

$4! = 4 \cdot 3 \cdot 2 \cdot 1 = 24$ ways

51. *Choosing Officers*   From a pool of 10 candidates, the offices of president, vice-president, secretary, and treasurer will be filled. In how many ways can the offices be filled if each of the 10 can hold any one of the offices?

Solution

$$_{10}P_4 = 10 \cdot 9 \cdot 8 \cdot 7 = 5040$$

53. *Number of Subsets*   List all the subsets with three elements that can be formed from the set of letters $\{A, B, C, D, E\}$.

Solution

$_5C_3 = 10$ subsets

$\{A, B, C\}, \{A, B, D\}, \{A, B, E\}, \{A, C, D\}, \{A, C, E\},$
$\{A, D, E\}, \{B, C, D\}, \{B, C, E\}, \{B, D, E\}, \{C, D, E\}$

55. *Identification Numbers*   In a statistical study, each participant is given an identification label consisting of a letter of the alphabet followed by a single digit. How many distinct identification labels are possible?

Solution

label	=	letter		number		
		26	$\cdot$	10	=	260 labels

57. *Committee Selection*   In how many ways can a committee of five be formed from a group of 30 people?

Solution

$$_{30}C_5 = \frac{30!}{25!\,5!} = \frac{30 \cdot 29 \cdot 28 \cdot 27 \cdot 26}{5 \cdot 4 \cdot 3 \cdot 2 \cdot 1} = 142{,}506 \text{ ways}$$

59. *Basketball Lineup*   A high school basketball team has 15 players. In how many different ways can the coach choose the starting lineup of five? (Assume each player can play each position.)

Solution

$$_{15}C_5 = \frac{15!}{5!\,10!} = 3003 \text{ ways}$$

61. *Group Selection*   Four people are to be selected from four couples. In how many ways can this be done if
(a) there are no restrictions?
(b) one person from each couple must be selected?

Solution

(a) $_8C_4 = \dfrac{8!}{4!\,4!} = \dfrac{8 \cdot 7 \cdot 6 \cdot 5}{4 \cdot 3 \cdot 2 \cdot 1} = 70$ (b) $_2C_1 \cdot {_2C_1} \cdot {_2C_1} \cdot {_2C_1} = 2 \cdot 2 \cdot 2 \cdot 2 = 16$

63. *Diagonals of a Polygon*   Find the number of diagonals of each polygon. (A line segment connecting any two nonadjacent vertices of a polygon is called a *diagonal* of the polygon.)

(a) Pentagon   (b) Hexagon   (c) Octagon

Solution

(a) Diagonals of Pentagon $= {_5C_3} - {_5C_1} = 5$ (b) Diagonals of Hexagon $= {_6C_4} - {_6C_1} = 9$
(c) Diagonals of Octagon $= {_8C_6} - {_8C_1} = 20$

10.6   Probability

5. *Sample Space*   Determine the sample space for the experiment.

   One letter from the alphabet is chosen.

Solution

{a, b, c, d, e, f, g, h, i, j, k, l, m
n, o, p, q, r, s, t, u, v, w, x, y, z}

7. *Sample Space*   Determine the sample space for the experiment.

   Two county supervisors are selected from five supervisors, A, B, C, D, and E.

Solution

{AB, AC, AD, AE, BC, BD, BE, CD, CE, DE}

9. You are given the probability $p = 0.35$ that an event will occur. Find the probability that the event will not occur.

Solution

$1 - 0.35 = 0.65$

11. *Coin Tossing*   A coin is tossed three times. Find the probability of the specified event.

   The event of getting two heads.

   Use the following sample space.

   HHH, HHT, HTH, THH, HTT, THT, TTH, TTT

Solution

$$P(E) = \frac{n(E)}{n(S)} = \frac{3}{8}$$

13. *Coin Tossing*   A coin is tossed three times. Find the probability of the specified event.

   The event of getting at least one head.

   Use the following sample space.

   HHH, HHT, HTH, THH, HTT, THT, TTH, TTT

Solution

$$P(E) = \frac{n(E)}{n(S)} = \frac{7}{8}$$

15. *Reading a Graph*   Use the circle graph in the textbook, which shows the number of people in the United States in 1990 with each blood type. A person is selected at random from the United States population. What is the probability that the person does not have blood type B?

Solution

$$P(E) = 1 - \frac{n(F)}{n(S)} = 1 - \frac{1}{10} = \frac{9}{10}$$

(*F* is event that person does have type B.)

17. *Random Selection*   Twenty marbles numbered 1 through 20 are placed in a bag, and one is selected. Find the probability of the specified event.

(a) The number is 12

(b) The number is prime.

(c) The number is odd.

(d) The number is less than 6.

Solution

(a) $P(E) = \dfrac{n(E)}{n(S)} = \dfrac{1}{20}$

(b) $P(E) = \dfrac{n(E)}{n(S)} = \dfrac{8}{20} = \dfrac{2}{5}$

(c) $P(E) = \dfrac{n(E)}{n(S)} = \dfrac{10}{20} = \dfrac{1}{2}$

(d) $P(E) = \dfrac{n(E)}{n(S)} = \dfrac{5}{20} = \dfrac{1}{4}$

19. *Meteorites*   The largest meteorite that ever landed in the United States was found in the Willamette Valley of Oregon in 1902. Earth contains 57,510,000 square miles of land and 139,440,000 square miles of water. What is the probability that a meteorite that hits the earth will fall onto land? What is the probability that a meteorite that hits the earth will fall into the water?

Solution

(a) $P(E) = \dfrac{n(E)}{n(S)} = \dfrac{57{,}510{,}000}{196{,}950{,}000} = \dfrac{1917}{6565}$

(a) $P(E) = \dfrac{n(E)}{n(S)} = \dfrac{139{,}440{,}000}{196{,}950{,}000} = \dfrac{4648}{6565}$

21. *Game Show*   On a game show, you are given five digits to arrange in the proper order to form the price of a car. If you arrange them correctly, you win the car. Find the probability of winning if you know the correct position of only one digit and must guess at the digits in all the other positions.

Solution

$P(E) = \dfrac{n(E)}{n(S)} = \dfrac{1}{1 \cdot 4 \cdot 3 \cdot 2 \cdot 1} = \dfrac{1}{24}$

23. *Defective Units*   A shipment of 10 food processors to a certain store contained two defective units. If you purchase two food processors as birthday gifts for friends, determine the probability that you get both defective units.

Solution

$P(E) = \dfrac{n(E)}{n(S)} = \dfrac{1}{{}_{10}C_2} = \dfrac{1}{\dfrac{10}{8! \, 2!}} = \dfrac{1}{\dfrac{10 \cdot 9}{2 \cdot 1}} = \dfrac{1}{45}$

25. *Card Selection*   Five cards are selected from a standard deck of 52 cards. Find the probability that the four aces are selected.

Solution

$P(E) = \dfrac{n(E)}{n(S)} = \dfrac{12}{{}_{52}C_5} = \dfrac{12}{\dfrac{52!}{47! \, 5!}} = \dfrac{12}{\dfrac{52 \cdot 51 \cdot 50 \cdot 49 \cdot 48}{5 \cdot 4 \cdot 3 \cdot 2 \cdot 1}} = \dfrac{12}{2{,}598{,}960} = \dfrac{1}{216{,}580}$

27. Describe the relationship between the graphs of f and g.

$$g(x) = f(x) - 4$$

Solution

g is vertical shift of f 4 units downward.

29. Describe the relationship between the graphs of f and g.

$$g(x) = -f(x)$$

Solution

g is a reflection of f in the x-axis.

31. Use a graphing utility to solve the system of equations.

$$5x - 2y = -25$$
$$-3x + 7y = 44$$

Solution

Solve each equation for y.

$$5x - 2y = -25 \qquad\qquad -3x + 7y = 44$$
$$-2y = -5x - 25 \qquad\qquad 7y = 3x + 44$$
$$y = 2.5x + 12.5 \qquad\qquad y = \frac{3}{7}x + \frac{44}{7}$$

Keystrokes:

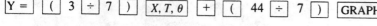

y_1 [Y =] 2.5 [X, T, θ] [+] 12.5 [ENTER]

y_2 [Y =] [(] 3 [÷] 7 [)] [X, T, θ] [+] [(] 44 [÷] 7 [)] [GRAPH]

Point of intersection $(-3, 5)$

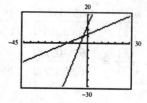

33. Determine the sample space for the experiment.

A taste tester must taste and rank three brands of yogurt, A, B, and C, according to preference.

Solution

{ABC, ACB, BAC, BCA, CAB, CBA}

35. Determine the sample space for the experiment.

A basketball tournament between two teams consists of three games. For each game, your team may win (W) or lose (L).

Solution

{WWW, WWL, WLW, WLL, LWW, LWL, LLW, LLL}

37. You are given the probability that an event will not occur. Find the probability that the event will occur.

$$p = 0.82$$

Solution

$$P(E) = 1 - p = 1 - 0.82 = 0.18$$

39. *Playing Cards* A card is drawn from a standard deck of playing cards. Find the probability of drawing the indicated card.

A red card

Solution

$$P(E) = \frac{n(E)}{n(S)} = \frac{26}{52} = \frac{1}{2}$$

41. *Playing Cards* A card is drawn from a standard deck of playing cards. Find the probability of drawing the indicated card.

A face card

Solution

$$P(E) = \frac{n(E)}{n(S)} = \frac{12}{52} = \frac{3}{13}$$

43. *Tossing a Die* A six-sided die is tossed. Find the probability of the specified event.

The number is a 5.

Solution

$$P(E) = \frac{n(E)}{n(S)} = \frac{1}{6}$$

45. *Tossing a Die* A six-sided die is tossed. Find the probability of the specified event.

The number is no more than 5.

Solution

$$P(E) = \frac{n(E)}{n(S)} = \frac{5}{6}$$

47. *Reading a Graph* Use the circle graphs in the textbook, which show for a certain college the numbers of incoming freshmen in each average high school grade category for the years 1970 and 1992.

A person is selected at random from the 1970 freshmen class. What is the probability that the person's high school average was an A?

Solution

$$P(E) = \frac{n(E)}{n(S)} = \frac{184}{1161} \approx 0.158$$

49. *Reading a Graph* Use the circle graphs in the textbook, which show for a certain college the numbers of incoming freshmen in each average high school grade category for the years 1970 and 1992.

What is the probability that a person selected from the 1970 freshman class did not have a high school average grade of C?

Solution

$$P(E) = \frac{n(E)}{n(S)} = \frac{851}{1161} \approx 0.733$$

51. *Geometry* A child uses a spring-loaded device to shoot a marble into the square box shown in the figure in the textbook. The base of the square is horizontal and the marble has an equal likelihood of coming to rest at any point on the base. Find the probability that the marble comes to rest in the red center.

Solution

$$P(E) = \frac{n(E)}{n(S)} = \frac{4\pi}{24^2} \approx 0.022$$

53. *Geometry* A child uses a spring-loaded device to shoot a marble into the square box shown in the figure in the textbook. The base of the square is horizontal and the marble has an equal likelihood of coming to rest at any point on the base. Find the probability that the marble comes to rest in the purple border.

Solution

$$P(E) = \frac{n(E)}{n(S)} = \frac{24^2 - 10^2\pi}{24^2} \approx 0.455$$

55. *Multiple Choice Test* A student takes a multiple-choice test in which there are five choices for each question. Find the probability that the question is answered correctly given the following conditions.

(a) The student has no idea of the answer and guesses at random.

(b) The student can eliminate two of the choices and guesses from the remaining choices.

(c) The student knows the answer.

Solution

(a) $P(E) = \dfrac{n(E)}{n(S)} = \dfrac{1}{5}$ (b) $P(E) = \dfrac{n(E)}{n(S)} = \dfrac{1}{3}$ (c) $P(E) = \dfrac{n(E)}{n(S)} = 1$

57. *Girl or Boy?* The genes that determine the sex of a human baby are denoted by XX (female) and XY (male). Complete the Punnett square shown in the textbook. Then use the result to explain why it is equally likely that a newborn baby will be a boy or a girl.

Solution

Female
	X	X
X	XX	XX
Y	XY	XY

(Male)

Probability of a girl $= \dfrac{2}{4} = \dfrac{1}{2}$

Probability of a boy $= \dfrac{2}{4} = \dfrac{1}{2}$

59. *Continuing Education* In a high school graduation class of 325 students, 255 are going to continue their education. What is the probability that a student selected at random from the class will not be furthering his or her education?

Solution

$P(E) = \dfrac{n(E)}{n(S)} = \dfrac{70}{325} = \dfrac{14}{65}$

61. *Lottery* You buy a lottery ticket inscribed with a five-digit number. On the designated day, five digits are randomly selected from the digits 0 through 9, inclusive. What is the probability that you have a winning ticket?

Solution

$P(E) = \dfrac{n(E)}{n(S)} = \dfrac{1}{10 \cdot 10 \cdot 10 \cdot 10 \cdot 10} = \dfrac{1}{100,000}$

63. *Preparing for a Test* An instructor gives her class a list of 10 study problems, from which she will select eight to be answered on an exam. If you know how to solve eight of the problems, find the probability you will be able to answer all eight question.

Solution

$P(E) = \dfrac{n(E)}{n(S)} = \dfrac{1}{{}_{10}C_8} = \dfrac{1}{45}$

65. *Defective Units*   A shipment of 12 compact disc players contains two defective units. A husband and wife buy three of these players to give their children as gifts.

(a) What is the probability that all three three are good players?

(b) What is the probability that they buy at least one defective player?

Solution

(a) $P(E) = \dfrac{n(E)}{n(S)} = \dfrac{{}_{10}C_3}{{}_{12}C_3} = \dfrac{120}{220} = \dfrac{6}{11}$

(b) $P(E) = \dfrac{n(E)}{n(S)} = \dfrac{{}_{2}C_1 \cdot {}_{10}C_2 + {}_{2}C_2 \cdot {}_{10}C_1}{{}_{12}C_3} = \dfrac{100}{220} = \dfrac{5}{11}$

Review Exercises for Chapter 10

1. Use sigma notation to write $[5(1) - 3] + [5(2) - 3] + [5(3) - 3] + [5(4) - 3]$.

Solution

$$\sum_{n=1}^{4} (5n - 3)$$

3. Use sigma notation to write $\dfrac{1}{3(1)} + \dfrac{1}{3(2)} + \dfrac{1}{3(3)} + \dfrac{1}{3(4)} + \dfrac{1}{3(5)} + \dfrac{1}{3(6)}$.

Solution

$$\sum_{n=1}^{6} \frac{1}{3n}$$

5. Simplify $\dfrac{20!}{18!}$.

Solution

$\dfrac{20!}{18!} = \dfrac{20 \cdot 19 \cdot 18!}{18!} = 380$

7. Simplify $\dfrac{n!}{(n-3)!}$.

Solution

$\dfrac{n!}{(n-3)!} = n(n-1)(n-2)$

9. Write the first five terms of $a_n = 132 - 5n$.

Solution

$a_1 = 132 - 5(1) = 127$

$a_2 = 132 - 5(2) = 122$

$a_3 = 132 - 5(3) = 117$

$a_4 = 132 - 5(4) = 112$

$a_5 = 132 - 5(5) = 107$

11. Write the first five terms of $a_n = \dfrac{3}{4}n + \dfrac{1}{2}$.

Solution

$a_1 = \dfrac{3}{4}(1) + \dfrac{1}{2} = \dfrac{5}{4}$

$a_2 = \dfrac{3}{4}(2) + \dfrac{1}{2} = 2$

$a_3 = \dfrac{3}{4}(3) + \dfrac{1}{2} = \dfrac{11}{4}$

$a_4 = \dfrac{3}{4}(4) + \dfrac{1}{2} = \dfrac{7}{2}$

$a_5 = \dfrac{3}{4}(5) + \dfrac{1}{2} = \dfrac{17}{4}$

13. Find the common difference of 30, 27.5, 25, 22.5, 20, ...

Solution

$d = -2.5$

15. Write the first five terms of $a_1 = 10$, $r = 3$.

Solution

$a_n = a_1 r^{n-1}$

$a_n = 10(3)^{n-1}$

$a_1 = 10(3)^{1-1} = 10$

$a_2 = 10(3)^{2-1} = 30$

$a_3 = 10(3)^{3-1} = 90$

$a_4 = 10(3)^{4-1} = 270$

$a_5 = 10(3)^{5-1} = 810$

17. Write the first five terms of $a_1 = 100$, $r = -\dfrac{1}{2}$.

Solution

$a_n = a_1 r^{n-1}$

$a_n = 100\left(-\dfrac{1}{2}\right)^{n-1}$

$a_1 = 100\left(-\dfrac{1}{2}\right)^{1-1} = 100$

$a_2 = 100\left(-\dfrac{1}{2}\right)^{2-1} = -50$

$a_3 = 100\left(-\dfrac{1}{2}\right)^{3-1} = 25$

$a_4 = 100\left(-\dfrac{1}{2}\right)^{4-1} = -12.5$

$a_5 = 100\left(-\dfrac{1}{2}\right)^{5-1} = 6.25$

19. Find the common ratio of $8, 12, 18, 27, \dfrac{81}{2}, \ldots$.

Solution

$r = \dfrac{3}{2}$

21. Find a formula for the nth term of the specified sequence.

Arithmetic sequence: $a_1 = 10$, $d = 4$

Solution

$a_n = dn + c$

$10 = 4(1) + c$

$6 = c$

$a_n = 4n + 6$

23. Find a formula for the nth term of the specified sequence.

Arithmetic sequence: $a_1 = 1000$, $a_2 = 950$

Solution

$a_n = dn + c$

$1000 = -50(1) + c$

$1050 = c$

$a_n = -50n + 1050$

25. Find a formula for the nth term of the specified sequence.

Geometric sequence $a_1 = 1$, $r = -\dfrac{2}{3}$

Solution

$a_n = a_1 r^{n-1}$

$a_n = 1\left(-\dfrac{2}{3}\right)^{n-1}$

27. Find a formula for the nth term of the specified sequence.

Geometric sequence $a_1 = 24$, $a_2 = 48$

Solution

$a_n = a_1 r^{n-1}$

$a_n = 24(2)^{n-1}$

29. Find a formula for the *n*th term of the specified sequence.

Geometric sequence: $a_1 = 12$, $a_4 = -\dfrac{3}{2}$

Solution

$a_n = a_1 r^{n-1}$

$a_n = 12\left(-\dfrac{1}{2}\right)^{n-1}$

31. Match the sequence with its graph in the textbook. [The graphs are labeled (a), (b), (c), (d), (e), and (f).]

$a_n = 5 - \dfrac{1}{n}$

Solution

(a)

33. Match the sequence with its graph in the textbook. [The graphs are labeled (a), (b), (c), (d), (e), and (f).]

$a_n = 5 - 2n$

Solution

(b)

35. Match the sequence with its graph in the textbook. [The graphs are labeled (a), (b), (c), (d), (e), and (f).]

$a_n = 6\left(\dfrac{2}{3}\right)^{n-1}$

Solution

(d)

37. Use a graphing utility to graph the first 10 terms of $a_n = \dfrac{3n}{n+1}$.

Solution

Keystrokes (calculator in sequence and dot mode):

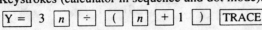

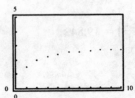

39. Use a graphing utility to graph the first 10 terms of $a_n = 5\left(\dfrac{3}{4}\right)^{n-1}$.

Solution

Keystrokes (calculator in sequence and dot mode):

$\boxed{Y=}\ 5\ \boxed{(}\ .75\ \boxed{)}\ \boxed{\wedge}\ \boxed{(}\ \boxed{n}\ \boxed{-}\ 1\ \boxed{)}\ \boxed{TRACE}$

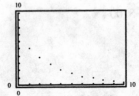

41. Evaluate the sum $\displaystyle\sum_{k=1}^{4} 7$.

Solution

$\displaystyle\sum_{k=1}^{4} 7 = 7 + 7 + 7 + 7 = 28$

43. Evaluate the sum $\displaystyle\sum_{n=1}^{4}\left(\dfrac{1}{n} - \dfrac{1}{n+1}\right)$.

Solution

$\displaystyle\sum_{n=1}^{4}\left(\dfrac{1}{n} - \dfrac{1}{n+1}\right) = \dfrac{1}{2} + \dfrac{1}{6} + \dfrac{1}{12} + \dfrac{1}{20} = \dfrac{30 + 10 + 5 + 3}{60} = \dfrac{48}{60} = \dfrac{4}{5}$

45. Evaluate the sum $\displaystyle\sum_{k=1}^{12}(7k-5)$.

Solution

$$\sum_{k=1}^{12}(7k-5) = 12\left(\frac{2+79}{2}\right) = 486$$

47. Evaluate the sum $\displaystyle\sum_{j=1}^{100}\frac{j}{4}$.

Solution

$$\sum_{j=1}^{100}\frac{j}{4} = 100\left(\frac{\frac{1}{4}+25}{2}\right) = 1262.5$$

49. Evaluate the sum $\displaystyle\sum_{n=1}^{12}2^n$.

Solution

$$\sum_{n=1}^{12}2^n = 2\left(\frac{2^{12}-1}{2-1}\right) = 8190$$

51. Evaluate the sum $\displaystyle\sum_{k=1}^{8}5\left(-\frac{3}{4}\right)^{k-1}$.

Solution

$$\sum_{k=1}^{8}5\left(-\frac{3}{4}\right)^{k-1} = 5\left(\frac{\left(-\frac{3}{4}\right)^8 - 1}{-\frac{3}{4}-1}\right) = 2.571106$$

53. Evaluate the sum $\displaystyle\sum_{i=1}^{8}(1.25)^{i-1}$.

Solution

$$\sum_{i=1}^{8}(1.25)^{i-1} = 1\left(\frac{1.25^8-1}{1.25-1}\right) = 19.842$$

55. Evaluate the sum $\displaystyle\sum_{n=1}^{120}500(1.01)^{n-1}$.

Solution

$$\sum_{n=1}^{120}500(1.01)^{n-1} = 500\left(\frac{1.01^{120}-1}{1.01-1}\right) = 115{,}019.34$$

57. Find the sum of the first 50 positive integers that are multiples of 4.

Solution

$$\sum_{n=1}^{50}4n = 50\left(\frac{4+200}{2}\right) = 5100$$

59. *Auditorium Seating*   Each row in a small auditorium has three more seats than the preceding row. Find the seating capacity of the auditorium if the front row seats 22 people and there are 12 rows of seats.

Solution

$$\sum_{n=1}^{12}(3n+19) = 12\left(\frac{22+55}{2}\right) = 462$$

61. *Population Increase*   A city of 85,000 people is growing at the rate of 1.2% per year. (That is, at the end of each year, the population is 1.012 times the population at the beginning of the year.)

(a) Find a formula for the *n*th term of the geometric sequence that gives the population *n* years from now.

(b) Estimate the population 50 years from now.

Solution

(a) $a_n = 85{,}000(1.012)^n$

(b) $a_{50} = 85{,}000(1.012)^{50} = 154{,}327.94 \approx 154{,}328$

63. Evaluate $_8C_3$.

Solution

$$_8C_3 = \frac{8!}{3! \, 5!} = \frac{8 \cdot 7 \cdot 6 \cdot 5!}{3 \cdot 2 \cdot 5!} = 56$$

65. Evaluate $_{12}C_0$.

Solution

$$_{12}C_0 = 1$$

67. Use a graphing utility to evaluate $_{40}C_4$.

Solution

Keystrokes:

40 MATH PRB 3 4 ENTER $_{40}C_4 = 91{,}390$

69. Use the Binomial Theorem to expand $(x + 1)^{10}$. Simplify your answer.

Solution

$$(x + 1)^{10} = 1x^{10} + 10x^9(1) + 45x^8(1)^2 + 120x^7(1)^3 + 210x^6(1)^4 + 252x^5(1)^5$$
$$+ 210x^4(1)^6 + 120x^3(1)^7 + 45x^2(1)^8 + 10x(1)^9 + 1(1)^{10}$$
$$= x^{10} + 10x^9 + 45x^8 + 120x^7 + 210x^6 + 252x^5 + 210x^4 + 120x^3 + 45x^2 + 10x + 1$$

71. Use the Binomial Theorem to expand $(y - 2)^6$. Simplify your answer.

Solution

$$(y - 2)^6 = 1y^6 - 6y^5(2) + 15y^4(2)^2 - 20y^3(2)^3 + 15y^2(2)^4 - 6y(2)^5 + 1(2)^6$$
$$= y^6 - 12y^5 + 60y^4 - 160y^3 + 240y^2 - 192y + 64$$

73. Use the Binomial Theorem to expand $\left(\dfrac{1}{2} - x\right)^8$.

Solution

$$\left(\frac{1}{2} - x\right)^8 = 1\left(\frac{1}{2}\right)^8 - 8\left(\frac{1}{2}\right)^7 x + 28\left(\frac{1}{2}\right)^6 x^2 - 56\left(\frac{1}{2}\right)^5 x^3 + 70\left(\frac{1}{2}\right)^4 x^4 - 56\left(\frac{1}{2}\right)^3 x^5 + 28\left(\frac{1}{2}\right)^2 x^6 - 8\left(\frac{1}{2}\right)x^7 + 1x^8$$

$$= \frac{1}{256} - \frac{1}{16}x + \frac{7}{16}x^2 - \frac{7}{4}x^3 + \frac{35}{8}x^4 - 7x^5 + 7x^6 - 4x^7 + x^8$$

75. Find the coefficient of the term x^5 in the expression $(x - 3)^{10}$. Simplify your answer.

Solution

$$_{10}C_5 = 252 \cdot (-3)^5 = -61{,}236$$

77. *Morse Code*   In Morse code, all characters are transmitted using a sequence of dots and dashes. How many different characters can be formed by using a sequence of three dots and dashes. (These can be repeated. For example, dash-dot-dot represents the letter *d*.)

Solution

$$2 \cdot 2 \cdot 2 = 8$$

79. *Committee Selection* Determine the number of ways a committee of five people can be formed from a group of 15people.

Solution

$$_{15}C_5 = \frac{15 \cdot 14 \cdot 13 \cdot 12 \cdot 11}{5 \cdot 4 \cdot 3 \cdot 2 \cdot 1} = 3003$$

81. *Rolling a Die* Find the probability of obtaining a number greater than 4 when a single six-sided die is rolled.

Solution

$$P(E) = \frac{n(E)}{n(S)} = \frac{5}{6}$$

83. *Book Selection* A child who does not know how to read carries a four-volume set of books to a book-shelf. Find the probability that the books are put on the shelf in the correct order.

Solution

$$P(E) = \frac{n(E)}{n(S)} = \frac{1}{4 \cdot 3 \cdot 2 \cdot 1} = \frac{1}{24}$$

85. *Hospital Inspection* As part of a monthly inspection at a hospital, the inspection team randomly selects reports from eight of the 84 nurses who are on duty. What is the probability that none of the reports selected will be from the 10 most experienced nurses on duty?

Solution

$$P(E) = \frac{n(E)}{n(S)} = \frac{_{74}C_8}{_{84}C_8} \approx 0.346$$

Test for Chapter 10

1. Write the first five terms of the sequence $a_n = \left(-\frac{2}{3}\right)^{n-1}$. (Begin with $n = 1$.)

Solution

$$a_n = \left(-\frac{2}{3}\right)^{n-1}$$

$$a_1 = \left(-\frac{2}{3}\right)^{1-1} = 1$$

$$a_2 = \left(-\frac{2}{3}\right)^{2-1} = -\frac{2}{3}$$

$$a_3 = \left(-\frac{2}{3}\right)^{3-1} = \frac{4}{9}$$

$$a_4 = \left(-\frac{2}{3}\right)^{4-1} = -\frac{8}{28}$$

$$a_5 = \left(-\frac{2}{3}\right)^{5-1} = \frac{16}{81}$$

2. Evaluate $\sum_{j=0}^{4}(3j + 1)$.

Solution

$$\sum_{j=0}^{4}(3j + 1) = 1 + 4 + 7 + 10 + 13 = 35$$

3. Evaluate $\sum_{n=1}^{5}(3 - 4n)$.

Solution

$$\sum_{n=1}^{5}(3 - 4n) = 5\left(\frac{-1 + -17}{2}\right) = -45$$

4. Using sigma notation to write the sum $\dfrac{2}{3(1) + 1} + \dfrac{2}{3(2) + 1} + \cdots + \dfrac{2}{3(12) + 1}$.

Solution

$$\sum_{n=1}^{12} \frac{2}{3n + 1}$$

5. Write the first five terms of the arithmetic sequence whose first term is $a_1 = 12$ and whose common difference is $d = 4$.

Solution

$a_n = a_1 + (n - 1)d$

$a_n = 12 + (n - 1)4 = 12 + 4n - 4 = 4n + 8$

$a_1 = 4(1) + 8 = 12$

$a_2 = 4(2) + 8 = 16$

$a_3 = 4(3) + 8 = 20$

$a_4 = 4(4) + 8 = 24$

$a_5 = 4(5) + 8 = 28$

6. Write a formula for the nth term of the arithmetic sequence whose first term is $a_1 = 5000$ and whose common difference is $d = -100$.

Solution

$a_n = a_1 + (n - 1)d$

$a_n = 5000 + (n - 1)(-100)$

$a_n = 5000 - 100n + 100$

$a_n = -100n + 5100$

7. Find the sum of the first 50 positive integers that are multiples of 3.

Solution

$$\sum_{n=1}^{50} = 50\left(\frac{3 + 150}{2}\right) = 3825$$

8. Find the common ratio of the geometric sequence $2, -3, \dfrac{9}{2}, -\dfrac{27}{4}, \ldots$.

Solution

$$r = -\frac{3}{2}$$

9. Find a formula for the nth term of the geometric sequence whose first term is $a_1 = 4$ and whose common ratio is $r = \dfrac{1}{2}$.

Solution

$a_n = a_1 r^{n-1}$

$a_n = 4\left(\dfrac{1}{2}\right)^{n-1}$

10. Evaluate: $\displaystyle\sum_{n=1}^{8} 2(2^n)$

Solution

$$\sum_{n=1}^{8} 2(2^n) = 4\left(\frac{2^8 - 1}{2 - 1}\right) = 1020$$

11. Evaluate $\displaystyle\sum_{n=1}^{10} 3\left(\frac{1}{2}\right)^n$

Solution

$$\sum_{n=1}^{10} 3\left(\frac{1}{2}\right)^n = \frac{3}{2}\left(\frac{\dfrac{1}{2}^{10} - 1}{\dfrac{1}{2} - 1}\right) = \frac{3069}{1024}$$

12. Fifty dollars is deposited each month in an increasing annuity that pays 8%, compounded monthly. What is the balance after 25 years?

Solution

$$A = P\left(1 + \frac{r}{n}\right)^{nt}$$

$$a_{300} = \left(1 + \frac{0.08}{12}\right)^{12(25)} = 50(1.0066667)^{300}$$

$$a_1 = 50(1.0066667)^1$$

$$\text{balance} = [50(1.0066667)^1]\left[\frac{1.0066667^{300} - 1}{1.0066667 - 1}\right] = \$47,868.09$$

13. Evaluate $_{20}C_3$.

Solution

$$_{20}C_3 = \frac{20 \cdot 19 \cdot 18}{3 \cdot 2 \cdot 1} = 1140$$

14. Explain how to use Pascal's Triangle to expand $(x - 2)^5$.

Solution

$$(x - 2)^5 = 1(x^5) - 5x^4(2) + 10x^3(2)^2 - 10x^2(2)^3 + 5x(2)^4 - 1(2)^5$$
$$= x^5 - 10x^4 + 40x^3 - 80x^2 - 80x - 32$$

15. Find the coefficient of the term x^3y^5 in the expansion of $(x + y)^8$.

Solution

The coefficient of x^3y^5 in expansion of $(x + y)^8$
is 56, since $_8C_3 = 56$.

16. How many license plates can consist of one letter followed by three digits?

Solution

$$\begin{array}{ccccccc}
\text{plates} = & \text{letter} & & \text{digit} & & \text{digit} & & \text{digit} \\
= & 26 & \cdot & 10 & \cdot & 10 & \cdot & 10 & = 26,000 \text{ plates}
\end{array}$$

17. Four students are randomly selected from a class of 25 to answer questions from a reading assignment. In how many ways can the four be selected?

Solution

$$_{25}C_4 = \frac{25!}{4! \, 21!} = \frac{25 \cdot 24 \cdot 23 \cdot 22}{4 \cdot 3 \cdot 2 \cdot 1} = 12,650$$

18. The weather report indicates that the probability of snow tomorrow is 0.75. What is the probability that it will not snow?

Solution

$1 - 0.75 = 0.25$

19. A card is drawn from a standard deck of playing cards. Find the probability that it is a red face card.

Solution

$$P(E) = \frac{n(E)}{n(S)} = \frac{6}{52} = \frac{3}{26}$$

20. Suppose two spark plugs require replacement in a four-cylinder engine. If the mechanic randomly removes two plugs, find the probability that they are the two defective plugs.

Solution

$$P(E) = \frac{n(E)}{n(S)} = \frac{1}{_4C_2} = \frac{1}{\dfrac{4!}{2!\,2!}} = \frac{1}{\dfrac{4 \cdot 3}{2 \cdot 1}} = \frac{1}{6}$$